A Companion to the Philosophy of Biology

Blackwell Companions to Philosophy

This outstanding student reference series offers a comprehensive and authoritative survey of philosophy as a whole. Written by today's leading philosophers, each volume provides lucid and engaging coverage of the key figures, terms, topics, and problems of the field. Taken together, the volumes provide the ideal basis for course use, representing an unparalleled work of reference for students and specialists alike.

Already published in the series:

1. The Blackwell Companion to Philosophy, Second Edition
   Edited by Nicholas Bunnin and Eric Tsui-James
2. A Companion to Ethics
   Edited by Peter Singer
3. A Companion to Aesthetics
   Edited by David Cooper
4. A Companion to Epistemology
   Edited by Jonathan Dancy and Ernest Sosa
5. A Companion to Contemporary Political Philosophy (2 Volume Set), Second Edition
   Edited by Robert E. Goodin and Philip Pettit
6. A Companion to Philosophy of Mind
   Edited by Samuel Guttenplan
7. A Companion to Metaphysics
   Edited by Jaegwon Kim and Ernest Sosa
8. A Companion to Philosophy of Law and Legal Theory
   Edited by Dennis Patterson
9. A Companion to Philosophy of Religion
   Edited by Philip L. Quinn and Charles Taliaferro
10. A Companion to the Philosophy of Language
    Edited by Bob Hale and Crispin Wright
11. A Companion to World Philosophies
    Edited by Eliot Deutsch and Ron Bontekoe
12. A Companion to Continental Philosophy
    Edited by Simon Critchley and William Schroeder
13. A Companion to Feminist Philosophy
    Edited by Alison M. Jaggar and Iris Marion Young
14. A Companion to Cognitive Science
    Edited by William Bechtel and George Graham
15. A Companion to Bioethics
    Edited by Helga Kuhse and Peter Singer
16. A Companion to the Philosophers
    Edited by Robert L. Arrington
17. A Companion to Business Ethics
    Edited by Robert E. Frederick
18. A Companion to the Philosophy of Science
    Edited by W. H. Newton-Smith
19. A Companion to Environmental Philosophy
    Edited by Dale Jamieson
20. A Companion to Analytic Philosophy
    Edited by A. P. Martinich and David Sosa
21. A Companion to Genethics
    Edited by Justine Burley and John Harris
22. A Companion to Philosophical Logic
    Edited by Dale Jacquette
23. A Companion to Early Modern Philosophy
    Edited by Steven Nadler
24. A Companion to Philosophy in the Middle Ages
    Edited by Jorge J. E. Gracia and Timothy B. Noone
25. A Companion to African-American Philosophy
    Edited by Tommy L. Lott and John P. Pittman
26. A Companion to Applied Ethics
    Edited by R. G. Frey and Christopher Heath Wellman
27. A Companion to the Philosophy of Education
    Edited by Randall Curren
28. A Companion to African Philosophy
    Edited by Kwasi Wiredu
29. A Companion to Heidegger
    Edited by Hubert L. Dreyfus and Mark A. Wrathall
30. A Companion to Rationalism
    Edited by Alan Nelson
31. A Companion to Ancient Philosophy
    Edited by Mary Louise Gill and Pierre Pellegrin
32. A Companion to Pragmatism
    Edited by John R. Shook and Joseph Margolis
33. A Companion to Nietzsche
    Edited by Keith Ansell Pearson
34. A Companion to Socrates
    Edited by Sara Ahbel-Rappe and Rachana Kamtekar
35. A Companion to Phenomenology and Existentialism
    Edited by Hubert L. Dreyfus and Mark A. Wrathall
36. A Companion to Kant
    Edited by Graham Bird
37. A Companion to Plato
    Edited by Hugh H. Benson
38. A Companion to Descartes
    Edited by Janet Broughton and John Carriero
39. A Companion to the Philosophy of Biology
    Edited by Sahotra Sarkar and Anya Plutynski

Forthcoming

40. A Companion to Hume
    Edited by Elizabeth S. Radcliffe
41. A Companion to Aristotle
    Edited by Georgios Anagnostopoulos

A Companion to the Philosophy of Biology

Edited by

Sahotra Sarkar

and

Anya Plutynski

BLACKWELL PUBLISHING
350 Main Street, Malden, MA 02148-5020, USA
9600 Garsington Road, Oxford OX4 2DQ, UK
550 Swanston Street, Carlton, Victoria 3053, Australia

First published 2008 by Blackwell Publishing Ltd

1 2008

Library of Congress Cataloging-in-Publication Data

A companion to the philosophy of biology / edited by Sahotra Sarkar and Anya Plutynski.
p. cm. – (Blackwell companions to philosophy ; 39)
Includes bibliographical references and index.
ISBN 978-1-4051-2572-7 (hardcover : alk. paper) 1. Biology–Philosophy. I. Sarkar, Sahotra. II. Plutynski, Anya.

QH331.C8423 2006
570.1–dc22

2007024735

A catalogue record for this title is available from the British Library.

Set in 10 on 12.5 pt Photina
by SNP Best-set Typesetter Ltd., Hong Kong
Printed and bound in Singapore
by Utopia Press Pte Ltd

The publisher's policy is to use permanent paper from mills that operate a sustainable forestry policy, and which has been manufactured from pulp processed using acid-free and elementary chlorine-free practices. Furthermore, the publisher ensures that the text paper and cover board used have met acceptable environmental accreditation standards.

For further information on
Blackwell Publishing, visit our website at
www.blackwellpublishing.com

Contents

Figures

Tables

Notes on Contributors

J. McKenzie Alexander is at the Department of Philosophy, Logic and Scientific Method at the London School of Economics and Political Science. His interests are in Evolutionary game theory, philosophy of social science, and rational choice theory. Recent publications include, "Follow the leader: local interactions with influence neighborhoods," *Philosophy of Science*, 72, 86–113 (2005), "Random Boolean networks and evolutionary games," *Philosophy of Science*, 70, 1289–1304 (2003),"Bargaining with neighbours: Is justice contagious?" (with Brian Skyrms), *Journal of Philosophy*, 96 (11), 588–98 (1999), and *The Structural Evolution of Morality* (Cambridge: Cambridge University Press, 2007).

Ron Amundson is Professor, Dept of Philosophy, University of Hawaii at Hilo. His research is in history and philosophy of biology, as well as bioethics and disability rights. Recent publications include: "Bioethics and disability rights: conflicting values and perspectives," (with Shari Tresky) in press, and *The Changing Role of the Embryo in Evolutionary Thought: Roots of Evo-Devo* (Cambridge: Cambridge University Press, 2005).

Stefan Artmann is at the Institute of Philosophy, Friedrich-Schiller-University Jena. In 2007, he finished his postdoctoral thesis on the philosophy of structural sciences (such as information theory, cybernetics, and decision theory). He has published several papers on questions related to the application of information theory and semiotics to biological systems, e.g., "Artificial life as a structural science," *Philosophia naturalis*, 40, 183–205 (2003), "Biosemiotics as a structural science: between the forms of life and the life of forms," *Journal of Biosemiotics*, 1, 183–209 (2005), and "Computing codes versus interpreting life. Two alternative ways of synthesizing biological knowledge through semiotics," in M. Barbieri (Ed.), *Introduction to Biosemiotics* (Dordrecht: Springer, 2007), pp. 209–33.

Mark A. Bedau is Professor of Philosophy and Humanities at Reed College in Portland, Oregon. He is also Editor-in-Chief of the journal *Artificial Life* (MIT Press); co-founder of the European Center for Living Technology, a research institute in Venice, Italy, that investigates theoretical and practical issues associated with living systems; and co-founder of ProtoLife SRL, a start-up company with the long-term aim of creating useful artificial cells. He has published and lectured around the world extensively on philosophical and scientific issues concerning emergence, evolution, life, mind, and the social and ethical implications of creating life from scratch.

Derek Bickerton is Professor Emeritus of Linguistics at the University of Hawaii. He is best known for his work on Creole languages, leading to the Language Bioprogram Hypothesis, and his subsequent work on the evolution of language. Among his books are *Roots of Language, Language and Species*, and *Language and Human Behavior*.

Mark Colyvan is Professor of Philosophy and Director of the Sydney Centre for The Foundations of Science at the University of Sydney in Sydney, Australia. He has published on the philosophy of mathematics, philosophy of science (especially ecology and conservation biology), philosophy of logic, and decision theory. His books include *The Indispensability of Mathematics* (OUP, 2001) and, with Lev Ginzburg, *Ecological Orbits: How Planets Move and Populations Grow* (OUP, 2004).

Michael R. Dietrich is an Associate Professor in the Department of Biology at Dartmouth College.

Marc Ereshefsky is Professor of Philosophy at the University of Calgary. He has written extensively on biological taxonomy and has published two books on the topic, *The Poverty of the Linnaean Hierarchy* and *The Units of Evolution*. His current research focuses on biological homology and its philosophical applications.

Justin Garson is a lecturer in Philosophy at the University of Texas at Austin. His interests are in the Philosophy of Science, with emphasis on the history and philosophy of neuroscience and psychiatry. He was one of the assistant editors of, and contributors to, *The Philosophy of Science: An Encyclopedia* (Routledge, 2005). Recent publications include "The introduction of information into neurobiology." *Philosophy of Science*, 70, 926–36 (2003).

Peter Godfrey-Smith is Professor of Philosophy at Harvard University, but also spends time at the Philosophy Program, Research School of Social Sciences, The Australian National University. Godfrey-Smith's primary research interests are in the philosophy of biology and philosophy of mind. He also has interests in other parts of philosophy of science, causation, and the philosophy of John Dewey. His books include: *Complexity and the Function of Mind in Nature* (Cambridge: Cambridge University Press, 1996) and *Theory and Reality: An Introduction to the Philosophy of Science* (Chicago: University of Chicago Press, 2003).

Paul E. Griffiths is a philosopher of science with a focus on biology and psychology, and was educated at Cambridge and the Australian National University. He taught at Otago University in New Zealand and was later Director of the Unit for History and Philosophy of Science at the University of Sydney, before taking up a professorship in the Department of History and Philosophy of Science at the University of Pittsburgh.

He returned to Australia in 2004, first as an Australian Research Council Federation Fellow and then as University Professorial Research Fellow at the University of Sydney. He is a Fellow of the Australian Academy of the Humanities.

Moira Howes is Associate Professor of Philosophy at Trent University. She specializes in metaphysics, philosophy of science (especially immunology), and feminist epistemology. Her current research concerns immunological models of the human female reproductive tract.

James Justus wrote a dissertation on "The Stability-Diversity-Complexity Debate of Theoretical Community Ecology: A Philosophical Analysis" at the University of Texas, Austin. He has published in the journals *Biology and Philosophy*, *Conservation Biology*, and *Philosophy of Science*. Recent publications include "Qualitative scientific modeling and crop analysis," *Philosophy of Science* (2005); and "Ecological and Lyapunov stability," *Philosophy of Science* (2007). His interests outside the philosophy of science include the history of analytic philosophy, environmental philosophy, logic and philosophy of mathematics, formal epistemology, and bird watching.

Jonathan M. Kaplan, an Associate Professor of Philosophy at Oregon State University, specializes in the Philosophy of Biology, Philosophy of Science and Political Philosophy. In addition to various articles and book chapters, he has published two books, most recently *Making Sense of Evolution: The Conceptual Foundations of Evolutionary Biology*, co-authored with evolutionary biologist Massimo Pigliucci (Chicago: Chicago University Press, 2006).

Marc Lange is Professor of Philosophy at the University of North Carolina at Chapel Hill. He is the author of *Natural Laws in Scientific Practice* (Oxford, 2000), *An Introduction to the Philosophy of Physics: Locality, Fields, Energy, and Mass* (Blackwell, 2002), and *Laws and Lawmakers* (Oxford, forthcoming).

James Lennox is Professor at the Department of History and Philosophy of Science at the University of Pittsburgh. Research specialties include Ancient Greek philosophy, science and medicine, and Charles Darwin and Darwinism. Lennox has published essays on the philosophical and scientific thought of Plato, Aristotle, Theophrastus, Boyle, Spinoza, and Darwin, especially focused on scientific explanation, and particularly teleological explanation, in the biological sciences. He is author of *Aristotle's Philosophy of Biology* (Cambridge 2000) and *Aristotle on the Parts of Animals I–IV* (Oxford, 2001), the first English translation of this work since 1937. He is co-editor of *Philosophical Issues in Aristotle's Biology* (Cambridge, 1987); *Self-Motion from Aristotle to Newton* (Princeton, 1995); and *Concepts, Theories, and Rationality in the Biological Sciences* (Pittsburgh and Konstanz, 1995).

Richard C. Lewontin is Alexander Agassiz Research Professor at the Museum of Comparative Zoology, Harvard University. He is an evolutionary biologist, geneticist, and pioneer in the application of techniques from molecular biology to questions of genetic variation and evolution. He is the author of *The Genetic Basis of Evolutionary*

Change and *Biology as Ideology*, and the co-author of *The Dialectical Biologist* (with Richard Levins) and *Not in Our Genes* (with Steven Rose and Leon Kamin).

Alan C. Love is Assistant Professor of Philosophy at the University of Minnesota. His current work focuses on the nature of conceptual change and explanation in the biological sciences, specifically within the dynamic of the discipline of evolutionary developmental biology, using a combination of historical and philosophical methodologies. Selected Publications include "Evolutionary morphology and evo-devo: hierarchy and novelty," *Theory in Biosciences*, 124, 317–33 (2006), "Evolvability, dispositions, and intrinsicality," *Philosophy of Science*, 70(5), 1015–27 (2003), and "Evolutionary morphology, innovation, and the synthesis of evolutionary and developmental biology", *Biology and Philosophy*, 18, 309–45 (2003).

Staffan Müller-Wille received his PhD in Philosophy from the University of Bielefeld for his dissertation on Linnaeus' taxonomy. He worked at the Max-Planck-Institute for the History of Science (Berlin) from 2000 to 2004 and is currently holding a post as Senior Research Fellow for Philosophy of Biology at the University of Exeter. He is author of the book *Botanik und weltweiter Handel* (1999) and has published articles on the history and epistemology of natural history, genetics, and anthropology.

Dominic Murphy is Assistant Professor of Philosophy at Caltech. He is the author of *Psychiatry in the Scientific Image* (MIT, 2006), as well as papers in the philosophy of mind and the philosophy of biology.

Bryan G. Norton is Professor in the School of Public Policy, Georgia Institute of Technology and author of *Why Preserve Natural Variety?* (Princeton University Press, 1987), *Toward Unity Among Environmentalists* (Oxford University Press, 1991), *Searching for Sustainability* (Cambridge University Press, 2003), and *Sustainability: A Philosophy of Adaptive Ecosystem Management* (University of Chicago Press, 2005). Norton has contributed to journals in several fields and has served on the Environmental Economics Advisory Committee of the US EPA Science Advisory Board, and two terms as a member of the Governing Board of the Society for Conservation Biology. His current research concentrates on sustainability theory and on spatio-temporal scaling of environmental problems. He was a member of the Board of Directors of Defenders of Wildlife from 1994 to 2005 and continues on their Science Advisory Board.

Jay Odenbaugh is a member of the Department of Philosophy and Environmental Studies Program at Lewis and Clark College. His main areas of research are in the philosophy of science (especially ecology and evolutionary biology) and environmental ethics. He is currently working on a book tentatively entitled *On the Contrary: A Philosophical Examination of the Environmental Sciences and their Critics* examining these issues especially in ecology, climatology, and environmental economics.

Samir Okasha received his doctorate from the University of Oxford in 1998 and is currently Reader in Philosophy of Science at the University of Bristol. He has published numerous research articles in journals such as *Philosophy of Science*, *Evolution*, *British*

Journal for the Philosophy of Science, Synthese, Biology and Philosophy, and others. He is currently completing a monograph on the units of selection debate to be published by Oxford University Press.

Kent A. Peacock is Associate Professor of Philosophy at the University of Lethbridge, Alberta, Canada. He has published in philosophy of physics and ecology, and is the editor of a text, *Living With the Earth: An Introduction to Environmental Philosophy* (Toronto: Harcourt Brace, 1996).

Anya Plutynski is an Assistant Professor of Philosophy at the University of Utah. Her research is in the history and philosophy of biology. She is the author of several recent articles and book chapters on the early synthesis and evolutionary explanation, in some of the following journals: *Proceedings of the Philosophy of Science, British Journal of Philosophy of Science, Biology and Philosophy*, and *Biological Theory*.

Hans-Jörg Rheinberger is Director at the Max Planck Institute for the History of Science in Berlin and a Scientific Member of the Max Planck Society. He studied philosophy (MA) and biology (PhD) at the University of Tübingen and the Free University of Berlin. He worked as a molecular biologist at the Max Planck Institute for Molecular Genetics in Berlin, and as a historian of science at the Universities of Lübeck and Salzburg. Among his books are *Toward a History of Epistemic Things* (1997), *Iterationen* (2005), *Epistemologie des Konkreten* (2006) and *Historische Epistemologie zur Einführung* (2007). He is also a co-editor of *The Concept of the Gene in Development and Evolution* (2000), and *The Mapping Cultures of Twentieth Century Genetics* (2004).

Alex Rosenberg is the R. Taylor Cole Professor of Philosophy, and co-Director of the Center for Philosophy of Biology, Duke University. His interests focus on problems in metaphysics, mainly surrounding causality, the philosophy of social sciences, especially economics, and most of all, the philosophy of biology, in particular the relationship between molecular, functional, and evolutionary biology. Recent publications include: *Darwinian Reductionism: Or, How to Stop Worrying and Love Molecular Biology* (University of Chicago, 2006).

Sahotra Sarkar is Professor of Integrative Biology, Geography and the Environment, and Philosophy at the University of Texas at Austin. His research is in history and philosophy of science, formal epistemology, philosophy of biology, and conservation biology. Recent books include *Doubting Darwin? Creationist Designs on Evolution* (Blackwell, 2007), *Biodiversity and Environmental Philosophy: An Introduction* (Cambridge, 2005), and *Molecular Models of Life* (MIT, 2005).

Christopher Stephens is Assistant Professor of Philosophy at University of British Columbia. Dr Stephens specializes in philosophy of biology, philosophy of science, and epistemology. Recent publications include "Modelling reciprocal altruism," *British Journal for the Philosophy of Science* (1996) and "When is it selectively advantageous to have true beliefs?" *Philosophical Studies* (2001). His current research interests include drift and chance in evolutionary biology, the evolution of rationality, and the relationship between prudential and epistemic rationality.

Marcel Weber received an MSc in molecular biology from the University of Basel, a PhD in philosophy at the University of Konstanz, and a *Habilitation* in philosophy at the University of Hannover. He is currently Swiss National Science Foundation Professor of Philosophy of Science at the University of Basel. His books include *Philosophy of Experimental Biology* (Cambridge University Press, 2005) and *Die Architektur der Synthese: Entstehung und Philosophie der modernen Evolutionstheorie* (Walter de Gruyter, Berlin, 1998).

Jon F. Wilkins is Research Professor at the Santa Fe Institute. His research interests include coalescent theory, genomic imprinting, human demographic history, altruism, and cultural evolution. Recent publications include: "Gene genealogies in a continuous habitat: a separation of timescales approach," *Genetics*, 168, 2227–44 (2004); (with J. Wakeley) "The coalescent in a continuous, finite, linear population," *Genetics*, 161, 873–88 (2002); "Genomic imprinting and methylation: epigenetic canalization and conflict," *Trends Genet.*, 21, 356–65 (2005); and (with D. Haig) "What good is genomic imprinting: the function of parent-specific gene expression," *Nat. Rev. Genet.*, 4, 359–68 (2003).

Acknowledgments

The editors gratefully acknowledge all the contributors and the editors at Blackwell, especially Jeff Dean, for their efforts in completing this volume.

Introduction

SAHOTRA SARKAR AND ANYA PLUTYNSKI

There are many different ways to do the philosophy of biology. At one end of a spectrum of possibilities would be works of general philosophical interest drawing on biological examples for illustration and support. At the other end would be works that deal only with conceptual and methodological issues that arise within the practice of biology. The strategy of this book is closer to the second way of approaching the subject. It aims to provide overviews of philosophical issues as they arise in a variety of areas of contemporary biology. Traditionally, evolution has been the focus of most philosophical attention. While it surely remains true that "nothing in biology makes sense except in light of evolution" (Dobzhansky, 1973), this tradition within the philosophy of biology is myopic insofar as it ignores much – if not most – of the work in contemporary biology. Intended primarily for students and beginning scholars, this book takes a wider perspective and addresses philosophical questions arising in molecular biology, developmental biology, immunology, ecology, and theories of mind and behavior. It also explores general themes in the philosophy of biology, for instance, the role of laws and theories, reductionism, and experimentation. In this respect, this book aims to break new ground in the philosophy of biology. Before we turn to what is new, let us briefly look at the background from which contemporary philosophy of biology emerged.

1. Background

When the logical empiricists reoriented the direction of philosophy of science in the 1920s and 1930s, the loci of their attention were mathematics (and within it, almost entirely mathematical logic) and physics (initially relativity theory, later also quantum mechanics). This not only set the agenda, but also the tone, for the philosophy of science. The relatively simple axiomatic structures of relativity theory and quantum mechanics – or, at least, how professional philosophers conceived those fields – became the yardstick of comparison for other disciplines. If these other disciplines were found to be less general in their intended domain, to be using different criteria of rigor (that is, using techniques different from the type of mathematics used in mathematical logic), or simply different, they were presumed to be wanting. This applied not only to biology or chemistry (or, for that matter, the social sciences) but even to other areas of physics. Biology thus suffered from a not always benign neglect throughout this period.

Yet, in spite of this limited attention, if the sophistication of the discussion is used as a standard, biology fared much better during the early decades of the logical empiricist regime (that is, from 1925 to 1945) than during the next 20 years. This is not only because many biologists – including Driesch (1929), J. S. Haldane (1929, 1931), Hogben (1930), and J. B. S. Haldane (1936, 1939) – explicitly debated philosophical positions, in particular, the relative roles of reductionism and holism in biology, during those decades. These debates within the biological community helped the development of philosophy of biology, but there were also significant attempts by philosophers to come to terms with the exciting developments that had taken place in biology, particularly in genetics and evolution, during the first three decades of the twentieth century. Woodger (1929) produced an exploration of traditional philosophical problems in biology, such as vitalism and mechanism, as well as a theory of biological explanation. In 1937 he went on to attempt to axiomatize parts of genetics.[1] By 1952 Woodger (1952) had clearly articulated what, after independent formulation and elaboration by Nagel (1949, 1951, 1961), became the standard model of theory reduction.[2] Nagel used this model in an attempt to explicate mechanistic explanation in biology. Less successfully, he attempted to provide a deflationary account of teleological explanation in biology (Nagel, 1961, 1977).

Arguably, until at least the 1960s, philosophers provided less philosophical insight about biology than theoretically oriented biologists. In the case of mechanistic explanation, for instance, as far as substantive biological questions are concerned, Nagel achieved little more than Hogben (1930). All he did was translate the simplest biological questions into the logical empiricists' framework and presumed that the result showed what was philosophically interesting about biology. Following the standard twentieth-century philosophical tradition, Nagel's writings on biology contributed little that scientists, even philosophically oriented biologists, found valuable. Nagel also displayed a strange refusal to follow contemporary developments in biology: between 1949 and 1961 he saw no reason to temper his bleak assessment of the state of mechanistic/reductionist explanation in biology – the events of 1953 either completely slipped by him, or failed to impress him. *The Structure of Science* from 1961 has several sections devoted to reductionism in biology but makes no mention of the double helix or, for that matter, any other development in molecular biology that had raised the potential for successful reduction in biology to an entirely different level (Nagel, 1961).

In the philosophy of biology, during the late 1950s and early 1960s only two notable exceptions stand out, Beckner's *The Biological Way of Thought* and, especially, Goudge's *The Ascent of Life*, the latter being a scientifically fairly sophisticated philosophical exploration of evolutionary theory (Beckner, 1959; Goudge, 1961; see also

---

1 Woodger (1937), under the sway of operationalism and skepticism about theoretical entities, attempted an axiomatization of genetics without "gene" as a term; Carnap (1958) developed some of Woodger's formal treatment in more interesting ways.

2 For a history, see Sarkar (1989).

Scriven, 1959). The situation changed for the better in the late 1960s and 1970s. Hull (1965, 1967, 1968) began to explore the conceptual structure of evolutionary biology. Wimsatt (1971, 1972) provided a detailed analysis of teleological explanation (and biological "feedback"), drawing extensively on contemporary work in theoretical biology. In a series of papers, Schaffner (1967a, b, 1969, 1976) began to argue the case for reductionism in molecular genetics while Hull (1972, 1974, 1976, 1981) questioned Schaffner's assessment. Ruse (1976) and Wimsatt (1976) were among those who joined this debate. A consensus emerged against reductionism (provided that reduction was construed in the fashion inherited from Nagel and the logical empiricists). Philosophy of biology also played its part, though rather late, in the rejection of logical empiricism in the 1960s and 1970s.

Since the early 1970s, the philosophy of biology has had a continuous and increasingly prominent presence in the philosophy of science. Occasional abuse of biology by philosophers has continued – as late as 1974, Popper would claim that Darwinism is not a scientific enterprise (Popper, 1974). Over the years, however, philosophy of biology has contributed to the development of the various alternatives to logical empiricism, including scientific realism, the semantic view of theories, and, in particular, naturalistic epistemology. Within the general context of the philosophy of biology, the last of these programs has been particularly natural and fecund presumably because philosophers of biology, because of their engagement with biology, are more likely than other philosophers to analyze how humans are evolutionarily produced, constrained, and challenged, as biological organisms. In fact, barring a very few exceptions, there is consensus among philosophers of biology of the great value of the naturalized perspective in philosophy where "naturalism" is very narrowly construed purely in evolutionary terms. Moreover, philosophers of biology have quite routinely begun to practice biology. If philosophy is to be done in continuity with science, as Quine once urged, no area in philosophy has followed that dictum more systematically than the philosophy of biology.

In the late 1970s, philosophy of biology became almost exclusively concerned with evolutionary theory. In some ways, this focus was productive; core philosophical questions were addressed about the foundations of evolutionary theory. For instance, Hull (1965a, b; see also Sober, 1988), advanced a discussion of different schools of phylogenetic analyses that has subsequently developed a rich literature on the methodological commitments of different schools of thought in systematics and phylogenetics. Philosophers including Wimsatt (1980), Brandon (1982), and Sober (1984) produced useful analyses of what constitutes the units of selection, while several prominent biologists, including Lewontin (1970) and Maynard Smith (1976), made important philosophical contributions. Sober's 1984 book, *The Nature of Selection*, advanced a clear analysis of the nature of laws and the structure of evolutionary theory, and particularly clarified related questions about the units-of-selection debate. Another 1984 book of equal merit was Flew's (1984) *Darwinian Evolution*. However, the almost exclusive focus on evolution in much of the literature of the late 1970s and 80s arguably hurt the development of the discipline. Many of the philosophical writings on biology from this period remained inattentive to molecular biology where, for better or for worse, most of biological research had become concentrated. Kitcher (1982, 1984) and Rosenberg (1985), however, are notable exceptions. Kitcher (1982) gave a thoughtful analysis of the transformation of biology after 1953, as well as a critical discussion of gene concepts (Kitcher, 1984), and Rosenberg (1985) advanced a perspective that treated genetics and molecular biology as being central to biology.

Given this state of the field, it is easy to understand the molecular biologists' lack of concern for philosophical critiques of their enterprise. This lack of concern was particularly noticeable during the debates over the initiation of the Human Genome Project in the late 1980s and early 1990s, a debate on which philosophers, unlike historians and social scientists, had no perceptible influence. (A notable exception to these generalizations is neurobiology which has always received considerable philosophical attention though usually in the context of the philosophy of mind.)

Since the early 1990s, in a very welcome development, philosophical writing on biology has extended its scope to cover many areas within biology beyond evolutionary theory.[3] There has been much recent interest in ecology, molecular and developmental biology. There has also finally been some attention to the role of experimentation in biology. In particular, Rheinberger (1993, 1997) has pioneered the use of techniques from the continental tradition of philosophy in the analysis of experimentation in molecular biology. Philosophers of biology have usually also paid ample attention to the history of biology. With intellectual and technical history gradually falling out of fashion in the professional history of science, philosophers of biology have done much to keep the history of the *science* of biology alive in contemporary research. This book reflects all these trends.

2. Structure of the *Companion*

Most of biology today is molecular biology, and the *Companion* begins with a section on molecular biology and genetics ("Molecular Biology and Genetics"). Rheinberger and Müller-Wille ("Gene Concepts") provide a historical review the various ways in which genes have been conceptualized, and how these have changed from the period of classical genetics to the post-genomic era in which we now find ourselves. Artmann ("Biological Information") explores the troubled question of whether and how biological information is susceptible to precise, quantitative measurement, an issue that has been hotly debated by philosophers (Godfrey-Smith, 2004; Sarkar, 2005). Contrary to many philosophers (Sarkar, 1996), he argues that there is more to informational talk in biology than mere metaphor.

Lewontin ("Heredity and Heritability") provides a philosophically sophisticated account of how classical genetics views heredity and adds a critique of the much-abused concept of heritability. Sarkar ("Genomics, Proteomics, and Beyond") speculates on where the study of heredity and development is going in the wake of the massive whole-genome sequencing projects. Both Lewontin and Sarkar emphasize the limitations of a gene-centered view of biology and argue for a more developmentally oriented approach to understanding the emergence of phenotypes.

The next section ("Evolution") turns to a number of classic issues addressed in the philosophy of biology, as well as some issues that have not perhaps received the atten-

3 The textbook by Sterelny and Griffiths (1999) is indicative of this trend.

tion they deserved. Reconciling Darwin's own views with the various ways in which "Darwinism" has been understood during the last 130 years has been a challenge for biologists, historians, and philosophers of biology. Lennox ("Darwinism and Neo-Darwinism") identifies the core principles of Darwin's original theory, and traces their empirical and conceptual development through the evolutionary synthesis, arguing that there is a meaningful set of commitments one can identify as "Darwinian." A further classic problem in evolutionary biology is how species should be defined and classified. Ereshefsky ("Systematics and Taxonomy") analyzes a variety of controversies that have arisen among biologists and philosophers of biology about the nature of species and their classification, ultimately defending a pluralist view of how species should be defined.

Population genetics has typically been viewed as the theoretical core of evolutionary biology. Stephens ("Population Genetics") recounts the history of the origins of population genetics, and reviews central debates in the history of the theory. He also considers a number of conceptual issues about representation and explanation that arise in the context of theoretical population genetics. Okasha ("Units and Levels of Selection") reviews the conceptual as well as empirical issues at stake in the debate over the units and levels of selection and gives a history of the debate from Darwin to the present day. He shows how this debate is tied to concerns about the evolution of altruism, the plausibility of group and kin selection, species selection and macroevolution, and concludes with a review of multilevel selection theory. Dietrich ("Molecular Evolution") describes the rise of the neutral theory of molecular evolution, and discusses how debates over drift versus selection in molecular evolution are exemplary of relative significance debates in biology.

One area that has received relatively little attention in philosophy of biology is the relationship between micro- and macro-evolution, and in particular, issues surrounding how hypotheses about change at and above the species level are tested. Plutynski ("Speciation and Macroevolution") addresses this question, and reviews recent empirical and theoretical work on speciation, the punctuated equilibrium debate, and questions about the disparity and evolvability. Finally, Godfrey-Smith and Wilkins ("Adaptationism") trace the history of the debate over "adaptationist" thinking, nicely demarcating different senses of adaptationism: empirical, explanatory, and methodological. In conclusion, they suggest a resolution to some of the controversy by illustrating how various alternatives might be resolved through careful attention to the grain at which evolutionary processes are being described.

The section on "Developmental Biology" contains three important contributions. Kaplan ("Phenotypic Plasticity and Reaction Norms") returns to the question of the relation between genotype and phenotype, already explored earlier by Lewontin. Once again the emphasis is on the complexity of this relation, which was largely ignored in classical genetics. Much of modern evolutionary theory was formulated at the genotypic level, ignoring the complexities of organismic development. The received view is that development can be put in a "black box" and phenotypic change tracked by recording changes at the genotypic level. However, it has long been recognized that, eventually, to understand the evolution of phenotypes, we must understand how developmental mechanisms have evolved. The past decade has seen a lot of excitement in evolutionary developmental biology, which many biologists now hold as finally successfully integrating evolutionary biology and studies of development. Amundson ("Development and Evolution") puts these studies in historical perspective, analyzing the long, sometimes idiosyncratic, and largely unsuccessful past attempts to integrate the two disciplines. It is an open question whether the near future will be much different from the past. In "Explaining the Ontogeny of Form: Philosophical Issues," Love provides a survey of issues surrounding the explanation of the ontogeny of form. He provides a philosophical framework for approaching different kinds of explanations in developmental biology, and addresses a variety of related epistemological and ontological issues; among them: representation, explanation, typology, individuality, model systems, and research heuristics.

The next section ("Medicine") takes up the relatively underexplored field of health and disease. One area that has received relatively little attention among philosophers of biology is immunology. Howes ("Self and Nonself") considers how philosophers can play a critical role in analyzing the conceptual foundations and empirical justifications of different models of self and nonself deployed in immunology. Murphy ("Health and Disease") considers "objectivist," "constructivist," and "revisionist" perspectives on health and disease, and focuses his discussion on the role of norms in judgments concerning mental illness.

The "Ecology" section summarizes much of the recent work on the philosophy of ecology, another area of the philosophy of biology that is receiving increased attention in recent years. Perhaps the most theoretically mature part of ecology is population ecology, and Colyvan ("Population Ecology") summarizes the philosophical work on the subject, showing how this is a fertile area to explore questions such as the role of laws and theories in biology. Justus ("Complexity, Diversity, and Stability") turns to a central issue in community ecology, whether there is any relation between diversity and stability. He shows how the concepts of diversity and stability (and, also, though to a lesser extent, complexity) can be interpreted in a variety of inconsistent ways, making it almost impossible to answer this question.

In the context of our increasing concern for the environment, Peacock ("Ecosystems") describes recent thinking on ecosystems, including work done within science, and philosophically intriguing ideas at the fringe of science such as the Gaia hypothesis. Turning to conservation biology, Norton ("Biodiversity and Conservation") shows how the concept of biodiversity is both descriptive (capturing some feature of habitats) and normative (reflecting the values people have which make them want to preserve nature). He also embeds philosophical discussions of biodiversity in the context of environmental policy.

The next section turns to mental and cultural life ("Mind and Behavior"), about which there is perhaps more scientific controversy than in any other area explored in depth by philosophers of biology. Griffiths ("Ethology, Sociobiology, and Evolutionary Psychology") gives a historical analysis that shows the deep connection between mid-twentieth-century ethology, human sociobiology, and contemporary Evolutionary Psychology. He notes that, while there is no reason to doubt that mental features are results of biological and cultural evolution, the research program of contemporary "Evolutionary Psychology" makes many controversial assumptions that should be scrutinized carefully. Alexander ("Cooperation") takes up recent approaches to the evolution of cooperative behavior including the many applications of game theory.

Finally, Bickerton ("Communication and Language") explores what we do and do not know about the emergence and evolution of human language and notes both the analogies and disanalogies between language and animal communication systems.

The final section ("Experimentation, Theory, and Themes") takes up a variety of general issues in the philosophy of biology, ranging from metaphysical issues about how to define life, or whether there are biological laws, to epistemological issues about how biologists investigate the living world. Bedau ("What is Life?") explores the variety of attempts to set out conditions for "life," and discusses how and why this question has become especially pressing with recent research into artificial life. Weber ("Experimentation") analyzes the special difficulties and characteristics of experimental work in biology. He considers the roles of model organisms, the limitations and advantages of laboratory work in biology, and the nature of evidence and objectivity in the biological sciences.

Many philosophers hold that biology is not at all like physics insofar as there are no "laws" of biology; however, Lange ("Is Biology Like Physics?") argues to the contrary. He considers the objection that laws of biology are not exceptionless and non-accidental, and argues, using a number of different examples, that lawful generalizations are an integral part of evolutionary biology. While it is uncontroversial that models and modeling are central to empirical and theoretical work in all branches of biology, philosophers do not agree on what a "model" is. Odenbaugh ("Models") reviews philosophical work on models, starting with the logical empiricists, explaining the subtle differences between the syntactic and semantic view of theories, and discusses a variety of historical and recent work on models and metaphors, and models as "mediators" between theory and data in the biological sciences.

It is hard to imagine biology without talk of functions but there is little philosophical agreement on what a function is. Garson ("Function and Teleology") gives a comprehensive review of the philosophical literature on functions, from etiological to consequentialist theories of function, and concludes with a defense of pluralist and context-dependent approaches to assignments of function. Yet another contentious issue in philosophy of biology has been the claim whether biological facts are reducible to molecular chemical or physical facts. Rosenberg ("Reductionism in Biology") takes a radical stance on this question, arguing that while the reducibility of theories, as the logical empiricists understood it, is implausible, generalizations in functional biology can and should be reduced, in the sense of being "completed, corrected, made more precise or otherwise deepened" by "fundamental explanations in molecular biology."

References

Beckner, M. (1959). *The biological way of thought*. New York: Columbia University Press.

Brandon, R. (1982). The levels of selection. In P. Asquith & T. Nickles (Eds). *Philosophy of Science Association Proceedings*, 1, 315–22.

Carnap, R. (1958). *Introduction to symbolic logic and its applications*. New York: Dover Publications.

Dobzhansky, T. (1973). Nothing in biology makes sense except in the light of evolution. *American Biology Teacher*, 35, 125–9.

Driesch, H. (1929). *Man and the universe*. London: Allen & Unwin.
Flew, A. (1984). *Darwinian evolution*. London: Paladin.
Godfrey-Smith, P. (2004). Genes do not encode information for phenotypic traits. In C. Hitchcock (Ed.). *Contemporary debates in the philosophy of science* (pp. 275–289). Oxford: Blackwell.
Goudge, T. A. (1961). *Ascent of life: a philosophical study of the theory of evolution*. Toronto: University of Toronto Press.
Haldane, J. B. S. (1936). Some principles of causal analysis in genetics. *Erkenntnis*, 6, 346–57.
Haldane, J. B. S. (1939). *The Marxist philosophy and the sciences*. New York: Random House.
Haldane, J. S. (1929). *The sciences and philosophy*. Garden City, NY: Doubleday, Doran and Co.
Haldane, J. S. (1931). *The philosophical basis of biology*. London: Hodder & Stoughton.
Hogben, L. (1930). *The nature of living matter*. London: Kegan Paul, Trench, Trubner.
Hull, D. (1965, a, b). The effect of essentialism on taxonomy – two thousand years of stasis, Parts 1, 2. *British Journal for the Philosophy of Science*, 15, 314–326, February; 16, 1–18, May.
Hull, D. (1967). Certainty and circularity in evolutionary taxonomy. *Evolution*, 21, 174–18.
Hull, D. (1968). The operational imperative: sense and nonsense in operationism. *Systematic Zoology*, 17, 438–57.
Hull, D. (1972). Reduction in genetics – biology or philosophy? *Philosophy of Science*, 39, 491–9.
Hull, D. (1974). *Philosophy of biological science*. Englewood Cliffs, NJ: Prentice Hall.
Hull, D. (1976). Are species really individuals? *Systematic Zoology*, 25, 174–91.
Hull, D. (1981). Reduction and genetics. *The Journal of Medicine and Biology*, 6, 125–40.
Kitcher, P. (1982). Genes. *British Journal for the Philosophy of Science*, 33, 337–59.
Kitcher, P. (1984). 1953 and all that: a tale of two sciences. *Philosophical Review*, 93, 335–74.
Lewontin, R. C. (1970). The units of selection. *Annual Review of Ecology and Systematics*, 1, 1–18.
Maynard Smith, J. (1976). Group selection. *Quarterly Review of Biology*, 61, 277–83.
Nagel, E. (1949). The meaning of reduction in the natural sciences. In R. C. Stauffer (Ed.). *Science and civilization* (pp. 99–135). Madison: University of Wisconsin Press,
Nagel, E. (1951). Mechanistic explanation and organismic biology. *Philosophy and Phenomenological Research*, 11, 327–38.
Nagel, E. (1961). *The structure of science*. New York: Harcourt, Brace and World.
Nagel, E. (1977). Teleology revisited. *Journal of Philosophy*, 74, 261–301.
Popper, K. (1974). Scientific reduction and the essential incompleteness of all science. In F. Ayala and T. Dobzhansky (Eds). *Studies in the philosophy of biology: reduction and related problems* (pp. 259–84). Berkeley: University of California Press.
Rheinberger, H.-J. (1993). Experiment and orientation: early systems in vitro protein synthesis. *Journal of the History of Biology*, 26, 443–71.
Rheinberger, H.-J. (1997). Experimental complexity in biology: some epistemological and historical remarks. *Philosophy of Science*, 64(4), S245–54.
Rosenberg, A. (1985). *The structure of biological science*. Cambridge/New York: Cambridge University Press.
Ruse, M. (1976). Reduction in genetics. *Boston Studies in the Philosophy of Science*, 32, 631–51.
Sarkar, S. (1989). *Reductionism and molecular biology: a reappraisal*. PhD Dissertation, Department of Philosophy, University of Chicago.
Sarkar, S. (1996). Biological information: a sceptical look at some central dogmas of molecular biology. In S. Sarkar (Ed.). *The philosophy and history of molecular biology* (pp. 187–201). Dordrecht: Kluwer.
Sarkar, S. (2005). *Molecular models of life: philosophical papers on molecular biology*. Cambridge, MA: MIT Press.
Schaffner, K. F. (1967a). Approaches to reduction. *Philosophy of Science*, 34, 137–47.

Schaffner, K. F. (1967b). Antireductionism and molecular biology. *Science*, 157, 644–7.

Schaffner, K. F. (1969). The Watson–Crick model and reductionism. *British Journal for the Philosophy of Science*, 20, 325–48.

Schaffner, K. F. (1976). Reductionism in biology: prospects and problems. *Philosophy of Science Association Proceedings, 1974*, 613–632.

Scriven, M. (1959). Explanation and prediction. *Science*, 130, 477–82.

Sober, E. (1984). *The nature of selection*. Cambridge: MIT Press.

Sober, E. R. (1988). *Reconstructing the past: parsimony, evolution, and inference*. Cambridge, MA: MIT Press, Bradford Books.

Sterelny, K. and Griffiths, P. E. (1999). *Sex and death: an introduction to philosophy of biology*. Chicago: University of Chicago Press.

Wimsatt, W. (1971). Function, organization, and selection. *Zygon: Journal of Religion and Science*, 6, 168–73.

Wimsatt, W. (1972). Teleology and the logical structure of function statements. *Studies in History and Philosophy of Science*, 3, 1–80.

Wimsatt, W. C. (1976). Reductive explanation: a functional account. *Boston Studies in the Philosophy of Science*, 32, 671–710.

Wimsatt, W. (1980). Reductionistic research strategies and their biases in the units of selection controversy. In T. Nickles (Ed.). *Scientific discovery: case studies* (pp. 213–59). Boston: Reidel.

Woodger, J. H. (1929). *Biological principles*. Cambridge: Cambridge University Press.

Woodger, J. H. (1937). *The axiomatic method in biology*. Cambridge: Cambridge University Press.

Woodger, J. H. (1952). *Biology and language*. Cambridge: Cambridge University Press.

Part I

Molecular Biology and Genetics

Chapter 1

Gene Concepts

HANS-JÖRG RHEINBERGER AND
STAFFAN MÜLLER-WILLE

1. Introduction

There has never been a generally accepted definition of the "gene" in genetics. There exist several, different accounts of the historical development and diversification of the gene concept. Today, along with the completion of the human genome sequence and the beginning of what has been called the era of post-genomics, genetics is again experiencing a time of conceptual change, with some even suggesting that the concept of the gene be abandoned altogether. As a consequence, the gene has become a hot topic in philosophy of science around which questions of reduction, emergence, or supervenience are debated. So far, however, all attempts to reach a consensus regarding these questions have failed. The concept of the gene emerging out of a century of genetic research has been and continues to be, as Raphael Falk has reminded us, a "concept in tension" (Falk, 2000).

Yet, despite this apparently irreducible diversity, "there can be little doubt that the idea of 'the gene' has been the central organizing theme of twentieth century biology," as Lenny Moss put it (Moss, 2003, p.xiii; see also Keller, 2001). The layout of this chapter will be largely historical. We will look at genes as epistemic objects. This means that we will not only relate established definitions of the gene, but rather analyze the processes in the course of which they became and still are being determined by changing experimental practices and experimental systems. After having thus established a rich historical panorama of gene concepts, some more general philosophical themes will be addressed, for which the gene has served as a convenient handle in discussion, and which revolve around the topic of reduction.

Before dealing with the historical stages of the gene concept's tangled development, it will be useful to have a short look at its nineteenth-century background. It was only in the nineteenth century that heredity became a major biological problem (Gayon, 2000; López Beltrán, 2004; Müller-Wille & Rheinberger, 2007), and with that the question of the material basis of heredity. In the second half of the nineteenth century, two alternative frameworks were proposed to deal with this question. The first one conceived of heredity as a force the strength of which accumulated over generations, and which, as a measurable magnitude, could be subjected to statistical analysis. This concept was particularly widespread among nineteenth-century breeders (Gayon &

Zallen, 1998) and influenced Francis Galton and the so-called "biometrical school" (Gayon, 1998, pp.105–46). The second saw heredity as residing in matter that was transmitted over the generations. Two major trends in this tradition are to be differentiated here. One of them regarded hereditary matter as particulate and amenable to breeding analysis. Charles Darwin called the presumed hereditary particles gemmules; Hugo de Vries, pangenes; Gregor Mendel, elements. None of these authors, however, associated these particles with a particular hereditary substance. They all thought that hereditary factors consisted of the stuff that the body of the organism is made of. A second category of biologists in the second half of the nineteenth century, to whom Carl Naegeli and August Weismann belonged, distinguished the body substance, the trophoplasm or soma, from a specific hereditary substance, the idioplasm, or germ-plasm, which was assumed to be responsible for intergenerational hereditary continuity. However, they took this idioplasmic substance as being not less particulate, but rather highly organized (Robinson, 1979; Churchill, 1987).

Mendel stands out among these biologists. He is generally considered as the precursor to twentieth-century genetics (see, however, Olby, 1979). As Jean Gayon has argued, his 1866 paper (Mendel, 1866) attacked heredity from a wholly new angle, interpreting it not as a measurable magnitude, as the biometrical school did at a later stage, but as a "structure in a given generation to be expressed in the context of specific crosses." This is why Mendel applied a "calculus of differences," that is, combinatorial mathematics, to the resolution of hereditary phenomena (Gayon, 2000, pp.77–8). With that, he also introduced a new formal tool for the analysis of hybridization experiments: the selection of discrete character pairs (Müller-Wille & Orel, 2007).

2. The Gene in Classical Genetics

The year 1900 is generally considered as the *annus mirabilis* that gave birth to a new discipline: genetics. During that year, three botanists, Hugo de Vries, Carl Correns, and Erich Tschermak, reported on their breeding experiments of the late 1890s and claimed to have confirmed the regularities that Mendel had already presented in his seminal paper (Olby, 1985, pp.109–37). In their experimental crosses with *Zea mays*, *Pisum*, and *Phaseolus*, they observed that the elements responsible for pairs of alternative traits segregated randomly, but in a statistically significant ratio, in the second filial generation (Mendel's law of segregation), and that different pairs of these elements were transmitted independently from each other (Mendel's law of independent assortment). The additional observation, that sometimes several elements behaved as if they were linked, contributed to the hypothesis soon promoted by Walter Sutton and by Theodor Boveri that these elements were located in groups on the different chromosomes of the nucleus. Thus the chromosome theory of inheritance assumed that the regularities of character transmission were grounded in the facts of cytomorphology (Coleman, 1965; Martins, 1999).

Despite initial resistance from the biometrical school (Provine, 1971; MacKenzie & Barnes, 1979) awareness rapidly grew that the possibility of independent assortment of discrete hereditary factors, based on the laws of probability, was to be seen as the very cornerstone of a new "paradigm" of heredity (Kim, 1994). This went together, after an initial period of conflation by the "unit-character fallacy" (Carlson,

1966, ch. 4), with the establishment of a categorical distinction between *genetic factors* on the one hand and *characters* on the other. The masking effect of dominant traits over recessive ones and the subsequent reappearance of recessive traits were particularly instrumental in stabilizing this distinction (Falk, 2001). Toward the end of the first decade of the twentieth century, after William Bateson had coined the term *genetics* for the emerging new field of transmission studies in 1906, Wilhelm Johannsen codified this distinction by introducing the notions of *genotype* and *phenotype*, respectively. In addition, for the elements of the genotype, he proposed the notion of *gene*.

Johannsen's distinction has profoundly marked all of twentieth-century genetics (Allen, 2002). We can safely say that it instituted the *gene as an epistemic object to be studied within its proper epistemic space*, and with that an "exact, experimental doctrine of heredity" (Johannsen, 1909, p.1) which concentrated on transmission only and not on the function and development of the organism in its environment. Some historians have spoken of a "divorce" of genetical from embryological concerns because of this separation (Allen, 1986; Bowler, 1989). Others hold that this separation was itself an expression of the embryological interests of early geneticists in their search for "developmental invariants" (Gilbert, 1978; Griesemer, 2000). Be that as it may, the result was that the relations between the two spaces, once separated by abstraction, were now experimentally elucidated in their own right (Falk, 1995). Michel Morange judged this "rupture to be logically absurd, but historically and scientifically necessary" (Morange, 1998, p.22).

Johannsen himself stressed that the genotype had to be treated as independent of any life history and thus as an "ahistoric" entity amenable to scientific scrutiny like the objects of physics and chemistry (Johannsen, 1911; see Churchill, 1974; Roll-Hansen, 1978a). Unlike most Mendelians, however, he remained convinced that the genotype would possess an overall architecture. He therefore had reservations with respect to its particulate character, and especially warned that the notion of "genes for a particular character" should always be used cautiously if not altogether be omitted (cf. Moss, 2003, p.29). Johannsen also clearly recognized that the experimental regime of Mendelian genetics neither required nor allowed any definite supposition about the material structure of the genetic elements. For him, the gene remained a concept "completely free of any hypothesis" (Johannsen, 1909, p.124).

On this account, genes were taken as the abstract elements of an equally abstract space whose structure, however, could be explored through the visible and quantifiable outcome of breeding experiments based on mutations of model organisms. This became the research program of Thomas Hunt Morgan and his group. From the early 1910s into the 1930s, the growing community of researchers around Morgan and their followers used mutants of the fruit fly *Drosophila melanogaster* in order to produce a map of the fruit fly's genotype in which genes, and alleles thereof, figured as genetic markers which occupied a particular locus on one of the four homologous chromosome pairs of the fly (Kohler, 1994). The basic assumptions that allowed the program to operate were that genes were located in a linear fashion on the chromosomes, and that the frequency of recombination events between homologous chromosomes gave a measure of the distance between the genes, at the same time defining them as units of recombination (Morgan et al., 1915). In this practice, identifiable aspects of the phenotype, assumed to be determined directly by genes, were used as indicators or "windows" for an outlook on the formal structure of the genotype. This is what Moss has termed the "Gene-P" (P standing for phenotype).

Throughout his career, Morgan remained aware of the formal character of his program (Morgan, 1935, p.3). In particular, it did not matter if one-to-one, or more complicated relationships reigned between genes and traits. Morgan and his school were well aware that, as a rule, many genes were involved in the development of a particular trait, and that one gene could affect several characters. To accommodate this difficulty and in line with their experimental regime, they embraced a differential concept of the gene. What mattered to them was the relationship between a change in a gene and a change in a trait, rather than the nature of these entities themselves. Thus the alteration of a trait could be causally related to a change in (or a loss of) a single genetic factor, even if it was plausible in general that a trait like eye-color was, in fact, determined by a whole group of variously interacting genes (Roll-Hansen, 1978b; Schwartz, 2000).

The fascination of this approach consisted in the fact that it worked, if properly conducted, like a precision instrument. Population geneticists like Ronald A. Fisher, J. B. S. Haldane, and Sewall Wright could make use of that same abstract gene concept in developing elaborate mathematical models describing the effects of evolutionary factors on the genetic composition of populations. As a consequence, evolution became re-defined as a change of gene frequencies in the gene pool of a population in what is commonly called the "evolutionary," "neo-Darwinian," or simply "modern synthesis" of the late 1930s (Dobzhansky, 1937) [SEE DARWINISM AND NEO-DARWINISM]. Considered as a "developmental invariant" (Griesemer, 2000), and solely obeying the Mendelian laws in its transmission from one generation to the next, the gene provided a kind of inertia principle against which the effects of both developmental (epistasis, inhibition, position effects, etc.) and evolutionary factors (selection, mutation, recombination, etc.) could be measured with utmost accuracy, assessed and accurately quantified (Gayon, 1995).

Nevertheless, it became the conviction of many geneticists in the 1920s, among them Morgan's student, Herman J. Muller, that genes had to be material particles. Muller saw genes as endowed with two properties: that of *autocatalysis* and that of *heterocatalysis*. Their autocatalytic function allowed them to reproduce as units of transmission and thus to connect the genotype of one generation to that of the next. Their heterocatalytic capabilities connected them to the phenotype, as functional units involved in the expression of a particular character. With his own experimental work, Muller added a significant argument for the materiality of the gene, pertaining to a third property of the gene, its susceptibility to mutations. In 1927, he reported on the induction of Mendelian mutations in *Drosophila* by using X-rays. He concluded that the X-rays must have altered some molecular structure in a permanent fashion. But the experimental practice of X-raying, which eventually gave rise to a whole "industry" of radiation genetics in the 1930s and 1940s, did not by itself open the path to the material characterization of genes as units of heredity (Muller, 1951, pp.95–6).

Meanwhile, cytological work had also added credence to the materiality of genes, residing on chromosomes. During the 1930s, the cytogeneticist, Theophilus Painter, correlated formal patterns of displacement of genetic loci on Morganian chromosome maps with visible changes in the banding pattern of giant salivary gland chromosomes of *Drosophila*. Barbara McClintock was able to follow with her microscope the changes

– translocations, inversions and deletions – induced by X-rays in the chromosomes of *Zea mays* (maize) Corn. Simultaneously, Alfred Sturtevant, in his experimental work on the Bar eye effect in *Drosophila* at the end of the 1920s, had shown what came to be called a *position effect*: the expression of a mutation was dependent on the position of the corresponding gene on the chromosome. This finding stirred wide-ranging discussions about the heterocatalytic aspect of a gene. If a gene's function depended on its position on the chromosome, it became questionable whether that function was stably connected to that gene at all, or as Richard Goldschmidt had assumed, whether physiological function was not determined by the organization of the genetic material (Goldschmidt, 1940; see also Dietrich, 2000).

Thus far, all experimental approaches in the new field of genetics had remained silent with respect to the two basic Mullerian aspects of the gene: its autocatalytic and its heterocatalytic function. Toward the end of the 1930s, Max Delbrück had the intuition that the question of autocatalysis, that is, replication, could be attacked through the study of phage. But the phage system, which he established throughout the 1940s, remained as formal as that of classical *Drosophila* genetics. Around the same time, Alfred Kühn and his group, as well as Boris Ephrussi and George Beadle, using organ transplantations between mutant and wild type insects, opened a window on the space between the gene and its presumed physiological function. Studying the pigmentation of insect eyes, they realized that genes did not directly give rise to physiological substances, but that they obviously first initiated what Kühn termed a "primary reaction" leading to ferments or enzymes, which in turn catalyzed particular steps in metabolic reaction cascades.

Kühn viewed his experiments as the beginning of a reorientation of what he perceived to be the preformationism of transmission genetics of his day. He pleaded for an epigenetics that would combine genetic, developmental, and physiological analyses to define heterocatalysis as the result of an interaction of two reaction chains, one leading from genes to particular ferments, and the other leading from one metabolic intermediate to the next by the intervention of these ferments, thus resulting in complex epigenetic networks (Kühn, 1941, p.258; Rheinberger, 2000a). On the other side of the Atlantic, George Beadle and Edward Tatum, working with cultures of *Neurospora crassa*, codified the first of these relations into the one-gene–one-enzyme hypothesis. But for Kühn, as well as to Beadle and Tatum, the material character of genes and the way these putative entities gave rise to primary products remained elusive and beyond the reach of experimental analysis.

The gene in classical genetics was already far from being a simple concept corresponding to a simple entity. Conceiving of the gene as a unit of transmission, recombination, mutation, and function, classical geneticists combined various aspects of hereditary phenomena. Well into the 1940s, only proteins were thought to be complex enough to perform these tasks. But owing to the lack of knowledge about the material nature of the gene, gene conceptions remained largely formal and operationalist, i.e., were substantiated indirectly by the successes achieved in explaining and predicting experimental results. This lack of a synthetic understanding of the gene notwithstanding, the mounting successes of the various research strands associated with classical genetics led to a "hardening" of the belief in the gene as a discrete, material entity (Falk, 2000, pp.323–6).

3. The Gene in Molecular Genetics

The enzyme view of gene function, as envisaged by Kühn and by Beadle and Tatum, gave the idea of genetic specificity a new twist and helped to pave the way to the molecularization of the gene. The same can be said about the findings of Oswald Avery and his colleagues in the early 1940s. They purified the deoxyribonuleic acid (DNA) of one strain of bacteria, and demonstrated that it was able to transmit the infectious characteristics of that strain to another, harmless one. Yet the historical path that led to an understanding of the nature of the molecular gene was not a direct follow-up of classical genetics. It was rather embedded in an overall molecularization of biology driven by the application of newly developed physical and chemical methods and instruments to problems of biology. Among these methods were ultracentrifugation, X-ray crystallography, electron microscopy, electrophoresis, macromolecular sequencing, and radioactive tracing. The transition also relied upon use of comparatively simple model organisms like unicellular fungi, bacteria, viruses, and phage. A new culture of physically and chemically instructed *in vitro* biology ensued, which in large part no longer rested on the presence of intact organisms in a particular experi8mental system (Rheinberger, 1997).

For the development of molecular genetics in the narrow sense, three lines of experimental inquiry proved to be crucial. They were not connected to each other when they gained momentum in the late 1940s, but they happened to merge at the beginning of the 1960s, giving rise to a grand new picture. The first of these developments was the elucidation of the structure of DNA as a macromolecular double helix by Francis Crick and James D. Watson in 1953. This work was based on chemical information about base composition of the molecule provided by Erwin Chargaff, on data from X-ray crystallography produced by Rosalind Franklin and Maurice Wilkins, and on mechanical model building as developed by Linus Pauling. The result was a picture of a nucleic acid double strand, the four bases (**A**denine, **T**hymine, **G**uanine, **C**ytosine) of which formed complementary pairs (A-T, G-C) that could be arranged in all possible combinations into linear sequences. At the same time, that molecular model suggested an elegant mechanism for the duplication of the molecule. Opening the strands and synthesizing two new strands complementary to each would suffice to create two identical helices from one. Thus, the structure of the DNA double helix had all the characteristics that were to be expected from a molecule serving as an autocatalytic hereditary entity (Chadarevian, 2002).

The second line of experiment that formed molecular genetics was the *in vitro* characterization of the process of protein biosynthesis to which many biochemical researchers contributed, among them Paul Zamecnik, Mahlon Hoagland, Paul Berg, Fritz Lipmann, Marshall Nirenberg, and Heinrich Matthaei. It started in the 1940s largely as an effort to understand the growth of malignant tumors. During the 1950s, it became evident that the process required a ribonucleic acid (RNA) template that was originally thought to be part of the microsomes on which the assembly of amino acids was seen to take place. It turned out that the process of amino acid condensation was mediated by a transfer molecule with the characteristics of a nucleic acid *and* the capacity to carry an amino acid. The ensuing idea that it was a linear sequence of ribonucleic acid derived from one of the DNA strands that directed the synthesis of a linear sequence of amino acids, or a polypeptide, and that this process was mediated by an adaptor molecule, was soon corroborated experimentally. The relation between these two classes of molecules was found to be ruled by a nucleic acid triplet *code*: three bases at a time specified one amino acid (Rheinberger, 1997; Kay, 2000). Hence, the *sequence hypothesis* and the *Central Dogma* of molecular biology, which Francis Crick formulated at the end of the 1950s:

> In its simplest form [the sequence hypothesis] assumes that the specificity of a piece of nucleic acid is expressed solely by the sequence of its bases, and that this sequence is a (simple) code for the amino acid sequence of a particular protein. [The central dogma] states that once 'information' has passed into protein *it cannot get out again*. In more detail, the transfer of information from nucleic acid to nucleic acid, or from nucleic acid to protein may be possible, but transfer from protein to protein, or from protein to nucleic acid is impossible. (Crick, 1958, pp.152–3)

With these two fundamental assumptions, a new view of biological specificity came into play (Sarkar, 1996). In its center stands the transfer of molecular order from one macromolecule to the other. In one molecule the order is preserved structurally; in the other it becomes expressed and provides the basis for a biological function carried out by a protein. This transfer process became characterized as molecular *information transfer* [SEE BIOLOGICAL INFORMATION]. Henceforth, genes could be seen as stretches of deoxyribonucleic acid (or ribonucleic acid in certain viruses) carrying the information for the assembly of a particular protein. Both molecules were thus thought to be colinear. In the end, both the fundamental properties that Muller had required of genes, namely autocatalysis and heterocatalysis, were perceived as relying on one and the same stereochemical principle, respectively: The base complementarity between nucleic acid building blocks C-G and A-T (U in the case of RNA) was responsible both for the faithful duplication of genetic information in the process of *replication*, and, via the genetic code, for the transformation of genetic information into biological function through *transcription* and *translation*. The code, as well as the mechanisms of transcription and translation, turned out to be nearly universal for all living beings. The genotype was thus reconfigured as a universal repository of genetic information, sometimes also addressed as a *genetic program*. Talk of DNA as embodying genetic "information," as being the "blueprint of life," which governs public discourse to this day, emerged from a peculiar conjunction of the physical and the life sciences during World War II, with Erwin Schrödinger's *What is Life?* as a source of inspiration (Schrödinger, 1944), and cybernetics, a discipline engaged in the study of complex systems and their self-regulation. It needs to be stressed, however, that initial attempts to "crack" the DNA code by purely cryptographic means soon ran into a dead end. In the end it was biochemists who unraveled the genetic code by the advanced tools of their discipline (Judson, 1996; Kay, 2000).

For the further development of the notion of DNA as a "program," we have to consider an additional third line of experiment, aside from the elucidation of DNA structure and the mechanisms of protein synthesis. This line of experiment came out of a fusion of bacterial genetics with the biochemical characterization of an inducible system of sugar metabolizing enzymes. It was largely the work of François Jacob and Jacques Monod and led, at the beginning of the 1960s, to the identification of messenger RNA as the mediator between genes and proteins, and to the description of a regulatory model of gene activation, the so-called operon model, in which two classes of genes became distinguished: One class was the *structural genes*. They were presumed to carry the "structural information" for the production of particular polypeptides. The other class was the *regulatory genes*. They were assumed to be involved in the regulation of the expression of structural information. A third element of DNA involved in the regulatory loop of an operon was a binding site, or *signal sequence*, that was not transcribed at all. These three elements, structural genes, regulatory genes, and signal sequences, provided the framework for viewing the genotype as an ordered, hierarchical system, as a "genetic program," as Jacob contended, not without adding that it was a very peculiar program, namely one that needed its own products for being executed (Jacob, 1976, p.297). If we take that view seriously, although the whole conception looks like a circle (Keller, 2000), it is in the end the organism which interprets or "recruits" the structural genes by activating or inhibiting the regulatory genes that control their expression.

The operon model of Jacob and Monod marked the precipitous end of the simple, informational concept of the molecular gene. Since the beginning of the 1960s, the picture of gene expression has become vastly more complicated (see Rheinberger, 2000b, and GENOMICS AND PROTEOMICS). Moreover, most genomes of higher organisms appear to contain huge DNA stretches to which no function can as yet be assigned. Finally, the "non-coding," but functionally specific, regulatory DNA-elements have proliferated: There exist promoter and terminator sequences; upstream and downstream activating elements in transcribed or non-transcribed, translated or untranslated regions; leader sequences; externally and internally transcribed spacers before, between, and after structural genes; interspersed repetitive elements and tandemly repeated sequences such as satellites, LINEs (long interspersed sequences), and SINEs (short interspersed sequences) of various classes and sizes (for an overview see Fischer, 1995).

As far as transcription, i.e., the synthesis of an RNA copy from a sequence of DNA, is concerned, overlapping reading frames have been found on one and the same strand of DNA, and protein coding stretches have been found to derive from both strands of the double helix. On the level of modification after transcription, the picture has become equally complicated. Soon it was realized that DNA transcripts such as transfer RNA and ribosomal RNA had to be trimmed and matured in a complex enzymatic manner to become functional molecules, and that messenger RNAs of eukaryotes underwent extensive post-transcriptional modification before they were ready to go into the translation machinery. In the 1970s, to the surprise of everybody, molecular biologists had to acquaint themselves with the idea that eukaryotic genes were composed of modules, and that, after transcription, *introns* were cut out and *exons* spliced together in order to yield a functional message. The gene-in-pieces was one of the first major scientific offshoots of recombinant DNA technology, and this technology has since continued to be useful for exploring unanticipated vistas on the genome. A spliced messenger sometimes may comprise a fraction as little as 10 percent or less of the primary transcript. Since the late 1970s, molecular biologists have become familiar with various kinds of *RNA splicing*: autocatalytic self-splicing, alternative splicing of one single transcript to yield different messages; and even trans-splicing of different primary transcripts to yield one hybrid message. Finally, yet another mechanism, or rather, class of mechanisms has been found to operate on the level of RNA transcripts. It is called *messenger RNA editing*. In this case, the original transcript is not only cut and pasted, but its nucleotide sequence is systematically altered after transcription. The nucleotide replacement happens before translation starts, and is mediated by various RNAs and enzymes that excise old and insert new nucleotides in a variety of ways to yield a product that is no longer complementary to the DNA stretch from which it was originally derived, and a protein that is no longer co-linear with the DNA sequence in the classical molecular biological definition.

The complications with the molecular biological gene continue on the level of translation, i.e., the synthesis of a polypeptide according to the sequence of triplets of the mRNA molecule. There are findings such as translational starts at different start codons on one and the same messenger RNA; instances of obligatory frame shifting within a given message; post-translational protein modification such as removing amino acids from the amino terminus of the translated polypeptide. Another phenomenon called *protein splicing* has been observed in the past few years. Here, portions of the original translation product have to be cleaved and joined together in a new order before yielding a functional protein. And finally, a more recent development from the translational field is that a ribosome can manage to translate two different messenger RNAs into one single polypeptide. François Gros, after a lifetime of research in molecular biology, has come to the rather paradoxically sounding conclusion that in view of this perplexing complexity, the "exploded gene" – *le gène éclaté* – could be specified, if at all, then only by "the products that result from its activity," that is, the functional molecules to which they give rise (Gros, 1991, p.297). But it appears difficult to follow Gros' advice of such a reverse definition, as the phenotype would then simply be used to define the genotype.

As Falk (2000) has argued, on the one hand, the autocatalytic property once attributed to the gene as a unit of replication has been relegated to the DNA at large. It can no longer be taken as being specific for the gene as such. After all, the process of DNA replication is not punctuated by the boundaries of coding regions. On the other hand, as many observers of the scene have remarked (Kitcher, 1982; Gros, 1991; Morange, 1998; Portin, 1993; Fogle, 2000), it has become ever harder to define clear-cut properties of a gene as a heterocatalytic entity. It has become a matter of choice as to which sequence elements are to be included and which ones excluded. There have been different reactions to this situation.

Scientists like Thomas Fogle and Michel Morange concede that there is no longer a precise definition of what could count as a gene. However, they continue to talk about genes in a contextual, generic, and pragmatic manner (Fogle, 2000; Morange, 2000). Elof Carlson and Petter Portin have also concluded that the present gene concept is abstract, general, and open, despite, or perhaps because of, present knowledge on the structure and organization of the genetic material having become so comprehensive and so detailed. But they, like Richard Burian (1985), take open concepts with a large reference potential not as a deficit to live with, but as a potentially productive tool in science. Such concepts offer options and leave choices open (Carlson, 1991; Portin, 1993). Philosopher Philip Kitcher, as a consequence of all the molecular data concerning the gene, some 20 years ago already drew the ultraliberal conclusion that "there is no molecular biology of the gene. There is only molecular biology of the genetic material" (Kitcher, 1982, p.357).

Consequently, there are those who take the heterocatalytic variability of the gene as an argument to treat genes no longer as fundamental units in their own right, but rather as a developmental resource. They claim that the time has come, if not to dissolve, then at least to embed genetics in development and even development in reproduction (Griesemer, 2000), and pick up the thread where Kühn and others left it half a century ago. Consequently, Moss defines "gene-D" as a "developmental resource (hence the D), which in itself is *indeterminate* with respect to phenotype. To be a gene-D is to be a transcriptional unit on a chromosome, within which are contained molecular template resources" (Moss, 2003, p.46). On this view, genetic templates constitute only one reservoir on which the developmental process draws and are not ontologically privileged as hereditary molecules.

With molecular biology, the classical gene "went molecular" (Waters, 1994). Ironically, the initial idea of genes as simple stretches of DNA coding for a protein dissolved in this process. Together with the material structure, which the classical gene acquired through molecular biology, biochemical mechanisms accounting for the transmission and expression of genes proliferated. The development of molecular biology itself, that enterprise so often described as an utterly reductionist conquest, has made it impossible to think of the genome any longer simply as a set of pieces of contiguous DNA co-linear with the proteins derived from them and each of them endowed with a specific function. When the results of the Human Genome Project were presented on the fiftieth anniversary of the double helix, molecular genetics seemed to have accomplished a full circle, readdressing reproduction and inheritance no longer from a purely genetic, but from an evolution *cum* development perspective.

4. The Gene in Evolution and Development

One of the more spectacular events in the history of twentieth-century biology as a discipline, triggered by the rise of genetics, was the so-called "modern evolutionary synthesis." In a whole series of textbooks, published by evolutionary biologists like Theodosius Dobzhansky, Ernst Mayr, and Julian S. Huxley, the results of population genetics were used to re-establish Darwinian, selectionist evolution. After the "eclipse of Darwinism," which had reigned around 1900 (Bowler, 1983), neo-Darwinism once again provided a unifying, explanatory framework for biology that also included the more descriptive, naturalist disciplines like systematics, biogeography, or paleontology (Provine, 1971; Mayr & Provine, 1980).

Scott Gilbert (2000) has singled out six aspects of the notion of the gene as it had been used in population genetics up to the modern evolutionary synthesis. First, it shared with the classical gene in the Morganian sense that it was an abstraction, an entity that had to fulfill formal requirements, but that did not need to be and indeed was not materially specified. Second, the evolutionary gene had to result in or had to be correlated with some phenotypic difference that could be "seen" or targeted by selection. Third, and by the same token, the gene of the evolutionary synthesis was the entity that was ultimately responsible for selection to occur and last between organisms. Fourth, the gene of the evolutionary synthesis was largely equated with what molecular biologists came to call "structural genes." Fifth, the gene was expressed in an organism competing for reproductive advantage. Sixth, and finally, the gene was seen as a largely independent unit. Richard Dawkins has taken this last aspect to its extreme by defining the gene as a "selfish" replicator competing with its fellow genes and using the organism as an instrument for its own survival (Dawkins, 1976).

Molecular biology, with higher organisms moving center stage during the past three decades, has made a caricature of this kind of evolutionary gene, and has presented to us genes and whole genomes as complex systems not only allowing for evolution to occur, but being themselves subjected to a vigorous process of evolution. The genome in its entirety has taken on a more and more flexible and dynamic configuration. The mobile genetic elements, characterized by McClintock more than half a century ago in *Zea mays*, have gained currency as *transposons* that can be regularly and irregularly excised and inserted all over bacterial and eukaryotic genomes. There are also other forms of shuffling that occur at the DNA level. A large amount of somatic gene tinkering and DNA splicing, for instance, is involved in organizing the immune response in higher organisms. This gives rise to the production of potentially millions of different antibodies. No genome would be large enough to cope with such a task if the parceling out of genes and a sophisticated permutation of their parts had not been invented during evolution. Gene families have arisen from duplication over time, containing silenced genes (sometimes called pseudogenes). Genes themselves appear to have largely arisen from modules by combination. We find jumping genes; and multiple genes of one sort giving rise to a genetic polymorphism on the DNA itself coding for different protein isoforms. In short, there appears to be a whole battery of mechanisms and entities that constitute what could be called a respiratory, or breathing, genome.

Molecular evolutionary biologists have barely started to understand this flexible genetic apparatus. It has become evident that the genome is a dynamic body of ancestrally tinkered pieces and forms of genetic iteration (Jacob, 1977). Genome sequencing combined with intelligent sequence data comparison may bring out more of this structure in the near future. If there is a chance to understand evolution beyond the classical, largely formal, evolutionary synthesis, it is from the perspective of learning more about the genome as a *dynamic* and *modular* configuration. The purported elementary events on which this complex machinery operates, such as point mutations, nucleotide deletions, additions, and oligonucleotide inversions, are no longer the only elements of the evolutionary process, but only one component in a much wider arsenal of *DNA tinkering*. The replication process, that is, the transmission aspect of genetics as such, has revealed itself to be a complicated molecular process whose versatility, far from being restricted to gene shuffling during meiotic recombination, constitutes a reservoir for evolution and is run by a highly complex molecular machinery including polymerases, gyrases, DNA binding proteins, repair mechanisms, and more. Genomic differences, targeted by selection, can be, but must not become, "compartmented into genes" during evolution, as Peter Beurton has put it (Beurton, 2000, p.303). Under this perspective, the gene is no longer to be seen as the unit of evolution, but rather as its late product, the eventual result of a long history of genomic condensation.

We have come a long way with molecular biology from genes to genomes. But there is still a longer way to go from genomes to organisms. The developmental gene, as described in the work of Ed Lewis and Antonio Garcia-Bellido, and from later work by Walter Gehring, Christiane Nüsslein-Volhard, Eric Wieschaus, Peter Gruss, Denis Duboule, and others, allows us possibly to go a step along on this way. As Gilbert (2000) argues, it is the exact counterpart to the gene of the evolutionary synthesis. But we need to be more specific and to direct attention to what has been termed "developmental genes." As it turned out, largely from an exhaustive exploitation of mutation saturation and genetic engineering technologies, fundamental processes in development such as segmentation or eye formation in such widely different organisms as insects and mammals are decisively influenced by the activation and inhibition of a class of regulatory genes that to some extent resemble the regulator genes of the operon model.

But in contrast to these long-known regulatory genes, whose function rests on their ability to be switched on and off according to the requirements of actual metabolic and environmental situations, developmental genes initiate irreversible processes. They code for so-called transcription factors which can bind to control regions of DNA and thus influence the rate of transcription of a particular gene or a whole set of genes at a particular stage of development. Among them are what we could call developmental genes of a second order which appear to control and modulate the units gated by the developmental genes of the first order. They act as a veritable kind of master switch and have been found to be highly conserved throughout evolution. An example is a member of the *pax*-gene family that can switch on a whole complex process such as eye formation from insects to vertebrates. Most surprisingly, the homologous gene isolated from the mouse can replace the one present in *Drosophila*, and when placed in the fruit fly, switch on, not mammalian eye formation, but insect eye formation. Many of these genes or gene families, like the *homeobox* family, are thought to be involved in the generation of spatial patterning during embryogenesis as well as in its temporal patterning.

Morange (2000) distinguishes two central "hard facts" that can be retained from this highly fluid and contested research field. The first is that the regulatory genes appear to play a central role in development as judged from the often drastic effects resulting from their inactivation. And second, it appears that not only have particular homeotic genes been highly conserved between distantly related organisms, but they tend to come in complexes which have themselves been structurally conserved throughout evolution, thus once more testifying to genomic higher-order structures. Another class of such highly conserved genes and gene complexes is involved in the formation of components of pathways that bring about intracellular and cell-to-cell signaling. These processes are of obvious importance for cellular differentiation and for embryonic development of multicellular organisms.

One of the big surprises of the extensive use of the technology of targeted gene knockout has been that genes thought to be indispensable for a particular function, when knocked out, did not alter or at least not significantly alter the organism's performance. This made developmental molecular biologists aware that the networks of development appear to be largely redundant. These networks are highly buffered and thus robust to a considerable extent with respect to changing external and internal conditions. Gene products are of course involved in these networks and their complex functions, but these functions are by no means specified by the genes alone. Another result, coming from embryonic gene expression studies with recently developed chip technologies, was that one and the same gene product can be expressed at different stages of development and in different tissues, and that it can be implicated in quite different metabolic and cellular functions.

These recent results seriously call into question the further applicability of straightforward "gene-for" talk. Highly conserved in evolution, yet highly redundant and variable in function, developmental genes rather look like molecular building blocks with which evolution tinkers in constructing organisms (Jacob, 1977; Morange, 2000) than like the pieces of DNA with a determinate function as originally envisioned by molecular genetics. The discovery of developmental genes throws light on the way in which the genome as a whole is organized as a dynamic, modular, and robust entity.

5. Conclusion: Genes, Genomics, and Reduction

As we argued in the preceding sections, the history of twentieth-century genetics is characterized by a proliferation of methods for the individuation of genetic components, and, accordingly, by a proliferation of gene definitions. These definitions appear to be largely technology-dependent. Major conceptual changes did not precede, but followed, experimental breakthroughs. Especially the contrast of the "classical" and the "molecular" gene, the latter succeeding the former chronologically, has raised issues of how such alternative concepts relate semantically, ontologically, and epistemologically. Understanding these relations might offer a chance to convey some order to the bewildering variety of meanings inscribed in the concept of the gene in the course of a long century.

In a now classical paper, Kenneth Schaffner argued that molecular biology – the Watson–Crick model of DNA in particular – effected a reduction of the laws of (classical) genetics to physical and chemical laws (Schaffner, 1969, p.342). The successes of molecular biology in identifying DNA as the genetic material – as Watson's and Crick's discovery of the DNA structure or the Meselson-Stahl experiment on the semi-conservative replication of DNA – lend empirical support, according to Schaffner, "for reduction functions involved in the reduction of biology as: $gene_1$ = DNA $sequence_1$." Schaffner's account was criticized by David Hull, who pointed out that relations between Mendelian and molecular terms are "many–many," not "one–one" or "many–one" relations as assumed by Schaffner, because "phenomena characterized by a single Mendelian predicate term can be reproduced by several types of molecular mechanisms [. . . and] conversely, the same type of molecular mechanism can produce phenomena that must be characterized by different Mendelian predicate terms" (Hull, 1974, p.39). "To convert these many–many relations," Hull concluded, "into the necessary one–one or many–one relations leading from molecular to Mendelian terms, Mendelian genetics must be modified extensively. Two problems then arise – the justification for terming these modifications 'corrections' and the transition from Mendelian to molecular genetics 'reduction' rather than 'replacement'" (Hull, 1974, p.43). To account for this difficulty and accommodate the intuition (which Hull shared) that there should be at least some way in which it makes sense to speak of a reduction of classical to molecular genetics, Alexander Rosenberg adopted the notion of supervenience (coined by Donald

Davidson and going back to George Edward Moore) to describe the relation of classical to molecular genetics. Supervenience implies that any two items that share the same properties in molecular terms also have the same properties in Mendelian terms, without, however, entailing a commitment that Mendelian laws must be deducible from the laws of biochemistry (Rosenberg, 1978). This recalls the way in which classical geneticists related gene differences and trait differences in the differential gene concept, where trait differences were used as markers for genetic differences without implying a deducibility of trait behavior, the dominance or recessivity of traits in particular, from Mendelian laws (Schwartz, 2000; Falk, 2001). Interestingly, Kenneth Waters has argued on this basis, and against Hull, that the complexity that was revealed by molecular genetics was simply the complexity already posited by some classical geneticists (Waters, 1994, 2000).

The literature on genetics and reductionism has meanwhile become almost as variegated and complex as the field of scientific activities it attempts to illuminate. In his book-length, critical assessment of that literature, Sahotra Sarkar made an interesting move by distinguishing five different concepts of reduction, of which he considers three to be particularly relevant to genetics: "weak reduction," exemplified by the notion of heritability; "abstract hierarchical reduction," exemplified by classical genetics; and "approximate strong reduction," exemplified by the use of "information"-based explanations in molecular genetics. The perhaps not so surprising result is that "reduction – in its various types – is scientifically interesting beyond, especially, the formal concerns of most philosophers of sciences" in that it constitutes a "valuable, sometimes exciting, and occasionally indispensable strategy in science" and thus needs to be acknowledged as being ultimately "related to the actual practice of genetics" (Sarkar, 1998, p.190). In a similar vein, Jean Gayon has expounded a "philosophical scheme" for the history of genetics which treats phenomenalism, instrumentalism, and realism not as alternative systems that philosophers have to decide upon, but as actual, historically consecutive strategies employed by geneticists in their work (Gayon, 2000).

We would finally like to address briefly two issues that are related to the problem of reduction and have occasioned repeated discussion in the philosophical literature. The first point concerns the notion of "information" in molecular genetics. The early molecular uses of the terms "genetic information" and "genetic program" have been widely criticized by philosophers and historians of science alike (Sarkar, 1996; Kay, 2000; Keller, 2000). No one less than Gunther Stent, one of the strongest proponents of what has been termed the "informational school" of molecular biology, warned long ago that talk about "genetic information" is best confined to its explicit and explicable meaning of sequence specification, that is, that it is best to keep it in the local confines of "coding" instead of scaling it up to a global talk of genetic "programming." "It goes without saying," he contends, "that the principles of chemical catalysis [of an enzyme] are not represented in the DNA nucleotide base sequences," and he concludes:

> After all, there is no aspect of the phenomena to whose determination the genes cannot be said to have made their contribution. Thus it transpires that the concept of genetic information, which in the heyday of molecular biology was of such great heuristic value for unraveling the structure and function of the genes, i.e., the explicit meaning of that information, is no longer so useful in this later period when the epigenetic relations which

remain in want of explanation represent mainly the implicit meaning of that information. (Stent, 1977, p.137)

However, it appears to us that one should remain aware of the fact that the molecular biological notion of a flow of information, both in terms of storage and expression in the interaction between two classes of macromolecules, has added a dimension of talking about living systems that helps to distinguish them specifically from chemical and physical systems characterized solely by flows of matter and flows of energy (Crick, 1958; Maynard Smith, 2000). Molecular biology, seen by many historians and philosophers of biology as a paragon of reductionism, not only introduced physics and chemistry into biology, or even reduced the latter to the former two, but – paradoxically – also helped to find a way of conceiving of organisms in a fundamentally non-reducible manner. In a broader vision, this implies "epigenetic" mechanisms of intracellular and intercellular molecular signaling and communication in which genetic information and its differential expression is embedded and through which it is contextualized. On this view, it appears not only legitimate, but heuristically productive to conceive of the functional networks of living beings in a biosemiotic terminology instead of a simply mechanistic or energetic idiom (Emmeche, 1999).

The second point concerns the already mentioned "gene-for" talk. Why has talk about genes coding for this and that become so entrenched? Why do genes still appear as the ultimate determinants and executers of life? As we have seen in the preceding two sections, the advances in conceptualizing processes of organismic development and evolution have thoroughly deconstructed the view of genes as it dominated classical genetics and the early phases of molecular genetics. Why is it, to use the formulation of Moss, that genetics is still "understood not as a practice of instrumental reductionism but rather in the constitutive reductionist vein" implying the "ability to account for the production of the phenotype on the basis of the genes" (Moss, 2003, p.50)? A recent empirical study by Paul Griffiths and Karola Stotz on how biologists conceptualize genes comes to the conclusion "that the classical molecular gene concept continues to function as something like a stereotype for biologists, despite the many cases in which that conception does not give a principled answer to the question of whether a particular sequence is a gene" (Stotz et al., 2004, and Griffiths, in press). Waters provides a surprising but altogether plausible epistemological answer to this apparent conundrum (Waters, 2004). He reminds us that in the context of scientific work and research, genes are first and foremost handled as entities of epistemological rather than ontological value. It is on the grounds of their epistemic function in research that they appear so privileged. Waters deliberately goes beyond the question of reductionism or antireductionism that has structured so much philosophical work on modern biology, especially on genetics and molecular biology over the past decades. He stresses that the successes of a gene-centered view of the organism are not due to the fact that genes are the major determinants of the main processes in living beings. Rather, they figure so prominently because they provide *highly successful entry points for the investigation of these processes*. The success of gene-centrism, according to this view, is not ontologically, but first and foremost epistemologically grounded.

From this, two major conclusions result: first, that it is the structure of investigation rather than an encompassing system of explanation that has grounded the scientific success of genetics; and second, that the essential incompleteness of genetic explanations, whenever they are meant to be located at the ontological level, calls for the promotion of a scientific pluralism (2006). Complex objects of investigation such as organisms cannot be successfully understood by a single best account or description, and any experimental science advances through the construction of successful models. Whether and how long these models will continue to be gene-based remains an open question. In any case, however, it will be contingent on the outcome of future research not on a presupposed, ontology of life.

Acknowledgement

This chapter is based on an entry we wrote for *The Stanford Encyclopedia for Philosophy* (Winter 2001 edition).

References

Allen, G. (1986). T. H. Morgan and the split between embryology and genetics, 1910–1926. In T. Horder, I. A. Witkowski, & C. C. Wylie (Eds). *A history of embryology* (pp. 113–44). Cambridge: Cambridge University Press.

Allen, G. (2002). The classical gene: its nature and its legacy. In L. S. Parker & R. A. Ankeny (Eds). *Mutating concepts, evolving disciplines: genetics, medicine and society* (pp. 13–42). Dordrecht: Kluwer Academic Press.

Beurton, P. (2000). A unified view of the gene, or how to overcome reductionism. In P. Beurton, R. Falk, & H.-J. Rheinberger (Eds). *The concept of the gene in development and evolution: historical and epistemological perspectives* (pp. 286–314). Cambridge: Cambridge University Press.

Bowler, P. (1983). *The eclipse of Darwinism*. Baltimore: Johns Hopkins University Press.

Bowler, P. (1989). *The Mendelian revolution: the emergence of hereditarian concepts in modern science and society*. Baltimore: Johns Hopkins University Press.

Burian, R. M. (1985). On conceptual change in biology: the case of the gene. In D. J. Depew & B. H. Weber (Eds). *Evolution at a crossroads: the new biology and the new philosophy of science* (pp. 21–42). Cambridge, MA: MIT Press.

Carlson, E. A. (1966). *The gene: a critical history*. Philadelphia: Saunders.

Carlson, E. A. (1991). Defining the gene: an evolving concept. *American Journal for Human Genetics*, 49, 475–87.

Chadarevian, S. de (2002). *Designs for life: molecular biology after World War II*. Cambridge: Cambridge University Press.

Churchill, F. (1974). William Johannsen and the genotype concept. *Journal of the History of Biology*, 7, 5–30.

Churchill, F. (1987). From heredity theory to Vererbung: the transmission problem, 1850–1915. *Isis*, 78, 337–64.

Coleman, W. (1965). Cell, nucleus and inheritance: a historical study. *Proceedings of the American Philosophical Society*, 109, 124–58.

Crick, F. (1958). On protein synthesis. *Symposium of the Society of Experimental Biology*, 12, 138–63.

Dawkins, R. (1976). *The selfish gene*. Oxford: Oxford University Press.

Dietrich, M. R. (2000). From gene to genetic hierarchy: Richard Goldschmidt and the problem of the gene. In P. Beurton, R. Falk, & H.-J. Rheinberger (Eds). *The concept of the gene in development and evolution: historical and epistemological perspectives* (pp. 91–114). Cambridge: Cambridge University Press.

Dobzhansky, T. (1937). *Genetics and the origin of species.* New York: Columbia University Press.
Emmeche, C. (1999). The Sarkar challenge: is there any information in a cell? *Semiotica*, 127, 273–93.
Falk, R. (1995). The struggle of genetics for independence. *Journal of the History of Biology*, 28, 219–46.
Falk, R. (2000). The gene – a concept in tension. In P. Beurton, R. Falk, & H.-J. Rheinberger (Eds). *The concept of the gene in development and evolution: historical and epistemological perspectives* (pp. 317–48). Cambridge: Cambridge University Press.
Falk, R. (2001). The rise and fall of dominance. *Biology and Philosophy*, 16, 285–323.
Fischer, E. P. (1995). How many genes has a human being? The analytical limits of a complex concept. In E. P. Fischer & S. Klose (Eds). *The human genome* (pp. 223–56). München: Piper.
Fogle, T. (2000). The dissolution of protein coding genes in molecular biology. In P. Beurton, R. Falk, & H.-J. Rheinberger (Eds). *The concept of the gene in development and evolution: historical and epistemological perspectives* (pp. 3–25). Cambridge: Cambridge University Press.
Gayon, J. (1995). Entre force et structure: genèse du concept naturaliste de l'hérédité. In J. Gayon & J.-J. Wunenburger (Eds). *Le paradigme de la filiation* (pp. 61–75). Paris: L'Harmattan.
Gayon, J. (1998). *Darwinism's struggle for survival: heredity and the hypothesis of natural selection.* Cambridge: Cambridge University Press.
Gayon, J. (2000). From measurement to organization: a philosophical scheme for the history of the concept of heredity. In P. Beurton, R. Falk, & H.-J. Rheinberger (eds.), *The concept of the gene in development and evolution: historical and epistemological perspectives* (pp. 69–90). Cambridge: Cambridge University Press.
Gayon, J., & Zallen, D. (1998). The role of the Vilmorin Company in the promotion and diffusion of the experimental science of heredity in France, 1840–1920. *Journal of the History of Biology*, 31, 241–62.
Gilbert, S. (1978). The embryological origins of the gene theory. *Journal of the History of Biology*, 11, 307–51.
Gilbert, S. (2000). Genes classical and genes developmental: the different use of genes in evolutionary syntheses. In P. Beurton, R. Falk, & H.-J. Rheinberger (Eds). *The concept of the gene in development and evolution: historical and epistemological perspectives* (pp. 178–92). Cambridge: Cambridge University Press.
Goldschmidt, R. (1940). *The material basis of evolution.* New Haven: Yale University Press.
Griesemer, J. (2000). Reproduction and the reduction of genetics. In P. Beurton, R. Falk, & H.-J. Rheinberger (Eds). *The concept of the gene in development and evolution: historical and epistemological perspectives* (pp. 240–85). Cambridge: Cambridge University Press.
Gros, F. (1991). *Les secrets du gène. Nouvelle édition revue et augmeentée.* Paris: Odile Jacob.
Hull, D. L. (1974). *Philosophy of biological science.* Englewood Cliffs, NJ: Prentice-Hall.
Jacob, F. (1976). *The logic of life.* New York: Vanguard (Originally published 1970).
Jacob, F. (1977). Evolution and tinkering. *Science*, 196, 1161–6.
Johannsen, W. (1909). *Elemente der exakten Erblichkeitslehre.* Jena: Gustav Fischer.
Johannsen, W. (1911). The genotype conception of heredity. *The American Naturalist*, 45, 129–59.
Judson, H. F. (1996). *The eighth day of creation: makers of the revolution in biology* (2nd edn). Plainview, NY: Cold Spring Harbor Laboratory Press.
Kay, L. (2000). *Who wrote the book of life? A history of the genetic code.* Stanford: Stanford University Press.
Keller, E. Fox (2000). Decoding the genetic program: or, some circular logic in the logic of circularity. In P. Beurton, R. Falk, and H.-J. Rheinberger (Eds). *The concept of the gene in development and evolution: historical and epistemological perspectives* (pp. 159–77). Cambridge: Cambridge University Press.
Keller, E. Fox (2001). *The century of the gene.* Cambridge, MA: Harvard University Press.

Kim, K-M. (1994). *Explaining scientific consensus: the case of Mendelian genetics*. New York: Guilford Press.

Kitcher, P. (1982). Genes. *British Journal for the Philosophy of Science*, 33, 337–59.

Kohler, R. E. (1994). *Lords of the fly: Drosophila genetics and the experimental life*. Chicago: University of Chicago Press.

López Beltrán, C. (2004). In the cradle of heredity: French physicians and *l'hérédité naturelle* in the early nineteenth century. *Journal of the History of Biology*, 37, 39–72.

MacKenzie, D., & Barnes, B. (1979). Scientific judgment: the biometry–Mendelism controversy. In B. Barnes & S. Shapin (Eds). *Natural order: historical studies of scientific culture* (pp. 191–210). Beverly Hills: Sage.

Martins, L. Al-Chueyer Pereira (1999). Did Sutton and Boveri propose the so-called Sutton–Boveri chromosome hypothesis? *Genetics and Molecular Biology*, 22, 261–71.

Maynard Smith, J. (2000). The concept of information in biology. *Philosophy of Science*, 67, 177–94.

Mayr, E., & Provine, W. (Eds). (1980). *The evolutionary synthesis: perspectives on the unification of biology*. Cambridge, MA: Harvard University Press.

Mendel, G. (1866). Versuche über Pflanzen-Hybriden. Verhand lungen des Naturforschenden Vereins Brünn, 4 (1865), 3–47. English translation: Mendel, G. (1965). Experiments in plant hybridization. Ed. by James Bennet. Edinburgh and London: Oliver and Boyd.

Morange, M. (1998). *La part des gènes*. Paris: Odile Jacob.

Morange, M. (2000). The developmental gene concept: history and limits. In P. Beurton, R. Falk, & H.-J. Rheinberger (Eds). *The concept of the gene in development and evolution: historical and epistemological perspectives* (pp. 193–215). Cambridge: Cambridge University Press.

Morgan, T. H. (1935). The relation of genetics to physiology and medicine. In *Les prix Nobel en 1933*, 1–16. Stockholm: Imprimerie Royale.

Morgan, T. H., Sturtevant, A. H., Muller, H. J., & Bridges, C. B. (1915). *The mechanism of Mendelian heredity*. New York: Henry Holt.

Moss, L. (2003). *What genes can't do*. Cambridge, MA: The MIT Press.

Muller, H. J. (1951). The development of the gene theory. In L. C. Dunn (Ed.). *Genetics in the 20th century: essays on the progress of genetics during its first 50 years* (pp. 77–99). New York: MacMillan.

Müller-Wille, S., & Rheinberger, H.-J. (2007). Heredity: the production of an epistemic space. In S. Müller-Wille & H.-J. Rheinberger (Eds). *Heredity produced: at the crossroads of biology, politics, and culture, 1500–1870* (pp. 3–34). Cambridge, MA: MIT Press.

Müller-Wille, S. & Orel, V. (2007). From Linnaean species to Mendelian factors: elements of hybridism, 1751–1870. *Annals of Science*, 64, 171–215.

Olby, R. C. (1979). Mendel no Mendelian? *History of Science*, 17, 53–72.

Olby, R. C. (1985). *Origins of Mendelism* (2nd edn). Chicago: University of Chicago Press.

Portin, P. (1993). The concept of the gene: short history and present status. *The Quarterly Review of Biology*, 68, 173–223.

Provine, W. B. (1971). *Origins of theoretical population genetics*. Chicago: University of Chicago Press.

Rheinberger, H-J. (1997). *Toward a history of epistemic things: synthesizing proteins in the test tube*. Stanford: Stanford University Press.

Rheinberger, H.-J. (2000a). Ephestia: the experimental design of Alfred Kühn's physiological developmental genetics. *Journal of the History of Biology*, 33, 530–76.

Rheinberger, H-J. (2000b). Gene concepts: fragments from the perspective of molecular biology. In P. Beurton, R. Falk, & H.-J. Rheinberger (Eds). *The concept of the gene in development and evolution: historical and epistemological perspectives* (pp. 219–39). Cambridge: Cambridge University Press.

Robinson, G. (1979). *A prelude to genetics: theories of a material substance of heredity, Darwin to Weismann*. Lawrence, KS: Coronado Press.
Roll-Hansen, N. (1978a). The genotype theory of Wilhelm Johannsen and its relation to plant breeding and the study of evolution. *Centaurus*, 22, 201–35.
Roll-Hansen, N. (1978b). Drosophila genetics: a reductionist research program. *Journal of the History of Biology*, 11, 159–210.
Rosenberg, A. (1978). The supervenience of biological concepts. *Philosophy of Science*, 45, 368–86.
Sarkar, S. (1996). Biological information: a skeptical look at some central dogmas of molecular biology. In S. Sarkar (Ed.). *The philosophy and history of molecular biology: new perspectives* (pp. 187–231). Dordrecht: Kluwer.
Sarkar, S. (1998). *Genetics and reductionism*. Cambridge: Cambridge University Press.
Schaffner, K. (1969). The Watson–Crick model and reductionism. *British Journal for the Philosophy of Science*, 20, 325–48.
Schrödinger, E. (1944). *What is life? The physical aspect of the living cell*. Cambridge: Cambridge University Press (Originally published 1943).
Schwartz, S. (2000). The differential concept of the gene: past and present. In P. Beurton, R. Falk, & H.-J. Rheinberger (Eds). *The concept of the gene in development and evolution: historical and epistemological perspectives* (pp. 26–39). Cambridge: Cambridge University Press.
Stent, G. S. (1977). Explicit and implicit semantic content of the genetic information. In R. E. Butts & J. Hintikka (Eds). *Foundational problems in the special sciences* (pp. 131–49). Dordrecht: Reidel.
Stotz, K., Griffiths, P. E., & Knight, R. D. (2004). How biologists conceptualize genes: an empirical study. *Studies in the History and Philosophy of Biological and Biomedical Sciences*, 35, 647–73.
Waters, K. C. (1994). Genes made molecular. *Philosophy of Science*, 61, 163–85.
Waters, K. C. (2000). Molecules made biological. *Revue Internationale de Philosophie*, 4, 539–64.
Waters, K. C. (2004). What was classical genetics? *Studies in History and Philosophy of Science*, 35, 783–809.
Waters, K. C. (2006). A pluralist interpretation of gene-centered biology. In S. Kellert, H. Longino, and K. C. Waters (Eds). *Scientific Pluralism* (pp. 190–214). Minneapolis: University of Minnesota Press.

Further Reading

Bateson, W. (1902). *Mendel's principles of heredity. A defense*. Cambridge: Cambridge University Press.
Beurton, P., Falk, R., & Rheinberger, H-J. (Eds). (2000). *The concept of the gene in development and evolution: historical and epistemological perspectives*. Cambridge: Cambridge University Press.
Burian, R. M. (1985). On conceptual change in biology: the case of the gene. In D. J. Depew & B. H. Weber (Eds). *Evolution at a crossroads. The new biology and the new philosophy of science* (pp. 21–42). Cambridge, MA: MIT Press.
Carlson, E. A. (1966). *The gene: a critical history*. Philadelphia: Saunders.
Falk, R. (1986). What is a gene? *Studies in the History and Philosophy of Science*, 17, 133–73.
Keller, E. Fox (2001). *The century of the gene*. Cambridge, MA: Harvard University Press.
Moss, L. (2003). *What genes can't do*. Cambridge, MA: The MIT Press.
Portin, P. (1993). The concept of the gene: short history and present status. *The Quarterly Review of Biology*, 68, 173–223.
Waters, K. C. (2004), What was classical genetics? *Studies in History and Philosophy of Science*, 35, 783–809.

Chapter 2

Biological Information

STEFAN ARTMANN

1. Introduction

The most remarkable property of living systems is their enormous degree of functional organization. Since the middle of the twentieth century, scientists and philosophers who study living complexity have introduced a new concept in the service of explaining biological functionality: the concept of information. The debate over biological information concerns whether it is appropriate to speak of the processing of information in biological systems. Proponents of biological information theory argue that many basic life processes, from the molecular foundations of inheritance to the behavior of higher organisms, are self-organizing processes of storing, replicating, varying, transmitting, receiving, and interpreting information. Opponents of biological information theory object that the concept of information does not have such explanatory power in biology.

Let us adduce some of the highly controversial theses that the proponents of biological information theory claim to be true:

- In molecular genetics, a set of rules for transmitting the instructions for the development of any organism has been discovered that is most appropriately described as a genetic code [SEE GENE CONCEPTS].
- The main research problem of developmental biology is how the decoding of these ontogenetic instructions depends upon changing biochemical contexts [SEE PHENOTYPIC PLASTICITY AND REACTION NORMS].
- Neurobiology cannot make decisive progress before neural codes that are needed for storing, activating, and processing simple features of complex cognitive representations are discovered.
- Ethology is a science of communication since it studies the astonishing variety of information-bearing signals, whose transmission can be observed, for example, in social insects, birds, and primates [SEE ETHOLOGY, SOCIOBIOLOGY, AND EVOLUTIONARY PSYCHOLOGY].
- Information-theoretical considerations are also of great importance to evolutionary biology: Macroevolutionary transitions – from co-operative self-replication of macromolecules, to sexual reproduction, to human language – established more and more complex forms of natural information processing [SEE SPECIATION AND MACROEVOLUTION].

- If all these claims prove true, the following answer must be given to the old problem of defining life: Life is matter plus information (Küppers, 2000, p.40; see WHAT IS LIFE?).

Such a wide appeal to the notion of information is a matter of much dispute. Does the application of the concept of information in biology impute consciousness or intentionality to biological systems? Or is it possible to use a formal concept of information to explain the structure, function, and history of organisms – without anthropomorphizing them unduly and without smuggling into biology an untenable metaphysics in mathematical disguise? If so, it must be shown that the usual paradigm of information transmission, communication between human beings through complex languages, is only a special case of a general theory of information.

There are two interrelated issues concerning the application of the term "information" to biological systems. First, what formal theory ought one to adopt? Second, which biological objects can be described as information-bearing? The construction of a theory of information is influenced by the objects we want to describe as bearers of information. Some authors try to restrict biological information to genes (Maynard Smith, 2000a, 2000b; Sterelny, 2000); others who want to be more inclusive (Godfrey-Smith, 2000; Sarkar, 2000; Griffiths, 2001) have also to decide whether there is just one concept of biological information (Sarkar, 2005, pp.270ff.) or a plurality of concepts (Godfrey-Smith, 2000, pp.206).

A discussion of such questions is made difficult by the fact that the word "information" is often used in an untechnical way. To prevent a further spread of the word "information" as an opaque metaphor in biology, the philosopher should work on two problems at the same time: first, define a technical concept of information that contributes directly to the explanation of concrete biological phenomena; second, develop an abstract understanding of this technical concept so that it can be fruitfully used in the development of a general theory of biological systems.

It is impossible here to give a complete account of every concept of information that has been proposed in biology, of all applications that have been attempted, and of the philosophical discussions that accompanied this research. Instead, there will be presented a general semiotic account of information, which will capture the different facets of biological information in one complex concept. First, a scenario of information transmission will be introduced and applied to genetics (Section 2). Note that the choice of this application does not mean that the concept of biological information must be restricted to genes. Second, the syntactic, semantic, and pragmatic dimension of transmitted information will be distinguished (Section 3). Focusing on heredity again, some basic aspects of the syntax and semantics of genetic information will be analyzed (Sections 4–6).

2. General Scenario for the Transmission of Information and Its Application to Genetics

The general scenario of information transmission, which will be used as a conceptual frame of reference for our discussion of biological information, comes from the birth certificate of information theory, Claude E. Shannon's *A Mathematical Theory of*

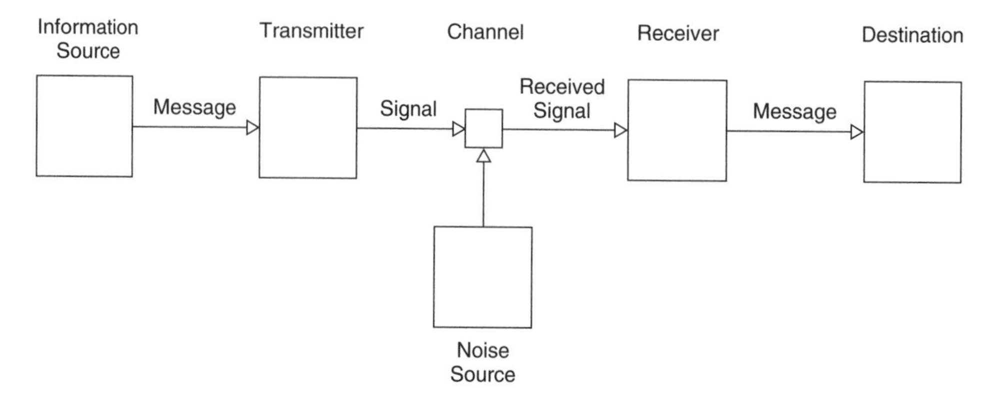

Figure 2.1 Schematic diagram of a general communication system (after Shannon, 1948)

Communication (1948). Shannon calls his scenario a "general communication system," since he generalizes it from telecommunication technologies like telephone and radio. The original point of view of information theory is the one of an engineer who tries to design a communication system that optimizes the conditions of information transmission to be both as economical and as reliable as possible. In a paper that started a new round in the debate on biological information, the late John Maynard Smith (2000a, pp.178ff., 183f.) emphasized the fundamental importance of the general analogy between engineered and evolved systems for our topic, and also that, if our analogies are not to mislead us from the beginning, they must be not merely intuitive, but formal. This recalls Shannon's warning that "the establishing of such applications [of information theory] is not a trivial matter of translating words to a new domain, but rather the slow tedious process of hypothesis and experimental verification." (Shannon, 1956, p.3)

Shannon's scenario consists of the following elements:

- *message*, the information to be transmitted;
- *information source*, the system that selects a message;
- *destination*, the system that receives a message;
- *channel*, the medium that is used for the transmission of a message;
- *transmitter*, the mechanism that encodes a message so that it can be transmitted over the channel;
- *receiver*, the mechanism that decodes a message which has been transmitted over the channel;
- *signal*, the encoded message that is transmitted over the channel;
- *noise source*, anything that possibly perturbs the signal during transmission.

To *transmit information* means to pass through the following procedure (see Figure 2.1). First, the information source selects a message to be transmitted. Then the transmitter encodes the chosen message. The resulting signal is transmitted via the channel. During the transmission noise can distort the signal. The receiver decodes the (possibly distorted) signal. Finally, the reconstructed message is delivered to the destination.

Now we must identify biological entities that are promising candidates for the different functional roles in Shannon's scenario. At the same time, we must take the implications of our assignments of these roles to biological entities into account.

Let us begin with the *signal* that is genetically transmitted. Any genetic signal we shall call a "gene." This simple nominal definition, which does not say anything about the material properties of genes, is a first example of the formal perspective of information theory.

Every path through which a genetic signal can be transmitted is a possible *channel* for the information the signal contains. This may happen by the division of a cell or by sexual reproduction. There are also other tracks of genetic transmission (e.g., horizontal gene transfer between plants), which we ignore for simplicity's sake. To fix a generic term for all genetic channels, we deploy August Weismann's concept of germ-track, though he exclusively used it to denote the course taken by the unaltered hereditary substance from ovum to reproductive cell (Weismann, 1893, p.184). Weismann's concept aptly suggests, however, the image of a continuous flow of genetic information, and he also discussed the connection between germ cells of two consecutive generations, so that it is just another step of generalization to call, from our formal perspective, a channel of genetic information transmission a "gene-track."

Where does a gene-track end? The *receiver* of a genetic signal must be a biochemical environment that makes the decoding of the transmitted signal possible. Maynard Smith (2000b, p.216) considers such an environment a channel condition, but if we use the complete Shannon scenario to capture genetic information processing, our assignment is more adequate. The biological term for the process of decoding a genetic signal is "gene expression," and the decoding system that controls how, and when, genes are expressed is the regulatory mechanism of gene expression. Without such a regulating biochemical context, a genetic signal could not even be recognized as a signal, i.e., as a biochemical structure that carries information (Küppers, 1996, p.142). If we emphasize the importance of the evolution of the receiver's decoding system (Jablonka, 2002, p.582), the "hermeneutical relativity" (Sarkar, 2000, p.212) of genetic information is one of its outstanding features. Consequently, genetic information does not exist in an absolute sense but only in a relative one: whether a biological object carries genetic information, and what information this object possibly carries, cannot be decided just by analyzing the structural properties of this object. Its functional properties – the causal roles it plays in different biochemical contexts – are of equal importance.

The *message*, which is reconstructed from a genetic signal by its receiver, consists of the primary structure of a protein. The decoding of so-called "structural genes" instructs the synthesis of proteins by organizing the linear sequence of their amino acid building blocks. If two signals are decoded as the same linear sequence of amino acid residues, they carry the same message. To put it formally: a message is the class of all signals that are equivalent to each other in respect to the result of their decoding (Sarkar, 2005, p.272). There are also parts of structural genes (called "regulatory sequences") and entire regulatory genes that fulfill a most important role for genetic decoding: they encode in which biochemical contexts the decoding mechanism of the receiver is applied to the structural genes (Maynard Smith, 2000b, p.216). As emphasized before, genetic information is relative: without regulatory signals, the structural genes would not carry genetic information – in fact, they would not count as genes. But there is (contrary to Sterelny [2000, p.199]) no need to interpret the hierarchical semantic structure of a genetic message as constituting an information-theoretical dualism between the message itself (structural genes) and its readers (regulatory sequences and genes).

We shall rather acknowledge the complexity of the message by developing a refined picture of genetic semantics (see Maynard Smith, 2000b, p.217 and Section 6).

Genetic information is not transmitted in a completely reliable way. What is the *noise source* that can distort the genetic signal and make the reconstruction of the original message difficult or impossible? It consists of all physical or chemical mechanisms that cause errors in any step of genetic information transmission by deleting, substituting, or inserting parts of the signal (Adami, 1998, p.61). Recognition errors during replication, transcription, or translation, mutagens, and ultraviolet radiation are some possible sources of such errors.

By decoding the genetic signal and by reconstructing its ontogenetic message, the *destination* of genetic information, the phenotype of an organism, is reached. A destination does not exist before the reception and decoding of the genetic signal, since its message consists in primary protein structures that are necessary for ontogenetic development (Maynard Smith, 2000b, p.216). Another functional role of (a subsystem of) the destination is to be the future gene-track of the same – or, due to noise, a very similar – signal.

The *information source* that selects the message to be transmitted shows another distinctive characteristic of genetic information transmission: a conscious designer of evolutionarily successful phenotypic features does not exist. Instead, the only possible source of genetic information lies in natural processes that bring about the origin, evolution, and development of organisms. Since evolutionary theory describes the working of these processes on the level of abstract mechanisms like natural selection, we can consider such mechanisms as information sources. Here Maynard Smith (2000a, p.190) uses a computer analogy: Natural selection "programmed" the information in a gene. This analogy – and other possible ones like tinkering (Artmann, 2004) – show that the information source is the element of Shannon's scenario where the analogy between the engineering of artifacts and the evolution of living systems can be easily misleading. In order to avoid presupposing a divine planning intentionality that guides the evolution of living systems, one appeals to a class of natural processes. These processes are formally studied by theoretical population genetics [SEE POPULATION GENETICS].

It would be a distorting restriction on information-theoretical research into heredity if we were to leave the evolutionary dimension of genetic information transmission out of the account. This would happen if the complete scenario of information transmission were situated in one organism, so that the channel had to be identified with protein synthesis (Gatlin, 1972, p.96f.), or if transcription of DNA in RNA were identified with encoding (Yockey, 1992, p.111). On the other hand, to equate the information source with the environment *in toto* (Adami, 1998, p.61) is too vague since the environment is an information source only due to an evolutionary mechanism like natural selection.[1] Whether, and how, features of the environment act as information sources depends upon the extent to which these features influence the transmission of genes across generations. The environment selects genetic messages that are transmitted

1 An analogous remark applies to identifying the destination of genetic information with the environment since the phenotypic organism is the receiving "environment" of the ontogenetic message, which is then evaluated by natural selection.

from one generation to the next only in so far as it has properties that are relevant to the survival or reproduction of the organisms living in this environment. Regarded as an information source, the environment evaluates organisms in terms of their reproductive success, i.e., their fitness [SEE THE UNITS AND LEVELS OF SELECTION].

Maynard Smith (2000a, p.189ff.) and Sterelny (2000, p.196) follow the same type of reasoning in their criticism of Dretske's (1981) concept of "causal information" that would, in case of biological information transmission, equate information with the causal covariance of changes in the environment and in the organisms living in this environment, so that any environmental change causing a change in the organism would automatically carry information about this organism – no evaluative mechanism is considered from this too simple perspective on the role of the environment in genetic information transmission.

Our discussion of the noise source and the information source in the transmission of genetic information can now be summarized. The source of genetic information consists of two subsystems, a generative and an evaluative one. The first subsystem is the noise source, which produces genetic variation, i.e., the elements of the set of possible genetic messages. In an evolutionary perspective, it is too simple to say that the environmental fluctuations cause only noise in the organism (Maynard Smith, 2000a, p.192), because noise can become information in evolution. How does this happen? All produced messages are evaluated by the second subsystem, e.g., natural selection. The destination of a genetic message, the phenotypic organism, transmits the decoded (or a very similar) message in as many signal copies as possible to other destinations. The phylogenetic information value of one and the same genetic message – its fitness – is thus measurable by the number of those of its destinations that are, in turn, able to transmit it further.

Is there a biological entity that fulfils the role of the *transmitter* of genetic information by encoding the selected message into a genetic signal? According to the so-called "Central Dogma" of molecular biology (see Section 5 and GENE CONCEPTS), genetic information flows only from nucleic acids to proteins – we can leave aside the reversal of the direction of information flow from DNA to RNA in case of retrotransposons and retroviruses, because we are only interested in the relation between a genetic signal and its protein message. If the Central Dogma is true, then natural selection cannot choose any message that is physically different from the signal to be transmitted, because there simply is no difference between signals and messages at the sending side of the channel. Maynard Smith (2000a, p.179) expresses this by identifying natural selection with the "coder," i.e., transmitter, without introducing any separate information source.

In comparison to the selectionist theory of evolution, the neutral theory of molecular evolution [SEE MOLECULAR EVOLUTION] defines no evaluative criteria on the natural dynamics of genetic information processing and renders thereby irrelevant, not only the encoding, but also the decoding of genetic information. This amounts to a complete withdrawal of the distinction between signal and message – which is not the case in the selectionist framework. Of course, the neutral theory is not inconsistent with the selectionist theory – nowadays most biologists who are selectionists accept that neutral evolution is very important on the molecular level. It is just completely superfluous to invoke information theory in a neutralist analysis of an evolutionary phenomenon.

3. Semiotic Dimensions of Biological Information

The application of Shannon's general scenario of information transmission to heredity has shown that we must analyze the syntax, the semantics, and the pragmatics of the genetic message (Cherry, 1966, p.219ff.; Küppers, 1990, p.31ff.). Syntax refers to the internal order of messages and signals: How are they structured? Semantics refers to the code mapping of transmitters and receivers: How do they encode messages and decode signals? Pragmatics refers to the generative history of information sources and destinations: How do mechanisms of selecting messages to be transmitted and acting upon messages received evolve? By integrating these semiotic dimensions into the general scenario of information transmission, we get a systematic picture of transmitted information.

The *syntactic* dimension of a signal (or message) is structured by its internal organization. For example, the change in frequency and amplitude of an analog radio signal over time constitutes its syntactic structure, which can be represented by the mathematical technique of Fourier analysis. Generally, the relational order of the elements of a signal (or message) constitutes its syntax. Any mapping of elements of a signal (or message) to elements of a message (and signal, respectively) must not be taken into account in syntactic descriptions.

The *semantic* dimension of information is structured by codes. Any code connects at least two syntactic orders with each other. The ASCII code associates, for example, each letter of the Latin alphabet, cipher, punctuation mark, special symbol, and control character with one, and only one, fixed sequence of eight binary digits. If we ignore the internal organization of messages and signals (their syntax), we can say that a code is the information-theoretical name of a mapping that relates, in case of encoding by transmitters, each of the possible syntactic elements of a message to a possible element of a signal. In case of decoding by receivers, each of the possible syntactic elements of a signal is mapped by the code to a possible element of a message (Cover & Thomas, 1991, p.79ff.). The semantics of a signal (or message) is thus given by the mapping from its syntactic elements to the syntactic elements of a message (and signal, respectively).

The *pragmatic* dimension of biological information is structured by boundary conditions on coding. In which contexts does it happen that a message is selected for being encoded, and in which contexts does it happen that a signal is decoded in a particular way? Moreover, how does the code itself originate? To describe the pragmatics of information involves at least two syntactic orders and one semantic mapping. Its analysis must include the information source, which selects a message to be encoded in a certain manner, and the destination, which is ready to get a message via a particular way of decoding. So the pragmatics of a message (or signal) is given by the generative history of its connection to a signal (and message, respectively). This definition generalizes the widespread idea that information is constituted pragmatically by the effect of a signal on its receiver (MacKay, 1969; Küppers, 1990; Jablonka, 2002).

Since the distinction between syntax, semantics, and pragmatics comes from semiotics, the science of signs (Morris, 1938), we call syntax, semantics, and pragmatics the "semiotic dimensions of information," so that we develop a theory of "semiotic information" (Sarkar, 2005, p.270ff.).

4. Syntactic Dimension I: Measuring the Statistical Entropy of Signals and Messages

A syntactic analysis of information investigates the internal organization of a message or a signal. Up to now, information-theoretical research into syntax has concentrated on two aspects, transmission and compression of information. In this section we discuss the first aspect, which is quantifiable from a probabilistic perspective by Shannon's statistical measure of entropy. The question of whether this measure is useful in biology has dominated the debate on biological information. Below we will analyze a typical example of a measurement of genetic signal entropy that may be statistically precise but turns out to be irrelevant for the explanation of the functional organization in living systems. Since the application of Shannon entropy in biology did not come up to the early high expectations of many biologists, its dominance over the debate on biological information is all the more distorting if one tries to extrapolate the general value of information theory for biology from a detailed history of this disappointment (as is done in Kay, 2000).

If transmitter and receiver share syntactic knowledge of the set of possible messages and semantic knowledge of the code, and if the destination is pragmatically interested in any message that is selected by the information source, then Shannon's theory of information transmission provides a statistical measure of syntactic information content (Shannon, 1948). A transmitter is able to send a selected message over the channel by encoding it. The code maps each letter, i.e., each element of the message, to a codeword, which consists of a finite sequence of elements from the code alphabet. To describe the set of possible messages statistically, we ascribe to each letter that is available to the information source a probability with which it will be selected as the next letter. In other words, we define a probability distribution over the range of the random variable called "information source." The probability that a particular message is selected can be calculated by multiplying the selection probabilities of the letters that occur in the message, if the selection of letter l at position x_i of a message does not alter the probability with which letter l' is selected at position x_k with $k > i$.

To define a statistical measure H for the syntactic information content of messages and signals, we identify the set of all letters l that can be selected by the information source S. Then we determine the probability $p(l)$ with which S selects l. Shannon's decisive idea was to equate the information content of a message selected by the information source with the amount of uncertainty that the destination loses by its reception. H will quantify the "surprise value" (Cherry, 1966, p.51), or "potential information" (Küppers, 1990, p.37), of a message. The less probably a letter l is selected, the less the destination expects this letter, and the more informative is its reception.

The probability distribution that assigns to each possible selection l by S the same probability $p(l) = 1/n$ (where n is the number of available letters) constitutes the most informative information source. Given such an equiprobability distribution, the measure H should moreover increase monotonously if n does likewise. Another property required of H is that its value should change continuously when the probability distribution of the information source does. A third demand on H turns up if we consider the selection of a letter as a sequence of selections between binary alternatives.

Let us assume, for example, that we have a set of three letters A, B, and C which are selected with probability $p(A) = 1/2$, $p(B) = 1/3$, and $p(C) = 1/6$. Then we can select a letter by two binary decisions. First we choose between A and $\{B, C\}$ with $p(A) = 1/2$ and $p(\{B, C\}) = 1/2$, and then between B and C with $p(B) = 2/3$ and $p(C) = 1/3$. Our measure H should be insensitive to such transformations in the representation of the selection process.

Given all three conditions on H, Shannon (1948: appendix 2) proved that H must be of the following form:

$$H(S) = -K \sum_{l} p(l) \log p(l), \tag{1a}$$

where K represents the value of a positive constant. Since we can freely choose a unit of measure, we select the binary digit (bit) – though "bit" as a unit of measure for $H(S)$ must not be confused with "bit" as a unit of measure for the space used when actually storing the possible messages, e.g., on the hard disk of a computer. In (1a) we thus set K equal to 1, take all logarithms to base 2, and get the formula

$$H(S) = -\sum_{l} p(l) \log_2 p(l), \tag{1b}$$

which can be understood as the expectation value of $\log_2 (1/p(S))$. Shannon called the information measure defined in (1b) "*entropy*," since (1a) is formally identical to the homonymous thermodynamic measure.

The entropy of DNA per nucleotide l,

$$H_1(S) = -\sum_{l=1}^{4} p(l) \log_2 p(l), \tag{1c}$$

is, for sizeable genomes, typically ≈ 1.9–2.0, i.e., very close to the maximum 2 (Percus, 2002, p.79). We can generalize (1c) to entropies of nucleotide words w of any length n,

$$H_n(S) = -\sum_{w=1}^{n^4} p(w) \log_2 p(w). \tag{1d}$$

For $n = 3$, (1d) gives the entropy of DNA per codon. In coding regions, it is typically ≈ 5.9, again close to the maximum 6 (Percus, 2002, p.79). Taking longer contexts into account, a more realistic measure of the unpredictability of the next nucleotide letter that is added to a DNA signal, the so-called "excess entropy," is defined as $H_n(S) = H_{n+1}(S) - H_n(S)$. For a variety of mammal genomes, an average value of $H_n(S) \approx 1.67$ has been estimated (Loewenstern & Yianilos, 1999, p.157).

Why is Shannon entropy – although it allows us to precisely measure syntactic properties of biological information – not as important to biology as it was hoped? We shall show the reason for this by giving a short technical analysis of a central section of Gatlin (1972), the first systematic account of biological information theory.

To define a biologically meaningful set of signals, Gatlin (1972, p.58ff.) shows that the DNA found in a specimen of some species can be considered, with certainty, as representative for all signals possibly selected by this species. Based on data about the relative frequency of the four DNA nucleotides in signals from many different species, ranging from phages to vertebrates, Gatlin (1972, p.73ff.) not only calculates the entropy H_1 of DNA per nucleotide. She is also interested, first, in the difference D_1 between the highest possible entropy of a message $H_{max} = \log_2 n$ (where n is the number of available letters) and its actual entropy H_{uncond} per nucleotide using unconditional probabilities $p(l_i l_{i+1}) = p(l_i)p(l_{i+1})$, and second, in the difference D_2 between H_{uncond} and the conditional entropy H_{cond} of DNA per nucleotide using nearest-neighbor frequencies, i.e., conditional probabilities $p(l_i l_{i+1}) = p(l_i)p(l_{i+1}|l_i)$. Gatlin wants to measure how much the probability distribution of an information source deviates, in case of D_1, from equi-probability and, in case of D_2, from independence. If the information density Id of a message is defined as the sum of D_1 and D_2, its redundancy R is equal to Id/H_{max}. Gatlin calculates also the relative contribution of D_1 and D_2 to redundancy, $RD_1 = D_1/(D_1 + D_2)$ and $RD_2 = D_2/(D_1 + D_2)$, respectively. Her result is that vertebrates have, in comparison to lower organisms, both higher R- and RD_2-values, and lower RD_1-values. Gatlin (1972, p.80) concludes that vertebrates have achieved their R-values by holding D_1 relatively constant and increasing D_2. Lower organisms that achieve R-values in the vertebrate range (or even higher) do so primarily by increasing D_1. Phenotypic complexity seems thus to correspond broadly to D_2, but there does not exist, according to Gatlin (1972, p.82), a more concrete correlation of this statistical measure for the internal order of genetic information with a classical taxonomic measure.

In addition, Gatlin (1972, p.79f.) must admit that her result is just an information-theoretical reformulation of the empirically well-known molecular-biological fact that bacteria and other lower organisms have wide variational range, from 20 to 80 percent, in respect to nucleotides C and G, whereas the nucleotide composition of vertebrates lies within a relatively restricted range, i.e., vertebrates have a restricted range of D_1- and D_2-values. Calculating entropies seems thus to be at best a new but rather complicated way of representing well-known biochemical data, because H- or D-values measure only the syntactic aspect of genetic information. They do not directly tell us anything about its semantic and pragmatic dimension. If the codes and histories of biological information are not taken into account, the most remarkable property of living systems, their enormous degree of evolved functional organization, is beyond the reach of information theory.

Gatlin (1972, p.191ff.) tries, indeed, to integrate her syntactic results into semantics and pragmatics. The reason for the higher redundancy in vertebrates is, according to her, that it is needed for error-correcting, so that there must be an RD_2-increase for the evolution of complexity, which does not necessarily possess a semantic consequence, since due to synonymity of codons two DNA signals could have different D_2-values without transmitting a different message. Gatlin draws the pragmatic consequence that there must exist some kind of coding selection which works at the input of the genetic channel and is distinct from Darwinian selection. Her argument would, however, amount to the thesis that the more redundant a signal is, the higher the complexity of its message – a counterintuitive and also obviously false consequence, since a totally redundant signal carries zero syntactic information content from an entropy point of view (Wicken, 1987, p.28).

5. Syntactic Dimension II: Estimating the Algorithmic Complexity of Signals and Messages

Francis Crick's original formulation of the hypothesis he ironically called the "Central Dogma" (see Section 2) implies a definition of the syntactic information content of signals and messages that is different from Shannon's. The Central Dogma

> [. . .] states that once "information" has passed into protein *it cannot get out again.* [. . .] Information means here the *precise* determination of sequence, either of bases in the nucleic acid or of amino acid residues in the protein. (Crick, 1958, p.153)

This suggests that the most important syntactical feature of a genetic signal or message is the exact arrangement ("precise determination") of the linear order ("sequence") in which its elements ("bases" or "amino acid residues") follow each other and which can be accurately transferred ("passed") from nucleic acids to proteins by means of a semantic code. The decisive idea behind Crick's perspective on syntax can be traced back to Erwin Schrödinger's famous speculation on the inner structure of the hereditary substance as an aperiodic crystal, whose internal organization links together building blocks of a few different types in very many possible combinations (Schrödinger, 1992, p.60ff.).

Shannon's entropy can measure a statistical aspect of the syntax of a set of possible messages, and even this only for stationary stochastic processes, which cannot capture the complex compositional heterogeneity in DNA sequences (Román-Roldán et al., 1998). To capture the specific internal order of single messages and signals, we can use a theory of information compression that is based upon an algorithmic approach (Kolmogorov, 1993, pp.184–93). The step from Shannon's statistical reasoning to Kolmogorov's algorithmic reasoning necessitates a change of perspective, which has not been adequately discussed in philosophy up to now – though the theory of algorithmic information was rigorously applied to biology for the first time in Küppers (1990). In the debate on Maynard Smith (2000a), only Winnie (2000) introduces the concept of algorithmic information, but he does not integrate it into Shannon's scenario of information transmission. Winnie argues that it is unnecessary to look for biological analogues of its elements, because he does not grasp the generality of this scenario, which can be applied to genetics without restricting biological information theory to the sole use of Shannon's measure of entropy.

Whenever the destination of a message asks to what extent a received message could be compressed, the destination can fall back on the idea of computation. "Compressing" means to encode a message so that the encoding is shorter than the message but nevertheless contains the same information. By "computation" we understand an ordered sequence of formally describable operations that is effective for solving a problem and can be executed by an automaton (preferably a universal Turing machine), if it is formulated as an algorithm in the automaton's programming language. The computational problem to be solved by this automaton consists in finding a minimal and lossless description of the syntactic structure of a message. The destination measures thus the syntactic information content of a message, its *algorithmic* (or Kolmogorov) *complexity*, by the bit length of the shortest program that as an input signal, can generate this message on the automaton and then stops. The more space is needed to store the compressed signal of the message, the more algorithmic syntactic information this message delivers to the destination.

Algorithmic complexity is, like Shannon's entropy, a purely syntactic measure insofar as only the internal order in which the elements of a message follow each other is relevant for its computation. Formally, the algorithmic complexity AC of a message m on an abstract automaton A is defined as the minimal element of the set of lengths L of all programs P generating m on A and then stopping.

$$AC_A(m) = \min\{L(P) : A(P) = m\}. \tag{2}$$

The program that is able to generate the message m and then stops can be used as an encoded version of m in order to transmit it very economically – on condition that both the transmitter and the receiver share knowledge of the employed automaton. Algorithmic information theory thus throws light on quantitative restrictions on the transmission of single messages. By measuring the entropy of an information source we can show how good the transmission codes can be, on average, for compressing any possible message. The statistical and the algorithmic theory of information describe therefore two syntactic aspects that are quantitatively closely related to each other if some well-specified formal conditions are fulfilled (Cover & Thomas, 1991, p.154), since both presuppose that the amount of information in a message has to do with the minimal length of an encoding signal, which is transmitted over a channel or fed as a program into an automaton.

The definition of algorithmic complexity can easily be used to express the syntactic aspect of Crick's information concept formally. Let us take, as an example, Jacques Monod's claim that the sequence of amino acids in the primary structure of a protein is random (1971, p.95): When the linear succession of 199 amino acids in a chain of 200 amino acids is well known, no rule exists to predict the last one. A message m is formally defined to be *random* on an automaton A if, and only if, its length L is approximately equal to its algorithmic complexity $AC_A(m)$,

$$AC_A(m) \approx \mathrm{L}(m). \tag{3}$$

A random message cannot (or only minimally) be compressed because it has (almost) no internal order. But is it possible to prove Monod's claim for a particular sequence? A formal axiomatic system *FAS* with $AC(FAS) > k$ is needed to generate the set of all theorems stating that a message has $AC(m) = n$ (for all $n \leq k$) and all theorems stating that $AC(m) > k$ (Chaitin, 1974, theorem 4.3). The impossibility of disproving that $AC(protein) \approx \mathrm{L}(protein)$ is thus itself not to be proven in the framework of a formal system *FAS* with $AC(FAS) < AC(protein)$. For great k, to find by chance in the set of more than 2^k possible sequences the one that proves, as an encoded version of *FAS*, the randomness of another message with algorithmic complexity k is very improbable and, for all practical purposes, negligible. To systematically search such an *FAS*, one must show that a candidate

FAS has the required complexity so that it is useful to test it; to prove this, we need an even more complex formal system, and so on (Küppers, 1990, p. 98ff.).

It is, for all practical purposes, impossible to prove the nonexistence of rule-governed regularities for reasonably long signals. If one does not invest as much algorithmic complexity as these signals have, one cannot prove that they have a particular algorithmic complexity. So one can only hope to make reasonable approximations by using powerful compression algorithms. Beyond such practical questions, the biological application of the theory of algorithmic complexity demonstrates that important theoretical questions lurk behind claims like Monod's, which seems to be purely empirical. Here information-theoretical research on a basic problem – how complex is genetic information? – encounters a fundamental limit of scientific explanation that cannot be crossed by simply accumulating new empirical knowledge. This does not mean that the study of biological information reached a dead end so that it should either be abandoned in favor of other scientific approaches to living systems or be replaced by metaphysical creeds in the inherent simplicity or randomness of nature. Instead, the rich conceptual results of using the theory of algorithmic complexity in the study of the syntactic dimension of genetic information suggest two positive consequences. First, it is necessary to develop sophisticated methods for the statistical approximation of the algorithmic complexity of genetic signals and messages so that empirical hypotheses about their algorithmic information content can be evaluated. Second, it may be sensible to use the theory of algorithmic complexity also in the study of the semantic dimension of genetic information (see next section).

6. Semantic Dimension: Classifying the Mutual Complexity of Transmitters and Receivers

A well-known biological problem is to identify function by structure. To infer, for example, the biochemical function of a protein from its tertiary structure, most often we need to know structural homologs, whose function has already been analyzed (Thornton et al., 2000). The relation between syntax and semantics of the genetic code exemplifies this problem, if we look at it from the perspective of the theory of "teleosemantic information," according to which the semantics of biological information lies in its function shaped by natural selection (Sterelny, 2000: 197f.; see FUNCTION AND TELEOLOGY).

It is controversial whether there is a stereochemical affinity between a codon and the amino acid it encodes. If this is the case, the spatial shape of a codon and the amino acid it encodes must be structured in a way that increases the probability that both interact chemically with each other; this interaction directly translates the genetic information. If it is not the case that there exists a stereochemical affinity between a codon and the amino acid it encodes, the semantic relation between both is chemically contingent (Monod, 1971, p.106; Maynard Smith, 2000a, p.185; Sarkar, 2005, p.274). This contingency does not appear in the transmission of the syntactic structure of the genetic signal (transcription and replication), but in the transmission of its semantic content (translation).

The genetic code does, of course, show internal regularities. Similar codons encode similar amino acids, so that transmitting and decoding errors are minimized. Moreover, we do not have to adopt Crick's hypothesis that the genetic code is a "frozen accident" – even Crick (1968) does not exclude that the code could have evolved. Selectional (adaptation for error minimization), historical (coevolution of metabolic relatedness of amino acids and the code), and stereochemical causes determining the probability that a particular codon encodes a particular amino acid in the RNA world (Knight et al., 1999) do not necessarily explain away the current contingent organization of the code. The unpredictability of semantics from syntax just means that the molecular biologist must also reconstruct the contingent physico-chemical boundary conditions in which the code originated and upon which its further evolution reacted. We can imagine that, though the semantics of the primeval genetic code may be stereochemically motivated, its evolution has led to semantic contingency, i.e., that these stereochemical affinities do not play any causal role in decoding anymore (Godfrey-Smith, 2000, p.204). For a code to be contingent it is sufficient that there could have originated and evolved another code which would be functionally equivalent (Beatty, 1995).

A well-known approach to the semantic dimension of information, which tries to transfer Shannon's statistical approach to semantics, was developed by Yehoshua Bar-Hillel and Rudolf Carnap (1964), but did not exercise any measurable influence on biology. Bar-Hillel and Carnap's theory works only in the abstract framework of a fixed formal language system based on a set of independent elementary sentences, whose combinations give a complete representation of all possible states of a system. The great problem, which is insurmountable in applications to an empirical science like biology, consists in finding these elementary sentences without having a finished theory of the system – and if we had it, information theory could not tell us anything new about the system.

Luciano Floridi (2003) developed another formal approach to the semantic dimension of information, according to which the standard definition of an information unit σ contains three components: first, σ consists of a nonempty set D of data d; second, the data d of D are well-formed according to a system of syntactic rules; third, these well-formed data d are meaningful. The definition of semantic information as data plus meaning is neutral in respect, not only to the formal type of data and to the material structure of the information carrier, but in particular to the truth value of an information unit. Floridi criticizes this alethic neutrality: False information and tautologies should not be accounted as information at all. Whatever the truth of this criticism, it is not very relevant for biology, since Floridi captures semantics from the perspective of information *about* reality (Floridi, 2004). Yet in biology the semantic aspect of information lies, above all, in the instructional nature of information *for* reality. The genome "[. . .] is a recipe, not a blueprint" (Maynard Smith, 2000a, p.187). Genetic information has a semantic aspect because the syntactic structure of a DNA signal instructs the construction of the syntactic structure of its protein message. Only through its information for molecular reality does the signal contain information about a cell. It is, therefore, not adequate to think that the concept of instruction adds something new to the concept of biological information and to call this novel quality, rather confusingly, "intentionality" (Maynard Smith, 2000a, p.192ff.). Instead, the concept of biological information analytically contains the idea of a process by which a syntactic structure instructs the assembly of another syntactic structure (Sarkar, 2000, p.211). Since the laws of physics and chemistry can explain all biochemical processes involved in this instructed assembly, there is no need to invoke any intentionality or final causation in order to explain how the syntax of nucleic acids is related to the syntax of proteins by means of the semantics of the genetic code.

Although biochemistry explains how structure and function of genetic information are connected to each other in today's living systems, the laws of physics and chemistry do not enforce the evolution of a particular connection between the syntax and the semantics of genetic information – otherwise, contingency in the genetic code could not have been evolved. Codes that were stereochemically motivated at their origin can be conventional nowadays (Griffiths, 2001, p.403). The instructional semantics of biological information has, therefore, to be analyzed also from the perspective of pragmatics. To describe the ontogenetic (or proximate) pragmatics of a signal (or message) means to narrate the process of its transmission: Why was a message selected for encoding, and why was a signal decoded in a certain way? To describe its phylogenetic (or ultimate) pragmatics means to narrate the evolutionary process of the establishment of the codes used for encoding and decoding (Jablonka, 2002).

Nevertheless, the instructional nature of biological information shows semantic aspects, which can be described separately since they are not directly deducible from pragmatics. Algorithmic information theory allows formulating precise research hypotheses about semantic information and its syntactic conditions. The compressibility of the primary structures of proteins, for example, was tested by various compression schemes relying on Markov dependence, and they seem to be random in the sense of (3) (Nevill-Manning & Witten, 1999). If we generalize this result, semantically functional sequences are syntactically almost incompressible (Küppers, 2000, p.37).

Furthermore, algorithmic information theory helps to describe semantic features of the genetic code, which are important for its comparison to other codes. A measure for semantic predictability based on (2) is the mutual algorithmic complexity $AC(s_1{:}s_2)$ of two sequences s_1 and s_2,

$$AC(s_1{:}s_2) = AC(s_1) + AC(s_2) - AC(s_1, s_2). \quad (4)$$

$AC(s_1{:}s_2)$ measures the algorithmic complexity shared by s_1 and s_2 when the program describing one of the two sequences can generate, at least partially, the other string as well. $AC(s_1, s_2)$ means, in (4), the joint algorithmic complexity of s_1 and s_2, i.e., the algorithmic complexity $AC(s_1)$ of the first sequence plus the conditional algorithmic complexity $AC(s_2|s_1)$ of the second sequence given the program to generate the first one.

Thanks to (4), we can speak objectively about degrees of semantic contingency (as demanded by Godfrey-Smith, 2000, p.206). A function between two sequences s_1 and s_2 for which $AC(s_1{:}s_2) \approx AC(s_1) \approx AC(s_2)$ has highest semantic predictability: If we know the internal order of s_1, we know also the one of s_2, and vice versa. This is the case in very simple sign systems like those used in symbolic logic. The genetic code is, however, much more complicated.

In a syntactic order, two elements should be identified as variants of a single semantic element – or, expressed more formally, as syntactic members of the same semantic equivalence class (Sarkar, 2005, p.272) – if they can substitute each other without causing changes in a semantically connected syntactic order. Six different DNA codons encode synonymously the amino acid serine, and there are synonymous codons also for other amino acids. This so-called "degeneracy" of the code mapping between the DNA signal and the protein message is, on the one hand, the reason why syntactic information is lost in decoding the signal – degenerate codes lead, thus, to a unidirectional information flow as stated by the Central Dogma (Yockey, 1992, p.105ff.). Due to degeneracy the semantically connected syntactic orders show, on the other hand, a different inner organization, so that semantic predictability decreases: $AC(s_1{:}s_2)$ is then both $<AC(s_1)$ and $<AC(s_2)$.

The genetic signal shows, moreover, an essential difference between the level of nucleotides and the level of codons. Whereas codons enter into semantic relations to amino acids, nucleotides do not do so. Codons, as amino acid-encoding units, are built of elements that are primarily just differentiating semantically between amino acids. On the level of such merely discriminating elements semantically connected syntactic orders do not have to be similar. So both the $AC(s_1)$- and the $AC(s_2)$-value can be high with respect to the program generating s_2 and s_1, respectively. In other words: $AC(s_1{:}s_2)$ can be very low compared to $AC(s_1)$ and $AC(s_2)$ because $AC(s_1, s_2)$ comes close to $AC(s_1) + AC(s_2)$.

In the DNA signal, three types of information-carrying syntactic units must be distinguished in order to capture the semantic relativity and complexity of genetic information (see Section 2). The first type consists of codons that encode amino acids. The syntactic units of the second type do not enter into a code mapping with a syntactic unit of the protein message; instead, they enter into code mappings with other syntactic units of the very same signal, e.g., codons that encode the termination of a cistron (the complete coding of a protein). The third type of information-carrying syntactic unit appears at the level of the operon, the transcriptional unit for mRNA. Its DNA sequence contains a promoter, which integrates the operator: the site where, in case of negative regulation, the repressor can bind DNA to prevent the transcription of the operon's protein-coding part. As a binary switch, the operator determines whether the genetic encoding of proteins is to be actually read, and constitutes thus an alternative, which consists of two mutually exclusive syntactic elements. Both enter into "meta-"code mappings to cistron encodings of proteins. The operon meta-code constitutes a set of semantic boundary conditions on the syntactic transcription of the genome and makes thereby pragmatic selections in the semantics of codewords possible.

Acknowledgment

I would like to thank the editors, Anya Plutynski and Sahotra Sarkar, for their very helpful comments on two preliminary versions of this entry.

References

Adami, C. (1998). *Introduction to artificial life*. New York: Springer.

Artmann, S. (2004). Four principles of evolutionary pragmatics in Jacob's philosophy of modern biology. *Axiomathes*, 14, 381–95.

Bar-Hillel, Y., & Carnap, R. (1964). An outline of a theory of semantic information. In Y. Bar-Hillel (Ed.). *Language and information: selected essays on their theory and application* (pp. 221–74). Reading, MA: Addison-Wesley and Jerusalem: Jerusalem Academic Press.

Beatty, J. (1995). The evolutionary contingency thesis. In G. Wolters & J. G. Lennox (Eds). *Concepts, Theories, and Rationality in the Biological Sciences*, 45–81. Pittsburgh: University of Pittsburgh Press.

Chaitin, G. J. (1974). Information-theoretic limitations of formal systems. *Journal of the ACM*, 21, 403–24.

Cherry, C. C. (1966). *On human communication* (2nd edn). Cambridge, MA: MIT Press.

Cover, Th. M., & Thomas, J. A. (1991). *Elements of information theory*. New York: Wiley-Interscience.

Crick, F. H. C. (1958). On protein synthesis. *Symposia of the Society for Experimental Biology*, 12, 138–63.

Crick, F. H. C. (1968). The origin of the genetic code. *Journal of Molecular Biology*, 38, 367–79.

Dretske, F. I. (1981). *Knowledge and the flow of information*. Cambridge, MA: MIT Press.

Floridi, L. (2003). From data to semantic information. *Entropy*, 5, 125–45.

Floridi, L. (2004). Open problems in the philosophy of information. *The Journal of Philosophy*, 35, 554–82.

Gatlin, L. L. (1972). *Information theory and the living system*. New York and London: Columbia University Press.

Godfrey-Smith, P. (2000). Information, arbitrariness, and selection: comments on Maynard Smith. *Philosophy of Science*, 67, 202–7.

Griffiths, E. (2001). Genetic information: a metaphor in search of a theory. *Philosophy of Science*, 68, 394–412.

Jablonka, E. (2002). Information: its interpretation, its inheritance, and its sharing. *Philosophy of Science*, 69, 578–605.

Kay, L. E. (2000). *Who wrote the book of life? A history of the genetic code*. Stanford: Stanford University Press.

Knight, R. D., Freeland, S. J., & Landweber, L. F. (1999). Selection, history and chemistry: the three faces of the genetic code. *Trends in Biochemical Science*, 24, 241–7.

Kolmogorov, A. N. (1993). *Selected works* (Vol. 3): *Information theory and the theory of algorithms*. Dordrecht, Boston, London: Kluwer.

Küppers, B-O. (1990). *Information and the origin of life*. (P. Woolley, Trans.). Cambridge, MA: MIT Press. (Original work published 1986).

Küppers, B-O. (1996). The context-dependence of biological information. In K. Kornwachs & K. Jacoby (Eds). *Information: new questions to a multidisciplinary concept* (pp. 137–45). Berlin: Akademie Verlag.

Küppers, B-O. (2000). The world of biological complexity: origin and evolution of life. In S. J. Dick (Ed.). *Many worlds: the new universe, extraterrestrial life, and the theological implications* (pp. 32–43). Radnor, PA: Templeton Foundation Press.

Loewenstern, D., & Yianilos, P. N. (1999). Significantly lower entropy estimates for natural DNA sequences. *Journal of Computational Biology*, 6, 125–42.

MacKay, D. M. (1969). *Information, mechanism and meaning*. Cambridge, MA: MIT Press.

Maynard Smith, J. (2000a). The concept of information in biology. *Philosophy of Science*, 67, 177–94.

Maynard Smith, J. (2000b). Reply to commentaries. *Philosophy of Science*, 67, 214–18.

Monod, J. (1971). *Chance and necessity*. (A. Wainhouse, Trans.). New York: Knopf (Original work published in 1970).

Morris, C. W. (1938). *Foundations of the theory of signs*. Chicago: University of Chicago Press.

Nevill-Manning, C. G., & Witten, I. H. (1999). Protein is incompressible. In J. A. Storer &M. Cohn (Eds). *Proceedings of data compressing conference* (pp. 257–66). Los Alamitos, CA: IEEE Press.
Percus, J. K. (2002). *Mathematics of genome analysis*. Cambridge: Cambridge University Press.
Ridley, M. (1996). *Evolution* (2nd edn). Cambridge, MA: Blackwell.
Román-Roldán, R., Bernaola-Galván, P., & Oliver, J. L. (1998). Sequence compositional complexity of DNA through an entropic segmentation method. *Physical Review Letters*, 80, 1344–7.
Sarkar, S. (2000). Information in genetics and developmental biology: comments on Maynard Smith. *Philosophy of Science*, 67, 208–13.
Sarkar, S. (2005). *Molecular models of life. Philosophical papers on molecular biology*. Cambridge, MA: MIT Press.
Schrödinger, E. (1992). *What is life?* Canto edition. Cambridge: Cambridge University Press. (Original work published 1944).
Shannon, C. E. (1948). A mathematical theory of communication. *Bell System Technical Journal*, 27, 379–423, 623–56.
Shannon, C. E. (1956). The bandwagon. *IRE Transactions on Information Theory*, IT-2 (March), 3.
Sterelny, K. (2000). The "genetic program" program: a commentary on Maynard Smith on information in biology. *Philosophy of Science*, 67, 195–201.
Thornton, J. M., Todd, A. E., Milburn, D., Borkakoti, N., & Orengo, C. A. (2000). From structure to function: approaches and limitations. *Nature Structural Biology*, 7, 991–4.
Weismann, A. (1893). *The germ-plasm: a theory of heredity*. (W. N. Parker and H. Rönnfeldt, Trans.). New York: Charles Scribner's Sons. (Original work published 1892).
Wicken, J. S. (1987). *Evolution, thermodynamics, and information: extending the Darwinian program*. New York: Oxford University Press.
Winnie, J. A. (2000). Information and structure in molecular biology: comments on Maynard Smith. *Philosophy of Science*, 67, 517–26.
Yockey, H. P. (1992). *Information theory and molecular biology*. Cambridge: Cambridge University Press.

Further Reading

Arndt, C. (2001). *Information measures: information and its description in science and engineering*. Berlin, Heidelberg, New York: Springer.
Avery, J. (2003). *Information theory and evolution*. Singapore: World Scientific.
Barbieri, M. (2003). *The organic codes: an introduction to semantic biology*. Cambridge: Cambridge University Press.
Barbieri, M. (Ed.). (2007). *Introduction to biosemiotics: the new biological synthesis*. Dordrecht: Springer.
Berlekamp, E. R. (Ed.). (1974). *Key papers in the development of coding theory*. New York: IEEE Press.
Biological Theory, 1(3), Summer (2006). Special issue on biological information. Cambridge, MA: MIT Press.
Harms, W. F. (2004). *Information and meaning in evolutionary processes*. Cambridge: Cambridge University Press.
Li, M., & Vitányi, P. M. B. (1997). *An introduction to Kolmogorov complexity and its applications* (2nd edn). New York, Berlin, Heidelberg: Springer.
Oyama, S. (2000). *The ontogeny of information: developmental systems and evolution* (2nd edn). Durham, NC: Duke University Press.
Slepian, D. (Ed.). (1974). *Key papers in the development of information theory*. New York: IEEE Press.
Weber, B., Depew, D., & Smith, J. (Eds.). (1988). *Entropy, information and evolution*. Cambridge, MA: MIT Press.

Chapter 3

Heredity and Heritability

RICHARD C. LEWONTIN

The problematic of the study of heredity derives from the observation that offspring tend to resemble their parents and other close relatives more than they resemble unrelated organisms. Lions give birth to lions, lambs to lambs, larger parents have offspring who are on the average larger, the children of rich people are usually richer than the average, and the offspring of English speakers know how to write in English, but seldom in Amharic. For single celled organisms that reproduce by division, the resemblance of the daughter cells to each other and to the cell from which they arose appears, at first sight, to be unproblematic. There seems to be simply a symmetrical division followed by growth of the duplicated objects. For multicellular organisms that begin life as a single fertilized egg, however, it is immediately obvious that there is a much deeper problem underlying the question of resemblance of parents and offspring. The fertilized eggs of lions and lambs look extremely similar to each other but neither has any resemblance at all to the adult organisms that will be produced. Parents do not pass on their characteristics to their offspring, but rather they instigate processes of development that have the property that they eventuate in a final state that is similar to the parental state. Thus, the problem of resemblance, which we usually think of as the problem of heredity, is in reality a problem of differential growth and development. Indeed, the problem of heredity was understood in the nineteenth and early twentieth centuries as a *part* of the problem of development. August Weisman, the originator of the theory of the germ-plasm, and T. H. Morgan, whose work elucidated the physical nature of Mendel's "factors," were established embryologists before they turned their interest to heredity, while E. B. Wilson, a founder of modern cell biology, made the connection explicit in his *magnum opus*, *The Cell in Development and Heredity*. Mendel was an informative exception. He was a physics student who was recruited by the Abbe Knapp, head of the Koeningensklöster at Brno, to try to establish a quantitative basis for understanding variation in plants, as an aid to Bohemian agriculture. This was an effort to create a set of formal predictors of the outcome of crosses, not an investigation of the material basis of those regularities. Yet Mendel's Laws turn out to describe the physical properties of hypothetical entities that link heredity and development. Mendel's "factors" relevant to a particular character are pairs of discrete particles possessed by parents, one member of each pair being passed to the next generation in each gamete. While these particles interact in determining the development of the mature organism

(Law of Dominance), they remain physically separate and uncontaminated either by the other member of the pair or by factors for other characters during development, so that they are passed on unchanged by the offspring to their offspring in turn (Law of Segregation and Law of Independent Assortment). Moreover, the Law of Dominance makes a specific claim about the nature of development, namely that one member of a factor pair will completely dominate development of a particular character, regardless of the presence of the other paired factor. While there is no Mendelian "Law of Independent Development," the analysis of crosses involving more than one variable character showed no interaction of the development of different characters, making it possible to infer a Law of Independent Assortment from these crosses. Mendel's Laws are as much claims about development as they are about heredity.

Until the rise of molecular genetics, which makes it possible to study the physical material of the inherited "factors," the genes, and to identify genetic variants by a direct observation of the molecular structure of the genic material itself, studies of heredity were necessarily studies of the outcome of developmental processes. The only exceptions were direct microscopic observations of a few cases of physically altered chromosomes, but even these were largely attempts to correlate such physical changes with altered development. Almost the entire history of the study of heredity since the rediscovery of Mendel's paper in 1900, including much of present-day studies of human genetics, is the study of the outcome of developmental processes and an attempt to make genetic inferences from the manifest appearance of organisms. But this raises the question of the relation between genotype and phenotype. While it is a fair inference that the distinction between lions and lambs is a consequence of the difference in their genomes, the way in which genetic differences influence differences in body size is more complex, involving nutrition as well as genes, the causal chains explaining the passage of social and economic power from parent to offspring yet more complex, and it is clear that the fact that the offspring of native speakers of English can write English is a social product in which the characteristics of both parents and the offspring come from the society at large. The investigation of the heredity of any character variation in any species must then begin with an investigation of the relationship between genetic and non-genetic influences on development.

1. The Relation of Genotype to Phenotype

The basic distinction that must be made in understanding heredity is that between *genotype* and *phenotype*. The genotype of an organism is the class to which it belongs, as specified by the physical specification of its genes. The organism's phenotype is the class of which it is a member, as specified by its biochemical, morphological, physiological, and behavioral characteristics. In practice, neither the entire genotype nor the entire phenotype of an organism can be specified, so actual observations, experiments, and inferences are based on partial genotypes and their relation to partial phenotypes. From their definitions it is clear that a genotype, being a physical description of the organism's genes, is also a partial phenotype. The fundamental phenomenological concept in understanding the influence of genotype on development of phenotype is the *norm of reaction*. The norm of reaction of a genotype is the mapping of environments

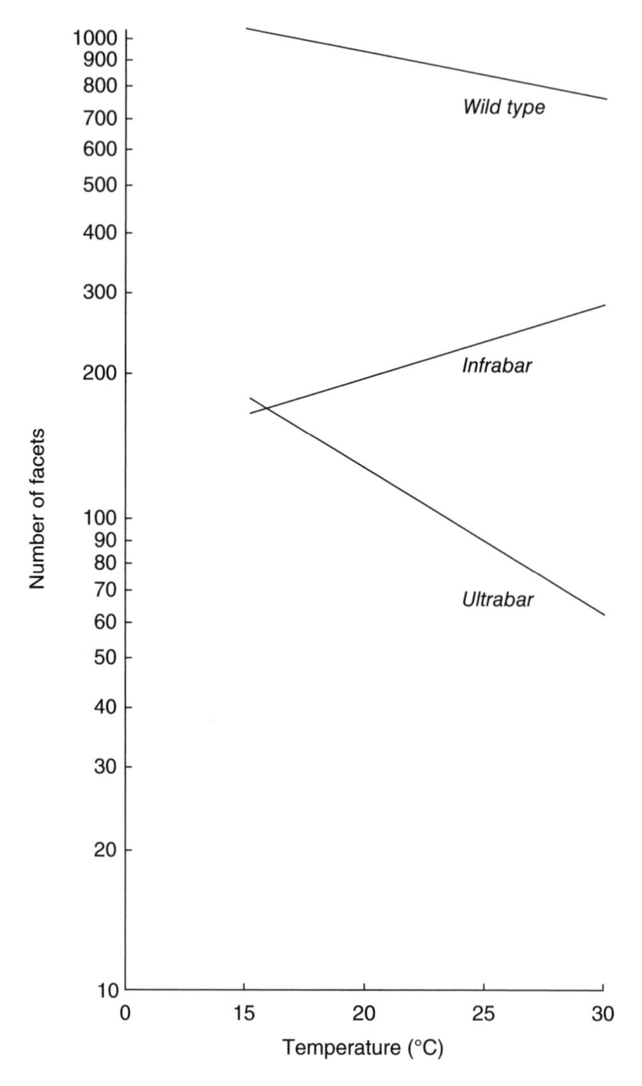

Figure 3.1 Norms of reaction of eye size as a function of temperature for three different genotypes of *Drosophila melanogaster* (from Lewontin, 2001)

in which the genotype may develop into the corresponding outcomes of development of that genotype. Typically, the norm of reaction is graphically represented as the measured phenotype plotted against some environmental variable. Figure 3.1 shows the norms of reaction of three different genotypes influencing the size of the eye of the fly, *Drosophila melanogaster*, as a function of the temperature at which the flies developed. In a genetically normal strain, eye size decreases with increasing temperature. In two mutant genotypes, the eye size is much smaller than normal at all temperatures, but for one the size decreases with temperature while for the other the size increases with temperature and they cross each other at about 15° C. One of the most extensive studies of norms of reaction is that of Clausen, Keck, and Hiesey (1958) on clones of the plant

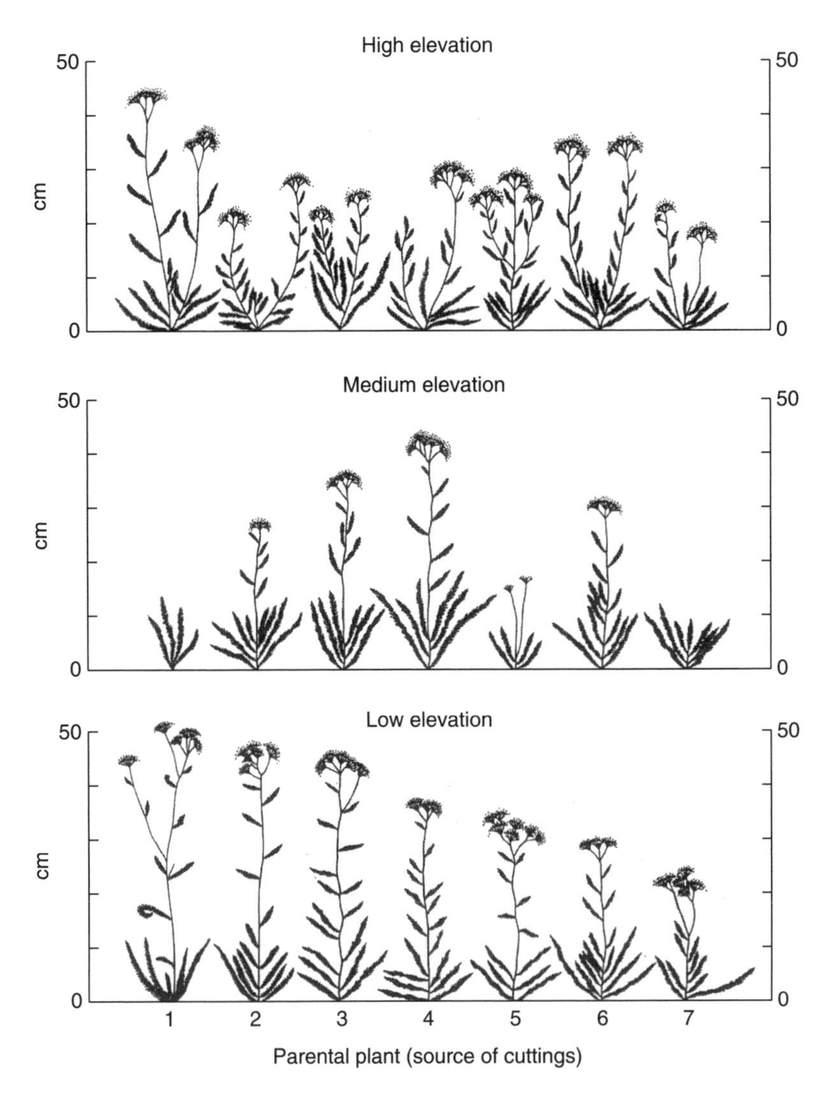

Figure 3.2 Growth of clones of *Achillea millefolium* at three different elevations. Plants arranged horizontally are seven different genotypes. Shown vertically for each plant are the growths of the three clones at three different elevations (from Clausen et al., 1958)

Achillea millefolium in California. Plants were collected from a natural population, each plant was cut into three pieces and the pieces were planted at a low, medium, or high elevation. Figure 3.2 shows a typical outcome of the experiment. The bottom row shows the different plants when they developed at the low elevation, arranged serially by decreasing size. The two rows above show the two other clones of each plant when a clone developed at medium or high elevation. The growth or flowering of a genotype at one elevation does not predict its relative growth at a different elevation. For example, the genotype with the best growth at low elevation is the poorest at medium elevation, while the second worst at low elevation is the second best at high elevation. There is, in fact, no correlation on the average between the growths in the different environments. Each genotype is characterized by its own pattern of interaction with the environment and each environment by its own pattern of interaction with different genotypes. This is not an exceptional outcome. It is characteristic of many studies in both plants and animals that development of a genotype in one environment is not a good predictor of its performance in other environments. Development is a consequence of a unique interaction between genotype and environment.

The degree to which variation in phenotype is affected by environmental or genetic variation differs widely. There are genotypes whose phenotypes are lethal or deformed in virtually all environments because some basic biochemical or physiological process necessary to life or development is missing. For example, an albino mutation in mice prevents the formation of any hair pigment irrespective of developmental environment, because an enzyme that is necessary for pigment formation is not produced in the absence of the normal gene. This accords with a model of development in which the environment provides the materials necessary for growth and development, while the conversion of these general materials into one specific structure rather than another is a consequence of the genetic "blueprint." The analogy is with bricks, mortar and wood which may be turned into any sort of structure, the specific form of which is determined by the prior plan. This model of development is illustrated in Figure 3.3a. At the other extreme is the case of the ability to pronounce the phonemes of English, as opposed to those of an African click language. In this model the genes determine the development of anatomical features that make any speech at all possible, but the particular features of that speech are acquired entirely from the particular social environment. This model is illustrated in Figure 3.3b. These two cases are the classical models of "genetic determination" and "environmental determination" that characterize the "nature versus nurture" debates. Such extreme cases are exceptions for organisms in general. Usually there is an interaction between genotype and environment as illustrated by the *Achillea* example and summarized in the model of Figure 3.3c.

There is yet a third factor, in addition to genotype and environment, that accounts for variation in development. The genes in the cells are the same on the right- and left-hand sides of a developing fruit fly. The pupa of the fly is about 3 mm long and in laboratory culture undergoes its development into an adult with its ventral surface stuck to the side of a glass vial so that the developmental environment of left and right sides are the same. Yet the number of sensory bristles that develop on the left and right sides are not the same. In one fly there will be six on the left and eight on the right, in another seven on the left and five on the right, and so on. The average number is the same on both sides but there is a fluctuating asymmetry from individual to individual. This fluctuating asymmetry is characteristic of bilaterally "symmetrical" organisms. It is a consequence of random events in development, a randomness that is traceable to the very small number of copies of biologically important molecules in cells. When cells divide they do not distribute these molecules exactly equally to the daughter cells. If some total number of copies of a molecule is necessary for further division then the daughter cells will require different times before they too can divide. Moreover, reactions between molecules require that they be in physical proximity and in the correct vibrational states, both of which are subject to random variation. When there are very few copies of such molecules there will then be a large variation in waiting time between

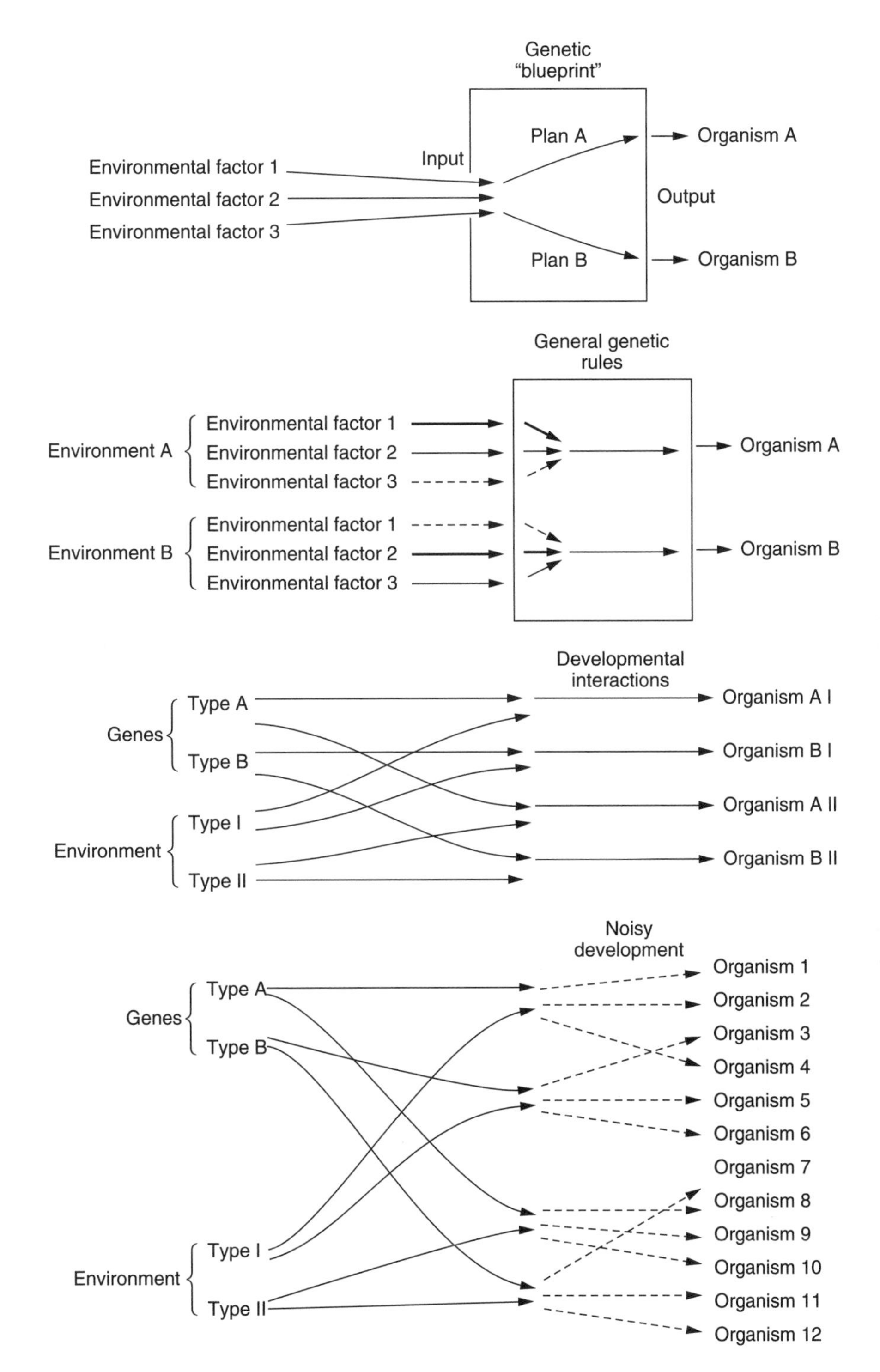

Figure 3.3 Four different models of the determination of phenotype. a) model emphasizing determination by genes; b) model emphasizing the role of the environment; c) interactive model showing the unique interactions of genes and environment; d) full model taking account of gene-environment interaction and of developmental stochasticity (from Lewontin, 2001)

successive biochemical reactions. To form a sensory bristle an original cell must divide twice to form a bristle-producing cell, a socket-producing cell and a nerve cell. But the time it takes to perform these cell divisions depends on how many copies of various molecules are included, by chance, after each division. This cluster of cells must then migrate from an interior position into an outer layer of the fly's developing integument. While this process of cell division and migration is occurring the outer layer of the integument is hardening so a cell cluster that arrives too late by chance will not succeed in producing a bristle. Stochastic events in cell metabolism and cell division explain, for example, why bacterial cells in a liquid culture initiated from a single bacterium do not divide simultaneously. Thus, the correct model of the determination of phenotype, shown in Figure 3.3d, contains not only the specification of both the genotype and the temporal order of environments during development, but also the effect of the "developmental noise" that is consequent on molecular stochasticity.

Environmental effects on development present two opposite problems for the study of heredity. The greater similarity of offspring to their parents as compared to their similarity to unrelated individuals is the basic observation on which an inference of biological inheritance is built. But if the environments of parent and offspring are different, then that similarity is reduced or even abolished. In the case of the *Achillea* experiment, clonally propagated individuals that developed in different environments were no more similar to each other than they were to other clones. There was no average correlation in height between individuals of the same clone. From such an observation no inference of heredity can be made, yet it is also clear that genotype has an influence on height since the different clones showed different norms of reaction across the same set of environments. Thus the evidence of heredity in this case comes not from similarity between relatives but from the diversity of reaction norms. But the determination of a reaction norm requires the duplication of the genotype as in clonal reproduction, which is rarely possible, or else the isolation of individuals of the same partial genotype identifiable either by the direct chemical characterization of some of their genes or by a characteristic phenotype that appears in some controlled environment, like the eye size mutations shown in Figure 3.1.

The opposite problem that arises from environmental effects is that similarity between related individuals may be increased by similar environments or even created in the absence of any genetic similarity. Biological heredity is not the only form of inheritance. In humans, money, education, nutrition, speech patterns, values, and attitudes are all inheritable by social mechanisms. Two characteristics in American populations that show consistently high correlation between parent and offspring are political party affiliation and religious sect, yet no one seriously proposes that there are genes for voting Republican or being a Methodist. Even morphological differences can be inherited entirely socially, as in groups that practice head-binding or circumcision in infants. A more general biological source of non-genetic similarity is the phenomenon of maternal effect. Fertilized seeds and eggs carry, in addition to the genes of the parents, nutrients and self-reproducing virus-like particles. Antigens are carried across placental membranes. More generally, maternal size and nutritional status influence the rate of growth of offspring so that large mothers may have large and vigorous progeny. The consequence of the various mechanisms of non-genetic inheritance is that a correct inference about genetic inheritance from an observed similarity between parent and offspring or any other comparison of related individuals can only be made when nongenetic sources of similarity have been accounted for or eliminated.

Because of the difficulties of genetic inference raised by the interaction of genotype and environment in development, geneticists have been biased toward using character differences for which environmental effects and developmental noise are small compared to the differences between genotypes, so that different genotypes can be unambiguously identified. The known mutations of the fruit fly, *Drosophila melanogaster*, the organism on which most of classical genetic work was done, are classified by "Rank." Flies carrying one or two copies of Rank I mutants differ from normal in every individual under the usual range of laboratory culture conditions so that the genotype of an individual can be unambiguously read from its phenotype. Rank II to Rank V mutants manifest their genotype phenotypically only in some variable fraction of individuals and under some restricted range of environmental conditions. No sensible geneticist will work with Rank V mutations. The alternative is to avoid the ambiguities raised by development and to study traits that are closer in the causal chain to the genes themselves. The first movement in this direction was biochemical genetics which used as phenotypic traits the enzymes and other proteins that were manufactured by cells as the direct product of reading and translating the genes. The principle of this school of genetics was "one gene, one enzyme." The problems of environmental and random effects are not entirely eliminated, however, because the timing and amounts of protein produced are under the influence of control systems in cells which, in turn, are influenced by environment. The final step in eliminating all developmental contingencies is the direct sequencing of the DNA, collapsing the genotype–phenotype distinction.

The long history of using Rank I mutants, biochemical traits, and DNA sequencing has resulted in a consciousness among geneticists that genes *determine* organisms. In this view one need only have the complete DNA sequence of an organism and a large enough computer and the organism can be computed. Genetics thus becomes DNA-centered. DNA is said to be self-replicating and the maker of proteins, which, in turn, produce all the structures and metabolic activities of the organism. In fact, DNA is not "self-reproducing" nor does it "make" anything. DNA is biochemically inert. New copies of DNA are manufactured by a cell machinery consisting of enzymatic proteins and a supply of small molecular materials. Proteins are the folded state of long chains of amino acids assembled by a protein machinery which reads the DNA sequence to determine the order of the amino acids. The folding of these chains into active proteins is only partly determined by the sequence of amino acids and partly by intracellular conditions. The timing and amount of production of various proteins by this machinery is controlled by complex feedback systems involving cellular environment, proteins, and RNA that bind to the genic DNA. The best metaphor for the genic DNA is that it is a library of recipes that is consulted by the cellular machinery, a library that is copied by the cell machinery and passed on to future generations.

A consequence of the removal of development from an integral role in the study of inheritance is the creation of a special discipline, Developmental Genetics, whose subject matter is the description of development in terms of the control circuits that determine the reading of particular genes at particular times and particular places in the life history of the organism. It then takes seriously the metaphor of "development," literally the "unfolding" (in Spanish, *desarollo* and in German *Entwicklung*, "unrolling"), of a preexistent program already contained in the fertilized egg. The entire process of development is seen as the result of a genetic machinery which, once set in motion, produces an end product independent of external contingencies. It accords then with the model of development in Figure 3.3a. It may, of course, be that when this program of description is completed the next step will be the introduction of environmental inputs and stochastic irregularities into the framework of the machinery to produce a more realistic model of phenotype production. Elements of such a move are already in place. One of the first discoveries of controlling circuits for the cell's reading of genes was the phenomenon of inducible enzymes. The gene coding for a particular enzyme is not read by the cell unless it is induced to do so by the appearance in the cell of the substrate on which the enzyme is to work. The classic example of such an induced enzyme formation is the galactosidase enzyme in bacteria which is only manufactured when there is lactose in the medium on which the bacteria are growing. The stage is thus set for an eventual incorporation of external conditions into the complete picture of the circuitry of development.

2. Statistical Approaches to the Study of Quantitative Characters

The picture of inheritance built up by the study of the effects on phenotype of single genes of large and unambiguous effect created, in the history of genetics, a major problem in understanding the genetics of characters like size and shape which, although clearly heritable, do not conform to the simple outcomes of crosses that Mendel found. One of the examples on which Mendel built his explanatory scheme was the inheritance of the difference between tall and short pea plants. A cross of tall with short produced a progeny generation that were all as tall as the tall parent. When these tall offspring were then crossed to each other they produced a generation in which the ratio of tall to short plants was the now classic 3 : 1. That is, plant height differences were the consequence of two alternative forms of a single gene, with Tall dominant to Short. But such a result is not generally characteristic of crosses between plants of different height. Mendel had the good fortune (or sense) to work with horticultural varieties characterized by mutations of large effect in a single gene. The more usual outcome is that the hybrids between tall and short plants are intermediate in height. When these intermediates are then crossed their offspring show a continuous distribution of heights spanning the difference between the original parents. While some of this variation is environmental or stochastic there is also heritability of the height differences. If two tall plants from the second generation are crossed with each other they will produce offspring with some variation but all will be at the tall end of the scale. What is to be made of such heritable variation? How can quantitative phenotypic traits be incorporated into an explanatory device designed to deal with qualitative phenotypic differences? In the early part of the twentieth century, before the hegemony of Mendelian genetics was established, such observations were the basis for alternative schemes of explanation for heredity. Continuous distributions of phenotypes suggested some underlying continuity of the mechanism of inheritance. In order to establish Mendelism as uncontroversial it was necessary to incorporate observations of continuously varying characters into the standard explanatory scheme based on discontinuous Mendelian factors and to provide methods for their analysis. Two landmarks of this project were R. A. Fisher's (1918) revealingly named paper, "On the correlation between relatives on the supposition of Mendelian inheritance" and his later book, *The Genetical Theory of Natural Selection* (1930). The scheme built by Fisher and elaborated thereafter in the discipline of "biometrical genetics" rested on two genetic and three developmental assumptions. The basic genetic claim was that continuously distributed character differences were not the consequence of variation in state of single genes of major developmental effect which could be identified and analyzed by Mendel's methods, but by a very large, but unspecified, number of Mendelian genes of equal but small individual effect which could not be separately investigated. Second, as suggested by Mendel's Law of Independent Assortment, the many Mendelian factors affecting the quantitative trait would be passed to gametes independently of each other so that a hybrid between two extremely different lines, say a very tall strain mated with a very short strain, would produce gametes containing variable numbers of factors with positive and negative effects on height. The first developmental claim was that the observed characters were the sum of the small effects of the individual gene pairs. Second, Mendel's Law of Dominance was relaxed so that for each gene pair a heterozygote, +/–, would make an intermediate contribution to the total phenotype, varying anywhere from complete dominance to complete recessivity of +. The average degree of dominance then became a parameter of the model to be determined from the data. Third, it was assumed that some of the observed variation was a consequence of random environmentally caused variation that was not specific to genotype but was added on to the genetic effects. This polygenic model became the standard apparatus for plant and animal breeding research in which intrinsically quantitative characters, such as number of seed set per plant or butterfat content of milk or degree of resistance to disease, were the focus of interest. It also became the model for human geneticists who are concerned with studying the inheritance of a variety of psychic and cognitive traits that are characterized by numerical scores rather than by typological classification into alternative states such as "schizophrenic" and "normal." At one time it was proposed that the "polygenes" postulated by the model were a special class of elements with a different location on the chromosomes and a different chemistry than Mendelian genes, but no clear evidence has been found for such an hypothesis and it has been abandoned as a last relic of the earlier belief that Mendelism was insufficient to explain continuous variation.

The introduction of the polygenic model as the basic scheme of explanation of quantitative variation had the effect of changing the questions to be asked of the data from those that characterize an investigation of simple Mendelian traits. Analysis of the classic Mendelian factors enumerates the various alternative forms that a particular gene may take by accumulating various mutations of the gene. The physical location of the gene on a chromosome and its DNA sequence are determined. The developmental effect of each of the alternate allelic forms both in homozygous state and when heterozygous with each of the other allelic states is observed. To the extent possible, the developmental effects are explained by the properties of the proteins specified by the gene and by the control apparatus that determines the place and timing of their production. In those cases where a distinct phenotype is the result of the interaction between the results of reading of two genes, the enumeration and explanation of the developmental effects include the combinations of alternative gene forms for both factors. The entire apparatus of investigation and explanation is at the level of a material description of genes and the physico-chemical apparatus of development to which those genes are relevant. In contrast, the explanatory structure of the polygenic model is concerned with the estimation of parameters of statistical distributions of gene effects without individual physical characterization of any of the unknown but large number of hypothesized genetic elements. The questions are framed in terms of standard statistical properties of distributions: means, variation as measured by the variance of the distribution, and average intensity of relationship between quantities as measured by covariances and correlations. Phenotypic measurements are made on a large number of individuals that result either from a cross between individuals of a specified phenotype or of a specified closeness of pedigree relationship (sibs, half-sibs, parent–offspring, etc.), or from a random mating among the individuals in a heterogeneous population. The possible questions that are asked from an analysis and manipulation of the data include:

(1) What difference in mean of the character is there between crosses?
(2) What is the correlation or covariance between the phenotype of parents and the phenotype of offspring or between the phenotypes of other relations of various degrees? The estimation of such correlations obviously requires experiments in which it is possible to identify family lines.
(3) How much of the correlation and covariation between relatives results from their genetic similarity and how much from the similarity of their individual environments?
(4) What proportion of the variance of the observations is estimated to arise from genetic variation among individuals (*genetic variance*) as opposed to variation in the environment (*environmental variance*)? Unless special experiments are devised, environmental variation includes random developmental noise.
(5) What proportion of the variance can be ascribed to interaction between particular genotypes and particular environments (*genotype-environment interaction variance*) as opposed to the general effects of the variable environment?
(6) How much genetic variance arises from the average effect of substituting + alleles for – alleles over all possible combinations of alleles in individuals (*additive genetic variance*)? How much genetic variance arises from the effects of dominance in heterozygotes (*dominance variance*) or from specific interactions among different genes (*epistatic variance*)?
(7) How many genes are estimated to be involved in producing variation of the phenotype? What fraction of the genetic variation can be assigned to specific chromosomes?

The purpose of these questions is to establish whether there is any evidence at all suggesting genetic effects on the character in question, to provide a quantitative estimate of the “importance” of genetic effects in causing variation, and, if there is evidence of such genetic effects, to give quantitative estimates of the “importance” of various kinds of interactions between genes and between genes and environment. These quantitative estimates of “importance” serve the double purpose of providing a heuristic picture of the role of genes in development and, in the case of breeding programs, to help in the choice of breeding and selection programs that will best achieve the goal of changing the properties of the organisms by selective breeding. Questions 1 and 2 are meant to tell whether there is any influence at all of one generation on another. Question 3 asks specifically to what extent intergenerational influence is genetic as opposed to environmental. So, the simple observation that there is a high correlation in religious doctrine between parents and offspring cannot be taken as a demonstration of the influence of genes on religiosity. One of the purposes of adoption and fostering studies is to separate direct parental environmental influence from genetic causation. If it were found in an adoption study that there was a high correlation of children with their biological parents but a low or no correlation with their adopting parents, this would be taken as evidence of genetic effects. It is important, however, that such a result cannot be interpreted as indicating the ease with which a character can be changed by environmental manipulation, as we will see below. Question 4, like question 3, is meant to provide a quantitative estimate of the relative "importance" of genes, using the currency of genetic variance rather than correlation. Question 5 is meant to provide guidance in designing programs of genetic and environmental improvement. Question 6 is of direct relevance to the design of selection programs in plants and animals. The additive variance is a predictor of how much change in the mean of a population can be made by choosing a biased sample of parents in each generation. No matter how much genetic variance may be present for a character, if there is little or no additive genetic variance, little or no progress in changing the mean of the population can be achieved by simply choosing a biased sample of parents. Finally, the estimate of the number of genes and the chromosomal location of the genetic variance is meant primarily to bring the abstract statistical model back into some contact with a physical reality. It must be noted that estimating the proportion of genetic *variance* that is associated with a particular chromosome is not the same as estimating the proportion of all the *genes* affecting the character that are on that chromosome unless the assumption that all genes have equal effect is taken seriously.

In the usual application of the model of quantitative genetics to the issue of how "important" genes are in explaining variation, a central is concept is *heritability*, which is defined in terms of variance. Let V_T be the total variance of a measured character in some population. Further let V_G be the total estimated genetic variance, V_A be the estimated additive genetic variance, V_{NA} be the estimated non-additive genetic variance, V_E be the estimated environmental variance and $V_{G{\times}E}$ the estimate of genotype–environment interaction variance.

Thus the genetic variance in the population is

$$V_G = V_A + V_{NA}$$

and the total variance in the population is

$$V_T = V_G + V_E + V_{G{\times}E} = V_A + V_{NA} + V_E + V_{G{\times}E}.$$

The methods for estimating these various components of variance are beyond the scope of this discussion (see Falconer, 1970). We simply observe here that correlations between relatives of various degrees are an important source of estimates of variance components. The heritability of a character is then defined as the proportion of total variance V_T of a character that is assigned to genetic causes. *Broad heritability*, H^2, is the proportion of the total variance that is made up of the total genetic variance:

$$H^2 = \frac{V_G}{V_T}$$

while narrow heritability, h^2, is defined as the proportion of all the variance that is additive genetic variance:

$$h^2 = \frac{V_A}{V_T}.$$

Narrow heritability is used for the prediction of changes that can be made in a population by a biased selection of parents. If the difference between the mean of the selected parents and the population as a whole from which they were selected is D, then the expected change in the mean by breeding the next generation from these selected parents is Dh^2. Narrow heritability is chiefly of use in agricultural applications. For making claims about the "importance" of genes in influencing characters it is broad sense heritability, H^2, referred to simply as "heritability," that is relevant. Heritability in this sense, the proportion of all variance in the character that is assigned to genetic variation, is used widely to provide a heuristic for claims about the biological basis of variation, about possible changeability of characters by environmental alterations, and for assignment of differences between groups as biological. When used in this way it is intended to have broad ideological and programmatic consequences, especially when applied to human individuals and groups. There are, however, serious methodological errors and conceptual misunderstandings that must be avoided in estimating and interpreting heritability.

3. Problems Raised by Statistical Methodologies

First, there are methodological problems. Because of the role of correlations between relatives in inferences about genetic variation, it is vital that non-genetic correlations between relatives be avoided. A difficult problem is posed by physiological maternal effects. Poorly nourished mothers have low birth-weight offspring who are then poorly nourished as nurslings. HIV positive mothers have HIV positive babies. Starved fruit flies lay smaller eggs which develop into smaller adults. Stunted plants set smaller seed with fewer nutrients for early stages of the growth of seedlings. Adoption or fostering studies must, if possible, be adoptions at birth or if that is not possible then the effect of age at adoption must be assessed and corrected for. The adopting parents must have no average correlation with the biological parents in environmental variables that might reasonably be considered as possible factors in affecting the measured characters. Comparison of the similarity between identical twins, who are genetically identical, with the similarity between fraternal twins or sibs, who share only half their genes on average, are valid only if it can be shown that parents do not treat identical twins more similarly than they do sibs. There is abundant evidence that this is not the case, identical twins often being given similar names, identical dress, haircuts, toys, schooling, and so on. There are even contests (the so-called Twin Olympics in Twinsburg, Ohio, for example) that award prizes for the greatest apparent similarity.

Second, even in methodologically perfect comparisons yielding unbiased estimates of correlations and heritability, there is a deep conceptual misunderstanding about the relationship between heritability and changeability, because correlation and variance estimates are missing important information relevant to how easily a character can be changed. The result typically seen in adoption studies in which IQ scores of parents and children are measured is encapsulated in the following IQ scores created for the purpose of illustration:

	IQ scores		
	Child	**Biological mother**	**Adoptive mother**
	101	91	109
	102	92	107
	103	93	110
	104	94	108
	105	95	102
	106	96	101
	107	97	105
	108	98	104
	109	99	106
	110	100	103
Mean	105	95	105

In this illustrative example correlation between biological mother and child is perfect (1.00) while it is essentially nonexistent between child and adoptive mother (.015). (Correlation between sets of values does not measure identity but the extent to which increases or decreases in one set of values is matched by decreases or increases in the other set.) So we judge, correctly, that there is essentially complete heritability of IQ score. The greater the IQ of the biological mother the greater the IQ score of the child, but no such relation is seen with the adoptive mother. Nevertheless, the IQ scores of the children are uniformly 10 points higher than their biological mothers and their mean IQ is equal to the mean IQ of the adopting mothers. These numbers illustrate the commonly seen phenomenon that on the average the IQ scores of adopting parents are higher than those of mothers who give up their children for adoption and that the adopted children resemble, on the average, their adoptive parents. The source of this discrepancy between heritability and changeability lies in the nature of the statistical structures, variance and correlation. These measure dispersion of values around a mean (variance) and the way the dispersion of one set of values is coupled to another set of values (correlation). But dispersion around a mean is completely unchanged if the mean is changed by adding or subtracting a constant amount to or from every value. Moreover, even if different amounts are added or subtracted from the values, the change that may occur in the dispersion has no fixed relation to the change in the mean. The mean may increase while the variance decreases. So, in concrete terms, if some change in environment occurs that changes all the values of a character in some direction, thus increasing the mean, there is no constraint on how this may affect the genetic variance of the character or the correlation between relatives. Conversely, a partitioning of the variance of a character between genetic and environmental components makes no prediction about how much the character can be changed by an environmental change because that partitioning only characterizes the components of variance in the original set of environments.

The underdetermination of the effect of mean changes on the variance of a character is an example of a generally misunderstood aspect of the basic statistical apparatus known as the *analysis of variance*. The technique is used to assay the relative role played by different causal factors in determining the distribution of some property in a collection of observations. The usual implication drawn by the assignment of different amounts of variance to different causal factors is that the effects of the factors have somehow been separated, but that is wrong. This error can be illustrated in the case of genotype and environment by Figures 3.4a and b showing hypothetical norms of reaction for some character. The two norms of reaction are linear and of negative slope, crossing at an intermediate value of the environmental variable. These norms map the distributions of an environmental variable, say temperature, shown on the abscissa, into distributions of phenotypes, say height, shown on the ordinate. In Figure 4a the distribution of environments centers around high temperatures with the result that the distribution of phenotypes is bimodal, each subdistribution having its own mean corresponding to one of the genotypes. The shape of the combined bimodal distribution of phenotypes depends on the relative numbers of each of the two component genotypes. For illustration we have supposed them to be equally frequent. There is a great deal of genetic variance in the population as measured by the square of the difference between the two means, in addition to the environmental variance which is the average of the variances of the two underlying phenotypic subdistributions. Now suppose that there is a general drop in temperatures, shown in Figure 4b as a downward shift of the distribution of environments. The result is that the phenotypic distributions of the two genotypes now lie nearly on top of one another and the total distribution of phenotypes becomes unimodal. The very small difference in the mean phenotypes of the two genotypes has the result that the amount of genetic variance has been greatly reduced. So a change in *environment* has resulted in a change in the *genetic* variance. (No change has occurred in environmental variance in this simple case because both norms are linear.) Conversely, suppose that instead of moving the environment in Figure 4a, the environmental distribution remains unchanged but the relative representation of the two genotypes in the population is changed so that Genotype II now comprises 90 percent of the population. The result is a reduction in genetic variance, but a large increase in environmental variance of the total population because the more environmentally sensitive genotype now makes up the largest fraction. Thus, a change in *genotype* frequencies has changed the *environmental* variance. The partitioning of the variance into genetic and environmental components does not correspond to a separation of the causal pathways. The genetic variance is the variance in phenotype contributed by differences in genotype conditional on the current distribution of environments. The environmental variance is the variance in phenotype

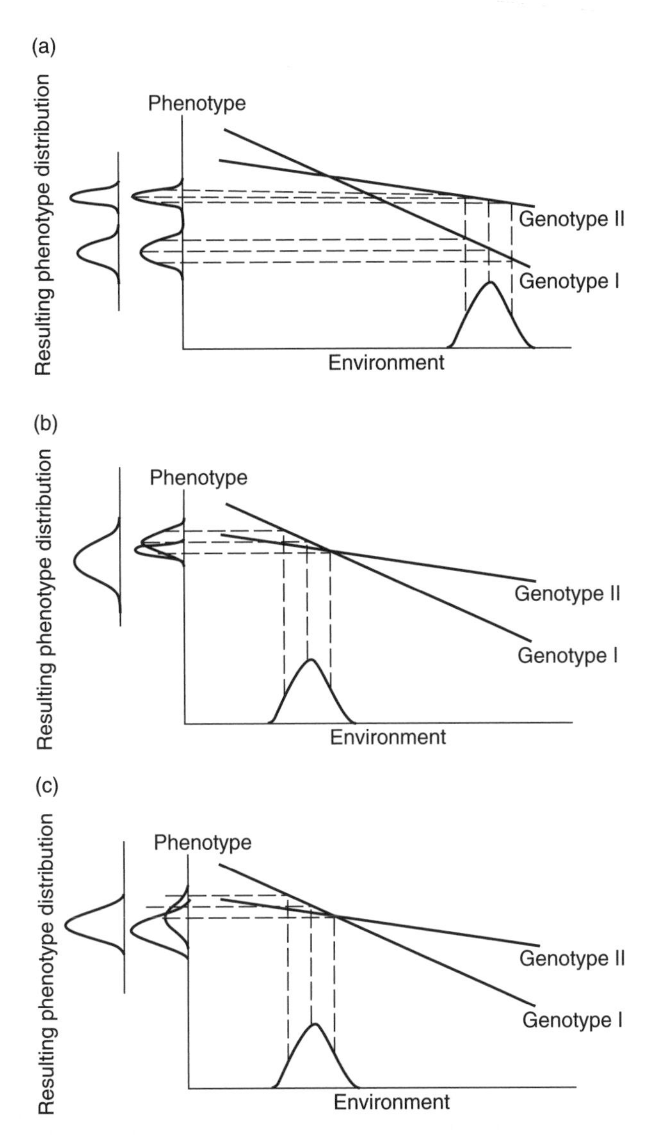

Figure 3.4 Hypothetical norms of reaction for two genotypes and the phenotypic distribution resulting from variation in the environment and genetic variation. a) An environmental distribution producing a bimodal phenotypic distribution with high genetic variance; b) An environmental distribution shifted so as to produce a unimodal phenotypic distribution with low genetic variance. (From Suzuki et al., 1981)

contributed by environmental differences conditional on the current distribution of genotypes. The analysis of variance is not an analysis of causes because the analysis is a local rather than global analysis.

Third, the heritability of a character within populations gives no information about the source of differences between populations. It is often argued, for example, that there is a reasonably high heritability of IQ scores within ethnic or racial groups so that it is probable that the differences in mean IQ scores between groups is largely genetic. But, irrespective of the correctness of the claim of high heritability of IQ scores within groups, the inference to between-group heritability is incorrect. The error is most easily seen by a hypothetical example. Suppose that one large handful of seeds from a genetically variable population is grown in a carefully prepared artificial culture solution, while another handful from the same source is grown in a similar solution from which an important nutrient has been left out. Among the plants that grow from the seeds there will be variation within each of the two lots that is entirely the consequence of the genetic variation from plant to plant, because there is no environmental variation within lots. There will also be a large difference in average growth between the two lots which is entirely the result of the environmental difference since the samples of genotypes are the same in the two lots. Thus there is a heritability of 1.00 within populations and a heritability of 0.0 between populations. Consider now, the opposite experiment. Large numbers of seeds from two different highly inbred lines are used in an experiment. Seed from each line is grown in individual pots, carelessly filled to different heights with potting soil taken from a poorly mixed combination of earth and fertilizer. The plants within lines will grow to different heights entirely because of random variation in potting conditions. There will be a difference in mean height between the two lines entirely because of the genetic difference between them. Thus, the heritability within lots is 0.0 while heritability between lines is 1.00. The observation of heritability of a characteristic within a population cannot be used as evidence about the heritability between populations. In general, the heritability of characteristics within populations contains, literally, no information about the heritability between populations and vice versa.

4. Making Quantitative Trait Genes Real

The statistical methods of investigation of the genetics of quantitative characters were, for a very long time, the only available attack on the problem. The assumption that such characters were based on large numbers of individual genes affecting development in fundamentally the same way as the classic Mendelian genes was consistent with the observations, but that consistency was largely the result of the freedom to make ad hoc assumptions about the numbers and physiological effects of the hypothetical genes of small effect. It is hard to see how any observations would have fatally contradicted what Fisher called "the supposition of Mendelian inheritance." The chief motivation for accepting the assumption was an appeal to parsimony. Nevertheless, genetics in general has been marked by a constant drive to link observed high-level regularities with detailed mechanical explanations based on low-level phenomenology. Genetics is a reductionist science par excellence. It was inevitable, then, that the development of molecular genetics would include an effort to make material the hypothesized polygenes. The result of this effort is a set of methods combining statistical, molecular, and classical techniques for the identification and physical and developmental characterization of what are now known as Quantitative Trait Loci (QTLs). The chief enabling tool has been the discovery of an immense store of previously undetected genetic variation, densely spread through the entire genome, that can be used as genetic markers for localizing and identifying QTLs influencing particular quantitative characters. These markers are single nucleotide variants with no apparent effect on the organism's phenotype, but their presence can be detected by DNA sequencing or other related methods of investigating changes at the level of DNA. Using the progeny of genetic crosses between lines that differ in some quantitative character, it is then possible ask what proportion of the quantitative trait difference is associated with a nucleotide variant at a particular spot in the genome. Entire genomes can be scanned and "hot spots" of association between nucleotide variants and significant differences in phenotypic score are identified. At first there is no suggestion that the nucleotide variant at such a spot is actually within a gene that is functionally connected to the phenotypic trait, but only that the DNA variant is a marker sufficiently close to a relevant gene that it segregates together with that unknown gene in crosses. The next step is to search for genes of known developmental influence in the immediate chromosomal vicinity of the nucleotide markers and thus identify *candidate genes* whose variation may turn out to be the causes of the phenotypic variation. The success of this last enterprise depends on how well known the genome of the organism is in terms of the developmental functions at the cellular level. While still in an early stage, this methodology brings quantitative characters into the main mechanistic scheme of genetics. Unfortunately the results so far obtained by QTL localization techniques have for the most part made the eventual mechanistic explanation seem very difficult. A typical result is a finding of a few dozen gene loci significantly associated with trait variation, none of which accounts for more than a small percent of the total variation. The number of relevant loci detected is typically smaller than was implied by the standard quantitative genetic model, but the results agree with that model in that the effects of individual genes are indeed quite small. Such small individual effects of multiple genes mean that it will be very difficult to provide an articulated developmental and physiological explanation of genetic variation in quantitative trait variation.

Other books by the author that are relevant to this chapter include *Not in Our Genes* with S. P. Rose and L. J. Kamin (1984, New York, Pantheon Books), *The Dialectical Biologist* with R. Levins (1985, Cambridge, MA, Harvard University Press), *Biology as Ideology* (1991, New York, Harper Perennial), and *The Triple Helix: Gene, Organism and Environment* (2001, Cambridge, MA, Harvard University Press).

Bibliography

Clausen, J., Keck, D. D., & Hiesey, W. (1958). *Experimental studies on the nature of species* (Vol. 3): *Environmental responses of climatic races of* Achillea millefolium. Carnegie Institution of Washington Publ. No. 581, pp. 1–129.

Falconer, D. S. (1970). *Introduction to quantitative genetics.* New York: Ronald Press.

Fisher, R. A. (1918). The correlation between relatives on the supposition of Mendelian inheritance. *Transactions of the Royal Society of Edinburgh*, 52, 399–433.

Fisher, R. A. (1930). *The genetical theory of natural selection.* Oxford: Clarendon Press.

Lewontin, R. C. (2001). *The triple helix; gene organism and environment.* Cambridge, MA: Harvard University Press.

Suzuki, D. T., Griffiths, A. J. F., & Lewontin, R. C. (1981). *Introduction to genetic analysis* (2nd edn). New York: W.H. Freeman.

Chapter 4

Genomics, Proteomics, and Beyond

SAHOTRA SARKAR

1. Introduction

The term "molecular biology" was introduced by Warren Weaver in 1938 in an internal report of the Rockefeller Foundation: "And gradually there is coming into being a new branch of science – molecular biology – . . . in which delicate modern techniques are being used to investigate ever more minute details of certain life processes."[1] Weaver probably only dimly foresaw that these new techniques would ultimately transform the practice of biology in a way comparable only to the emergence of the theory of evolution in the previous century. By the beginning of the twenty-first century molecular biology has become most of biology, either *constitutively*, insofar as biological structures are characterized at the molecular level as a prelude for further study, or at least *methodologically*, as molecular techniques have become a preferred mode of experimental investigation of a domain. Recent biological work at the organismic and lower levels of organization – cytology, development, neurobiology, physiology, etc. – increasingly fall under the former rubric. Work in demography, epidemiology, and ecology falls under the latter, with ecology perhaps being the sub-discipline within biology which has most resisted molecularization. Work in evolution falls under both: constitutively, when the evolution of molecules and molecular structures forming organisms is studied for its own sake, and methodologically, when molecular techniques (most notably, DNA sequencing) are used to reconstruct evolutionary history.

This chapter traces the conceptual shifts that have marked the development of molecular biology during the past half-century with an emphasis on epistemological issues raised by the more recent changes. Section 2 provides the background of classical molecular biology. Section 3 moves on to the genomic and post-genomic era. Section 4 analyzes the prospects for proteomics. Section 5 turns to the nascent project of systems biology. Finally Section 6 turns to the philosophical implications of these developments, namely, the status of reductionism, of the informational interpretation of molecular biology, and the prospect that systems biology will finally reintroduce dynamical considerations in molecular biology. Section 7 invites readers to pursue

1 As quoted by Olby (1974, p.442).

more philosophical exploration of the issues raised by molecular biology which have, until recently, often been ignored by philosophers.

2. Classical Molecular Biology

During the decade following Weaver's introduction of "molecular biology" experimental work showed that the hereditary substance – specifying "genes" [SEE GENE CONCEPTS] – was deoxyribonucleic acid (DNA). Attention then focused on deciphering the physical structure of DNA, a problem that was solved by Watson and Crick (1953) with their double helix model from 1953. The construction of this model and its subsequent confirmation was a development of signal importance for modern biology.[2] It ushered in the "classical" age of molecular biology with an intriguing informational interpretation of biology [SEE BIOLOGICAL INFORMATION]. Important conceptual innovation also came from Monod and Jacob in the early 1960s, who constructed the "allostery" model to explain cooperative behavior in proteins and the "operon" model of gene regulation.[3] Genes were interpreted as DNA sequences either specifying proteins (the *structural* genes) or controlling the action of other genes (the *regulatory* genes). Perhaps the most important development in classical molecular biology was the establishment of a genetic "code" delineating the relation of DNA sequences to amino acid residue sequences in proteins.[4] Gene *expression* took place by the *transcription* of DNA to ribonucleic acid (RNA) at the chromosomes (in the nucleus), and the *translation* of these transcripts into protein at the ribosomes (in the cytoplasm). The one gene–one enzyme credo of classical genetics was transformed into the one DNA segment–one protein chain credo of molecular biology.

Crucial to the program of molecularizing biology was the expectation – first explicitly stated by Waddington (1962) – that gene regulation explained tissue differentiation and, ultimately, morphogenesis in complex organisms. Genetic reductionism, the thesis that genes alone can explain organismic features, long predates molecular biology (Sarkar, 1998). However, the molecular interpretation of the gene allowed the general explanatory success of molecular biology to be co-opted as a success of molecular genetics. In such a context, Waddington's thesis was positively received and helped usher in an era dominated by *developmental genetics* according to which organismic development was to be understood through the action of genes. Mayr (1961) and Jacob and Monod (1961) independently introduced the metaphor of the genetic program to characterize the putative relation between genomic DNA and organismic development. As molecular genetics began to dominate the research agenda of molecular biology in the 1970s, the emergence of organismic features came to be viewed as determined by "master control genes" (Gehring, 1998). This view was initially supported by the demonstration that some DNA sequences (such as the "homeobox") were conserved across a wide variety of species. DNA came to be viewed as the molecule "defining" life, a view that

---

2 Sarkar (2005, ch. 1) argues this point in detail.

3 See below, and Monod (1971) and Jacob (1973).

4 Both DNA and protein are linear molecules in the sense that they consist of units connected in a chain through strong (covalent) chemical bonds.

helped initiate the massive genome sequencing projects of the 1990s, which were supposed to produce a gene-based complete biology that delivered on all the promises of molecular developmental genetics. In general, because of the presumed primacy of DNA in influencing organismic features, starting in the early 1960s, molecular genetics began to dominate research in molecular biology.

Thus, genetics and development were the earliest biological sub-disciplines to be reconstituted by molecular biology. In the case of evolutionary biology, as early as the 1950s, Crick (1958) pointed out that the genotype–phenotype relation could be reinterpreted as the relation between DNA and protein, with proteins constituting the subtlest form of the expression of a phenotype of an organism. Consequently, the evolution of proteins (and, later, DNA sequences), especially the question of what maintained their diversity within a population, became a topic of investigation. In the 1960s, these studies led to the neutralist challenge to the received view of evolution [SEE MOLECULAR EVOLUTION]. More importantly, changes at the level of DNA sequences, provided that these were selectively neutral, permitted the construction of a "molecular clock" that can arguably be used to reconstruct evolutionary history more accurately than what can be achieved by traditional morphological methods (even though such reconstructions have on occasion proved to be controversial).

Meanwhile, biochemistry and immunology also fell under the spell of the new molecular biology. That enzyme interactions and specificity would be explained in molecular terms was no surprise. However, immunological specificity was also believed to be explainable by the same mechanism. This model of immune action was coupled to a selectionist theory of cell proliferation to generate the clonal theory of antibody formation, which combined molecular and cellular mechanisms in a novel fashion [SEE SELF AND NONSELF]. In both biochemistry and immunology, what was largely at stake was the development of models that could explain the observed specificity of interactions: enzymes reacted only with very few substrates; antibodies were highly specific to their antigens.

Classical molecular biology can be viewed in continuity with both the genetics and the biochemistry of the era that preceded it. From biochemistry – in particular, the study of enzymes in the 1920s and 1930s – it inherited the proposed mechanism that the function or behavior of biological molecules is "determined" by its structure.[5] In the 1950s, structural modeling of biological macromolecules, especially proteins, was pioneered by Pauling and his collaborators using data from *x*-ray crystallography (e.g., Pauling & Corey, 1950). By the early 1960s a handful of such structures were fully solved. These structures, along with the structure of DNA, seemed to confirm the hypothesis that structure explains behavior. Perhaps more surprisingly, it was found that structural interactions seemed to be mediated entirely by the shape of active sites on molecules and that the sensitive details of structure and shape were maintained by very weak interactions.

These experimental observations led to four seemingly innocuous rules about the behavior of biological macromolecules which, in the 1960s and 1970s, formed the theoretical core of molecular biology:[6]

---

5 This idea is of earlier vintage, going back to Ehrlich's "side-chain" theory in the late nineteenth century.

6 For details, see Sarkar (1998, pp.149–50).

(i) the *weak interactions* rule – the interactions that are critical in molecular interactions are very weak;
(ii) the *structure-function* rule – the behavior of biological macromolecules can be explained from their structure as determined by techniques such as crystallography;
(iii) the *molecular shape* rule – these structures, in turn, can be characterized entirely by molecular size and, especially, external shape, and some general properties (such as the hydrophobicity) of the different regions of the surfaces;
(iv) the *lock-and-key fit* rule – in molecular interactions, molecules interact only when there is a lock-and-key fit between the two molecular surfaces. There is no interaction when these fits are destroyed.

Such a lock-and-key fit, based on shape, achieves what is called "stereospecificity," thus resolving the critical problem for classical molecular biology, which was to explain how structure specified behavior. Of the four rules introduced above, the molecular shape and lock-and-key fit rules are the most important because they are the ones that are most intimately involved in the explanation of specificity. In what follows, these four rules will be called the rules of classical molecular biology.

In the 1960s and 1970s these rules were deployed with remarkable success. As noted earlier, enzymatic and immunological interactions were among those that were immediately brought under the molecular aegis. Two other cases are even more philosophically interesting: (i) the allostery model explains why some molecules such as hemoglobin show *cooperative* behavior. In the case of hemoglobin, there is a nonlinear increase in the binding of oxygen after binding is first initiated. This is explained by conformational – shape – changes in the molecular subunits of hemoglobin as the first oxygen molecules begin to bind to them; and (ii) the operon model of gene expression explains *feedback*-mediated gene regulation in prokaryotes. This model explains how the presence of a substrate activates the production of a protein that interacts with it, and its absence inhibits that production.[7] Section 6 will emphasize the philosophical significance of the success of such structural explanation in molecular biology.

However, the 1950s also saw the elaboration of a radically different model of biological specificity, one based on the concept of information, which was only introduced in genetics in 1953 (Sarkar, 1996). This concept soon came to play a foundational role in molecular genetics. DNA was supposed to be the repository of biological information, a genetic "program" was supposed to convert this information into the adult organism, and new information was supposed to result from random mutation (when such mutations were maintained by selection). Information was never incorporated from the environment into the genome. Crick (1958, p.153) enshrined these assumptions in what he called the "Central Dogma" of molecular biology: "This states that once 'information' has passed into protein *it cannot get out again*. In more detail, the transfer of information from nucleic acid to nucleic acid, or from nucleic acid to protein may be possible, but transfer from protein to protein, or from protein to nucleic acid is impossible." Information, in Crick's model, was defined by the sequence of nucleotide bases

7 See Monod (1971) for an accessible accurate account of these two examples and a conceptual summary of theoretical reasoning in early molecular biology.

in DNA or the sequence of amino acid residue in protein molecules. Note the contrast here with the stereospecific physical model of specificity: specificity comes from the combinatorial order or arrangement of subunits in DNA and protein, and not from the physical shape. The Central Dogma has continued to be an important regulative principle of molecular biology in the sense that it is presumed for further theoretical reasoning. Whether it survives recent developments will be discussed later in this chapter.

By the late 1970s it became clear that the simplicity of the picture of genetics inherited from the 1960s was being lost. The initial picture was generated from an exploration of the genomes of prokaryotes (single-celled organisms without a nucleus), especially the bacterium, *Escherichia coli*. In prokaryotes, every piece of DNA has a structural or regulatory function. In the 1970s, it was discovered that the genetics of eukaryotes (organisms with cells with nuclei) turns out to have an unexpected complexity. In particular, large parts of the genomic DNA sequences apparently had no function: these segments of "junk" DNA were interspersed between genes on chromosomes and also within genes. After RNA transcription, non-coding segments within genes were *spliced* out before translation. Gene regulation in eukaryotes was qualitatively different and more complicated than in prokaryotes. Some organisms used nonstandard genetic codes, etc.[8]

Subsequent work in molecular biology has only enhanced the complexity of this picture, so much so that it is reasonable to suggest that the classical picture is breaking down. RNA transcripts are subject to *alternative splicing*, with the same DNA gene corresponding to several proteins. RNA is edited, with bases added and removed, before translation at the ribosome, sometimes to such an extent that it is difficult to maintain that some gene codes for a given protein. There is also no obvious relation between the amount of DNA in an organism and its morphological or behavioral complexity, an observation that is sometimes called the C-value paradox (Cavalier-Smith, 1978). Most importantly, it now appears that a fair amount of the so-called junk DNA is transcribed into RNA, though not translated. Thus, presumably, much of the so-called junk DNA is functional, though the nature of these functions remains controversial (see Section 3).

The complexities of eukaryotic genetics, as discovered in the 1970s and 1980s, already begin to challenge the Central Dogma.[9] Much of this work was made possible by the development of technologies based on the polymerase chain reaction (PCR) in the 1980s. There were five salient discoveries that challenged the simple picture inherited from prokaryotic genetics:[10]

(i) the genetic code is not fully universal, the most extensive variation being found in mitochondrial DNA in eukaryotes. However, there is also some variation across taxa (Fox, 1987);

(ii) DNA sequences are not always read sequentially in blocks. There are overlapping genes, genes within genes, and so on (Barrell et al., 1976). Thus, two or more

8 See Sarkar (1996) for a detailed account.

9 Thiéffry and Sarkar (1998) give a history of several earlier challenges. Even in the 1960s there was no unanimity about the status of the Central Dogma.

10 For details, see Sarkar (2005), chapter 8.

different proteins could be specified by the same "gene." Once again the Central Dogma is under challenge since the genome alone does not seem to contain all the information necessary to determine which protein is encoded by the "gene" in question;

(iii) as noted earlier, not all DNA in the genome is functional. Intervening sequences – within and between structural genes – must be spliced out from transcripts (Berget et al., 1977; Chow et al., 1977). This discovery helped resolve the *C*-value paradox mentioned earlier, that is, the absence of any obvious correlation between the size of the genome and morphological and behavioral complexity of an organism;

(iv) the same transcript may be spliced in different ways (Berk & Sharp, 1978). One consequence of such *alternative splicing* is that, as with overlapping genes, two or more different proteins could be specified by the same "gene";

(v) besides splicing, RNA is sometimes subject to extensive editing before translation at the genome (reviewed by Cattaneo, 1991).

Both points (iv) and (v) challenge the Central Dogma for the same reason as point (ii), These developments have led to increasing skepticism of the relevance of the coding model of the DNA–protein relationship and, especially, of the informational model of specificity (see Sarkar, 2005, and Section 6). It is no longer even clear that there is a coherent concept of information in molecular biology (see, however, BIOLOGICAL INFORMATION). Though philosophers – and some biologists – have been slow to recognize this, the one DNA segment–one protein chain credo has long become irrelevant in molecular biology. These developments in eukaryotic genetics paved the way to a reconceptualization of heredity in the emerging field of genomics.

3. Genomics and Post-Genomics

Genomics was ushered in by the decision to sequence the entire human genome as an organized project (the Human Genome Project [HGP]), involving a large number of laboratories in the late 1980s. Subsequently, similar projects were established to sequence the genome of many other species. To date, genomes of over 150 species have been sequenced. Almost every month sees the announcement of the completion of sequencing for a new species. The sheer volume of sequence information that has been produced has spawned a new discipline of "bioinformatics" dedicated to computerized analyses of biological data.

When the HGP was first proposed, there was considerable controversy among biologists about its wisdom (Tauber & Sarkar, 1992; Cook-Deegan, 1994). There were: (i) doubts about its ability to deliver on the bloated promises made by proponents of its scientific and, especially, its medical benefits; (ii) questions whether such organized "Big Biology" projects were wise science policy because of their potential effect on the ethos of biological research; and (iii) worries that society would be legally and medically ill-prepared to cope with the results of rapid sequencing, rather than the normal slower accumulation of human genomic sequence information. It was feared that legislation protecting genetic privacy and preventing genetic discrimination would not be in place; there would be a shortage of genetic counselors; and so on.

In one important aspect, the critics were correct: there have been few immediate medical benefits from the HGP and no significant such innovation seems forthcoming. Instead, recent work underscores the importance of gene–environment interactions that critics had routinely invoked to criticize the claims of the HGP [SEE HEREDITY AND HERITABILITY]. However, in another sense, even the most acerbic critics should now accept that the scientific results of the sequencing projects, taken together, have been breathtaking.

Contrary to the expectations of the HGP's proponents, few successful and interesting predictions about organismic development have come from sequence information alone (Stephens, 1998). However, as the following list shows, genomic research is persistently throwing up surprises:

(i) the most important surprise from the HGP was that there are probably only about 30,000 genes in the human genome compared to an estimate of 140,000 as late as 1994 (Hahn & Wray, 2002).[11] In general, plant genomes are expected to contain many more genes than the human genome. Morphological or behavioral complexity is not correlated with the number of genes that an organism has. This has been called the G-value paradox (Hahn & Wray, 2002);

(ii) the number of genes is also not correlated with the size of the genome, as measured by the number of base pairs. The fruit-fly, *Drosophila melanogaster*, has 120 million base pairs but only 14,000 genes; the worm, *Caenorhabditis elegans* has 97 million base pairs but 19,000 genes; the mustard weed, *Arabidopsis thaliana* has only 125 million base pairs and 26, 000 genes, while humans have 29,000 million base pairs and 30,000 genes (Hahn & Wray, 2002);

(iii) at least in humans, the distribution of genes on chromosomes is highly uneven. Most of the genes occur in highly clustered sites. Most genes that occur in such clusters are those that are expressed in many tissues – the so-called "housekeeping" genes (Lercher et al., 2002). However, the spatial distribution of cluster sites appears to be random across the chromosomes. (Cluster sites tend to be rich in *C* and *G*, whereas gene-poor regions are rich in *A* and *T*.) In contrast, the genomes of arguably less complex organisms, including *D. melanogaster*, *C. elegans*, and *A. thaliana*, do not have such pronounced clustering;

(iv) only 2 percent of the human genome codes for proteins while 50 percent of the genome is composed of repeated units. Coding regions are interspersed by large areas of non-coding DNA. However, some functional regions, such as *HOX* gene clusters, do not contain such intervening sequences;

(v) scores of genes appear to have been horizontally transferred from bacteria to humans and other vertebrates, though apparently not to other eukaryotes. However, this issue remains highly controversial;

(vi) once attention shifts from the genome to the proteome (the protein complement of a cell – see Section 4), a strikingly different pattern emerges. The human

11 If past trends are at all indicative of the future, all estimates of the number of genes in "higher" animals will decline even further.

proteome is far more complex than the proteomes of the other organisms for which the genomes have so far been sequenced. According to some estimates, about 59 percent of the human genes undergo alternative splicing, and there are at least 69,000 distinct protein sequences in the human proteome. In contrast, the proteome of *C. elegans* has at most 25,000 protein sequences (Hahn & Wray, 2002);

(vii) it now appears that non-coding DNA is routinely transcribed into RNA but not translated in complex organisms (Mattick, 2003). It seems that these RNA transcripts form regulatory networks that are critical to development. Interestingly, the amount of non-coding DNA sequences in organisms appears to grow monotonically with the morphological complexity of organisms;

(viii) at least in *A. thaliana*, there is evidence of genome-wide non-Mendelian inheritance during which specifications from the grandparental, rather than parental, generation are transmitted to descendants (Lolle et al., 2005).

An important task of modern molecular biology is to make sense of these disparate unexpected discoveries. One conclusion seems unavoidable: any concept of the gene reasonably close to that in classical genetics will be irrelevant to the molecular biology of the future [See Gene Concepts].

4. Proteomics

The term "proteome" was introduced only in 1994 to describe the total protein content of a cell produced from its genome (Williams & Hochstrasser, 1997). Unlike the genome, the proteome is not even approximately a fixed feature of a cell (let alone an organism) because it changes over time during development as different genes are expressed. Deciphering the proteome, and following its temporal development during the life cycle of each tissue of an organism, has emerged as the major challenge for molecular biology in the post-genomic era. This project has been encouraged by the discovery of unexpected universality of developmental processes at the level of cells and proteins (Gerhart & Kirschner, 1997). For instance, even though hundreds of genes are known to specify molecules involved in transport across cellular membranes, there are only about twenty transport mechanisms in all living systems. The emergence of proteomics in the wake of the various sequencing projects signals an acceptance of the position that studying processes entirely, or even largely, at the DNA level will not suffice to explain phenomena at the cellular and higher levels of biological organization, including organismic development. Even genomics did not go far enough; a sharper break with the past will be necessary.

Nevertheless, in one very important sense, the emergence of proteomics recaptures the spirit of early molecular biology, when all molecular types, but especially proteins, were foci of interest, and the deification of DNA had not replaced a pluralist vision of the molecular basis for life. In the late 1960s, Brenner and Crick proposed "Project K" which was supposed to be "the complete solution of *E. coli*." *E. coli* (strain K-12) was selected as a model organism because of its simplicity (as a unicellular prokaryote) and ease of laboratory manipulation. Project K included: (i) a "detailed test-tube study of the structure and chemical action of biological molecules (especially proteins)" (Crick, 1973); (ii) completion of the models of protein synthesis; (iii) work on the structure and function of cell membranes; (iv) the study of control mechanisms at every level of organization; and (v) the study of the behavior of natural populations, including population genetics. Once *E. coli* was solved, biology was supposed to move on to more complex organisms.

Notice that in this project: (i) DNA receives no preferential attention at the expense of other molecular components; and (ii) the centrality of proteins as the most important active molecules in a cell is recognized. Project K accepts that there is much more to the cell than DNA; it accepts that no simple solution of the cell's behavior can be read from the genomic sequence. After a generation of infatuation with DNA and genetic reductionism, the aims of proteomics return in part to the vision of biology incorporated in Project K. However, at least in one important way, that project went beyond proteomics as currently understood: it emphasized all levels of organization whereas the explicit aims of proteomics are limited to the protein level. To understand the biology of organisms, the future will probably require even further expansion – see Section 5.

Meanwhile, work on proteins has also generated unexpected challenges. In particular, the four rules of classical molecular biology have not survived intact and at least the last three will require some modification. It now appears – though the essential idea goes back to the 1960s – that the fit between interacting sites of protein molecules is more dynamic than in the classical model, with the active site often "inducing" an appropriate fit.[12] It also appears that a more complicated model than the original allostery model will be required to account for many cases of cooperativity.

5. Towards a Systems Biology?

Over a half-century ago, Wiener (1948) suggested that living organisms be viewed as systems governed by feedback control. Wiener attempted to found a new discipline – "cybernetics" – for the study of such systems. In spite of Wiener's proselytization on behalf of the new discipline, cybernetics did not amount to much. It generated some excitement in the social sciences in the 1950s and then fizzled out (Heims, 1991). Engineers occasionally referred to cybernetic concepts (especially feedback) but, by the 1980s, that was about all the attention it received. In biology, especially in the emerging field of molecular biology, cybernetics contributed nothing of substance in spite of many attempts to use it (Sarkar, 1996).

Unexpectedly, at the beginning of the twenty-first century, Wiener's vision has returned to the forefront of attention in contemporary molecular biology. The context of Wiener's return is the new "systems biology" approach to the organism. As one of the proponents of the new approach, Kitano (2002), puts it: "Since the days of Norbert Wiener system-level understanding has been a recurrent theme in biological science." Kitano is partly right: ecosystem ecology, also going back to the 1950s, and large-scale studies of the immune system, starting in the 1960s, have both been important parts of biology even though Wiener's direct influence is hard to discern. But, in the new

12 See, for example, Koshland and Hamadani (2002).

molecular biology that came to dominate most of biological research, starting in the 1960s (as discussed in the earlier sections of this chapter), systems thinking was irrelevant. Research was dominated by what will be called "reductionism" in Section 6: trying to explain wholes by constructing them out of smaller and smaller parts (Sarkar, 1998).

Systems biology claims to be the culmination of the move from genetics to genomics to proteomics. Its aim is to study cells and larger units within organisms as composite systems described in terms of both the structures within them and the processes that occur in these structures (Ideker et al., 2001; Weston & Hood, 2004). Almost all advocates of systems biology endorse a collaborative technology-driven enterprise. Biologists, engineers, and computer scientists (among others) are supposed to collaborate to set up the necessary technological infrastructure to track all relevant processes within the cell and record the massive amounts of data that are produced. Integration at all levels – intellectual disciplines, conceptual frameworks, technology creation, and research culture – is expected to be critical to the success of this approach.

The most important innovation of systems biology is its explicit reintroduction of considerations of time into molecular biology – see Section 6 for further reflection on this point. One of the peculiar characteristics of molecular biology has been its avoidance of explicit reference to time: flows of information between nucleic acids, and from them to proteins, control of gene expression through negative feedback and switches – these mechanisms all replace explicit discussion of how the chemical composition of cells change over time. This is one of the salient features that make molecular biology look so different from the biochemistry that preceded it. Systems biology seems to be returning to the older biochemical view, worrying about processes, and how they change over time, but with a radical expansion of scale: in systems biology, thousands of reactants are potentially tracked over time rather than the ten or so which were the limit of classical biochemistry. Systems biology presents a much more dynamic view of biology than traditional molecular biology or even genomics. It promises both conceptual and technological innovations. If it leads to a successful model of even a single cell, it will already have justified the massive spending of the genome sequencing projects.

6. Philosophical Implications

It is time to draw some philosophical implications, first about reductionism which has long been of interest to philosophers, next about the notion of biological information which has recently seen a rapid growth of philosophical attention,[13] and finally about the return of temporal considerations in molecular biology.

6.1. *Beyond reductionism?*

One of the few philosophical issues in molecular biology that have routinely been discussed is that of reductionism [See Reductionism]. Here, reduction will be construed as

13 That is, relative to other issues in molecular biology; no area of molecular biology has received the philosophical attention it deserves, as Section 7 will note.

the explanation of wholes by parts, that is, reductionist explanations are those in which the weight of a putative explanation is borne by properties of the parts alone.[14] The wholes are biological entities, from cellular organelles to entire organisms. The parts are macromolecular and other components of the cell (and the extra-cellular matrix). Reductionism is the (empirical) thesis that explanations in some discipline will continue to be reductionist. The four rules of classical molecular biology embrace such reductionism and the remarkable success of classical molecular biology marks one of the most important triumphs of reductionism in the history of science (Sarkar, 1998). From the perspective of a reductionist, perhaps the most satisfactory aspect of this success is that cooperative behavior (in the case of allostery) and feedback regulation (in the case of the operon) were accommodated under the reductionist rubric in spite of being important exemplars from the traditional holists' repertoire.[15]

Moreover, the fact that the four rules of classical molecular biology are being challenged (recall the end Section 4), at least to some extent, is not reason enough to generate any new skepticism about the reductionist interpretation of explanation in molecular biology. They do not bring the physical explanation of wholes by parts into question. Rather, they show that the physical rules needed to explain macromolecular behavior are more complicated than previously thought, for instance, by an enzyme's active site inducing a fit with a reactant rather than merely responding to it. In contrast, if RNA-based (or other) regulatory networks turn out to be crucial to explaining development (and evolution, as Mattick [2003] argues – see Section 4), the reductionist interpretation *may* be in trouble. If network-based explanations are ubiquitous, it is quite likely that what will often bear the explanatory weight in such explanations is the topology of the network rather than the specific entities of which it is composed.[16]

Topological explanations have not received the kind of attention from philosophers they deserve even though networks have lately entered the center stage of scientific attention (Mattick & Gagen, 2005). Here "topology" refers to the connectivity properties of systems such as networks which, without loss of generality, can be modeled as directed graphs. The vertices of such a graph represent components of a system, and edges (between vertices), with appropriate directionality and weights, represent interactions between such vertices. How topological an explanation is becomes a matter of degree: the more an explanation depends on individual properties of a vertex, the closer an explanation comes to traditional reduction. The components matter more than the structure. Conversely, the more an explanation is independent of individual properties of a vertex, the less reductionist it becomes. In the latter case, if explanations invoke properties of a graph that measure its connectivity, then these are topological explanations. Such connectivity measures include the number of edges in the graph, the distribution of edge degree between vertices (the "degree" of a vertex being the number of edges incident on it), and so on.[17]

---

14 This is what Sarkar (1998) has called "strong" reduction – for a more carefully characterized treatment of varieties of reduction and reductionism, consult that work.

15 Recall the discussion of Section 2; for more detail, see Sarkar (1998).

16 Some classical phenomena such as dominance have already been interpreted to resist straightforward reductionist explanation (Sarkar, 1998).

17 For a review of network theory, see Newman (2003).

If topological explanations become necessary in molecular biology, it will mark a serious philosophical break with the reductionist classical era, though one that is not completely unexpected. Sarkar (1998) noted how the phenomenon of dominance had no straightforward structural explanation at the molecular level. Rather, the best molecular explanation of dominance involved complex reaction networks, the topological structure of which accounted for why one allele rather than the other was expressed at the phenotypic level.[18] This model predicts that dominance would be ubiquitous because such networks are common. Such an explanation depends very little on exactly what molecules comprise a network. If such network-based models begin to thrive in the post-genomic era, the reductionist interpretation of molecular biology will be seriously threatened.

Finally, systems biologists also reject reductionism – see, for instance, Aderem (2005) – even though the project of system biology emerged from the large-scale genome sequencing projects that had taken reductionism to its limits within biology (Tauber & Sarkar, 1992). As noted earlier (Section 3), contrary to most expectations, the results of sequencing only showed how little functional biology can be read off from sequences alone. Some systems biologists explicitly abandon reductionism to endorse philosophical doctrines such as emergence, according to which properties of wholes cannot be predicted or explained from the properties and organization of parts (Aderem, 2005). Few philosophers who defend reductionism will accept emergence easily, but the question can only be decided when the holists have specific examples in which properties of composite systems have deep explanations but none in terms of their parts. It will be a while before systems biology models get to that stage.

6.2. *Beyond DNA information?*

As noted earlier (Section 3), it is no longer clear that an informational account is appropriate for molecular biology. Even in the context of an informational account, the developments within eukaryotic genetics and, especially, genomics strongly suggest the view that DNA is the *sole* carrier of information. However "information" is explicated, such a view of DNA cannot be sustained for organisms more complicated than prokaryotes. Most of the critical interactions that determine the future behavior of a cell seem to occur at the level of RNA: splicing, RNA editing, and so on. Because of this feature of cellular interactions, Sarkar (2005, ch. 14) has speculated that the DNA genome consists of a relatively static set of sequestered modular templates (resulting in the "SMT" model of the genome), far from the classical view of the genome coding a program for development. The failure of the sequence hypothesis for many proteins only increases skepticism about the classical picture.

The routine generation of untranslated RNA transcripts from the genome also suggests that, should cellular processes be viewed informationally, RNA networks form a parallel information-processing system partly independent from the genomic DNA (Mattick, 2003). At present, it is unclear whether such information must be viewed *semiotically*, as in the case of DNA, where there is a symbolic coding relation. Similarly,

---

18 The original model goes back to Kacser and Burns (1981).

the discovery of ubiquitous non-Mendelian genetic specification in *A. thaliana* (Lolle et al., 2005) also suggests that there is yet another parallel system of heredity that can also perhaps be viewed informationally and, once again, is not specified through DNA. However, it is also possible that all such phenomena are best interpreted not informationally but using the more traditional – generally structural – conceptual apparatus of physics and chemistry. However, the distinction between the two frameworks becomes blurred in the case of RNA because the relation between the sequence and three-dimensional conformation seems to be relatively straightforward, at least much more so than in the case of proteins.

Note, however, that in these discussions of biological information, two issues should be distinguished: (i) whether an informational framework for molecular biology is of any use; and (ii) whether, within any such framework, DNA (or, more restrictively, genomic DNA) is the sole repository of that information. The problems mentioned here provide an argument against the second claim, leaving open the status of the first.

6.3. The return of time?

One of the peculiar characteristics of molecular biology has been its avoidance of explicit reference to the temporal dimension of the biological processes going on inside the cell and at other levels. The problem with informational interpretations of molecular biology is that these have always been static: flows of information between nucleic acids, and from them to proteins, control of gene expression through negative feedback and switches – these mechanisms all replace explicit discussion of how the chemical composition of cells change dynamically. Time does not enter explicitly into these accounts of biology though, implicitly, such transfer must take place during some time interval. Systems biology seems to be returning to the older biochemical view, worrying about processes and how they change over time. Systems biology thus presents a much more dynamic view of biology than traditional molecular biology. If systems biology lives up to its promise, the end result will be radically different from the classical molecular biology (discussed in Section 1).

However, even if the nascent project of systems biology fails to develop into anything substantive, proteomics also brings back considerations of time to molecular biology. Recall that the proteome is not a static feature of the cell, let alone the organism: proteomics requires a commitment to the characterization of cellular and organismic change over time. Moreover, the recent discoveries of potentially ubiquitous RNA network-based regulation also underscore the importance of dynamic accounts explicitly taking time into account. Moreover, new micro-array techniques and their extensions are increasingly making temporal stages of cellular changes empirically accessible. The challenge remains to develop a theoretical framework to interpret the empirical information. Any such framework can begin with either a physicalist or an informational characterization of cellular processes or a mixture of both, though prospects for a physicalist account do not seem particularly promising because of the sheer complexity of the molecular networks involved (Sarkar, 2005, ch. 10). But a dynamic informational account also leads to uncharted territory.

In retrospect, what seems surprising is how successful the static framework for classical molecular biology has been given that organisms are obviously dynamic entities undergoing development over time. It is hard not to predict a future in which molecular biology has an explicit temporal dimension in its models.

7. Conclusions: An Invitation

With perhaps the exception of the question of reductionism, molecular biology has not received the extent of philosophical attention it deserves, and the little that it has received has been limited to the classical period. There are at least two reasons why philosophers should invest more work on the subject: (i) without at least a partial methodological commitment to molecular concepts and techniques, any sub-discipline within biology will likely soon be relegated to irrelevance. Philosophy of biology that does not take molecular biology fully into account will remain incomplete; and (ii) modern molecular biology raises fundamentally new epistemological questions, especially about the relevance of physical versus semiotic or informational accounts that have both dominated discussions of biology for the last century and lived in uneasy tension with each other. The deployment of philosophical techniques – particularly formal techniques – may contribute significantly to the advancement of the field.

The most important task in the philosophy of biology for the next few decades will be to conceptualize the functional role of DNA within the cell so as to explain the surprising organization and other properties of the genome that were discussed earlier. Philosophers will also probably be faced with new problems that arise as molecular biology becomes a dynamic discipline (that is, one in which models have a temporal component to them), whether or not the program of systems biology flourishes. The extent to which the biological sciences are similar to and different from the physical sciences will then have to be reassessed. It also remains an open question whether the new molecular biology will finally be able to explain most, preferably all, facets of organismic development and perhaps help to integrate development with evolution [See Development and Evolution]. In all these areas physical and informational accounts will probably have to interact in order to create a consistent satisfactory picture. As Section 6 indicates, any such attempt must necessarily begin with a clearer account than what is currently available of what "information" must mean in a biological context. This is probably where philosophers have most to contribute to the future of molecular biology [See Biological Information]. Perhaps techniques from formal epistemology or semantics will enable progress where traditional biological tools have largely failed.

References

Aderem, A. (2005). Systems biology: its practice and challenges. *Cell*, 121, 511–13.

Barrell, B. G., Air, G. M., & Hutchison III, C. A. (1976). Overlapping genes in bacteriophage PhiX174. *Nature*, 264, 34–41.

Berget, S., Moore, C., & Sharp, P. (1977). Spliced segments at the 5′ terminus of Adenovirus 2 Late mRNA. *Proceedings of the National Academy of Sciences, USA*, 74, 3171–75.

Berk, A., & Sharp, P. (1978). Structure of the Adenovirus 2 Early mRNAs. *Cell*, 14, 695–711.

Cattaneo, R. (1991). Different types of messenger RNA editing. *Annual Review of Genetics*, 25, 71–88.

Cavalier-Smith, T. (1978). Nuclear volume control by nucleoskeletal DNA, selection for cell volume and cell growth rate, and the solution of the DNA C-value paradox. *Journal of Cell Science*, 34, 247–78.

Chow, L., Gelinas, R., Broker, T., & Roberts, R. (1977). An amazing sequence arrangement at the 5' ends of Adenovirus 2 messenger RNA. *Cell*, 12, 1–18.

Cook-Deegan, R. (1994). *The gene wars*. New York: Norton.

Crick, F. H. C. (1958). On protein synthesis. *Symposia of the Society for Experimental Biology*, 12, 138–63.

Crick, F. H. C. (1973). Project K: "The Complete Solution of *E. coli*." *Perspectives in Biology and Medicine*, 17, 67–70.

Gehring, W. (1998). *Master control genes in development and evolution: the homeobox story*. New Haven: Yale University Press.

Gerhart, J., & Kirschner, M. (1997). *Cells, embryos, and evolution*. Oxford: Blackwell Science.

Fox, T. D. (1987). Natural variation in the genetic code. *Annual Review of Genetics*, 21, 67–91.

Hahn, M. W., & Wray, G. A. (2002). The G-value paradox. *Evolution & Development*, 4, 73–5.

Ideker, T., Galitski, T., & Hood, L. (2001). A new approach to decoding life: systems biology. *Annual Review of Genomics and Human Genetics*, 2, 343–72.

Jacob, F. (1973). *The logic of life: a history of heredity*. New York: Pantheon.

Jacob, F., & Monod, J. (1961). Genetic regulatory mechanisms in the synthesis of proteins. *Journal of Molecular Biology*, 3, 318–56.

Kitano, H. (2002). Systems biology: a brief overview. *Science*, 295, 1662–4.

Koshland, D. E., Jr., & Hamadani, K. (2002). Proteomics and models for enzyme cooperativity. *Journal of Biological Chemistry*, 277, 46841–4.

Lercher, M. J., Urrutia, A. O., & Hurst, L. D. (2002). Clustering of housekeeping genes provides a unified model of gene order in the human genome. *Nature Genetics*, 31, 180–3.

Lolle, S. J., Victor, J. L., Young, J. M., & Pruitt, R. H. (2005). Genome-wide non-Mendelian inheritance of extra-genomic information in *Arabidopsis*. *Nature*, 434, 505–9.

Mattick, J. (2003). Challenging the dogma: the hidden layer of non-protein-coding RNAs in complex organisms. *BioEssays*, 25, 930–9.

Mattick, J., & Gagen, M. J. (2005). Accelerating networks. *Science*, 307, 856–7.

Mayr, E. (1961). Cause and effect in biology. *Science*, 134, 1501–6.

Monod, J. (1971). *Chance and necessity: an essay on the natural philosophy of modern biology*. New York: Knopf.

Newman, M. E. J. (2003). The structure and function of complex networks. *SIAM Review*, 45, 167–256.

Olby, R. C. (1974). *The path to the double helix*. Seattle: University of Washington Press.

Pauling, L., & Corey, R. B. (1950). Two hydrogen-bonded spiral configurations of the polypeptide chains. *Journal of the American Chemical Society*, 71, 5349.

Sarkar, S. (1996). Biological information: a skeptical look at some central dogmas of molecular biology. In S. Sarkar (Ed.). *The philosophy and history of molecular biology: new perspectives* (pp. 187–231). Dordrecht: Kluwer.

Sarkar, S. (1998). *Genetics and reductionism*. New York: Cambridge University Press.

Sarkar, S. (2005). *Molecular models of life: philosophical papers on molecular biology*. Cambridge, MA: MIT Press.

Stephens, C. (1998). Bacterial sporulation: a question of commitment? *Current Biology*, 8, R45–8.

Tauber, A. I., & Sarkar, S. (1992). The Human Genome Project: has blind reductionism gone too far? *Perspectives on Biology and Medicine*, 35(2), 220–35.

Thiéffry, D., & Sarkar, S. (1998). Forty years under the Central Dogma. *Trends in Biochemical Sciences*, 32, 312–16.

Weston, A. D., & Hood, L. (2004). Systems biology, proteomics, and the future of health care: towards predictive, preventative, and personalized medicine. *Journal of Proteome Research*, 3, 179–96.

Wiener, N. (1948). *Cybernetics.* Cambridge, MA: MIT Press.

Watson, J. D., & Crick, F. H. C. (1953). Molecular structure of nucleic acids – a structure for desoxyribose nucleic acid. *Nature*, 171, 737–8.

Waddington, C. H. (1962). *New patterns in genetics and development.* New York: Columbia University Press.

Williams, K. L., & Hochstrasser, D. F. (1997). Introduction to the proteome. In M. R. Wilkins, K. L. Williams, R. D. Appel, & D. F. Hochstrasser (Eds). *Proteome research: new frontiers in functional genomics* (pp. 1–12). Berlin: Springer.

Yockey, H. P. (1992). *Information theory and molecular biology.* Cambridge: Cambridge University Press.

Part II

Evolution

Chapter 5

Darwinism and Neo-Darwinism

JAMES G. LENNOX

1. Introduction

Scientific theories are historical entities, and like every historical entity, they undergo change through time. Indeed, a scientific theory might undergo such significant changes that the *only* point of continuing to name it after its source is to identify its lineage and ancestry. This may seem obviously true in the case of the theory of evolution by natural selection, still often referred to as Darwinism or neo-Darwinism. For when one looks at an advanced text on evolutionary biology today, especially one that stresses the centrality of mathematical population genetics to the theory, one might wonder what the point would be of applying to such a theory the name of a confessed mathematical illiterate with no clear ideas about the mechanisms of variation and inheritance. Nevertheless, there is merit to the view recently expressed by Jean Gayon, one of Darwinism's most thoughtful narrators:

> The Darwin–Darwinism relation is in certain respects a causal relation, in the sense that Darwin influenced the debates that followed him. But there is also something more: a kind of isomorphism between Darwin's Darwinism and historical Darwinism. It is as though Darwin's own contribution has constrained the conceptual and empirical development of evolutionary biology ever after. (Gayon, 2003, p.241)

Darwinism identifies a core set of concepts, principles, and methodological maxims that were first articulated and defended by Charles Darwin and which continue to be identified with a certain approach to evolutionary questions.[1] This is so *despite* the radical changes that this approach has undergone since the 1920s. One very important reason for this continuity has to do with the fact that most of its concepts, principles, and methods have been continuously challenged, not by those opposed to evolution,

1 So described, Darwinism denotes not so much a theory as a "research tradition" (Laudan, 1976) or a "scientific practice" (Kitcher, 1993); that is, at any given time in its history Darwinism consists of a family of theories related by a shared ontology, methodology, and goals; and through time, it consists of a lineage of such theories. I am using "theory" above in the very broad sense in which, from early on in his notebooks, Darwin kept referring to "my theory."

but by evolutionary biologists who portray themselves as *non-Darwinian* in one or more ways.[2]

For that reason it is worthwhile to begin with Darwin's Darwinism as formulated in *On the Origin of Species* in 1859. Charles Darwin was not, as we use the term today, a philosopher, though he was often so described during his lifetime.[3] If the concept of Darwinism has legitimate application today, it is due to a set of principles, both scientific and philosophical, that were articulated by Darwin and that are still widely shared by those who identify with "Darwinism."

2. Darwin's Life

Charles Darwin was born on February 12, 1809 and died on April 18, 1882. It was a time of radical changes in British culture, and his family background put him in the midst of those changes. Both of his grandfathers, physician/poet/philosopher Erasmus Darwin and pottery manufacturer Josiah Wedgwood, were members of an informal group of free thinkers that met regularly in Birmingham to discuss everything from the latest philosophical and scientific ideas to the latest advances in technology and industry. The members of the self-styled Lunar Society,[4] which included James Watt and Joseph Priestly, shared a "non-conforming" religious inclination. Robert Darwin, Charles's father, followed in his father's footsteps and became a doctor, and married Josiah Wedgwood's favorite offspring, Susannah. Charles was the youngest of five children she bore, but she died when he was but eight years old, and much of his upbringing he owed to his three sisters and brother, Erasmus, with whom he shared an early passion for chemistry, and with whom, at the age of 16, he went off to Edinburgh for the best medical education Great Britain had to offer.

Privately, Charles early on decided he could not practice medicine. But his already serious inclination toward science was considerably strengthened both by some fine scientific lectures in chemistry, geology, and anatomy, and by the mentoring of Dr Robert Grant, a Lamarckian who introduced Darwin to marine invertebrates and the use of the microscope in their study. This interest became a lifelong obsession, climaxing in his massive study of fossil and living Cirripedia or "barnacles" (Barrett & Freeman, 1988, vols. 11–13).

Eschewing medicine, he enrolled to take a degree in Divinity at Christ College, Cambridge University, from which he graduated in January of 1831. While in Cambridge, he befriended two young men attempting to institute a serious program of natural science at Cambridge, Rev. John Henslow, who was trained in botany and mineralogy, and Rev. Adam Sedgwick, a leading member of the rapidly expanding community of geologists. Through Henslow, to whom he shipped all his collections

---

2 Some of those biologists considered "non-Darwinian," such as Stephen Jay Gould, insist that in some respects they are closer to Darwin than defenders of the Synthesis. (Cf. Gould's forward to Dobzhansky, 1937/1982, p.xix.)

3 The word "scientist" was coined by William Whewell during Darwin's lifetime, but very few of Darwin's contemporaries owned up to it.

4 An entertaining account of the culture of the key members of this group can be found in Uglow (2002).

during the *Beagle* voyage, Darwin was introduced to leading figures in geology and natural history, as well as to Sir John Herschel and Rev. William Whewell, both serious students of the history and philosophy of science. Adam Sedgwick took Darwin on extended geological tours of England and Wales. Darwin's cousin, William Darwin Fox, a year ahead of him at Cambridge, helped convert his amateur passion for bug collecting into serious entomology. All of these influences built on those of Robert Grant, so that despite the lack of science required by his Divinity degree, Darwin graduated a very well-trained naturalist.[5]

3. Darwin's Darwinism

Darwin's mentors decisively shaped his philosophical attitudes and scientific career. Henslow was the final link in securing his position on the H. M. S. *Beagle*. The combination of meticulous field observation, collection, experimentation, note taking, reading, and thinking during that five-year journey through a wide cross-section of the earth's environments was to set the course for the rest of his life. During the voyage, he read and reread Charles Lyell's newly published *Principles of Geology*, which articulated a philosophical vision of rigorously empirical historical science, oriented around four key ideas:

(1) Geology includes the study of the history of life as evidenced by the fossil record and the past and present geographic distribution of species.
(2) It must also search for the causes of the extinction, introduction, or changing distribution of species.
(3) That search must be limited to causes of the same kind and intensity as those "now in operation."
(4) Lamarck's attempt to explain the introduction of new species by the hypothesis of "indefinite modification" of their ancestors fails on both methodological and empirical grounds.

Lyell's vision influenced Darwin profoundly. By the time of Darwin's return to England, likely influenced by conversations with Sir John Herschel in South Africa, he was convinced that the fossil record and current distribution of species were best explained by some form of species transformation. He set out to articulate a causal theory that measured up to Lyell's standards. He struggled to formulate a theory that would account for such transformations by referring only to "causes now in operation," causes that could be investigated empirically. The problem and the methodological constraints were those established by Lyell and received their philosophical defense from Herschel.

Darwin, of course, expected, and got, outraged reactions from religiously conservative colleagues, such as his old geology teacher Sedgwick, who, in a review, expressed his "deep aversion to the theory; because of its unflinching materialism; – because it has deserted the inductive track, – the only track that leads to physical truth; – because

5 For an expanded sketch of Darwin's early years see my entry, "Darwinism," for the *Stanford Encyclopedia of Philosophy* at http://plato.stanford.edu/entries/darwinism. The best biography is that by Janet Browne (Browne, 1995, 2002).

it utterly repudiates final causes, and therby [sic] indicates a demoralized understanding on the part of its advocates." What he had not expected was Lyell's refusal to openly endorse his theory and Herschel's decisive (if polite) rejection of its key elements. After setting out the theory in its Darwinian form, we can consider these reactions from those who apparently shared Darwin's philosophical norms about scientific theory, explanation, and confirmation.

The theory can be set out as three fundamental truths about species (1–3); four consequences of these truths that give rise to "natural selection" (4–7); and then three extrapolations from these consequences that will result in the origin and extinction of species (8–10).

(1) Species are comprised of individuals that vary ever so slightly from each other with respect to their many traits.
(2) Species have a tendency to exponentially increase their numbers over generations.
(3) This tendency is held in check by limited resources – as well as disease, predation, and so on – which creates a constant struggle for survival among the members of a species.
(4) Some individuals will *by chance* have variations that give them a slight advantage in this struggle, variations that allow more efficient or better access to resources, greater resistance to disease, greater success at avoiding predation, and so on.
(5) These individuals will *tend* to survive better and leave more offspring.
(6) Offspring *tend* to inherit the variations of their parents.
(7) Therefore, favorable variations will *tend* to be passed on more frequently than others, a tendency Darwin labeled "Natural Selection."
(8) Over time, especially in a slowly changing environment, this process will cause species to change.
(9) Given a long enough period of time, the descendant populations of an ancestor species will differ enough to be classified as different species, a process capable of indefinite iteration.
(10) There are, in addition, forces that encourage both divergence among descendant populations and the elimination of intermediate varieties.

Clearly every aspect of the mechanism of natural selection is capable of empirical investigation – indeed the published confirmatory studies of this process would fill a small library.[6] One can understand why devout and orthodox Christians would have problems; but why did Darwin's philosophical and scientific mentors? It would seem to be the model of Herschelian/Lyellian orthodoxy.

---

6 A more recent phenomenon than is usually appreciated. In Dobzhansky (1937/1982), after describing Ronald Fisher's "extreme selectionism," he quotes, as a "good contrast," the following remark of selection skeptics G. C. Robson and O. W. Richards (1936): "We do not believe that natural selection can be disregarded as a possible factor in evolution. Nevertheless, there is so little positive evidence in its favor . . . that we have no right to assign to it the main causative role in evolution."

4. Philosophical Problems with Darwin's Darwinism

The answer lies in five philosophically problematic elements of the theory.

[i] *Probability and Chance.* Note the language of "tendencies" and "frequencies" in the above principles. Privately, Darwin learned, Herschel had referred to his theory as "the Law of higgledy-piggledy," likely a reference to the probabilistic character of Darwin's claims. His theory is, as we would say today, a "statistical" theory, about what *tends* to happen due to clearly articulated causes. It allows us to make accurate predictions about *trends*, at the level of populations, but not to predict with certainty what will happen in each and every case. The proper philosophical understanding of this aspect of Darwinism is still elusive.

[ii] *The Nature, Power, and Scope of Selection.* For many people, natural selection is the core of Darwin's theory. And yet, even Darwin's strongest supporters and closest allies had problems with it. Some saw it as an "intermediate cause" instituted and sustained by God, others as a purely materialist and aimless process, and thus utterly incapable of dealing with adaptation. Some denied that it could originate species, seeing selection as a negative force eliminating what has already been created by mutation. Many felt that "selection" inappropriately imported into natural history an anthropomorphic vision of Nature choosing purposefully between variants. In a devastating review of *On the Origin of Species*, Fleeming Jenkin happily accepted the principle of natural selection but argued that it must be limited in scope to the production of varieties. [See Population Genetics]. All of these issues re-emerge during the resurgence of Darwinian principles in the creation of the evolutionary synthesis.

[iii] *Selection, Adaptation, and Teleology.* Because Darwin was fond of describing natural selection both as a natural process and one that worked for the good of each species, Darwin's followers seemed to have diametrically opposed views as to whether his theory *eliminated* final causes from natural science or breathed new life into them. In either case, there was serious disagreement on whether this was a good thing or a bad thing.[7]

[iv] *Nominalism and Essentialism.* There is a fundamental philosophical problem with the idea that a species can undergo a series of changes that will cause it to become one or more other species. The problem is well illustrated by the first question faced in the second volume of the *Principles of Geology*:

> . . . first, whether species have a real and permanent existence in nature; or whether they are capable, as some naturalists pretend, of being indefinitely modified in the course of a long series of generations. (Lyell, 1831, II, p.1)

Lyell assumes that a "real" species must have "permanent existence in nature," or ". . . fixed limits beyond which the descendants from common parents can never deviate from a certain type . . ." (Lyell, 1831, II, p.23). For Lyell, evolutionism implies a variety of nominalism about species, i.e., it implies that species names do not refer to types or kinds but only to collections of similar individuals. Darwin sometimes seems to agree.[8]

---

7 On which see Beatty (1990) and Lennox (1993).

8 Darwin was examined as an undergraduate on John Locke's *Essay on Human Understanding*. As far as I know he never discusses whether this had any impact on his willingness to articulate the views expressed in this quote.

> . . . I look at the term species, as one arbitrarily given for the sake of convenience to a set of individuals closely resembling each other, and that it does not essentially differ from the term variety, which is given to less distinct and more fluctuating forms. (Darwin, 1859/1964, p.52)

Given enough time, the individual differences found in all populations can give rise to stable varieties, these to subspecies, and these to populations that systematists will want to class as distinct species. Moreover, Darwin concludes the *Origin* with very strong words on this topic, words bound to alarm his philosophical readers:

> In short, we will have to treat species in the same manner as those naturalists treat genera, who admit that genera are merely artificial combinations made for convenience. This may not be a cheering prospect; but we shall at least be freed from the vain search for the undiscovered and undiscoverable essence of the term species. (Darwin, 1859/1964, p.485)

Lyell, Herschel, Whewell, and Sedgwick certainly would not find this a cheering prospect, since they were unrepentant essentialists about species.[9] Members of a species possess a "type" established in the original parents, and this type provides "fixed limits" to variability. Lyell provided evidence for this view in *Principles* Vol. II; and it was canvassed again in Jenkin's review of the *Origin*. Such fixed limits to a species' ability to track environmental change easily explain extinction. But a naturalistic account of species origination is more difficult, since those "fixed limits" must somehow be transgressed.

Yet, adopting the sort of nominalism advocated above by Darwin has undesirable consequences as well. How are we to formulate objective principles of classification? What sort of a science of organisms is possible without fixed laws relating their natures to their characteristics and behaviors? In chapter 2 of the *Origin*, Darwin sought to convince the reader that, in practice, botanists and zoologists accepted a natural world organized as he described:

> It must be admitted that many forms, considered by highly competent judges as varieties, have so perfectly the character of species that they are ranked by other highly competent judges as good and true species. (Darwin, 1859/1964, p.49)

This is a predictable consequence of the fact that the organisms we wish to classify are products of a slow, gradual evolutionary process. In a given genus some naturalists may see ten species with a few varieties in each; others may rank some of the varieties as species and see twenty species. [See Systematics and Taxonomy]. Both classifications may be done with the utmost objectivity and care by skilled observers. Some systematists are "lumpers," some are "splitters." Reality is neither.

[v] *Tempo and Mode of Evolutionary Change*. Whether or not Darwin's views entailed nominalism about natural kinds, they seem to reflect a belief that the evolutionary process is slow and gradual. I stress slow *and* gradual, for it is clear that one could have a *slow but non-gradual* evolutionary process (perhaps the geologically rapid periods of speciation postulated by Eldridge and Gould's "punctuated equilibrium model" are such

9 There is a very important, and underexplored, tension here, at least in Lyell and Herschel, both of whom seem to be in many other respects orthodox followers of Scottish empiricism.

[SEE SPECIATION AND MACROEVOLUTION]), and one could have a *rapid but gradual* one (for example the process George Gaylord Simpson labeled "adaptive radiation").

One of the strongest arguments for insisting that "Darwinism" and "neo-Darwinism" as they are used today are isomorphic to Darwin's Darwinism, as Gayon puts it, is that each of these questions is still hotly debated, and has been throughout the theory's history. Despite the changes wrought by the genetic, biochemical, and molecular revolutions [SEE MOLECULAR EVOLUTION]; the development of mathematical population genetics and ecology; and cladistic analysis in systematics, many evolutionary biologists still adhere to Darwinism, and are recognized as doing so by both themselves and their critics. We may thus organize the discussion of the "evolution" of Darwin's Darwinism into "neo-Darwinism" around these themes.[10]

5. The Core Problems and Darwinism

The philosophical problems of *Darwin's* Darwinism arise from questions concerning: [i] the role of chance as a factor in evolutionary theory and the theory's apparently probabilistic nature; [ii] the nature of selection; [iii] whether selection/adaptation explanations are teleological; [iv] the ontological status of species and the epistemological status of species concepts; and [v] whether evolutionary change is invariably slow and gradual. One dominant approach to evolutionary biology, represented by the so-called "neo-Darwinian Synthesis," sides with Darwin on these issues (and on many less fundamental ones, besides). That in itself is remarkable, given the radical transformations that the theory has undergone since the infusion of mathematics and Mendelian genetics that took place in the period from 1915 to 1930. [SEE POPULATION GENETICS]. But, it is the more remarkable because the Darwinian position on each issue has been continuously under pressure from non-Darwinian evolutionary biologists from Darwin's death to the present.

A full understanding of the underlying philosophical disagreements on these questions requires a historical study of how the "Synthesis" positions on these various issues, and those of their critics, arose. That cannot be done here; but it will be helpful to have a historically accurate summary of that theory.

The use of the term "synthesis" seems to have been suggested by the title of Julian Huxley's account, *Evolution: The Modern Synthesis*. What he intended is not entirely clear. In his chapter on natural selection he emphasizes the need for "facts and methods" from virtually every domain of biology as well as a number of related disciplines. But he immediately admits that most of these disciplines have developed in relative isolation. The synthesis he discusses is in the future and will be greatly aided by a "re-animation of Darwinism" (cf. Huxley, 1942, p.13).

10 I will use "neo-Darwinism" to refer to an explanatory framework created by the founders of the evolutionary synthesis of natural selection and population genetics and who hoped to bring a wide spectrum of biological subdisciplines within that explanatory framework. It was, of course, used much earlier to characterize a related framework defended by August Weissman in the 1880s and 1890s.

Among the fields he mentions, embryology and comparative anatomy played no significant part in the "neo-Darwinian synthesis." Huxley focuses most of his attention on the synthesis of Mendelian genetics and Natural Selection forged by R. A. Fisher, J. B. S. Haldane, and Sewall Wright; he also discusses its empirical support in both laboratory and ecological genetics (his book is dedicated to T. H. Morgan).

Thus, in standard accounts of the synthesis one can discern two stages: in the first stage (say, 1912–31) we see the growth of the laboratory genetics associated with T. H. Morgan, H. J. Muller, and A. H. Sturtevant, and the formulation of the mathematical theory of the genetics of populations developed by Fisher, Haldane, and Wright; in the second stage we see the publication of the books of Theodosius Dobzhansky, Ernst Mayr, George Simpson, and Huxley. In these latter works, all published between 1937 and 1944, the implications for paleontology, systematics, and *natural* selection of experimental laboratory genetics and the theoretical models of population genetics were explored.

This picture leaves out a number of important elements, two of which will be briefly noted. The "ecological genetics" exemplified in the work of E. B. Ford, A. J. Cain, P. M. Sheppard, and H. B. D. Kettlewell was critical both to the understanding and the acceptance of the power of selection in natural populations. And, while it is true that many evolutionary biologists tended to ignore development as irrelevant to their interests, two embryologists, C. D. Darlington and G. De Beer, were considered serious contributors to the synthesis.[11]

Nevertheless, after allowance is made for these and a number of other corrections, there is a profound truth in the claim that "the Evolutionary Synthesis" is, at its core, a brilliant integration. Experimental and mathematical genetics are wedded to those subjects that dominate *On the Origin of Species*: natural selection acting on chance variation as the principal mechanism of evolutionary change; the fossil record as the principal historical evidence of the evolutionary process; and biogeographic distribution providing overwhelming evidence that current populations are the products of an evolutionary process. A few key quotations make this clear:

> Since evolution is a change in the genetic composition of populations, the mechanisms of evolution constitute problems of population genetics. (Dobzhansky, 1937/1982, p.12)

> The paleontological record is consistent with the usual genetical opinion that mutations important for evolution, of whatever eventual taxonomic grade, usually arise singly and are small, measured in terms of structural change. (Simpson, 1944/1984, p.58)

> . . . the variability within the smallest taxonomic units has the same genetic basis as the differences between the subspecies, species and higher categories. . . . selection, random gene loss, and similar factors, together with isolation, make it possible to explain species

11 See, for example, the extensive citations of Darlington's work in cytology in Huxley (1942), Dobzhansky (1937/1932), and Simpson (1953). De Beer's *Embryos and Ancestors* is Simpson's primary source on the subject of how developmental genetics can play a role in determining the extent of a mutation's effect on the phenotype (Simpson, 1953, p.97). Darlington is cited as often as Fisher in Huxley's *Evolution: the Modern Synthesis* and in Dobzhansky's *Genetics and the Origin of Species*.

formation on the basis of mutability, without any recourse to Lamarckian forces. (Mayr, 1942/1982, p.70)

The element of the synthesis that, in the minds of all three men, makes it *Darwinian* is the central role of natural selection on the small, genetically based variations studied by the geneticists and modeled mathematically by Fisher, Haldane, and Wright. At the time Dobzhansky, Mayr, and Simpson were writing their seminal works, it was easy for them to cite a large body of evidence skeptical of any significant role for natural selection in the production of evolutionary change. And even within this group, Wright's papers written between 1930 and 1932, which had a significant impact on Dobzhansky, restricted selection's importance to small, relatively isolated populations (cf. Dobzhansky, 1937/1982, p.191). In fact Dobzhansky closes his chapter on "Selection" by quoting Wright's 1932 statement of the view that evolution is due to a "shifting balance" of mutation, selection, inbreeding within colonies, and cross-breeding between them.

We will now turn to some philosophical problems the theory faced during its elaboration between 1930 and 1960. I will discuss only the first four of the five I mentioned at the beginning of this section.

5.1. *The roles of chance in neo-Darwinism*

In evolutionary theory, "chance" plays a key role both in discussing the *generation* of variation and the *perpetuation* of variation (a distinction I owe to John Beatty; see also Sober, 1984, ch. 4). Consider the following *variation grid*, created by asking whether the *contribution to fitness* of a variation does or does not bias its chances in favor of being *generated* or of being *perpetuated*:

	Variations	
	Generation	**Perpetuation**
Fitness biased	Lamarck	Darwin
Not fitness biased	Darwin	Lamarck
	Neutralism	Neutralism

The uniquely Darwinian position is that a variation's future contribution to fitness does not produce a bias in favor of its generation (as it would for Lamarckian theories), but contribution to fitness does produce a bias in favor of its perpetuation. Neutralism, to be discussed shortly, claims that a significant amount of evolutionary change, particularly at the molecular level, is due to randomly generated variation that is also perpetuated by chance. [See Molecular Evolution].

The above grid might lead one to conclude that both in the case of the generation of variation and the perpetuation of variation, "chance" will refer to the absence of a bias created by fitness differences. We get to the heart of the problem of the concept of "chance" within neo-Darwinism by seeing why that conclusion is, at best, misleading.

As we have seen, it was Darwin's view that advantageous variations occasionally arise "by chance," and have a "better chance" of being perpetuated than those that are not advantageous (cf. Darwin, 1859/1964, pp.80–1).

On this issue, orthodox neo-Darwinism agrees whole-heartedly with Charles Darwin. Fisher, Wright, and Haldane all start with the Hardy–Weinberg equilibrium principle [SEE POPULATION GENETICS] that represents the current state of a biological population in terms of the *relative frequencies* of alleles which "in a relatively large, closed population remains constant in the absence of any unbalanced pressure due to mutation or selection" (Wright, 1939/1986, p.285; he cites Haldane as the first to put the issue of evolutionary change in these terms; cf. Fisher, 1930/1999, pp.9–10). It is the *presence* of such pressures that is viewed as the principal mechanism of evolutionary change.

Thus understood, fitness differences must be understood in terms of increasing or decreasing the likelihood of the perpetuation of a trait (or gene) above (or below) what might be called "chance" levels. To take a simple case: if there are three possible combinations of alleles at a given locus in a population, we can characterize the outcome of a reproductive cycle as "chance" if, given a certain frequency distribution, each of the three possible combinations occurs at a frequency determined strictly by the laws of probability. Neo-Darwinism conceives of natural populations as "gene pools," and thinks of evolution as long-run changes in the frequencies of different combinations of genes from generation to generation. Thus, even when one factors in natural selection, being relatively better adapted merely increases an organism's "chances," i.e., its probability of leaving viable offspring; it does not guarantee it. Since natural selection is itself a stochastic process, Darwinians from Darwin to the present rightly characterize it in terms of selection influencing the "chances" of a given outcome, in interaction with other variables such as population size, population structure, or mutation rate.

Conceptual confusion arises from the fact that neo-Darwinians often, even typically, contrast the generation of variation due to "chance" and "randomness" with alternative theories that claim the generation of variation is "guided along beneficial lines" (to borrow a phrase from Asa Gray). Darwin defined natural selection as the preservation of variations that *happen* to be beneficial. This was in sharp contrast with the view of variation both of his botanist friend Asa Gray, who at least hoped it was due to design, and of Lamarck and his followers, who saw variation as a direct response to adaptive demands. Against this background, "chance" or "random" variation contrasts with variations arising by design or in response to a need.

The concept of "random variation" is today often used as a synonym for "chance variation" in precisely this latter sense. One of the founding fathers of the Synthesis puts it this way:

> ... mutation is a random process with respect to the adaptive needs of the species. Therefore, mutation alone, uncontrolled by natural selection, would result in the breakdown and eventual extinction of life, not in the adaptive or progressive evolution. (Dobzhansky, 1970, p.65)

At least a significant amount of confusion concerning the role of chance in evolution can be avoided by determining whether, in a given case, "chance" or "randomness" is being used to characterize the *origins* or *generation* of variation or the *perpetuation* or *spread* of a variation.

Because of the stochastic character of natural selection, neo-Darwinians occasionally characterize it so as to make it almost indistinguishable from random drift. (For a presentation of the problem and various solutions cf. Beatty, 1984; Brandon, 1990, 2005; Lennox & Wilson, 1994; Millstein, 2002, 2005). The fitness of a genotype is characterized as its relative contribution to the gene pool of future generations – the genotype increasing in frequency being the fitter. But, of course, that could easily be the result of a "random" – non-fitness biased – sampling process; which organisms would be declared "fitter" by this method might have nothing to do with natural selection.

In order to provide a proper characterization of the role of chance in evolutionary change, then, we need a more robust and sophisticated account of fitness. But even with such an account there remains a substantial empirical question of what role indiscriminate sampling of genotypes (or phenotypes) plays in evolutionary change. Sewall Wright's work in the 1930s defended the possibility that genes neutral with respect to fitness could, due to the stochastic nature of population sampling, increase their representation from one generation to the next, with the likelihood increasing as effective population size decreases. Wright believed that species were typically subdivided into relatively small, relatively isolated, populations (or "demes") with significant in-breeding, and thus that it was likely that "neutral genotypes" becoming fixed at relatively high levels was significant. Though he gradually toned down this aspect of his work, a significant school of mathematical population geneticists in the 1960s and 70s developed these ideas into the "Neutralist" approach to evolutionary change mentioned earlier. Whether or not such a process plays a significant role in evolution is not a philosophical issue, but it is highly relevant to whether evolutionary biology is seen as predominantly Darwinian. For if any view is central to Darwinism, it is that the evolutionary process is guided predominantly by natural selection preserving randomly generated variation. It is to natural selection and related concepts that we now turn.[12]

5.2. The nature, power, and scope of selection

Darwin consistently refers to natural selection as a *power of preserving* advantageous, and eliminating harmful, variations. As noted in the last section, whether an advantageous variation *arises* is, in one sense of that term, a matter of chance; and whether an advantageous variation is actually *preserved* by selection is, in another sense of the term, also a matter of chance, but selection increases the chances of some variations relative to others. For Darwinism, selection is the force or power that favors advantageous variations, or to look ahead to the next section, of adaptations. It is this that distinguishes selection from drift.

As Darwinism developed in the mid-twentieth century, the expression "survival of the fittest" has essentially been eliminated from any serious presentation of the theory.

12 This is in fact the Synthesis view: see Mayr in Mayr and Provine (1980, p.3); Simpson (1984, p.xvii); Eldridge in Mayr (1982, p.xvi); Huxley (1942, pp.26–7).

On the other hand, the concept of "fitness" has played a prominent, and problematic, role. How that came about is a puzzle.[13]

R. A. Fisher's famously perplexing "Fundamental Theorem of Natural Selection" states that "the rate of increase in fitness of any organism at any time is equal to its genetic variance in fitness at that time" (Fisher, 1930/1999, p.35). However, none of the four classic proponents of the "Synthesis" we have discussed (Dobzhansky, Huxley Mayr, Simpson) even mention the term "fitness." Even in Dobzhansky's long discussion of Fisher's work, he uses "differences in viability" where Fisher will use "fitness differences," and he uses mathematical formulae borrowed from Wright's work rather than Fisher's. How and why the concept of fitness becomes central to textbook presentations from the 1950s on is an interesting question for which I have no answer. Nevertheless, from that point on, the mathematical models used in population genetics use "fitness" to refer either to the abilities of the different genotypes in a population to leave offspring, or to the measures of those abilities, represented by the variable ***W***. Here is a rather standard textbook presentation of the relevant concepts:

> In the neo-Darwinian approach to natural selection that incorporates consideration of genetics, fitness is attributed to particular genotypes. The genotype that leaves the most descendants is ascribed the fitness value $W = 1$, and all other genotypes have fitnesses, relative to this, that are less than 1. . . . Fitness measures the relative evolutionary advantage of one genotype over another, but it is often important also to measure the relative penalties incurred by different genotypes subject to natural selection. This relative penalty is the corollary of fitness and is referred to by the term **selection coefficient**. It is given the symbol *s* and is simply calculated by subtracting the fitness from 1, so that: $s = 1 - W$. (Skelton, 1993, p.164)

The dual senses of fitness (as capacity and measure) are instructively conflated in this quotation. When fitnesses are viewed as differential *abilities* (or propensities) of organisms with different genotypes to leave different numbers of offspring, we are encouraged to suppose that "fitness" refers to the relative selective advantages of genotypes. But if "fitness" refers to a *measure* of reproductive success, it is a quantitative representation of small-scale evolutionary change in a population, and it leaves entirely open the question of the *causes* of the change – in which case the assumed connections among the concepts of fitness, adaptation, and natural selection are severed. "Selection coefficients" may have nothing to do with selection; what *W* represents may have nothing to do with selective advantage.

Fisher would have been unhappy with treating "fitness" as a measure. In a fascinating comparison between his fundamental theorem and the second law of thermodynamics he notes that both are statistical laws, dependent upon measurable constants, ranging over populations (Fisher, 1930/1999, p.36). Nevertheless, he goes on to note five "profound" differences, including that, though there is a standardized method for *measuring* fitness, *what is measured* is qualitatively different in every population; whereas entropy is presumed to be a measure of the same property for all physical systems (Fisher, 1930/1999, p.37).[14]

---

13 The pre-history of this puzzle is interestingly explored in Gayon (1992, 1995).

14 It is also likely that Fisher, as well as Haldane, saw these models as experimental or as ways of demonstrating possibilities (cf. Lennox, 1991; Plutynski, 2004).

For Fisher then, fitness, the "measurable property," though always measured in terms of relative increases and decreases in gene frequencies, must not be identified with this measure. Fitness is a relationship between population members and their environments, and that relationship will differ depending on the nature of the population and the nature of the environment.

Following out Fisher's insight, we can formulate the theory in its "synthesis" guise without collapsing the common method of *measuring* fitness with the *heterogeneity* of instantiations of fitness. Since there are a number of confirmed ways in which natural populations can evolve in the absence of natural selection, and since stabilizing selection may prevent a population from evolving in its presence, measuring changes in the genetic make-up of a population does not establish natural selection and failing to detect such changes does not establish its absence. Population genetics and its associated models provide ways of establishing that a population either is or is not in equilibrium, and sophisticated tools for predicting subtle differences in expected trajectories depending on the values of the various variables in the models. Moreover, like the kinematics of any physical theory, if we see cross-generational change in a population of the sort predicted by a certain population model, it not only suggests that there are causes to be found – the detailed contours of those measures may suggest what sorts of causes to look for. What such models cannot do *on their own* is provide knowledge of the actual forces at work. To use language introduced by Elliott Sober, fitness, unlike natural selection, is *causally inert*. As I understand it, this is simply recognition of Fisher's point that the uniformity of the fitness measure hides the very different causal interactions that underlie it.

If we suppose that the standard neo-Darwinian view shares with Darwin a view of natural selection favoring certain organisms in virtue of their phenotypic variations, we can see two challenges to today's Darwinism with respect to *levels* of selection. There are those, such as G. C. Williams and Richard Dawkins, who argue that selection is always and only of genes. Here is a clear statement:

> These complications [those introduced by organism/environment interactions] are best handled by regarding individual [organismic] selection, not as a level of selection in addition to that of the gene, but as the primary mechanism of selection at the genic level. (Williams, 1992, p.16)

Dawkins refers to organisms – or interactors – as the *vehicles* of their genes, in fact, as vehicles constructed by the genome for its own perpetuation.

This view has been extensively challenged by philosophers of biology on both methodological and conceptual grounds, though there are, among philosophers, enthusiastic supporters (cf. Dennett, 1995). Oddly, defenders of this view claim to be carrying the Darwinian flag (an oddity noted by Gayon, 1998; Gould, 2002). Dawkins, for example, regularly refers to himself as a neo-Darwinian (e.g., Dawkins, 1982, pp.50–1). Yet, advocates of the "neo-Darwinian synthesis" invariably gave causal primacy to the interaction between organisms in populations and ever-variable ecological conditions; changes in the gene pools of those populations are viewed as the quantifiable and measurable *effects* of natural selection. On the other hand, both Dawkins and Williams are defenders of the adaptationist program; and at least part of their defense of genic selectionism is that it seems like a plausible interpretation of kin-selection explanations of so-called "altruistic" behavior. After all, if an animal behaves in a way that slightly lowers its individual fitness while increasing its "inclusive fitness," does that not suggest that it is the genes that are in the driver's seat?

Darwinism also faces challenges from the opposite direction. In the 1970s a number of biologists working in the fields of paleontology and systematics challenged the neo-Darwinian dogma that you could account for "macroevolution" by simple, long-term extrapolation from microevolution. Gould, in particular, opens Part II of *The Structure of Evolutionary Theory* (*Towards a Revised and Expanded Evolutionary Theory*), with a chapter entitled "Species as Individuals in the Hierarchical Theory of Selection." That chapter title combines two conceptually distinct theses: first, the thesis defended by Michael Ghiselin (Ghiselin, 1997) and championed and refined by David Hull (Hull, 2001), that species are, in a robust sense of the term, "individuals"; and second, that there may well be selection among groups of organisms, *qua* groups. [SEE SYSTEMATICS AND TAXONOMY; SPECIATION AND MACROEVOLUTION]. Gould's title exemplifies one approach to group selection – the unit of selection is always the individual, but there are individuals at various ontological levels, any of which may be subject to selection. A very different result emerges if one assumes that groups of organisms such as demes, kin-groups, or species, though not individuals, are nevertheless, under tightly specified conditions, subject to selection. Adding to the conceptual complexity, some researchers propose that "group selection" be restricted to the process whereby group-level traits provide advantages to one group over another, in which case there are strict conditions delimiting cases of group selection. Others define group selection primarily in terms of group level *effects*. Thus, a debate analogous to that earlier discussed regarding the definitions of "fitness" emerges here – by group selection do we mean a distinct type of causal process that needs to be conceptually distinguished from selection at the level of individual organism or gene, or do we mean a tendency within certain populations for some well-defined groups to displace others over time? (For further discussion, see Sterelny & Griffiths, 1999, pp.151–79; Hull, 2001, pp.49–90.)

5.3. *Selection, adaptation, and teleology*

Early in the introduction to *On the Origin of Species*, Darwin observes that the conclusion that each species had descended from others "even if well founded, would be unsatisfactory, until it could be shown how the innumerable species inhabiting this world have been modified so as to acquire that perfection of structure and co-adaptation which most justly excites our admiration" (Darwin, 1859/1964, p.3). One might say that this was the central promise of Darwinism – to account for both phylogenic continuity *and* adaptive differentiation by means of the same principles.

The nature of "selection explanations" is a topic to which much philosophical attention has been devoted in recent years. (Distinctive book-length treatments can be found in Brandon, 1990, and Sober, 1984.) Here, I want to focus on only one important question – to what extent is the teleological appearance of such explanations simply an appearance masking a causal process in which goals play no role?

The *appearance* of teleology is certainly present in Darwinian explanations, and has been since Darwin spoke of natural selection *working solely for the good* of each being (Darwin, 1859/1964, p.84). The appearance of teleology stems from the ease with which both evolutionary biology and common sense take it for granted that animals and plants have the adaptations they do *because* of some benefit or advantage to the organism provided by those adaptations.

Virtually every biologist identified with the neo-Darwinian synthesis has felt the need to address this issue. Haldane is reported to have compared teleology to the biologist's mistress: he cannot live without it but he doesn't want to be seen in public with it (Mayr, 1976, p.392). Dobzhansky stated that "some modern biologists seem to believe that the word 'adaptation' has teleological connotations, and should therefore be expunged from the scientific lexicon," a view with which he "emphatically disagreed" (Dobzhansky, 1937/1982, p.150). In a collection of papers edited by G. G. Simpson and A. Roe, C. S. Pittendrigh acknowledged that the evolutionary biologist cannot get along without references to ends and functions, but recommended replacing the word "teleology" with "teleonomy," a recommendation sometimes endorsed by Simpson, Mayr, and G. C. Williams (Williams, 1966, p.258). Perhaps the best survey of Synthesis views on this topic is to be found in Ernst Mayr's "Teleological and Teleonomic: A New Analysis," which includes a footnote in which a letter from Pittendrigh is quoted at length on why he coined the term "teleonomy." The clearest analysis from an "orthodox" neo-Darwinian of the teleological nature of selection explanations is that by Francisco Ayala (Ayala, 1970).

Whatever term one uses, the serious philosophical issue is whether the functions provided by adaptations (i.e., selected traits) play a central and irreducible role in their explanation. Only if the answer is "yes" are the explanations teleological.[15] [SEE FUNCTIONS AND TELEOLOGY].

Let us begin with a simple, yet realistic, example. In research carried out over many years, John Endler was able to demonstrate that the color patterns of males in the guppy populations he studied resulted from a balance between mate selection and predator selection. To take one startling example, he was able to test and confirm a hypothesis that a group of males with a color pattern that matched that of their river beds except for bright red spots have that pattern because a common predator in those rivers, a prawn, is color blind for red. Red spots provided no selective disadvantage and attracted mates (Endler, 1983, p.173–90). This pattern of coloration is a complex *adaptation* that serves the functions of predator avoidance and mate attraction (Williams, 1966, p.261; Brandon, 1985; Burian, 1983). Do those functions explain why these male guppies have the coloration they do?

---

15 I need to stress here that this discussion is restricted to explanations of adaptation within the Darwinian framework, i.e., by reference to natural selection. Whether other sorts of explanation in other aspects of biology are teleological or not, and whether, if they are, the explanation would take the same form, I leave entirely open. For a good survey of this question, and a defense of a distinct understanding of biological function in the domain of comparative morphology, see Amundson and Lauder (1998).

In order for it to be a product of natural selection, there must be an array of color variation available in the genetic/developmental resources of the species wider than this particular pattern but including this pattern. In popular parlance (and the parlance favored by Darwin), this color pattern is present in the population because it is *good for* the male guppies that have it, and for their male offspring (Binswanger, 1990; Brandon, 1985; Lennox, 2002). That is why natural selection favors this coloration. The analysis offered here is more robust than standard accounts in terms of "selected effects" or "consequence etiologies" in stressing that selection ranges over *value* variation. The reason for one among a number of color patterns having a higher fitness value has to do with the *value* of that pattern *relative to the survival and reproductive success* of its possessors (Lennox, 1993, 1999, 2002).

A commitment to a strong role for natural selection in the evolution of life is certainly central to neo-Darwinism, a commitment sometimes referred to as "adaptationism" or the "adaptationist program." Explanations by reference to selection are a particular kind of teleological explanation, an explanation in which a trait's *adaptive functions, its valuable consequences*, account for its differential increase or maintenance in the population. Given neo-Darwinism's commitment to selection as the source of adaptation, then, it is not surprising that all the central figures in the Synthesis felt it necessary to address this question. Their ambivalence is also understandable. Teleology was closely associated with two discredited biological research programs, natural theology and vitalism. A great deal of work by philosophers of biology over the past 30 years has obviated the need for such ambivalence.

5.4. *Species and the concept of "species"*

In listing the topics under which I would discuss neo-Darwinism, I distinguished the question of the ontological status of species from the epistemological status of the species *concept*. Though they are closely related questions, it is important to keep them distinct. As will become clear as we proceed, this distinction is rarely honored. Moreover, it is equally important to distinguish the *species concept* from the categories of features that belong in their *definitions*. Advances in our theoretical understanding may lead us to reconsider the sorts of attributes that are most important for determining whether a group of organisms is a species, and thus whether it deserves to be assigned a name at that taxonomic level. It should not be assumed that such changes constitute a change in the species concept, though at least some such changes may lead us to restrict or expand the taxa within that category.

In his contribution to the Synthesis, *Systematics and the Origin of Species*, Ernst Mayr titled chapter five "The Systematic Categories and the New Species Concept." Recall that Darwin made a point of treating the species category as continuous with "well-marked variety" and "sub-species," and made the radical suggestion that its boundaries would be just as fluid. Without explicitly acknowledging Darwin, Mayr takes the same tack, discussing "individual variants" and "sub-species" as a preliminary to discussing the species concept. Mayr notes that for someone studying the evolutionary process, speciation is a critical juncture; ". . . his interpretation of the speciation process depends largely on what he considers to be the final stage of this process, the species" (Mayr, 1942/1982, p.113). With this in mind, he offers the following definition, the now infamous "biological species concept" (BSC):

> Species are groups of actually or potentially interbreeding natural populations, which are reproductively isolated from other such groups. (Mayr, 1942/1982, p.120; 1976, p.518)

Mayr was well aware of the limitations of this definition, and treated it somewhat as a "regulative ideal." Dobzhansky in 1937 gave what he claimed to be a definition of species, but which seems, as Mayr noted (Mayr, 1976, p.481) much more a definition of *speciation*:

> . . . that stage of evolutionary process at which the once actually or potentially interbreeding array of forms becomes segregated in two or more separate arrays which are physiologically incapable of interbreeding. (p.312)

Simpson (1944/1984) and others built even more historicity into the concept. These are all, of course, intended as *definitions* of the species *category*, and they attempt to provide a test (or a "yardstick": Mayr, 1976, p.479) that in principle will permit a researcher to decide whether a group of individuals should all be identified by a single species-level concept such as "homo sapiens." The test for species membership is the *capacity* to interbreed; the test distinguishing two species is *incapacity* to interbreed. Dobzhansky makes the importance of this test transparent – the transition from a single interbreeding population to two reproductively isolated ones is the process of speciation. [See Speciation and Macroevolution].

Now in each of these cases, little attention is paid to the actual methods used by taxonomists and systematists in differentiating between varieties of a species and distinct species, something to which Darwin gave a great deal of attention. Darwin's nominalism regarding the species concept likely stemmed from his close attention to his own taxonomic practices and those of other specialists. But nominalism typically combines a view about the *ontology* of species with one about the epistemological status of the species *concept*. On the first question, the nominalist insists that there are no species – there are more or less similar individuals. On the second question, the nominalist typically insists that the species *concept* is, at best, a useful or convenient grouping of similar individuals or, at worst, an *arbitrary* grouping of similar individuals.

In his work, Mayr relates different approaches to the species concept to the philosophical distinction between essentalism and nominalism. He associates essentialism with the view that a species *concept* refers to a universal or type. This view of the referent of the concept leads to the Typological Species Concept, which he traces from Linnaeus back to Plato and Aristotle and claims "is now universally abandoned" (1976, p.516). At the opposite extreme is nominalism, which combines the view that only individuals exist in nature and that species are concepts invented for the purpose of grouping these individuals collectively.

Mayr claims that his Biological Species Concept (BSC) is an advance on both; individual species members are objectively related to one another not by a shared relation to a type but by causal and historical relationships to one another. Notice, however, that this is, from an ontological perspective, nominalism. Mayr's position can be understood as arguing for a new way of understanding the epistemological grounds for grouping individuals into species. This new way of grouping stresses historical, genetic, and various ecological relationships among the individuals as the grounds for determining species membership. His claim is that this is more reliable and objective than similarities of phenotypic characteristics. This makes sense of the importance he eventually places on the fact the BSC defines species relationally:

> . . . species are relationally defined. The word species corresponds very closely to other relational terms such as, for instance, the word *brother*. . . . To be a different species is not a matter of degree of difference but of relational distinctness. (Mayr, 1976, p.518)

Brothers may or may not look alike; the question of whether two people are brothers is determined by their historical and genetic ties to a common ancestry. Notice, however, that this is a claim about which, among the many characteristics that they have, should be taken most seriously in determining the applicability to them of the concept "brother." That is, it is a defense of a sort of essentialism.

A number of critics have pointed out that essentialism need not be committed to "types" understood as *universalia in re*; and on certain accounts of essences any species taxon that meets the standards of BSC does so in virtue of certain essential (though relational and historical) properties. At one extreme Michael Ghiselin and David Hull (and Mayr [1987] acknowledges this as an extension of his ideas) have argued that this causal/historical structure of species provides grounds, at least within evolutionary biology, for considering species to be individuals. Organisms are not members of a class or set, but "parts" of a phylogenetic unit.

A critical issue in this debate over the account of the species concept most appropriate for Darwinism is the extent to which the process of biological classification – taxonomy – should be informed by advances in biological theory. Besides those already discussed, the moderate pluralism associated with Robert Brandon and Brant Michler or the more radical pluralism defended by Philip Kitcher argue that different explanatory aims within the biological sciences will require different criteria for determining whether a group constitutes a species. Cladists, on the other hand, employ strictly defined phylogenetic tests to determine species rank.

Unlike many of the other topics that define the history of Darwinism, there is no clear-cut position on this question that can be identified as "Darwinian" or "neo-Darwinian." In a recent collection of papers defending most of the viable alternatives (Ereshefsky, 1992), my suspicion is that virtually every author would identify himself as Darwinian. This may be because many of the positions defended could plausibly be traced to roots in Darwin's own theory and practice (see Beatty, 1985; reprinted in Ereshefsky, 1992).

6. Conclusion

In this essay I have built a case for the claim that a certain stance within evolutionary biology today is legitimately referred to as "Darwinism" or "neo-Darwinism," despite the remarkable changes that the theory of evolution by natural selection has undergone since *On the Origin of Species* was first published. The case consists of identifying core principles of Darwin's original theory (with their associated philosophical problems) and tracing the development of those principles through the neo-Darwinian synthesis. I have argued that, despite the radical changes brought about by the fusion of the theory with Mendelism via mathematical population genetics, those core principles survive, and serve to differentiate a "Darwinian" approach to evolutionary biology from other approaches. Moreover, the development of the theory has resulted from a continuous history of philosophical pressure on each of those principles.

References

Amundson, R., & Lauder, G. (1994). Function without purpose: the uses of causal role function in evolutionary biology. *Biology and Philosophy*, 9, 443–69.

Ayala, F. (1970). Teleological explanation in evolutionary biology. *Philosophy of Science*, 37, 1–15.

Barrett, P. H., & Freeman R. B. (Eds). (1988). *The works of Charles Darwin*, vols. 1–29. New York: New York University Press.

Beatty, J. (1984). Chance and natural selection. *Philosophy of Science*, 51, 183–211.

Beatty, J. (1990). Teleology and the relationship between biology and the physical sciences in the nineteenth and twentieth centuries. In F. Durham & R. Purrington (Eds). *Some truer method: reflections on the heritage of Newton* (pp. 113–44). New York: Columbia University Press.

Binswanger, H. (1990). *The biological basis of teleological concepts*. Los Angeles: ARI Press.

Brandon, R. (1985). Adaptation explanations: are adaptations for the good of replicators or interactors? In B. Weber & D. Depew (Eds). *Evolution at a crossroads: the new biology and the new philosophy of science* (pp. 81–96). Cambridge, MA: MIT Press.

Brandon, R. (1990). *Adaptation and environment*. Princeton: Princeton University Press.

Brandon, R. (2005). The difference between selection and drift: a reply to Millstein. *Biology and Philosophy*, 20, 153–70.

Brandon, R., & Burian, R. (Eds). (1984). *Genes, organisms, populations: controversies over the units of selection*. Cambridge, MA: Harvard University Press.

Browne, J. (1995). *Charles Darwin: voyaging*. Princeton: Princeton University Press.

Browne, J. (2002). *Charles Darwin: the power of place*, New York: Alfred A. Knopf.

Burian, R. (1983). Adaptation. In M. Grene (Ed.). *Dimensions of Darwinism* (pp. 287–314). Cambridge: Cambridge University Press.

Darwin, C. (1859/1964). *On the Origin of Species: a facsimile of the first edition*. Cambridge, MA: Harvard University Press.

Dawkins, R. (1982). *The extended phenotype*, Oxford: Oxford University Press.

Dennett, D. (1995). *Darwin's dangerous idea: evolution and the meanings of life*. New York: Simon and Schuster.

Dobzhansky, T. (1937/1982). *Genetics and the Origin of Species* (Columbia Classics in Evolution, with an introduction by Stephen Jay Gould). New York: Columbia University Press.

Dobzhansky, T. (1970). *Genetics of the evolutionary process.* New York: Columbia University Press.

Durham, F., & Purrington, R. (Eds). (1992). *Some truer method: reflections on the heritage of Newton.* New York: Columbia University Press.

Endler, J. (1983). Natural and sexual selection on color patterns in poeciliid fishes. *Environmental Biology of Fishes*, 9, 173–90.

Ereshefsky, M. (Ed.). (1992). *The units of evolution: essays on the nature of species.* Cambridge, MA: Harvard University Press.

Fisher, R. (1930/1999). *The genetical theory of natural selection: a complete variorum edition* (Henry Bennett, ed.). Oxford: Oxford University Press.

Gayon, J. (1998). *Darwinism's struggle for survival: heredity and the hypothesis of natural selection.* Cambridge: Cambridge University Press.

Gayon, J. (2003). From Darwin to today in evolutionary biology. In J. Hodge & G. Radick (Eds). *The Cambridge Companion to Darwin* (pp. 240–64). Cambridge: Cambridge University Press.

Ghiselin, M. (1997). *Metaphysics and the origin of species.* Albany, NY: SUNY Press.

Gould, S. (2002). *The structure of evolutionary theory.* Cambridge, MA: Harvard University Press.

Grene, M. (Ed.). (1983). *Dimensions of Darwinism.* Cambridge: Cambridge University Press.

Herbert, S. (Ed.). (1980). *The red notebook of Charles Darwin.* Ithaca, NY: Cornell University Press.

Hodge, J., & Radick, G. (Eds). (2003). *The Cambridge Companion to Darwin.* Cambridge: Cambridge University Press.

Horowitz, T., & Massey, G. (Eds). (1991). *Thought experiments in science and philosophy.* Savage, MD: Rowman and Littlefield.

Hull, D. (2001). *Science and selection: essays on biological evolution and the philosophy of science.* Cambridge: Cambridge University Press.

Huxley, J. (1942). *Evolution: the modern synthesis.* London: George Allen & Unwin.

Kitcher, P. (1993). *The advancement of science: science without legend, objectivity without illusions.* Oxford: Oxford University Press.

Laudan, L. (1976). *Progress and its problems.* Berkeley: University of California Press.

Lennox, J. (1991). Darwinian thought experiments: A function for just-so stories. In T. Horowitz & G. Massey (Eds). *Thought experiments in science and philosophy* (pp. 223–46). Savage, MD: Rowman and Littlefield.

Lennox, J. (1993). Darwin *was* a teleologist. *Biology and Philosophy*, 8, 409–422.

Lennox, J. (1999). The philosophy of biology. In M. Salmon (Ed.). *Introduction to the philosophy of science* (pp. 269–309). Indianapolis: Hackett Publishing.

Lennox, J. (2002). Che bene è un adattamento? *Iride*, XV, n. 37, 521–35.

Lennox, J., & Wilson, B. (1994). Natural selection and the struggle for existence. *Studies in History and Philosophy of Science*, 25, 65–80.

Lyell, C. (1831–3/1991). *Principles of geology* (1st edn, vols. I–III). Chicago: University of Chicago Press.

Mayr, E. (1942/1982). *Systematics and the Origin of Species* (Columbia Classics in Evolution, with an introduction by Niles Eldredge), New York: Columbia University Press.

Mayr, E. (1976). *Evolution and the diversity of life: selected essays.* Cambridge, MA: Harvard University Press.

Mayr, E. (1987). The ontological status of species: scientific progress and philosophical terminology. *Biology and Philosophy*, 2, 145–66.
Millstein, R. (2002). Are random drift and natural selection conceptually distinct? *Biology and Philosophy*, 17, 33–53.
Millstein, R. (2005). Selection vs. drift: a response to Brandon's reply. *Biology and Philosophy*, 20, 171–5.
Plutynski, A. (2004). Explanation in classical population genetics. *Philosophy of Science*, 71, 1201–14.
Robson, G. C., & Richards, O. W. (1936). *The variation of animals in nature*. London: Longmans.
Salmon, M. et al. (1992). *Introduction to philosophy of science*. Indianapolis: Hackett.
Simpson, G. G. (1944/1984). *Tempo and mode in evolution* (Columbia Classics in Evolution, with a new introduction by George Gaylord Simpson). New York: Columbia University Press.
Simpson, G. G. (1953). *The major features of evolution*. New York: Columbia University Press.
Skelton, P. (Ed.). (1993). *Evolution: a biological and palaeontological approach*. London: Pearson.
Sober, E. (1984). *The nature of selection*. Cambridge, MA: The MIT Press.
Sterelny, K., & Griffiths, P. (1999). *Sex and death: an introduction to philosophy of biology*. Chicago: Chicago University Press.
Uglow, J. (2002). *The lunar men: five friends whose curiosity changed the world*. New York: Farrar, Strauss and Giroux.
Weber, B., & Depew, D. (Eds). (1985). *Evolution at a crossroads: the new biology and the new philosophy of science*. Cambridge, MA: MIT Press.
Williams, G. (1966). *Adaptation and natural selection: a critique of some current evolutionary thought*. Princeton: Princeton University Press.
Williams, G. (1992). *Natural selection: domains, levels, and challenges*. Oxford: Oxford University Press.

Further Reading

Three excellent, but quite different, perspectives on the historical development of Darwinism are:

(1) Depew, D. J., & Weber, B. H. (1995). *Darwinism evolving: systems dynamics and the genealogy of natural selection*. Cambridge, MA: MIT Press.
(2) Gayon, J. (1998). *Darwinism's struggle for survival: heredity and the hypothesis of natural selection*. Cambridge: Cambridge University Press.
(3) Shanahan, T. (2004). *The evolution of Darwinism: selection, adaptation, and progress in evolutionary biology*. Cambridge: Cambridge University Press.

Three studies focused on different aspects of the "modern synthesis" are:

(4) Mayr, E., & Provine, W. B. (Eds). *The evolutionary synthesis: perspectives on the unification of biology*, Cambridge, MA: Harvard University Press.
(5) Provine, W. B. (1971). *The origins of theoretical population genetics*. Chicago: Chicago University Press.
(6) Sarkar, S. (Ed.). (1992). *The founders of evolutionary genetics: a centenary reappraisal*. Dordrecht: Kluwer.

And finally, the classic that coined the term "modern synthesis" for the attempt to base diverse biological disciplines on a "genetical" theory of evolution by natural selection:

(7) Huxley, J. (1942). *Evolution: the modern synthesis*. London: George Allen & Unwin.

A website devoted to making all of Darwin's published works and unpublished notebooks available online, and with links to many other valuable sites is: http://darwin-online.org.uk.

Chapter 6

Systematics and Taxonomy

MARC ERESHEFSKY

1. Introduction

Biological taxonomy may seem like a simple science – biologists merely observe similarities among organisms and construct classifications according to those similarities. But biological taxonomy is not so simple. Consider an obvious type of similarity referred to as "morphological similarity": when organisms have a similar body shape and structure. Dogs have a different morphology than coyotes, and dogs and coyotes are more similar to one another than either is to foxes. Mammals come in neat morphological packages. However, morphology is an inadequate marker for classifying many organisms, especially insects, molds, fungi, and bacteria. For example, the fruit flies *Drosophila persimilis* and *Drosophila pseudoobscura* have nearly identical morphologies. It took years for biologists to determine that many organisms thought to be *Drosophila persimilis* are in fact members of a different species, *Drosophila pseudoobscura*. Matters get worse in bacteria. Some bacteriologists have thrown up their hands in classifying parasitic bacteria. The morphological differences between such bacteria grade into one another, resulting in a continuum of organisms. Bacteria are not an exceptional case. Most of life on Earth, in terms of both biomass and biodiversity, is bacterial.

Perhaps a better foundation for biological classification can be found in genetics. We live in the heady days of the Human Genome Project and other genome projects. Perhaps the organisms of one species are genetically more similar to one another than they are to organisms in other species. If this is true, then classification can be based on genetic similarity. There are, however, strong challenges to this suggestion; one being that genes are insufficient for distinguishing species. Turning to fruit flies again, there can be more genetic variation between different populations of a single fruit fly species than there is between two such species (Ferguson, 2002). In other words, two organisms in different species can be more similar to one another genetically than either is to the members of its own species.

Alternatively, one might think that species are distinguished in terms of sexual reproduction. Introductory biology texts tell us that the members of the same species can interbreed and produce viable offspring. Classification, then, should be based on the relations between organisms – in this case, interbreeding relations – rather than on similarities. Interbreeding relations do provide clean divisions among mammals and birds.

Nevertheless, the interbreeding approach to classification runs aground of a glaring biological fact: the vast majority of organisms on Earth do not reproduce by interbreeding. Most organisms reproduce asexually by cloning, self-fertilization, or by other means. So the interbreeding approach does not apply to most of life on this planet.

Which type of trait should be used for classifying organisms? This is a thorny issue. To complicate matters further, there are a number of philosophical controversies within biological taxonomy. Four of these controversies are the focus of this chapter. One controversy concerns the ontological nature of species. Are species natural kinds akin to the chemical elements whose members share theoretically significant similarities, or are they "individuals" analogous to particular organisms whose parts are connected by casual relations? Another controversy concerns the unity of science. Is there a single correct way to sort organisms into species, or are there multiple correct ways to classify the organic world? This debate pits monists against pluralists. A third philosophical controversy concerns phylogenetic inference. The majority of taxonomists would like biological classification to reflect branching on the tree of life, but how should information about organismic traits be used to infer such branching? A fourth controversy concerns the framework of biological classification, the Linnaean hierarchy. The Linnaean hierarchy was developed in the eighteenth century, well before the advent of evolutionary theory. We now live in a Darwinian age, and many biologists believe that the Linnaean hierarchy is theoretically outdated and should be replaced.

The resolution of the above philosophical issues within biological taxonomy has implications outside of taxonomy. For example, decisions concerning the nature of species affect how biological conservation should be conducted. If we consider species the basic units for assessing biodiversity, then the approach to species we choose will affect our choice of biological entities to preserve. Philosophical issues in taxonomy also affect our conception of human nature. If an account of human nature has a biological basis, then our approach to species affects what it means to be a human. Is there a genetic or other sort of biological essence to *Homo sapiens*, or is each one of us a human because we share a common evolutionary history? If the latter is true, then little can be said about what is normal or natural for humans.

Before turning to fuller discussion of these issues, some terminological clarification is in order. The terms "classification," "taxonomy," and "systematics" are often used in taxonomic discussions. It is important to be clear about their meanings. Biological taxonomy provides the principles and methods for constructing classifications. Biological taxonomy tells us how to sort organisms into species, and it provides the principles for classifying taxa into more inclusive taxa. Classifications themselves are the products of taxonomy. Biological systematics is more foundational. Systematics is the study of how organisms and taxa are related in the natural world. Ideally, the results of systematics determine the principles of taxonomy, which in turn tell us how to construct classifications of the organic world.

2. The Ontological Nature of Species

Most but not all philosophers believe that species are natural kinds. And even among those philosophers who agree that species are natural kinds, there is disagreement about the nature of those kinds. This section will review two approaches to the idea that species are natural kinds as well as the thesis that species are individuals rather than natural kinds.

2.1. *Species essentialism*

The standard philosophical account of natural kinds assumes that the members of a kind share a common essential property or essence. This essentialist approach to natural kinds traces back to Aristotle and is found in the work of Hilary Putnam and Saul Kripke. Stated simply, kind essentialism has two main tenets: (1) All and only the members of a kind share a kind-specific essential property; and (2) a kind's essential property is causally responsible for other properties typically found among the members of that kind. The essence of the natural kind gold, for example, is gold's atomic structure, which occurs in all and only gold and is used for predicting and explaining other properties associated with pieces of gold, such as their ability to conduct electricity.

If species are essentialist kinds, what are their essences? Linnaeus thought that the essence of a plant species was its genus' fructification system and whatever traits distinguish that species from the other species in its genus. Locke thought that the essence of a species was its unique microstructure, although he did not know the nature of such microstructures. Some have speculated that the essence Locke was looking for was none other than DNA. In the past fifty years, a number of philosophers and biologists have argued that species are not natural kinds with essences (Mayr, 1959; Hull, 1965; Ghiselin, 1974; Sober, 1980; Dupré, 1981). They maintain that species essentialism is inconsistent with evolutionary theory and therefore should be abandoned. Anti-essentialists offer many arguments against species essentialism (see Ereshefsky, 2001, for a review). We will focus here on the argument that biological forces work against the existence of biological essences.

The first tenet of essentialism requires that there be a biological property in all and only the members of a particular species. Biologists have been hard-pressed to find such properties. Evolutionary biology explains why. In order for a property to be a species' essence, it must be present in *all* the members of a species. However, processes such as mutation work against a trait occurring in all members of a species. Suppose a trait is universal among the members of a species. A mutation can eliminate that trait in an organism in the next generation. If a trait fails to occur in one member of a species, then that trait is not the essence of that species. Recombination can have the same effect. Recombination does not alter DNA, but reshuffles it such that a trait universal in one generation of a species may fail to appear in a member in the next generation. In general, genetically based traits are vulnerable to the forces of mutation and recombination, which makes the universality of a trait in a species fragile.

Suppose, nevertheless, that a trait occurs in all members of a species. Essentialism also requires that this trait be *unique* to the members of the species. This constraint rules out many traits as species essences because those traits occur in other species. Evolutionary theory explains why similar traits frequently occur in different species. Organisms in closely related species inherit common genes and developmental programs from their shared ancestors. These common genetic and developmental resources cause the members of different species to be similar. Another source of similarity across taxa is parallel evolution. Similar adaptive needs cause similar traits in different species. For example, the eye of the octopus and the human eye are functionally similar, but each type of eye has a different evolutionary origin.

It is an empirical claim that evolutionary forces work against species having essences. So it is possible that a trait could occur in all and only the members of a species. But consider the stringent requirements of essentialism. A trait is the essence of a species only if it occurs in all members of that species for the entire duration of that species. Furthermore, a trait is unique among the members of a species as long as it does not occur in any other organism for the *entire* history of life in the universe. If the trait occurs just once in another species, then that trait is not the essence of the species in question. Recall, also, that the second tenet of essentialism places a further requirement on essentialism. A trait might occur in all and only the members of a species, but this occurrence would be insufficient for it being a species' essence unless it also caused the other traits typically associated with that species. Given the high standards of essentialism and the confounding forces of evolution, species essentialism is probably false.

2.2. Species as individuals

If species are not natural kinds with essences, then what are they? Some philosophers and biologists believe that species are not natural kinds but individuals. The two most prominent advocates, Ghiselin (1974) and Hull (1978), contrast natural kinds and individuals in terms of space-time locality. Natural kinds, they suggest, are spatiotemporally unrestricted entities: a member of the kind gold is gold regardless of its location in space and time. The motivation for this requirement is that laws of nature refer to natural kinds and laws of nature are not restricted to particular space-time regions. If "All water boils at 100 degrees Celsius" is a law of nature, then water will boil at that temperature anywhere in space and time. In contrast to natural kinds, Ghiselin and Hull suggest that individuals are spatiotemporally restricted entities. Consider an analogy. In geology there are various kinds of rocks (granite, shale, and so on), and there are individual rocks (the granite rock in a garden). Granite, the kind, can have members across the universe, but the parts of the granite rock in the garden must be located in a restricted space-time region to be parts of that rock. Ghiselin and Hull argue that species are also spatiotemporally restricted entities, akin to the rock in the garden, hence species are individuals rather than kinds.

What is their argument for species being spatiotemporally restricted entities? Hull writes that species are units of evolution and as units of evolution species must be spatiotemporally restricted. Suppose that selection causes species to evolve. For evolution by selection to occur, the selected traits must be passed down through the generations of species. Traits are not inherited unless some causal connection exists between the members of a species. In particular, sex and reproduction require that organisms or their parts (gametes, DNA) come into contact. Evolution, thus, requires that the organisms of a species be connected genealogically. Just as the parts of an individual organism must be appropriately connected causally, so must the members of a particular species. The organisms of a species cannot be scattered throughout the universe. Hence, species are individuals.

Hull (1978) and others have drawn many implications from the thesis that species are individuals. One implication is that there is no biological essence to being a human. From an evolutionary perspective, humans are merely parts of the evolving lineage *Homo sapiens*. There is no qualitative property that all and only humans must have. Having a certain cognitive ability, social ability, even being able to communicate with language is not required for being a human. Being part of a particular evolving lineage is all that matters. Traditional accounts of human nature require that all humans have a distinctive human quality. If species are individuals, then such accounts of human nature lack a biological basis.

2.3. *Species as homeostatic property cluster kinds*

The debate over the ontological status of species does not end with the claim that species are individuals. A handful of philosophers argue that we should not reject the view that species are natural kinds (Boyd, 1999a; Griffiths, 1999; Wilson, 1999). The problem, they suggest, is the standard essentialist account of natural kinds. Adopt a better approach to natural kinds and species will be returned to their proper place as natural kinds. The approach to natural kinds they advocate is Boyd's Homeostatic Property Cluster (HPC) Theory. The members of an HPC kind share a cluster of similar properties, but none of these properties is essential for membership in an HPC kind. Nevertheless, these properties must be stable enough to allow for successful induction. That is, they must be stable enough to allow us to predict with better than chance probability that a member of an HPC kind will have certain properties. The members of the kind *Canis familaris* share many similar properties such that we can reasonably predict that, if an organism is a dog, it will have certain properties. According to HPC theory, the co-occurrence of properties among the members of an HPC kind is due to a kind's homeostatic mechanisms. Such homeostatic mechanisms include interbreeding, shared ancestry, and common developmental constraints.

HPC theory provides a more promising account of species as natural kinds than traditional essentialism. HPC theory allows for variation among the members of a species, and it does not require that the members of a species share a common essence. All that is required is that the members of a species share a cluster of co-occurring properties. HPC theory also recognizes the importance of genealogy. Shared ancestry and reproductive relations are homeostatic mechanisms that maintain similarities among the members of a species.

Does HPC theory provide an adequate account of species as natural kinds? Some argue that it does not (Ereshefsky & Matthen, 2005). While it is undoubtedly true that the members of a species have many similarities, it is also true that species are characterized by dissimilarities. Polymorphism – variation within a species – is an important feature of nearly every species. For example, the males and females of a species can vary dramatically, and the members of a species can vary in their life stages, as exemplified by the caterpillar and butterfly stages of a single organism. Stable polymorphism is an essential feature of nearly all species, yet HPC theory gives no account of this feature. HPC theory focuses only on explaining those similarities that exist within a species. So the first problem with HPC theory is that it provides an impoverished account of species.

A second problem with HPC theory turns on the requirement that species are lineages. Hull and others argue that from an evolutionary perspective species must be genealogical entities. HPC theory allows that species are genealogical entities, but HPC theory does not require that all species be genealogical entities (Boyd, 1999b, p.80). HPC kinds are first and foremost kinds whose members have sufficient similarity to underwrite successful predictions. Yet, as Boyd recognizes, genealogy and similarity can part company. When genealogy and similarity conflict, Boyd prefers similarity to genealogy and posits species that are not genealogical lineages. Evolutionary theory requires that all species be genealogical lineages; HPC theory does not. In sum, HPC theory is inconsistent with an evolutionary account of species. Moreover, it fails to explain the occurrence of stable polymorphism in species. The claim that species are individuals fares better on both counts. The individuality thesis is premised on the assumption that species are genealogical lineages. Furthermore, the individuality thesis provides a more robust account of the nature of species. The individuality thesis appeals to the genealogy of a species to explain the similarities *and* dissimilarities among the members of a species.

3. Taxonomic Pluralism

A common assumption in biology and philosophy is that one true classification of the organic world exists. That is, if we had a god's eye perspective, we would see that each organism belongs to a particular species, that each species belongs to a particular genus, and so on up the Linnaean hierarchy. This view, called "monism," also assumes that there is one correct definition of "species" and there is one correct method for classifying taxa into more inclusive taxa. In contrast, pluralism is the view that there are multiple correct definitions of "species" (Kitcher, 1984; Ereshefsky, 1992; Dupré, 1993). According to pluralists, there are different kinds of species and different but legitimate Linnaean classifications of the organic world.

What is the argument for taxonomic pluralism? It begins with the observation that biologists provide various definitions of the term "species" – what biologists refer to as "species concepts." The dozen or so species concepts in the current biological literature are not fringe concepts, but have widespread support among biologists. The most prominent species concepts fall into three types: interbreeding, ecological, and phylogenetic. According to interbreeding concepts, species are groups of organisms that can interbreed and produce fertile offspring. Interbreeding species are distinct gene pools, bound and maintained by sexual reproduction. Ecological species concepts also focus on the forces that maintain species. An ecological species is a lineage of organisms that live in a particular ecological niche. The selection forces in a species' niche cause a lineage to be a distinct species. Interbreeding and ecological species concepts stem from work in evolutionary biology, whereas phylogenetic species concepts are derived from the school of taxonomy called "Cladism" (see Section 4). According to cladists, organisms should be classified by their shared ancestry. Each taxon should contain all and only the descendants of a common ancestor. Such taxa are labeled "monophyletic." According to phylogenetic species concepts, species are the smallest monophyletic taxa within the Linnaean hierarchy.

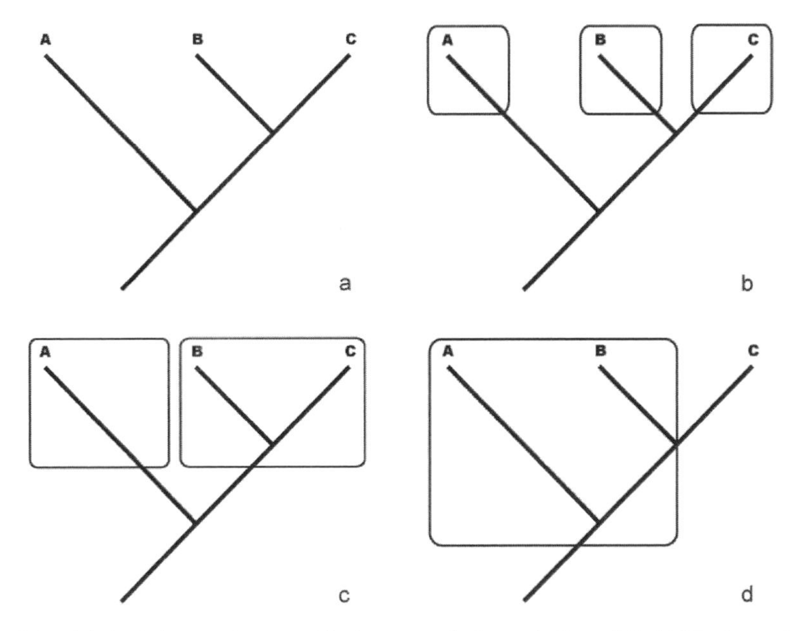

Figure 6.1 (a) A phylogenetic tree with three populations, A, B, and C. (b) The tree with three phylogenetic species, A, B, and C. (c) The tree with two ecological species, A and B + C. (d) The tree with one interbreeding species, A + B

These three approaches to species – interbreeding, ecological, and phylogenetic – assume that species are genealogical lineages. Nevertheless, these approaches highlight different types of lineages as species. As a result, they give rise to different classifications of a single group of organisms. Consider a hypothetical example, which is based on empirical studies showing that interbreeding, ecological, and phylogenetic species concepts often pick out different groups of organisms in nature (Ereshefsky, 2001). Suppose that three insect populations, A, B, C, live on the side of a mountain (Figure 6.1a), and each population forms a single basal monophyletic taxon. The organisms in B and C share a common ecological niche, while the organisms in A occupy their own distinct niche. The organisms in A and B can successfully produce fertile offspring, whereas the organisms in C reproduce asexually. Given these biological considerations, how should we classify the insects in question? According to the phylogenetic approach, there are three species: A, B, and C (Figure 6.1b). According to the ecological approach, there are two species: A and B + C (Figure 6.1c). According to the interbreeding approach, there is only one species: the species consisting of A + B (Figure 6.1d). These different approaches to species provide three different classifications of the same group of insects.

When we apply these approaches to all of life, the result is three different classifications of the organic world. Different species concepts give rise to a plurality of classifications. A monist might respond that this situation is due to our lack of biological knowledge and is merely temporary. One of the species concepts discussed, or one to be discovered, is the correct approach to species. Once biologists have settled on that correct concept, we will have a single classification of the world's organisms. However, species pluralists maintain that the case for pluralism is not our lack of information about the organic world. Quite the contrary. We have substantial information from evolutionary biology that the tree of life is segmented by various evolutionary forces into different types of species (interbreeding, ecological, phylogenetic). Taxonomic pluralism is a result of a fecundity of biological forces rather than a paucity of scientific information.

Monists offer many responses to taxonomic pluralism (Sober, 1984; Hull, 1999). We will consider two recent monist responses. De Queiroz (1999) and Mayden (2002) argue that among the species concepts found in the literature, one concept should be considered the primary species concept. They observe that, despite their differences, all species are lineages. Thus, de Queiroz and Mayden offer a lineage account of species. The lineage account of species, according to Mayden (2002, p.191), "serves as the logical and fundamental over-arching conceptualization of what scientists hope to discover in nature behaving as species. As such, this concept . . . can be argued to serve as the primary concept of diversity." De Queiroz and Mayden suggest that the various species concepts in the literature provide criteria for discovering species, but only the lineage account properly defines "species." De Queiroz and Mayden believe that although their approach to species is monistic, it captures what is correct in species pluralism. They offer a single correct species concept – the lineage account. At the same time, de Queiroz and Mayden allow that the world is populated by different types of lineages, namely, interbreeding, ecological, and phylogenetic species.

While pluralists appreciate the recognition of different types of species, they do not believe that the lineage approach to species provides a unified (monist) account of the organic world. Phylogenetic and interbreeding concepts identify different species. Consider the example of classifying insects. A phylogenetic species concept identifies three species (A, B, C), while an interbreeding species identifies one species (A + B). On Mayden and de Queiroz's lineage approach, both answers are correct. So even when one recognizes that all species are lineages, there remains a plurality of conflicting classifications. If monism is the view that there is a single correct classification of the organisms in the world, then de Queiroz and Mayden's response to pluralism fails.

A second monist response to pluralism is inspired by advances in molecular sequencing. Perhaps a single correct species concept should be based on genetic similarity. As more molecular studies are performed we may discover the distinctive genome of each species. We can then use that information to construct a single classification of the organic world. Despite its initial appeal, molecular data is not the answer to taxonomic pluralism. Molecular data provides yet another classification of the organic world. Ferguson (2002) provides examples where overall genetic similarity and the ability to interbreed do not coincide. The result is two different classifications: one that sorts organisms according to interbreeding, and another classification based on overall genetic similarity. Add to these classifications a third classification based on ecological adaptedness. Wu (2004) cites cases where a classification based on genes for ecological adaptedness fails to coincide with a classification based on overall genetic similarity. Moreover, neither of these classifications coincides with a classification based on interbreeding behavior. Bringing molecular data to the table does not reduce the number of classifications but increases their number.

A promoter of a molecular approach to classification may respond that classifications based on genetic similarity should be preferred over all other types of classifications. An argument would then need to be made for why classifications based on genetic similarity are more fundamental than classifications based on interbreeding or ecological adaptations. Some biologists doubt that such an argument is forthcoming. Molecular data faces many of the same problems that confront traditional data, in addition to its own problems (Maddison, 1997; Mayden, 2002). Furthermore, the pressing problems of biological taxonomy, such as phylogenetic inference (see Section 4), apply to molecular and nonmolecular data alike. Molecular data is not the antidote for pluralism.

4. Two Major Schools of Biological Taxonomy

We have seen that biologists disagree over the proper approach to classifying organisms into species. We now turn to the task of classifying species into genera, genera into families, and so on up the Linnaean hierarchy. As is the case with species, biologists disagree on the proper way to classify taxa into more inclusive taxa. Those differences arise from biologists subscribing to different schools of biological taxonomy, where each school provides its own principles and methods for constructing classifications.

The twentieth century saw three major schools of biological taxonomy: Evolutionary Taxonomy, Pheneticism (Numerical Taxonomy), and Cladism. Cladism is currently the most popular school among taxonomists, although many still subscribe to the tenets of Evolutionary Taxonomy. Pheneticism is no longer considered a viable taxonomic school by the vast majority of taxonomists. This section will introduce Evolutionary Taxonomy and Cladism and the philosophical issues surrounding these schools. Pheneticism will not be discussed here (for a philosophical introduction to pheneticism see Sober [1993]).

Evolutionary Taxonomy is a product of evolutionary thinking in the early twentieth century. In the 1930s, a handful of biologists developed a Mendelian framework for Darwinian evolutionary theory. The result of their work was the "evolutionary synthesis": the integration of Mendelian genetics and Darwinian theory [See Population Genetics]. Theodore Dobzhansky, Ernst Mayr, and Gaylord Simpson used the insights of the evolutionary synthesis to forge the school Evolutionary Taxonomy. That school has two main tenets. First, the members of a taxon must be descendants of a common ancestor; that is, all taxa must be genealogical lineages. Second, as Mayr (1981 [1994], p.290) writes, "evolutionary taxonomists . . . aim to construct classifications that reflect both of the two major evolutionary processes, branching and divergence (cladogenesis and anagenesis)." In cladogenesis, a single lineage is split into two branches (Figure 6.2a). Suppose a population of a species becomes isolated from the rest of the species. If that population is exposed to new selection forces, it may undergo a "genetic revolution" and become a new species. In anagenesis, speciation occurs in a single lineage (Figure 6.2b). Suppose a species enters a new environment and acquires a radically new suite of adaptations. If that change is drastic enough, then the lineage has evolved into a new species.

Given that speciation can occur through either cladogenesis or anagenesis, evolutionary taxonomists believe that classifications should highlight the two types of taxa

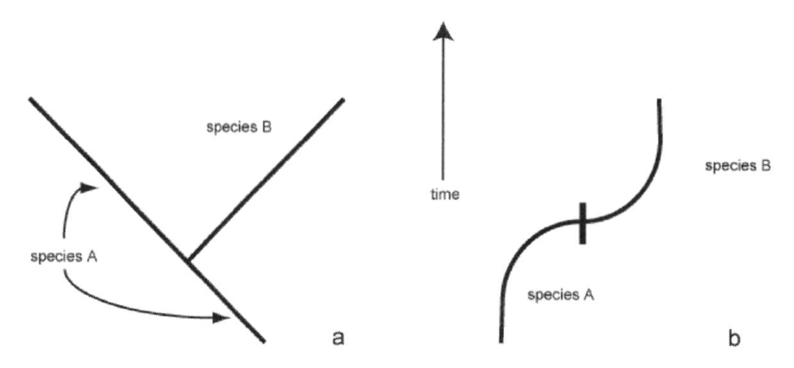

Figure 6.2 (a) Speciation by cladogenesis. (b) Speciation by anagenesis

that arise from these processes: monophyletic taxa and paraphyletic taxa. A monophyletic taxon contains an ancestor and all and only its descendants. In Figure 6.3, the group containing crocodiles and birds is monophyletic, as is the group containing lizards and snakes, and the group containing lizards, snakes, crocodiles, and birds. Monophyletic taxa are the result of cladogenesis or branching events. A paraphyletic taxon contains an ancestor and some but not all of its descendants. The group Reptilia, which contains lizards, snakes, crocodiles, but not birds, is paraphyletic. Paraphyletic taxa are the result of anagenesis. The lineage leading to birds has diverged significantly from lizards, snakes, and crocodiles, so evolutionary taxonomists exclude birds from the taxon Reptilia. In brief, evolutionary taxonomists believe that classifications should highlight only genealogical taxa, and those taxa can be either monophyletic or paraphyletic.

In the second half of the twentieth century the taxonomic school Cladism was introduced by Willi Hennig. The word "Cladism" is based on the Greek word for branch. Hennig believed that only those taxa that are the result of cladogenesis should be classified. His aim was to construct classifications that reflect common ancestry. If two taxa originate in the same branching event, then they have a common ancestor that is not shared by any other taxon. Crocodiles and birds have a common ancestor that is not shared by lizards and snakes (Figure 6.3). So a cladistic classification of those taxa places crocodiles and birds in a taxon that excludes lizards and snakes. Cladists believe that classifications should be based on common ancestry and nothing else. This view of classification has implications for the types of taxa that cladists represent in their classifications. Monophyletic taxa are defined in terms of common ancestry: a monophyletic taxon contains all and only the descendants of a common ancestor. So only monophyletic taxa are represented in cladistic classifications. Paraphyletic taxa are excluded from such classifications. A paraphyletic taxon does not contain all the descendants of a common ancestor. Because the taxon Reptilia excludes birds, this taxon does not contain all the descendents of its most recent ancestor (Figure 6.3). Thus cladists do not recognize the taxon Reptilia.

We can now see a major difference between Cladism and Evolutionary Taxonomy. Cladists only cite monophyletic taxa in their classifications because such taxa are the result of common ancestry. Evolutionary taxonomists represent both paraphyletic and

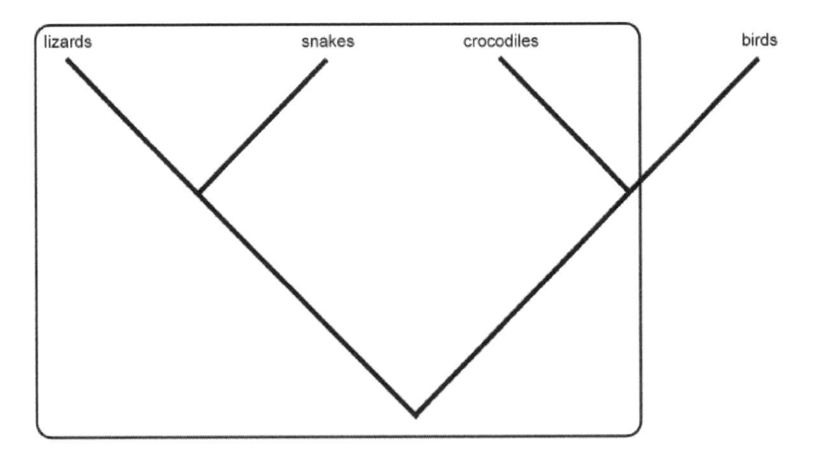

Figure 6.3 A phylogenetic tree of lizards, snakes, crocodiles, and birds

monophyletic taxa in their classifications because they believe that two types of information should be represented in classifications: common ancestry, and how much a taxon has diverged from its neighbors. This difference in taxonomic thought causes cladists and evolutionary taxonomists to construct opposing classifications. For example, evolutionary taxonomists posit the taxon Reptilia while cladists do not.

Cladists have two main complaints with Evolutionary Taxonomy (Hennig, 1966; Eldridge & Cracraft, 1980). We have already seen the first, namely that evolutionary taxonomists allow the existence of paraphyletic taxa. For cladists, such taxa are incomplete lineages: they do not contain all the descendants of a common ancestor. Cladists believe that placing crocodiles in a taxon that excludes birds ignores the unique common ancestor shared by birds and crocodiles. Another problem that cladists see with evolutionary taxonomy concerns the meaning of "significant divergence." When evolutionary taxonomists maintain that birds and reptiles have diverged significantly they cite the adaptive and phenotypic differences between those organisms. Birds, they suggest, live in a very different adaptive zone than reptiles. Furthermore, birds have significantly different traits than reptiles, such as wings and feathers. Cladists respond that the concepts of phenotypic difference and adaptive zone are ambiguous and are applied inconsistently to different types of taxa (Hennig, 1966; Eldridge & Cracraft, 1980). Cladists believe that the concepts of phenotypic diversity and adaptive zone are too malleable and reject them as grounds for classifying taxa.

Evolutionary taxonomists, for their part, think that cladists are wrong for not recognizing the existence of paraphyletic taxa (Mayr, 1981 [1994]). According to evolutionary taxonomists, paraphyletic taxa are not limited to a few marginal cases but occur throughout the organic world. Consider the case of ancestral species. Speciation frequently begins when a small population becomes isolated from the main body of a species. That "founder population" is exposed to different selection factors and becomes an incipient species. Meanwhile the main body of the old species – the ancestral species – continues to live. In Figure 6.2a, species b is the new species and species a is the ancestral species. Species a is paraphyletic because it contains some but not all of the descendants of species a's founder population: some of species a's descendants are members of species b. Cladists deny the existence of ancestral species because such species are not monophyletic. Evolutionary taxonomists respond that ancestral species abound in the world and cladists should not deny the existence of such a frequent type of phenomena.

Despite such criticisms, Cladism has become the prominent school of taxonomy. This is largely due to its precise and unambiguous methods for constructing classifications. Cladists aim to use only evidence of common ancestry to infer classifications. That evidence comes in the form of traits called "homologies." A homology occurs in two (or more) organisms and has been passed down from a common ancestor. Eyes in humans and dogs are homologous. Cladists attempt to avoid constructing classifications using traits called "homoplasies." A homoplasy is a similar trait in two (or more) organisms that has been passed down from different ancestors. Octopus eyes are similar to mammalian eyes but they evolved in different lineages. So octopus eyes and human eyes form a homoplasy and are not evidence of common ancestry. A challenge for cladists is distinguishing those similarities that are homologies from those similarities that are homoplasies. This is an important distinction for cladists because only homologies serve as evidence for cladistic classifications.

Cladists disagree over the criteria for distinguishing homologies from homoplasies (Hall, 1994). Here are two proposed criteria. According to one criterion, a homology must be a fundamental similarity rather than a superficial similarity between two traits. Bird wings and bat wings violate this criterion. They look similar on the surface, but they are supported by different digits and made of different materials. A second criterion demands that a homology be similar in both the adult form and in the embryonic stages that lead to adulthood. Barnacles and limpets have similar adult traits, such as a hard external armor and the ability to feed through a hole in that armor. However, the embryonic stages of these traits are dissimilar. So these traits are considered homoplasies.

Once a cladist determines which traits are homologies, the cladist constructs a cladogram. A cladogram represents the branching relations among a group of taxa and provides the basis for cladistic classification. The move from data concerning homologies to positing a cladogram is called "phylogenetic inference." Unfortunately the inference from putative homologies to a single cladogram is not straightforward. Cladists use dozens of traits to infer the correct cladogram for a set of taxa, and more often than not, the traits used to construct a cladogram provide conflicting evidence: some traits support one cladogram, while other traits support a different cladogram. Consider an example. Suppose that a biologist wants to classify three taxa, A, B, and C, and she has information about three traits, x, y, and z. Suppose also that each trait comes in one of two states, 0 and 1, where 0 is ancestral and 1 is derived. The distribution of traits found in the three taxa is the following:

	A	B	C
x	1	1	0
y	1	1	0
z	0	1	1

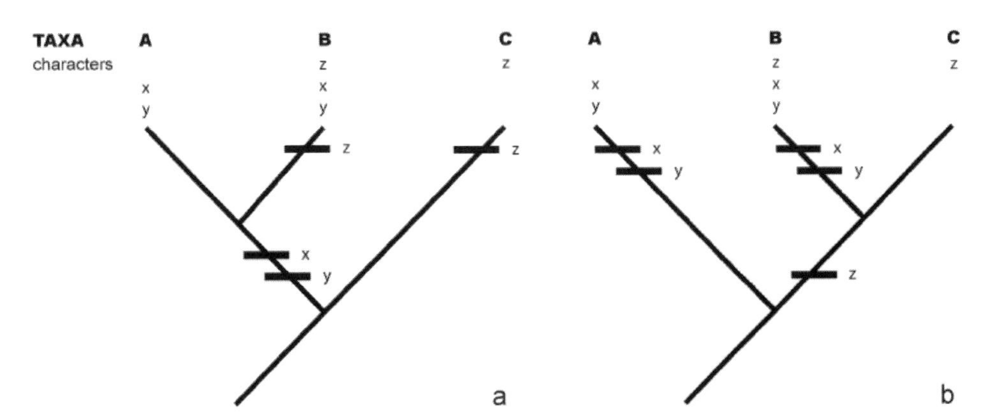

Figure 6.4 Two cladograms of the same taxa. (a) and (b) represent different evolutionary scenarios

This distribution gives rise to two conflicting cladograms (Figures 6.4a and 6.4b). According to one cladogram, A and B are more closely related to each other than either is to C (Figure 6.4a). According to the other cladogram, B and C are more closely related (Figure 6.4b). Which cladogram should a cladist posit?

Cladists agree about the source of confusion in such cases: some of the putative homologies are actually homoplasies. If the first cladogram (Figure 6.4a) is correct, then z is a homoplasy and has evolved at least twice: once in the branch leading to B and once in the branch leading to C. On the other hand, if the second cladogram (Figure 6.4b) is correct, then x and y are homoplasies. If that is the case, then x and y have each evolved at least twice: both on the branch leading to A and both on the branch leading to B. The task for cladists is to determine which of the putative homologies are homoplasies. To do this, cladists employ the principle of parsimony: choose the phylogeny that requires the minimal number of changes to arrive at a given trait distribution. In the example under consideration, the principle of parsimony counsels choosing the phylogeny represented by the first cladogram. The phylogeny captured by the first cladogram requires a minimum of four changes (they are represented by slash-marks on the cladogram), whereas the phylogeny represented by the second cladogram requires a minimum of five changes.

Cladists offer various justifications for their reliance on the principle of parsimony. Some suggest that evolution itself is parsimonious. Ridley (1986) reasons that because it is unlikely for a mutation to be selected in a species, it is even more unlikely for similar mutations to occur and to be selected in multiple species. Ridley's justification for parsimony turns on general assumptions about evolution. Other cladists argue that assumptions about evolution can justify the use of parsimony, but only on a case-by-case basis (Felsenstein, 1978). In some instances evolution is parsimonious, in others it is not, depending on local mutation rates and selection coefficients. So in some situations the use of parsimony is empirically justified, in other situations it is not.

A third group of cladists justifies the use of parsimony on more philosophical grounds. They suggest that the preference for the more parsimonious cladogram need not depend on any assumptions about evolution. Instead, we should prefer the more parsimonious cladogram because it is more falsifiable (see Farris, 1983). Cladists in this group follow Karl Popper's philosophy of science: all scientific hypotheses must be falsifiable; those hypotheses that are unfalsifiable – that cannot possibly be shown to be false with empirical evidence – are unscientific. Some cladists argue that to posit a homoplasy without empirical evidence is to posit an unfalsifiable hypothesis. The more homoplasies a cladogram requires, the greater the number of unfalsifiable hypotheses posited. Thus, the more parsimonious cladogram is preferred for the methodological reason that it is more falsifiable. (Popper might disagree with this application of his philosophy because he does not think that falsifiability comes in degrees.) Cladists have written extensively on the proper justification of parsimony. The issue is far from settled.

5. The Linnaean Hierarchy

Having discussed the philosophical issues surrounding the nature of species and the principles for classifying taxa, we now turn to the framework for constructing classifications – the Linnaean hierarchy. Although many aspects of biological taxonomy are under debate, one might hope that the Linnaean hierarchy is universally accepted. Unfortunately this is not the case. The continued use of the Linnaean hierarchy has been challenged. Some biologists and philosophers believe that the Linnaean hierarchy has outlived its usefulness and should be replaced (de Queiroz & Gauthier, 1992; Ereshefsky, 1994). Other biologists believe that the Linnaean hierarchy is still the best system available and should be retained (Forey, 2002). The debate over the Linnaean hierarchy is an important issue because much of biological theory employs the Linnaean ranks, from prey–predator relations in ecology to hypotheses concerning the tempo and mode of macroevolution.

The current Linnaean hierarchy contains 21 ranks, from subspecies to kingdom. Linnaeus posited a hierarchy of 5 ranks: variety (subspecies), species, genus, order, and class. Evolutionary taxonomists believed that Linnaeus's 5 ranks were insufficient for representing life's diversity, so they posited the 21 ranks used today. From Linnaeus's time to the advent of Cladism, taxonomists have offered various definitions of Linnaean ranks that aim to highlight a common biological factor among the taxa of a particular rank. For example, taxonomists have tried to find a biological factor that is common to families and that distinguishes families from tribes and orders. Evolutionary taxonomists and cladists have offered various suggestions for defining the higher Linnaean ranks (those ranks above the rank of species); none of those definitions has withstood criticism.

Evolutionary taxonomists have suggested that such factors as phenotypic diversity and ecological breadth indicate the rank of a taxon. The greater the phenotypic diversity within a taxon, or the greater the size of a taxon's adaptive zone, the more inclusive a taxon. For example, the adaptive zone of a tribe will be greater than the adaptive zone of a family. As discussed in the previous section, cladists consider the concepts "adaptive zone" and "phenotypic diversity" to be ambiguous and applied inconsistently across phyla. Hennig playfully asks "whether the morphological divergence between an earthworm and a lion is more or less than between a snail and a chimpanzee?" (1966, p.156). Most taxonomists now believe that the concepts of phenotypic diversity and adaptive zone are too malleable to serve as measures of a taxon's rank.

Hennig (1965) offered an alternative way of defining the higher Linnaean ranks by suggesting that taxa of the same rank originate in the same time period. Classes, for example, should be defined as all and only those taxa that originated during the Late Cretaceous. Orders would be defined as all and only those taxa that originated during a more recent time period. Hennig's suggestion for defining the higher Linnaean ranks is problematic as well. Taxa that originate in the same period often have different phylogenetic structures. Some taxa that originated during the Late Cretaceous are quite successful and contain a number of orders and genera; such taxa have extensive phylogenetic branching. Other taxa that originated during the same period are monotypic and contain only a single basal taxon; they are phylogenetic twigs. From a phylogenetic perspective, Hennig's criterion places different types of taxa under a single rank. Cladists, including later Hennig, abandoned the idea of correlating the rank of a taxon with time of origin.

Neither evolutionary taxonomists nor cladists have established a universal criterion for defining the higher Linnaean ranks. Instead, they use a patchwork of criteria for determining the ranks of such taxa. As a result, taxa of the same rank can vary dramatically. Families can vary in their age, their phylogenetic structure, their phenotypic diversity, and the breadth of their adaptive zone. Calling a taxon a "family" merely means that *within* a particular classification that taxon is more inclusive than a genus and less inclusive than a class. There is no ontological commonality among all taxa we call "families." Some have generalized this conclusion and questioned whether the higher Linnaean ranks correspond to any categories in nature (de Queiroz & Gauthier, 1992; Ereshefsky, 1994).

Thus far we have discussed the meaning of the higher Linnaean ranks, but what of the rank of species? While many taxonomists question whether there are higher Linnaean categories in nature, most continue to believe in the existence of the species category. However, species pluralism poses a threat to the claim that "species" refers to a unified category in nature. Recall that biologists offer a myriad of species concepts. Some biologists define a species as a group of organisms that successfully interbreed and produce fertile offspring. Cladists assert that a species is a group of organisms bound by a unique phylogeny. Still other biologists suggest that a species is a group of organisms that share a unique ecological niche. Each proposal highlights a different biological feature for defining "species." Species pluralists believe that each of these concepts is theoretically legitimate. Some species are groups of interbreeding organisms; others consist of asexual organisms. Some species are monophyletic, that is, good phylogenetic species; others are not. These different types of taxa that we call "species" are real, yet they lack a common significant feature.

If the above arguments concerning the Linnaean ranks are correct, then the Linnaean ranks, from species up, refer to heterogeneous collections of taxa. There is no unique and universal biological feature found in all taxa called "species," just as there is no common and distinct feature among those groups of organisms referred to as "families." This result undermines the reality of the species and other Linnaean categories. In the end, the Linnaean hierarchy may be a fictitious grid that we place on nature.

The heterogeneity of the Linnaean categories has practical implications, especially for biodiversity studies. The units biologists use for measuring biodiversity are Linnaean:

biologists count the number of species present in a location, or the number of genera or families. However, the Linnaean ranks can mask important biological differences. Suppose that we want to measure the biodiversity of a class of organisms by the number of families present. Suppose further that the comparison is between snail families and mammalian families. Snail families have much denser phylogenetic structures than mammalian families. That is, snail families contain many more species than mammalian families. If we measure biodiversity by number of families, then we are not measuring comparable units. Because some families will have many species and other families will have few species, "family" does not refer to a consistent biological unit. In general, the Linnaean ranks do not correspond to categories in nature, and they should not be employed in biodiversity studies. Instead, these studies should use parameters that capture such biological phenomena as phylogenetic structure or ecological breadth [SEE BIODIVERSITY: ITS MEANING AND VALUE].

Thus far we have talked about the Linnaean ranks. Often when biologists talk about the Linnaean hierarchy they mean more than just the Linnaean ranks. They also have in mind the Linnaean rules of nomenclature for naming taxa. Some of these rules were introduced by Linnaeus, other rules were introduced by evolutionary taxonomists in the twentieth century. To avoid confusion, let the "Linnaean system of classification" refer to both the Linnaean hierarchy and the Linnaean rules of nomenclature.

The centerpiece of the Linnaean naming rules is the requirement that the name of a taxon should indicate a taxon's rank and classification. For species this is achieved with Linnaeus's binominal rule. The names of all species contain two parts: a generic name and a specific name. The generic name refers to a species' genus. For example, the generic name of *Homo sapiens* is *Homo*. The specific name of a species distinguishes that species from all other species in its genus. *Sapiens* is the specific name of our species. Binomial names clearly indicate which taxa are species: all and only species have binomial names, while other taxa have singular names. The generic name of a binomial indicates the classification of a species: the name *Homo* shows that our species is part of the genus *Homo*.

Similar Linnaean rules require that the names of higher taxa display a higher taxon's rank and classification. The names of most higher taxa have rank-specific endings showing the ranks of those taxa. For example, the rank of the family Hominidae is represented by the suffix *–idae*, and rank of the tribe Hominini is indicated by the suffice *–ini*. The name of a higher taxon is formed from the name of that taxon's type genus – the genus contained in that taxon. "Hominidae" and "Hominini" and are formed from the root *Homin*, which stands for the type genus *Homo*. All taxa whose names include the root *Homin* form a hierarchy of taxa containing the genus *Homo*.

Representing the ranks and classifications of taxa in their names seems like a good idea. One just needs to read a taxon's name to know that taxon's rank and classification. Since its inception, the popularity of the Linnaean system has been attributed to its practical features, such as Linnaeus's binomial rule. But some question the practicality of the Linnaean rules of nomenclature (de Queiroz & Gauthier, 1992; Ereshefsky, 1994).

Consider the activity of taxonomic revision. Such revision occurs when a taxon is assigned a new rank or given a new position in a classification. Taxonomic revision is the norm not the exception in biological taxonomy and can occur for many reasons. New DNA evidence may imply that a species should be reassigned to a different genus.

Or a shift in taxonomic theory, such as cladists eliminating paraphyletic taxa from classifications, may require changing a taxon's rank from tribe to family. Taxonomic revision causes instability in classification: new evidence or new theoretical considerations give us reason to revise our classifications. The Linnaean rules of nomenclature are also a source of instability because the names of taxa reflect the rank and classification of a taxon. As such, these rules make the job of the taxonomist harder than need be. Not only must biologists revise a taxon's classification, they must rename the taxon as well. This may not sound like much of an inconvenience, but it is. A case of taxonomic revision can involve renaming hundreds of taxa. The Linnaean rules of nomenclature themselves are a source of instability.

The Linnaean rules cause other practical problems. For instance, when taxonomists disagree on the rank of a taxon they must assign that taxon different names containing different rank-specific endings. For example, one biologist may think that a taxon is a family and another biologist may consider the same taxon to be a tribe. Following the Linnaean rules, the first biologist must name the taxon "Hominidae" while the second biologist must name it "Hominini." Even though the biologists agree they are talking about the same taxon, the Linnaean rules require the taxon to have multiple names. Another problem with the Linnaean rules of nomenclature is that they cause hasty classification. Recall that a species must be given a binomial name that includes the name of a species' genus. Often biologists do not know the genus of a newly discovered species. Yet if a biologist wants to name a new species, she must first assign that species to a genus. According to some biologists, the binomial rule often causes the assignment of a species to a genus on inadequate grounds. Thus, the binomial rule is a cause of inaccurate classification.

Supporters of the Linnaean system are well aware of the problems facing the Linnaean system of classification. But they argue that the Linnaean system is still the best system available. Detractors of the Linnaean system have constructed alternative systems of classification. The most prominent one to date is the Phylocode, which was developed by a group of cladists (Cantino et al., 2001). Supporters of the Linnaean system believe that the Phylocode is an inferior system (Forey, 2002). This judgment may be hasty because proponents of the Phylocode are just starting to develop their system. Whether biologists should continue using the Linnaean system or adopt an alternative is hotly debated by taxonomists. It is too early to predict the outcome of this debate.

One other challenge to the Linnaean hierarchy is worth mentioning. This challenge is not only to the Linnaean hierarchy, but to any system of classification that assumes life is hierarchically arranged. The Linnaean hierarchy and rival systems of classification assume that there is a single hierarchical tree of life, a tree with a single origin, and speciation events that give rise to non-overlapping lineages. Each species is a part of one and only one genus, each genus is a part of one and only one family, and so on up the hierarchy. However, recent molecular studies of bacteria challenge the assumption that life forms a single hierarchical tree (Doolittle, 1999).

Whether or not life forms a single hierarchical tree depends on how genetic information is transferred. The common assumption is that the vast majority of genes are passed down from parent to offspring. For the most part species are closed gene pools. Hybridization may occur between closely related species, but organisms in different genera and families rarely exchange genes. Bacteria and archaea (ancient bacteria), however, do not play by the same reproductive rules. Bacteria do not reproduce sexually. Molecular studies reveal that considerable amounts of bacterial DNA are transferred laterally among organisms of the same generation. Consequently, bacteria evolution is not one of a branching tree but of an intertwined bush. Ford Doolittle (1999, p.2124) suggests "molecular phylogeneticists will have failed to find the 'true tree,' not because their methods are inadequate or because they have chosen the wrong genes, but because the history of life cannot properly be presented as a tree."

One might question the significance of Doolittle's suggestion given that it is based on information concerning bacteria and archaea, but his suggestion should be taken very seriously. There are more types of bacteria and archaea than all other types of organisms, and the combined weight of all bacteria and archaea is greater than the combined weight of all other organisms (Tudge, 2000, p.107). We are not talking about a few isolated cases here. Molecular studies of bacteria and archaea indicate that *most* of life does not form a phylogenetic tree.

In summary, we have seen that the nature of species, the general principles of biological taxonomy, and the soundness of the Linnaean hierarchy are controversial. We now see that even the assumption that life is hierarchically arranged has been challenged. The philosophical problems facing biological taxonomy are foundational. How these problems are resolved will have widespread implications both inside and outside of biology. Biological taxonomy is rife with conceptual problems and fertile ground for philosophical analysis.

References

Boyd, R. (1999a). Homeostasis, species, and higher taxa. In R. Wilson (Ed.). *Species: new interdisciplinary essays* (pp. 141–85). Cambridge, MA: MIT Press.

Boyd, R. (1999b). Kinds, complexity and multiple realization: comments on Millikan's "Historical Kinds and the Special Sciences." *Philosophical Studies*, 95, 67–98.

Cantino P. D. et al. (2001). Phylocode: a phylogenetic code of biological nomenclature. Available from: http:/www.ohiou.edu/phylocode/.

de Queiroz, K. (1999). The general lineage concept of species and the defining properties of the species category. In R. Wilson (Ed.). *Species: new interdisciplinary essays* (pp. 49–90). Cambridge, MA: MIT Press.

de Queiroz, K., & Gauthier, J. (1992). Phylogenetic taxonomy. *Annual Review of Ecology and Systematics*, 23, 480–99.

Doolittle, W. F. (1999). Phylogenetic classification and the universal tree. *Science*, 284 (June), 2124–8.

Dupré, J. (1981). Natural kinds and biological taxa. *Philosophical Review*, 90, 66–90.

Dupré, J. (1993). *The disorder of things: metaphysical foundations of the disunity of science*. Cambridge, MA: Harvard University Press.

Eldridge, N., & Cracraft, J. (1980). *Phylogenetic patterns and the evolutionary process.* New York: Columbia University Press.

Ereshefsky, M. (1992). Eliminative pluralism. *Philosophy of Science*, 59, 671–90.

Ereshefsky, M. (1994). Some problems with the Linnaean hierarchy. *Philosophy of Science*, 61, 186–205.

Ereshefsky, M. (2001). *The poverty of the Linnaean hierarchy: a philosophical study of biological taxonomy*. Cambridge: Cambridge University Press.

Ereshefsky, M., & Matthen, M. (2005). Taxonomy, polymorphism and history: an introduction to population structure theory. *Philosophy of Science*, 72, 1–21.
Farris, J. S. (1983). The logical basis of phylogenetic analysis. In N. I. Platnick & V.A. Funk (Eds). *Advances in cladistics II* (pp. 7–36). New York: Columbia University Press.
Felsenstein, J. (1978). Cases in which parsimony or compatibility methods will be positively misleading. *Systematic Zoology*, 27, 401–10.
Ferguson, J. (2002). On the use of genetic divergence for identifying species. *Biological Journal of the Linnean Society*, 75, 509–19.
Forey, P. (2002). Phylocode – pain, no gain. *Taxon*, 51, 43–54.
Ghiselin, M. (1974). A radical solution to the species problem. *Systematic Zoology*, 23, 536–44.
Griffiths, P. (1999). Squaring the circle: natural kinds with historical essences. In R. Wilson (Ed.). *Species: new interdisciplinary essays* (pp. 209–28). Cambridge, MA: MIT Press.
Hall, B. K. (Ed.). (1994). *Homology: the hierarchical basis of comparative biology*. San Diego: Academic Press.
Hennig, W. (1965). Phylogenetic systematics. *Annual Review of Entomology*, 10, 97–116.
Hennig, W. (1966). *Phylogenetic systematics*. Urbana: University of Illinois Press.
Hull, D. (1965). The effect of essentialism on taxonomy: two thousand years of stasis. *British Journal for the Philosophy of Science*, 15, 314–26.
Hull, D. (1978). A matter of individuality. *Philosophy of Science*, 45, 335–360.
Hull, D. (1999). On the plurality of species: questioning the party line. In R. Wilson (Ed.). *Species: new interdisciplinary essays* (pp. 23–48). Cambridge, MA: MIT Press.
Kitcher, P. S. (1984). Species. *Philosophy of Science*, 51, 308–33.
Maddison, W. (1997). Gene trees in species trees. *Systematic Biology*, 46(3), 523–536.
Mayden, R. (2002). On biological species, species concepts and individuation in the natural world. *Fish and Fisheries*, 3, 171–96.
Mayr, E. (1959). Typological versus population thinking. In *Evolution and anthropology: a centennial appraisal*. Washington: The Anthropological Society of Washington; In E. Mayr, *Evolution and the diversity of life* (pp. 26–9). Cambridge, MA: Harvard University Press (1976).
Mayr, E. (1981 [1994]). Biological classification: toward a synthesis of opposing methodologies. *Science*, 214, 510–16. Reprinted in E. Sober (Ed.). *Conceptual issues in evolutionary biology* (2nd edn, pp. 277–94). Cambridge, MA: MIT Press.
Ridley, M. (1986). *Evolution and classification: the reformation of cladism*. London: Longman.
Sober, E. (1980). Evolution, population thinking and essentialism. *Philosophy of Science*, 47, 350–83.
Sober, E. (1984). Sets, species, and natural kinds. *Philosophy of Science*, 51, 334–41.
Sober, E. (1993). *Philosophy of biology*. Boulder, CO: Westview.
Tudge, C. (1999). *The variety of life: a survey and a celebration of all the creatures that have ever lived*. Oxford: Oxford University Press.
Wilson, R. (1999). Realism, essence, and kind: resuscitating species essentialism? In R. Wilson (Ed.). *Species: new interdisciplinary essays* (pp. 187–208). Cambridge, MA: MIT Press.
Wu, C., & Ting, C. (2004). Genes and speciation. *Nature Genetics*, 5, 114–22.

Further Reading

Ereshefsky, M. (Ed.). (1992). *The units of evolution: essays on the nature of species*. Cambridge, MA: MIT Press.
Ghiselin, M. (1997). *Metaphysics and the origin of species*. Albany, NY: SUNY Press.

Hull, D. (1988). *Science as a process: an evolutionary account of the social and conceptual development of science.* Chicago: University of Chicago Press.

Mayr, E. (1988). *Toward a new philosophy of biology.* Cambridge, MA: Harvard University Press.

Panchen, A. (1992). *Classification, evolution, and the nature of biology.* Cambridge: Cambridge University Press.

Sober, E. (1988). *Reconstructing the past: parsimony, evolution, and inference.* Cambridge. MA: MIT Press.

Wheeler, Q., & Meier, R. (Eds). (2000). *Species concepts and phylogenetic theory: a debate.* New York: Columbia University Press.

Chapter 7

Population Genetics

CHRISTOPHER STEPHENS

Population genetics is the study of processes that influence gene and genotype frequencies. It has been obsessed with two related questions: what is the extent of the genetic variation between individuals in nature and what are the factors that are responsible for this variation? Much of the historical, methodological, and philosophical interest in population genetics results from the fact that these two central questions – the *extent* and *explanation* of genetic variation – have proved extraordinarily difficult to answer. It is impossible to know the complete genetic structure of any species, and there are significant underdetermination problems in figuring out which factors are the relevant causes of evolutionary change, even if one knows a lot about the genetic structure of a population. Despite these difficulties, population genetics has had remarkable successes, and is widely viewed as the theoretical core of evolutionary biology. Significant evolutionary changes often occur over thousands or millions of years. Because of this, it is impossible to observe these changes directly. As a result, understanding the causes of evolution depends crucially on theoretical insights that flow from the mathematical models of population genetics.

This essay will proceed as follows. It begins with an historical overview of the problems that led to the development of population genetics in its modern form. Many of the controversies that motivate contemporary discussion about population genetics began in the immediate aftermath of the publication of Darwin's *On the Origin of Species* (Darwin, 1859). One of the main problems that neo-Darwinian theory faced was to show how evolutionary change could occur in the available time based on small fitness differences. One of the most significant changes in evolutionary theory from Darwin's day compared to contemporary evolutionary theory is the quantification and formalization of evolutionary change. After examining this history, some of the basic models of contemporary population genetics will be explained, including the famous "Hardy–Weinberg" law. Next, four major controversies that have involved population genetics will be briefly considered. First, the debate between two of the founders of modern population genetics, R. A. Fisher and Sewall Wright will be addressed. After an introduction to contemporary population genetics, a discussion of fitness and the tautology problem follows. Fourth, the so-called "classical/balance" debate between Dobzhansky and Muller will be examined. This will give an opportunity to discuss some of the social dimensions of population genetics. Next will follow a discussion of how the classical/

balance debate was transformed into the neutralism debate by the development of molecular techniques for detecting genetic variation as well as by subsequent theoretical innovations. Finally, what biologists such as S. J. Gould view as a challenge to population genetics in its traditional form – the saltationism–gradualism debate will be examined. Each of these debates will be examined with an eye to thinking about how they have transformed population genetics.

1. Historical Overview

Questions about the extent and nature of individual variation have been with evolutionary biology since Darwin (1859) and Wallace (1859). Indeed, people have wondered about these issues long before Darwin and Wallace came along, but their development and defense of evolution by natural selection gave these issues a new theoretical setting. Darwin and his contemporaries did not have a modern understanding of genetics, but analogous questions about the amount of variation at the phenotypic level and whether natural selection was a sufficient explanation of population changes were controversial as soon as *On the Origin of Species* was published. Fleeming Jenkin (1867) raised the most important initial problems for the development of population genetics in his review of the *Origin*. Jenkin (1867), an engineer, raised three major objections, two of which are most relevant to our purposes. First, he expressed skepticism about the idea that a species can indefinitely vary in a given direction, and second, he argued that under Darwinian assumptions about inheritance, natural selection alone will not be able to lead to substantial evolutionary change.

Jenkin's related objections focus on the connection between selection and variation. With respect to variation, Jenkin's concern was that if one assumes that inheritance occurs by blending, then the offspring of an individual will tend toward a species mean rather than toward a novel or more extreme value. He agreed with Darwin that many flora and fauna show considerable variation, and that natural and artificial selection could modify these populations. However, Jenkin had a traditional view that each species had a mean or "type" to which it would tend to revert.

Jenkin raised questions and doubts about whether natural selection could be as powerful as Darwin believed it to be. His questions were of three sorts. First, Jenkin wondered whether the system of inheritance was a "blending" theory in which an offspring's traits are intermediate between the parental values, or a "particulate" theory, in which a given trait might be dominant. Second, he wondered about whether the variation was slight or large (large variations were called "sports"). Third, regardless of whether the variation is small or large, is it present in a large number of individuals in a population, or just one or a few individuals?

While Darwin did not believe that evolution proceeded by large jumps ("sports"), other contemporaries of Darwin's such as Huxley did defend such a process of evolution by macromutation. Darwin, on the other hand, thought that natural selection could lead to large evolutionary change over a long period of time by acting simply on small, continuous variation. Because of this, and his acceptance of a "blending" theory of heredity, Darwin's theory had a major problem. Jenkin wondered how a beneficial character could get established in a population, if natural selection acted on just one or a few individuals. First, why couldn't the trait easily be lost due to chance? Second, even if the variation in traits were not eliminated by chance, under the hypothesis of blending inheritance, any advantage will be blended away.

We can illustrate the problem with blending inheritance using a few simple ideas from contemporary biology. Darwin didn't know about genes, but knew that there had to be some underlying mechanism that caused organisms to have the traits they did and that enabled them to pass their traits on. Here is a contemporary gloss on the problem that Jenkin raised. Suppose that we have a simple model of blending where *AA* codes for the tallest phenotype, and *aa* codes for the shortest, and *Aa* for an intermediate height. If *AA* (tall) mate with *aa* (short), there will be *Aa* (medium height offspring). The outcome here is not any different from that in a Mendelian model. The difference comes in the next generation. In a case of blending, the *A* and *a* gametes come together to form a new "blended" *Aa* type. If *Aa* mates with another *Aa*, these parents "blend" the gametes they inherited from their parents into a new gamete – call it *B*, which is then passed on to their offspring. So it is blended *B* that codes for medium height. Now if a *B* type mates with another *B* type, these parents will pass on different gametes from those they inherited. Instead of producing half *A* and half *a* gametes, they would produce only *B*. And so now all the offspring would be of medium height, and the initial variation in the population has been lost.

Darwin was unable to deal with these difficulties. Although many biologists immediately accepted that evolution was true, there was much continued controversy about the mechanism, in part for the reasons that Jenkin raised in his review. Right from the beginning, there was a tension between Darwinian selection, in which natural selection acts primarily on small continuous variations, and the theory of blending inheritance, in which small variations would seem to be "blended away" and lost over time.

In light of these problems, many biologists adopted alternative mechanisms to natural selection that might explain evolutionary change. Directed mutation theories, such as the Lamarckian inheritance of acquired characteristics, were proposed as a way to overcome the problem of how acting on small changes could add up to large-scale evolutionary change (Gayon, 1998). Another option was to find a new theory of inheritance so that variation was not lost in every generation. Mendel's (1865) theory was just such a theory and was rediscovered around the turn of the twentieth century.

Despite the fact that it avoided the problems of a blending theory of inheritance, Mendelism was controversial for several years (Olby, 1966). In the late nineteenth and early twentieth centuries, there was a heated debate between two groups. On the one hand were the biometricians led by Pearson and Weldon, who emphasized a statistical approach to measuring continuous variation, with the hope that Darwin's gradualism about natural selection would be vindicated. In the other camp were the Mendelians, who emphasized experimental results from breeding in which offspring seemed to inherit traits in the traditional Mendelian ratios. The Mendelians, though, were not (at least initially) gradualists – they thought that natural selection worked on major mutations, and that evolution proceeded by leaps. Both camps were inspired by Galton – the biometricians by his statistical work and the Mendelians by his rejection of gradualism (Provine, 1971).

Although mathematically sophisticated, one reason that biometricians such as Pearson didn't have much success in developing a theory was their methodological conservatism, influenced by their attraction to positivism. Unlike Darwin, who relied on something like inference to the best explanation, Pearson was much more conservative, and thought that the proper aim of science should be to develop laws that are mere "summaries" or "descriptions" of our observations, rather than attempt to postulate underlying explanations that are not themselves directly observable.

Most of the early Mendelians, such as Hugo de Vries (one of the rediscoverers of Mendel's laws), were fans of the view that mutation, rather than selection, was the primary mechanism behind evolutionary change. According to mutationists, it is only discontinuous change that is a source of evolution. Jenkin's objection that natural selection has limited variation on which to act was a problem for the biometricians, who, like Darwin, favored a gradualist approach to selection. Mutationists, in contrast, were motivated in part by this worry, and had an alternative explanation to selection to account for the source of variation. A typical mutationist defended the view that while selection might be relevant to whether a species *survives*, it fails as an explanation for the *origin* of species. Debates about different possible explanations of the survival and origin of species continue to this day.

Mendelian genetics merged with Darwinian views on selection in the 1910s and 1920s. Mendelian genetics offered a theory of heredity that showed that populations could be stable – heredity was not a factor that varied in intensity. Hardy (1908) and Weinberg (1908) independently derived this result theoretically, stating what is known as the Hardy–Weinberg law. Because Mendelian genetics offered a kind of baseline for how to think about changes, one could subsequently think more clearly about how other possible factors such as selection, mutation, and drift would affect the evolutionary process.

Fisher's (1918) paper is one of the best candidates for the start of the so-called modern synthesis. In it, Fisher showed that the statistical results about continuous variation of various characteristics due to the biometricians could be reconciled with Mendelian inheritance. Another important early result was in R. C. Punnett's 1915 book, *Mimicry in Butterflies*, which included the mathematician H. T. J. Norton's table in an appendix (pp. 154–6). In this table, Norton calculated the number of generations that it would take for a Mendelian population subject to selection to go from one distribution to another. What was striking about the results in Norton's table is that even with relatively weak selection intensities, one form can completely replace another form in a relatively small number of generations. For example, even with a selection intensity of 1 percent in favor of the dominant form, a population can go from being about 2 percent of the dominant form (including both heterozygotes and homozygotes) to about 97 percent of the population being of the dominant form in about 1,100 generations. Although we do not have the papers that demonstrate the algorithms that Norton used, they are apparently similar to those that Haldane (1924) later developed. Norton's table and Haldane's later work convinced biologists that Mendelism and Darwinism were compatible – natural selection could work in a Mendelian framework. Thus, it was not necessary to employ Lamarckian principles such as inheritance of acquired characteristics to explain evolutionary change in natural populations. Natural selection, working on small differences, could, in a Mendelian framework, explain the observed results.

In its mature form, population genetics began with the so-called modern synthesis of the 1920s and 30s, largely due to the theoretical work of three people: R. A. Fisher, J. B. S. Haldane, and Sewall Wright. Although important developments occurred both before and afterwards, the modern synthesis did the most to shape the structure of what we think of now as population genetics. Fisher (1930) and Haldane (1932) wrote books and Wright (1931) an important summary paper, each of which, in different ways, used mathematics to merge Mendelian genetics with evolutionary theory. All three of these authors wanted to answer the following question: Given the existence of a number of possible factors such as selection, mutation, migration, and random drift all acting on genetic variation, how can one predict the frequencies of genes at many loci in many organisms over many generations?

The problem is similar to that faced by physicists in developing theories of statistical mechanics to try to explain the behavior of particles in a gas without being able to track the behavior of each individual particle. All three of them developed models that were simplified in various ways, e.g., they assumed that selection rates and population size were constant and ignored the possibility of frequency dependent selection. Even so, the mathematical achievement was significant.

In his book, *The Genetical Theory of Natural Selection*, Fisher (1930) wanted to show that selection, operating on very large populations, was the primary mechanism of evolutionary change. His theory showed that even with very slight selection pressures, a rare individual allele could replace a less advantageous alternative. He also argued that evolution was determined largely by what affects individuals. Fisher developed what he called the "fundamental theorem of natural selection," which Fisher compared to the second law of thermodynamics, although his law is supposed to lead to a kind of increase, rather than decrease, of order. Sewall Wright (1931), on the other hand, focused more on the issue of what circumstances would best favor the origin and spread of a novel adaptation. Wright put more emphasis on the interactive effects of genes, and argued that population structure and random "drift" played more important roles. Unlike Wright and Fisher, Haldane focused more on the issue of modeling what could really be measured, and so often concentrated on high selection rates because they could be more easily observed.

2. Population Genetics Models

Richard Lewontin (2000, p.5) refers to population genetics as the "auto mechanics of evolutionary theory." The main task of population genetics models is to predict frequencies of genotypes in one generation based on their frequencies in a prior generation. Population genetics usually recognizes natural selection, mutation, migration, and random genetic drift as possible causes of evolutionary change. The system of mating can also affect the genotype frequencies.

The Hardy–Weinberg law is a good place to start in introducing population genetics. In diploid organisms such as humans, chromosomes come in pairs. Suppose that in a population of infinite size (no drift) there is random mating, and no selection, mutation, or migration. In such a case, the Hardy–Weinberg equilibrium law determines that the genotype frequencies will evolve to a stable equilibrium regardless of where the frequencies begin.

In the so-called "standard selection model" (Nagylaki, 1992, p.51; Sarkar, 1994, p.5), several additional assumptions are made. One assumes that there are two sexes, each with the same initial frequency and with each individual organism being exactly one of the two sexes. In addition, one assumes that the locus at issue is not on a sex-determining chromosome that is not fully diploid (e.g., the XY male sex chromosomes). One supposes further that there is exactly one locus with two alternate alleles, *A* and *a*, which are segregated in gametes. The standard model also assumes that the alleles assort independently, that there are discrete, non-overlapping generations, and that the fitnesses are frequency and time independent.

In such a model there are three possible diploid genotypes: *AA* and *aa* (homozygotes) and *Aa* (the heterozygote). Let the frequencies of *AA*, *aa*, and *Aa* be P, Q, and R, respectively (where $P + Q + R = 1$). If we let p be the frequency of *A* and q be the frequency of *a*, then we can calculate the allele frequencies from the genotype frequencies as follows:

$$p = P + 1/2R \tag{1}$$

$$q = Q + 1/2R \tag{2}$$

where $p + q = 1$. If we have random mating and a very large population (so that picking out an organism of one type does not effectively change the probability of picking another organism of the same type), there is a P chance of picking out a *AA* homozygote at random, a P chance of picking out a second homozygote, and hence the probability of $AA \times AA$ matings is P^2. Similarly, the chance of $aa \times aa$ matings is Q^2, and that of $Aa \times Aa$ matings is R^2. There are two ways of getting each of the other combinations such as $AA \times Aa$. One is to pick *AA* first and *Aa* second, and the other way is pick the *Aa* first and *AA* second. So the frequencies of $AA \times Aa$, $AA \times aa$, and $Aa \times aa$ matings are $2PR$, $2PQ$, and $2RQ$, respectively. The frequency of each mating and the proportions of the offspring genotypes are given in the following table:

Mating	**Frequency of mating**	**Offspring genotype frequencies**		
		AA	**Aa**	**aa**
$AA \times AA$	P^2	P^2	0	0
$AA \times Aa$	$2PR$	PR	PR	0
$AA \times aa$	$2PQ$	0	$2PQ$	0
$Aa \times Aa$	R^2	$R^2/4$	$R^2/2$	$R^2/4$
$Aa \times aa$	$2RQ$	0	RQ	RQ
$aa \times aa$	Q^2	0	0	Q^2

As you can see from the table, only three combinations of matings will lead to *AA* offspring. The frequency of *AA* in the next generation (P′) is:

$$P' = P^2 + PR + R^2/4 \tag{3}$$

which is equivalent to $P' = (P + 1/2R)^2$. Using equation (1), we can substitute and get the frequency of *AA* in the new generation as:

$$P' = p^2 \tag{4}$$

Using similar reasoning for the remaining two genotypes, we can calculate that after one generation, the frequencies of the genotypes, *AA*, *Aa*, and *aa* will be p^2, $2pq$, and q^2 respectively, regardless of what the initial genotype frequencies *P*, *R*, and *Q* were. Furthermore, as long as mating is random, the population size is sufficiently large and there is no selection, mutation, or migration, the population will remain at the Hardy–Weinberg equilibrium. This means that the variation in the three genotypes will be preserved, unlike in blending inheritance.

The Hardy–Weinberg equilibrium provides us with the baseline or (zero-force) model of what we should expect about genotype frequency change over time (Roughgarden, 1979; Sober, 1984). How could an additional factor such as selection be combined with the Hardy–Weinberg law? Here is a simple case of viability fitness, which is a measure of an organism's chance of surviving to adulthood. Suppose that *A* is dominant to *a* (this means that *AA* and *Aa* genotypes will produce the same phenotypes). Suppose further that *AA* and *Aa* are fitter than *aa*. We can then describe their relative chances for survival in the following way: we assign *AA* and *Aa* a relative fitness of 1 (the fittest genotypes) and *aa* a fitness value of $1 - s$, where s is known as the selection coefficient. This is the amount *aa* is less fit relative to the fittest genotypes. So if $s = .01$, this means that aa individuals have a 99 percent chance of survival compared to a 100 percent chance for the fittest genotypes. The fittest individual is given a 100 percent chance of survival merely as a matter of algebraic convenience; what matters is the chance of survival relative to the chance of survival of the fittest members of the population.

We can now see how to combine selection with the Hardy–Weinberg principle stated above. Each of the genotypes *AA*, *Aa*, and *aa* has its initial frequencies of p^2, $2pq$, and q^2 at birth. We can then multiply each one of these H–W frequencies by their relevant fitness values – in this case they are 1, 1, and $1 - s$. So the adult frequencies of *AA*, *Aa*, and *aa* become $p^2/(1 - sq^2)$, $2pq/(1 - sq^2)$, and $q^2(1 - s)/(1 - sq^2)$.

In each case we must divide by $(1 - sq^2)$ because this is the mean fitness of the population and the frequencies must sum to 1:

$$p^2 + 2pq + q^2(1 - s) = (1 - sq^2), \text{ since we know that } p^2 + 2pq + q^2 = 1.$$

We can then calculate the new frequency of a particular allele such as *A* by noting that the frequency of *A* at a given time is equal to the frequency of *AA* plus $^1/_2$ the frequency of *Aa*. Hence, the new frequency of the *A* allele is: $p' = p/(1 - sq^2)$.

The Hardy–Weinberg law can be extended to cover three or more alleles at a locus. In order to deal with two or more loci, however, one must take into account linkage disequilibrium, which occurs when the loci do not combine independently (Moran, 1968). In two or more locus models, one is concerned with haplotype frequencies, which are the set of alleles at more than one locus inherited from one of the parents. If *A* and *a* are the alleles at one locus and *B* and *b* the alleles at another, the offspring will have possible haplotypes of *AB*, *Ab*, *aB* and *ab*. Recombination is one factor that tends to eliminate linkage disequilibrium over time. Linkage equilibrium, which occurs when each of the two alleles at one locus are equally likely to be associated with alleles at another locus, is a kind of two-locus analog to the Hardy–Weinberg equilibrium for one locus. This is because it describes the equilibrium that is achieved in the absence of selection in a randomly mating, infinite population.

3. The Tautology Problem

Darwin eventually adopted Spencer's (1864) phrase "survival of the fittest" as a synonym for natural selection. Biologists (e.g., Peters, 1976), philosophers (e.g., Popper, 1963; retracted in Popper, 1978) and various creationists (Gish, 1979; ReMine, 1993) have sometimes alleged that natural selection is a tautology, and that evolutionary theory is problematic, metaphysical, or somehow otherwise flawed as a result. If natural selection is the survival of the fittest, and the fittest are those who survive, isn't natural selection merely a tautology? And if so, doesn't this render fitness unexplanatory? It hardly seems explanatory to say that the reason various organisms survived is because they were fitter if by definition the fittest are those who survive.

Philosophers and biologists have responded to this charge in two ways. First, they point out that even if some *component* of evolutionary theory were a tautology, this does not mean that evolutionary theory as a *whole* is "untestable" or "true by definition." There is a lot more to evolutionary theory besides the definition of fitness. Even if it is true by definition that "bachelors are unmarried" this doesn't mean there aren't empirical ways of finding out about other facts concerning bachelors, such as which people are bachelors and whether they eat mac and cheese. Similarly, if a claim about fitness is true by definition, this doesn't mean there aren't empirical ways of finding out about fitness (Sober, 1984). So even if some notion of fitness were tautological, this wouldn't have the dire consequences for evolutionary theory that these critics sometimes suggest.

The second kind of response to the tautology problem is found in philosophers such as Mills and Beatty (1979), who argue that fitness should be understood as a propensity, so that it is defined in terms of *expected*, rather than *actual*, survival and reproductive success. According to the propensity interpretation, fitness is a probabilistic dispositional property analogous to dispositions such as solubility. Just as a lump of sugar has a certain probabilistic disposition to dissolve when immersed in water, fitness can be understood as a probabilistic tendency to survive to adulthood (viability fitness) or to have a certain number of offspring (fertility fitness). If one organism's viability fitness is .7 and another's is .2, this means that we should expect that the first organism is more likely to survive than the second. However, it is not a tautology that the first will survive and the second will die. The actual survivorship of these individuals can be distinct from the expected survivorship.

Although the propensity interpretation avoids defining fitness as a tautology, it is not without its problems. First, one might worry that explaining why organisms with one trait survived and organisms with another trait did not by saying that the former trait was fitter than the latter isn't especially illuminating. This is because it amounts to saying that what happened did so because it was more probable than its alternative (Sober, 1984).

Other objections have been raised to the propensity interpretation of fitness (Brandon, 1978; Rosenberg, 1985; Beatty & Finsen, 1989; Sober, 2001). One puzzle is that there are several different ways of making sense of fitness as a propensity. For example, a particular trait might have a high short-term fitness but a low long-term fitness or vice versa. Also, if offspring contribution varies solely *between* generations then the geometrical mean is the best measure of fitness, but if there is variation solely *within* generations then the arithmetical mean is a better measure (Gillespie, 1973; Beatty & Finsen, 1989).

4. The Wright–Fisher Debate

Soon after the publication of their major works in the early 1930s, Wright and Fisher began a debate that would last until Fisher's death in 1962 and was carried on by their followers (see Provine, 1986). There were two major sources of disagreement – one about the phenomena to be explained, and the other about how to explain them. Two questions that they answered differently were:

(1) How much genetic diversity is there?
(2) What processes of evolutionary change are most important in explaining the relevant phenomena in (1)?

Wright tended to think that there was more variation and placed a greater emphasis on drift, migration, interdemic selection, and, most importantly, population structure than Fisher, who tended to think that selection was the only important factor. Wright put more emphasis on the role of many small, subdivided populations in evolution whereas Fisher emphasized large, unstructured populations with random mating. Similarly, they disagreed about the process of speciation. Wright thought it was the by-product of local adaptation in epistatic systems whereas Fisher thought it was a result of disruptive or locally divergent selection. Fisher emphasized the additive effects of genes whereas Wright emphasized gene interactions and the possibility of pleiotropy.

The debate centered on Sewall Wright's so-called Shifting Balance Theory. Wright wanted a theory that could explain how natural selection could modify a population so that it goes from a state of lower to higher fitness when the intermediate stages are even less fit than the initial state. Wright claims that the idea of his theory came from various observations about guinea pigs and domestic livestock (Provine, 1986). In guinea pigs, Wright's primary animal of experimental choice, he noticed that the effect of various combinations of genes did not lead to what would be predicted by thinking only about their individual effects (e.g., the rosette hair pattern). He also observed that inbreeding in guinea pigs led to greater variability. The third observation was a kind of group-level phenomenon where breeders of cattle would get better bulls not by focusing on and breeding *individual* bulls that had the relevant properties, but instead by picking bulls from herds where the entire herd had the desired property.

The three-phase shifting balance theory can be summarized as follows (Wright, 1931, 1932; Provine, 1986). Wright was fond of using adaptive landscape diagrams that became very popular in part due to the fact that they were easier to understand than the algebra. Originally, Wright interpreted the axes as representing different traits and the height of hills and depth of valleys would represent fitness levels of various trait combinations. Later, however, Wright used the same diagrams but interpreted each point on the surface as an entire population (Wright, 1978).

Suppose a population is at one fitness peak on a Wrightian adaptive landscape. In order to pass over a valley to get to a higher peak, the population goes through three phases. In the first phase, a large population is broken up into partially isolated subpopulations, each of which is small enough that the allele frequencies within them are affected primarily by genetic drift. At some point, one of the subpopulations may drift into the combination of allele frequencies that is more fit, and is within the "domain of attraction" of a higher peak, a peak that was previously unreachable. In this first phase, the population actually must decline (initially) in fitness before drifting over into the basin of attraction of a potentially higher peak.

In the second phase, the gene frequencies will lead this subpopulation to "climb" the higher peak. This phase depends on selection, and occurs relatively rapidly. In the final phase, the subpopulation that is now at the higher peak sends migrants to the other subpopulations that are back at the lower peak. These migrants allow the rest of the subpopulations to move to the higher fitness level. The process can then begin again. In summary, the first phase must be one in which drift dominates, the second, one in which selection dominates, and the third, one that favors directional migration.

Wright's theory has been criticized mostly with respect to the frequency with which it applies (Coyne, Barton, & Turelli, 1997, 2000; Ewens, 2000), although it also has contemporary defenders (Peck, Ellner, & Gould, 1998; Wade & Goodnight, 1991, 1998). Because of its strict requirements on population structure, it has been argued that the theory has limited scope. For example, it is a process whereby an increase in fitness can occur only by temporarily reducing the fitness of a population. Wright's theory also requires there to be a number of distinct subpopulations – these must not be too distinct (there must be some migration between the subpopulations), but at the same time, distinct enough so that they aren't just one big population. Furthermore, the populations must be isolated enough during the first phase that drift can do its work but not so isolated in the third phase so that the relevant migration can occur (see also (Haldane, 1957; Crow, Engels, & Denniston, 1990, for discussion).

Wright's shifting balance theory sounds superficially like various group selection models, but it is important not to confuse his theory with the evolution of altruism by group selection, which also requires a particular population structure (Wright, 1945; Sober & Wilson, 1998). In the case of the evolution of altruism, the altruistic trait is selected against within every group, whereas the traits that Wright is thinking of, though their fitness depends on what other genes are common in the group, are stable within the group. Wright himself developed a separate model of the evolution of altruism (1945) but, unlike his shifting balance theory, he was skeptical that the particular conditions necessary in this model were likely to hold in the world.

Provine (1986) argues that the Fisher–Wright debate was good for population genetics, but Ewens (2000) disagrees. Fisher and Wright disagreed about, e.g., the importance of drift (the effective population size), with Wright giving it a significant role, arguing that in natural populations the effective population size was much smaller than the entire species, and Fisher more inclined to think that in most cases the effective population size could be understood as the entire species – it is a useful idealization to assume that the population size is infinite and hence that drift plays no role. Their dispute began with Wright objecting to various aspects of Fisher's theory of dominance. That is, even though they came to largely agree about the quantitative features of one another's models, they still had deep disagreements over empirical issues. One reason to think, with Ewens (2000), that the debate was not particularly good for population genetics is because of the amount of misunderstanding between the two camps, especially in the later years of the dispute. For example, Fisher and Ford (1947) presented Wright's view of evolution as one where random genetic drift is the only important factor, despite Wright's explicit denial of this.

5. Classical/Balance Hypothesis Debate

As with many of the major disputes in population genetics, the classical vs. balance hypothesis debate was about how best to explain a certain sort of genetic variation. One of the main reasons that the dispute was difficult to resolve is that it was a relative significance dispute where the two sides weren't always exactly clear about the hypotheses they wanted to defend (Beatty, 1987). Because neither held an extreme position, each could accommodate some exceptions.

Besides its biological interest, the debate also had interesting political dimensions, relating to both eugenics and nuclear testing. This section begins with the biological debate and then examines the socio-political context.

The two major disputants in this debate were Th. Dobzhansky and H. J. Muller. Dobzhansky (1955) coined the terms "classical" and "balance" and defended the more modern-sounding balance hypothesis, and stuck Muller with the "classical" view even though Muller protested that it was really Dobzhansky's view that was backward and archaic. Despite Muller's initial protestations about the accuracy of the names, they eventually became widely used.

The main empirical disagreement between the two sides was over the extent and importance of variation and heterozygosity. The amount of heterozygosity refers to the proportion of heterozygous loci in an average individual in natural populations. Muller believed that heterozygosity was rare, whereas Dobzhansky believed it was common. More specifically, according to the classical view selection generally reduces allelic variation and tends to promote uniformity at most loci in most organisms. This happens because there are optimal states for most traits, and corresponding optimal alleles at most loci. Also, except in rare cases, no allele is completely dominant with respect to the other alleles at its locus. Consequently, homozygotes for the optimal allele are selectively favored over heterozygotes. Muller did recognize a number of possible causes of variation in a population, including mutation, frequency-dependent selection, multiple-niche polymorphism, and migration. In addition, he acknowledged that heterozygote superiority, though he believed it to be rare, could play a significant role – for instance, in the case of sickle-cell anemia. He put the most emphasis, however, on the role of mutation in explaining variation in a population.

According to the classical view, it makes sense to talk about the "normal" or "typical" allele at a locus. This is why Dobzhansky caricatured the view as a kind of "Platonic ideal" theory. In contrast, according to the balance view, it makes less sense to talk about the normal allele at a locus. Heterozygotes are often fitter than either homozygote, so selection often preserves, rather than eliminates, variation. Heterozygotes are supposed to be more plastic and capable of dealing with more variable environments because they have two different alleles at a locus. The name "balance" refers to the fact that according to the balance hypothesis, selection favors a "balance" of the alternative allele frequencies – a mix of alleles, rather than one allele being dominant, as the "classical" view suggests.

Before molecular sequencing techniques were introduced in the 1960s, biologists didn't have a direct way of figuring out the amount of heterozygosity in a natural population. Two main issues divided the classical and balance hypotheses. First, how much heterozygosity is in natural populations? Though both sides usually stayed away from precise estimates, the classical side thought that around 5–10 percent of the loci in the average individual in a population were heterozygous, whereas defenders of the balance hypothesis thought that the number was greater than 50 percent.

Second, how much overdominance is there in natural populations? *Overdominance* occurs when the phenotype of the heterozygote (*Aa*) is outside the range of the phenotypes of the homozygotes (*AA*, *aa*). Consider a trait such as height. If we imagine that one loci is responsible for the height of some plant (an oversimplification), it might be that the *Aa* heterozygote plants are taller than both *AA* and *aa* homozygotes when all three types are exposed to the same environmental conditions. Overdominance is *one* way for heterozygotes to be more fit then either homozygote, and if the heterozygote is the most fit, then this is *one* way in which variation can be maintained in the population. Crow (1987) usefully distinguishes between two claims, one of which both sides in the classical/balance debate could accept, the other of which is controversial. One might claim more modestly that overdominance makes a major contribution to the existence of variance in natural populations, or one might make the stronger claim that the *majority* of individual loci in natural populations are overdominant. It is this latter, much stronger claim that Dobzhansky (1955) and Wallace (Wallace & Dobzhansky 1958) eventually held and that Muller (1950, 1955) and his allies (Muller & Kaplan, 1966) resisted.

It was difficult to determine the answer to the question about overdominance because many early experiments were not able to distinguish between real and apparent overdominance. Apparent overdominance can be caused by the close linkages of favorable dominants with deleterious recessives. In such cases, a homozygote is actually more fit than the heterozygote, but the recessive alleles are maintained in a population because of linkage. As a result, it appears that the heterozygote is more fit than it really is. Also, it was discovered that mutation rates in certain populations of corn, which had previously been taken to be evidence of overdominance, were underestimated, thus the observations could be explained by the higher mutation rates (Crow, 1987).

One of the factors that gave the classical vs. balance debate its interest was the social dimensions of the debate. In addition to what might appear to be a narrow debate about the extent and explanation of genetic variation in a population, these two sides disagreed about eugenics and nuclear testing (see Beatty, 1987). Both were ultimately concerned with evolution of humans. Muller was clearly a eugenicist – he thought it would be good for humans if we reduced the variation that artificial, cultural practices were preserving. Dobzhansky, on the other hand, thought it would be bad for humans if we reduced the amount of variation, and pointed to experiments on fruit flies where low levels of radiation increased the variation of alleles without causing chromosomal damage. Dobzhansky was not necessarily against eugenics per se, just the form that Muller advocated which involved reducing variation.

For similar reasons, Muller was against nuclear testing because he thought it would increase mutations and lead to more variation. Dobzhansky did not think variation was in general a bad thing; he and his allies were less opposed to nuclear testing as a result. Dobzhansky appealed to the dangers in prematurely resolving the controversy in Muller's favor – when the evidence seemed to be counting in favor of the classical view and against the balance position, Dobzhansky advocated higher evidential standards for resolving the dispute. Both sides occasionally criticized the other because of the perceived pernicious social consequences – Muller was concerned that Dobzhansky's defense of the balance view would lead people to underestimate the hazards of nuclear testing, whereas Dobzhansky and his followers criticized Muller's defense of the classical view as leading to an objectionable eugenics.

Lewontin (1987) argues that the original classical/balance debate underwent a transformation in the 1960s and 1970s with the introduction of the sophisticated neutralist alternative. The original issue was the extent to which natural populations were homozygous at most loci or whether a substantial fraction of the alleles within a population were polymorphic. According to Lewontin, one of the interesting motivations behind this debate was whether natural selection had significant variation to act on – if there was a lot of heterozygote superiority, for instance, this would mean that there would already be the variation needed for selection to work quickly for evolutionary change.

With the introduction of the neutralist hypothesis, however, the polymorphism that exists might be for variation that has no connection with natural selection, because these genetic differences might not have any connection with differences in the organisms' phenotypes. So, even though, with the work of biologists such as Lewontin (summarized in Lewontin, 1974), there was a lot of evidence in favor of polymorphism at many loci, this does not mean that the role of selection had been resolved, because with the introduction and development of neutralist hypotheses, we now have two different kinds of competing explanations of genetic polymorphism. This leads us to our next debate.

6. The Neutralism Controversy

The two major factors that affect the relative importance of selection and drift are the selection coefficients of the different genotypes and the effective population size. If the selection coefficients are large, natural selection will tend to dominate in a population, and if the effective population size is small, random genetic drift will tend to dominate the population.

Kimura (1968, 1983), along with King and Jukes (1969), developed and defended the neutral theory, with the new data that were appearing as a result of using gel electrophoresis to estimate the amount of genetic variation in fruit flies and other organisms. There is a chance that two randomly drawn alleles will differ at an average locus, and there is the overall percentage of polymorphic loci. The evidence that Hubby and Lewontin (1966) and Harris (1966) provided suggested that in natural populations about 10–20 percent of the loci at the protein level are polymorphic. Kimura argued that these numbers were too high if proteins had evolved primarily by natural selection, because there is a limit to the amount of genetic load a population can endure. Roughly, genetic load is a measure of how the fitness of an average individual in a population compares to the fitness of the fittest member of the population. Different factors can cause the genes that result in genetic load. For example, genetic load that results from mutation is known as mutation load, whereas load that results from genes that are favorable in heterozygotes but not in homozygotes is known as segregation load (for discussion see Crow, 1992). Any process or factor that leads to fitness variability can generate a genetic load. Haldane (1957) introduced what he called the cost of natural selection (and what is now usually known as substitutional load). The main idea is that there is a kind of cost to natural selection that makes high genetic loads unlikely. If selection is extremely strong, it will drive a population extinct. Kimura thought that the allelic replacement rates suggested by the data were too rapid for natural selection to be the primary cause.

Although important in helping formulate neutralism, Kimura himself later recognized that the argument is inconclusive because of the response one can give by distinguishing between hard and soft selection. The Haldane (1957)–Kimura argument assumes that natural selection is primarily "hard" selection, in which selection occurs on top of the background mortality. In soft selection, however, the selective deaths are deaths that would have been a result of background mortality. In such a case, rapid evolution can occur as a result of natural selection if the selection is soft, rather than hard. The population size need not decrease dramatically as a result of this kind of selection.

Another early argument in favor of the neutral theory was based on the rates of molecular evolution, which appeared to be too high if natural selection was the primary cause of such change. Kimura argued that if natural selection operated as strongly as necessary to explain these rates of molecular evolution, most organisms would have to be killed off by the intense selection pressure. With neutral drift, however, there is no such cost in substituting one allele for another. Kimura's argument is indecisive, however, because he made unrealistic assumptions about the amount of genetic load.

Other tests have been used to examine the controversy, such as evidence about the comparative rates of evolution in functionally more and less important areas of proteins. Biologists have found that the rate of evolution in the less important areas tends to be faster than that in the slower areas. Once again, however, it appears that both natural selection and neutral drift can explain this tendency. One reason that the controversy has continued is that neutralist theories have changed over time from being purely neutral to being "nearly neutral" (see Ohta, 1973; Ohta & Gillespie, 1996). Nearly neutral theories allow for the possibility that there is random fixation of slightly deleterious mutations, which allowed defenders of the neutral theory to explain certain phenomena that were otherwise problematic for their approach.

Perhaps the main reason that the debate will continue for some time is that the values for the selection coefficients, the effective population size, and the mutation rates are usually unknown. Depending on what values these variables are assigned, it will be possible to give a selection or a (nearly) neutralist explanation of almost any evolutionary rate or frequency of polymorphism (see Gillespie, 1991 for a review of many of these issues).

7. Saltationism vs. Gradualism

Darwin is usually understood as a gradualist in the sense that he thought that natural selection typically acted on variations that had small effects – complex adaptations do not appear all at once, but only gradually. This view was controversial as soon as the *Origin* was published. This is one of the few significant criticisms that Huxley (1859) raised to Darwin. Huxley thought that Darwin was unnecessarily restricting himself by claiming that saltations were insignificant.

It is worth pointing out that saltation is not necessarily an alternative to natural selection. This is because saltationists, just like gradualists, can invoke natural selection as the primary mechanism for explaining how a trait gets established in a population. The debate between gradualism and saltationism is primarily one about how a complex trait *first appears* in an individual, not about how it spreads.

Some biologists, e.g., Gould and Eldredge (1977), have argued that certain features of the fossil record are a problem for the traditional neo-Darwinian modern synthesis theory of evolution. In particular, it does not account for the *origin* of species or for long-term evolutionary trends. The fossil record, according to them, involves alternating periods of stasis followed by sudden and rapid change. Gould and Eldredge argue that this provides support for their new theory of "punctuated equilibrium," which requires various macroevolutionary processes in addition to traditional gradualist natural selection. These macroevolutionary changes are sudden and lead to new species and forms in a relatively few number of generations; consequently, a process of species selection plays a major role in the long-term trends in evolution, whereas traditional natural selection plays a lesser role because species don't change that much in between these large-scale macroevolutionary changes.

Thus, it would seem that this pattern of stasis alternating with periods of rapid change threatens to challenge the explanatory and predictive scope of traditional population genetics. Turner (1986) argues convincingly through a detailed example of the evolution of various butterfly wing patterns that these patterns likely were static for long periods of time before undergoing relatively rapid changes. If they were the kind of organism that is easily fossilized, they would demonstrate the punctuated equilibrium pattern. This is because in cases of Batesian mimicry, once a few significant genes change there is considerable selective pressure on the rest of the population to mimic the new successful patterns. Turner shows that depending on the selection coefficient, we might expect genes for new patterns to sweep through the population in just a few centuries, a relatively short time period in the evolutionary scheme. Turner discovered that a relatively few alleles need to change in order for dramatic changes to occur in the pattern of these butterfly wings – "whole red patches and yellow bars are taken out or put in at one go, or in two or three rather large chunks, and their shapes are likewise altered by single mutations of comparatively large effect, breaking a solid yellow patch into a group of dots." He concludes, appropriately enough, by saying that "Evolution appears to have been much less gradual than one might expect" (Turner, 1986, p.189).

The basic process is that one species of butterfly ends up mimicking another toxic species when it is exposed to its population by changing one or a few genes. It turns out, if Turner's argument can be generalized, that population genetics should actually lead us to *expect* the kind of punctuated patterns that Gould and Eldredge emphasize (see, e.g., Maynard Smith, 1983, and Newman et al., 1985).

Eldredge and Gould originally argued that phyletic gradualism – the view that evolution has a fairly constant rate, and that new species tend to arrive by the gradual transformation of other species – is false. The general question of phyletic gradualism vs. punctuated equilibrium is an open question – some cases seem to support their theory, others don't. Macromutations result from genes that control developmental processes, and so new species might arise by this sort of change. The central negative claim is that natural selection and ordinary variation only lead to small, microevolutionary change.

Eldrege and Gould are making two basic claims. First, that evolution proceeds in an alternating pattern of stasis and rapid change. Second, the rapid change will tend to be accompanied by speciation events. Evolutionary change tends to happen in geographically isolated populations. Turner's example is one where there is rapid change without speciation. [See Speciation and Macroevolution].

8. Conclusions

Most of these debates about population genetics concerned both the frequency and explanation of certain kinds of variation. These debates have been transformed in various ways by the discovery of new techniques, but continue because (1) they are usually "relative frequency" debates and so can only be properly evaluated in the long run, and (2) there are significant underdetermination problems.

Acknowledgments

Thanks to Anya Plutynski and Sahotra Sarkar for their patience and for very helpful comments on earlier drafts.

References

Beatty, J. (1987). Weighing the risks: stalemate in the classical/balance controversy. *Journal of the History of Biology*, 20, pp. 289–319.

Beatty, J., & Finsen, S. (1989). Rethinking the propensity interpretation: a peek inside Pandora's box. In M. Ruse (Ed.). *What the philosophy of biology is* (pp. 17–30). Dordrecht: Kluwer Academic Publisher.

Brandon, R. (1978). Adaptation and evolutionary theory. *Studies in History and Philosophy of Science*, 9, 181–206.

Coyne, J., Barton, N., & Turelli, M. (1997). A critique of Sewall Wright's shifting balance theory of evolution. *Evolution*, 51, 643–71.

Coyne, J., Barton, N., & Turelli, M. (2000). Is Wright's shifting balance process important in evolution? *Evolution*, 54(1), 306–17.
Crow, J. (1987). Muller, Dobzhansky and overdominance. *Journal of the History of Biology*, 20, 351–80.
Crow, J. (1992). Genetic load. In E. Fox Keller & E. A. Lloyd (Eds). *Keywords in evolutionary biology* (pp.132–6). Cambridge, MA: Harvard University Press.
Crow, J. F., Engels, W. R., & Denniston, C. (1990). Phase three of Wright's shifting-balance theory. *Evolution*, 44(2), 223–47.
Darwin, C. (1859). *On the origin of species by means of natural selection, or the preservation of favoured races in the struggle for life.* London: Murray.
Dobzhansky, Th. (1955). A review of some fundamental concepts and problems of population genetics. *Cold Spring Harbor Symposium on Quantitative Biology*, 20, 1–15.
Ewens, W. (2000). The mathematical foundations of population genetics. In R. S. Singh & C. B. Krimbas (Eds). *Evolutionary genetics: from molecules to morphology* (vol. 1, pp. 24–40). Cambridge: Cambridge University Press.
Fisher, R. A. (1918). The correlation between relatives on the supposition of Mendelian inheritance. *Transactions of the Royal Society of Edinburgh*, 52, 399–433.
Fisher, R. A. (1930). *The genetical theory of natural selection.* Oxford: Clarendon Press.
Fisher, R. A., & Ford, E. B. (1947). The spread of a gene in natural conditions in a colony of the moth *Panaxia dominula*. *Heredity*, 1, 143–7.
Gayon, J. (1998). *Darwinism's struggle for survival: heredity and the hypothesis of natural selection.* Cambridge: Cambridge University Press.
Gish, D. (1979). *Evolution? The fossils say no!* San Diego: Creation Life Publishers.
Gillespie, J. H. (1973). Natural selection for within-generation variance in offspring number. *Genetics*, 76, 601–6.
Gillespie, J. H. (1991). *The causes of molecular evolution.* New York: Oxford University Press.
Haldane, J. B. S. (1924). A mathematical theory of natural and artificial selection. Parts I and II. *Transactions of the Cambridge Philosophical Society*, 23, 19–41.
Haldane, J. B. S. (1932). *The causes of evolution.* London: Longman.
Haldane, J. B. S. (1957). The cost of natural selection *Journal of Genetics*, 55, 511–24.
Hardy, G. H. (1908). Mendelian proportions in a mixed population. *Science*, 28, 49–50.
Harris, H. (1966). Enzyme polymorphisms in man. *Proceedings of the Royal Society of London B*, 164, 298–310.
Hubby, J. L., & Lewontin, R. C. (1966). A molecular approach to the study of genic heterozygosity in natural populations. I. The number of alleles at different loci in *Drosophila pseudoobscura*. *Genetics*, 68, 235–52.
Jenkin, F. (1867). The origin of species. *North British Review*, 44, 277–318.
Kimura, M. (1968). Evolutionary rate at the molecular level. *Nature*, 217: 624–6.
Kimura, M. (1983). *The neutral theory of molecular evolution.* Cambridge: Cambridge University Press.
King, L., & Jukes, T. (1969). Non-Darwinian evolution. *Science*, 164, 788–9.
Lewontin, R. C. (1974). *The genetic basis of evolutionary change.* New York: Columbia University Press.
Lewontin, R. C. (1987). Polymorphism and heterosis: old wine in new bottles and vice versa. *Journal of the History of Biology*, 20, 337–49.
Lewontin, R. C., & Hubby, J. L. (1966). A molecular approach to the study of genic heterozygosity in natural populations. II. Amount of variation and degree of heterozygosity in natural populations of *Drosophila pseudoobscura*. *Genetics*, 54, 595–609.
Lewontin, R. C. (2000). The problems of population genetics. In R. Singh & C. B. Krimbas (Eds). *Evolutionary genetics: from molecules to morphology* (pp. 5–23). Cambridge: Cambridge University Press.

Maynard Smith, J. (1983). The genetics of stasis and punctuation. *Annual Review of Genetics*, 17, pp. 11–25.

Mendel, G. (1865). Experiments on plant hybrids. Translated and reprinted in *The origin of genetics: a Mendel source book* (Ed. Stern and Sherwood, 1966; pp. 1–48). New York: W. H. Freeman and Company.

Mills, S., & Beatty, J. (1979). The propensity interpretation of fitness. *Philosophy of Science*, 46, 263–88.

Moran, P. A. P. (1968). On the theory of selection dependent on two loci. *Annals of Human Genetics*, 32, 183–90.

Muller, H. J. (1950). Evidence of the precision of genetic adaptation. *Harvey Lectures*, 18, 165–229.

Muller, H. J. (1955). Comments on the genetic effects of radiation on human populations. *Journal of Heredity*, 46, 199–220.

Muller, H. J., & Kaplan, W. D. (1966). The dosage compensation of drosophila and mammals as showing the accuracy of the normal type. *Genetic Research Cambridge*, 8, 41–59.

Ohta, T. (1973). Slightly deleterious mutant substitutions in evolution. *Nature*, 246, 96–8.

Ohta, T., & Gillespie, J. (1996). Development of neutral and nearly neutral theories. *Theoretical Population Biology*, 49, 128–42.

Olby, R. (1966). *Origins of Mendelism*. Chicago and London: The University of Chicago Press.

Peck, S. L., Ellner, S. P., & Gould, F. (1998). A spatially explicit stochastic model demonstrates the feasibility of Wright's shifting balance theory. *Evolution*, 52, 1834–39.

Peters, R. (1976). Tautology in evolution and ecology. *American Naturalist*, 110, 1–12.

Popper, K. (1963). *Conjectures and refutations*. London: Hutchinson.

Popper, K. (1978). Natural selection and the emergence of mind. *Dialectica*, 32, 339–55.

Provine, W. B. (1971). *The origins of theoretical population genetics*. Chicago and London: The University of Chicago Press.

Provine, W. B. (1986). *Sewall Wright and evolutionary biology*. Chicago and London: University of Chicago Press.

ReMine, W. J. (1993). *The biotic message: evolution versus message theory*. St. Paul, MN: St. Paul Science Publishers.

Rosenberg, A. (1985). *The structure of biological science*. Cambridge: Cambridge University Press.

Roughgarden, J. (1979). *Theory of population genetics and evolutionary ecology*. New York: Macmillan.

Sober, E. (1984). *The nature of selection*. Cambridge, MA: MIT Press.

Sober, E., & Wilson, D. S. (1998). *Unto others: the evolution and psychology of unselfish behavior*. Cambridge, MA: Harvard University Press.

Sober, E. (2001). The two faces of fitness. In R. Singh, D. Paul, C. Krimbas, & J. Beatty (Eds). *Thinking about evolution: historical, philosophical and political perspectives* (pp. 309–21). Cambridge: Cambridge University Press,

Spencer, H. (1864). *The principles of biology*. London: William and Norgate.

Turner, J. R. G. (1986). The evolution of mimicry: a solution to the problem of punctuated equilibrium. In D. M. Raugh & D. Jablonski (Eds). *Patterns and processes in the history of life* (pp. 42–66). Chichester: John Wiley.

Wade, M. J., & Goodnight, C. J. (1991). Wright's shifting balance theory: an experimental study. *Science*, 253, 1015–18.

Wade, M. J., & Goodnight, C. J. (1998). The theories of Fisher and Wright in the context of metapopulations: when nature does many small experiments. *Evolution*, 52, 1537–53.

Wallace, A. R. (1859). On the tendency of varieties to depart indefinitely from the original type. *Journal of the Proceedings of the Linnean Society (Zoology)*, 3.

Wallace, B., & Dobzhansky, Th. (1958). *Radiation, genes, and man.* New York: Henry Holt.

Weinberg, W. (1908). On the demonstration of heredity in man. Translated and reprinted in S. H. Boyer (Ed.). *Papers on human genetics* (1963; pp. 4–15). Englewood Cliffs, NJ: Prentice-Hall.

Wright, S. (1931). Evolution in Mendelian populations. *Genetics*, 16, 97–159.

Wright, S. (1932). The roles of mutation, inbreeding, crossbreeding and selection in evolution. *Proceedings of the Six International Congress of Genetics*, 1, 356–66.

Wright, S. (1945). Tempo and mode in evolution: a critical review. *Ecology*, 26, 415–19.

Wright, S. (1978). *Evolution and the genetics of populations* (Vol. 4): *Variability within and among natural populations.* Chicago: University of Chicago Press.

Further Reading

Crow, J., & Kimura, M. (1972). *Introduction to population genetics theory.* New York: Harper & Row.

Ewens, W. (2004). *Mathematical population genetics: I. Theoretical introduction.* New York: Springer.

Gayon, J. (1998). *Darwinism's struggle for survival: heredity and the hypothesis of natural selection.* Cambridge: Cambridge University Press.

Gillespie, J. (2004). *Population genetics: a concise guide* (2nd edn). Baltimore: Johns Hopkins University Press.

Hartl, D., & Clark, A. (1997). *Principles of population genetics* (3rd edn). Sunderland, MA: Sinauer.

Chapter 8

The Units and Levels of Selection

SAMIR OKASHA

1. Introduction

The "units of selection" question is one of the most fundamental in evolutionary biology. Though the debate it has generated is multifaceted and complex, the basic issue is straightforward. Consider a paradigmatic Darwinian explanation – of why the average running speed in a zebra population has increased over time, for example. The explanation might go as follows: "in the ancestral population, zebras varied with respect to running speed. Faster zebras were better at avoiding predators than slower ones, so on average left more offspring. And running speed was heritable – the offspring of fast zebras tended to be fast runners themselves. So over time, average running speed in the population increased." In this explanation, the "unit of selection" is the individual organism. It is the differential survival and reproduction of *individual zebras* that causes the evolutionary change from one generation to the next. We could also express this by saying that natural selection "acts at the level of the individual organism."

Traditional Darwinian theory treats the individual organism as the basic unit of selection. But in theory at least, there are other possibilities. For the principle of natural selection can be formulated wholly abstractly – it involves no essential reference to organisms or any other biological units. The principle tells us that if a population of entities vary in some respect, and if different variants leave different numbers of offspring, and if offspring entities resemble their parents, then over time the composition of the population will change, ceteris paribus. In Lewontin's famous formulation, natural selection will operate on any entities that exhibit "heritable variation in fitness" (Lewontin, 1970). Entities at many levels of the biological hierarchy could satisfy these conditions – including genes, chromosomes, organelles, cells, multicellular organisms, colonies, groups and even whole species. Since each of these entities undergoes reproduction, or multiplication, the notion of fitness, and thus heritable variation in fitness, applies to each. The hierarchical nature of the biological world, combined with the abstractness of the principle of natural selection, means that there is a range of candidate units on which selection can act.

From this brief sketch, the units of selection question might seem purely empirical. Given the multiplicity of possible levels at which selection *can* act, surely it is just a matter of finding out the levels at which it *does* act? With enough empirical data, surely the question can be conclusively answered? In fact matters are not quite so simple. As many authors have noted, the units of selection debate comprises a curious amalgam of empirical, theoretical, and conceptual questions, often not sharply distinguished from one another. (This is why philosophers of science have written so much about it.) The debate is of course responsible to empirical facts, but this cannot be all there is to it. For quite frequently, one finds authors in agreement about the basic biological facts in a given case but in disagreement about what the "true" unit of selection is. Disagreements of this sort are conceptual or philosophical in nature, rather than straightforwardly empirical.

A brief remark about terminology is needed. The expressions "units of selection" and "levels of selection" can both be found in the literature. Some authors treat these expressions as effective synonyms. On this usage, if the unit of selection is the individual organism, for example, then selection can be said to act at the organismic level. So it is possible to translate freely between talk of units and levels of selection. However, there is another usage, associated with the "replicator/interactor" view of evolution discussed below, which severs the close link between units and levels (e.g., Reeve & Keller, 1999). On this alternative usage, "unit of selection" refers to the replicators, typically genes, that transmit hereditary information across generations, while "level of selection" refers to the hierarchical level(s) at which there is variation in fitness. The former usage will be adopted here unless otherwise indicated.

2. Historical Remarks

The units of selection question traces back to Darwin himself. For the most part Darwin treated the individual organism as the unit of selection, but he recognized that not all biological phenomena could be interpreted as products of organism-level selection. Worker sterility in the social insect colonies was one such phenomenon, and it puzzled Darwin considerably. Sterile workers forgo reproduction, instead devoting their whole lives to assisting the reproductive efforts of the queen – by foraging for food, feeding the young, and protecting the colony. Such behavior does not benefit the workers themselves, so it is hard to see how it could evolve by selection at the organismic level. Worker sterility is a classic example of an *altruistic* trait: it reduces the fitness of the organism that expresses the trait but increases the fitness of others. (By an organism's fitness we mean the expected number of offspring that it leaves; this quantity depends on the probability that the organism survives to reproductive age, and the reproductive success it will enjoy if it does survive.)

The problem of how altruistic traits can evolve is intimately linked to the units of selection question, historically and conceptually.

Darwin's most explicit assault on the problem of altruism occurred in *The Descent of Man* (1871). Discussing the evolution of self-sacrificial behavior among early humans, Darwin wrote: "he who was ready to sacrifice his life, as many a savage has been, rather than betray his comrades, would often leave no offspring to inherit his noble nature" (1871, p.163). Darwin then argued that self-sacrificial behavior, though disadvantageous at the individual level, might be beneficial at the *group* level: "a tribe including many members who . . . were always ready to give aid to each other and sacrifice themselves for the common good, would be victorious over most other tribes; and this would be natural selection" (1871, p.166). Darwin's suggestion is that the behavior in question may have evolved by a process of between-group selection. Groups containing many altruists (self-sacrificers) might do better than groups containing fewer, even though *within* any group, altruists do less well than their selfish counterparts. So Darwin was open to the idea that at least sometimes, groups as well as individual organisms can function as units of selection.

August Weismann, the famous German evolutionist whose work on inheritance discredited Lamarckism, also saw that selection can operate at multiple hierarchical levels, as Gould (2002) has emphasized. While Darwin had toyed with the idea that selection could occur at levels above the organism, Weismann was interested in the possibility of sub-organismic levels of selection. His doctrine of "germinal selection" described a selection process between variant "determinants" (hypothetical hereditary particles) that occurred during the lifespan of a developing organism (Weismann, 1903). Though Weismann's theory of development has not stood the test of time, his idea that selection can operate on variant units within the lifespan of a complex organism has endured. Selection between different cell lineages within multicellular organisms plays a major role in the vertebrate immune response, in neuronal development, and also, tragically, in carcinogenesis (Edelman, 1987; Cziko, 1995; Frank, 1996). This process is sometimes referred to as "somatic" or "developmental" selection.

The units of selection debate in its modern form owes much to G. C. Williams' iconoclastic book *Adaptation and Natural Selection* (Williams, 1966). Williams' stated aim was to bring some "discipline" to the study of adaptation. His concern was with a growing trend in biology, particularly among ecologists and ethologists, to think of adaptation in terms of "benefit to the species" rather than "benefit to the individual." Thus for example Konrad Lorenz, the Nobel Prize-winning ethologist, would routinely explain an observed animal behavior by citing a benefit that the behavior confers on the species as a whole. If the Darwinian process one has in mind is ordinary organismic selection, this is a fallacious argument. For organismic selection produces adaptations that benefit *individual organisms*, and it is an open question whether such adaptations will on aggregate benefit any larger units (as both Fisher (1930) and Haldane (1932) had previously pointed out). Williams stressed that only a process of between-group selection would produce genuine group-level adaptations, and he regarded group selection as a weak evolutionary force, which would only rarely have significant effects. His main argument was that the generation time of groups is typically much longer than that of individual organisms, so the effects of group selection would be swamped by individual selection. The fragility of group selection as an evolutionary mechanism was also emphasized by Maynard Smith (1964).

As a result of Williams' and Maynard Smith's work, evolutionists in the 1960s and 70s increasingly came to see the importance of the units of selection question, and in particular to view the concept of group selection with great suspicion. This period also witnessed the rise of two crucial theoretical developments: the theory of kin selection, stemming from the seminal work of William Hamilton (1964) on the evolution of social behavior, and the "gene's eye view of evolution," stemming from the work of Hamilton and Williams and popularized by Dawkins (1976); see Section 3 below. Though no one could doubt the importance of these new developments, they complicated the units of selection issue considerably, generating a certain amount of conceptual confusion and a proliferation of terminology. The relationships between "individual selection," "kin selection," "genic selection," "frequency-dependent selection," "group selection," and "species selection" were not always perspicuous; nor was it clear whether these types of selection were strict alternatives to each other at all. Unsurprisingly, it was at this stage that philosophers of science started to take a serious interest in the debate.

3. The Gene's Eye View of Evolution

In *The Selfish Gene* (1976), Dawkins defends a gene-centric view of the evolutionary process. Ordinarily we think of natural selection as a competition between individual organisms, the winners surviving and reproducing, the losers dying. But Dawkins argues that organisms are mere epiphenomena of the evolutionary process – the real competition takes place between individual genes. Genes are engaged in a perpetual struggle to bequeath as many copies of themselves to future generations as possible, and organisms are simply "vehicles" that genes have built to assist them in this task. So the phenotypic adaptations we see all around us are not there because they benefit the organisms that display them, less still the groups or species to which the organisms belong. Rather, adaptations are there for the benefit of the underlying genes that produce them, Dawkins argues. Genes "program" their host organisms to express phenotypes – behavioral, morphological, and physiological – which help the organisms survive and reproduce, thus ensuring that copies of the genes will be found in future generations. The ultimate beneficiary of the evolutionary process, and thus the true unit of selection, is the individual gene, Dawkins claims.

This so-called "gene's eye view of evolution" has its roots in the work of Hamilton (1964), mentioned above. Hamilton was concerned with the very problem Darwin had puzzled over – altruism. As we have seen, an animal that behaves altruistically will have lower fitness than its selfish counterparts, so altruism, and the genes which cause it, should be disfavored by natural selection. But Hamilton realized that if the altruistic behavior is directed at relatives, rather than at unrelated members of the population, then the situation is immediately changed. For relatives share genes, so there is a certain probability that the beneficiary of the altruistic act will *itself* carry the gene for altruism. So to determine whether the altruism-causing gene (and thus the altruistic behavior itself) will spread, we need to take into account not just the effects of the gene on the fitness of its bearer, but also on the fitness of the bearer's relatives. Hamilton's achievement was to express this insight in precise mathematical form. The condition required for the spread of an altruistic gene in a population, Hamilton showed, was $b/c > 1/r$, where c denotes the cost incurred by the altruist, b denotes the benefit enjoyed by the recipient, and r is the *coefficient of relatedness* between donor and recipient, which measures how closely related they are. This inequality is known as *Hamilton's rule*; it tells us that altruism will be favored by natural selection so long as the cost to the altruist is offset by a sufficient amount of benefit to sufficiently closely related relatives, where the costs and benefits are measured in units of reproductive fitness. For obvious reasons, this idea came to be known as "kin selection."

Hamilton's work revolutionized the way biologists study animal behavior. But what matters for the moment is the *way* Hamilton arrived at his idea. He did so by employing the gene's eye view of evolution. Hamilton realized that in trying to determine whether a given trait (e.g., altruism) will evolve, it is not enough to ask whether the trait benefits the individual organism that expresses it. The real test is whether the net effect of the trait leads the gene underlying the trait to increase or decrease in frequency; only that tells us whether the trait will spread. So to explain why a given trait has evolved, we need to show that the trait confers a selective advantage on the gene that causes the trait, rather than on the organism that expresses the trait. Looked at from this gene's eye view, the phenomenon of altruism makes perfect sense. Causing its host organism to behave altruistically to relatives is simply a strategy devised by a "selfish" gene to ensure its future propagation, and so long as the costs and benefits satisfy Hamilton's rule, the strategy will work.

The gene's eye view is a powerful heuristic for thinking about evolution, particularly where social behaviors are involved. Another phenomenon that looks anomalous from the traditional organismic viewpoint but makes sense from the gene's eye view is *intra-genomic conflict*. Usually the genes within a single organism behave cooperatively, because they have a common interest in ensuring the organism's survival and reproduction; that is why genes generally have phenotypic effects that benefit their host organism. But in some cases an individual gene can promote its own interests at the expense of the rest of the genome. Segregation-distorter (SD) genes, which violate the rules of Mendelian inheritance to secure a greater than 50 percent representation in the gametes of heterozygotes, are an example. SD genes often have adverse phenotypic effects on the organism itself, so from the organism's point of view, and from the point of view of all other genes in the genome, the SD gene is a liability. But from the gene's eye view, the behavior of the SD gene makes perfect sense – it has simply devised an unusual strategy for ensuring its transmission to future generations. Recent research has revealed intra-genomic conflict to be more common than was originally thought, and it constitutes one of the best arguments in favor of the gene's eye view of evolution (Pomiankowski, 1999; Hurst, Allan, & Bengston, 1996; Burt & Trivers, 2006).

Dawkins offered another, quite different argument for treating the gene as the true unit of selection. (The argument had been hinted at, but not systematically articulated, by Williams.) Genes are what Dawkins calls *replicators*: entities which leave copies of themselves in subsequent generations. Thanks to the fidelity of DNA replication, the members of a gene lineage are usually perfect or near-perfect copies of one another. Entities such as organisms, colonies, and species also stand in ancestor–descendant relations, hence form lineages, but in no case does the fidelity of reproduction approach that found in gene lineages. This is especially true of sexually reproducing organisms, where offspring contain a mixture of genetic material from two parents. DNA replication is thus qualitatively different from organismic reproduction for Dawkins, for genes in existence today are descended unchanged or nearly unchanged from genes that existed hundreds of thousands ago; the same is obviously not true of whole organisms. Only genes have sufficient permanence to qualify as units of selection, Dawkins argues; organisms and their properties are mere temporary manifestations.

In the light of the gene's eye view of evolution, what becomes of the traditional units of selection debate? Prior to Dawkins, the debate had generally pitched group selectionists against organismic selectionists. But Dawkins (1976) argues that both are wrong, for the true unit of selection is the gene. This suggests that the claim "the gene is the unit of selection" is logically incompatible with the claim "the organism is the unit of selection" or "the group is the unit of selection." G. C. Williams (1966) also contrasts genic selection with group selection, again implying that these are incompatible views of how evolution proceeds (1966, p.55).

However, in his later work Dawkins (1982) adopts a different line, arguing that genic selection is not really an alternative to traditional organismic selection at all. Rather, the gene's eye view is simply a different *perspective* on the process of evolution that is heuristically valuable in certain contexts. So we can think of evolution either in the traditional way, in terms of selection between organisms, or in the gene's eye way, in terms of selection between genes. There is no fact of the matter about which is right – both are valid perspectives on one and the same set of facts. Central to this argument is Dawkins' distinction between replicators and vehicles, or replicators and interactors in the more widely used terminology of Hull (1981). As we have seen, genes are the paradigmatic replicators – they leave copies of themselves in future generations. However, natural selection does not operate on genes "directly" but only indirectly, via the effect the genes have on their host organisms. For it is whole organisms that survive, reproduce, and die, not individual genes. Organisms are thus interactors – entities that interact directly with their environment and are thus the direct target of selection. Both replicators and interactors are involved in the evolutionary process, according to Dawkins and Hull.

Dawkins and Hull argue that the expression "unit of selection," as it occurred in the early discussions, was often ambiguous between replicators and interactors. Arguments about whether the gene or the organism is the unit of selection typically traded on this ambiguity (though not always – see below). In retrospect this was a bad question to ask, for it commits a category mistake, pitting a replicator against an interactor. (Similarly, Williams' contrast between "genic selection" and "group selection" was a category mistake.) Arguments about whether the organism or the *group* is the unit of selection are different, however; this is a question about interactors, and does not commit a category mistake. It is an empirical question that can only be resolved by looking at the empirical facts, and may receive a different answer in different cases. So the Dawkins/Hull conceptualization permits a neat separation of the conceptual from the empirical aspects of the units of selection debate. "Group versus organism" is an empirical issue, but "organism versus gene" is not; rather, it is "an argument about what we ought to *mean* when we talk about a unit of natural selection," in Dawkins' words (1982, p.82).

This is a compelling analysis, but it raises certain questions. If the gene's eye view is ultimately equivalent to the orthodox organismic view, what becomes of phenomena such as intra-genomic conflict and junk DNA, which *don't* appear explicable in terms of advantage to the individual organism? The existence of such phenomena, which formed part of Dawkins' original case for genic selection, sits badly with the idea that the gene's eye view is merely a heuristic perspective, rather than an empirical thesis about the course of evolution. One response to this problem, favored by a number of commentators, is to allow that the gene is *sometimes* the unit of selection in the *same* sense as that in which the individual organism is the unit of selection, i.e., the unit of interaction (Sober & Wilson, 1998; Reeve & Keller, 1999). On this view, if the genes within a single organism differ in fitness, as in cases of intra-genomic conflict, then "genic selection" takes place, but if, as is usually the case, the genes within any single organism have identical fitness, then all the selection must occur at a higher level, e.g., the organismic level, or the group level.

This means that we must sharply distinguish the *process* of genic selection, which is relatively infrequent, from the changes in gene frequency that are the *product* of selection at other hierarchical levels, which are ubiquitous (Okasha, 2004a, 2006). Organismic, kin, and group selection all will in general lead to changes in gene frequency; so a gene's eye perspective is always going to be available on selection processes that occur at these levels. But in addition, there are selection processes that take place at the genic level itself – as in cases of intra-genomic conflict. The expression "genic selection" should be reserved for such processes. Thus we should not confuse the gene's eye viewpoint, which is a heuristic tool for thinking about selection processes that may occur at many different hierarchical levels, with genic selection itself, which is a specific level of selection that is logically distinct from individual, kin, or group selection. Increasingly, this is how the label "genic selection" is in fact being used in the literature, e.g., by Maynard Smith and Szathmary (1995) and Okasha (2006).

In retrospect, it is clear that Dawkins' arguments in *The Selfish Gene* failed to distinguish sharply enough between the units of selection and of inheritance. The distinction between selection and inheritance is conceptually straightforward: selection concerns which variants survive best/reproduce the most, while inheritance concerns the transmission of genotypic and phenotypic characters across generations. Thus quantitative geneticists typically distinguish selection itself from the evolutionary response to selection – where the latter depends on the heritability of the trait selected for. But Dawkins and Williams used facts about *inheritance*, e.g., that genes are faithfully replicated across generations while whole genotypes and organismic characters are not, to privilege the gene as the unit of *selection*. Had the distinction between selection and inheritance (or transmission) been kept clearly in mind, there probably would have been no need to introduce the terminology of replicators and interactors at all.

Indeed there are reasons for thinking that the replicator/interactor framework, though valuable for certain purposes, does not provide a fully general account of evolution by natural selection, despite what its advocates have thought (Griesemer, 2000). One such reason is that Lewontin's "heritable variation in fitness" formulation arguably *does* provide a fully general account, and it involves no distinction between replicators and interactors, thus undermining the Dawkins/Hull idea that *any* selection process must involve entities of both these types. (Similarly, Maynard Smith's (1988) abstract account of the conditions required for Darwinian evolution – multiplication, variation, and heredity – involves only one type of entity.) This suggests that the original Lewontin formulation of the units of selection question – "which are the entities that possess heritable variation in fitness?" – is superior to the replicator/interactor formulation (Okasha, 2006). Of course, rejecting the Dawkins/Hull framework as a *general* way of thinking about the units of selection does not mean abandoning the gene's eye view of evolution; the latter has proved invaluable for understanding a whole host of evolutionary phenomena.

4. Group Selection and Kin Selection

The group selection question is one of the most intriguing, and polemical, chapters in the units of selection debate. As we saw in Section 2, group selection fell out of favor among evolutionary biologists in the 1960s, due mainly to the work of Williams and Maynard Smith. The essence of their argument was that group selection is a weak evolutionary force compared to individual selection, for the turnover of groups will generally be much slower than that of individuals, thus permitting individual selection to accumulate adaptations at a faster rate. Moreover, the phenomena which group selection had originally been invoked to explain, such as altruism, could be explained in other more parsimonious ways, they argued, such as kin selection or the evolutionary game theory of Maynard Smith and Price (1973). So not only was the hypothesis of group selection implausible, it was also explanatorily superfluous.

Something like this is probably still the majority view in evolutionary biology, but it has not gone unchallenged. D. S. Wilson has vigorously opposed the orthodox rejection of group selection for many years, both alone and in collaboration with Elliott Sober (Wilson, 1975, 1980, 1989; Sober & Wilson, 1998). Wilson argues that group selection was wrongly rejected by biologists in the 1960s and 1970s, and is in fact a potent evolutionary force after all. The early mathematical models, which purported to show the impotence of group selection, relied on unrealistic and maximally unfavorable assumptions, Wilson holds. More controversially, he claims that the supposed alternatives to group selection, such as kin selection and evolutionary game theory, are not in fact alternatives at all; rather, they are *versions* of group selection theory, but presented in a formal framework which tends to obscure this fact.

The precise relation between kin and group selection has long been a point of controversy. Some authors insist that these modes of selection are of a piece, while others see a sharp distinction between them (cf. Uyenoyama & Feldman, 1980). Hamilton's own views on the matter underwent an interesting evolution, as Sober and Wilson (1998) have documented. Initially Hamilton treated group selection with suspicion, but later he came round to the view that his own models for the evolution of altruism did actually involve a component of group selection after all (Hamilton, 1996). Despite this change of heart by Hamilton, many biologists continue to regard kin selection as an alternative to group selection, not an instance of it. The issue here is in partly terminological – must "group" mean group of unrelated organisms? – but it runs deeper than this. To focus the issue, let us recall the basic problem of altruism, then contrast Darwin's group selectionist solution with Hamilton's solution.

The basic problem is simply that in any group containing both altruists and selfish organisms, the latter will be at an advantage – they will enjoy the benefits of others' altruism but without incurring any of the costs. So within any one group the frequency of altruists will always decline. Darwin suggested that in a multi-group scenario the accounting may change, for groups containing many altruists, all engaged in mutual assistance, may out-reproduce groups containing predominantly selfish types; in this way, group selection in favor of altruism may counteract individual selection against. Hamilton suggested that if altruists preferentially direct their altruism towards relatives, rather than towards unrelated members of the population, then altruism may spread, owing to the fact that relatives share genes.

Although Darwin and Hamilton may *seem* to have offered quite different solutions to the problem of altruism, there is actually a deep underlying commonality. In both cases, what permits the spread of altruism is that *the beneficiaries of altruistic actions have a better than random chance of being altruists themselves*; as Hamilton (1975) himself said, this is the "crucial requirement" for altruism to evolve. Darwin's scenario, involving a population subdivided into groups, which differ in their frequencies of altruists, and Hamilton's scenario, involving organisms which behave altruistically towards kin, are simply two different ways of satisfying this fundamental requirement. This is why Sober and Wilson (1998) maintain that kin-directed altruism, far from constituting an alternative to group selection, is actually group selection in disguise, an argument that Hamilton (1975) also endorsed.

Opponents of this argument point out that group selection, as traditionally conceived, involved discrete multi-generational groups reproductively isolated from other such groups, but kin-directed altruism may occur within a single population whether or not it contains such groups (Maynard Smith, 1976, 1998). However, Sober and Wilson (1998) reply that in the relevant sense of "group," a group exists whenever a number of organisms interact in a way that affects their fitnesses, whether or not the group is reproductively isolated, spatially discrete, or multi-generational. So in the limit, two organisms that engage in a fitness-affecting interaction just once in their lifetime constitute a group. This concept was first developed by Wilson (1975) in his well-known "trait group" model for the evolution of altruism, in which the trait groups are simply temporary alliances of organisms that break up and re-form every generation. The transitory nature of these alliances in no way prevents them from qualifying as groups, Wilson insists, for groups must be defined by the criterion of fitness interaction.

The trait-group model and similar models of "intra-demic" selection have generated an interesting philosophical discussion. Sober and Wilson (1998) insist that these models involve a component of group selection, for the trait-groups exhibit differential productivity. Different trait-groups contribute different numbers of offspring to the subsequent generation, so there is selection between groups as well as selection within them. (This is why the trait-group models permit altruism to evolve.) However, other authors argue that these models involve only individual selection in a group-structured population (Maynard Smith, 1998). On this view, an organism's trait-group is simply a part of its overall selective environment, so all the selection is at the level of the individual organism; the trait-groups are relevant only in that they partially determine individual fitnesses. Still others, including Dugatkin and Reeve (1994) and Sterelny and Griffiths (1998), have defended a pluralistic line. They argue that trait-group models *can* be construed as involving a component of group selection as per Sober and Wilson, but can equally be regarded as individual selection in a structured environment as per Maynard Smith. There is no fact of the matter as to which is right, according to these authors – we are faced with a choice of perspective, not empirical fact.

One notable contribution to this debate comes from Kerr and Godfrey-Smith (2002), who offer a sophisticated defense of pluralism. They construct a simple mathematical model of selection in a group-structured population and show that the model's dynamics can be fully described by two sets of parameter values, one of which ascribes fitness values only to individuals, the other of which ascribes fitnesses to groups *and* individuals. The former is called a "contextual" parameterization, for the fitness of an individual depends on its group context, while the latter is called a "multilevel" parameterization, for both individuals and groups are ascribed fitnesses. Kerr and Godfrey-Smith demonstrate that the two parameterizations are mathematically equivalent – each set of parameter values can be derived from the other. This does not *prove* that pluralism is the correct position – for it might be argued that that only one of the parameterizations correctly captures the causal facts, even though the two are mathematically interchangeable, hence computationally equivalent. But Kerr and Godfrey-Smith certainly make a strong case for a pluralistic interpretation of the trait-group models.

It is obvious that the group selection controversy is partly fuelled by disagreement about what exactly the process of group selection amounts to. Damuth and Heisler (1988) argue that there are two distinct concepts of group selection (or multilevel selection more generally), which have often been conflated in the literature. The distinction hinges on the meaning of "group fitness" and its relation to organismic fitness. In group selection type 1 (GS1), the fitness of a group is defined as the *average fitness of its constituent organisms*, so there is a definitional relationship between group and organismic fitness. The fittest groups, in this sense, are the ones that contribute the most offspring organisms to the next generation of organisms (per capita). In group selection type 2 (GS2), the fitness of a group is defined as the expected number of offspring groups that it leaves, rather than the average fitness of its constituent organisms. The fittest groups, in this sense, are those that contribute the most offspring *groups* to the next generation of *groups*. Although in many situations the groups that are fittest by the GS1 criterion will also be fittest by the GS2 criterion, and vice versa, the two concepts are logically distinct. So there are two quite different things that "group selection" can mean.

The essence of the difference between GS1 and GS2 concerns the "focal" level, i.e., the level we are interested in. In GS1 the focal level is the individual organism, while in GS2 it is the group. This means that GS1 and GS2 have different explanatory targets. The former can explain the changing frequency of different types of *individual* in a group-structured population, while the latter can explain the changing frequency of different types of *group* in a metapopulation of groups. (Put differently, in GS1 we count individuals while in GS2 we count groups.) As Damuth and Heisler (1988) note, most of the literature on group selection has dealt with GS1: the aim has been to understand the evolution of an *individual* phenotype, often altruism, in a population subdivided into groups. So group fitness, in models for the evolution of altruism, has usually been defined as average organismic fitness. By contrast, the literature on species selection has had a GS2 focus: the aim has been to understand the changing frequency of different types of *species*, not their component organisms (see Section 5 below). So species fitness is usually defined as expected number of offspring species, rather than as average organismic fitness. It follows that species selection is *not* simply a higher-level analog of group selection, as the latter has traditionally been understood, for it is of a different logical type (Arnold & Fristrup, 1982; Okasha, 2001, 2006).

The GS1/GS2 distinction is relevant to the debate over pluralism and trait-group selection. As we saw above, Kerr and Godfrey-Smith (2002) argue for pluralism by showing the interdefinability of the multilevel and contextual parameterizations of their model. However, this interdefinability result holds *only* in cases where group fitness is defined as average organismic fitness, i.e., GS1. If group fitness were defined in the GS2 way, as expected number of offspring groups, it would not be possible to switch between a multilevel and an individualist parameterization (Okasha, 2006). This means that group selection of the GS2 variety cannot be re-analyzed as organismic selection in a structured population. GS2 is thus an *irreducibly* group-level process, in one legitimate sense of the word "reducible." This indicates a limitation on the types of selection process for which the pluralist thesis – that there is "no fact of the matter" about the true level of selection – will be tenable. One *might* take this to show that only GS2 is "real" group selection, as authors such as Vrba (1989) have argued, but this inference is not mandatory; it would have the unwelcome implication that much of the work purporting to be about group selection does not really deal with that topic at all.

The distinction between GS1 and GS2 goes a long way towards clarifying the group selection question, but certain outstanding issues remain. One such issue concerns causality. Virtually everybody agrees that the theory of natural selection is a causal theory – it aims to provide a causal-historical explanation for changes in gene/trait frequency over time. Therefore, where multiple levels of selection are in play, it follows that causes must be operating at more than one hierarchical level. Sober's (1984) book contained a detailed attempt to use philosophical ideas about causality to address questions about the levels of selection. Recent work by Okasha (2004c, 2006) also addresses the issue of causality, though from a somewhat different angle. Most approaches to the levels of selection have addressed a purely *qualitative* question, namely, what are the level(s) of selection in a given situation? But this fails to address an important *quantitative* question, namely, given the levels of selection that are in play, what fraction of the total evolutionary change can be attributed to each? For example, suppose both group and organismic selection are in operation in a given situation. How do we tell how *much* of the resulting evolutionary change is due to selection at each level? Okasha (2004c) explores three different statistical techniques designed to address this question, and finds that they yield incompatible results – each decomposes the total change into different components, allegedly corresponding to distinct levels of selection. This raises an overarching philosophical issue: how do we choose between the techniques? Or is there perhaps "no fact of the matter" about which is correct? Focusing on the quantitative rather than just the qualitative question brings new conceptual problems to the fore.

5. Species Selection and Macroevolution

The concept of species selection was developed in the 1970s by Stanley (1975) and Eldredge and Gould (1972) as part of their attempt to "decouple" macroevolution from microevolution. The long-term evolutionary patterns revealed in the fossil record are not simply the cumulative upshot of the microevolutionary forces that adapt local populations to their environments, these authors argued. Phenomena such as the origins of new species and higher taxa, long-term phylogenetic trends, and the greater diversification of some clades compared to others need to be studied "at their own level," not treated as incidental effects of microevolution. This requires us to recognize the existence of autonomous macroevolutionary forces, of which species selection is a potential example.

The basic idea of species selection is that a selective force operates on whole species, analogous to but distinct from ordinary organismic selection, favoring those species that are fittest and disfavoring the least fit. Organismic death is analogous to species extinction, and organismic reproduction to speciation. So just as an organism's fitness is its expected number of offspring organisms, so a species' fitness is its expected number of offspring species. It is obvious that species vary in their characters, or traits. Some species are more geographically widespread than others, some are ecological generalists while others are specialists, some are more genetically diverse than others, some are composed of larger-bodied organisms than others, and so on. Conceivably, these species-level traits could affect fitness – either by affecting a species' probability of extinction or of speciation. If so, and if the traits in question are inherited by offspring species, then species selection could in theory have a significant effect on long-term evolutionary trends.

Most though not all biologists accept that species selection is possible, but there is substantial disagreement over its empirical significance. Additionally, there is disagreement about what exactly the concept of species selection amounts to, what type of evolutionary phenomena it is capable of explaining, and how the relation between species selection and lower-level selection should be understood. These conceptual issues require resolution before the empirical case for species selection can be adequately assessed.

In a series of publications, Elisabeth Vrba has argued that genuine species selection is extremely rare; most of the alleged examples involve only "species sorting," she claims (Vrba, 1984a, 1984b, 1989). The idea behind Vrba's selection/sorting distinction is that even if differential extinction or speciation rates correlate with species-level characters, this does not necessarily mean that an autonomous higher-level selection process exists. The trend may instead be a by-product of lower-level causal forces, such as organismic selection. For example, if red and grey squirrels compete for the same resources and the former are driven to extinction, it would be inappropriate, intuitively, to attribute this to species selection. Grey squirrels had higher individual fitness than red ones, and as a consequence the latter all died, hence the species went extinct. But no causal forces were acting on the species *as units*. So the higher-level trend, i.e., the survival of the one species and the extinction of the other, is not the product of species selection. Rather, it is the by-product of selection at the *organismic* level, the effects of which "percolate up" the biological hierarchy. In Vrba's terms, this is a case of species sorting but not species selection.

Most biologists agree with Vrba that genuine species selection involves more than mere differential extinction/speciation, but there is disagreement over exactly what the missing ingredient is. Vrba herself argues that true species selection requires the existence of "emergent" species-level characters that causally influence species fitness. Emergent characters are usually contrasted with "aggregate" or "sum of the parts" characters such as "average height" or "average running speed" that are produced by combining measurements on individual organisms. Intuitively such characters are statistical artifacts rather than real species-level traits. Emergent characters, by contrast, are not mere statistical summations of organismic characters. Vrba cites "characteristic population size, spatial and genetic separation between populations, and the nature of a species periphery" as possible examples of emergent characters of species (1984a, p.325). Genuine species selection only occurs, Vrba holds, where emergent properties lead to differences in species fitness.

The significance of the distinction between aggregate and emergent characters has proved controversial. One problem is that the distinction itself, while intuitively clear, is difficult to characterize in general terms; Vrba herself offers several non-equivalent characterizations. Another problem is that the emergent character requirement represents a substantial metaphysical thesis, which surely requires further explanation. For emergent characters of species, no less than aggregate ones, supervene on underlying organismic characters. Characters such as species range or spatial separation between populations are ultimately dependent on organismic characters and behaviors, e.g., dispersal distance. Vrba's requirement implies that a genuine species-level causal process occurs only when species fitness is affected by emergent characters. But since aggregate and emergent characters are *both* determined by underlying organismic characters, some explanation of this alleged difference in causal potential is surely needed. Alternative approaches to distinguishing "real" species selection from its surrogates are explored by Williams (1992), Gould (2002), Gould and Lloyd (1999), Sterelny (1996), and Okasha (2006).

A quite different challenge to species selection comes from Damuth (1985), who argues that species are not the right *type* of entity to function as units of selection in the first place. Most species are divided into many partially isolated populations, each subject to different local conditions, Damuth stresses. So there are unlikely to be selection pressures acting on a whole species as a unit; rather, different populations within the species will be subject to different selection pressures. In short, species are not ecologically localized the way that individual organisms are, and thus not the sorts of thing to which Darwinian fitness can be ascribed. Damuth thus proposes to replace the concept of species selection with "avatar" selection. Avatars are local populations of species that are ecologically localized, hence capable of competing and interacting with local populations of other species. This move is required to preserve the analogy with organismic selection that motivated the idea of species selection in the first place, Damuth argues.

Even if the concept of species selection can overcome the conceptual and empirical challenges it faces, there is still a fundamental reason for regarding the species as a relatively unimportant unit of selection. For species are not functionally organized the way other paradigmatic units of selection, such as cells, organisms, and insect colonies, are. These entities exhibit a division of labor between their constituent parts, the hallmark of true functional organization. The different proteins in a cell, the different tissues and organs in an organism, and the different castes in an insect colony each perform distinct roles in the functioning of the larger entity. The same is not true of the organisms that make up a species. (For this reason, the species should probably not be thought of as a level of biological *organization* at all.) Though this disanalogy does not invalidate the concept of species selection altogether, if only because many rounds of cumulative selection are required to produce functionally integrated entities, it does suggest that species selection has been much less important than selection at lower hierarchical levels.

6. Multilevel Selection Theory and The Major Transitions in Evolution

The expression "multilevel selection theory" is increasingly common in the biological literature. The basic idea of this theory – that natural selection may operate simultaneously at more than one hierarchical level – is not new; indeed, it is implicit in the very earliest discussions of the levels of selection, including Darwin's. What is new is the *use* to which multilevel selection is currently being put. Increasingly, biologists interested in explaining what Maynard Smith and Szathmary call the "major transitions in evolution" have made use of ideas from multilevel selection theory (Buss, 1987; Michod, 1997, 1999; Maynard Smith & Szathmary, 1995; Frank, 1997; Queller, 2000). The work of these authors extends the traditional units of selection question in an important new way.

The "major transitions in evolution" refer to the transitions from solitary replicators to networks of replicators enclosed within compartments, from independent genes to chromosomes, from prokaryotic cells to eukaryotic cells containing organelles, from unicellular to multicellular organisms, and from solitary organisms to colonies. Some of these transitions occurred in the distant evolutionary past, others much more recently. In each case a number of smaller units, originally capable of surviving and reproducing on their own, became aggregated into a single larger unit, thus generating a new level of biological organization. The challenge is to understand these transitions in Darwinian terms. Why was it advantageous for the lower-level units to sacrifice their individuality, cooperate with one another, and form themselves into a larger corporate body? And how could such an arrangement, once first evolved, be evolutionarily stable?

This is where multilevel selection enters the picture. As Buss, Michod, and Maynard Smith and Szathmary all stress, to understand the major transitions we need to know why lower-level selection did not disrupt the formation of the higher-level unit. In the transition to multicellularity, for example, we need to know why selection between competing cell lineages did not disrupt the integrity of the emerging multicellular organism. One possibility is that selection acted on the higher-level units themselves, leading them to evolve adaptations that minimize conflict and increase cooperation among their constituent parts. Thus in the case of multicellularity, Buss and Michod argue that early sequestration of the germ-line may be one such adaptation, for it reduces the probability that mutant cells, arising during ontogeny, will find their way into the next generation. Another idea is that passing the life cycle through a single-celled stage, as occurs in most animal and plant species, is an adaptation for minimizing within-organism conflict, for it increases the relatedness, hence decreases the competition, between the cells within an organism. These particular examples have both been contested, but the general idea that the major transitions involve an interaction between selection at different levels is very widely accepted.

Though still in their infancy, these theoretical developments suggest that the traditional way of posing the units of selection question was somewhat inadequate. For as Griesemer (2000) notes, the traditional formulations of the question, including Lewontin's "heritable variation in fitness" formulation employed above, generally take the existence of the biological hierarchy for granted, as if hierarchical organization were simply an exogenously given fact about the biotic world. But of course the biological hierarchy is *itself* the product of evolution – entities further up the hierarchy, such as multicellular organisms, have obviously not been there since the beginning of life on earth. The same is true of cells and chromosomes. So ideally, we would like an evolutionary theory which explains how the biological hierarchy came into existence, rather than treating it as a given. From this perspective, the units of selection question is not simply about identifying the hierarchical level(s) at which selection *now* acts, which is how it was traditionally conceived, but about identifying the mechanisms which led the hierarchy to evolve in the first place (Okasha, 2005).

This new "diachronic" perspective gives the units of selection question a renewed sense of urgency. Some biologists were inclined to dismiss the traditional debate as a storm in a teacup – arguing that in practice, selection on individual organisms is the only important selective force in evolution, whatever about other theoretical possibilities. But as Michod (1999) stresses, multicellular organisms did not come from nowhere, and a complete evolutionary theory must surely try to explain how they evolved, rather than simply taking their existence for granted. So levels of selection other than that of the individual organism *must* have existed in the past, whether or not they still operate today. From this expanded point of view, the argument that selection on individual organisms is "all that matters in practice" is clearly unsustainable. Moreover, this lends further weight to the view that group selection was prematurely dismissed in the 1960s. For multicellular organisms are themselves groups of cooperating cells, and chromosomes are groups of cooperating genes. Since multi-cellular organisms and chromosomes obviously have evolved, the efficacy of group selection cannot be denied (Michod, 1999; Sober & Wilson, 1998).

The attempt to understand the major transitions has thrown up a number of interesting questions. One concerns the extent to which the different transitions are thematically similar, and thus explicable in similar terms. For example, is the transition from unicellularity to multicellularity relevantly similar to the transition from solitary insects to eusocial insect colonies? If so, then can the theoretical principles needed to understand the former be extrapolated to the latter and vice versa? More generally still, can concepts such as kin selection and the gene's eye view of evolution, originally developed to help explain social behavior in animals, shed light on the major transitions? Theorists take different stands on these questions. Most agree that the principle of kin selection is of fundamental importance at all hierarchical levels, especially in the evolution of multicellularity, though Buss (1987) accords much less explanatory weight to this principle than others. Maynard Smith and Szathmary (1995) explicitly advocate a Williams/Dawkins gene-centered approach to the major transitions, but Michod (1999) describes Dawkins' gene-centric view of evolution as a "mistake" (p.139). These disagreements show that the application of multilevel selection theory to the major transitions raises substantial, and as yet unresolved, conceptual issues.

7. Conclusion

In some ways it is surprising that the units of selection question has engendered so much conceptual and foundational discussion, for the principle of natural selection is essentially straightforward and can be formulated very simply. Nonetheless, as the forgoing survey has hopefully made clear, the myriad of conflicting opinions among evolutionary biologists about the units of selection are not the "ordinary" scientific disagreements of opinion that arise from lack of empirical data. Rather, they are disagreements about which concepts to employ, which questions to ask, and which explanatory strategies to pursue. It is hard to predict what direction the debate will take in the twenty-first century, though it is likely that the flurry of interest in the major evolutionary transitions will continue. It remains to be seen whether the ensuing biological discussions will provide as fertile a ground for philosophy of science as did the units of selection discussions of the twentieth century.

References

Arnold, A. J., & Fristrup, K. (1982). The theory of evolution by natural selection: a hierarchical expansion. *Paleobiology*, 8, 113–29.
Burt, A., & Trivers, R. (2006). *Genes in conflict*. Harvard: Harvard University Press.
Buss, L. (1987). *The evolution of individuality*. Princeton: Princeton University Press.
Cziko, G. (1995). *Universal Darwinism*. Cambridge, MA: MIT Press.
Damuth, J. (1985). Selection among "species": a formulation in terms of natural functional units. *Evolution*, 39(5), 1132–46.
Damuth, J., & Heisler, L. (1988). Alternative formulations of multilevel selection. *Biology and Philosophy*, 3, 407–430.
Darwin, C. (1871). *The descent of man*. New York: Appleton.
Dawkins, R. (1976). *The selfish gene*. Oxford: Oxford University Press.
Dawkins, R. (1982). *The extended phenotype*. Oxford: Oxford University Press.
Dugatkin, L. A., & Reeve, H. K. (1994). Behavioural ecology and levels of selection: dissolving the group selection controversy, *Advances in the Study of Behaviour*, 23, 101–33.
Edelman, G. M. (1987). *Neural Darwinism*. New York: Basic Books.
Eldredge, N., & Gould, S. J. (1972). Punctuated equilibria: an alternative to phyletic gradualism. In T. J. M. Schopf (Ed.). *Models in paleobiology* (pp. 82–115). San Francisco: Freeman.
Fisher, R. A. (1930). *The genetical theory of natural selection*. Oxford: Clarendon Press.
Frank, S. A. (1997). Models of symbiosis. *The American Naturalist*, 150, S80–S99.
Frank, S. A. (1996). The design of natural and artificial systems. In M. R. Rose & G. V. Lauder (Eds). *Adaptation* (pp. 451–505). San Diego: Academic Press.
Gould, S. J. (2002). *The structure of evolutionary theory*. Cambridge, MA: Harvard University Press.
Gould, S. J., & Lloyd, E. A. (1999). Individuality and adaptation across levels of selection. *Proceedings of the National Academy of Sciences*, 96(21), 11904–9.
Griesemer, J. (2000). The units of evolutionary transition. *Selection*, 1, 67–80.
Haldane, J. B. S. (1932). *The causes of evolution*. New York: Cornell University Press.
Hamilton, W. D. (1964). The genetical evolution of social behaviour I and II. *Journal of Theoretical Biology*, 7, 1–16, 17–32.
Hamilton, W. D. (1996). *Narrow roads of gene land* (Vol. 1). New York: W. H. Freeman.
Hamilton, W. D. (1975). Innate social aptitudes in man: an approach from evolutionary genetics. In R. Fox (Ed.). *Biosocial anthropology* (pp. 133–53). London: Malaby Press.
Hull, D. (1981). Units of evolution: a metaphysical essay. In R. Jensen & R. Harre (Eds). *The philosophy of evolution* (pp. 23–44). Brighton: Harvester Press.

Hurst, L., Atlan, A., & Bengtsson, B. (1996). Genetic conflicts. *Quarterly Review of Biology*, 71, 317–64.

Kerr, B., & Godfrey-Smith, P. (2002). Individualist and multi-level perspectives on selection in structured populations. *Biology and Philosophy*, 17(4), 477–517.

Lewontin, R. C. (1970). The units of selection. *Annual Review of Ecology and Systematics*, 1, 1–18.

Maynard Smith, J. (1964). Group selection and kin selection. *Nature*, 201, 1145–7.

Maynard Smith, J. (1976). Group selection. *Quarterly Review of Biology*, 51, 277–83.

Maynard Smith, J. (1983). Models of evolution. *Proceedings of the Royal Society of London B*, 219, 315–25.

Maynard Smith, J. (1998). The origin of altruism. *Nature*, 393, 639–40.

Maynard Smith, J., & Price, G. R. (1973). The logic of animal conflict. *Nature*, 146, 15–18.

Maynard Smith, J., & Szathmary, E. (1995). *The major transitions in evolution*. San Francisco: Freeman.

Michod, R. (1997). Evolution of the individual. *American Naturalist*, 150, S5–S21.

Michod, R. (1999). *Darwinian dynamics*. Princeton: Princeton University Press.

Okasha, S. (2001). Why won't the group selection controversy go away? *British Journal for the Philosophy of Science*, 51, 25–50.

Okasha, S. (2003). Does the concept of "clade selection" make sense? *Philosophy of Science*, 70, 739–51.

Okasha, S. (2004a). The "averaging fallacy" and the levels of selection. *Biology and Philosophy*, 19, 167–84.

Okasha, S. (2004b). Multi-level selection, covariance and contextual analysis. *British Journal for the Philosophy of Science*, 55, 481–504.

Okasha, S. (2004c). Multi-level selection and the partitioning of covariance: a comparison of three approaches. *Evolution*, 58(3), 486–94.

Okasha, S. (2005). Multi-level selection and the major transitions in evolution. *Philosophy of Science*, 72, 1013–28.

Okasha, S. (2006). *Evolution and the levels of selection*. Oxford: Oxford University Press.

Pomiankowski, A. (1999). Intra-genomic conflict. In L. Keller (Ed.). *Levels of selection in evolution* (pp. 121–52). Princeton: Princeton University Press.

Queller, D. C. (2000). Relatedness and the fraternal major transitions. *Philosophical Transactions of the Royal Society of London B*, 355, 1647–55.

Reeve, H. K., & Keller, L. (1999). Levels of selection: burying the units-of-selection debate and unearthing the crucial new issues. In L. Keller (Ed.). *Levels of selection in evolution* (pp. 3–14). Princeton: Princeton University Press.

Sober, E. (1984). *The nature of selection*. Cambridge, MA: MIT Press.

Sober, E., & Wilson, D. S. (1998). *Unto others: the evolution and psychology of unselfish behavior*. Cambridge, MA: Harvard University Press.

Stanley, S. M. (1975). A theory of evolution above the species level. *Proceedings of the National Academy of the Sciences*, 72, 646–50.

Sterelny, K. (1996). Explanatory pluralism in evolutionary biology. *Biology and Philosophy*, 11, 193–214.

Sterelny, K., & Griffiths, P. (1998). *Sex and death: an introduction to the philosophy of biology*. Chicago: University of Chicago Press.

Uyenoyama, M., & Feldman, M. W. (1980). Theories of kin and group selection: a population genetics perspective. *Theoretical Population Biology*, 17, 380–414.

Vrba, E. (1984a). What is species selection? *Systematic Zoology*, 33, 318–28.

Vrba, E. (1984b). Patterns in the fossil record and evolutionary processes. In M. W. Ho & P. Saunders (Eds). *Beyond neo-Darwinism* (pp. 115–39). London: Academic Press.

Vrba, E. (1989). Levels of selection and sorting with special reference to the species level. *Oxford Surveys in Evolutionary Biology*, 6, 111–68.
Weismann, A. (1903). *The evolution theory*. London: Edward Arnold.
Williams, G. C. (1966). *Adaptation and natural selection*. Princeton: Princeton University Press.
Williams, G. C. (1992). *Natural selection: domains, levels, challenges*. Oxford: Oxford University Press.
Wilson, D. S. (1975). A theory of group selection. *Proceedings of the National Academy of Science*, 72, 143–6.
Wilson, D. S. (1980). *The natural selection of populations and communities*. Menlo Park, CA: Benjamin Cummings.
Wilson, D. S. (1989). Levels of selection: an alternative to individualism in biology and the human sciences. *Social Networks*, 11, 357–72.

Further Reading

Alexander, R. D., & Borgia, G. (1978). Group selection, altruism and the levels of organization of life. *Annual Review of Ecology and Systematics*, 9, 449–74.
Bonner, J. T. (1988). *The evolution of complexity*. Princeton: Princeton University Press.
Brandon, R., & Burian, R. (Eds). (1984). *Genes, organisms and populations*. Cambridge, MA: MIT Press.
Eldredge, N. (1985). *Unfinished synthesis: biological hierarchies and modern evolutionary thought*. Oxford: Oxford University Press.
Frank, S. A. (1998). *Foundations of social evolution*. Princeton: Princeton University Press.
Godfrey-Smith, P., & Lewontin, R. (1993). The dimensions of selection. *Philosophy of Science*, 60, 373–95.
Gould, S. J. (1999). The evolutionary definition of selective agency. In R. Singh, C. Krimbas, D. Paul, & P. Beatty (Eds). *Thinking about evolution* (pp. 208–34). Cambridge: Cambridge University Press.
Grafen, A. (1984). Natural selection, kin selection and group selection. In J. Krebs & N. Davies (Eds). *Behavioural ecology: an evolutionary approach* (pp. 62–84). Oxford: Blackwell Scientific Publications.
Hammerstein, P. (Ed.). (2003). *Genetic and cultural evolution of cooperation*. Cambridge, MA: MIT Press.
Hull, D. (1980). Individuality and selection. *Annual Review of Ecology and Systematics*, 11, 311–32.
Keller, L. (Ed.). (1999). *Levels of selection in evolution*. Princeton: Princeton University Press.
Maynard Smith, J. (1987). How to model evolution. In J. Dupré (Ed.). *The latest on the best: essays on evolution and optimality* (pp. 119–31). Cambridge, MA: MIT Press.
Maynard Smith, J. (1988). Evolutionary progress and levels of selection. In M. H. Nitecki (Ed.), *Evolutionary progress* (pp. 219–30). Chicago: University of Chicago Press.
Price, G. R. (1995). The nature of selection. *Journal of Theoretical Biology*, 175, 389–96.
Seeley, T. D. (1997). Honey bee colonies are group-level adaptive units. *American Naturalist*, 150, S22–S41.
Sober, E., & Lewontin, R. (1982). Artifact, cause and genic selection. *Philosophy of Science*, 49, 157–80.
Sterelny, K., & Kitcher, P. S. (1988). The return of the gene. *Journal of Philosophy*, 85, 339–61.

Trivers, R. L. (1971). The evolution of reciprocal altruism. *Quarterly Review of Biology*, 46, 35–57.

Wade, M. (1978). A critical review of the models of group selection. *Quarterly Review of Biology*, 53, 101–14.

Wilson, E. O. (1975). *Sociobiology: the new synthesis*. Cambridge, MA: Harvard University Press.

Wilson, D. S. (1997). Altruism and organism: disentangling the themes of multi-level selection theory. *American Naturalist*, 150, S122–34.

Wright, S. (1980). Genic and organismic selection. *Evolution*, 34, 825–43.

Chapter 9

Molecular Evolution

MICHAEL R. DIETRICH

Molecular evolution emerged as a hybrid discipline in the 1960s. Blending theoretical and experimental traditions from evolutionary genetics, molecular biology, biochemistry, systematics, anthropology, and microbiology, molecular evolution represented a significant reconsideration of several key features of the preceding neo-Darwinian evolutionary synthesis. Where neo-Darwinians articulated a unified understanding of the evolutionary process dominated by selection, by the 1970s most molecular evolutionists recognized that the domain of evolutionary biology was divided into molecular and morphological levels. Where neo-Darwinians advocated variable rates of evolution driven by environmental change, molecular evolutionists advocated a molecular clock that approximated a constant rate of change in proteins and nucleic acids. Where systematics had been based on morphological features, it now had a vast new array of molecular data and the challenge of reconciling sometimes divergent phylogenetic inferences.

The changes introduced by molecular evolution created enormous controversy during the 1960s and 1970s. While these disputes have tended to ease over time, controversy remains one of the persistent features of the history of molecular evolution. As such, molecular evolution provides a very rich history for the analysis of scientific controversy, testing, experimentation, and methodology.

1. The Neutral Theory of Molecular Evolution

When molecular evolution emerged as a field in the early 1960s, biochemists, molecular biologists, and some evolutionary biologists began to consider that some changes in proteins and nucleic acids were not selected. The possibility of neutral mutations was widely acknowledged by evolutionary biologists, such as Theodosius Dobzhansky (1955), but the existence of a significant number of neutral mutations was not taken seriously by most evolutionary geneticists.

Attitudes began to change in 1968 when Motoo Kimura argued that many substitutions at the molecular level were not subject to natural selection, but were instead governed by random drift (Kimura, 1968; also see Dietrich, 1994, and Suarez & Barahona, 1996). Using protein sequence data generated by biochemists such as Emile Zuckerkandl and Emmanuel Margoliash, Kimura and his colleague Tomoko Ohta compared mammalian protein sequences and used the number of detected differences across species to calculate a rate of molecular evolution. Kimura then reasoned that if most mutations were in fact harmful, then the rate of evolution calculated for mammals would create an intolerable genetic load (an accumulation of too many harmful alleles). Since mammals were not extinct or staggering under an enormous genetic load, Kimura concluded that most detected molecular variants were in fact neutral (Kimura, 1968).

Kimura's conclusion and argument were controversial, but the dispute between neutralists and selectionists was guaranteed in 1969 when Tom Jukes and Jack King wrote their neutralist manifesto under the provocative title of "Non-Darwinian Evolution." King and Jukes brought a large variety of evidence to bear in favor of large numbers of neutral mutations (King & Jukes, 1969). By using evidence from the growing field of molecular evolution to support the idea of neutral mutations and the importance of random drift, they spelled out the molecular consequences of the neutral hypothesis more clearly than Kimura had. King and Jukes built their case using phenomena such as synonymous mutations, the Treffors mutator, the relationship between amino acid frequencies and the genetic code, and the growing body of data on specific proteins such as cytochrome c.

Although many biologists were extremely skeptical of the neutral theory, Kimura and his colleague Tomoko Ohta pursued the neutral theory vigorously. In 1969, Kimura used the constancy of the rate of amino acid substitutions in homologous proteins to argue powerfully for neutral mutations and the importance of random drift in molecular evolution (Kimura, 1969). At the same time, Kimura was also calling on his earlier work on stochastic processes in population genetics to forge a solid theoretical foundation for the neutral theory. Kimura's diffusion equation method provided the theoretical framework he needed to formulate specific models which in turn allowed him to address issues such as the probability and time to fixation of a mutant substitution as well as the rate of mutant substitutions in evolution (Kimura, 1970). Working in collaboration with Tomoko Ohta, Kimura also extended the neutral theory to encompass the problem of explaining protein polymorphisms. This was a central concern of population genetics, and Kimura and Ohta were able to show that protein polymorphisms were a phase in mutations' long journey to fixation (Kimura & Ohta, 1971).

In 1971 the Sixth Berkeley Symposium on Mathematical Statistics and Probability devoted a session to Darwinian, neo-Darwinian, and non-Darwinian evolution. Selectionist responses to King and Jukes' paper had created a full-blown controversy (Clarke, 1970; Richmond, 1970). Although the positions were becoming well articulated, there had only been a handful of empirical tests proposed. James Crow was charged with giving a review of both sides of the debate to start the conference session. Crow had been Kimura's advisor and remained a close friend and colleague. He was disposed toward the neutral theory, but was more skeptical than Kimura, Ohta, King, or Jukes. Like many others at the time, Crow believed that there was a continuum of fitness values for new mutations ranging from extremely detrimental through neutral to slightly beneficial. The conflict between neutralists and selectionists was thus a matter of the *relative importance* of neutral alleles and random drift relative to selection. In general, the neutralists had two battles to win: they had to prove that neutral alleles exist, and they had to prove that they play a significant role in evolution.

Crow's review was sympathetic to Kimura's position and as such answered a number of criticisms and provided several important arguments for the value of the neutral theory. Among the reasons *not* to accept neutralism or non-Darwinian evolution listed by Crow was the idea that "a random theory may discourage a search for other explanations and thus may be intellectually stultifying" (Crow, 1972, p.2) and that neutral changes were not as interesting as adaptive changes. Since adaptive change was a central concern, the neutral theory was not considered relevant. However, Crow notes that the neutral theory leads to a formulation of the important factors in evolution that has both new ideas and quantitative predictions. Moreover, "it is directly concerned with the gene itself, or its immediate products, so that the well-developed theories of population genetics become available. It produces testable theories about the rates of evolution" (Crow, 1972, p.2). Clearly Crow thought that the neutral theory was not intellectually stultifying; it was instead a source of innovation because of its testability and its new integration with molecular biology.

Tapping into the data and techniques of molecular biology was an important source of innovation for population genetics in the early 1970s. For population biology, the 1950s and 60s had been marked by a dispute over the type of genetic variation in natural populations and the forces responsible for maintaining that variation. Extreme positions advocating large amounts of homozygosity and purifying selection (the classical position) or large amounts of heterozygosity and balancing selection (the balance position) divided the community. H. J. Muller and Crow both advocated versions of the classical position, while Theodosius Dobzhansky and many of his students advocated versions of the balance position. By the mid-1960s, however, the controversy had stalemated – traditional experiments using radiation induced mutations were proving to be indecisive and extremely controversial (Beatty, 1987a; Lewontin, 1991). As Richard Lewontin, a student of Dobzhansky's, puts it, "population genetics seemed doomed to a perpetual struggle between alternative interpretations of great masses of inevitably ambiguous data" (Lewontin, 1991, p.658). What was needed was a way of breaking this deadlock. In 1964, Richard Lewontin thought he had found it in Jack Hubby's work using electrophoresis. Electrophoresis is a biochemical technique for separating proteins by charge and size. When applied to proteins from *Drosophila*, Hubby and Lewontin detected higher than expected levels of heterozygosity (Hubby & Lewontin, 1966; Lewontin & Hubby, 1966). This level was high enough to tilt the dispute toward the balance position, if only for a short while. Kimura's proposal that much of the variability that Hubby and Lewontin had detected was in fact neutral shifted the conceptual foundations of the classical balance dispute, but the technique of electrophoresis itself shifted the debate in terms of experimental practice (Suarez & Barahona, 1996).

Electrophoresis brought experimental population genetics down to the molecular level. In Lewontin's words, "Here was a technique that could be learned easily by any moderately competent person, that was relatively cheap as compared with most physiological and biochemical methods, that gave instant gratification by revealing before one's eyes the heritable variation in unambiguously scoreable characters, and most important, could be applied to *any* organism whether or not the organism could be genetically manipulated, artificially crossed, or even cultivated in the laboratory greenhouse. It is little wonder that there was a virtual explosion of electrophoretic investigations" (Lewontin, 1991, p.658). The introduction of electrophoresis to population genetics opened up the possibility of routine experimentation at the molecular level. It was in this context that Crow had advocated the molecularization of population genetics at the 12th International Congress of Genetics in 1968. There he wrote that, "What molecular biology is now doing so elegantly for population genetics is to provide a greatly improved opportunity to study the actual quantities – the gene frequencies and gene substitutions – to which the theory applies most directly. This is especially true for alleles that have small selective differences; until recently these have been largely outside the realm of experimental inquiry" (Crow, 1969, pp.106–7). The value of this kind of experimental access and the quantitative predictions that result from it is in part derived from the immediate context of population genetics: quantitative theory in population genetics, according to Crow, has "mainly centered around the individual gene and its rate of replacement" (Crow, 1969, p.107). To population geneticists used to problematic predictions and ambiguous data, molecular biology seemed to offer a way to sharpen both their predictions and data in such a way as to allow decisive tests to be made.

At the Berkeley Symposium, G. L. Stebbins and Richard Lewontin advocated a selectionist position. According to Stebbins and Lewontin, the neutral theory is so permissive that it is weak as a testable hypothesis (Stebbins & Lewontin, 1972, p.35). For instance, the neutral theory in its simplest form predicts that allele frequencies will vary from population to population, but in *Drosophila pseudoobscura* and *willistoni*, widely separate populations show very similar allele frequencies. A migration rate as low as one migrant per generation, however, is enough to account for the similarity. Armed with these assumptions about migration rate, Stebbins and Lewontin charge that no observation could contradict the prediction. Appealing to Karl Popper's philosophy of science, they labeled the neutral theory "'empirically void' because it has no set of potential falsifiers" (Stebbins & Lewontin, 1972, pp.35–6). Despite their arguments, Stebbins and Lewontin do not reject the idea of neutral mutation and the effects of random drift. Instead they see the nature of evolutionary processes as unresolved and even encourage the pursuit of both neutralist and selectionist explanations (Stebbins & Lewontin, 1972, p.40).

Concerns about testing continued to haunt the neutralist–selectionist controversy for the next decade. While the popularity of electrophoresis meant that plenty of new data on genetic variability was being produced, devising the statistical tests that relied on that data was difficult. Warren Ewens, for instance, created a test for neutrality derived from his sampling formula (Ewens, 1972). When this test was applied to electrophoretic data, however, it did not have sufficient statistical power to distinguish drift from selection. In 1977, Geoff Waterson refined Ewens' test, but could not eliminate the problems with low statistical power (Watterson, 1977; Lewontin, 1991). The results of other tests were similarly indecisive or actively disputed. Francisco Ayala's group, for instance, tested neutralists' predictions about heterozygosity with data on the electrophoretic variability detected in natural populations of *Drosophila*. Ayala and his coworkers predicted that the distribution of heterozygous loci should cluster around a value of 0.177. The observed distribution, however, was fairly even except that it had many loci with very little heterozygosity. Ayala argued that the detected excess of rare alleles was evidence against the neutral theory (Ayala et al., 1974, p.378). Jack King responded by questioning the assumptions of the model that Ayala had used; the infinite alleles model, King asserted, was the source for the rare alleles discrepancy. Moreover, King noted that the predictions generated with an infinite alleles model should not be compared to data from electrophoresis, since the differences detected by electrophoresis did not necessarily correspond to allelic differences (King, 1976). As a result, Ayala and his coworkers adapted their tests to use the charge ladder model of mutation that was designed for electrophoretic data. The excess of rare alleles remained, however.

As Ayala's results were debated and refined, Tomoko Ohta articulated the Nearly Neutral Theory that proposed a larger proportion of slightly deleterious mutants that while selected were so weakly selected that they acted as if they were neutral (Ohta, 1973, 1992; Ohta & Gillespie, 1996). One of the chief benefits of the Nearly Neutral Model was that it could accommodate the large number of rare alleles observed by Ayala. At the same time, Masatoshi Nei looked to population dynamics such as the possibility of population bottlenecks as a means of explaining the excess of rare alleles (Nei, 2005). In the end, Ayala's test was very influential and created the impetus for significant revisions of the neutralist position, but Ayala's tests did not settle the controversy. Instead, the results of Ayala's and other tests led the neutralists to put more stock in the molecular clock as a source of supporting evidence.

2. The Molecular Clock

The idea that the rate of change in biological molecules was constant over time was christened the molecular clock in 1965 by Emile Zuckerkandl and Linus Pauling (Zuckerkandl & Pauling, 1965; Morgan, 1998). Zuckerkandl and Pauling based this claim on their comparison of similarities and differences in the amino acid sequences of hemoglobins from different species. When different hemoglobins were compared, the number of differences seemed to be proportional to the length of time that the species in question had been separated evolutionarily. Zuckerkandl and Pauling were interested in using molecular characteristics to infer evolutionary relationships and immediately saw the value of the molecular clock for not only inferring relationships, but the times of divergence.

The molecular clock was not perfect, however. Like clocks based on radioactive decay, the molecular clock was stochastic. Differences did not emerge at a perfectly constant rate. The constancy of the clock was instead an average of sometimes highly variable substitution events. Thus, from its beginnings, the clock was understood to have some variability in its rate. For the clock's many critics, however, one of the key questions at hand revolved around how much variability the clock could have and still remain a clock.

The controversies over the variability of the molecular clock were compounded by its role in the neutralist–selectionist controversy. Zuckerkandl and Pauling had initially invoked both selection and drift to explain the mechanism of the clock (Morgan, 1998). When Kimura, King, and Jukes began to advocate the neutral theory, they recognized that neutrality provided an elegant explanation for the observed constancy (see Dietrich, 1998, and Morgan, 1998). The neutral theory predicted that for neutral sites or alleles the rate of mutation would be the same as the rate of substitution. Substitutions were the observed differences between molecules. These detected differences did not represent all of the changes produced by mutation. They represented those changes remaining after selection had eliminated the more harmful mutants and fixed the most beneficial. The rate of substitution for a mutation subject to selection would depend on the factors that normally affected selection processes, such as population size and environment. Selection should produce a highly variable rate of substitution. For a neutral allele or site, however, the process of moving from origination as a mutant to fixation was a process of random drift. The rate of substitution should then depend on the rate at which new mutants are introduced. For neutral changes, if the rate of mutation was approximately constant, then the rate of substitution would be as well.

When neutralists championed their explanation of the molecular clock's constancy, they inherited the problem of also explaining its variability. As soon as differences between molecules began to be compared, researchers noted that different molecules seemed to have different rates of change. Neutralists explained these different clocks in terms of the distribution of selected and neutral sites within each type of molecule. Hemoglobins, for instance, have sites that never change across species. These highly conserved sites were understood to be strongly selected; changing them would render the molecule less functional or non-functional and so were selected against. Other regions in hemoglobins show numerous differences among different species. These variable regions were interpreted as being neutral or weakly selected. A molecule such as histone IV was observed to have a large number of constrained sites and a low rate of substitution, whereas fibrinopepetide A was much less constrained and had a much faster molecular clock (King & Jukes, 1969, p.792). The problem of variability across types of molecules could thus be explained away, but rate variability within a molecule type was another matter. Very early in the history of the molecular clock, speedups and slowdowns were observed for the same molecule. Comparisons of insulin sequences, for instance, revealed that insulins in the guinea pig lineage seemed to have evolved faster than insulins in other mammalian lineages (King & Jukes, 1969; Ohta & Kimura, 1971, p.19). Primates, in contrast, seem to have experienced a slower rate of evolution for some proteins. Even as the neutralists defended the idea that types of molecules possessed intrinsic rates of change, they had to explain these deviations.

In 1971, Tomoko Ohta and Motoo Kimura compared sequence differences from alpha and beta hemoglobins, and for cytochrome c. Ohta and Kimura's statistical analysis of the variability in the rate of substitution for these proteins confirmed that both beta hemoglobin and the cytochrome c had significantly more variability than expected (Ohta & Kimura, 1971, p.21). Ohta and Kimura tried to explain away this high variability in terms of the effects of the influence of the positively selected regions in each molecule. Variability was the result of selection, but need not detract from the overall constancy of the molecule (Ohta & Kimura, 1971, p.23). The problem of variability of rates across lineages was not so easily resolved, however. In 1974, Walter Fitch and Charles Langley produced a new statistical analysis that demonstrated even greater variability (Langley & Fitch, 1974).

Additional evidence of slowdowns and speedups from various lineages produced by Morris Goodman and others reinforced doubts about the clock's constancy (Goodman, Moore, & Matsuda, 1975). Kimura responded by emphasizing the constancy of the intrinsic rate of each type of molecule. Emphasizing "local fluctuations" was, in his mind, "a classic case of 'not seeing the forest for the trees' " (Kimura, 1983). Selectionists did not share Kimura's vision. Indeed, growing evidence of rate variability fueled selectionist criticisms.

In 1984, John Gillespie proposed an episodic molecular clock with a selectionist mechanism that explained both the constancy and variability evident in the patterns of substitution (Gillespie, 1984, 1991). Neutralists, such as Naoyuki Takahata and Tomoko Ohta, revised their models of the molecular clock in order to explain both the observed constancy and variability (Takahata, 1987). At the same time, Francisco Ayala used sequence comparisons for molecules such as superoxide dismutase (SOD) to demonstrate that the clock was erratic and unreliable (Ayala, 1986). The variability of rates across genera and families continued to render other molecules useless as clocks and reinforced Ayala's calls for skepticism of the clock as evidence in support of neutrality (Ayala, 1997, 1999, 2000).

3. The Neutral Null Model

The availability of DNA sequence data in the mid-1980s transformed the neutralist–selectionist controversy. While electrophoresis allowed evolutionary biologists access to variability at the molecular level, its resolution was limited. DNA sequencing promised direct access to genetic variability. Indeed, as DNA sequences became available, new tests of neutrality and selection made it possible to distinguish drift from selection.

DNA sequencing was introduced into evolutionary genetics by Martin Kreitman in 1983 (Kreitman, 1983). As Richard Lewontin's graduate student at Harvard, Kreitman used the sequencing techniques he learned in Walter Gilbert's laboratory to analyze the sequences of alcohol dehydrogenase (ADH) genes in *Drosophila melanogaster*. ADH had a well-known polymorphism for fast- and slow-moving electrophoretic variants. Kreitman's investigation of the DNA sequences of the fast/slow ADH polymorphism revealed many differences between the DNA sequences of eleven different alleles, but only one DNA difference that corresponded to an amino acid difference. This non-synonymous DNA substitution was at the site of the fast–slow protein polymorphism. The striking difference between synonymous changes (which cause no change in amino acid sequence) and non-synonymous changes (which cause a change in amino acid sequence) led Kreitman and his collaborators to devise new statistical tests for selection.

Kimura, King, and Jukes had proposed that synonymous changes, which occur mainly in the third position in the triplet of DNA bases (a codon) that code for an amino acid, should be neutral because they do not lead to changes in amino acid composition. If synonymous changes are neutral, they should evolve at a higher rate than amino acid changes that are more likely to be subject to negative selection (assuming that most amino acid changes would be deleterious). The rate of synonymous changes should only be surpassed if positive selection is accelerating the substitution process by driving nucleotide changes to fixation at a higher rate. Using the rate of synonymous substitutions as a measure of the neutral rate of change, Kimura proposed that comparisons of synonymous and non-synonymous rates could provide a test for positive selection (Kimura, 1983). Kreitman extended Kimura's idea of comparing synonymous and non-synonymous substitutions by contrasting changes within and between species. The resulting McDonald–Kreitman test compares the ratio of non-synonymous to synonymous changes within a species and between two species. If the sequences are neutral, the ratios should remain the same. If there is positive selection, then non-synonymous changes should have accumulated over time, so there would be more non-synonymous changes between species than within a species. The McDonald–Kreitman test and many other statistical tests that followed allow evolutionary biologists to detect balancing selection, adaptive protein evolution, and population subdivision (McDonald & Kreitman, 1991; Kreitman, 2000). Where earlier statistical tests using electrophoretic data had been stalled by low power, these comparisons using DNA sequence data succeeded in distinguishing the effects of drift and selection.

The success of tests of selection did not tip the balance of the neutralist–selectionist controversy in favor of the selectionists. Instead, it supported an important shift in attitude toward the neutral theory that cast it as the methodological starting place for molecular evolutionary analysis. The neutral theory emerged in a climate of panselectionism – most evolutionary biologists understood natural selection to be the most important factor in biological evolution and as a result assumed that searching for selection and its effects was the method of choice (see Kimura, 1983). Indeed in response to Stephen Jay Gould and Richard Lewontin's famous attack on panselectionism in their "The Spandrals of San Marco and the Panglossian Paradigm," Ernst Mayr argued that biologists should give selectionist explanations priority, because random drift could not be demonstrated (Mayr, 1983). Mayr's confidence in selection was the result of earlier efforts that reinterpreted supposed cases of random drift governing the fate of morphological traits as actually the result of natural selection. As a result, for Mayr and many others, drift became equated with an admission of ignorance of how selection was in fact operating (Beatty, 1987b). Indeed part of the initial hostility toward the neutral theory undoubtedly was a result of its equation with these earlier, discredited attempts (Provine, 1990). Ernst Mayr would have been hard pressed to hold such a stringent denial of drift only a few years later as statistical tests using DNA data became accepted tools in molecular evolution. By the late 1980s, both proponents and critics of the neutral theory recognized that neutrality, not selection, was a useful starting hypothesis when analyzing DNA sequences (Kreitman, 2000; Beatty, 1987b).

The methodological shift toward neutrality as a starting assumption is frequently expressed by referring to the neutral theory as a null hypothesis. In Roger Selander's words, "All our work begins with tests of the null hypothesis that variation in allele frequencies generated by random drift is the primary cause of molecular evolutionary change" (Selander, 1985, p.87). Selander notes that beginning with a neutral null hypothesis does not exclude selection as a possibility or predispose him toward neutrality. He starts with neutrality because he prefers "to begin with the simplest model" because it allows him to determine "a baseline for further analysis and interpretation" (Selander, 1985, p.88).

However, not every drift hypothesis has the form of a standard null hypothesis. If the standard null hypothesis proposes that there is no difference between two populations, there may be many cases where hypotheses of drift do not conform to a claim of no difference (Beatty, 1987b). That said, predictions generated by the statistical tests of selection and neutrality using DNA data do resemble no-difference null hypotheses. The methodological shift toward neutrality, however, involves more than its usefulness as a null hypothesis in statistical testing. In his review of methods to detect selection, Kreitman argues that "Kimura's theory of neutrally evolving mutations is the backbone for evolutionary analysis of DNA sequence variation and change" because a "substantial fraction" of the genome is best modeled as selectively neutral, because selective neutrality is a "useful null hypothesis," and because "statistical analysis of (potentially) neutral variation in a gene (or other region of the genome) can be informative about selection acting at linked sites" (Kreitman, 2000, pp.541–2). Kreitman's view accepts both that there is a substantial amount of neutral variation and that the neutral theory is essential for detecting selection at the DNA level.

The acceptance of neutrality as a starting place for molecular evolutionary research might be viewed as an important weakening of panselectionism in evolutionary biology. However, the impact of neutralism can be lessened if the rise of molecular evolution is interpreted as a diversification of the levels of biological phenomena. In other words, molecular techniques introduced information about a new level of biological organization: the molecular level where drift plays a significant role. On this view, panselectionism could be alive and well when it comes to morphological traits, but a non-starter when DNA sequence evolution is considered. Molecular evolutionists helped create the divide between the molecular and morphological levels as a way of culling out space where their research could develop independently of the selectionist agenda of the architects of the neo-Darwinian synthesis (Dietrich, 1998; Aronson, 2002; Hagen, 1999). The same molecular evolutionists also sought to find ways to integrate molecular and morphological evolution. Allan Wilson, for instance, proposed that the constant rate of change at the molecular level and the erratic rate of change at the morphological level might be explained by mutations in regulatory genes that produce relatively large phenotypic changes from small molecular changes (Wilson, Maxson, & Sarich, 1974). In a similar fashion, Tomoko Ohta has turned to evolutionary developmental interpretations of heat shock proteins, like Hsp90, to explain how the accumulation of neutral or nearly neutral changes could act as a capacitor for future morphological evolution (Ohta, 2002, 2003). As more integrative explanations link molecular and morphological evolution, morphological panselectionism will continue to weaken, although it will probably never undergo the kind of shift that grants neutralism primacy.

4. Controversy in Molecular Evolution

Controversies are a prominent feature of the history of evolutionary genetics and molecular evolution (Dietrich, 2006). While controversies are by definition disputes extended in time, they need not be disagreements between alternate positions such that resolution would be equated with the triumph of one position over the other. Indeed, controversies in molecular evolution, like those in evolutionary genetics, are "relative significance" disputes (Beatty, 1997). Within its proposed domain of application, the relative significance of a theory is "roughly the proportion of phenomena within its intended domain that the theory correctly describes" (Beatty, 1997, p.S432). For instance, in the neutralist–selectionist controversy the dispute concerns the relative significance of both selection and drift. Selectionists advocate a strong role for selection, but do not deny the possibility of drift at the molecular level. Neutralists acknowledge an important role for selection, but argue that most detected molecular differences are neutral. In part the dispute is over the proportion of the domain of molecular evolution explained by selection or drift.

Where a binary controversy may proceed through the accumulation of evidence in favor of one position over another or conversely the accumulation of a greater number of anomalies by one position when compared to its rival, relative significance controversies tend to have a different dynamic and pattern of resolution. Controversies such as the classical–balance controversy in evolutionary genetics or the neutralist–selectionist controversy in molecular evolution rapidly polarized into extreme positions early in both disputes. Over time, however, these disputes depolarized, meaning that most of the biologists engaged in the controversy moved from advocating large differences in relative significance to smaller differences or a range of differences. For instance, in 1968, Kimura advocated that most detected molecular differences were neutral (Kimura, 1968), while, in 1973, Christopher Wills asserted that "virtually any change in amino acid composition of any protein molecule produces a molecule of slightly different properties and therefore of slightly different selective value from the original" (Wills, 1973, p.23). By contrast, DNA sequencing and successful statistical testing depolarized the dispute by admitting significant roles for both neutrality and selection, while providing a means to empirically detect selection on a case-by-case basis. Depolarized controversies, such as the neutralist–selectionist controversy today, are not closed or settled. Instead they are characterized by a kind of pluralism – both selection and drift are accepted as probable influences on the evolution of a molecule. As a result, the need to declare a winner in the controversy is fading in the face of explanatory diversification.

Acknowledgment

John Beatty provided helpful guidance and discussion for which I am very grateful.

References

Aronson, J. (2002). Molecules and monkeys: George Gaylord Simpson and the challenge of molecular evolution. *History and Philosophy of the Life Sciences*, 24, 441–65.

Ayala, F. (1986). On the virtues and pitfalls of the molecular evolutionary clock. *The Journal of Heredity*, 77, 226–35.

Ayala, F. (1997). Vagaries of the molecular clock. *Proceedings of the National Academy of Sciences*, 94, 7776–83.

Ayala, F. (1999). Molecular clock mirages. *Bioessays*, 21, 71–5.

Ayala, F. (2000). Neutralism and selectionism: the molecular clock. *Gene*, 261, 27–33.

Ayala, F., Tracey, M., Barr, L., McDonald, J., & Perez-Salas, S. (1974). Genetic variation in natural populations of five Drosophila species and the hypothesis of the selective neutrality of protein polymorphisms. *Genetics*, 7, 343–84.

Beatty, J. (1987a). Weighing the risks: stalemate in the classical/balance controversy. *Journal of the History of Biology*, 20, 289–319.

Beatty, J. (1987b). Natural selection and the null hypothesis. In J. Dupré (Ed.). *The latest on the best* (pp. 53–75). Cambridge, MA: MIT Press.

Beatty, J. (1997). Why do biologists argue like they do? *Philosophy of Science*, S64: 231–42.

Clarke, B. (1970). Darwinian evolution of proteins. *Science*, 168, 1009–11.

Crow, J. F. (1969). Molecular genetics and population genetics. *Proceedings of the Twelfth International Congress of Genetics*, 3, 105–13.

Crow, J. F. (1972). Darwinian and non-Darwinian evolution. In L. LeCam et al. (Eds). *Proceedings of the sixth Berkeley symposium on mathematical statistics* (Vol. V): *Darwinian, neo-Darwinian, and non-Darwinian evolution* (pp. 1–22). Berkeley: University of California Press.

Dietrich, M. R. (1998). Paradox and persuasion: negotiating the place of molecular evolution within evolutionary biology. *Journal of the History of Biology*, 31, 85–111.

Dietrich, M. R. (2006). From Mendel to molecules: a brief history of evolutionary genetics. In C. W. Fox & J. B. Wolf, (Eds). *Evolutionary genetics: concepts and case studies* (pp. 3–13). New York: Oxford University Press.

Dobzhanksy, Th. (1955). A review of some fundamental concepts and problems in population genetics. *Cold Spring Harbor Symposium on Quantitative Biology*, 20, 1–15.

Ewens, W. (1972). The sampling theory of selectively neutral alleles. *Theoretical Population Biology*, 3, 87–112.

Gillespie, J. (1984). The molecular clock may be an episodic clock. *Proceedings of the National Academy of Sciences*, 81, 8009–13.

Gillespie, J. (1991). *The causes of molecular evolution*. New York: Oxford University Press.

Goodman, M., Moore, G., & Matsuda, G. (1975). Darwinian evolution in the genealogy of hemoglobin. *Nature*, 253, 603–8.

Hagen, J. (1999). Naturalists, molecular biologists, and the challenges of molecular evolution. *Journal of the History of Biology*, 32, 321–41.

Hubby, J. L., & Lewontin, R. C. (1966). A molecular approach to the study of genic heterozygosity in natural populations. I. The number of alleles at different loci in *Drosophila pseudoobscura*. *Genetics*, 54, 546–95.

Kimura, M. (1968). Evolutionary rate at the molecular level. *Nature*, 217, 624–6.

Kimura, M. (1969). The rate of molecular evolution considered from the standpoint of population genetics. *Proceedings of the National Academy of Sciences*, 63, 1181–8.

Kimura, M. (1970). The length of time required for a selectively neutral mutant to reach fixation through random frequency drift in a finite population. *Genetical Research*, 15, 1131–3.

Kimura, M. (1983). *The neutral theory of molecular evolution*. Cambridge: Cambridge University Press.

Kimura, M., & Ohta, T. (1971). Protein polymorphism as a phase in molecular evolution. *Nature*, 229, 467–9.

King, J. (1976). Progress in the neutral mutation random drift controversy. *Federation Proceedings*, 35, 2087–91.

King, J., & Jukes, T. (1969). Non-Darwinian evolution. *Science*, 164, 788–98.

Kreitman, M. (1983). Nucleotide polymorphism at the alcohol dehydrogenase locus of *Drosophila melanogaster*. *Nature*, 304, 412–17.

Kreitman, M. (2000). Methods to detect selection in populations with application to the human. *Annual Review of Genomics and Human Genetics*, 1, 539–59.

Langley, C., & Fitch, W. (1974). An examination of the constancy of the rate of molecular evolution. *Journal of Molecular Evolution*, 3, 161–77.

Lewontin, R. C. (1991). Twenty-five years ago in *Genetics*: Electrophoresis in the development of evolutionary genetics: milestone or millstone? *Genetics*, 128, 657–62.

Lewontin, R. C., & Hubby, J. L. (1966). A molecular approach to the study of genic heterozygosity in natural populations. II. Amount of variation and degree of heterozygosity in natural populations of *Drosophila pseudoobscura*. *Genetics*, 54, 595–609.

Mayr, E. (1983). How to carry out the adaptationist program, *American Naturalist*, 121, 324–34.

McDonald, J. H., & Kreitman, M. (1991). Adaptive protein evolution at the Adh locus in Drosophila. *Nature*, 351, 652–4.

Morgan, G. (1998). Emile Zuckerkandl, Linus Pauling, and the molecular evolutionary clock, 1959–1965. *Journal of the History of Biology*, 31, 155–78.

Nei, M. (2005). Selectionism and neutralism at the molecular level. *Molecular Biology and Evolution*, 22, 2318–42.

Ohta, T. (1973). Slightly deleterious mutant substitutions in evolution. *Nature*, 246, 96–8.

Ohta, T. (1992). The nearly neutral theory of molecular evolution. *Annual Review of Ecology and Systematics*, 23, 263–86.

Ohta, T. (2002). Near-neutrality in evolution in genes and gene regulation. *PNAS*, 99, 16134–7.

Ohta, T. (2003). Evolution by gene duplication revisited: Differentiation of regulatory elements versus proteins. *Genetica*, 118, 209–16.

Ohta, T., & Gillespie, J. (1996). Development of neutral and nearly neutral theories. *Theoretical Population Biology*, 49, 128–42.

Ohta, T., & Kimura, M. (1971). On the constancy of the evolutionary rate of cistrons. *Journal of Molecular Evolution*, 1, 18–25.

Provine, W. (1990). The neutral theory of molecular evolution in historical perspective. In N. Takahata & J. Crow (Eds). *Population biology of genes and molecules* (pp. 17–31). Tokyo: Baifukan.

Richmond, R. (1970). Non-Darwinian evolution: a critique. *Nature*, 225, 1025–8.

Selander, R. K. (1985). Protein polymorphism and the genetic structure of natural populations of bacteria. In T. Ohta & K. Aoki (Eds). *Population genetics and molecular evolution* (pp. 85–106). Tokyo: Japan Scientific Societies.

Stebbins, G. L., & Lewontin, R. C. (1972). Comparative evolution at the level of molecules, organisms, and populations. In L. LeCam et al. (Eds). *Proceedings of the sixth Berkeley symposium on mathematical statistics* (Vol. V): *Darwinian, neo-Darwinian, and non-Darwinian evolution* (pp. 23–42). Berkeley: University of California Press.

Suarez, E., & Barahona, A. (1996). The experimental roots of the neutral theory of molecular evolution. *History and Philosophy of the Life Sciences*, 18, 55–81.

Takahata, N. (1987). On the overdispersed molecular clock. *Genetics*, 116, 169–79.

Watterson, G. (1977). Heterosis or neutrality? *Genetics*, 85, 789–814.

Wills, C. (1973). In defense of naïve panselectionism. *American Naturalist*, 107, 23–34.

Wilson, A., Maxson, L., & Sarich, V. (1974). Two types of molecular evolution: evidence from studies of interspecific hybridization. *Proceedings of the National Academy of Sciences*, 71, 2834–47.

Zuckerkandl, E., & Pauling, L. (1965). Evolutionary divergence and convergence in proteins. In V. Bryson& H. Vogel (Eds). *Evolving genes and proteins* (pp. 97–166). New York: Academic Press.

Further Reading

Dietrich, M. R. (1994). The origins of the neutral theory of molecular evolution. *Journal of the History of Biology*, 27, 21–59.

Chapter 10

Speciation and Macroevolution

ANYA PLUTYNSKI

1. Introduction

Speciation is the process by which one or more species[1] arises from a common ancestor, and "macroevolution" refers to patterns and processes at and above the species level – or, transitions in higher taxa, such as new families, phyla, or genera. "Macroevolution" is contrasted with "microevolution,"[2] evolutionary change within populations, due to migration, selection, mutation, and drift. During the 1930s and 40s, Haldane (1932), Dobzhansky (1937), Mayr (1942), and Simpson (1944) argued that the origin of species and higher taxa were, given the right environmental conditions and sufficient time, the product of the same microevolutionary factors causing change within populations. Dobzhansky reviewed the evidence from genetics, and argued, "nothing in the known macroevolutionary phenomena would require other than the known genetic principles for causal explanation" (Dobzhansky, 1951, p.17). In sum, genetic variation between species was not different in kind from the genetic variation within species. Dobzhansky concluded that one may "reluctantly put an equal sign" between micro- and macroevolution. This view was not accepted by all, however. Richard Goldschmidt, for instance, argued that microevolution does not, by the sheer accumulation of small, adaptive changes, lead to novel species. In his words, "the facts of microevolution do not suffice for macroevolution" (Goldschmidt, 1940, p.8).

Goldschmidt's position was regarded by many during the synthesis as implausible. However, similar arguments, questioning the sufficiency of microevolutionary processes for macroevolutionary change, were offered up at different stages subsequent to the 1940s. In this same vein but based on very different arguments, Gould and Eldredge (1977) argued that there are causal processes operating at and above the species level which are not reducible to, or explainable in terms of, change within populations. They claim that patterns of extinction or survival through periods of mass extinction might

1 For further discussion of species concepts, see Ereshefsky, this volume.

2 The terms were coined in 1927 by the Russian entomologist Iuri'i Filipchenko in *Variabilität und Variation* (according to Bowler, 1983).

involve species or clade selection. [SEE THE UNITS AND LEVELS OF SELECTION]. In other words, there are features of species, or perhaps higher clades, that render them more or less likely either to go extinct, or to survive and diversify. Species-level traits that have been suggested are broad geographic range, or broad habitat tolerance. At the level of whole clades, certain body types or developmental features may render clades more likely to diversify. Clade or species selectionists argue that such traits are properties of whole taxa, not reducible to properties of individual members. Needless to say, questions about what counts as "individual" or "species" or perhaps "clade"-level traits complicates the question of whether and how frequently species selection drives macroevolutionary change.

Opponents of species and higher clade selection argue that explaining change at higher taxonomic levels does not require appeal to higher-level processes. In other words, radical revision of the theoretical framework defended by the founders of the early synthesis (Fisher, 1930; Wright, 1931; Haldane, 1932) is not necessary; selection, drift, etc., on individual organisms is sufficient to explain speciation, etc. This is not to say that there have not been new and important insights since 1932 that are in the process of being integrated into that theoretical framework. Comparative cellular and developmental biology has identified deep homologies in signaling pathways (Halder, Callaerts, & Gehring, 1995), which has illuminated a good deal about the constraints on body plans and their evolutionary trajectories (Gerhart & Kirschner, 1997; Raff, 1996). Just as theoretical population genetics provides an account of how evolution in populations is possible, so too, developmental biology provides an account of how characters can vary, as well as which body plans may evolve from others.

The view that evolution below and above the species level is not distinct in kind is often called "neo-Darwinism," insofar as Darwin (1859) did not view microevolution and macroevolution as distinct problems requiring distinct solutions. Darwin viewed speciation as a by-product of adaptive divergence; the diversity of life today is the product of a series of branching processes. The branching process is not qualitatively different as one ascends the Linnaean hierarchy. [SEE DARWINISM AND NEO-DARWINISM].

The structure of this essay will be as follows. First, there will be a review of some of the key episodes in the history of speciation research, focusing on one controversy: the debate over founder effect. The last half of this essay will review the literature on evolutionary rates, and then turn to definitions of and explanations for disparity, continuity, and stasis in the fossil record. These will illustrate some of the central epistemological issues that arise in the context of research into speciation and macroevolution. For discussion of the metaphysics of species, see [SYSTEMATICS AND TAXONOMY].

2. Speciation: Studying and Classifying Modes of Speciation

Speciation occurs (for the most part) in "geological time," or time scales that span many scientists' lifetimes. One can rarely observe speciation "in action" (excepting perhaps polypoloid speciation in plants (Soltis & Soltis, 1999)). The waiting time for speciation[3] ranges from 100,000 years (in Malawi Cichlids) to hundreds of millions of years (300 million in the crustaceans of the order Notostraca, and 120 million in the Ginko) (Coyne & Orr, 2004). Unlike studies of change within interbreeding populations, genetic analysis of reproductive isolation is difficult. Lewontin (1974) called the problem of studying the genetics of speciation a "methodological contradiction" at the heart of speciation research, insofar as one by definition cannot do genetics between species, or interbreed members of reproductively isolated groups.

Deciding among competing hypotheses about patterns and processes of speciation involves assessing a variety of indirect evidence, and thus, there has been a great deal of dispute about the major mechanisms involved in speciation. In particular, one dispute concerns the relative significance of selection versus drift in speciation.[4] This debate has been just as heated in the biological literature as parallel debates about change within populations. Not coincidentally, some of the same authors are involved in both disputes (e.g., Charlesworth, Lande, & Slatkin, 1982). In a review of both theoretical and empirical work on speciation, Coyne and Orr (2004) argue that selection is the major mechanism in most cases of speciation. More precisely, indirect selection, or reproductive isolation evolving as a pleiotropic side effect, or byproduct of selection on other characters, is the major mechanism of speciation. They and others (Turelli, Barton, & Coyne, 2001) argue that the evidence suggests that drift plays a relatively minor role in speciation; however, the debate is not over, as new models of speciation and empirical case studies are being developed all the time (Gavrilets, 2004).

The standard way to classify modes of speciation is with respect to biogeography. That is, whether reproductive isolation arose with or without geographic isolation determines the major categories of speciation. For instance, "allopatric" speciation refers to speciation following geographical isolation, "parapatric" speciation occurs with semi-isolation, and speciation in "sympatry" occurs within the ancestral population, or with the possibility of gene flow. The choice of categorizing modes of speciation with respect to biogeographic factors is a matter of historical accident; one might better categorize speciation by its genetic basis or by the evolutionary forces producing reproductive isolation (Kirkpatrick & Ravigne, 2002). The question of whether the first stage of speciation requires geographic isolation emerged in the nineteenth century, and remains contentious today (Berlocher, 1998). The extent to which the role of biogeog-

---

3 There are several different measures of speciation rates, each with advantages and limitations (for a review, see: Coyne & Orr, 2004). The BSR (*biological speciation rate*) is the average rate at which one species branches to produce two reproductively isolated groups (this averages about a million years). The BSI (*biological speciation interval*) is the mean time elapsing between the origin of a lineage and the next branching event. The NDI (net diversification interval) is the reciprocal of the NDR (net diversification rate), which is simply the change in the number of surviving lineages per unit time. (The above estimates are in NDI.)

4 See Baker, J. M. (2005). Adaptive speciation: the role of natural selection in mechanisms of geographic and non-geographic speciation. *Studies in History and Philosophy of Biological and Biomedical Sciences*, 36, 303–26.

raphy has historically served as a polarizing factor in speciation research explains, but does not justify, this emphasis in categorizing modes of speciation.

In the vicariant or "dumbbell" allopatric model, two large subpopulations are subdivided by some external cause – a geographical barrier like a mountain range, river, island, or glacier. After the subpopulations have remained isolated for sufficiently long, drift or adaptation to local environmental conditions results in reproductive incompatibility. When the two incipient species come into secondary contact, they cannot mate, or, if mating is still possible, the hybrids are inferior. Further evolution of premating or postmating isolation eventuates in two discrete species.[5] Theoretical modeling of this process demonstrates that geographic isolation can lead to complete reproductive isolation, given sufficient time, and strong enough selection (Orr, 1995; Orr & Orr, 1996; Orr & Turelli, 2001). Drift may lead to speciation in such cases, but theory indicates that drift alone is much less effective than selection or a combination of drift and selection. There is a great deal of laboratory evidence in favor of reproductive isolation evolving as a pleiotropic byproduct of selection on other factors (reviewed in Rice & Hostert, 1993). A vast number of instances of concordance of species borders with existing geographic or climatic barriers suggest that vicariant speciation is common (reviewed in Coyne & Orr, 2004).

Another form of speciation in allopatry, "peripatric" speciation, involves the isolation of a small founder population. A "founder event" is when one or a few individuals colonize a distant habitat, such as an island or a lake. The "founder effect" is a form of genetic drift induced by population size restriction. This is said to cause speciation during a founder event. Mayr (1942, 1954, 1963) placed special emphasis on the role of founder effect, or population bottlenecks and drift during founder events. He argued that loss of heterozygosity via genetic drift in the founder population would cause a change in the genetic background of the species, and thus a change in the net fitness of genotypes under selection. This would lead to what Mayr called a "genetic revolution" – or, a radical shift in the genetic constitution of the species. While founder events followed by adaptive radiations are ubiquitous in nature, the evidence that founder effect is a major mode of speciation is slim. A variety of special conditions need to be met for this kind of speciation to go forward. There are very few plausible cases of peripatric speciation via founder effect in the wild (Coyne & Orr, 2004); this will be discussed in greater detail below.

Speciation in sympatry is speciation within the "cruising range" of the ancestral species. There has been a resurgence of interest in speciation in sympatry (Via, 2001). Stickleback, cichlid fishes, and the apple maggot fly, *Rhagoletis pomonella*, show evidence of speciation in sympatry, though these cases are contentious (Schleiwen, Tautz, & Paabo, 1994; Albertson et al., 1999; Bush, 1994). In these cases, behavioral isolation may be followed by reproductive isolation, a byproduct on selection for genes with pleiotropic effects associated with host or niche specialization (Bush, 1994; Schilthuizen,

5 Of course, in nonsexual or uniparental populations, populations may become genetically distinctive and diverge due to isolation, mutation, selection, and drift, but not due to reduced gene flow. This chapter will deal exclusively with speciation in sexual organisms. Unfortunately, however, despite the fact that most of the diversity of living fauna is microbial, the literature on speciation deals almost exclusively with sexual species.

2001). Maynard Smith (1966), Kondrashov (1983a, 1983b, 1986; Kondrashov & Kondrashov, 1999) and others have developed a number of models of speciation in sympatry, due to habitat shift or behavioral isolation. The conditions necessary for speciation in sympatry to go forward are rather restrictive. Disruptive selection needs to be fairly intense in order to overcome interbreeding. Most models include a number of loci which influence reproduction, and which are strongly linked, at least one of which is subject to disruptive selection.

Recent theoretical work (Kirkpatrick, 1982, 1987, 2000; Kirkpatrick & Ryan, 1991; Pomiankowski & Iwasa, 1998) suggests that sexual selection can play a significant role in speciation. For instance, a flashy trait in males and the preference for it in females will become associated (the alleles for the male's flashy trait and the alleles for females choosing this trait come into linkage disequilibrium). This is most likely to occur where there is avid competition for mates, as in polygamous species. Indeed, it has been found that ornamented polygamous species are twice as speciose as plain, monogamous species (Moller & Cuervo, 1998).

The consensus developed during the evolutionary synthesis led many to assume that geographical isolation was required for speciation, because simple disruptive selection could not possibly be enough to overpower the effects of interbreeding. However, new work suggests that intraspecific variation, such as plasticity, or variation governed by developmental switches, might lead to incipient speciation and eventual divergence either in allopatry or sympatry (West-Eberhard, 2005). West-Eberhard has defended what she calls the "developmental plasticity hypothesis of speciation," according to which intraspecific differences in the form of alternative phenotypes can contribute to the evolution of reproductive isolation. For instance, dimorphisms, such as mites with normal versus "phoretic" reduced segment body types, might become fixed, due to either selection or chance, and lead (for instance, via sexual selection) to reproductive isolation. West-Eberhard calls this process "phenotypic fixation." This new synthesis of evo-devo and micro-macro is a potentially promising avenue of research that is only now being explored (see also, Kirschner & Gerhard, 2005).

2.1. Founder effect

"Founder effect" is the means by which, following a founder event, novel allele combinations are generated. That is, a "genetic bottleneck" leads to radical changes in the genetics of a population, so that new gene combinations would be exposed to selection. Mayr (1954) argued that species possess "genetic homeostasis" and "unity of the genotype," so that, without geographical isolation or genetic bottlenecks, it would be difficult if not impossible for new adaptive combinations of genes to come about. There was an "evolutionary inertia" in large populations that required either geographical isolation or population bottlenecks for what Mayr called a "genetic revolution" to be possible – i.e., the generation of a novel "homeostatic gene complex."[6] Mayr's arguments had a lasting influence in the evolutionary literature; the question of whether founder effect occurred, and how, became a major problem in much of the literature on speciation from the

6 See Provine (1989) for a review and discussion.

1960s until the 1980s. If one could understand how peak shifting (or the shift to a novel adaptive gene combination) via drift was possible, one could understand how founder effect worked at a genetic level.

Carson and Templeton (1984) built on Mayr's work and argued that while Mayr's notion of the genetic revolution was vague, they could supplement it with a robust mechanistic explanation, "founder flush." Small populations of founders, or single individuals, would occasionally "flush" or increase dramatically in size, due to one of several proposed mechanisms. Carson thought that relaxed selection due to decreased competition among members of the founder population would lead the population to expand in size. Templeton (1981) argued that the effects of founder events might lead to novel selection pressures for some alleles on otherwise homogeneous genetic backgrounds. This could trigger changes at other loci, with effects cascading through the "epistatic genetic system," eventually leading to reproductive isolation (Templeton, 1980, p.1015). He called this "transilience."

Theoretical work has demonstrated the implausibility of speciation via peak shifting via drift (Lande, 1985; Barton & Charlesworth, 1984; Barton, 1989; Gavrilets, 2004), and empirical work in both lab and field has demonstrated that speciation via founder flush is implausible (Coyne & Orr, 2004), though, of course, this is controversial. First, Barton (1989) argued that even for very small populations, with relatively shallow valleys, the chance of a peak shift is very small. This is because the chance of such a shift occurring decreases with population size and depth of valley, but the waiting time to a peak shift grows exponentially with the product of the population size and the depth of valley. In other words, the conditions for peak shifting via drift are very restrictive. They summarize:

> Perhaps the most important objection to peak shift models is that the chances of such shifts are small and, even if they do occur, they yield only trivial reproductive isolation . . . the probability of a peak shift is proportional to the size of population and depth of valley . . . the deeper the valley, the smaller the chances of a peak shift . . . [and] the less gene flow there is. The lesson is clear, while deeper valleys yield greater reproductive isolation, they are less likely to be crossed. (Coyne & Orr, 2004, p.395)

In other words, the population genetic scenario that Mayr envisioned is implausible. Small populations are more likely to go extinct than to drift into the vicinity of nearby adaptive peaks.

Classic empirical examples of founder flush have been challenged with molecular data. For example, Templeton's cases of island Hawaiian *Drosophila* may be just as genetically variable as mainland species (Bishop & Hunt, 1988), indicating that their rapid radiation may not be due to founder flush, as was previously supposed (see also, Coyne & Orr, 2004, pp.402–3). Instead, there is also evidence that their radiation was a product of divergence under sexual selection (Kambysellis et al., 1995). Further, analysis of molecular variation in Darwin's finches (*Geospiza*) on the Galapagos suggests that the most recent common ancestor of the group is about 15 million years old, far too long ago for the single founder model to be plausible (Vincek et al., 1997). Further, a multi-generation experiment of fifty populations, over 14 generations, that attempted to reproduce bottleneck effects in *Drosophila* (Moores, Rundle, & Whitlock,

1999) was unsuccessful at generating reproductive isolation, though some contest that experimental work such as this is not decisive (Carson, 2003). In short, both theoretical and empirical work on founder-flush models seems to show that this particular mode of speciation is far less significant than other modes of speciation.

The debates over founder effect illustrate a variety of epistemological issues that arise in speciation research. One such question is how hypotheses about speciation can be subject to test, and whether and when such tests are decisive. Support for different views comes from three kinds of considerations: arguments drawing upon theoretical models, experimental studies of speciation in the laboratory, and natural history, or biogeographical and ecological studies of species distribution in nature. There are limits to the value of each of these sources of evidence. First, theoretical models, while they may demonstrate that a proposed mechanism of speciation depends upon more or less restrictive conditions, also necessarily oversimplify a process that involves a complex of factors. Second, experimental work on speciation can focus on only one aspect of the evolution of reproductive isolation at a time, isolating other factors of potential relevance. Finally, biogeography may only occasionally serve to rule some processes out. Much of the speciation literature is taken up with plausibility arguments and relative significance debates. One way to test hypotheses is to examine the background assumptions, e.g., of theoretical models. While much of the evidence is lost in the distant geological past, the molecular revolution has transformed this area of evolutionary biology, as it has other areas. Molecular data, for instance, has proven decisive in some debates about founder effect. It appears that the best evidence to date, both theoretical and empirical, suggests that founder flush, and more generally, peak shifting via drift, is unlikely as a mechanism of speciation.

3. Rates of Evolution and Punctuated Equilibrium

In 1972, Eldredge and Gould published a controversial paper, defending the theory of punctuated equilibrium. They made two central claims; first, that the fossil record showed periods of rapid change, or "punctuation" followed by relative stasis, and second, the process by which this takes place is not gradual transformation within ancestral populations, but rapid speciation in small, peripherally isolated populations. They claimed that this pattern challenged the neo-Darwinian consensus on the major mechanisms of evolutionary change, which they deemed "phyletic gradualism."

Gould and Eldredge observed that a common pattern in the fossil record was for a species to appear relatively suddenly, persist for a period without a great deal of morphological change, and then go extinct. This might be explained (1) by appeal to the peripheral isolate model of speciation, (2) by constraints on the possible trajectory of different body plans, or (3) by what they later (1977) called "species selection." First, if species arise in small isolated populations, at the periphery of the main breeding group, then the fossil record will most likely not reveal the speciation event. Since small populations are not likely to leave fossils, transitional forms would not be recorded. After a peripheral isolate population speciated, it would reinvade the ancestral population, outcompeting its ancestor, and thus leave a record of a sudden appearance of a new type. Second, they argued that developmental or genetic constraints could explain the patterns of relative stasis. And third, they claimed that this whole process involves a higher-level sorting process; entire species are selected as units having their own group fitness. Differential diversification, they thought, could not be explained by mere population genetic change within species. Rather, there was selection at the species level for whatever trait (e.g., large home range) would lend itself to higher rates of diversification.

Eldredge and Gould generated controversy over three questions. First, is it in fact counter to the tenets of the synthesis that there should be patterns of stasis and relatively abrupt change in the fossil record? Second, are the rates of evolution indeed as they suggest, or is there a diversity of evolutionary rates? Third, does the pattern they describe necessarily rule out explanation in terms of ordinary population genetic mechanisms of selection, drift, etc.?

Eldredge and Gould claim that Darwin and the founders of the synthesis were "phyletic gradualists." Phyletic gradualists endorse the view that new species arise by transformation of an ancestral population into modified descendants, the transformation is even and slow, involves the entire ancestral population, and occurs over all or a large part of the ancestral species' geographic range. Moreover, the fossil record for the origin of a new species should consist of a long sequence of continuously graded, intermediate forms, and morphological breaks in postulated phyletic sequences are due to imperfections in the fossil record (Eldredge & Gould, 1972, p.89).

While it is true that many proponents of the synthesis emphasized, and perhaps overemphasized, gradualism (Mayr, 1942, 1963; Dobzhansky, 1942), it is not clear that any evolutionary biologist, living or dead, actually accepts *all* of these claims. Highly variable rates of evolution were recognized by Darwin, as well as by paleontologists both long before and during the synthesis. Darwin wrote, "the periods, during which species have undergone modification, though long as measured by years, have probably been short in comparison with the periods during which they retain the same form" (cf. Charlesworth et al., 1982, p.475). Simpson's *Tempo and Mode in Evolution* (1944), one of the central texts in the synthesis, closely examined the variety of evolutionary rates, noting that rates vary between taxa, character, and times. Haldane (1949) developed a quantitative measure of evolutionary rate within lineages, the darwin. Thus, the historical claim that proponents of the synthesis were naive phyletic gradualists is, at best, overstated, and at worst, false. It is perhaps better to view phyletic gradualism and punctuated equilibrium as extremes along a continuum. Some biologists may take punctuated change followed by stasis to occur more often than others.

Since Eldredge and Gould's 1972 article, a huge empirical literature on evolutionary rates has accumulated (for a review, see Vrba & Eldredge, 2005). Estimating rates of evolution is complicated by the fact that the fossil record is incomplete, and so does not provide (except in some rare cases) documentation of the evolution of entire families and higher taxa. The entire geological range of a species, as well as at least 100,000 years of its evolutionary history, would have to be well documented in the fossil record for one to accurately assess the pattern and rate of species change, but these conditions are rarely if ever met (Carroll, 1997). So, a test of punctuated equilibrium (the pattern hypothesis) is a difficult matter, requiring a complete stratigraphic record and careful biometrical measurements. What evidence that is available suggests a variety of different patterns, along a continuum from some cases of punctuated equilibrium, to cases of gradual change (reviewed in Levinton, 2001, and Gingerich, 1983, 1993).[7]

Gingerich (1983) did an exhaustive survey of evolutionary rates within and between lineages; he showed that rates vary over time and across taxa. For instance, gradual change is relatively common in vertebrates (about .08 darwins), though some rapidly evolving vertebrates lineages show rates as high as 10 darwins, over short periods. A darwin is the difference between the natural log of the average measures of some character (say, the height of a fossilized molar from base to crown) taken at two times, divided by the total time interval, or $r = (\ln x_2 - \ln x_1/\Delta t)$ (Haldane, 1949). These changes in the fossil record appear consistent with rates achieved in microevolutionary contexts. Indeed, experimental selection has produced rates of change orders of magnitude faster than the fossil record (Lenski & Travisano, 1994). In experimental and some field populations, biologists have been able to generate rates of evolution as high as 10,000 darwins (Papadopoulos et al., 1999). Reznick has been able to generate very rapid rates of evolution in experimental manipulations of guppy populations (Reznick et al., 1997); and Hendry has done the same with introduced populations of salmon (Hendry et al., 2000). It seems that Gould and Eldredge's claims to the effect that patterns of speciation in the fossil record are inconsistent with ordinary population genetic mechanisms of selection, mutation, migration, and drift are overstated. Maynard Smith (1983) theoretically demonstrated that appearance of punctuated change could result from the ordinary processes (mutation, migration, selection, drift, etc.) of population genetics.

Eldredge and Gould claimed that major phenotypic change, when it does occur, is often concentrated at times of speciation. Gould's favored example is that of the fossils found in the Burgess shale at the Cambrian; this appears to be an example of very rapid and unusually diverse proliferation of body types. However, contra Gould, it does not appear that this example requires exceptional speciation mechanisms. There is evidence that the Cambrian explosion was preceded by a long period of cladogenesis in which many modern phyla diversified (Fortey, Briggs, & Wils, 1996, 1997; Knoll & Carroll, 1999; Valentine, Jablonski, & Erwin, 1999). So, the "explosion" was not so explosive as some had thought; some studies date the early origins of the explosion at a much younger date of 630 mya, leaving an additional 100 million years for cladogenesis via standard modes of speciation before the radiation appears in the fossil record (Lynch, 1999; cf. Leroi, 2001).

There are several studies of punctuated fossil sequences; Cheetham's (1986) work on the Miocene to Pliocene bryozoans is a well-worn example. Cheetham shows almost

7 Simpson (1953) distinguished two kinds of evolutionary rates – taxonomic frequency rates – or, the rate at which new taxa or genera replace previous ones – and, phylogenetic (or, phyletic) rates – rates of change in single characters or complexes of characters. Phylogenetic rates are easier to measure and describe in quantitative terms than are taxonomic rates. One can either measure number of standard deviations by which the mean of a character changes per unit time, or take average measures of some character (say, the height of a fossilized molar from base to crown) at two times, and take the natural log of each. The evolutionary rate in darwins, (r), is the difference between the two divided by total time interval ($\ln x_2 - \ln x_1/\Delta t$) (Haldane, 1949). Using this measure, biologists have asked a number of descriptive questions about evolutionary rates. What is the average rate of change within lineages? Do different taxonomic groups have different rates of change?

static lineages coexisting with lineages that appear, from phylogenetic analysis, to be their descendants. Almost no intermediates were found, suggesting that new species arose relatively rapidly. While this is clearly a punctuated pattern, it is not clear that such a pattern must be explained by speciation in peripherally isolated populations, or, for that matter, that the appearance of stasis cannot be explained by standard microevolutionary processes, e.g., of stabilizing selection.

While the inception of higher taxa is frequently marked by rapid evolution of many characteristics, after which the rate of morphological evolution is much slower, the evidence for the role of founder effect in speciation is fairly slim (see above). Moreover, over long periods of time, though individual features appear to evolve very slowly, Gingerich (1983) found that there is an inverse relation between evolutionary rate and the time interval over which it is measured. That is, the shorter the time scale, the more likely one is to find evidence of rapid evolution, perhaps due to patterns of fluctuating selection. In other words, once one looks at shorter time scales, stasis turns into rapid, fluctuating change.

Does punctuated equilibrium challenge the neo-Darwinian view of evolution? As for the descriptive claim, the observation that there is a variety of rates, and that these rates vary over time, was well known to paleontologists long before Eldredge and Gould (1973). So, it is not clear that this requires a radical revision of neo-Darwinian theory. There is abundant evidence that populations can respond quickly to selection, and that this has occurred in the fossil record with or without speciation. So the claim that change at the species barrier is somehow qualitatively different from microevolutionary change, or that rapid change only occurs in speciation, is false. Moreover, there are several well-studied lineages where gradual change has occurred (Gingerich, 1986, 1987; Levinton, 2001). In sum, Eldredge and Gould's hypothesis does not seem so revolutionary after all; it is not inconsistent with the theoretical framework of evolution articulated by the founders of the synthesis.

4. Diversity and Disparity: Definition and Causes

Gould (1989) argued that while diversity of life has increased, disparity has decreased since the Cambrian. More precisely, while the total number of species in the history of life, or species richness, continues to grow, *disparity* among different lineages, or the "degree of morphological differentiation among taxa," has decreased (McNamara & McKinney, 2005). There are a variety of different definitions of disparity, more and less precise. Some refer rather vaguely to the "differences among body plans" (Carroll, Grenier, & Weatherbee, 2001), or a measure of "how fundamentally different organisms are" (Raff, 1996, p.61). There have been some attempts to make this more precise and quantitative (Eble, 2002; Zelditch et al., 2003), where the measure taken is of "distance in a state space," average spread and spacing of forms in "morphospace," where one takes relative measures of adult forms. Others have suggested measures of developmental disparity, or "ontogenetic disparity" – the extent to which organisms change over the period of ontogenesis (Eble, 2002).

However, some have argued that disparity is a vague measure (Ridley, 1990). They doubt that there is a principled way to measure degree of morphological disparity.

Choice and measure of characters, and decisions about what to compare in terms of similarity and difference, they argue, are subjective. They contend that deciding what counts as the dimensions of morphospace, and determining measures along these dimensions, such that one can compare oysters and brachiopods, is difficult if not impossible. This remains a serious challenge to those who see disparity as a fact of the history of life to be explained.

However, it seems that the discussion of how or whether disparity has decreased in the history of life has gone forward absent a univocal definition of disparity. Some have argued that certain body plans evident in the Burgess Shale, Gould's exemplary case of a proliferation of disparity, possessed "key innovations" that enabled them to diversify. Whether or not one views disparity as an objective measure, it seems clear that certain body plans were eminently successful, while others went by the wayside. What capacity do such lineages have that others lack? One of the most noted features shared by the most diverse phyla is modularity (Schlosser & Wagner, 2004). Modularly organized animals, put most simply, have parts – "integrated" or relatively "autonomous" parts – that yet function together in the system as a whole. Modularity can occur at the genetic, developmental, or organismic level, and can be a property of a process (e.g., ontogenesis) or an entity (e.g., a genetic regulatory network). Moreover, modularity comes in degrees; modular features of an organism may be more or less autonomous or "decomposable." A modular organism may have repeated, serially homologous parts, or modular genetic regulatory or developmental systems.

Some argue that modular organisms are more evolvable, where "evolvability" is defined as "the capacity to generate heritable, selectable phenotypic variation" (Kirschner & Gerhart, 1998, p.8420). Sometimes evolvability is referred to as "the space of evolutionary possibility to which [lineages] have access" (Sterelny, in press). The greater the space, the more "evolvable" a particular lineage is.

Differences in the evolutionary potential of different lineages can be traced to features that either generate or constrain the variation on which selection acts. Such features cannot simply be genetic; developmental features of the organisms in question surely play a role, as does population structure. Some organisms may have more "entrenched" mechanisms of development than others, and, in turn, are less flexible evolutionarily. Modularity in development may be an important feature enabling the evolution of novelty. Hierarchical organization of development by genetically complex switches is one example of modularity, and phenotypic plasticity may play a role in enabling organisms to evolve (West-Eberhard, 2005). Gerhart and Kirschner (1997) argue that evolvability is importantly connected to what they call "flexibility" of developmental mechanisms. More flexible mechanisms have "greater capacity to change in response to changing conditions, to accommodate change" (Ibid., p.445).

The best example of modularity is the family of Homeobox genes. Homeotic genes control differentiation of body segments; such genes were first found in Drosophila. The critical DNA-binding region of the homeotic gene is called the "Homeobox," and "Hox" genes are those genes that control the patterning of gene expression along the Anterior–Posterior (A-P) axis in development. Hox genes have been found in all animal phyla, including higher vertebrates. All phyla have multiple Hox genes, with very similar Homeobox sequences, suggesting that a gene family has replicated serially and can be traced to the common ancestor of all metazoans, more than 550 million years ago.

The significance of the Hox genes is not simply their shared ancestry, but their common regulatory functions in development. The same genes are associated with regulation of body plan development in frogs, mice, and humans. As many as 59 to 60 amino acid residues are shared across these gene complexes in different animals. Hox genes regulate axial morphology and development of body segments in these vastly different organisms. And, they most likely evolved in a "modular" fashion, by replication of these gene complexes (Carroll et al., 2001).

5. Conclusions

The above discussion reviews only a few of the many advances in the study of speciation and macroevolution in the past fifty years. However, the view defended here is that this fact should not require a new "paradigm" for evolutionary biology. Speciation and the origin of higher taxa do not require mechanisms distinct in kind from those operating at the level of populations. Microevolutionary processes, in particular indirect selection, most likely plays the major role in most speciation events. And, patterns of stasis and rapid change in the fossil record do not require an overhaul of neo-Darwinism.

Work in experimental evolution, in both the lab and field, has shown that selection can change the genetic constitution of a population extremely rapidly. Lenski et al.'s (1991) study with 12 replicate populations of *E. coli* demonstrated that evolution can go extremely fast. Recent work (Travisano et al., 1995), suggests that evolving strains can continue to adapt to novel conditions. In natural populations, Reznick et al.'s (1997) study of guppies transplanted to pools with novel predation regimes demonstrates that selection can change a population extremely quickly (evolving at rates from 3,700 to 45,000 darwins). In addition, work on sticklebacks and cichilid species flocks in African lakes (discussed above) demonstrates that competition and sexual selection can very quickly bring about rapid morphological divergence (Schulter, 1996; Albertson et al., 1999; Coyne & Orr, 2004).

Advances in developmental and molecular biology have not overturned the insights of the synthesis, but supplemented and indeed supported many of them. Nor does it appear that micro- and macro-evolution are fundamentally different kinds of process requiring different explanatory resources. Micro- and macro-evolution are continuous, both governed by the same processes, though often operating at different scales and at different levels of organization.

References

Albertson, R. C., Markert, J. A., Danley, P. D., & Kocher, T. D. (1999). Phylogeny of a rapidly evolving clade: the cichlid fishes of Lake Malawi, East Africa. *Proceedings of the National Academy of the Sciences*, 96, 5107–10.

Barton, C. (1989). Founder effect speciation. In D. Otte & J. A. Endler (Eds). *Speciation and its consequences* (pp. 229–56). Sunderland, MA: Sinauer Associates.

Barton, N. H., & Charlesworth, B. (1984). Genetic revolutions, founder effects, and speciation. *Annual Review of Ecology and Systematics*, 15, 133–64.

Berlocher, S. (1998). Origins: a brief history of research on speciation. In D. J. Howard & S. H. Berlocher (Eds). *Endless forms: species and speciation* (pp. 3–18). New York: Oxford University Press.

Bishop, J. G., & Hunt, J. A. (1988). DNA divergence in and around the Alcohol Dehydrogenase locus in five closely related species of Hawaiian Drosophila. *Molecular Biological Evolution*, 5, 415–41.

Bush, G. (1994). Sympatric speciation in animals – new wine in old bottles. *Trends in Ecology and Evolution*, 9, 285–8.

Carroll, S. B. (1997). *Pattern and process in vertebrate evolution*. Cambridge: Cambridge University Press.

Carroll, S. B., Grenier, J. K., & Weatherbee, S. D. (2001). *From DNA to diversity: molecular genetics and the evolution of animal design*. Malden, MA: Blackwell.

Carson, H. L. (2003). Mate choice theory and the mode of selection in sexual populations. *Proceedings of the National Academy of Sciences USA*, 100(11), 6584–7.

Carson H. L., & Templeton, A. R. (1984). Genetic revolutions in relation to speciation phenomena: the founding of new populations. *Annual Review of Ecology and Systematics*, 15, 97–131.

Charlesworth, B., Lande, R., & Slatkin, M. (1982). A neo-Darwinian commentary on macroevolution. *Evolution*, 36, 474–98.

Cheetham, A. H. (1986). Tempo of evolution in a Neogene bryozoan: rates of morphologic change within and across species boundaries. *Paleobiology*, 12, 190–202.

Cole, L. J. (1940). The relation of genetics to geographic distribution and speciation; speciation I. Introduction. *American Naturalist*, 74, 193–7.

Cook, O.F. (1906). Factors of species-formation. *Science*, 23, 506–7.

Coyne, J. A., & Orr, H. A. (2004). *Speciation*. Sunderland, MA: Sinauer Associates.

Darwin, C. (1859). *On the origin of species*. Facsimile of 1st edn, Harvard University Press. Images (Writings of Charles Darwin, British Library, http://pages.britishlibrary.net/charles.darwin2/diagram.jpg).

DeVries, H. (1901). *Die Mutationstheorie* (Vol. 1). Leipzig: von Veit Verlag.

Dobzhansky, Th. (1937; later edn, 1942, 1951). *Genetics and the origin of species*. New York: Columbia University Press.

Eble, G. J. (2002). Multivariate approaches to development and evolution. In N. Minugh-Purvis, & K. J. McNarmara (Eds). *Human evolution through developmental change* (pp. 51–78). Baltimore: Johns Hopkins University Press.

Eldredge, N., & Gould, S. J. (1972). Punctuated equilibria: an alternative to phyletic gradualism. In T. J. M. Schopf (Ed.). *Models in paleobiology* (pp. 82–115). San Francisco, Freeman, Cooper.

Fisher, R. A. (1930, 1958). (J. H. Bennett, Ed., 2000 valorium edition). *Genetical theory of natural selection*. Oxford: Oxford University Press.

Fortey, R. A., Briggs, D. E. G., & Wils, M. A. (1996). The Cambrian evolutionary explosion: decoupling cladogenesis from morphological disparity. *Biological Journal of the Linnean Society*, 57, 13–33.

Fortey, R. A., Briggs, D. E. G., & Wils, M. A. (1997). The Cambrian evolutionary explosion recalibrated. *Bioessays*, 19, 429–433.

Futuyama, D. (1983). Mechanisms of speciation. *Science*, 219, 1059–60.

Gavrilets, S. (2004). *Fitness landscapes and the origin of species*. Monographs in Population Biology, 41. Princeton: Princeton University Press.

Gerhart, J., & Kirschner, M. (1997). *Cells, embryos and evolution: toward a cellular and developmental understanding of phenotypic variation and evolutionary adaptability*. Oxford: Blackwell Science.

Gingerich, P. D. (1983). Rates of evolution: effects of time and temporal scaling. *Science*, 222, 159–161.

Gingerich, P. D. (1993). Quantification and comparison of evolutionary rates. *American Journal of Science*, 293A, 453–78.

Goldschmidt, R. (1940). *The material basis of evolution*. New Haven: Yale University Press.

Gould, S. J. (1989). *Wonderful life: the Burgess Shale and the nature of history*. New York: Norton.

Gould, S. J., & Eldredge, N. (1977). Punctuated equilibria: the tempo and mode of evolution reconsidered. *Paleobiology*, 3, 115–51.

Haldane, J. B. S. (1932). *The causes of evolution*. Princeton: Princeton University Press.

Haldane, J. B. S. (1949). Suggestions as to a quantitative measurement of rates of evolution. *Evolution*, 3, 51–56.

Halder, G. P., Callaerts, P., & Gehring, W. J. (1995). Induction of ectopic eyes by targeted expression of the *eyeless* gene in *Drosophila*. *Science*, 267, 1788–92.

Hendry, A. P., Wenburg, J. K., Bentzen, J., Volk, E. C., & Quinn, T. P. (2000). Rapid evolution of reproductive isolation in the wild: evidence from introduced salmon. *Science*, 290(5491), 516–18.

Howard, D. J., & Berlocher, S. H. (1998). *Endless forms: species and speciation*. New York: Oxford University Press.

Huxley, J. (1942). *Evolution: the modern synthesis*. New York: Wiley & Sons.

Kambysellis, M. P., Ho, K. F., Craddock, E. M., Piano, F., Parisi, M., & Chohen, J. (1995). Patterns of ecological shifts in the diversification of Hawaiian Drosohila inferred from a molecular phylogeny. *Current Biology*, 5, 1129–1139.

Klein, S. A. (2003). Explanatory coherence and empirical adequacy: the problem of abduction, and the justification of evolutionary models. *Biology and Philosophy*, 18, 513–27.

Kirschner, M. W., & Gerhart, J. C. (2005). *The plausibility of life: resolving Darwin's dilemma*. New Haven: Yale University Press.

Kirkpatrick, M. (1982). Sexual selection and the evolution of female mate choice. *Evolution*, 36, 1–12.

Kirkpatrick, M. (1987). Sexual selection by female choice in polygynous animals. *Annual Review of Ecology and Systematics*, 18, 43–70.

Kirkpatrick, M. (2000). Reinforcement and divergence under assortative mating. *Proceedings of the Royal Society of London B*, 267, 1649–55.

Kirkpatrick, M., & Ravigne, V. (2002). Speciation by natural and sexual selection: models and experiments. *American Naturalist*, 159, S22–35.

Kirkpatrick, M., & Ryan, M. J. (1991). The evolution of mating preferences and the paradox of the lek. *Nature*, 350, 33–8.

Kirschner, M., & Gerhart, J. (1998). Evolvability. *Proceedings of the National Academy of Sciences, USA*, 95, 8420–7.

Knoll, A. H., & Carroll, S. B. (1999). Early animal evolution: emerging views from comparative biology and geology. *Science*, 284, 2129–37.

Kondrashov, A. S. (1983a). Multilocus model of sympatric speciation. I. One character. *Theoretical Population Biology*, 24, 121–135.

Kondrashov, A. S. (1983b). Multilocus model of sympatric speciation II. Two characters. *Theoretical Population Biology*, 24, 136–144.

Kondrashov, A. S. (1986). Multilocus model of sympatric speciation III. Computer simulation. *Theoretical Population Biology*, 29, 1–15.

Kondrashov, A. S., & Kondrashov, F. A. (1999). Interactions among quantitative traits in the course of sympatric speciation. *Nature*, 400, 351–4.

Lande, R. (1985). Expected time for random genetic drift of a population between stable phenotypic states. *Proceedings of the National Academy of Sciences, USA*, 82, 7641–5.

Lenski, R. E., & Travisano, M. (1994). Dynamics of adaptation and diversification: a 10,000 generation experiment with bacterial populations. *Proceedings of the National Academy of Sciences, USA*, 91, 6808–14.

Lenski R. E., Rose, M. R., Simpson, S. C., & Tadler, S. C. (1991). Long term experimental evolution in Escherichia coli 1. Adaptation and divergence during 2000 generations. *American Naturalist*, 138, 1315–41.

Leroi, A. M. (2000). The scale independence of evolution. *Evolution and Development*, 2, 267–77.

Levinton, J. S. (2001). *Genetics, paleontology, and macroevolution* (2nd edn). Cambridge: Cambridge University Press.

Lewontin, R. (1974). *The genetic basis of evolutionary change*. New York: Columbia University Press.

Maynard-Smith, J. (1966). Sympatric speciation. *American Naturalist*, 100, 637–50.

Maynard-Smith, J. (1983). The genetics of stasis and punctuation. *Annual Review of Genetics*, 17, 11–25.

Mayr, E. (1942). *Systematics and the origin of species*. New York: Columbia University Press.

Mayr, E. (1954). Change of genetic environment and evolution. In J. Huxley & E. B. Ford (Eds). *Evolution as a process* (pp. 157–80). London: George Allen & Unwin, Ltd.

Mayr, E. (1963). *Animal, species and evolution*. Cambridge, MA: Belknap Press.

Mayr, E. (1982). *The growth of biological thought*. Cambridge, MA: Harvard University Press.

McNamara, K. J., & McKinney, M. L. (2005). Heterochrony, disparity and macroevolution. *Paleobiology*, 31(2), 17–26.

Moller, A. P., & Cuervo, J. J. (1998). Speciation and feather ornamentation in birds. *Evolution*, 52, 859–69.

Moores, A. O., Rundle, H. D., & Whitlock, M. C. (1999). The effects of selection and bottlenecks on male mating success in peripheral isolates. *The American Naturalist*, 153, 437–44.

Orr, H. A. (1995). The population genetics of speciation: the evolution of hybrid incombatibilities. *Genetics*, 139, 1805–13.

Orr, H. A. (1998). The population genetics of adaptation: the distribution of factors fixed during adaptive evolution. *Evolution*, 52, 935–49.

Orr, H. A., & Orr, L. H. (1996). Waiting for speciation: the effect of population subdivision on the time to speciation. *Evolution*, 51, 1742–9.

Orr, H. A., & Turelli, M. (2001). The evolution of postzygotic isolation: accumulating Dobzhansky–Mueller incompatibilities. *Evolution*, 55, 1085–94.

Papadopoulos, D., Schneider, D., Meier-Eiss, J., Arber, W., Lenski, R. E., & Blot, M. (1999). Genomic evolution during a 10,000-generation experiment with bacteria. *Proceedings of the National Academy of Sciences, USA*, 96, 3807–12.

Pomiankowski, A., & Iwasa, Y. (1998). Runaway ornament diversity caused by Fisherian sexual selection. *Proceedings of the National Academy of the Sciences, USA*, 95, 5106–11.

Provine, W. B. (1989). Founder effects and genetic revolutions in microevolution and speciation: a historical perspective. In L. V. Giddings, K. Kaneshiro, & W. Anderson (Eds). *Genetics, speciation, the founder principle* (pp. 43–78). New York: Oxford University Press.

Raff, R. A. (1996). *The shape of life: genes, development and the evolution of animal form*. Chicago: University of Chicago Press.

Reznick, D. N., Shaw, F. H., Rodd, F. H., & Shaw, R. G. (1997). Evaluation of the rate of evolution in natural populations of guppies. *Science*, 275, 1934–7.

Ridley, M. (1990). *Dreadful beasts. The London Review of Books*, 28 June, pp. 11–12.

Rice, W. H., & Hostert, E. E. (1993). Laboratory experiments on speciation: what have we learned in 40 years? *Evolution*, 47, 1637–53.

Schilthuizen, M. (2001). *Frogs, flies and dandelions: the making of species*. Oxford: Oxford University Press.

Schleiwen, U. K., Tautz, D., & Paabo, S. (1994). Sympatric speciation suggested by monophyly of crater lake cichlids. *Nature*, 368, 629–32.

Schlosser, G., & Wagner, G. P. (2004). *Modularity in development and evolution*. Chicago: University of Chicago Press.

Schulter, D. (1996). Ecological causes of adaptive radiation. *American Naturalist*, 148, S40–S64.

Simpson, G. G. (1944). *Tempo and mode in evolution*. New York: Columbia University Press.

Simpson, G. G. (1953). *The major features of evolution*. New York: Columbia University Press.

Soltis, D. E., & Soltis, P. S. (1999). Polyploidy: recurrent formation and genome evolution. *Trends in Ecology and Evolution*, 14, 348–52.

Sterelny, K. (in press). Evolvability in Matthen and Stephens. *Elsevier Handbook to Philosophy of Biology*.

Templeton, A. R. (1980). The theory of speciation via the founder principle. *Genetics*, 94, 1011–38.

Travisano, M., Mongold, J. A., Bennett, A. F., & Lenski, R. E. (1995). Experimental tests of the roles of adaptation, chance, and history in evolution. *Science*, 267, 87–90.

Turelli, M., Barton, N. H., & Coyne, J. A. (2001). Theory and speciation. *Trends in Ecology and Evolution*, 16, 330–43.

Valentine, J. W., Jablonski, D., & Erwin, D. H. (1999). Fossils, molecules and embryos: new perspectives on the Cambrian explosion. *Development*, 126, 551–859.

Via, S. (2001). Sympatric speciation in animals: the ugly duckling grows up. *Trends in Ecology and Evolution*, 16, 381–90.

Vincek, V., O'Huigin, C., Satta, Y., Takahata, N., Boag, P. T., Grant, P. R., Grant, B.R., & Klein, J. (1997). How large was the founding population of Darwin's finches? *Proceedings of the Royal Society of London, B*, 264, 111–18.

Vrba, E. S., & Eldredge, N. (Eds). (2005). *Macroevolution: diversity, disparity, contingency: essays in honor of Stephen Jay Gould*. Supplement to *Paleobiology*, 31(2).

West-Eberhard, M. (2003). *Developmental plasticity and evolution*. Oxford: Oxford University Press.

Wright, S. (1931). Evolution in Mendelian populations. *Genetics*, 16, 97–159.

Zelditch, M. L., Sheets, H. D., & Fink, W. L. (2003). The ontogenetic dynamics of shape disparity. *Paleobiology*, 29, 139–56.

Further Reading

Carroll, R. L. (1997). *Patterns and processes of vertebrate evolution*. Cambridge Paleobiology Series. Cambridge: Cambridge University Press. This is a balanced, accessible introduction to the study of vertebrate paleontology.

Coyne and Orr's (2004) *Speciation* is the best recent survey textbook introduction to speciation studies.

Schilthuizan's (2002) *Frogs, flies and dandelions* is an accessible and fun read on speciation, suitable for undergraduates.

Maynard Smith, J., & Szathmáry, E. (1997). *The major transitions in evolution*. New York: Oxford University Press. This is an overview of the main stages of evolution and attempts to explain them.

Presgraves, D. C. (2007). Speciation genetics: epistasis, conflict and the origin of species. *Current Biology*, 17, R125–7. This is an excellent review of recent work on the genetics of speciation.

Elisabeth S. Vrba and Niles Eldredge's (2005) *Macroevolution: diversity, disparity, contingency: essays in honor of Stephen Jay Gould* is an interesting, eclectic collection of essays pro and con recent controversies in paleontology and macroevolution.

Chapter 11

Adaptationism

PETER GODFREY-SMITH AND JON F. WILKINS

1. Introduction

The "adaptationism" debate is about the role of natural selection in relation to other evolutionary factors. The term "adaptationist" is used for views that assert or assume the primacy, or central importance, of natural selection in the project of explaining evolutionary change. This "central importance," however, can take a variety of forms. The debate can also involve questions about *how* natural selection operates, and what sorts of outcomes it tends to produce. But most discussion of adaptationism is about the relative significance of selection, in comparison with the various other factors that affect evolution.

The term "adaptationism" is only a few decades old, but the debate itself is an extension of long-running debates that reach back to the early days of evolutionary theory in the late nineteenth century. Darwin himself constantly fine-tuned his claims about the relations between natural selection and other evolutionary factors, especially in successive editions of the *Origin of Species*. Many of the topics covered in recent debates can also be recognized in debates about gradualism, the role of mutation, and the significance of Mendelism to evolutionary theory in the early twentieth century (Provine, 1971). During the early years of the "evolutionary synthesis," the debate between R. A. Fisher and Sewall Wright was in large part a debate about the role of subtle non-selective factors such as population structure and random drift (Fisher, 1930; Wright, 1932).

So the debate about the relative importance of selection is old, but it was transformed by a famous 1979 paper by Stephen Jay Gould and Richard Lewontin. They used the term "adaptationism" for one set of views about the primacy of selection. They then attacked that view, and defended a "pluralist" position in which many evolutionary factors are explicitly taken into account. Selection is then seen as constrained by a range of developmental and architectural factors, and evolutionary outcomes reflect accidents of history as much as ecological demands. Gould and Lewontin also attacked poor methodological practices that they saw as common within the "adaptationist" camp.

Although the debate initially appeared to be primarily biological and empirical, it came to occupy the attention of philosophers as well as biologists. In part this can be attributed to philosophers' keen interest in theoretical debates in evolutionary theory. But as the debate developed it became entangled in abstract issues in the philosophy of science. These include questions about idealization, teleological thinking, and the overall role of evolutionary theory in the scientific world view (Dupre, 1987; Dennett, 1995). The debate is now transforming and, to some extent, subsiding.

Our discussion here will have three parts. First, we discuss the development of the debate in more detail, focusing especially on recent transformations. Then we discuss distinctions between several different kinds of adaptationist position. Within "the" problem of assessing adaptationism, at least three distinct problems are often mixed together. This distinction enables us to sort the more empirical from the more non-empirical aspects of the problem.

Once this has been done, in the final section we present a novel treatment of some of the more empirical aspects of the debate. This analysis will be partly deflationary; we suggest that some (though not all) conflicts in this area are not as real as they seem. They arise from paying insufficient attention to some crucial differences in the "grain" of evolutionary analysis.

2. The Development of the Debate

We will not trace deep history of debates about the role of selection and adaptation, but will start from the specific discussion initiated by Gould and Lewontin's "spandrels" paper.

Gould and Lewontin argued for several claims. First, they argued that evolutionary thinking had become far too focused on natural selection as a determinant of evolutionary change. A more subtle line of critique concerned how natural selection itself should be understood. Gould and Lewontin argued that organisms had come to be seen as patchworks of traits that had each been selected as a "solution" to some "problem" posed by the organism's environment. Gould and Lewontin saw two errors in this picture of organisms and environments. One error was a reductionist picture of organisms as collections or amalgams of distinct traits. We can call this, more specifically, an "atomistic" view of the organism. The other is what Lewontin has elsewhere (1983) called an "alienated" conception of the organism in relation to its environment. This second error can more simply be called an "externalist" conception of evolution. In this view, the environment is taken as a preexisting condition to which the organism must respond.

In their critique, Gould and Lewontin put a lot of weight on the etymology and metaphorical loading of the terms "adaptation" and "adaptive." They saw mainstream evolutionary theory as beholden to a picture of organisms that is in some ways pre-Darwinian and pre-scientific. Organisms were seen as fitting their environments' demands as a key fits a lock. Although evolutionists invoke natural mechanisms to explain this "fit" between organism and circumstances, the conception of this relationship itself is, for Gould and Lewontin, too close to the tradition of natural theology, in which God has designed every organism to be ideally fitted for its circumstances and role.

The atomism and externalism of mainstream English-speaking evolutionary biology should be replaced, Gould and Lewontin argued, by a view that recognizes the integrated nature of organisms, and also recognizes the reciprocal or two-way interaction of organisms and environments. The argument was not that *no* traits are solutions to environmental problems in the standard sense, but that a great many traits are not.

Further, according to Gould and Lewontin, the focus on adaptive explanation had led to careless and biased methodological habits in much of evolutionary biology. The aim, allegedly, had become that of finding *some* adaptive rationale for every trait that could be described. Explanation was incomplete until an adaptive story had been found, and the biologist's work was done once an adaptive explanation had been found that had reasonable fit to available data. Some parts of evolutionary thinking were turning into an exercise in concocting "just-so stories." There was, according to Gould and Lewontin, little willingness to seriously consider different kinds of explanation, or to raise the standard of proof for an adaptive explanation to a level appropriate for science.

In sketching such alternative explanations, Gould and Lewontin co-opted an architectural term, "spandrel." Spandrels are features of a structure that were not directly shaped by natural selection or deliberate design, but are *byproducts* of selection (or design) operating on other features. Though this term achieved wide currency via Gould and Lewontin's paper, it does not capture with much accuracy the shape of the alternative explanatory program that Gould and Lewontin were trying to describe. The core of this alternative program is the idea that evolutionary processes are subject to a long list of influences, many of them quite well understood in isolation, but interacting in very complex ways. For example, evolutionary biology had focused largely on the features of adult organisms, neglecting the fact that adults are the outcomes of developmental sequences that start with a single cell. A possible adult phenotype with very high fitness is evolutionarily irrelevant if it cannot feasibly be produced by the developmental trajectory characteristic of that kind of organism. In indicating the structure of alternative explanations Gould and Lewontin also cited constraints on evolution deriving from the genetic systems of organisms, constraints imposed by an organism's "bauplan" or basic layout, and various roles for accident and happenstance (Kitcher (1987) gives a good survey of all these factors. See Pigliucci and Preston (2004) for a collection of work that focuses on the integrated nature of phenotypes.)

The argument in Gould and Lewontin's paper was expressed generally, but a crucial target both here and in subsequent discussion was the evolutionary study of behavior, especially human behavior. Sociobology had arisen as a specific research program a few years earlier (Wilson, 1975), and Gould and Lewontin saw the problem of rampant adaptive speculation as especially acute and harmful in this area. Special criticism was also focused on the then-novel strategy of "optimality analysis," a set of formal tools that embody the assumption that selection will generally produce the best-possible solution to an adaptive problem (Maynard Smith, 1978; Parker & Maynard Smith, 1990).

Gould and Lewontin's critique generated a heated discussion. Some biologists – perhaps most – thought that Gould and Lewontin had caricatured the selection-oriented style of biological work. So some responses took the form of arguing that a reasonable sensitivity to non-selectionist factors was already present in mainstream biology and no corrective was needed. For instance, Maynard Smith (1982) points out that one part of setting up any optimality analysis is definition of the set of alternative phenotypes that are to be compared, and that this is equivalent to a description of developmental constraints. While it may be true that insufficient attention has been paid to exactly how those constraints should be formulated, Maynard Smith argues, it is not fair to claim that these constraints are absent from this type of analysis. Others argued that a strong focus on selection was both real and warranted, either by theoretical considerations or the successful track record of this approach (Mayr, 1983; see also the next section below).

The intensity of this debate has subsided in recent years, but it would be wrong to say that the debate was "won" by either the adaptationists or the anti-adaptationists. For most evolutionary biologists, natural selection continues to play a privileged explanatory role, but no longer a solitary one. To varying degrees, the criticisms leveled by Gould and Lewontin have been internalized by the field, and are reflected in contemporary methodologies. (For reviews of these developments, see Pigliucci and Kaplan (2000) and various essays in Walsh (forthcoming).)

Evolutionary Psychology (EP), which is commonly viewed as the modern reincarnation of Sociobiology, is perhaps the field that has been most resistant to the anti-adaptationist critique. However, while this field is still primarily concerned with the identification of adaptive explanations for particular human behaviors, the approach is generally less naïve than many of the analogous efforts of the pre-"spandrels" era. For instance, one of the standard components of contemporary adaptive explanations of human behavior is the "Environment of Evolutionary Adaptation" (EEA). This concept acknowledges that the perceived adaptive value of a trait in a contemporary cultural context is irrelevant to an explanation of the evolutionary origin of that trait (Barkow, Cosmides, & Tooby, 1992). An adaptive explanation must refer to selective value in an environment like the one in which most of human evolution is thought to have occurred (e.g., small groups, hunter-gatherer lifestyle). While many would still describe EP as a field with an adaptationist bent (in the pejorative sense), the EEA incorporates at least some sense of a historical constraint.

Another sign of the integration of the anti-adaptationist critique into mainstream evolutionary biology is the explicit and widespread use of phylogenetics. At one point, there was significant debate within systematics over whether the most appropriate mode of taxonomic categorization was based on shared features or shared ancestry. That debate has largely been settled in favor of shared ancestry. It is now common to pursue the construction of taxonomic relationships among species in parallel with the study of the evolution of particular traits. Trait changes are explicitly mapped onto phylogenetic trees. In this view, selection always occurs in a historical context.

Another area where it is possible to see this integration is in the study of the evolution of development ("evo-devo"). Here the entity that is evolving is not a "trait" in the traditional sense, but rather a developmental trajectory (Raff, 1996). Selection may still be the prime mover in changes in these developmental trajectories, but it is impossible to formulate a question about selection in this framework without explicitly considering developmental, "bauplan" constraints, the integrated form of the organism as a whole, and the possibility that changes in one trait may result in changes in other traits. It is natural, if not unavoidable, in this framework to assume that some traits have been the subject of direct selection and others have not.

The idea that organisms reshape their environments, rather than just adapting to them, is not new. However, there has been a recent renewal of interest in explicitly considering these processes (Odling-Smee, Laland, & Feldman, 2003). The term "niche construction" is often now used for this process, specifically in an effort to undermine the concept of the "niche" as a preexisting thing that an organism must fit itself into.

As a final example, it is interesting to consider the development of molecular evolution and population genetics over the past three decades. One of the major research agendas in this area has been the development and application of statistical methods for identifying signs of selection from molecular data (e.g., DNA sequence data). This work is interesting in that its entire premise implies a selectionist perspective tempered by the type of caution urged by Gould and Lewontin. The idea that it is possible to identify particular genes that have recently been subject to a particular type of selection – and the idea that this is a worthwhile thing to do – implies that selection is of particular interest, and that if we can develop the right tools, we can find it. The idea that we have to develop powerful laboratory and statistical methods to find it implies that in many cases, selection may not be the most useful description of what is going on. The focus on statistical methods also takes on board the idea that it is appropriate to require a rigorous standard of evidence when making assertions about the role of selection.

3. Varieties of Adaptationism

One role for philosophical work in this area is distinguishing several different *kinds* of commitment that have been tangled together in adaptationism debates. That will be the focus of this section.

First, it is worth noting the gap between a commitment to a strong form of *adaptationism* and what might better be called *selectionism*. In the Gould and Lewontin critique, and in Lewontin's other work (e.g., 1983), the focus is not just the primacy of natural selection, but a particular conception of how selection works and what it produces. For Lewontin, as noted above, mainstream evolutionary thinking has operated with a strongly asymmetric picture of organism/environment relations. The organism is seen as responding to structure in the environment that exists independently of what organisms are like and how they change. Not all of the biological work focused on natural selection has this character, though. In game-theoretic models of evolution, the "environment" encountered by any organism is constituted primarily by the behaviors of other organisms in the same population. These models tend to place great emphasis on selection, but they do not see populations as adapting to independently existing environmental features. So in a sense, game-theoretic work is selectionist without being adaptationist. In most discussion of adaptationism, however, this sort of distinction has not been made. Below we will follow the more familiar practice of using "adaptationist" for work that asserts the primacy of selection, whether or not the explanatory pattern is strongly externalist with respect to organism/environment relationships.

A more pervasive problem in the debate over adaptationism has been the mixing together of different senses in which selection might be said to be the "primary" or "most important" evolutionary factor. One sense is empirical: selection might be seen as the strongest force in evolution, or most efficacious causal factor. Another sense is not straightforwardly empirical at all, and has more to do with the role of evolutionary theory within science as a whole.

This second kind of position is illustrated especially by Richard Dawkins (1986). Dawkins is often associated with an extreme form of adaptationism. But this commitment is of a special kind. For Dawkins, the central importance of natural selection does not involve a claim about how *much* of what we see in the biological world has been shaped by selection. A huge amount of what we see might be due to other factors. Selection can in a sense still be "the most important" evolutionary factor, because only selection can answer the most important *questions* faced by biology.

Accordingly, Godfrey-Smith (2001) distinguishes three kinds of adaptationism.

Empirical adaptationism: Natural selection is a powerful and ubiquitous force, and there are few constraints on the biological variation that fuels it. To a large degree, it is possible to predict and explain the outcome of evolutionary processes by attending only to the role played by selection. No other evolutionary factor has this degree of causal importance.

Explanatory adaptationism: The apparent design of organisms and the relations of adaptedness between organisms and their environments are the *big questions* for biology. Explaining these phenomena is the core intellectual mission of evolutionary theory. Natural selection is the key to solving these problems. Because it answers the biggest questions, selection has unique explanatory importance among evolutionary factors.

Methodological adaptationism: The best way for scientists to approach biological systems is to look for features of adaptation and good design. Adaptation is a good "organizing concept" for evolutionary research.

Strictly speaking, all three of these views are logically independent. Any combination of "yeses" and "nos" is possible in principle. There are relations of *support* between them, but not relations of implication. And further, the relations of support between them are quite complicated. Evidence that supports one of the three may not support others. (See also Lewens (forthcoming) for a more fine-grained categorization of possible views here, and Sterelny and Griffiths (1999) for further discussion of the key distinctions.)

Let us look more closely at the relations between empirical and explanatory adaptationism. Empirical adaptationism, as outlined above, is intended to be a contingent claim about the causal role of selection in the actual biological world. Explanatory adaptationism, in contrast, is a claim about the role of selection in the total edifice of scientific knowledge. It is a claim about the role of selection in solving what would otherwise be an insoluble scientific problem. Selection can play this role *even if it is rare*, even if most of what we see is the product of other evolutionary factors.

A useful illustration is provided by some responses to the rise of the "neutral theory" of molecular evolution (Kimura, 1983). The neutral theory holds that most genetic variation observed at the molecular level is not to be explained in terms of selection; it is a consequence of mutation and random genetic drift. Neutralism is clearly a denial of the omnipresence of selection. It is a denial of some forms of empirical adaptationism. Recent decades have seen a lively debate between neutralists and their "selectionist" opponents [See Molecular Evolution]. But some others who see themselves as "selectionist" or "adaptationist" in their orientation to evolution see neutralism as no threat to their position at all. Dawkins (1986) is an example. This is because the processes described by neutralism are agreed on all sides to have no direct role in the explanation of well-adapted phenotypes that exhibit "apparent design." The neutralists are not trying to answer questions about apparent design in nature; they are trying to describe genetic variation considered as a whole. So Dawkins sees himself as having nothing directly invested in the neutralism debate. To the extent that the neutralists win, he gains a useful tool (a reliable molecular clock), but the neutralist denial of "selectionism" does not even touch on his core claims. For a pure explanatory adaptationist, selection might only explain 1 percent of all molecular genetic change, but if this is the 1 percent that is responsible for highly adapted phenotypes that give the appearance of design, then this is the 1 percent that counts.

Assessing the explanatory adaptationist position involves two stages. One is the assessment of whether it is really true that apparent design is the "big question" for biology. Is focusing on this question no more than a personal preference, or even a misguided concession to a pre-Darwinian, creationist point of view? The other stage is the assessment of whether selection is really the *answer* to the question, in the strong sense seen in the explanatory adaptationist tradition.

Some biologists have directly criticized the view that selection has a primary role in the explanation of apparent design (Kauffman, 1987; Goodwin, 1994), and have tried to develop more "internalist" explanations for roughly the same class of phenomena that adaptationists focus on. A more subtle and promising view of this matter might be extracted from more mainstream evolutionary thinking, however. According to this view, which can be associated with Wright's position in his debate with Fisher, we should see the evolutionary mechanism that can result in highly adapted phenotypes as comprising a *much more complex machine* than adaptationism envisages. For mutation and selection to produce highly adapted phenotypes, they must operate in a context in which many other evolutionary parameters take suitable values. Wright, for example, argued that partial subdivision of the population is needed for selection to avoid getting stuck on (what is now called) a mere "local optimum," a state of moderately high fitness that is inferior to the state that the population could in principle achieve via a more thorough exploration of the space of possibilities. (For assessments of the shifting balance theory, see Coyne et al. (1997) and follow-up discussions.)

Here we use Wright's appeal to population subdivision as an illustration of a role that could be played by a number of different evolutionary factors, any or all of which might be needed to enable evolution to produce highly adapted states of organisms. Mere mutation and selection alone, on this view, is too blunt an instrument. So some evolutionists would hold that even if explanatory adaptationism's view on the "big questions" in biology is accepted, it is an error to see selection as having primary importance in biology's answer. So this second part of the assessment of explanatory adaptationism depends very much on empirical questions, while the first part, concerning the alleged "big question" for biology, is much less empirical.

We will make only a few comments here about the third adaptationist position, methodological adaptationism. There are several distinct ways by which such a view might be motivated. First, an argument might be given on the basis of a prior commitment to empirical adaptationism. Another style of argument, often made informally but expressed explicitly by Mayr (1983), is an argument from simple induction. Some biologists hold that regardless of what we might make of the other two forms of adaptationism, the track record of strongly selectionist thinking in biology has been so impressive that this approach should be continued. If this argument is made, an interesting counter to it can be given on the basis of Lewontin's other work (1983), which expresses a historicist view of the matter. For Lewontin, it was productive during an earlier stage in the development of biology to apply an adaptationist mindset, but we have now passed that stage. What was once a useful organizing framework has now become an impediment to further progress.

We will look in more detail at the problem of assessing empirical adaptationism. First, it is worth noting a way in which the standard vocabulary is misleading here. Questions about empirical adaptationism are often described as questions about the "power of selection" to determine the course of evolution and to produce highly adapted states of organisms. But as Sober (1987) has noted, these questions are as much about the "power of mutation" as the power of selection. Often, the crucial question to ask in these cases is a question about the supply of *variation* in an evolutionary process, rather than a question about the size of fitness differences. This is one reason why our formulation of empirical adaptationism above includes a statement about the abundance and unconstrained nature of the supply of variation.

In the section following this one, we will present a new way of thinking about the problem of empirical adaptationism. According to this view, some of the apparent oppositions in this area can be dissolved via a more careful treatment of the "grain" of evolutionary analysis. In this section, though, we will first discuss an earlier attempt to describe a direct "test" of empirical adaptationism. This test was offered by Steven Orzack and Elliot Sober (1994). Orzack and Sober did not employ distinctions of the kind used in this section, but what they call "adaptationism" is basically a version of empirical adaptationism. Their "adaptationism" is the view that natural selection is the most powerful evolutionary force, and able to create near-optimal phenotypes.

They propose that we test this view by asking the following question: are predictions about evolution that are based only upon information about forces of natural selection just as good, or nearly as good, as predictions based on consideration of the entire range of evolutionary factors? The way to answer that question, in turn, is to investigate a large range of specific biological phenomena, and work out how adequate a purely selection-based model is for explaining each. In each case we ask whether an account of the phenomenon that considers *only* the role of selection fits the data as well as a richer model that considers a wide range of factors. If this approach is vindicated in the great majority of cases, then adaptationism is vindicated as a general claim about the biological world.

We think that the Orzack and Sober proposal was quite useful as a first attempt to make questions around empirical adaptationism more concrete and tractable. In particular, it is useful as an attempt to make empirical sense of the idea of selection as "the most powerful force" – an idea that is often intuitively attractive but is, at least in part, metaphorical. However, their proposal has some internal problems, and can also be seen to omit a factor that is crucial for making overall sense of the situation. Internal problems are discussed in this section; the missing factor is discussed in the next.

First, we note that the test that Orzack and Sober propose involves a comparison of simpler with more complex models of the same phenomenon. Given that the richer model includes all the factors that the simpler one has, the richer one cannot do worse than the simpler one. It can only do as well or better. So how *much* better does the complex model have to do before it is to be preferred? Such questions are hard to assess because they require that we make quantitative comparisons between two very dissimilar things: the complexity of a particular model and the goodness of fit to observed data. In practice, formal tools such as the Akaike Information Criterion or Bayes Information Criterion are often employed in making this comparison. These criteria impose a penalty for each model parameter. The favored explanation is the one that provides the best fit to the data, but only after this penalty has been applied. These methods often contain arbitrary features, however. One interesting recent approach that avoids some problems has been proposed by Rissanen (2005), and uses the concept of "normalized maximum likelihood." In this approach the goal is to find the model that permits the shortest possible description of the system, where the "system" includes both the model and the data being modeled. In principle, one would determine the number of bits required to specify the model, and the number of additional bits required to describe the data within this model. Each bit of information is equivalent, whether that bit is applied to the model or the data. However, even this refinement still does not adequately account for other, more human, aspects of the problem. Prior to any analysis one must determine which aspects of the data need to be explained. While this approach could be used to adjudicate among explanations for particular phenomena, it will not address disagreements among biologists about what features are most deserving of explanation. Likewise, a given model must be specified within the context of many unspecified or implicit assumptions. It may often be the case that the disagreements over the role of selection reflect different – perhaps implicit – views on what should be assumed prior to studying a particular problem.

These considerations suggest that a test of adaptationism should ideally focus on a contest between models of comparable complexity. If we are constrained to include in our model some specific number of parameters, and a specific level of tractability, then should we "invest" only in a very detailed specification of the selective forces relevant to the situation, or should we use a less complete specification of the selective forces along with some information about other factors as well? This comparison might be one between adaptationism and pluralism, but it also could be one between an adaptationist model and a model in which some single non-selective factor is described in great detail and made to carry all the predictive weight. The non-selective factor in question might be drift, or perhaps the "laws of biological form" described by a modern-day rational morphology. Empirical adaptationism as a general claim would be vindicated, on this proposal, if in the majority of cases a better fit to the data is achieved by a selection-based model than is achieved by any other model of comparable complexity. That is, empirical adaptationism is vindicated if a description of the relevant forces of selection is more informative than any other description at a similar level of detail.

In some respects this proposal derives from an application of Richard Levins' views about models (Levins, 1966; see also Wimsatt, 1987; Weisberg, forthcoming). Models can have a range of virtues and goals. Different levels of tractability and understandability are sought in different types of investigation, and great generality may or may not be desired. A precise fit to particular phenomena can be traded off against generality. The contest between models described above is designed to take these facts about model building into account. Relative to the scientific goals at hand, and the general style of model which is suitable for the occasion, which type of information is more useful: information about selection, or information about something else?

We think that something like this trade-off is often on the table in contemporary modeling of behavioral evolution. Game theory has become an important tool in this area (Maynard Smith, 1982). Game-theoretic approaches to behavior choose to "invest" heavily in a detailed specification of the fitnesses of different strategies, and how they change with frequencies and circumstances. As a consequence, however, game-theoretic models must make radical simplifications about other evolutionary factors, especially the role of the genetic system. Often they even abstract away from different ways in which a stable distribution of behaviors might be realized in a population of individuals (Bergstrom & Godfrey-Smith, 1998). The game-theoretic strategy embodies the idea that it is more informative to give a detailed specification of selective forces, and a minimal treatment of everything else, rather than "investing" some of the complexity of the model in a careful treatment of other factors.

When we envisage a contest between models of comparable complexity, we also avoid a problem that Brandon and Rausher (1996) found in Orzack and Sober's approach. Brandon and Rausher claim that Orzack and Sober's proposal is biased in favor of adaptationism. This is because in Orzack and Sober's proposal, if a simplified selectionist model succeeds predictively then it is said to be vindicated – even if some *other* simplified model would do just as well in a similar test. As Brandon and Rausher say, if there is to be an unbiased test between simpler and richer models, then a range of different kinds of simpler model should be included. This point is right, and the problem is avoided under the revised proposal in which the comparison to be made is always between comparably simple models.

We see this "contest between models" scenario as itself a simplified way of thinking about a more complex set of empirical questions. Not all aspects of the problem of empirical adaptationism can be assessed by a direct comparison of models in their dealing with data (Godfrey-Smith, 2001; Sterelny & Griffiths, 1999) In this chapter, though, we will continue to operate within a somewhat formal and idealized approach to the problem. But in the next section, we will introduce a richer framework than the ones used so far.

There is a simple way of motivating the shift from the framework used in this section to the one assumed below. In both the original proposal of Orzack and Sober, and the modified one outlined just above, there was no distinction made between different "grains" at which evolutionary processes can be described. In the "contests between models" discussed above, any biological phenomenon could be chosen and made subject to an instance of the test. The *same* approach was employed at *all* levels of grain. But perhaps the key to the problem, or a large part of it, lies in distinguishing between several different grains at which evolutionary processes can be described. This is a straightforward idea, but in the next section we will make it more precise with the aid of a formal framework that is popular, although controversial, within evolutionary biology itself.

4. The Role of Zoom and Grain

One of the most prominent metaphors in evolutionary biology is that of the adaptive landscape (Wright, 1932; Gavrilets, 2004). In its simplest form, the landscape represents a mapping from an organism's genotype or phenotype to its fitness; natural selection is the tendency for populations of organisms to move "uphill" in this landscape – that is, towards regions associated with higher fitness. Populations move locally on the landscape because mutations are assumed to have small effects.

There are many concerns about the validity of the adaptive landscape metaphor, and some biologists favor discarding it altogether (Moran, 1964). Most biologists are still comfortable with the idea of a fitness being associated with a particular phenotype. However, it is now commonly accepted that this fitness is context dependent. That is, the shape of the landscape – the locations of fitness peaks and valleys – may be extremely sensitive to the distribution of organisms on the landscape, in both the past and the present. Furthermore, the relationship between an organism's genotype and its phenotype is increasingly seen to be a complex one, mediated through developmental pathways and environmental interactions.

Some who reject the whole idea of the adaptive landscape claim that the image of individual organisms or populations climbing fitness peaks suggests the sort of intentionality that often haunts sloppy evolutionary reasoning. The assumption of continuity of movement on a phenotypic landscape also seems to disregard the disjoint nature of phenomena like Mendelian inheritance and recombination. So for some critics, the adaptive landscape metaphor implicitly reinforces some of the same kinds of problematic simplifying assumptions that Gould and Lewontin's critique targeted.

With these caveats in mind, we suggest that this metaphor may nonetheless be useful for understanding how seemingly contradictory scientific approaches to evolutionary questions can, in fact, be complementary. We also suggest that one's perception of the extent to which evolution is characterized by adaptation shapes – and is shaped by – the level of resolution at which one considers evolutionary processes.

When one begins to consider an evolutionary problem, one must first choose a scale, or grain, of analysis. We can think of this as choosing how large a region of the adaptive landscape we want to include in the analysis. Choosing a larger region means considering a wider variety of alternatives (alleles, genotypes, phenotypes, strategies, etc.), but does not permit the same depth of analysis that could be performed on a smaller region. Also, if most evolutionary processes (e.g., mutation and selection) result in local movement on the landscape, then inclusion of a larger region implicitly considers these processes over a longer timescale.

To see the importance of grain in relation to attitudes towards adaptation, we will consider three different scales of analysis. To make the application of this framework simple and clear, we will suppose that these are three scales at which the same very large landscape is being viewed. The landscape itself is phenotypic; it represents fitness (height) as a function of phenotypic variables describing individuals. In some ways, what we are imagining here is something that is not fully coherent, because we are supposing that the same measure of fitness can be applied very different organisms (elephants and jellyfish, for example). But we think that the landscape idea has genuine heuristic power here nonetheless.

At the highest, most "zoomed out" level, evolutionary analyses consider broad patterns of occupancy across large regions of the landscape. At this level of analysis, populations are represented by single points. (This creates problems in the case of extremely sexually dimorphic species, like some barnacles, but we will idealize away from that problem.) At this level of grain, the natural questions to ask include the following: What portion of conceivable peaks is occupied by populations in the real world? To what extent are the occupied peaks the highest peaks? How thoroughly and how predictably does natural selection explore the adaptive landscape?

To most biologists working at this highest level of analysis, what is most striking is the emptiness of the landscape. In the history of life on earth, organisms have explored a vanishingly small fraction of the conceivable ways of making a living, and the idea that the modes of life that we see today somehow represent the "best" of these possible forms has been broadly rejected. Rather, populations have been restricted to a small subset of local peaks by chance, as well as by historical and developmental constraints. From this vantage point, the power of natural selection to produce adaptation appears quite limited.

The situation changes if we zoom in on a particular region of the landscape, perhaps containing only one or a few peaks. At this intermediate level of grain, whole populations still tend to occupy single points, but they are vague or smudged ones. These analyses also typically focus on variation in a small number of dimensions, assuming (often implicitly) that the traits represented by the other dimensions are invariant over the timescale of the analysis, and evolve independently of the trait or traits under consideration.

A question asked in this second context might take the form, "Given that there is a population that is in this region of the landscape, where, within the region, should we expect to find that population?" The answer that many biologists would give is that we expect to find the population at or near one of the local peaks. Given enough time and a local topography that is conducive to a thorough evolutionary exploration of the region, we might even expect to find the population at or near the highest of those local peaks.

Perhaps more commonly, however, research at this scale starts with the empirical observation that a population occupies a particular location within the region. The task is then to uncover whether and why there should be a peak in that location. At this intermediate scale, many will hold that the salient feature of evolution is local adaptation. Populations tend to be found near peaks, as opposed to in the adjacent valleys. Some biologists may disagree with that claim, but the crucial point is that selectionist conclusions drawn at this level of analysis do not contradict non-selectionist conclusions drawn at the higher level. The fact that the population is in this region of the landscape, as opposed to some very distant one, is simply one of the background assumptions made when working at this second level of analysis.

Now let us consider a third level of "zoom" with which we might view the landscape. We now focus in great detail on a very small region. When we do this, certain features of evolution that are typically ignored at the higher levels become critically important. Rather than thinking of a population as occupying a single location, or a small, diffuse area, in the landscape, we explicitly consider the distribution of individuals that make up the population, and the very complex processes of change to which this distribution is subject. Analysis of the evolutionary dynamics of the system must account for drift and mutation, as well as selection, and, in the case of diploid organisms, Mendelian inheritance and recombination. The way the population moves on the landscape from generation to generation is not continuous, at this level of grain. Points appear suddenly, some distance from their parents.

As we focus more on the details of evolution at this lowest level, the adaptationist features that were prominent at the intermediate level of analysis recede in importance. Other evolutionary processes involving random fluctuations, interactions among alleles or loci, and constraints inherent in the mechanism of inheritance become more important aspects of our understanding. Here again, there will be some disagreement among biologists about which factors are in fact most significant. But many will hold that at this lowest level, as we also saw at the highest level, what is often most salient are those features of evolution that frustrate adaptation. And once again, the crucial point here is that de-emphasizing selection at this lowest level is not at all inconsistent with applying a strongly selectionist approach at the second, middle level.

To construct a specific example, consider the case of sickle-cell disease, which afflicts individuals who inherit two copies of a mutant allele of the Hemoglobin alpha chain gene. Despite the devastating effects of this condition, this mutant allele is present at an appreciable frequency in certain populations, because heterozygous individuals, who carry one copy of the mutant allele and one copy of the normal allele, have an increased resistance to malaria. It is instructive to consider how different scales of analysis of this system drive different relationships to adaptationism.

We can first analyze this system in the context of population genetics – considering how the selective effects of malaria and of sickle-cell anemia alter the relative frequencies of the mutant and normal alleles. This perspective highlights the way in which Mendelian inheritance undermines humans' capacity to adapt to the presence of malaria. A population composed entirely of "adapted" heterozygotes is inherently unstable: our mechanism of inheritance re-creates both classes of maladapted homozygote every generation.

At a slightly higher level, we begin to see signs of adaptation. If we expand our view to encompass the mutation event that created the mutant allele, we see that human evolution has found the beginnings of a solution to the problem of malaria. From an adaptationist point of view, heterozygous resistance to the disease certainly represents an advance over a population where everyone is susceptible. If we expand, hypothetically, to an even longer timescale, we can imagine solutions to the problem that are not disrupted by the diploid genetic architecture. For example, the right mistake in recombination could create a chromosome that carried both the normal and mutant alleles. This new allele might confer malaria resistance to its carriers without producing sickle-cell anemia in homozygotes.

As it happens, natural populations in Africa do, in fact, contain an allele (C) that confers resistance in homozygous state without harmful sickling, but this allele has low fitness in heterozygous state with each of the prevalent alleles (A and S), so it has been unable to advance from very low frequency (Templeton, 1982; Gilchrist & Kingsolver, 2001). Here we see the role of a different kind of fine-grained constraint, the requirement that a new favorable allele be advantageous when appearing as a heterozygote with locally common alleles. Again, when operating within a coarser-grained perspective we can expect new alleles to eventually arise that overcome this constraint.

If we continue to zoom out, the appearance of adaptation begins to recede again. Why, for instance, do we not have a fundamentally different immune system that would make malaria a non-issue? Our susceptibility to malaria suggests a whole other class of features limiting our adaptation. We are subject to numerous historical and developmental constraints. Perhaps most importantly, however, is that our capacity to adapt to our environment is limited by the fact that many aspects of our environment – such as the malaria parasite – are simultaneously adapting to us.

Let us note explicitly some of the features of the three levels of grain that make adaptive features of evolution more, or less, prominent. At the finest level of grain, existing features of the genetic architecture are treated as fixed constraints against which selection must act. At the intermediate level, we suppose that some of these can be altered by such things as modifier alleles. At the intermediate level, timescales are also longer than they are at the finer level. So at the intermediate level, we can suppose that there is a constant steady flow of new variants arising in the population. So the likely fate of any *particular* mutation is not important. (The distinction between shorter timescales in which the set of available variants is fixed, and longer ones in which the set of variants is not fixed, has also been treated in formal modeling framework by Eshel and Feldman (2001) and Nowak and Sigmund (2004), with conclusions that complement the present analysis.)

So our suggestion is that some of the apparent oppositions between adaptationist approaches and their alternatives might be resolved through careful attention to the grain at which evolutionary processes are being described. It is important that making a suggestion of this kind involves taking a stand on some substantive biological issues. In particular, some will contest our claim that at the finest of our three levels of grain, non-selective factors have great importance. Some would say that even at this fine-grained level selection tends to dominate, and that the role of other factors has been over-sold by theoreticians who are enamored of the subtleties of complex population genetics models. Our main aim here is not so much to rule out such a position, as to present a better framework with which these questions can be assessed. One way to represent different kinds of adaptationist commitment is in terms of size of the region, within the range of possible levels of "zoom," in which evolutionary change tends to have an adaptive character. Some will say the region is large, others will insist that it is small. But we think that without this conceptualization of the situation, or something akin to it, it is very difficult to *frame* the debate in a way that makes it tractable. Some attention to these questions of grain is already present, often implicitly, in much biological practice. Our suggestion is that a more explicit and systematic treatment of this factor could be of considerable use in clarifying, and then resolving, fundamental questions about the role of selection in evolution.

References

Barkow, J., Cosmides, L., & Tooby, J. (Eds). (1992). *The adapted mind: evolutionary psychology and the generation of culture.* New York: Oxford University Press.

Bergstrom, C., & Godfrey-Smith, P. (1998). On the evolution of behavioral heterogeneity in individuals and populations. *Biology and Philosophy*, 13, 205–231.

Brandon, R. N., & Rausher, M. D. (1996). Testing adaptationism: A comment on Orzack and Sober. *American Naturalist*, 148, 189–201.

Coyne, J. A., Barton, N. H., & Turelli, M. (1997). A critique of Sewall Wright's shifting balance theory of evolution. *Evolution*, 51, 643–71.

Dawkins, R. (1986). *The blind watchmaker*. New York: Norton.

Dennett, D. C. (1995). *Darwin's dangerous idea*. New York: Simon & Schuster.

Dupre, J. (ed.). (1987). *The latest on the best: essays on evolution and optimality*. Cambridge, MA: MIT Press.

Fisher, R. A. (1930). *The genetical theory of natural selection*. Oxford: Clarendon Press.

Eshel, I., & Feldman, M. (2001). Optimality and evolutionary stability under short-term and long-term selection. In S. Orzack & E. Sober (Eds). *Adaptationism and optimality* (pp. 161–90). Cambridge: Cambridge University Press.

Gavrilets, S. (2004). *Fitness landscapes and the origin of species*. Princeton: Princeton University Press.

Gilchrist, G., & Kingsolver, J. (2001). Is optimality over the hill? The fitness landscapes of idealized organisms. In S. Orzack & E. Sober (Eds). *Adaptationism and optimality* (pp. 219–41). Cambridge: Cambridge University Press.

Godfrey-Smith, P. (2001). Three kinds of adaptationism, in S. Orzack & E. Sober (Eds). *Adaptationism and optimality* (pp. 335–57). Cambridge: Cambridge University Press.

Gould, S. J., & Lewontin, R. C. (1979). The spandrels of San Marco and the Panglossian paradigm: a critique of the adaptationist program. *Proceedings of the Royal Society, London*, 205, 581–98. Reprinted in Sober (1994) pp. 73–90.

Kimura, M. (1983). *The neutral theory of molecular evolution*. Cambridge: Cambridge University Press.

Kitcher, P. S. (1987). Why not the best? In J. Dupre (Ed.). *The latest on the best: essays on evolution and optimality* (pp. 77–102). Cambridge, MA: MIT Press.

Levins, R. (1966). The strategy of model-building in population biology. *American Scientist*, 54, 421–31.

Lewens, T. (forthcoming). Seven types of adaptationism. In D. M. Walsh (Ed.). *Twenty-five years of spandrels*. Oxford: Oxford University Press.

Lewontin, R. C. (1983). The organism as the subject and object of evolution. Reprinted in R. Levins & R. C. Lewontin, *The dialectical biologist* (pp. 85–106). Cambridge, MA: Harvard University Press, 1985.

Maynard Smith, J. (1978). Optimization theory in evolution. *Annual Review of Ecology and Systematics*, 9, 31–56.

Maynard Smith, J. (1982). *Evolution and the theory of games*. Cambridge: Cambridge University Press.

Mayr, E. (1983). How to carry out the adaptationist program? *American Naturalist*, 121, 324–33. Reprinted in E. Mayr, *Towards a New Philosophy of Biology* (pp. 149–59). Cambridge, MA: Harvard University Press, 1988.

Moran, P. A. P. (1964). On the non-existence of adaptive topographies. *Annals of Human Genetics*, 27, 383–93.

Nowak, M., & Sigmund, K. (2004). Evolutionary dynamics of biological games. *Science*, 303, 793–9.

Odling-Smee, J., Laland, K., & Feldman, M. (2003). *Niche construction: the neglected process in evolution*. Princeton: Princeton University Press.

Orzack, S. H., & Sober, E. (1994). Optimality models and the test of adaptationism. *American Naturalist*, 143, 361–80.

Orzack, S., & Sober, E. (2001). *Adaptationism and optimality*. Cambridge: Cambridge University Press.

Parker, G. A., & Maynard Smith, J. (1990). Optimality theory in evolutionary biology. *Nature*, 348, 27–33.

Pigliucci, M., & Kaplan, J. (2000). The rise and fall of Dr Pangloss: adaptationism and the spandrels paper 20 years later. *Trends in Ecology and Evolution*, 15, 66–70.

Pigliucci, M., & Preston, K. (Eds). (2004). *Phenotypic integration: studying the ecology and evolution of complex phenotypes*. Oxford: Oxford University Press.

Provine, W. (1971). *The origins of theoretical population genetics*. Chicago: University of Chicago Press.

Rissanen, J. (2005). Complexity and information in modeling. In K. Velupillai (Ed.). *Computability, complexity and constructivity in economic analysis* (pp. 85–104). London: Blackwell.

Raff, R. (1996). *The shape of life*. Chicago: University of Chicago Press.

Sober, E. (1987). What is adaptationism? In J. Dupre (Ed.). *The latest on the best: essays on evolution and optimality* (pp. 105–18). Cambridge, MA: MIT Press.

Sober, E. (Ed.). (1994). *Conceptual issues in evolutionary biology* (2nd edn). Cambridge MA: MIT Press.

Sterelny, K., & Griffiths, P. (1999). *Sex and death. An introduction to the philosophy of biology*. Chicago: University of Chicago Press.

Templeton, A. R. (1982). Adaptation and the integration of evolutionary forces. In R. Milkman (Ed.). *Perspectives on evolution* (pp. 15–31). Sunderland, MA: Sinauer.

Walsh, D. M. (Ed.). (forthcoming). *Twenty-five years of spandrels*. Oxford: Oxford University Press.

Weisberg, M. (forthcoming). Who is a modeler? *British Journal for the Philosophy of Science*.

Williams, G. C. (1966). *Adaptation and natural selection.*_Princeton: Princeton University Press.

Wilson, E. O. (1975). *Sociobiology: the new synthesis*. Cambridge, MA: Harvard University Press.

Wimsatt, W. C. (1987). False models as means to truer theories. In M. Nitecki & H. Hoffman (Eds). *Neutral models in biology* (pp. 23–55). Oxford: Oxford University Press.

Wright, S. (1932). The roles of mutation, inbreeding, crossbreeding and selection in evolution. *Proceedings of the Sixth International Congress of Genetics*, 1, 356–66.

Part III

Developmental Biology

Chapter 12

Phenotypic Plasticity and Reaction Norms

JONATHAN M. KAPLAN

1. Introduction: What is Phenotypic Plasticity?

Even organisms that are genetically identical will often develop very different adult phenotypic traits when exposed to different environments during development; in other words, the relationship between the genotype and the phenotype is *plastic* – capable of varying based on developmental environment. This fact was recognized even before the so-called "modern synthesis" [SEE DARWINISM AND NEO-DARWINISM], and indeed, some early geneticists were actively studying plasticity soon after the rediscovery of Mendel's laws (see Sarkar, 2004). Research pursued since at least the early 1960s has made it increasingly obvious that phenotypic plasticity is in fact a basic feature of many adaptive responses; this realization has been responsible, in part, for the increased attention garnered by research into phenotypic plasticity (see Sarkar, 2004; Schlitching & Pigliucci, 1998).

Consider, for example, the semi-aquatic plant *Sagittaria sagittifolia*. When grown in dry conditions, the plant produces arrow-shaped ("saggitate") leaves; when grown submerged, the plant produces linear leaves, similar in appearance to seaweed (see Schlichting & Pigliucci, 1998, pp.38–41; Schmalhausen, 1949, pp.195–8). Both sorts of leaves are produced on plants that are grown partially submerged. For any individual plant, growing under a particular set of conditions, one form or the other (or a particular combination of leaf types) will be better – saggitate leaves are well adapted to terrestrial photosynthesis and carbon uptake, whereas linear leaves are well adapted to photosynthesis and carbon uptake in aquatic conditions. But the ability of any particular plant to produce different kinds of well-adapted leaves, based on the location it happens to find itself in, is itself an adaptive response – a response to the *variability* of the environment in which plants of this type evolved in. What evolved – what is adaptive – is not just the particular leaf-structures, but rather the ability to produce different leaf-structures based on the local conditions.

The adaptive nature of plasticity is closely related to Dobzhansky's claim that, because the phenotype of an organism "is the outcome of a process of organic development," and hence influenced both by the genotype of the organism and by the environment the organism finds itself in, that "the genotype of an individual" organism determines *not* its phenotype, but rather the *norm of reaction* of that organism

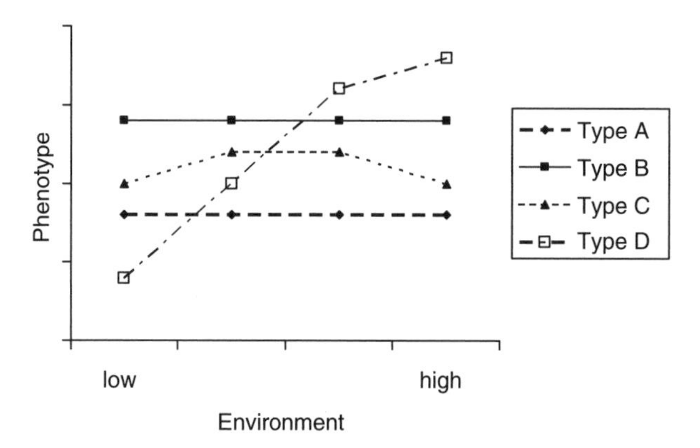

Figure 12.1 Phenotypic plasticity and reaction norms. In this example, organisms of genotype A and genotype B do not display plasticity with respect to variation in the environment tested; while these examples are hypothetical, one can imagine that for example the environmental variable in question is something like temperature (from low to high temperatures, say) and the phenotype in question something like leaf-size (from small to large, say). The variation between organisms of Type A and B is associated with the different genetic (and perhaps epigenetic) resources available. Organisms of Type C and D, on the hand, both display plasticity with respect to the environment tested; the difference between the performance of organisms of Type C in low environments and those of Type C in the middle environments is associated not with the availability of different genetic resources (since they have the same genotype), but with the availability of different environmental resources. Note that over the range of environments tested, organisms of Type D display a more plastic response than those of Type C

(Dobzhansky, 1955, p.74). The *norm of reaction* (aka *reaction norm*) of a genotype is a graphical representation of the phenotypic plasticity of that genotype. Consider, again, the case of semi-aquatic plants such as *Sagittaria sagittifolia*. These plants have genotypes that permit several distinct *kinds* of leaf development depending on the environment they find themselves in; so a particular genotype of *Sagittaria sagittifolia* develops differently in different environments. In this case, we might generate a reaction norm in which we graph the proportion of leaves of different sorts on a plant as a function of the extent to which the plant is submerged.

Reaction norms are best generated by raising genetically identical "clones" of the organisms in question in a variety of different environments; often, one environmental condition is varied and attempts are made to hold the rest of the environmental conditions constant (when it is not possible to use genetically identical clones of organisms, the reaction norms cannot represent the genotype per se, but only the aspect of the genotype that the researcher has managed to hold constant). When clones are used, the same genetic resources (genes, gene complexes, etc.) are available to be used in development (whether the same genetic and/or epigenetic resources are in fact used is another matter); the only developmental resources that differ systematically, then, are the differences in the environmental resources available. Where the phenotype in question does not vary within the different environments tested, there is no phenotypic plasticity for that trait associated with that genotype in the environments tested (see Figure 12.1, Type A and Type B); where the phenotype in question varies with

Box 12.1 Differences in Plasticity Associated with Non-Genetic Heritable Differences

Despite the history of the concept, there is no reason to conceive of phenotypic plasticity as being primarily about the different responses that identical genotypes have to different environments, or, relatedly, to think of different reaction norms as representing the differing responses of different genotypes of organisms. Indeed, unless the organisms in question are identical not just in genotype, but in all the other (more or less) reliably inherited developmental resources (except those being varied within the study in question), studies of plasticity will be missing one potential source of phenotypic differentiation, and hence one potential source of differences in the developmental response to the environment (differences in plasticity). (For discussion of the idea of developmental resources, see for example Oyama, Griffiths, & Gray, 2001; Moss, 2003; and Robert, 2004).

It is certainly possible to investigate the reaction norms not just of organisms that differ in genotype, but of those that, while sharing the same genotype, differ with respect to particular epigenetic mechanisms (such as DNA methylation or chromatin structure, or, to extend the epigenetic concept further, particular membrane states or intracellular diffusion gradients; see, e.g., Robert, 2004, and cites therein; Odling-Smee, Laland, & Feldman, 2003, and cites therein). So far, however, there has been vastly more empirical research into the differences in phenotypic plasticity associated with differences in genotype than in differences associated with other (heritable) developmental resources (see Sollars et al., 2003, for *hints* of what research into plasticity associated with heritable epigenetic resources might be like). In what follows, general points about differences in plasticity will often be framed in the more traditional way, with reference to differences genotype. Keep in mind, however, that the same points could be made with reference to any differences in available *developmental resources* generally.

variation in the environments, there is phenotypic plasticity for the trait in question in that range of environments (see Figure 12.1, Type C and Type D). In the usual graphs of reaction norms, distinct lines represent genetically distinct lineages of organisms (genetically different sets of clones) (but see Box 12.1 on epigenetic resources). These different genotypes will often display differing degrees of phenotypic plasticity (see Figure 12.1, Type C versus D). In these cases, differences in development in the sample organisms *in a particular environment* are associated with the different genetic resources (genes, gene complexes, etc.). In contrast, where different genotypes of a type of organism are grown in similar environments, systematic differences in development are the result of the different genotypes.

Studies of plasticity can be undertaken with a variety of different aims, including at least: elucidating the functional association between particular genes and particular traits; exploring the role played by particular environmental, genetic, or other resources in development; testing particular adaptive hypotheses; and exploring the fitness consequences of genetic variation within a population. While these different research efforts all aim at exploring phenotypic plasticity, and often generate reaction norms, the details of the studies are of course different. In studies that focus on the functional association between particular genes and phenotypes, comparisons of organisms that differ with respect to a single locus will tend to be preferred; in contrast, in those studies that focus on the maintenance of genetic diversity, there will often be less concern with identifying *particular* genetic differences and more interest in genetic differentiation per se. Or again, in those projects that focus on the sensitivity of particular developmental pathways to environmental variation, biologists will tend toward controlling the environment so that only a single environmental condition can be varied, whereas in those studies that focus on the evolution of locally adapted populations (sometimes referred to as ecotypes) biologists will instead examine range(s) of environmental conditions that the populations in question actually encountered during their recent evolutionary history. In the latter case, as well as in studies analyzing the significance of differences in plasticity at the species (or higher) level, researchers are often interested not in the particular norm of reaction for a genotype, but in the average (or "generalized") norm of reaction for the population in question (see Sarkar & Fuller, 2003).

Studies of plasticity tend to be limited by the logistical difficulties of experimental design. Each environment tested requires a separate experimental unit of sufficient size to generate statistically significant results for the size of the effects expected; thus, a study of four temperatures in *Drosophila* will need to be substantially larger (twice as big, ceteris paribus) than one of two temperatures. And because different environmental conditions can (and in many cases, verifiably do) interact in non-additive ways, each *kind* of environmental variable added effectively multiplies the number of discrete experimental set-ups by the number of variations in the new environment one wishes to test. When the desirability of running simultaneous replicates of each set-up is taken into account, it becomes clear why studies that attempt to discover the plasticity of more than two interacting environmental variables are quite rare, and even studies of two environmental variables tend to be restricted to at most three distinct values for each environmental variable (for nine discrete experimental set-ups for each genotype). For similar reasons, plasticity studies (like so much of experimental biology) are best performed on small, short-lived organisms. Far more is known about the plasticity of annual plants than about the plasticity of large deciduous trees, for example.

Over the past twenty years or so there has been increased interest in the mechanisms that make such adaptive plasticity responses possible (see Scheiner & DeWitt, 2004). As noted above, development requires many different sorts of resources (environmental, genetic, epigenetic, etc.) – variation in particular kinds of resources may or may not be associated with phenotypic variation. Further, only some kinds of phenotypic variation will be adaptive, and only a subset of that variation will actually be the result of developmental adaptations (developmental mechanisms selected by natural selection because of their association with adaptive variation). It is not enough to ask, as some researchers have, whether phenotypic plasticity is an *adaptation*, or whether there are genes whose *function* is to permit plastic responses. Rather, we must recognize that there are different types of plasticity, associated with different developmental mechanisms and different evolutionary histories. It is to these distinctions that we now turn.

2. Developmental Conversion and Developmental Sensitivity: Two Forms of Phenotypic Plasticity

Some of the confusion regarding the evolutionary significance of phenotypic plasticity can no doubt be traced to researchers not taking seriously enough the distinction between *developmental sensitivity* and *developmental conversion*. This distinction was first explored in some detail by Schmalhausen (1949), who noted that some kinds of environmental influences on phenotype were likely the passive result of the action of the environment on aspects of the developmental pathways producing the trait, as when, for example, increased temperature increases the rate of a chemical reaction, and thus changes the resulting phenotype. This is *developmental sensitivity* – the sensitivity of the developmental process to environmental variation. In other cases, Schmalhausen noted, the response to environmental variation was more obviously a complex adaptive response that directed phenotypic development down one pathway rather than another (Schmalhausen, 1949, p.6); these are cases of *developmental conversion*. In both cases, the same genotype of an organism will be associated with different phenotypes in different environments – developmental sensitivity and developmental conversion are therefore both kinds of phenotypic plasticity. But these different kinds of plasticity are associated with different sorts of developmental mechanisms and can be associated with different evolutionary histories.

In the case of *developmental sensitivity* (aka *developmental modulation*, see Schlichting & Pigliucci, 1998, pp.72–3), the environment influences the trait through what are essentially passive mechanisms. If, for example, a particular kind of plant responds to increased available soil nutrients by growing larger and more rapidly, various traits will show plasticity with respect to the nutrient levels; however, the explanation for this plasticity may simply be the increased availability of resources for development in the plants grown under favorable conditions (see Dudley, 2004, pp.163–5). In these cases, the development of a particular trait in an organism will vary (more or less) continuously based on the environment experienced during development (see Figure 12.2, Genotypes A and B) (note, however, that the influence of other genes on the trait in question or the developmental links between the trait in question and other traits can mask what would otherwise be a straightforward response to the environmental variable). There is, in these cases, a *quantitative* response by the trait in question to *quantitative* variation in the environment. There need have been no selection in this particular kind of plant for more rapid development in favorable conditions and less rapid development in unfavorable conditions; rather, development in both conditions may rely on the same developmental pathways utilizing available nutrients. Such plasticity may or may not be adaptive, and in any event does not require any particular adaptive mechanisms to be present – it is *often* simply the result of the same causal pathways utilizing different amounts of the same kinds of available resources.

At the molecular level, this can occur when particular genetic resources are utilized in the same way, but to differing degrees, based on the available resources (examples of this have been referred to as *allelic sensitivity* in the literature, in reference to the sensitivity of the variant of the gene in question to particular kinds of environmental variation; see Schlitching & Pigliucci, 1998, pp.72–3). Note that this description makes

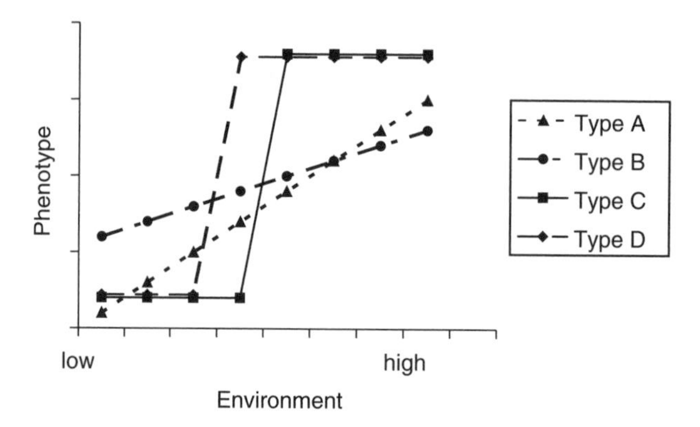

Figure 12.2 Developmental sensitivity versus developmental conversion. Genotypes Type A and Type B show phenotypic plasticity that is (likely) an example of developmental sensitivity; note that Type A is more plastic than Type B in this range of environments. Genotypes Type C and Type D show phenotypic plasticity via developmental conversation; in this case, Type C undergoes conversation at a lower environmental value than do those of Type D

it clear how norms of reaction associated with phenotypic sensitivity *can* (but again, need not) evolve in adaptive ways. Where, for example, phenotypic plasticity is associated with allelic sensitivity, the developmental resources that make the use of (for example) larger quantities of particular resources may be selected for in environments in which such large quantities are sometimes available, and hence the plasticity associated with variation in the amounts of the resource available could increase through selection on the ability to make use of larger quantities of the resource (this might occur via, say, the duplication of promoter regions and subsequent selection of those organisms with the more active promoters that are therefore able to utilize more effectively large amounts of the environmental resource available).

When phenotypic plasticity is the result of *developmental conversion*, some aspect of the environment associated with the environmental variation in question acts as a "signal" or "cue" which results in the organism (or more narrowly, the specific trait) in question taking one developmental pathway rather than another. Developmental conversion is, therefore, essentially an *active* response requiring at the very least a mechanism for detecting the environmental cue during development, a regulatory system that "switches" development toward one distinct kind of phenotype rather than another (activates one developmental cascade rather than another), and the developmental resources (at the molecular and other levels) to produce alternative phenotypes during development (see Figure 12.2, Genotypes C and D). Consider, for example, the shade-avoidance responses in the plant *Impatiens capensis*. Various phytochrome molecules are responsible for detecting the ratio of red to far-red light, which serves as a cue to vegetative density (in practice, the plants competing for light are often members of the same species). Other receptors then interpret the signal generated from the phytochrome molecules and activate a developmental cascade that produces one of two general responses – a plant with an elongated stem, or a plant that remains closer to the ground. The former has higher fitness in areas of dense growth, as it produces leaves higher up where they can be more successful in competition for light, whereas the latter has higher fitness in areas of less-dense growth, as the shorter stem is more resistant to physical damage (see Dudley & Schmitt, 1996). Given the complexity of the systems necessary for developmental conversion, and the costs of maintaining those regulatory pathways, it is reasonable to suggest that developmental conversion will *usually* be an adaptation – that is, an evolved response to environmental variability. Hence, understanding phenotypic plasticity that results from developmental conversion will usually require an evolutionary account of the developmental process in a way that understanding phenotypic plasticity that results from developmental sensitivity often will not.

Since, in the case of developmental conversion, there must be particular developmental resources that can detect some environmental variable and convert it into a biological signal, regulatory mechanisms for converting that biological signal into the activation and deactivation of various particular developmental pathways, and developmental systems in place for the generation of the developmentally distinct phenotypes, one can reasonably hope to find a variety of molecular resources involved specifically in such tasks (sometimes referred to "plasticity genes"; see Schlitching & Pigliucci, 1993, 1998; Windig, De Kovel, & de Jong, 2004; but see Box 12.2). While

Box 12.2 What Are "Plasticity Genes" and Do They Exist?

The concept of "plasticity genes" has a somewhat troubled history; different researchers have proposed different definitions, and then proceeded to argue for the existence or non-existence of "plasticity genes" on the basis of those definitions (see Windig et al., 2004, pp.36–8, and Sarkar, 2004, pp.25–7, for discussion). As noted in the main text, phenotypic plasticity of the developmental sensitivity form is often associated with *allelic sensitivity*. While the developmental pathways that produce those traits (the traits that show developmental sensitivity) will of course involve genes (as well as various epigenetic mechanisms and environmental conditions), the causal involvement of those genes is not different in any particularly significant way from the way that similar genes (and epigenetic mechanisms) are used in the development of less-plastic traits. Calling these "plasticity genes" would therefore be to stretch the concept in a particularly unhelpful way.

Similarly, the existence of genes that are associated (often causally) with *variation* in phenotypic plasticity within particular populations is not in question; but whether or not these should be properly called "plasticity genes" can (and should) be seriously questioned. Like all complex traits, adaptive phenotypic plasticity requires the presence of a complete complex of developmental resources (genes, gene complexes, intracellular systems, and environmental factors, to name a few); those genes that happen to be associated with *variation* in the plastic response may or may not be a particularly interesting aspect of the developmental system that produces the adaptive response (what Schlichting and Pigliucci refer to the necessary "infrastructure" for the evolution of plasticity: 1998, pp.74–7).

Recently, Pigliucci (forthcoming) has suggested that even the best candidates for plasticity genes – those genetic resources (genes and suites of genes) involved in

Continued

"switching" development down one developmental pathway rather than another (those genes most critically wrapped up in developmental conversion) – are likely too diverse a group of genetic resources to be properly thought of as a single *type* or *kind* of gene (that is, as "plasticity genes"). Rather, he suggests that any such genes (and/or gene complexes) should be analyzed in terms of the specific uses to which they are put in development, and their particular evolutionary histories within the populations in question. We should think, in other words, in terms of "the specific molecular underpinnings of particular kinds of plasticities" (Pigliucci, forthcoming) rather than of "plasticity genes" in general.

there will of course be genes involved in every kind of phenotypic plasticity, in the case of developmental sensitivity these genes will be the same (kinds of) genes that are involved in the development of the phenotypes themselves – not genes whose function is to actively control the phenotypic response.

Phenotypic plasticity is adaptive whenever the fitnesses of the different phenotypes generated in the different environments regularly encountered by organisms of that type are higher in the environments in which they appear than they would be in the other environments, and where no "intermediate" phenotype would do as well in all the environments regularly encountered. In some cases, of course, phenotypes that develop in one kind of environment simply could not develop in other kinds; the case discussed above, of annual plants that are larger and have higher fruit output in high-quality environments (nutrient rich, proper water, high sunlight, etc.) than in low-quality environments is a case in point – the "large plant" phenotype may simply be unobtainable in low-quality environments due to a lack of environmental resources. However, the "large plant" phenotype cannot be fitter than the "small plant" phenotype in low-quality environments, because the resources to produce the "large plant" phenotype are simply unavailable in that environment; the "small plant" phenotype, then, has, by default, the highest fitness in that low-quality environment.

Phenotypic plasticity is not always adaptive. For example, tragically, various aspects of the human phenotype are sensitive to the presence of thalidomide during early development – limb structure, for example, is *plastic* with respect to the presence of thalidomide. In this case, though, development is *derailed* by certain kinds of environments. Nor is adaptive phenotypic plasticity present in every case where one might imagine it to be useful; adaptive developmental conversion, the ability of a particular genotype to produce different well-adapted phenotypes in response to an environmental variable, is not ubiquitous.

3. Environmental Heterogeneity, Cues, and Plasticity

As the above suggests, it is possible to generalize somewhat about the relationship between environmental heterogeneity and phenotypic plasticity. First, and most obviously, environmental homogeneity will in general be associated with reduced adaptive phenotypic plasticity. Insofar as the environment is homogeneous, adaptive phenotypic plasticity will be unnecessary to maintain reasonably high fitness, and developmental sensitivity will be minimized by the limited range of developmental environments encountered. Of course, outside the range of environments normally encountered, the organisms in question may well display extensive *non-adaptive* plasticity (see Section 4, below). But in homogeneous environments (keeping in mind that, for example, other members of the same population can themselves be part of the environment), adaptive plasticity will be neither selected for nor present.

Environmental heterogeneity may or may not be associated with adaptive phenotypic plasticity, depending upon the particular features of the environmental heterogeneity (and the particular developmental pathways involved in the production of the trait). Generally, if the environment is heterogeneous in a way that can impact fitness, such that different phenotypes will be better adapted (than other specialized phenotypes or any "generalist" phenotype; see Box 12.3) to the different environments, if there are

Box 12.3 Life-History Strategies

Below are sketches of several possible life-history strategies. In fact, of course, most actual life-history strategies are a compromise or a combination of several of these, and different strategies may be adopted with respect with respect to different traits within a particular kind of organism.

(1) Specialist. The "specialist" life-history strategy is focused on the production of a phenotype that is adapted to one particular environment, to which it is particularly well suited. It is likely to have rather low relative fitness in other environments. Specialist strategies are particularly likely to evolve in relatively homogeneous environments.
(2) Adaptive Plasticity by Developmental Conversion. Where there are environmental cues that can signal the environment to be encountered, and where different phenotypes would have high fitness in the different environments, developmental conversion can permit an organism to develop the appropriate phenotype for the environment encountered.
(3) Generalist. The "generalist" life-history strategy produces a phenotype that can be expected to be reasonably successful in many of the environments likely to be encountered, but which is not particularly well suited to any particular environment. If there are no environmental cues that can be used to generate an adaptive developmental conversion response, and if a "middling" phenotype trait yields reasonable fitness across a range of environments, this may be a successful strategy.
(4) Adaptive Coin-Flipping. In cases where no "generalist" strategy is likely to generate a reasonably successful response to the environmental variation (where, for example, a "middle" phenotype would have very low fitness in the environments likely to be encountered), where there are no environmental cues that can be used, and where the various combinations of "right" and "wrong" phenotypes for the particular variant of the environment encountered have roughly equal high and low fitnesses, the best strategy for

Continued

organisms may be simply to adopt one of a number of alternate forms "randomly" and hope for a match.

(5) Bet-Hedging. This strategy is similar to adaptive coin-flipping, but is favored where the fitness of one combination is strikingly different. For example, where getting it "wrong" in one direction is lethal but getting it wrong in the other direction only moderately lowers fitness, adopting the phenotype that is unlikely to yield the lethal result may be the "safest bet."

no costs to plasticity, and if the plastic phenotype is able to take on any of the "specialized" phenotypes fittest in those environments, then, in the presence of useful (timely and accurate) environmental cues to the environmental condition the organism will be facing, "perfect" plasticity will be superior to other strategies (generalization, "coin-toss" specialization, etc; see Box 12.3) (DeWitt & Langerhans, 2004, p.99). It should be obvious that these conditions are never actually (exactly) met, but the degree to which they are approximated in particular circumstances, or the particular ways in which they fail to obtain, may suggest the likelihood of plasticity being selected over other strategies (DeWitt & Langerhans, 2004, p.100ff).

Unless there are environmental cues that reliably signal the environmental condition in time for development to respond with the production of an appropriate phenotype, adaptive phenotypic plasticity cannot evolve. In some cases, the development of the phenotype will be triggered by the same (or closely related) environmental cues that the phenotype is responding to; in other cases, the cue will be an environmental condition that is reliably correlated with the environmental condition to which the phenotype is an adaptive response. For example, many species of semi-aquatic plants (such as *Sagittaria sagittifolia* and *Ranunculus aquatilis*) produce different forms of leaves and stems above and below water, and yet a third sort at the transition area. The development of the different leaf and stem form can be triggered by the presence of moisture during the development of that leaf or stem, producing an appropriate "above" or "below" water form on the basis of the actual presence or absence of water. However, the association of "above" and "below" water leaves and water level is not perfect, and it can be shown that other cues (temperature, day-length, light quality, etc.) influence whether a particular leaf or stem forms into a "below" water or an "above" water sort (see Wells & Pigliucci, 2000). It is possible that by using these other environmental conditions as cues, some semi-aquatic plants are able to "anticipate" (predict with some reasonable accuracy) the likely conditions that leaves and stems not yet developed will find themselves in, and hence generate an overall phenotype with a higher fitness than could be achieved by leaf and stem development following *current* water conditions only. Similarly, various species of the common crustacean *Daphnia* (though *Daphnia pulex* is perhaps the best studied) respond during development not to the presence or absence of predators per se, but to the presence or absence of a chemical produced by the larvae of a common predator; *Daphnia* develop defensive neck spines when the chemical is present but not otherwise (see Windig et al., 2004, p.40).

Not only must there be timely environmental cues, but in order for adaptive phenotypic plasticity to evolve, the environmental variation in question has to be of a "grain" appropriate to an adaptive developmental response. If the heterogeneity is too fine-grained, either spatially or temporally, a plastic response could not be helpful. When, for example, the environment is likely to change far more rapidly than the phenotype in question can be produced, attempting to match the phenotype to the environment will fail, and a "generalist" strategy will be necessary (or, if some environments are far more stressful than others, perhaps a "bet-hedging" strategy will be most successful; see Box 12.3). Similarly, if the heterogeneity is too coarse-grained, again either spatially or temporally, adaptive phenotypic plasticity will be unlikely to evolve and "ordinary" selection for fixed specialist phenotypes will be necessary – for example, even nicely cyclic changes in climate over geological timescales would not generate plasticity in short-lived annual plants!

Since adaptive plasticity by developmental conversion requires the biological mechanisms necessary to detect cues and signal a response, as well as the developmental systems necessary to produce the alternative phenotypes, it is reasonable to suppose that adaptive plasticity must often accrue costs absent from "fixed" strategies (see DeWitt, Sih, & Wilson, 1998, pp.79–80). However, measuring the costs of plasticity in particular cases is difficult, and results have been mixed, with some studies pointing toward some kinds of plasticity being relatively costly, and others failing to detect significant costs (see DeWitt et al., 1998, for review). A further difficulty is that current methods of attempting to measure the costs of plasticity rely primarily on comparisons of the fitness between (relatively) "fixed" and (relatively) "plastic" genotypes in a variety of environments; these comparisons may not be reliable, as both the environments and the genotypes tested may or may not be sufficiently similar to those found in natural populations (see DeWitt et al., 1998, esp. Box 2; and DeWitt, 1998). In any event, the benefits of plastic responses must outweigh the costs for plastic strategies to be selected.

The fact that not every trait displays adaptive phenotypic plasticity in response to environmental variation suggests that phenotypic plasticity is not cost-free, that not every aspect of environmental heterogeneity provides appropriate cues or is appropriately grained for the development of plastic responses, that appropriate heritable phenotypic variation has not been generated in the recent history of the lineage, and/or that the developmental pathways responsible for the phenotypes preclude the spread of adaptive plasticity in the population (DeWitt & Langerhans, 2004, 99). Non-adaptive phenotypic plasticity – especially non-adaptive phenotypic sensitivity – does not require that the above conditions be met, as it is a byproduct of development. It is, then, the fact that mal-adaptive plasticity *is not* completely ubiquitous that requires explanation, and it is to the (adaptive) buffering of developmental systems that we now turn.

4. Phenotypic Plasticity and Developmental Buffering

Phenotypic plasticity is usually thought of as the ability of organisms with identical genotypes to produce distinct phenotypes under different environmental conditions; in these cases, differences in the environment are associated with different developmental results (see Box 12.1 for more on other heritable development resources). However, the ability of organisms to recover from environmental (or other) insults during

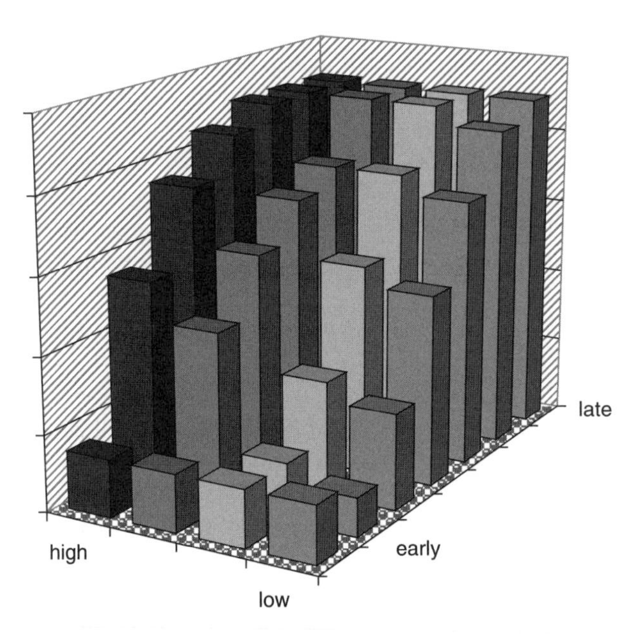

Figure 12.3 A developmental reaction norm visualization of buffering. In this case, there is substantial phenotypic variation associated with environmental variation (from low to high environmental values) early in development, but that variation is not maintained through development (at the 'late' developmental stage, the phenotypic values associated with different developmental environments are all approximately equal). Note that different 'pathways' towards the 'final' phenotype are taken depending upon the starting environment. Development is plastic, but the final phenotype does not show a plastic response to the environment. Representing a variety of different types of organisms in a single DRN in a way that is clear graphically is difficult, and I do not attempt to do so here, relying rather on comparing DRNs of single developmental types of organisms

development and produce similar phenotypes *despite* the different environments encountered is equally a form of phenotypic plasticity. Plasticity that is focused on development rather than on the final phenotypic form is usually referred to as *developmental phenotypic plasticity*, and it is this kind of plasticity that is responsible for *buffering* organisms from environmental variation and *canalizing* development toward particular phenotypic outcomes. Developmental processes that act to prevent environmental variation from influencing final phenotype reduce the apparent phenotypic plasticity of the genotype; however, the pathways necessary to produce similar outcomes despite dissimilar environmental inputs themselves represent a kind of *developmental* plasticity.

Developmental phenotypic plasticity can be represented graphically as a *developmental reaction norm (DRN)*; in a DRN, the phenotype of the organism in question is followed through a variety of developmental stages of the organism in a variety of different environments. Where the development of a particular phenotype is *canalized*, different developmental pathways produce the same resultant phenotype in a range of different developmental environments (see Figure 12.3); in the environmental ranges in which an organism's phenotypic development is canalized, one can speak of development being *buffered* from that environmental variation. In cases where plasticity is the

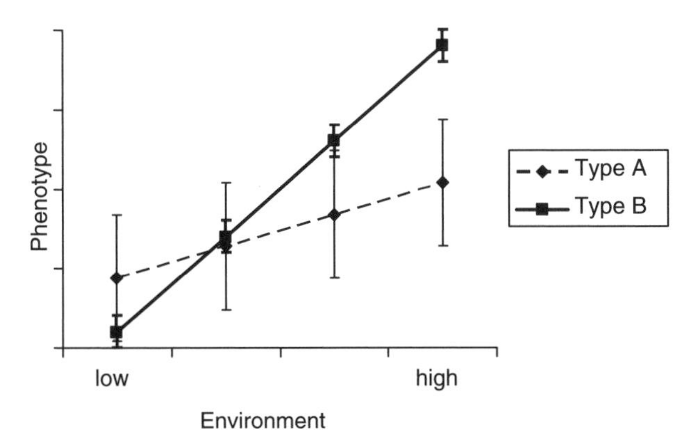

Figure 12.4 Developmental stability versus phenotypic plasticity. In this example, the lines represent the mean phenotypic value of the genotype given the environment; the error bars measure the degree of phenotypic variation within that environment for each genotype. While organisms of genotype Type B display a more plastic response to the environmental variable, those of Type A show greater developmental instability

result of developmental sensitivity with respect to the environment in question, the phenotype in question is not canalized; in these cases, phenotypic plasticity is the result of a *lack* of buffering (see Figure 12.2, Genotypes Type A and Type B). By contrast, in the case of phenotypic plasticity that results from developmental conversion, the development of the phenotype in question can be canalized with respect to the environments in question. Development in these cases is channeled in one direction rather than another on the basis of the presence or absence of an environmental cue, and the development of the phenotype in question may be canalized around two different developmental pathways, and hence toward two distinct, but individually well buffered, phenotypic results.

It is important not to confuse *phenotypic plasticity* with *developmental instability*. Again, phenotypic plasticity is the ability of organisms with identical genotypes (again, and/or other heritable developmental resources) to develop different phenotypes in different environments. Developmental instability, on the other hand, is a measure of the reliability with which particular developmental pathways in a particular organism will produce particular phenotypic outcomes *within* a particular environment; the result of developmental instability is sometimes referred to as *developmental noise*. The sensitivity of a plant's overall size (say) to available nitrogen in the soil is an example of plasticity – variation in the availability of nitrogen is associated with variation in the phenotype. On the other hand, the different number of bristles on either side of an individual fruit fly is not, so far as one can tell, associated with environmental variation; the difference in bristle number on the left- and right-hand sides of a fly is, therefore, an example of developmental noise (see Lewontin, 1974). So the mechanisms that produce canalization (buffering) are not necessarily the same as the mechanisms that ensure developmental stability; the former produce similar (mean) phenotypes in *different* environments; the latter reduces the variability of the resulting phenotype *within* a particular environment (see Figure 12.4). While some researchers have argued that developmental instability is the result of our inability to completely eliminate or control for variations in the developmental environments (Kitcher 2001), others have appealed to the complexity of the developmental process and to the possibility that development is not wholly deterministic (see Lewontin, 1974, for examples; see Dupré, 1993, for the argument in favor of non-determinism). It is currently impossible to discriminate empirically between these views of the causes of developmental instability, and in any event the distinction is irrelevant for most practical purposes.

Some researchers have claimed that increased plasticity might be associated with increased developmental instability (see DeWitt et al., 1998). The ability to produce multiple phenotypes might reduce the reliability with which any particular phenotype develops. There is, however, little or no empirical support for this contention, and given that the mechanisms responsible for buffering and the canalization of particular developmental pathways are at least partially independent of those responsible for the plastic responses that channel development down one pathway rather than another, there are good reasons to think that in fact there is no such association between developmental instability and adaptive plasticity (see Schlitching & Pigliucci, 1998, pp.64–6), though further research on this topic is no doubt desirable.

Developmental plasticity that buffers organismal development can result not only in reduced sensitivity to variation in the external environment, but also reduced sensitivity to genetic variation and to variations in epigenetic mechanisms. That is, the mechanisms that channel development toward particular phenotypic outcomes can "suppress" the expression of genetic and epigenetic variation that would otherwise result in phenotypic variation. This can be considered a kind of developmental buffering; and the canalization of resulting phenotypes is similar to, and likely uses some of the same developmental mechanisms as, the buffering of phenotypes with respect to (external) environmental variation (see Pigliucci, 2002 and 2003, for review).

Breakdowns in buffering systems, caused either by external environmental variation that overwhelms the buffering system or by breakdowns in the internal systems responsible for buffering, are therefore associated with increased *non-adaptive* phenotypic plasticity. The novel phenotypes that result from such breakdowns will have been shielded from recent natural selection by the buffering systems. Once released, however, such variation will of course be subject to selection. It is for this reason that some researchers have spoken of such buffering systems as maintaining a "reserve" of genetic and epigenetic variation in natural populations. It should be noted that despite the occasional sloppy language surrounding such claims (see Rutherford & Lindquist, 1998, p.341), most researchers are not seriously proposing that buffering systems have been *selected* for their ability to maintain such a reserve and permit rapid phenotypic evolution by breaking down under stress. Schlichting perhaps comes closest when he suggests that some lines of research have hinted that lineages with suppressed plastic responses may be more resistant to extinction, and hence that lineage selection may play a role in the way that canalizing systems break down (2004, p.199). But it seems more likely that the breakdown of the buffering systems that permits the exposure of such variation is a broadly non-adaptive feature of the limits on the ability of those buffering systems to respond to novel or overwhelming environmental or genetic variation. Nevertheless, the existence of such unexpressed variation can have significant evolutionary consequences. Breakdowns in the buffering system can permit the very rapid exploration of novel phenotypes; some of these phenotypes may be selected for and then maintained by the very buffering systems the breakdown of which permitted their expression (see Pigliucci, 2002 and 2003, for review).

5. The Future of Phenotypic Plasticity Research

The importance of research into phenotypic plasticity is currently well recognized, and there are many researchers and research programs focused on plasticity (see Scheiner & DeWitt, 2004). Phenotypic plasticity is an important aspect of the development of many, if not most, complex phenotypic traits. Despite the progress made over the past several decades, however, there are still a number of important questions that remain to be answered. These include at least: how plasticity responses are developmentally instantiated; how and under what conditions phenotypic plasticity evolves; how and under what conditions variation in plastic responses is maintained in populations; how and under what conditions phenotypic plasticity will be adaptive; and what kinds of factors limit the evolution of plasticity in particular lineages. In working toward answering these questions, researchers will no doubt uncover more examples of, more kinds of, and more variation within, plasticity in particular populations.

While clearly related to more general problems in developmental biology, understanding how phenotypic plasticity of different sorts is developmentally instantiated, and what implications the different kinds of instantiation have for the kinds of plasticity observed, carries its own set of difficulties. Even restricting the question to development in the case of adaptive phenotypic plasticity via developmental conversion, it has become clear that different species, and different plasticity responses, use different (kinds of) developmental resources in the singling process that leads to the development of one phenotypic response rather than another, and different kinds of developmental processes are evoked in the generation of the phenotypes (see Pigliucci, forthcoming). There may, indeed, be no one way in which the development of plastic responses differs from the development of fixed responses; nevertheless, understanding the developmental pathways that lead from particular environmental cues through the generation of alternative phenotypes in particular populations will be key to understanding plasticity within those populations.

The evolution of phenotypic plasticity has been explored by a number of methods, including modeling, artificial selection experiments, regression analyses, and phylogenetic analyses. Despite criticisms of some of these methods, this remains an active avenue for research, and further research, including the further refinement of these techniques (and perhaps the development of new techniques), will no doubt begin to shed more light on the conditions under which adaptive phenotypic plasticity can be expected to evolve, as well as the conditions under which adaptive developmental plasticity leading to well-canalized/well-buffered traits can be expected to evolve. This will no doubt include the exploration of the history of plasticity, and the genes and epigenetic mechanisms associated with plasticity, in various lineages.

The role of phenotypic plasticity in maintaining genetic diversity is another area of contemporary research that will likely continue to yield interesting results. There are a number of different phenomena that might be included under this general rubric. As noted above, developmental plasticity that results in canalized phenotypic outcomes can suppress genetic variation, and hence the systems responsible for buffering development can maintain a hidden reserve of genetic variation. Heritable variation in the reaction norms of a particular population can, in principle, maintain genetic diversity; however, the extent to which this phenomenon is in fact responsible for maintaining genetic diversity is still controversial, and more research, both theoretical and empirical, will be needed before any answers can be regarded as more than very tentative. Finally, as plasticity may reduce the degree of specialization within populations, it may result in additional genetic variation being maintained as a byproduct of, for example, permitting a larger effective population size. Again, this hypothesis has not been adequately tested empirically nor modeled effectively, and further research will be needed to establish whether this has any real impact on population-level genetic variation.

Finally, despite recent interest in modeling and empirically testing for the possible costs of phenotypic plasticity (see, e.g., Berrigan & Scheiner, 2004; DeWitt & Langerhans, 2004; DeWitt et al., 1998), the exploration of the limits of plasticity and the causes of those limits is still in its relative infancy. It is currently not known why some kinds of plasticity seem relatively common, whereas other kinds, of plausibly high fitness value, seem relatively unusual. For example, there are many examples of adaptive shade avoidance in plants and of predator avoidance in crustaceans, but relatively few examples of plants that can take on either vine-like, shrubby, or tree-like forms based on local conditions. Scheiner and DeWitt go further, and wonder why "there are so many species, rather than just a few highly plastic ones" (2004, p.202). Alas, they go on to note that this will be a hard question to answer, as "it is much harder to demonstrate why something does not evolve than to demonstrate why it has evolved" (Scheiner & DeWitt, 2004, p.202). But while exploring the costs and limits of plasticity will no doubt be difficult, this view suggests that it will also be critical to the understanding of such phenomena as speciation, the formation of ecotypes, and biological diversity more generally.

The diverse phenomena grouped under the broad heading of "phenotypic plasticity" are at the heart of many, if not most, of the central issues in current evolutionary and developmental biology. For this reason, we should expect that studies of phenotypic plasticity will continue to move to the center stage of evolutionary and developmental biology, and that the "plasticity perspective" will be an essential part of the growing interest in the intersection between developmental and evolutionary biology.

Acknowledgments

Sharyn Clough, Massimo Pigliucci, Anya Plutynski, and Sahotra Sarkar provided useful feedback and comments on previous drafts of this piece; it is much improved thanks to their help. My work with Massimo on these topics, now stretching to almost a decade, has deeply influenced my thinking on these issues in innumerable ways. Any confusions or oversights that remain are of course my own.

References

Berrigan, D., & Scheiner, S. M. (2004). Modeling the evolution of phenotypic plasticity. In T. J. DeWitt & S. M. Scheiner (Eds). *Phenotypic plasticity: functional and conceptual approaches* (pp. 82–9). New York: Oxford University Press.

DeWitt, T. J., Sih, A., & Wilson, D. S. (1998). Costs and limits of phenotypic plasticity. *Trends in Ecology and Evolution*, 13(2), 77–81.

DeWitt, T. J. (1998). Costs and limits of phenotypic plasticity: tests with predator-induced morphology and life history in a freshwater snail. *Journal of Evolutionary Biology*, 11, 465–80.

DeWitt, T. J., & Langerhans, R. B. (2004). Integrated solutions to environmental heterogeneity: theory of multimoment reaction norms. In T. J. DeWitt & S. M. Scheiner (Eds). *Phenotypic plasticity: functional and conceptual approaches* (pp. 98–111). New York: Oxford University Press.

Dobzhansky, T. (1955). *Evolution, genetics, and man*. New York: John Wiley & Sons, Inc.

Dudley, S. A., & Schmitt, J. (1996). Testing the adaptive plasticity hypothesis: density-dependent selection on manipulated stem length in *Impatiens capensis*. *American Naturalist*, 147, 445–65.

Dudley, S. A. (2004). The functional ecology of phenotypic plasticity in plants. In T. J. DeWitt & S. M. Scheiner (Eds). *Phenotypic plasticity: functional and conceptual approaches* (pp. 151–72). New York: Oxford University Press.

Dupré, J. (1993). *The disorder of things: metaphysical foundations of the disunity of science*. Cambridge, MA: Harvard University Press.

Kitcher, P. (2001). Battling the undead: how (and how not) to resist genetic determinism. In R. Singh, C. Krimbas, D. Paul, & J. Beatty (Eds). *Thinking about evolution: historical, philosophical and political perspectives* (pp. 396–414). Cambridge: Cambridge University Press.

Lewontin, R. C. (1974). The analysis of variance and the analysis of causes. *American Journal of Human Genetics*, 26, 400–11.

Moss, L. (2003). *What genes* can't *do*. Cambridge, MA: The MIT Press.

Odling-Smee, F. J., Laland, K. N., & Feldman, M. W. (2003). *Niche construction: the neglected process in evolution*. Princeton and Oxford: Princeton University Press.

Oyama, S., Griffiths, P. E., & Gray, R. (Eds). (2001). *Cycles of contingency: developmental systems and evolution*. Cambridge, MA: MIT Press.

Pigliucci, M. (2002). Buffer zone. *Nature*, 417 (June), 598–9.

Pigliucci, M. (2003). Epigenetics is back! Hsp90 and phenotypic variation. *Cell Cycle* 2 (1, January/February), 34–5.

Pigliucci, M. (forthcoming). The genetics of phenotypic plasticity: where are we, and where are we going? *New Phytologist*.

Robert, J. S. (2004). *Embryology, epigenesis, and evolution: taking development seriously*. Cambridge: Cambridge University Press.

Rutherford, S. L., & Lindquist, S. (1998). Hsp90 as a capacitor for morphological evolution. *Nature*, 396 (November), 336–42.

Sarkar, S. (2004). From the *Reaktionsnorm* to the evolution of adaptive plasticity: a historical sketch, 1901–1999. In T. J. DeWitt & S. M. Scheiner (Eds). *Phenotypic plasticity: functional and conceptual approaches* (pp. 10–30). New York: Oxford University Press.

Sarkar, S., & Fuller, T. (2003). Generalized norms of reaction for ecological developmental biology. *Evolution & Development*, 5, 106–15.

Scheiner, S. M., & DeWitt, T. J. (2004). Future research directions. In T. J. DeWitt & S. M. Scheiner (eds.), *Phenotypic plasticity: functional and conceptual approaches* (pp. 201–6). New York: Oxford University Press.

Schlichting, C. D., & Pigliucci, M. (1993). Control of phenotypic plasticity via regulatory genes. *American Naturalist*, 142, 366–70.

Schlichting, C. D. (2004). The role of phenotypic plasticity in diversification. In T. J. DeWitt & S. M. Scheiner (Eds). *Phenotypic plasticity: functional and conceptual approaches* (pp. 191–200). New York: Oxford University Press.

Schlitchting, C. D., & Pigliucci, M. (1998). *Phenotypic evolution: a reaction norm perspective.* Sunderland, MA: Sinauer Associates.

Schmalhausen, I. I. (1949). *Factors of evolution: the theory of stabilizing selection.* (T. Dobzhansky, Trans.). Chicago and London: Chicago University Press (Original work published 1946).

Sollars, V., Lu, X., Xiao, L., Wang, X., Garfinkel, M. D., & Ruden, D. M. (2003). Evidence for an epigenetic mechanism by which Hsp90 acts as a capacitor for morphological evolution. *Nature Genetics*, 33 (January), 70–4.

Wells, C. L., & Pigliucci, M. (2000). Adaptive phenotypic plasticity: the case of heterophylly in aquatic plants. *Perspectives in Plant Ecology, Evolution and Systematics*, 3(1), 1–18.

Windig, J. J., De Kovel, C. G. F., & de Jong, G. (2004). Genetics and the mechanics of plasticity. In T. J. DeWitt & S. M. Scheiner (Eds). *Phenotypic plasticity: functional and conceptual approaches* (pp. 31–49). New York: Oxford University Press.

Further Reading

There have been two recent book-length reviews of phenotypic plasticity. DeWitt and Scheiner's edited collection, *Phenotypic plasticity: functional and conceptual approaches* (2004, New York: Oxford University Press) provides an excellent snapshot of the current state of the art in several key areas of plasticity studies. A more systematic introduction to the field is provided by Pigliucci's *Phenotypic plasticity: beyond nature and nurture* (2001, Baltimore: John Hopkins University Press). West-Eberhard's *Developmental plasticity and evolution* (2003, Oxford University Press) explores in some detail the possible relationships between phenotypic plasticity and evolution. "Phenotypic plasticity and evolution by genetic assimilation," by Pigliucci, Murren, and Schlitchting (2006), briefly reviews some of the issues involved in linking plasticity to evolutionary change in populations (*The Journal of Experimental Biology*, 209, 2362–7).

Chapter 13

Explaining the Ontogeny of Form: Philosophical Issues

ALAN C. LOVE

The aim of this article is to survey philosophical issues that arise in offering scientific explanations of the ontogeny of form. Section 1 presents a conceptual framework from which to understand these explanations as responses to many distinct but related questions in developmental research. The second section identifies and describes the biological content of these questions, both in terms of the phenomena to be explained and current preferences for molecular genetic approaches. Each subsequent section focuses on an area of epistemology relevant to explaining the ontogeny of form (representation, explanation, and methodology). Topics discussed include typology, individuation, model systems, reduction, and research heuristics. In closing, I draw attention to several metaphysical topics that deserve further scrutiny.

1. The Old Problem (Agenda) of the Ontogeny of Form

Explaining the ontogeny of form, that is, discerning the processes and causes that generate the different shapes, size, and structural features of an organism as it develops from embryo to adult, is an old problem domain in the life sciences. The basic issues surrounding these explanations go back to ancient Greece. Aristotle rejected purely "efficient" causal explanations for the developmental origination of morphological features. "Formal" and "final" causation were necessary to adequately explain the ontogeny of form. A clear lesson from Aristotle is that philosophical commitments about scientific explanation permeate questions about what is required to explain how macroscopic complex "form" features of organisms emerge from seemingly simpler features of the embryo (Lennox, 2001).

One of the most persistent dichotomies in explanatory projects directed at these questions is epigenesis versus preformation (Maienschein, 2005). Epigenesis is the claim that heterogeneous, complex features of form emerge from homogeneous, less complex embryonic structures through interactive processes. Thus an explanation of the ontogeny of these form features requires attention to how these interactions occur. Preformation is a claim to the contrary that complex form preexists in the embryo and "unfolds" via ordinary growth processes. An adequate explanation involves detailing how growth occurs. Although preformation has a lighter explanatory burden in accounting for how form emerges during ontogeny (on the assumption that growth is easier to explain than process interactions), it must also address how the starting point of the next generation is formed with the requisite heterogeneous complex features. This was sometimes accomplished by embedding smaller and smaller miniatures *ad infinitum* inside the organism. Though nothing prevents mixing these two outlooks in explaining different aspects of the ontogeny of form, polarization into dichotomous positions has occurred frequently (Maienschein, 2005; Roe, 1981; Smith, 2006).

Attending to only preformation and epigenesis is a drastic oversimplification of the historical dimensions of explaining the ontogeny of form (see, e.g., Lenoir, 1989; Oppenheimer, 1967). For many of the issues discussed here, one key aspect of recent history is the molecularization of experimental (as opposed to comparative) embryology (Fraser & Harland, 2000), with the concomitant stress on the explanatory power of genes. Although the emphasis on genes has been controversial among developmental biologists (Berrill, 1961), one point of commonality is that explaining the ontogeny of form consists of many interrelated questions rather than a single problem. These questions have been manifested with differing frequency and vigor through history. The ability to answer any of them, as well as the nature of the questions themselves, is contingent on different research strategies and methodologies.

We can observe this multiplicity of questions in philosophical commentary on the problem of explaining development. For example, Sober refers to just two questions of interest on the agenda of problems surrounding ontogeny. "There are problems in biology that remain unsolved. The area of development (ontogeny) is full of unanswered questions. How can a single-celled embryo produce an organism in which there are different specialized cell types? How do these cell types organize themselves into organ systems?" (Sober, 2000, p.24). Moss claims that "the real question concerning metazoan ontogeny is just how a single cell gives rise to the requisite number of differentiated cell lineages with all the right inductive developmental interactions required to reproduce the form of the mature organism" (Moss, 2003, p.97). There are clearly *many* questions lurking in Moss's description of "*the* real question," including but not limited to features of cellular differentiation and inductive interactions.[1] The central *problem* of development is actually composed of *many* different but related scientific questions, each of which can be seen as requiring answers to obtain an adequate explanatory framework.[2] Claims that developmental research has shown a lack of "erotetic progress" because of an inability to decompose its central question are unsubstantiated,[3] which will become clear as these questions are identified and characterized in detail (Section 2).

---

1 Other descriptions are susceptible to a similar analysis. "The central problem of developmental biology is to understand how a relatively simple and homogeneous cellular mass can differentiate into a relatively complex and heterogeneous organism closely resembling its progenitor(s) in relevant aspects" (Robert, 2004, p.1).

2 This can be observed among biologists as well. "Vertebrate mesoderm induction is *one* of the classical problems in developmental biology" (Kimelman, 2006, p.360, emphasis mine). In his textbook, Gilbert speaks of the "general problems of developmental biology" (Gilbert, 1997, p.2) or "general questions scrutinized by developmental biologists" (Gilbert, 2003, p.4).

3 "In contemporary developmental biology, there is . . . uncertainty about how to focus the big, vague question, How do organisms develop?" (Kitcher, 1993, p.115).

Although there are a number of independent reasons for preferring an analytical strategy in philosophy of science focused on problems rather than theories, it is more profitable to move directly to the idea of a "problem agenda" before using it to interpret attempts to explain the ontogeny of form. "Problem agenda" refers to any distinguishable set of related phenomena that pose a suite of intertwined research questions. These questions are investigated with the aim of providing a satisfactory explanatory framework capable of addressing all of the component phenomena. *Problem* highlights the emphasis on that which is unknown, uncertain, or perplexing – questions rather than answers. *Agenda* denotes the multifaceted nature of the unit. What is unknown is not one thing, but many, a sort of "list of things to be done" by a group of scientific researchers. Researchers address the problem agenda through the ongoing development of a satisfactory explanatory framework, as well as articulating new questions and reframing old ones. Problem agendas are larger units of analysis than individual empirical or theoretical problems and can be thought of as "big" questions (abstractly framed) concerning a particular domain of inquiry. Most individual researchers focus their attention on concrete *research questions* ("empirical problems") within the context of specific biological systems, tackling them theoretically or experimentally using a variety of different formal and laboratory techniques. Answering research questions contributes to a greater understanding of the problem agenda phenomena. Problem agendas are a combination of domains of phenomena with the cognitive activity of asking questions about these domains (cf. Bechtel, 1986). Formally, they can be seen as analogous to individual questions in philosophical discussions of scientific explanation (e.g., van Fraassen, 1980, ch. 5)

This necessarily truncated discussion generates several indicators for teasing apart what is involved in the project of explaining the ontogeny of form. We can expect to isolate and characterize problem agenda features such as the phenomena to explained, interrelated questions about those phenomena (with particular presuppositions), proposed explanations of phenomena, and implicit or explicit reasons for seeing specific explanations as adequate answers to member questions of the problem agenda.[4]

2. Explaining the Ontogeny of Form

Although there are many questions in the problem agenda of the ontogeny of form, philosophers of biology have turned to development over the past decade because of its promise to provide help in rethinking evolutionary theory (e.g., developmental systems theory; Oyama, Griffiths, & Gray, 2001) and deflate overstated claims about the causal power of genes (Keller, 2002; Neumann-Held & Rehmann-Sutter, 2006). Seemingly, many biologists have given up explaining *development* in favor of explaining *the role of genes in development*, while tacitly maintaining that the latter task is equivalent to the former (Robert, 2004). Whether this is in fact true needs to be investigated because it would imply a reduction in the number of research questions associated with the

4 There is no implicit commitment that the interrelated questions of problem agendas must exhibit hierarchical relationships, as others have argued for with respect to structural relationships among questions in a domain of inquiry (e.g., Kitcher, 1993, ch. 4).

ontogeny of form, as well as a negative evaluation of alternative, non-genetic explanations. Contemporary textbooks are a natural place to begin. "Developmental biology is at the core of all biology. It deals with the process by which the genes in the fertilized egg control cell behavior in the embryo and so determine its pattern, its form, and much of its behavior" (Wolpert et al., 1998, p.v). Besides the central role of genes offered, this description highlights that there is more to developmental biology than explaining the origin of form.[5] Wolpert distinguishes pattern and behavior, although it is natural to include pattern *form*ation in the category of "form." This conceptual slipperiness arises from the fact that "form" is not so straightforwardly characterized.

Some have cast form in terms of the production of "shape" (Davies, 2005), where the key process of "morphogenesis" is flagged etymologically ("morph" ≈ form; "genesis" ≈ coming to be). This excludes differentiation and signaling, which are often included in discussions of morphogenesis because cellular differentiation can lead to changes in cell *shape* (Minelli, 2003, ch. 6) and cell death (apoptosis) can sculpt morphology (Lohmann et al., 2002). A broader account can be culled from morphological investigation where form has been defined in terms of the material composition and arrangement, shape, or appearance of organic materials (Bock & Wahlert, 1965). Understanding form in this way recovers Wolpert's distinguishing of behavior from other aspects of development. The ontogeny of function, at all levels of organization, is a critical component for understanding ontogeny, but it is often bracketed because of the visibility (both past and present) of questions surrounding the ontogeny of form.

Most textbooks (e.g., Gilbert, 2003; Slack, 2006; Wolpert et al., 1998) describe a canonical set of events that occur in metazoan ontogeny. The first of these is *fertilization* (in sexually reproducing species), where an already structured egg (upper surface, animal pole; lower region, vegetal pole) is penetrated by sperm followed by the fusion of the nuclei to generate the appropriate complement of genetic material. Second, the fertilized egg undergoes several rounds of *cleavage*, which are mitotic divisions without cell growth that subdivide the zygote into many distinct cells. After a number of rounds of cleavage this spherical conglomerate of cells (now called a *blastula*) begins to exhibit some specification of the germ layers (endoderm, mesoderm,[6] and ectoderm), and then proceeds to invaginate at the vegetal pole, a process referred to as *gastrulation*, eventually generating a through-gut. (All three germ layers become established during or shortly after gastrulation is complete.) *Organogenesis* refers to the production of tissues and organs through the interaction and rearrangement of cell groups. Events confined to distinct taxonomic groups include *neurulation* in chordates, whereas others correlate with mode of development (*metamorphosis* from a larval to adult stage).

Several key processes underlie these distinct developmental events and the resulting features of form that emerge (the *through-gut* formed subsequent to gastrulation or the *heart* formed during organogenesis). These processes are critical to the ontogeny of form

---

5 Other textbooks see development primarily in terms of form: "Developmental biology is the science that seeks to explain how the structure of organisms changes with time. Structure, which may also be called morphology or anatomy, encompasses the arrangement of parts, the number of parts, and the different types of parts" (Slack, 2006, p.6).

6 Cnidarians (such as jellyfish and coral) do not have a mesodermal germ layer. They are sometimes referred to as "diploblastic" in contrast to metazoans with three germ layers ("triploblasts").

and mediate the types of research questions posed in the problem agenda. First, cellular shapes change during ontogeny. This is largely a function of cellular differentiation whereby cells adopt specific fates that include shape transformations.[7] Second, regions of cells in the embryo are designated through arrangement and composition alterations to generate different axes (dorsal–ventral, anterior–posterior, left–right, and proximal–distal). The successive establishment of these regions[8] is referred to as pattern formation. Third, cells translocate and aggregate into layers (e.g., endoderm and ectoderm, followed by the mesoderm in many lineages) and later tissues (aggregations of differentiated cell types). Fourth, cells and tissues migrate and interact to generate new arrangements and shapes composed of multiple tissue layers with novel functions (i.e., organs). These last two sets of processes are usually termed morphogenesis (Davies, 2005; Hogan, 1999) and include many distinct mechanisms (Figure 13.1). Fifth, there is growth in the size of different form features in the individual, remarkably obvious when comparing zygote to adult, although proportional changes between different forms (termed *allometry*) are often of primary interest (Richtsmeier, 2003).

None of these processes occur in isolation and explanations of particular form features usually draw on all of them simultaneously, often presuming form features that originated earlier in ontogeny by different instantiations and combinations of the processes. These core processes capture the broad contours of what kinds of questions are asked about "form" arising during development: how do various iterations and combinations of these processes generate form features during ontogeny? There is a shared presupposition that the phenomena (e.g., shape of the heart) are in need of explanation and not artifacts. A related presupposition is that these processes are routinely involved in the ontogeny of form.

A particular case of form origination illustrates the multiplicity of research questions in the problem agenda. How does the vertebrate heart, with its internal and external shape and structure (as well as location) originate during ontogeny (Harvey, 2002; Harvey & Rosenthal, 1999)? This particular phenomenon poses a number of interrelated questions related to the core processes. How does the heart come to exhibit left/right asymmetry in the body cavity, and be in that particular location? How do muscle cells migrate to, aggregate in, and differentiate at this location? How does the interior of the heart adopt a particular tubular structure with various chambers (that differ among vertebrate species)? How does the heart grow at a particular rate and achieve a specific size? How do different tissues interact to progressively generate the form of the heart? Answers to these questions entail characterizing the operation of the core processes. But cellular differentiation alone does not explain why the heart has particular cell types rather than others. Solutions relevant to explaining the ontogeny of form characterize causal factors that drive these core processes, especially the *specificity* of their outcomes. What causes cells to adopt a *muscle* cell fate? What causes certain tissues to interact in the prospective location of the heart? What causes the *arrangement* of the internal tubular shape of the heart? What causes growth in size to occur in the

7 "Totipotent" cells can adopt any fate whereas "pluripotent" cells are able to adopt many but not all fates.

8 Metaphorically termed "embryo geography" (Carroll, 2005) or "compartment maps" (Kirschner & Gerhart, 2005).

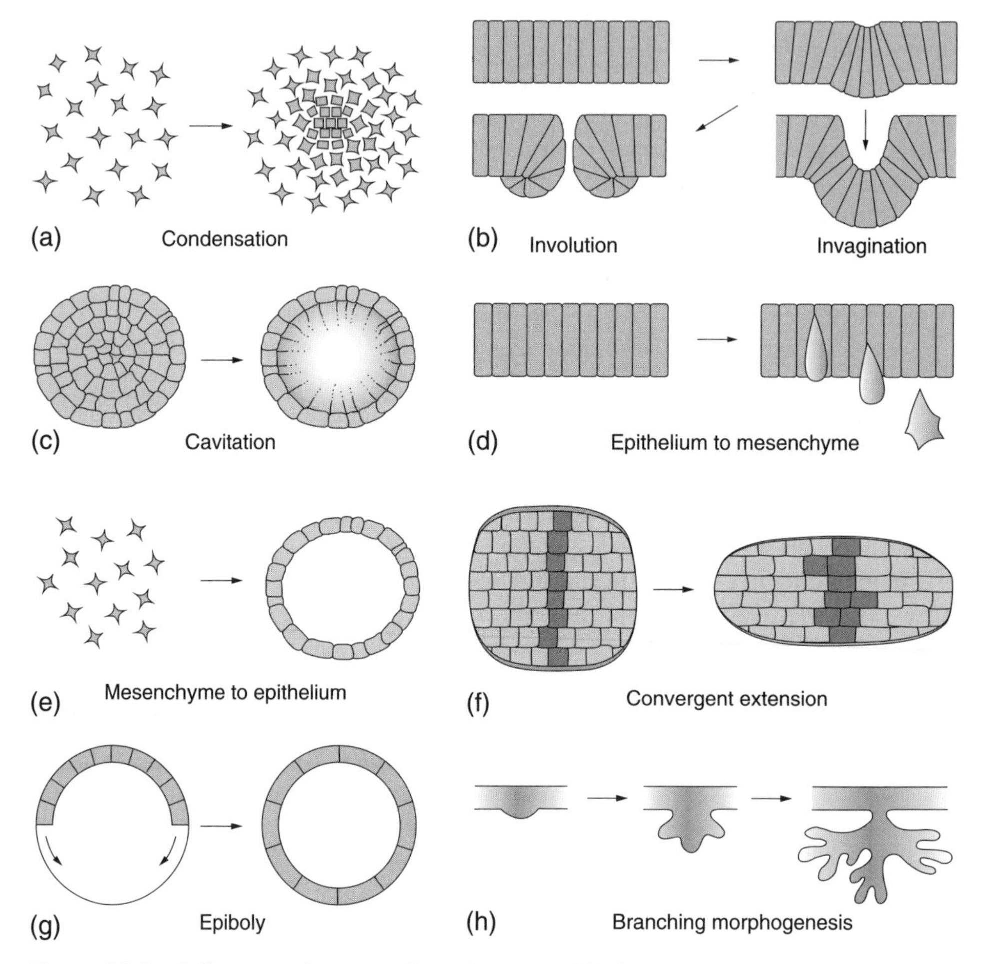

Figure 13.1 Different mechanisms of morphogenesis (Slack 2006, 17)

heart? Proposed solutions to the problem agenda of the ontogeny of form must appeal to causal factors relevant to questions such as these that pertain to the nature of the core processes.

It is no secret that the primary candidates for causal factors involved in answers to these questions are *genes*.[9] One primary rationale for this privileging (in the sense of holding genetic explanations more adequate than alternatives) is that the specificity of outcomes produced by the core processes is thought to lie in genetic "information" [SEE BIOLOGICAL INFORMATION]. This encourages the use of "blueprint," "program," and other linguistic metaphors in developmental investigations (Keller, 2002; Moss, 2003):

9 In fact, spatiotemporally regulated gene expression is taken as a complete solution to the origin of form by some researchers: "We now understand how complexity is constructed from a single cell into a whole animal" (Carroll, 2005, p.10). Many would evaluate this sentiment as premature.

"How is Form Encoded in the Genome?" (Carroll, 2005, p.34). Robert identifies a "consensus" around these metaphors that underwrites a blending of preformation and epigenesis themes according to three core theses: *genetic informationism* ("genes contain the entirety of the preformed, species-specific developmental 'information'"), *genetic animism* ("a genetic programme in the zygotic DNA controlling the development of an organism"), and *genetic primacy* ("the gene is the unit of heredity, the ontogenetic prime mover, and the primary supplier and organizer of material resources for development, such that the phenotype is the secondary unfolding of what is largely determined by the genes") (Robert, 2004, 39). This consensus is a mixture of themes from preformation and epigenesis because a preformed genetic program (passed along by inheritance) contains all the information determining the epigenetic outcomes observed during ontogeny.

However, there are reasons for thinking there might not be a consensus on development. Take an incriminating textbook example.

> How are the organizing principles of development embedded within the egg and in particular within the genetic material – DNA? . . . Genes control development mainly by determining which proteins are made in which cells and when. . . . The differences between cells must therefore be generated by differences in gene activity. Turning the correct genes on or off in the correct cells at the correct time becomes the central issue in development. All the information for embryonic development is contained within the fertilized egg. (Wolpert et al., 1998, pp.1, 13)

These loaded statements are often redacted or qualified.[10] For present purposes we only need evaluate the prospects of gene privileging for explanations of the ontogeny of form, not its actual distribution among current researchers. Robert argues against the privileging of genes by illustrating that they do not have the favored status attributed to them, either causally during ontogeny or transgenerationally via inheritance (Robert, 2004). The role of genes in development is only a subset of what is required to explain the reliable causal production of phenotypic features from generation to generation. This conclusion is synthesized from a variety of arguments offered by philosophers of biology to demonstrate that genetic informationism, genetic animism, and genetic primacy are all problematic (Jablonka & Lamb, 2005; Keller, 2002; Moss 2003; Oyama et al., 2001; Sarkar, 2000).

Instead of rehearsing these arguments, we can observe the abstract conclusion against privileging genetic explanations by returning to vertebrate cardiogenesis. Are there problems with claiming that genes contain all of the developmental "information" to form vertebrate hearts? Is there a genetic program in the DNA controlling heart development? Are genes the primary supplier and organizer of material resources for heart development, largely determining the phenotypic outcome? Existing studies of

10 "As all the key steps in development reflect changes in gene activity, one might be tempted to think of development simply in terms of mechanisms for controlling gene expression. But this would be highly misleading. For gene expression is only the first step in a cascade of cellular processes that change cell behavior and so direct the course of embryonic development. To think only in terms of genes is to ignore crucial aspects of cell biology, such as change in cell shape, . . ." (Wolpert et al., 1998, p.15).

heart development have identified a role for fluid forces in specifying the internal form of the heart (Hove et al., 2003) and its left/right asymmetry (Nonaka et al., 2002). Additionally, biochemical gradients of extracellular calcium are responsible for activating the asymmetric expression of the regulatory gene *Nodal* (Raya et al., 2004) and inhibition of voltage gradients scrambles normal asymmetry establishment (Levin et al., 2002). A number of genes are also critical to these processes (Hamada et al., 2002) but the conclusion seems to be that genes do not carry all the "information" needed to generate form features of the heart. And if there is a genetic program for these features, it is difficult to assign it "control" since an extragenetic feature is the initial cue for asymmetric spatiotemporal gene expression (Raya et al., 2004; cf. Farge, 2003). Also, genes do not "determine" the outcome because the experimental manipulation of fluid forces causes severe phenotypic malformations in the heart (Hove et al., 2003).[11]

Another pivotal reason for being wary of gene privileging is "phenotypic plasticity," the phenomenon of phenotypic differences arising from variation in development due to environmental factors (Hall, Pearson, & Müller, 2004; Pigliucci, 2001; Schlichting & Pigliucci, 1998; West-Eberhard, 2003) [SEE PHENOTYPIC PLASTICITY AND REACTION NORMS]. If the same set of genetic resources produces very different phenotypic outcomes due to diversity in the environmental factors present, then the specificity of form features originates from more than gene expression. Relevant "information" or determining causes required to explain the ontogeny of form reside "outside" of the organism. Intrinsic "environment" dynamics are also relevant, such as developmental selection (Kirschner & Gerhart, 2005), whereby competition among components (e.g., neurons) leads to the preferential preservation of one array of components rather than others.

It may be unsurprising that a concrete example reveals many of the difficulties identified by others regarding the privileging of genetic explanations in development. Claims about the "hardwiring of development" (Arnone & Davidson, 1997) lack support on several fronts but should be treated as philosophically interesting in their own right. Continued attempts to privilege genes in explanations of the ontogeny of form are clues to epistemological issues. For example, *modeling* genetic regulatory interactions in terms of input/output network wiring diagrams encourages the "hardwiring" metaphor analogous to the control attributed to an electronic circuit board (cf. Keller, 2002). Part of the rationale is an increased generalization of the explanatory apparatus purchased through abstraction. Abstract "network" models are applicable to very diverse phenomena (Shiffrin & Börner, 2003). Having jumped ahead to some of these philosophical concerns in looking at the emphasis on genetic explanations, it is now time to cast our net more widely.

---

11 Similar comments can be observed from researchers working on different form features, such as avian feathers. "The genetic control provides transcription and translational control of molecules. Specific sets of cell surface molecules and intra-cellular signaling are produced for particular cell types. The molecular information endows cells and their micro-environment with particular properties. Based on these properties, cells interact in accordance to physical-chemical rules, and there are competition, equilibrium, randomness, and stochastic events, at this cellular level. Epigenetic events appear to play important roles at the cellular level. The integument pattern we observe is the sum of these cell behaviors" (Jiang et al., 2004, pp.131–2).

3. Epistemological Issues: Representation

The first representational decision made in explaining the ontogeny of form concerns what constitutes the system of investigation ("intrinsic") and what is the outside or environment ("extrinsic"). In most cases this is implicitly determined by the intuitive inside/outside epithelial boundary exhibited by organisms studied in the laboratory. This does not prevent appeals to "extrinsic" causal factors in explanations but distinguishes the labeling ("representation") of those factors as either intrinsic (e.g., gene expression) or extrinsic (e.g., nutrition) to the organism. As with intuitive conceptions of biological individuality, a number of reasons can be marshaled to question a privileged circumscribing of developmental systems (cf. Keller, 2001).

A second key epistemological issue is how continuous ontogenetic trajectories are to be discretely represented. Often ontogenies are partitioned into developmental "stages" consisting of a numbered sequence. For example, chick ontogeny is divided up into 45 stages (Bellairs & Osmond, 1998), which were originally established over fifty years ago (Hamburger & Hamilton, 1951). The practice of dividing ontogeny into stages has only recently begun to be systematically investigated historically (Hopwood, 2005). There is agreement that "chronological age" is of little use, in part because of variability despite homogeneous environments; the same stage in the same system under identical control conditions is reached at different "ages." But how exactly these representational decisions are made is largely unique to each model system (because it involves discernible characters as indices) and is a function of several factors including the ability to communicate results among researchers unambiguously, replicate experimental results, and coordinate stages with other taxa. These decisions are contingent on the historical period in which the stages were set forward. Closely connected with the determination of stages are fate maps meant to show features of later stages prospectively in an earlier stage embryo (e.g., cleavage), such as where heart cells will originate prior to their migration and differentiation.

Decisions about how to stage development naturally provoke questions about how time itself is represented for ontogeny, especially since stages do not straightforwardly correlate with hours (or days). The changes that occur in ontogeny are all physically continuous and thus the measures of time utilized must connect the "stages" represented. Several basic distinctions about time can be recognized (Reiss, 2003; cf. Minelli, 2003, ch. 4). The first is between sequence and duration. Sequence concerns event ordering, such as gastrulation occurring prior to organogenesis, whereas duration concerns a succession of defined intervals, which may or may not map onto sequences of events. For any sequence we can ask about the relative duration of the events (for interval definition d, A to B occurs over $3d$ in one species whereas in another species it occurs over $4d$), and whether they exhibit reliable transformation ordinality (A always precedes B; B always precedes C: or, A always precedes B; B sometimes precedes C). Relative timing of one set of sequences to another can also be assessed using an "intrinsic" interval definition. For two event sequences ($A \rightarrow B \rightarrow C$; $D \rightarrow E \rightarrow F$), the timing of $D \rightarrow E \rightarrow F$ can be measured with respect to the interval occurrences defined by $A \rightarrow B \rightarrow C$. Alternatively, one or more event sequences or intervals can be measured according to extrinsic time measures. ("The transition from event A to event B occurs in 2–3 hours.") These choices are usually relative to explanatory aims but not necessarily explicitly justified.

Another critical issue is the recognition of sameness for units and similarity of mechanisms in different species. This is a necessary prerequisite for making generalizations outside of the model used for laboratory investigation. It is a manifestation of the problem of homology and is unavoidable in answering questions pertaining to representation of units across taxa.[12] In the attempt to assess whether a particular explanation of form origination in one species can be generalized to another, an assessment of the sameness of causal factors and phenomena in other taxa must be presumed (if not established). Thus, to claim that factor *x* (protein) causally explains the form feature *y* (heart shape) that occurs in the event *E* (organogenesis of the heart) in vertebrates requires that *x*-type factors, *y*-type form features, and *E*-type events are instantiated in vertebrate taxa. We can exemplify this as a research question: are genes, cardiac cells, and "hearts" of *Drosophila* relevantly homologous (Bodmer & Venkatesh, 1998)? Homology judgments concerning the individuation and sameness of these different aspects of ontogeny must be made prior to assessments of generalization, such as the behavior of particular genes in heart development or what counts as a segment (Minelli, 2003, ch. 9). Representational issues surrounding time and stage are directly pertinent to this question.

The factor of time alongside homology allows us to see another issue in a different light: typology. Although typological thinking and its ignoring of variation have a history of being disparaged because of *metaphysical* incompatibility with population thinking in evolutionary theory (Mayr, 1976),[13] type concepts may be necessary for explanatory purposes (Amundson, 1998, 2005). Variations of "typological thinking" are manifested in explaining the ontogeny of form as a consequence of conceptualizing continuous ontogenies in terms of discrete partitions and generalizing processes (morphogenesis), events (organogenesis), and form features (heart) across all of the instances within an organism kind, as well as to other developmental systems. These explanatory practices require that particular kinds of variations be disregarded. This is not to say that they are unbiased, as is the case for all representational decisions made in scientific investigation, and developmental generalizations are fraught with difficulty (Alberch, 1985; Minelli, 2003, ch. 4). Developmental stages can be questioned with respect to what counts as "typical" ontogeny.[14] But the reasons why researchers adopt different

12 Formally, homology concerns sameness ("correspondence") rather than similarity but representational claims about similarity of mechanisms are usually predicated on sameness of mechanism components and their activities.

13 "Population thinking" usually refers to the ontological claim that only individual organisms are real as a consequence of the variations they exhibit and any statistical terms used to describe them collectively are abstractions and not objective features of the world. "Typological thinking" is supposed to represent a contrary (metaphysical) position, whereby the "types" used to collectively describe organisms are objectively real (often equated with "essences") and, in some sense, downplay the reality of variations exhibited by individuals.

14 For example, in the original paper establishing stages for the chick embryo, the authors claim "we have tried to establish average or 'standard' types by comparing a considerable number of embryos in each stage, and we have selected for illustrations those embryos which appeared typical" (Hamburger & Hamilton, 1951, p.52).

kinds of typology should be sought in the epistemic context of explaining the ontogeny of form, not by way of contrast with a metaphysical essentialism that is in conflict with population thinking in evolutionary biology.

4. Epistemological Issues: Explanation

Causal explanations of the ontogeny of form can be distilled out of our earlier discussion of questions in the problem agenda. For example, within the domain of organogenesis, questions can be asked about what causal factors active in the core processes of development produce the specific form features of organs, such as the heart. Researchers are seeking to isolate and identify developmental causes that bring about specific form feature "effects." But not all explanations appeal to material causal factors, such as particular proteins. We can distinguish another related set of explanations that identify *structural* aspects of causal explanations, such as mathematical relations between features of developing organisms due to physical rules or constraints. Two historically famous examples are Thompson's use of geometrical shape transformations to show that specific form features arise solely from proportional changes in the growth of parts (Thompson, 1992 [1942]), and Turing's use of gradient equations to show how the diffusion of molecules can produce patterns (Turing, 1952; cf. Keller, 2002). These approaches causally explained the ontogeny of form without the invocation of specific genes. Structural and material explanatory strategies need not be in competition but, as in the case with epigenesis and preformation, there has been a widespread perception of mutual exclusivity.

More recent instantiations of these approaches include shape analysis of form features during ontogeny using geometric morphometrics (Zelditch et al., 2004) and "embryo physics" (Forgacs & Newman, 2005). Physical rules (e.g., surface area to volume ratios) are often used to generate models of core processes such as morphogenesis (Takaki, 2005) and specific events such as gastrulation or neurulation (Schiffman, 2005). Often there are several material explanations that could fit within the structural constraints (Davidson et al., 1995). This is taken by some as a motivation for the prioritization of material explanatory strategies because the structural aspects are necessary but not sufficient for the specification of form during ontogeny. But a number of researchers have argued that explanations appealing to physical features of biological "matter" are sufficient to explain specific form features, especially early in evolutionary history (Newman, 1994; Newman & Müller, 2000). Segments, tubes, hollow spheres, and layers of cells are generic structures attributable to biomechanical forces (Minelli, 2003, ch. 3) and can be multiply realized by different material components (e.g., proteins or cells). Related phenomena include the wrinkling of an elastic sheet under tension (Sharon et al., 2002) or the elasticity of biological gels (Storm et al., 2005). Studying these mechanical properties of biological materials that are responsive to stress and strains experienced during development is a strategy for explaining the ontogeny of form that utilizes a different set of causal factors. A philosophical motivation for this approach is that generalizations based on physical principles have a wider scope in the sense of operating in all ontogenies, whereas appeals to particular material factors may not be instantiated widely. Explanatory trade-offs are also conditional upon the degree to which structural explanatory strategies can account for the origination of specific forms as adequately as material-based strategies.

4.1. *Model systems and generalizations*

Explanations of form's ontogeny focus on form feature *types* (kinds) rather than form feature *tokens* (instances). Although some authors have stressed the explanatory value of token reductionism in developmental biology (Delehanty, 2005; Weber, 2005, chs. 1, 8), a central feature of current research is the search for generalizations across organism instances and different species relevant to the origination of form. These generalizations can be assessed along at least three dimensions: abstraction (how much a generalization is able to ignore particular details or variation), stability (how resilient the generalization is to changes in causal structures and relations), and strength (how frequently the generalization holds) (Mitchell, 2000). In general, strength and stability are the focus of developmental biologists utilizing material explanatory modes, whereas abstraction is also critical to structural ones.

One of the most significant features affecting these different properties of generalizations is the use of model organisms. The National Institutes of Health (NIH) primarily sponsor developmental research on a small number of animal models: round worms (*C. elegans*), fruitflies (*Drosophila*), zebrafish, frogs (*Xenopus*), and mice (http://www.nih.gov/science/models/). Most observations and analyses of core developmental processes are made in these systems, as well as in the historically important chicken (*Gallus*) (cf. Slack, 2006, section 2). Many explanations of the ontogeny of form are predicated on the assumption that these species can serve as models for the developmental processes extant (and extinct) in the diversity of life. There are many reasons to question this assumption because the models were chosen for *non*-representative reasons: small body size, rapid embryonic development/short gestation period, early sexual maturation (shorter generation time), optical translucency of the embryo, and ease of laboratory cultivation (Ankeny, 2001; Burian, 1993; Bolker, 1995; Schaffner, 1998). These are largely aspects of highly derived (and therefore "atypical") ontogenies (Hedges, 2002).

One explanation for the optimism of developmental researchers and pessimism of evolutionary researchers can be seen through the lens of different hierarchical levels of developmental organization (such as protein, cell, tissue, organ, etc.). Some developmental researchers are confident in the generalization potential of model systems because characters at lower levels (such as gene network components) are widely instantiated across a diversity of taxa.[15] This has led to unprecedented experimental manipulation, such as the expression of fruit-fly genes in mice. But alongside this success has been a growing body of evidence indicating that higher levels of organization (tissues, organs, and anatomical parts) can be multiply realized by different lower-

---

15 "The mechanisms of development are very similar for all animals, including humans. This fact has only been known since it has become possible to examine the *molecular basis* of developmental processes" (Slack 2006: 3, emphasis mine). The expectation underwrites the motivation for studying model systems, as in this *Drosophila* paper: "We expect that similar mechanisms may specify pattern formation in vertebrate developmental systems that involve intercellular communication" (Flores et al., 2000, p.75).

level constituents. In part this is because these higher levels emerge from combinations of compositional and procedural hierarchies during ontogeny not widely instantiated in other species; molecular level generality is *not* transitive (McShea, 2001; Salthe, 1985). A generalization that holds across model organisms ("gene *x* plays the same causal role during cardiogenesis in *Drosophila* and vertebrates") does not necessarily yield a generalization about higher levels of organization ("epithelial–mesenchymal interactions, in which gene *x* is expressed, play the same causal role during cardiogenesis in *Drosophila* and vertebrates"). Evidence for this non-transitivity includes the dissociation of homologous gene expression from homologous structures (Wray, 1999), co-option and convergence of gene expression (True & Carroll, 2002), self-organization dynamics (Camazine et al., 2001), and epigenetic interactions occurring during ontogeny (Müller, 2003). Cardiogenesis in vertebrates involves neural crest cells, which are not present in *Drosophila*. But many of the same genes are expressed during cardiogenesis in both organisms. Strong and stable molecular-level generalizations that hold across many species do not translate into generalizations that obtain at all hierarchical levels for those species.[16]

This empirical situation serves as another plank in the argument against gene privileging: a *solitary* explanatory strategy of decomposition and localization of developmental components (genes) and their interactions (gene networks) is insufficient for explaining the ontogeny of form apart from further, distinct evidential support (cf. Bechtel & Richardson, 1993). A model organism may represent a lower hierarchical level in other taxa quite accurately while simultaneously being a poor model for other (higher) levels. Caution is necessary when explanations of form origination gleaned from one level of biological organization are applied to another level in different species. Studies of cellular differentiation in bacteria (Iber et al., 2006) are relevant but insufficient for comprehending higher-level form feature origination.

4.2. Reductionism

Model systems and the non-transitivity of molecular generalizations also raise problems related to reductionism. A tendentious discussion in recent philosophy of biology comes from Rosenberg (Rosenberg, 1997; see REDUCTIONISM), where he sets out two different principles putatively at work in antireductionist approaches to developmental phenomena:

Principle of Autonomous Reality: The levels, units, kinds identified in functional biology are real and irreducible because they reflect the existence of objective explanatory generalizations that are autonomous from those of molecular biology.

Principle of Explanatory Primacy: At least sometimes, processes at the functional level provide the best explanation for processes at the molecular level.

16 A related issue is making generalizations across different anatomy *within* the same model, such as developmental mechanisms underlying the establishment of nerve and blood vessels (Carmeliet & Tessier-Lavigne, 2005). These generalizations are motivated by the exhibition of shared form features, such as stereotypical branching.

Rosenberg takes a dim view of both principles, holding that molecular developmental biology rejects them: "there are no explanatory generalizations at higher levels of organization" (Rosenberg, 1997, p.447). Many have challenged his account. Keller is concerned that Rosenberg misreads contemporary developmental biology (Keller, 1999), especially its metaphors, whereas Wagner and Laubichler claim he is not sufficiently sensitive to the many–many relations between developmental outcomes and molecular constituents (Laubichler & Wagner, 2001; cf. Frost-Arnold, 2004), highlighted above in terms of the non-transitivity of molecular-level generality and the role of biomechanical forces in the origin of specific form features.

An important aspect of this discussion is that what is meant by reductionism varies tremendously (Sarkar, 1998, chs. 2–3). "Reductionism" is rejected by some *cell* biologists,[17] which should at least lead us to pause about "reductionism" in developmental biology. One distinction of crucial importance is the difference between genetic and physical reductionism (Sarkar, 1998). Genetic reductionism is the project of explaining the phenotype in terms of abstract genes in an abstract (non-spatial) hierarchical relationship between genotype and phenotype. Physical reductionism is the explanation of biological phenomena using the physical properties of constituent molecules and macromolecules, usually conceptualized in a spatial hierarchy. Considerations of spatial hierarchy highlight the relevance of part/whole relations (Hüttemann, 2004; Sarkar, 1998, ch. 3; Wimsatt, 1976). Rosenberg's position is a conflation of genetic and physical reductionism that prefers certain kinds of macromolecules (DNA, RNA, proteins) to explain the ontogeny of form features in a presumed spatial hierarchy. Some difficulties with this position include an inability to defend a preferential treatment of particular macromolecules, especially since others (phospholipids, fatty acids, cholesterols, and carbohydrates) play key developmental roles (e.g., Hsu et al., 2006), and not having a explicit articulation of the hierarchical relationships involved. Developmental phenomena are heterogeneous and "developmental biology" is multidisciplinary as a consequence. Ignoring this diversity of research programs facilitates missing the heterogeneity of explanatory aims directed at different core processes in ontogeny and their characterization at multiple levels of organization (cf. Keller, 2002). Generalizations relevant to explaining the ontogeny of form are diverse and higher-level generalizations in particular can be objectively identified (cf. Gilbert & Sarkar, 2000).

One feature not routinely recognized for reductionism concerning part–whole relations is *temporality*. Supervenience is an atemporal notion, capturing relations of dependence at a particular time (Rueger, 2000; Sober, 1999). But causation is inherently diachronic, which is especially applicable to ontogeny. Given the representational dimension of time and the focus on causal explanation, understanding "reductionism" along a temporal axis is critical. Are higher-level form features (such as hearts) causally produced by the activity of their component parts (e.g., proteins) at earlier times? Further work is required to turn any synchronic realizations into diachronic dependencies between parts and wholes in biological hierarchies. Temporality opens up a broader space of alternatives for explanations of the ontogeny of form not captured by

17 E.g., "Our results suggest that the cellular responses . . . may be an emergent property that cannot be understood fully considering only the sum of individual . . . interactions" (Kung et al., 2005, p.3587).

synchronic ideas of reduction. This can be seen through attention to explanatory norms.

One norm for causal explanations is that more fine-grained explanations are preferable (ceteris paribus) (Jackson & Petit, 1992). But "fine-grain" can mean either "small grain" (prefer micro to macro causal information) or "close grain" (prefer proximate to distal causal information). Almost all discussion surrounding reduction in philosophy of biology has concerned "small grain." Consider an argument for the "small grain" preference.

(1) To explain is to provide information on the causal history of the *explanandum* phenomena.
(2) Better causal information is obtained at the micro-level ("small grain").
(3) Therefore, micro-level explanations are better.

A parallel argument is obtained by substituting "close grain" for "small grain" with the conclusion that proximate causal information is preferable. But the "small grain" preference is problematic because the second premise is not supported; there are times to prefer "large grain" because better causal information is available (Jackson & Petit, 1992). Since the close grain premise is similarly problematic, especially in embryogenesis where distal causal factors are sometimes highly relevant, a form of explanatory pluralism seems warranted even when temporality is emphasized.

But what are the consequences of preferring proximate causal information in developmental explanations? One possibility is that proximate causes constrain or channel earlier causal factors. Another is that wholes may "bring about" other wholes or parts (temporally), both of which are composed of (and maybe even "reducible" to) parts (spatially). Biologists have recognized something akin to this: "The unidirectional flow from genes to shape is being modified to include cell movements that cause 'physical stress' in neighbouring cells inducing specific gene expression. This causal chain, from a molecular event to physical stress inducing the next molecular event appears as an emergent acting as a downward cause" (Soto & Sonnenschein, 2005, p.115). Diachronic considerations are largely orthogonal to most discussions of reductionism (e.g., Rosenberg, 1997). Proximate factors may include entities favored by both "reductionism" and "antireductionism" because the main issue concerns relative location of the processes in a temporal sequence regardless of their level of organization. The close-grain preference allows higher levels of organization to causally explain lower levels of organization even if synchronic supervenience holds (Sober, 1999). Candidates for these kinds of explanations include the role of mechanical loading of muscle in shaping the form of bones (Rot-Nikcevic et al., 2006), cellular and tissue mechanosensation from compression leading to gene expression (Farge, 2003; Tschumperlin et al., 2004), and fluid forces in proper cardiac development or vascular remodeling (Hove et al., 2003; Tzima et al., 2005).

All of this bears on Rosenberg's two principles. It is patently false that "in developmental molecular biology there is no room for downward explanation, in which some regularity at the level of cell physiology plays a role in illuminating the molecular processes that subserve development" (Rosenberg, 1997, p.455) once the temporality of developmental processes is absorbed into the explanatory project of understanding the ontogeny of form. Generalizations about higher levels of organization compose the *explanans* of developmental biology, not just a halfway house of *explananda*, and no implicit teleological claims are involved.[18]

Even if we set aside the issue of temporality, difficulties remain. On the assumption that a particular lower level of explanation is preferred, there are questions about types of entities at that level and how many of them are explanatorily relevant. Physical reductionism does not inherently decide between macromolecular types. Much of the excitement in recent developmental biology arose from the discovery of conserved transcription factors and signaling proteins (from "regulatory" genes) that spatiotemporally modulate transcriptional activity during ontogeny (Carroll, 2005; Davidson, 2001). But structural genes also play a critical role in producing form features (Sakai, Larsen, & Yamada, 2003). How does one evaluate the contribution of genes, spontaneous electrical activity, fatty acids, and competition (*inter alia*) to neuronal morphology arising during ontogeny? There is no accepted currency for comparing these different causal factors to establish their relative role in the ontogeny of a form feature, either in term of causal contribution or difference making (Sober, 1988). This also holds for the structural aspects derived from physical rules. Answers to these questions have an impact on the kinds of generalizations available, which are not solved even if one accepts a physical reductionism that favors molecular explanations.

5. Epistemological Issues: Methodology

Many of the methodological questions that emerge in the problem agenda for the ontogeny of form can be extracted from our earlier discussion. Why choose a particular staging of an organism's ontogeny? Why preferentially investigate factors deemed intrinsic to the system versus extrinsic variables? Instead of teasing each of these out, it is useful to turn to research heuristics (or simplifying assumptions) utilized in explanations of the ontogeny of form. Following earlier analyses on the role of research heuristics in scientific investigation (Wimsatt, 1980, 1986), Robert has reconstructed an argument for a (genetic) reductionist research heuristic that explains development (and thus the ontogeny of form) in terms of the role of gene activity during ontogeny (Robert, 2004).

(1) Simplifying strategies and assumptions, as such, are absolutely necessary in biological science.
(2) Simplifying the context of a system is advantageous if we want to learn about intrasystemic causal factors.
(3) Genes by themselves are not causally efficacious, as genes and environments (at many scales) interact (differentially, over time) in the generation of any phenotypic trait.

18 Rosenberg raises this specter: "Cellular structures only come into existence through molecular processes that precede them. There is . . . no scope for claims about the indispensable role of cellular structures in these molecular processes. The future cannot cause the past" (Rosenberg, 1997, p.455).

(4) We decide to focus on the causal agency of genes against a constant background of other factors, for pragmatic or heuristic reasons.
(5) A trait *x* is caused by a gene *y* only against a constant background of supporting factors (conditions), without which *x* would not be present (even if *y* is present).
(6) Therefore, organismal development is a matter of gene action and activation, as particular alleles have their specific phenotypic effects against standard environmental background conditions.

The second premise is subject to two alternate readings according to Robert. The first pragmatically ignores the biases, such as a tendency to concentrate on lower-level intrasystemic factors or underestimate the impact of intersystemic factors (Wimsatt, 1980, 1986) and generates the hedgeless hedge heuristic (HHH).[19] HHH encourages proceeding *as if* genes are sufficient to explain developmental processes. When objections arise one admits their insufficiency for explaining the ontogeny of form while continuing to prosecute a gene-focused methodology. But the isolation of a genetic causal factor against a fixed background shows that this gene activity is a *relevant* factor, not the only or most important causal factor (or type of factor). Because the HHH does not experimentally explore the role of any extragenetic factors, using it alone involves researchers in a methodological fallacy.

From a second reading of premise (2) Robert generates a different strategy, the constant factor principle heuristic (CFPH): "Against standard background conditions, aspects of organismal development may be partially a matter of gene action and activation, and it remains to be determined whether (and how) extragenetic factors make a specific causal contribution to ontogenesis" (Robert, 2004, p.17). CFPH prevents an unlicensed inference from pragmatic choices about methodology to claims about gene activity as the best explanation. If we return to the study of fluid forces in cardiogenesis, something similar to the CFPH seems to have motivated the investigative strategy.

> The formation of a functional heart is regulated by the coordinated interplay between a genetic programme, fluid mechanical stimuli, and the inter- and intracellular processes that link them. While the genetics of cardiogenesis are being analysed intensely, studies of the influence of epigenetic factors such as blood flow on heart development have advanced more slowly owing to the difficulty of mapping intracardiac flow in vivo. (Hove et al., 2003, p.172)

The authors readily admit that genetic factors have received the most scrutiny for practical reasons and that technical difficulties were a major hurdle.

But CFPH leaves a key question unanswered: what heuristic do we use to isolate and characterize "standard background conditions" and the causal role of extragenetic factors during ontogeny? CFPH protects us from drawing illicit inferences about development from the role of genes in development but it does not guide us toward experiments that identify extragenetic factors in ontogeny. Even if CFPH produces a compulsion to execute different experiments, it does not by itself tell us what *kind* of experiments

19 Hedge$_{df}$ = a word or phrase used to allow for additional possibilities or to avoid overly precise commitment. Thus, the HHH seemingly recognizes additional possibilities but in fact does not.

these are or how to establish appropriate simplifying assumptions. To isolate causes relevant to explaining the ontogeny of form in terms of something other than genes, new positive research heuristics need to be articulated that are responsible to the conceptual arguments made against privileging genetic explanations of development.

One strategy for analyzing "standard background conditions" involves comparing the ontogeny of form in a model system with a closely related non-model system. For example, developmental stage 10 for *Xenopus* used to have dorsal mesoderm originating only from the deep mesenchymal layer. Two studies of a related anuran (*Hymenochirus boettgeri*) alongside *Xenopus* demonstrated that dorsal mesoderm also originated from surface cells in both species (Minsuk & Keller, 1996, 1997). Any gene that was expressed in the surface cells would not have been considered as a mesodermal contributor prior to this reevaluation. Basic descriptive and manipulative embryology evaluating "standard background conditions" is still required in order to interpret gene expression patterns. Another result of these investigations was that the contribution of surface cells to mesoderm varies between spawnings for *Xenopus*, ranging from nearly absent to almost ubiquitous, and that surface epithelial cells invade the notochord and somites via a novel developmental mechanism not previously described. The standardized background conditions presumed for the model system were problematic and required revision.

The seeds of one alternative positive heuristic are available in our discussion of temporality and a latent aspect of the previously discussed example of left/right asymmetry in cardiogenesis. How did researchers identify crucial extragenetic causal factors if they were focusing on the role of genes in left/right asymmetry origination *as a substitute* for left/right asymmetry origination? A glance at the investigative motivations show that they were driven to find the symmetry breaking event that initiates asymmetrical gene expression (Raya et al., 2004). They were led to extracellular Ca^{2+} because of prior work identifying a voltage gradient across the midline (Levin et al., 2002). The reasoning takes the form of following a causal chain backwards, seeking earlier and earlier antecedent causal factors in the ontogenetic trajectory. This suggests a different kind of heuristic strategy, one not fundamentally focused on reductionism. Following a causal chain involves seeking the next most proximate cause in a temporally extended causal sequence. A proximate cause heuristic (PCH) makes a simplifying assumption that focuses on the causal agency of proximate factors against a constant background of distal factors (for pragmatic or heuristic reasons), despite the recognition that distal causes play important roles in producing form features during ontogeny. PCH illustrates a potential *method* for finding higher-level explanatory generalizations, even under strong commitments favoring reductionism. The proximate cause of a particular form feature can be a higher-level entity without having to deny that gene expression and cellular dynamics are critical for generating the entity in the first place.

The application of PCH will be methodologically complex because of different conceptualizations of developmental time. What counts as proximate and distal will be relative to the sequences or durations specified. PCH also naturally transgresses the intrinsic/extrinsic boundary in searching for causal factors (Gilbert, 2001; Van der Weele, 1999). Whereas reductionist research heuristics are biased toward localization of causal factors within a system as opposed to its environment (Wimsatt, 1980), tracing causal chains and looking for proximal (or distal) causes are not. Following a sequence of events in time might lead to extrinsic causal factors that are relevant to particular processes underlying form origination, such as limitations on growth from precocious hatching due to vibrational cues from predators (Warkentin, 2005) or diet induced transformations of morphology (Greene, 1996).

6. Unexplored Issues and Summary

Though we have ranged widely over a variety of issues pertinent to explaining the ontogeny of form, we have left many untouched. One worth mentioning is the experimental utilization of developmental trajectories conceived of in terms of fertilized egg to adult from sexually reproducing species. Prior to molecularization, embryological studies concerned with form origination often concentrated on asexually reproducing species, specifically choosing asexual budding to understand the ontogeny of form (Berrill, 1961; cf. Minelli, 2003). Regenerative developmental phenomena have also received less attention (Alvarado, 2003). This nexus of issues touches directly on representational preferences, the scope of generalizations, and methodological biases.

Metaphysical issues have also been largely ignored here, in part to keep the focus on explanations. Some points of contact include: (a) reduction, emergence, physicalism, and concepts of supervenience, especially once temporality is included (Rueger, 2000); (b) causation, both in terms of concepts relevant to preformation and epigenesis such as "production" and "propagation" (Salmon, 1998) and whether *probabilistic* causation (Hitchcock, 2002) is useful for articulating a common currency to assess multiple causal contributions during ontogeny in the production of form features; and (c) dispositional properties, especially as they bear on transient "potentiality" in development and whether causal powers are intrinsically located (cf. Love, 2003). Questions about individuation and identity through time are also salient. Canonical events in form origination (such as gastrulation or organogenesis) direct us to consider the status of events in relation to other entities (Macdonald, 2005), especially whether "event" or "aspect" is more appropriate for developmental causes (Paul, 2000).

The idea of a problem agenda set forth earlier can also be applied to philosophical questions. Investigations of epistemological and metaphysical issues attending the attempt to causally explain the developmental origin of the material composition, arrangement, shape, and appearance of organismal features are interpretable as part of a *philosophical* problem agenda. It should be transparent that this agenda of philosophical issues affiliated with the ontogeny of form contains more than its fair share of outstanding questions, many of them distinct from evolutionary theory and the causal power of genes. Developmental phenomena have been persistent provocateurs of intellectual reflection for two millennia. In addition to constituting a multifaceted problem agenda for ongoing empirical research in developmental biology, the associated philosophical questions warrant increased scrutiny from philosophers of biology.

Acknowledgment

I am grateful to Jim Lennox, Ric Otte, Anya Plutynski, Sahotra Sarkar, and Ken Waters for timely and constructive feedback on an earlier version of this chapter. Thanks also to comments from attendees of the "Pizza Munch" discussion group at the California Academy of Sciences in San Francisco (June 2006).

References

Alberch, P. (1985). Problems with the interpretation of developmental sequences. *Systematic Zoology*, 34(1), 46–58.

Alvarado, A. S. (2003). Regeneration in the Metazoa. In B. K. Hall & W. M. Olson (Eds). *Keywords and concepts in evolutionary developmental biology* (pp. 318–25). Cambridge, MA: Harvard University Press.

Amundson, R. (1998). Typology reconsidered: two doctrines on the history of evolutionary biology. *Biology and Philosophy*, 13, 153–77.

Amundson, R. (2005). *The changing role of the embryo in evolutionary thought: structure and synthesis*. New York: Cambridge University Press.

Ankeny, R. (2001). The natural history of *C. elegans* research. *Nature Reviews Genetics*, 2, 474–78.

Arnone, M. I., & Davidson, E. H. (1997). The hardwiring of development: organization and function of genomic regulatory systems. *Development*. 124, 1851–64.

Bechtel, W. (1986). The nature of scientific integration. In W. Bechtel (Ed.). *Integrating Scientific Disciplines*. Dordrecht: M. Nijhoff. 3–52.

Bechtel, W., & Richardson, R. (1993). *Discovering complexity: decomposition and localization as strategies in scientific research*. Princeton: Princeton University Press.

Bellairs, R., & Osmond, M. (1998). *The atlas of chick development*. San Diego, CA: Academic Press.

Berrill, N. J. (1961). *Growth, development, and pattern*. San Francisco: W.H. Freeman & Company.

Bock, W. J., & Wahlert, G. V. (1965). Adaptation and the form–function complex. *Evolution*, 19, 269–299.

Bodmer, R., & Venkatesh, T. V. (1998). Heart development in *Drosophila* and vertebrates: conservation of molecular mechanisms. *Developmental Genetics*, 22, 181–6.

Bolker, J. A. (1995). Model systems in developmental biology. *BioEssays* 17(5), 451–5.

Burian, R. M. (1993). How the choice of experimental organism matters: epistemological reflections on an aspect of biological practice. *Journal of the History of Biology*, 26(2), 351–67.

Camazine, S., Deneubourg, J.-L., Franks, N. R., Sneyd, J., Theraulaz, G., & Bonabeau, E. (2001). *Self-organization in biological systems*. Princeton and Oxford: Princeton University Press.

Carmeliet, P., & Tessier-Lavigne, M. (2005). Common mechanisms of nerve and blood vessel wiring. *Nature*, 436, 193–200.

Carroll, S. B. (2005). *Endless forms most beautiful: the new science of evo-devo*. New York: W.W. Norton.

Davidson, E. H. (2001). *Genomic regulatory systems: development and evolution*. San Diego: Academic Press.

Davidson, L. A., Koehl, M. A. R., Keller, R., & Oster, G. F. (1995). How do sea urchins invaginate? Using biomechanics to distinguish between mechanisms of primary invagination. *Development*, 121, 2005–18.

Davies, J. A. (2005). *Mechanisms of morphogenesis: the creation of biological form*. San Diego, CA: Elsevier Academic Press.

Delehanty, M. (2005). Emergent properties and the context objection to reduction. *Biology and Philosophy*, 20(4), 715–34.

Farge, E. (2003). Mechanical induction of twist in the *Drosophila* foregut/stomodeal primordium. *Current Biology*, 13, 1365–77.

Flores, G. V., Duan, H., Yan, H., Nagaraj, R., Fu, W., Zou, Y., Noll, M., & Banerjee, U. (2000). Combinatorial signaling in the specification of unique cell fates. *Cell*, 103, 75–85.

Forgacs, G., & Newman, S. A. (2005). *Biological physics of the developing embryo*. New York: Cambridge University Press.

Fraser, S. E., & Harland, R. M. (2000). The molecular metamorphosis of experimental embryology. *Cell*, 100, 41–55.

Frost-Arnold, G. (2004). How to be an anti-reductionist about developmental biology: response to Laubichler and Wagner. *Biology and Philosophy*, 19, 75–91.

Gilbert, S. F. (1997). *Developmental biology* (5th edn). Sunderland, MA: Sinauer Associates, Inc.

Gilbert, S. F. (2001). Ecological developmental biology: developmental biology meets the real world. *Developmental Biology*, 233, 1–12.

Gilbert, S. F. (2003). *Developmental biology* (7th edn) Sunderland, MA: Sinauer Associates, Inc.

Gilbert, S. F., & Sarkar, S. (2000). Embracing complexity: organicism for the 21st century. *Developmental Dynamics*, 219, 1–9.

Greene, E. (1996). Effect of light quality and larval diet on morph induction in the polymorphic caterpillar *Nemoria arizonaria* (Lepidoptera: Geometridae). *Biological Journal of the Linnean Society*, 58(3), 277–85.

Hall, B. K., Pearson, R. D., & Müller, G. B. (Eds). (2004). *Environment, development and evolution*. Cambridge, MA: The MIT Press, A Bradford Book.

Hamada, H., Meno, C., Watanabe, D., & Saijoh, Y. (2002). Establishment of vertebrate left–right asymmetry. *Nature Reviews Genetics*, 3, 103–13.

Hamburger, V., & Hamilton, H. L. (1951). A series of normal stages in the development of the chick embryo. *Journal of Morphology*, 88, 49–92.

Harvey, R. P. (2002). Patterning the vertebrate heart. *Nature Reviews Genetics*, 3, 544–56.

Harvey, R. P., & Rosenthal, N. (Eds). (1999). *Heart development*. San Diego, CA: Academic Press.

Hedges, S. B. (2002). The origin and evolution of model organisms. *Nature Reviews Genetics*, 3, 838–49.

Hitchcock, C. (2002). Probabilistic causation. In E. N. Zalta (Ed.). *The Stanford encyclopedia of philosophy* (Fall 2002 edn). Available from: http://plato.stanford.edu/archives/fall2002/entries/causation-probabilistic/.

Hogan, B. (1999). Morphogenesis. *Cell*, 96, 225–33.

Hopwood, N. (2005). Visual standards and disciplinary change: normal plates, tables and stages in embryology. *History of Science*, 43, 239–303.

Hove, J. R., Köster, R. W., Forouhar, A. S., Acevedo-Bolton, G., Fraser, S. E., & Gharib, M. (2003). Intracardiac fluid forces are an essential epigenetic factor for embryonic cardiogenesis. *Nature*, 421, 172–7.

Hsu, H.-J., Liang, M.-R., Chen, C.-T., & Chung, B.-C. (2006). Pregnenolone stabilizes microtubules and promotes zebrafish embryonic cell movement. *Nature*, 439, 480–3.

Hüttemann, A. (2004). *What's wrong with microphysicalism?* London: Routledge.

Iber, D., Clarkson, J., Yudkin, M. D., & Campbell, I. D. (2006). The mechanism of cell differentiation in *Bacillus subtilis*. *Nature*, 441, 371–4.

Jablonka, E., & Lamb, M. J. (2005). *Evolution in four dimensions: genetic, epigenetic, behavioral, and symbolic variation in the history of life*. Cambridge, MA: A Bradford Book, The MIT Press.

Jackson, F., & Petit, P. (1992). In defense of explanatory ecumenism. *Economics and Philosophy*, 8, 1–21.

Jiang, T. X., Widelitz, R. B., Shen, W.-M., Will, P., Wu, D.-Y., Lin, C.-M., Jung, H.-S., & Chuong, C.-M. (2004). Integument pattern formation involves genetic and epigenetic controls: feather arrays simulated by digital hormone models. *International Journal of Developmental Biology*, 48, 117–36.

Keller, E. F. (1999). Understanding development. *Biology and Philosophy*, 14(3), 321–30.

Keller, E. F. (2001). Beyond the gene but beneath the skin. In S. Oyama, P. E. Griffiths, & R. D. Gray (Eds). *Cycles of contingency: developmental systems and evolution* (pp. 299–312). Cambridge, MA: A Bradford Book, The MIT Press.

Keller, E. F. (2002). *Making sense of life: explaining biological development with models, metaphors, and machines*. Cambridge, MA: Harvard University Press.

Kimelman, D. (2006). Mesoderm induction: from caps to chips. *Nature Reviews Genetics*, 7, 360–72.

Kirschner, M. W., & Gerhart, J. C. (2005). *The plausibility of life: resolving Darwin's dilemma*. New Haven and London: Yale University Press.

Kitcher, P. (1993). *The advancement of science: science without legend, objectivity without illusions*. New York: Oxford University Press.

Kung, C., Kenski, D. M., Dickerson, S. H., Howson, R. W., Kuyper, L. F., Madhani, H. D., & Shokat, K. M. (2005). Chemical genomic profiling to identify intracellular targets of a multiplex kinase inhibitor. *Proceedings of the National Academy of Sciences USA*, 102, 3587–92.

Laubichler, M. D., & Wagner, G. P. (2001). How molecular is molecular developmental biology? A reply to Alex Rosenberg's reductionism redux: computing the embryo. *Biology and Philosophy*, 16, 53–68.

Lennox, J. G. (2001). *Aristotle's philosophy of biology: studies in the origin of life science*. Cambridge: Cambridge University Press.

Lenoir, T. (1989). *The strategy of life: teleology and mechanics in nineteenth century German biology*. Chicago: University of Chicago Press.

Levin, M., Thorlin, T., Robinson, K. R., Nogi, T., & Mercola, M. (2002). Asymmetries in H^+/K^+ –ATPase and cell membrane potentials comprise a very early step in left–right patterning. *Cell*, 111, 77–89.

Lohmann, I., McGinnis, N., Bodmer, M., & McGinnis, W. (2002). The *Drosophila Hox* gene *deformed* sculpts head morphology via direct regulation of the apoptosis activator *reaper*. *Cell*, 110, 457–66.

Love, A. C. (2003). Evolvability, dispositions, and intrinsicality. *Philosophy of Science*, 70(5), 1015–27.

Macdonald, C. (2005). *Varieties of things: foundations of contemporary metaphysics*. Oxford: Blackwell Publishers Ltd.

Maienschein, J. (2005). Epigenesis and preformationism. In E. N. Zalta (Ed.). *Stanford encyclopedia of philosophy* (Winter 2005 edn). Available from: http://plato.stanford.edu/archives/win2005/entries/epigenesis/.

Mayr, E. (1976). Typological versus populational thinking. In E. Mayr (Ed.). *Evolution and the diversity of life* (pp.26–9). Cambridge, MA: Harvard University Press.

McShea, D. W. (2001). Parts and integration: consequences of hierarchy. In J. B. C. Jackson, S. Lidgard, & F. K. McKinney (Eds). *Evolutionary patterns: growth, form, and tempo in the fossil record* (pp. 27–60). Chicago and London: University of Chicago Press.

Minelli, A. (2003). *The development of animal form: ontogeny, morphology, and evolution*. Cambridge: Cambridge University Press.

Minsuk, S. B., & Keller, R. E. (1996). Dorsal mesoderm has a dual origin and forms by a novel mechanism in *Hymenochirus*, a relative of *Xenopus*. *Developmental Biology*, 174, 92–103.

Minsuk, S. B., & Keller, R. E. (1997). Surface mesoderm in *Xenopus*: a revision of the Stage 10 fate map. *Development Genes and Evolution*, 207, 389–401.

Mitchell, S. D. (2000). Dimensions of scientific law. *Philosophy of Science*, 67, 242–65.

Moss, L. (2003). *What genes can't do*. Cambridge, MA: MIT Press, A Bradford Book.

Müller, G. B. (2003). Embryonic motility: environmental influences and evolutionary innovation. *Evolution & Development*, 5(1), 56–60.

Neumann-Held, E. M., & Rehmann-Sutter, C. (Eds). (2006). *Genes in development: re-reading the molecular paradigm.* Durham and London: Duke University Press.

Newman, S. A. (1994). Generic physical mechanisms of tissue morphogenesis: a common basis for development and evolution. *Journal of Evolutionary Biology*, 7(4), 467–88.

Newman, S. A., & Müller, G. B. (2000). Epigenetic mechanisms of character origination. *Journal of Experimental Zoology (Mol Dev Evol)*, 288, 304–17.

Nonaka, S., Shiratori, H., Saijoh, Y., & Hamada, H. (2002). Determination of left–right patterning of the mouse embryo by artificial nodal flow. *Nature*, 418, 96–9.

Oppenheimer, J. M. (1967). *Essays in the history of embryology and biology.* Cambridge, MA: MIT Press.

Oyama, S., Griffiths, P. E., & Gray, R. D. (Eds). (2001). *Cycles of contingency: developmental systems and evolution.* Cambridge, MA: MIT Press.

Paul, L. A. (2000). Aspect causation. *Journal of Philosophy*, 97, 235–56.

Pigliucci, M. (2001). *Phenotypic plasticity: beyond nature and nurture.* Baltimore and London: The Johns Hopkins University Press.

Raya, Á., Kawakami, Y., Rodríguez-Esteban, C., Ibañes, M., Rasskin-Gutman, D., Rodríguez-León, J., Büscher, D., Feijó, J. A., & Belmonte, J. C. I. (2004). Notch activity acts as a sensor for extracellular calcium during vertebrate left–right determination. *Nature*, 427, 121–8.

Reiss, J. O. (2003). Time. In B. K. Hall & W. M. Olson (Eds). *Keywords and concepts in evolutionary developmental biology* (pp. 359–68). Cambridge, MA: Harvard University Press.

Richtsmeier, J. T. (2003). Growth. in B. K. Hall & W. M. Olson (Eds). *Keywords and concepts in evolutionary developmental biology* (pp. 161–9). Cambridge, MA: Harvard University Press.

Robert, J. S. (2004). *Embryology, epigenesis, and evolution: taking development seriously.* New York: Cambridge University Press.

Roe, S. A. (1981). *Matter, life, and generation: 18th century embryology and the Haller–Wolff debate.* Cambridge: Cambridge University Press.

Rosenberg, A. (1997). Reductionism redux: computing the embryo. *Biology and Philosophy*, 12, 445–70.

Rot-Nikcevic, I., Reddy, T., Downing, K. J., Belliveau, A. C., Hallgrímmson, B., Hall, B. K., & Kablar, B. (2006). *Myf5*$^{-/-}$: *MyoD*$^{-/-}$ amyogenic fetuses reveal the importance of early contraction and static loading by striated muscle in mouse skeletogenesis. *Development Genes and Evolution*, 216, 1–9.

Rueger, A. (2000). Physical emergence, diachronic and synchronic. *Synthese*, 124, 297–322.

Sakai, T., Larsen, M., & Yamada, K. M. (2003). Fibronectin requirement in branching morphogenesis. *Nature*, 423, 876–81.

Salmon, W. C. (1998). Causality: production and propagation, in *Causality and Explanation.* New York: Oxford University Press, 285–301.

Salthe, S. N. (1985). *Evolving hierarchical systems: their structure and representation.* New York: Columbia University Press.

Sarkar, S. (1998). *Genetics and reductionism.* Cambridge: Cambridge University Press.

Sarkar, S. (2000). Information in genetics and developmental biology: comments on Maynard Smith. *Philosophy of Science*, 67(2), 208–13.

Schaffner, K. F. (1998). Genes, behavior and developmental emergentism: one process, indivisible? *Philosophy of Science*, 65(2), 209–52.

Schiffman, Y. (2005). Induction and the Turing–Child field in development. *Progress in Biophysics and Molecular Biology*, 89, 36–92.

Schlichting, C. D., & Pigliucci, M. (1998). *Phenotypic evolution: a reaction norm perspective.* Sunderland, MA: Sinauer Associates, Inc.

Sharon, E., Roman, B., Marder, M., Shin, G.-S., & Swinney, H. L. (2002). Buckling cascades in free sheets. *Nature*, 419, 579.

Shiffrin, R. M., & Börner, K. (2003). Mapping knowledge domains. *Proceedings of the National Academy of Sciences USA*, 101(Suppl. 1), 5183–5.

Slack, J. M. W. (2006). *Essential developmental biology* (2nd edn). Malden, MA: Blackwell Publishing.

Smith, J. E. H. (Ed.). (2006). *The problem of animal generation in early modern philosophy*. New York: Cambridge University Press.

Sober, E. (1988). Apportioning causal responsibility. *Journal of Philosophy*, 85, 303–18.

Sober, E. (1999). Physicalism from a probabilistic point of view. *Philosophical Studies*, 95, 135–74.

Sober, E. (2000). *Philosophy of biology* (2nd edn). Boulder, CO: Westview Press.

Soto, A. M., & Sonnenschein, C. (2005). Emergentism as a default: cancer as a problem of tissue organization. *Journal of Biosciences*, 30(1), 103–19.

Storm, C., Pastore, J. J., MacKintosh, F. C., Lubensky, T. C., & Janmey, P. A. (2005). Nonlinear elasticity in biological gels. *Nature*, 435, 191–4.

Takaki, R. (2005). Can morphogenesis be understood in terms of physical rules? *Journal of Biosciences*, 30, 87–92.

Thompson, D. A. W. (1992 [1942]). *On growth and form* (Complete rev. edn). New York: Dover Publications, Inc.

True, J. R., & Carroll, S. B. (2002). Gene co-option in physiological and morphological evolution. *Annual Review of Cell and Developmental Biology*, 18, 53–80.

Tschumperlin, D. J., Dai, G., Maly, I. V., Kikuchi, T., Laiho, L. H., McVittie, A. K., Haley, K. J., Lilly, C. M., So, P. T. C., Lauffenburger, D. A., Kamm, R. D., & Drazen, J. M. (2004). Mechanotransduction through growth-factor shedding into the extracellular space. *Nature*, 429, 83–6.

Turing, A. M. (1952). The chemical basis of morphogenesis. *Philosophical Transactions of the Royal Society of London B Biological Sciences*, 237, 37–72.

Tzima, E., Irani-Tehrani, M., Kiosses, W. B., Dejana, E., Schultz, D. A., Engelhardt, B., Cao, G., DeLisser, H., & Schwartz, M. A. (2005). A mechanosensory complex that mediates the endothelial cell response to fluid shear stress. *Nature*, 437, 426–31.

Van der Weele, C. (1999). *Images of development: environmental causes in ontogeny*. Buffalo, NY: State University of New York Press.

van Fraassen, B. (1980). *The scientific image*. New York: Oxford University Press.

Warkentin, K. M. (2005). How do embryos assess risk? Vibrational cues in predator-induced hatching of red-eyed treefrogs. *Animal Behaviour*, 70, 59–71.

Weber, M. (2005). *Philosophy of experimental biology*. New York: Cambridge University Press.

West-Eberhard, M. J. (2003). *Developmental plasticity and evolution*. New York: Oxford University Press.

Wimsatt, W. C. (1976). Reductive explanation: a functional account. In R. S. Cohen (Ed.). *Proceedings of the Philosophy of Science Association, 1974* (pp. 671–710). Dordrecht: D. Reidel Publishing Company.

Wimsatt, W. C. (1980). Reductionistic research strategies and their biases in the units of selection controversy. In T. Nickles (Ed.), *Scientific discovery: case studies* (pp. 213–59). Boston Studies in the Philosophy of Science. Dordrecht: D. Reidel Publishing Company.

Wimsatt, W. C. (1986). Forms of aggregativity. In A. Donagan, A. N. Perovich, Jr., & M. V. Wedin (Eds). *Human nature and natural knowledge* (pp. 259–91). Dordrecht: D. Reidel Publishing Company.

Wolpert, L., Beddington, R., Brockes, J., Jessell, T., Lawrence, P. A., & Meyerowitz, E. M. (1998). *Principles of development*. New York: Oxford University Press.

Wray, G. A. (1999). Evolutionary dissociations between homologous genes and homologous structures. In G. R. Bock & G. Cardew (Eds). *Homology* (pp. 189–206). Chichester: John Wiley & Sons.

Zelditch, M. L., Swiderski, D. L., Sheets, H. D., & Fink, W. L. (2004). *Geometric morphometrics for biologists: a primer*. San Diego, CA: Academic Press.

Chapter 14

Development and Evolution

RON AMUNDSON

1. Introduction

The relation between embryological development and evolution has become a lively topic in recent years. During the 1990s, a wide range of molecular genetic discoveries showed that the basic regulatory genes of virtually all metazoa (multicellular organisms) were shared. Bodies that had seemed to show almost no similarities, like insects and vertebrates, were discovered to be sculpted during their development by shared genes. This was a shocking discovery (for reasons we will soon discuss). It gave rise to the new field of *evolutionary developmental biology* or evo-devo. For most of the twentieth century, most evolutionary biologists (and the philosophers who worked with them) considered development to have little or no relevance to evolutionary biology. These thinkers were neo-Darwinian, in the sense that they regarded natural selection as responsible for the great majority of evolutionary phenomena. Those who insisted on the evolutionary importance of development were often criticized as being *typological thinkers*. The accusation of typology stems from the fact that developmental evolutionists had very little interest in the variation that exists within a species. They concentrated their attention on patterns of commonality at high taxonomic levels. They studied, for example, aspects of body structure that were shared by all mammals, or all vertebrates. These were not just the characters (the backbone, for example) that taxonomists used to group species together. In the early nineteenth century it was discovered that the limbs of bats, horses, porpoises, and humans all have the same internal patterns of bones, even though they looked and functioned very differently. Developmental thinkers hypothesized *archetypes* and *bauplans* (body plans) to represent the common structures within a group. This fondness for abstract types was regarded as unscientific and almost mystical by most mid-twentieth-century evolutionists. The accusation of typological thinking aligned them with pre-Darwinian (and possibly pre-evolutionary) thought. Their lack of interest in within-species variation also justified this label, because variation within populations was the raw material for natural selection. Developmental thinkers were certainly not population thinkers, and population thinking was (and is) held to be the core of modern evolutionary biology by many. Only recently has developmental evolutionary thought shed the stigma of typological thinking. This chapter will first discuss the serpentine history of the relation between development and evolution up to and including the recent inauguration of evo-devo. It will then discuss the debates that remain concerning the relation between evolutionary and developmental biology.

2. The Nineteenth Century: Evolution Intertwined with Development

Organisms change. Individual organisms change as they develop, from the moment of their first individuality as a zygote through their adult life. Populations of organisms change, as the frequencies of traits in descendent populations vary from those in ancestral populations. On a larger scale, species change. Ancestral species give rise to descendant species, and those to others in a branching pattern. The result, over eons of time, is the current diversity of life. Every metazoon alive today is the result of these two processes of change. The more recent process is the individual organism's own development from a zygote to an adult: its ontogeny. The ancient process is phylogeny: the evolution of the organism's lineage from remote ancestors, through gradual populational change, successive speciation events, and the evolutionary origins of new traits and the losses of old ones.

The tremendous diversity of metazoan life has within it patterns of commonality. Diversity is not chaotic, but patterned. Species with more recent common ancestors are more similar to each other than those with remote common ancestors. Some of the similarities are obvious. Hawks are all similar, and different from other birds; birds are all similar, and different from other vertebrates. However, one set of especially intriguing similarities cannot be seen in adult organisms. We must look at the ontogenies of organisms, the processes of their embryological development. Karl Ernst von Baer in the 1820s showed that the organisms of related species are more similar in their embryonic forms than in their adult forms. This was the beginning of comparative embryology. All embryos begin as a single cell, then proceed through early generalized and homogeneous embryonic forms, until they reach their specialized and heterogeneous adult forms. The early embryo appears to be an unformed lump, but its parts gradually become distinct from each other until they become the various body parts and organs of the juvenile and adult organism. This process is called differentiation. Patterns of differentiation can be compared in different species, and these patterns closely reflect the taxonomic relatedness of the various species. Remotely related embryos begin to diverge from each other with the first patterns of cell division. Closely related organisms share each other's ontogenetic changes until late in embryonic development, when they begin to diverge. The divergence in embryonic form follows a tree-like pattern – a pattern very much like phylogeny itself. In fact, if we look closely at the ontogeny of an individual organism, it can be read as a recapitulation of its ancestors' evolution. In the 1870s Ernst Haeckel proposed this as the *biogenetic law*: ontogeny recapitulates phylogeny. Each organism, during its development, traces the pathway of its ancestors through evolutionary time.

But this picture is too simple. Von Baer's laws of development are only approximately true, and Haeckel's biogenetic law has almost as many exceptions as confirmations. In fact, von Baer's laws had been invented in order to refute an earlier, pre-evolutionary version of Haeckel's biogenetic law. Von Baer had insisted that embryonic organisms are poorly organized and homogeneous at the start, and become more organized and heterogeneous as they develop. Early embryos only resemble the *generalized embryos* of related organisms. They never resemble the *adult forms* of other species, as the biogenetic law requires. In point of fact, neither von Baer nor Haeckel is wholly correct. The shared patterns of embryological development are far too complex to be captured in such simple models. Some of the early embryonic stages of complex species do resemble the adults of ancestral species, and not merely a generalized version of adults of the same species. The embryological precursors of adult mammalian jaws and inner ears look very much like the gill support structures of our fishy ancestors. As complex as these correspondence patterns were, by the middle of the nineteenth century several people were beginning to see them as evidence for evolution.

One such person was Charles Darwin, whose *Origin of Species* was published in 1859. Darwin used the embryological evidence for evolution in the *Origin*, and considered it his "pet bit" (quoted in Ospovat, 1981, p.165). He convinced the scientific world of the fact of common ancestry, the Tree of Life. Darwin was unsuccessful, however, in convincing most of his contemporaries that natural selection had been the driving force behind evolutionary change. Only a minority of scientists accepted natural selection as the primary evolutionary cause. Selection was often seen as the cause of adaptation within a species, but it was harder to conceive of selection producing new species. In retrospect we can find several reasons why natural selection was disfavored during the nineteenth century. Two of them are of relevance here, because they touch on the complex relation between development and evolution. The first problem for natural selection was heredity. Many theories of heredity were proposed during the nineteenth century; at least thirty have been studied. However, none of these theories could be demonstrated to be consistent with natural selection as a cause of continuous evolutionary change in species. One reason was that almost all of the theories shared one feature: heredity was seen as an aspect of embryological development. The word *heredity* did not just name the similarity between parents and offspring. Instead, heredity was thought to be the construction in the embryo of parent-resembling features. In other words, heredity was a part of embryology. As long as the causes of embryological development were still obscure, evolutionists were not able to explain how natural selection could operate *through* them to yield continuing evolutionary change in a species. The universal acceptance of natural selection would have to wait for a new theory of heredity. It was 1915 before that theory appeared.

The second reason for the unpopularity of natural selection was that a separate research tradition dominated biological thought in the nineteenth century, a tradition that included both von Baer and Haeckel. This was morphology, the science of organic form. Morphology included embryology, and it had provided Darwin with crucial evidence for the Tree of Life. After 1859 morphologists rapidly converted to evolution, and became what one historian has termed "the first generation of evolutionary biologists" (Bowler, 1996, p.14). The goal of the program was the explanation of organic form, how it arises in ontogeny and how the processes of ontogeny are modified through phylogenetic time. Natural selection was accepted as a cause of adaptation within species, but adaptation was not an important topic within the morphological research tradition. Haeckel, Carl Gegenbaur, and many others believed that the careful study of comparative embryology would allow both.

(1) a reconstruction of the history of life, and
(2) the causal explanation of how changes in the processes of ontogeny gave rise to changes in the forms of adult organisms.

When morphologists thought about the "mechanism of evolution" they thought of (2), not of natural selection.

The biogenetic law had been the simplest and most dramatic explanation of how evolution worked through embryology. As we saw, Haeckel believed that successive embryological stages of modern species represented the adult forms of their ancestors in phylogenetic time. But the law was known to have exceptions. It would have worked perfectly if evolutionary changes had only occurred in adults. When evolutionary change occurs by the addition of new traits onto adult organisms, the *newest* traits (in evolutionary terms) appear in the *latest* stages of ontogeny, and the oldest traits are in the earliest stages. This is what the biogenetic law says should happen. But not all evolutionary changes happen in adults. Sometimes an evolutionary innovation happens in an embryo, and is inherited by its descendants. When this happens, the descendants of that organism have their *newest* trait (the new innovation) occurring *early* in their ontogenies. The biogenetic record becomes scrambled, and ontogeny no longer represents an accurate phylogenetic history. Because of this possibility, embryologists who observed an early embryological trait of an advanced organism could not be sure how to interpret it. Did it represent an adult trait of a very ancient ancestor, or is it a recent innovation that was inserted into early development? These two origins must be distinguished if the biogenetic law is to be useful in understanding evolution. If the trait we are considering is the mammalian placenta, clearly it must have been an innovation that occurred in early ontogeny, not a trait added onto to an ancestral adult. (The simple reason is that no adult could survive wrapped in a placenta!) Comparative studies revealed more and more ambiguities; embryonic traits simply could not be "sequenced" into a neat phylogenetic order merely by comparing the embryos of different species.

It is important to remember that comparative embryology at this stage was an observational science, not an experimental science. Its data were careful observations of embryological stages in different but related species of organisms. When the problems (ancient adult versus recent insertion) became clear, one possible solution was recognized. If we could discover the internal causes, in the embryo, of the changes it went through, we might be able to decipher which embryonic traits were those of ancient adults and which were recent insertions. (We might be able to tell, for example, how difficult it would be for a particular kind of trait to become inserted in an early embryo.) But how do we discover those internal causes? Gegenbaur had thought that careful observation would lead to the discovery of internal causes, but hope was fading. Others believed that only experimental manipulation would permit the discovery of the direct, local, proximate causes that propelled the embryo through its successive modifications towards adulthood. Experimentalists such as Wilhelm Roux claimed that experimental embryology could in this way be of service to evolutionary morphology by discovering how proximate causes controlled embryonic development.

Experimental method was rejected by Haeckel, however, and for an intriguing methodological reason. He claimed that proximate developmental causation was irrelevant to evolutionary ("ultimate") origins (Nyhart, 1995, p.189). (The semantic distinction between proximate and ultimate causes was popularized much later by Ernst Mayr, but the distinction applies quite clearly to Haeckel's reasoning; Mayr, 1961.) In effect, Haeckel wanted to black-box ontogenetic causation. He claimed that phylogeny was itself the mechanical cause of ontogeny (Gould, 1977, pp.76–85). It sounds bizarre today to claim that evolutionary history *causes* the growth of an individual embryo. But Haeckel insisted that the complex proximate causes that operate during ontogeny were merely irrelevant details that distracted from the big picture. The big picture was phylogeny (Amundson, 2005, p.121).

It is important to recognize the difference between Haeckel's program and others of his era. Haeckel denied the relevance of proximate embryological causation for the understanding of evolution. Other evolutionary morphologists, such as Gegenbaur, believed that proximate causation must be understood in order to distinguish between embryonic traits of ancient ancestors and those of recent insertion. Roux and other experimentalists expanded Gegenbaur's critique of Haeckel, and urged that proximate causation must underpin any developmental understanding of evolution. Surprisingly, in this sense, Haeckel can thus be seen as an *opponent* of the developmental understanding of evolution – at least of proximate-causal developmental understanding. The biogenetic law declared that embryological *patterns alone* – and not the details of embryological causation – would explain evolution. The failure of the biogenetic law was originally seen not as a refutation of the importance of development to evolution. Instead it was seen as a proof that the proximate causes of embryonic development must be understood before development would shed its light on evolution.

The experimentalists prevailed. The early experimentalists did not directly reject phylogeny. They hoped to contribute to its understanding. Even though the biogenetic law had failed, evolutionary changes in adult form were still seen as products of changes in embryonic development. When the proximate causes of ontogeny were finally understood (it was hoped), the changes in ontogeny that constitute phylogeny could be deciphered. This hope was premature. The proximate causes of ontogeny proved immensely complex. (Indeed we are still working them out, and we are nowhere near a final answer.) Long before embryologists knew enough to return to evolutionary morphology, a new and different evolutionary theory had sprung up. The new theory, called the Evolutionary Synthesis, considered ontogenetic development to be virtually irrelevant to the process of evolution. For the first time in history, development and evolution were seen as completely distinct phenomena, ships passing in the night. Only in the late twentieth century, after radical advances in developmental biology (the successor to experimental embryology), was development again seen as crucial to understanding evolution.

3. The Twentieth Century: A New Heredity Gives Rise to a New Evolution

Thomas Hunt Morgan began his academic life as an evolutionary morphologist. He and many colleagues abandoned evolutionary morphology in favor of experimental embryology. Early in the twentieth century, while he was studying how the sex of an individual organism was determined during its embryological development, Morgan became convinced that the inherited material substance that controlled the traits of the embryo lay in its chromosomes. In the year 1900 several students of heredity had independently rediscovered the work of one of the rare nineteenth-century heredity theorists who had *not* considered heredity as an aspect of embryonic development. That theorist was Gregor Mendel, who hypothesized unobserved *factors* (later called genes) that somehow carried adult similarities between the parents and offspring (never mind the embryological processes that produced those similarities).[1] Morgan and his colleagues incorporated Mendel's idea of hereditary factors into their theory of the chromosomal location of heredity. The result was the Mendelian chromosomal theory of heredity (MCTH). This was the basis of modern genetics. This theory would have many important influences on twentieth-century biology. One would be to enable, at long last, the construction of a detailed evolutionary theory that had natural selection at its core. Another influence, ironically, would be to prohibit the relevance of embryological development to that new evolutionary theory. Not the failure of the biogenetic law, but the success of the MCTH drove a wedge between evolution and development. To understand this effect, we must appreciate the differences between nineteenth-century concepts of heredity and the new MCTH.

Recall our discussion of nineteenth-century theories of heredity. Almost all of them regarded embryonic development as the action of heredity; to understand heredity we must understand development. To understand why traits are similar between parent and offspring we must first understand how those traits arise in ontogeny. Then we may be able to understand how they arise similarly in parent and in offspring. Heredity named the process by which a parent passed on to an offspring *the ability to develop* its characteristics – its spinal column, its limbs, and eventually the characteristics that made it resemble its parents rather than other members of its species. Development was an expression of heredity. Almost no one except Mendel considered heredity distinct from development, and Mendel's innovative work on heredity was virtually unknown among evolutionary thinkers until 1900. Until at least 1910, Morgan himself accepted this embryological view of heredity. When he developed the MCTH, his views changed. The MCTH made a very radical assertion: development is irrelevant to heredity. Genes are the hereditary causes of the adult traits even though geneticists had no idea how the possession of a gene contributes to the embryological development of the trait in the adult.

---

1 Besides Mendel, the only other nineteenth-century non-developmental heredity theorist I am aware of is Karl Pearson. Pearson based his views on the kind of epistemological phenomenalism that will be discussed in Section 3.

Embryologists were horrified at this theory (Lillie, 1927; Hertwig, 1934; Sapp, 1987, pp.17–28; Burian, 2005, pp.183–9; Amundson, 2005, pp.175–88). It was known by this time that (with a few exceptions) all of the cells in the body contained the same genes. One of the most basic facts of embryonic development was differentiation, in which all of the distinct body parts and tissues of the adult organism are produced within an originally formless embryo. This produces a paradox: How can the *same* genes (in each cell) be the causes of *differentiated* body parts (arms and legs, bone and muscle)? The *cause of* an adult character was, to an embryologist, the developmental process that built that character and differentiated it from other characters within the developing embryo. To talk about the cause of an adult trait while ignoring development was to speak nonsense. Did the gene in the zygote magically reach through time and space to insert the trait into the adult? Surely not. Then what sense can it make to speak of hereditary *causes* of traits without taking account of the intervening mechanical steps by which the inherited trait is brought into being?

In reply to this challenge, Morgan and his colleagues distinguished between *transmission genetics* and *developmental genetics*. Transmission genetics was heredity. Heredity was understood as a probabilistic correlation between the traits of adults and offspring, assuming certain facts about the segregation and independent assortment of genes (the carriers of traits) but making no assertions at all about how genes acted within the embryo. Developmental genetics (a study for the future) had the job of explaining how genes acted during development. The important point was this: Transmission genetics *is* heredity, and therefore development is irrelevant to heredity. Morgan had black-boxed embryological causation for the purposes of heredity, just as Haeckel had black-boxed it for the purposes of phylogeny forty years earlier. "The theory of the gene is justified without attempting to explain the nature of the causal processes that connect the gene and the characters. . . . the sorting out of characters in successive generations can be explained at present without reference to the way in which the gene affects the developmental process" (Morgan, 1926, pp.26–7). Embryologists continued to resist this co-option of the term *heredity*, but they gradually lost the battle.

Our discussion at this point will be aided by a distinction between two kinds of scientific methodology, realism and phenomenalism. Although philosophers often interpret the two doctrines as universally applied throughout science, in actual practice the doctrines are selectively applied. A given scientist might be a phenomenalist about some areas of science, but a realist about others. A given scientific theory might receive a phenomenalist interpretation during some period of time, and a realist interpretation later (or earlier). The phenomenalist/realist contrast concerns the proper interpretation of scientific theories. Phenomenalists and realists agree that observation is extremely important to science, but they differ on what can be legitimately inferred from a set of observations. The difference is this: Realists believe that it is proper, appropriate, and productive to infer the existence of entities and processes that are not directly observed by the scientists (entities that are therefore called "theoretical entities"). Phenomenalists do not. Phenomenalists believe that the goal of science (or at least the goal of the particular branch of science under discussion) is to discover the laws that account for variations in the observed phenomena. These so-called phenomenal laws make no reference to unobserved theoretical entities or processes. Once a set of phenomenal laws has been discovered, phenomenalists are satisfied with that achievement and might turn their attention toward discovering how those laws relate to other known laws. A realist, on the other hand, would ask an additional question: What unobserved processes explain why the phenomenal law operates as it does?

Phenomenalist and realist views are favored under a number of different circumstances. Sometimes a given realistic theory is just too speculative for the phenomenalist to swallow, and no better realist theories are available. At other times, a remarkably good theory may seem to conflict with widely accepted principles of science, or even metaphysics. Phenomenalism can rescue scientists from that uncomfortable position by allowing them to say "The conflict is only apparent; I am not making assertions about reality, but only predictions about observations." This allows a scientist to continue research without worrying about the underlying conflicts. Newton himself made this phenomenalist claim about the law of gravity. Newton's concept of gravity is a force that acts between bodies that have no contact with each other. In his day, this violated the metaphysical principle of "no action at a distance"; it was believed that force could only be conveyed between objects that were in contact. Phenomenalism about gravitation allowed Newton to ignore this problem.

Morgan may have found himself in a position somewhat similar to Newton's. Prior to 1910 he believed, with most of his colleagues, that heredity was a matter of embryology. But the causal understanding of embryology was proceeding very slowly. If heredity could be given a phenomenalist interpretation that divorced it from embryology, progress might be faster. By 1915 Morgan was ready to divorce heredity from embryology, and use the MCTH to link parental traits with offspring traits by correlation alone, with no explanation of the development of the offspring's traits. Development was irrelevant to transmission genetics. Morgan's claim about heredity is in fact very similar to Newton's claim about gravity. Transmission-genetic causation is literally *action at a distance*: the genes in the zygote cause traits in the adult, and the intervening embryological processes are black-boxed and ignored.

The phenomenalist nature of transmission genetics is seldom recognized by philosophers (but see Sarkar, 1998, ch. 5), possibly because developmental genetics has always been in the background with the promise of a realistic explanation of *how* genes contribute to the development of traits. Transmission genetics was tremendously successful, but developmental genetics was very slow in producing results. There was scarcely a glimmer of how genes *could* produce embryonic differentiation until the 1960s. Then Francois Jacob and Jacques Monod proposed a simple model of how bacteria could respond to the nutrition in their environment by modifying the expression of their genes. The very concept of the *expression* of genes was beyond the reach of transmission genetics. For most of the twentieth century "genetics" meant transmission genetics, a field of study that was carefully, and phenomenalistically, defined to exclude development from its purview.

The MCTH separated embryological development from heredity. How does that affect the relation between development and evolution? The answer is simple. The Evolutionary Synthesis of the 1930s and 1940s was based on the MCTH, together with a mathematical analysis of the distribution of genes – *transmission* genes of course – in evolving populations of organisms. The MCTH was the very first theory of heredity that was proven to be consistent with natural selection as a cause of long-term evolutionary change. The proof took the form of the equations of mathematical population genetics.

These equations modeled populations of organisms and their genes. As generations passed, the gene frequencies could be shown to be affected by natural selection as well as several other factors such as migration, mutation, and random drift. However – and here is the important thing – embryological development had no place within the models of population genetics. Population genetics was based on transmission genetics, which was *defined* in terms of the Mendelian patterns of correlation of phenotypic traits between generations. Embryological development had been black-boxed by transmission genetics. When transmission genetics was incorporated into the Evolutionary Synthesis, development remained in its black box. Just as in Haeckel's interpretation of the biogenetic law, proximate causes of embryological development were considered irrelevant to the understanding of evolution.

In recent years the advocates of evo-devo have sometimes argued that the resistance of mainstream Evolutionary Synthesis theorists to development was a sort of conspiracy from the start. However, careful study of the historical record reveals no evidence of this. Morgan and his coworkers were actively interested in developmental genetics, although they produced few results. The leaders (the so-called architects) of the Evolutionary Synthesis did react harshly toward some developmentally inclined adversaries such as Richard Goldschmidt, who opposed the Darwinian principle that evolution was a smooth and gradual process. But others were tolerated and sometimes even encouraged. C. H. Waddington and I. I. Schmalhausen were among the developmental advocates who were regarded as relatively friendly to the Synthesis. However, they had no lasting effect on Synthesis theorizing. The importance of development continued to be advocated by a minority of theorists throughout the century, including some comparative anatomists and paleontologists as well as embryologists. The real, open conflicts between mainstream Synthesis evolution and the advocates of development only arose around 1980. [SEE POPULATION GENETICS].

4. The Nature of Developmentalist Explanation: 1920–80

We see that the developmental view of evolution – the view that understanding evolution requires understanding the causal processes of development – was historically confronted with two black boxes. The first was Ernst Haeckel's declaration that the proximate causes of ontogeny were irrelevant because phylogeny itself was the cause of ontogeny. Haeckel is usually remembered as a friend of ontogeny. In fact he was a friend only of ontogenetic *pattern*; not ontogenetic causation. Haeckel's black box died when the biogenetic law died. The proximate causes of ontogeny (for example, the interactions among the developing body parts in the embryo) must be understood in order to decipher the tangled web of evolutionary changes that had occurred at various embryological stages in different lineages. But the task of understanding embryological causation was barely begun when the second black box appeared. It was constructed by the MCTH; development was now irrelevant to heredity. The new theory of heredity was parlayed into a new theory of evolution by the Evolutionary Synthesis. So again the proximate causes of ontogeny were black-boxed with respect to evolution. The logic behind the second black box is one step more complex than the first. Haeckel had said that because phylogeny is the direct cause of ontogeny, the intervening proximate causes of ontogeny are irrelevant to phylogeny. The later version, due to the Evolutionary Synthesis, states that phylogeny (evolutionary history) is the cause of contemporary genotypes, and contemporary genotypes are the direct cause of phenotypes. Therefore the intervening proximate causes of ontogeny – those that build the phenotype out of the genotype – are irrelevant to phylogeny. As long as genotypes were conceived as the direct causes of phenotypes, evolutionary biology had no causal room, or explanatory need, for ontogeny.

The developmental evolutionist's challenge is the same for either black box. It involves two interrelated tasks. The first task is to argue convincingly that some features of evolution cannot be explained in the absence of a proximate understanding of development. In the case of Haeckel's black box, this was shown by the continued inability to distinguish between traits of ancient adult ancestors and recent insertions. The second task is to show that an understanding of developmental causation can explain those aspects of evolution that non-developmental theories could not explain. Roux, Gegenbaur, and others hoped that this could be accomplished with the aid of knowledge about proximate causation within embryos; we might discover (for example) that certain traits were easy to insert into early embryonic stages, while others could only have gotten there by inheritance from ancient adults. This would allow us to separate the ancient-adult traits from those inserted into early embryology. Unfortunately, the causal structure of ontogeny was far too complex for this program to succeed in the early twentieth century.

These two tasks were the same for the twentieth-century developmentalists who were confronted by the black box produced by the MCTH and the Evolutionary Synthesis. First, find a phenomenon that is unexplained by non-developmental theories of evolution. Second, explain it as a consequence of the facts of development. The twentieth-century developmentalists' intuitions were the same as those of their predecessors: evolutionary changes were changes in ontogeny. But as we have seen, progress was very slow in experimental embryology and its successor developmental biology. The developmentalists were left with few explanatory resources. The problem was exacerbated by the fact that the linked disciplines of transmission genetics and the Evolutionary Synthesis were making great progress. The Evolutionary Synthesis explained evolved traits mostly as adaptations produced by natural selection from genetic variations in ancestral populations. These so-called adaptationist explanations were in competition with any explanation proposed by developmentalists. Some of the adaptationist explanations were extremely well confirmed, both by experiment and by observation of natural populations. But others were quite speculative, at least as seen by developmentalists. The problem for developmentalists was that their understanding of ontogenetic causes lacked the detail needed to offer alternative explanations to the adaptationists. They were forced to construct explanations out of mid-level ontogenetic patterns instead of genuine proximate causes.

Two examples of mid-level patterns are allometry and heterochrony. Allometry refers to correlations, sometimes expressed in complex equations, in the relative sizes of body parts during growth. These correlations are presumed to be produced by the mechanisms of development. When the modifications in relative size of two body parts during the growth of an individual are seen to correspond to size comparisons between two related species, the phylogenetic ratios between the species are said to reflect the same developmental causes as the ontogenetic ratios during development. It has been argued, for example, that body size increases faster than tooth size in the growth of individual mammals, and that therefore the relatively small teeth of gorillas as compared to chimpanzees is to be explained by the mere fact that gorillas are larger, as adults, than chimpanzees. This is allometry. The problem is that adaptationists could easily claim that natural selection caused the smaller tooth size, and so argued in opposition to the allometric explanation.

Heterochrony is a modification in the relative timing of different developmental events during embryogenesis. For example, if sexual maturation were selected to occur earlier than the maturation of body form within a lineage, the adult forms of descendants might retain juvenile bodily traits. This is *paedomorphosis*, one particular heterochronic pattern. The process is often said to have been involved in the evolution of humans, because juvenile chimpanzees show greater similarity to human adults than do adult chimpanzees. Even though natural selection is invoked in the paedomorphosis explanation, the traits that are explained developmentally are not directly selected for. Instead they are linked by developmental mechanisms to the selected-for traits. The adaptationist alternative is to argue that the juvenile traits of the descendant population were individually selected for – they did not piggy-back along, because of developmental linkage, on the single selected-for trait of early sexual maturity.

Heterochrony and allometry both refer to proximate causation during embryonic development: how form is generated in the body. However, proximate causation is inferred from observable morphological patterns rather than studied directly through experiment. Because embryological causation was so difficult to trace, many developmental evolutionists of the mid-twentieth century were comparative morphologists and paleontologists, rather than embryologists. Morphologists and paleontologists had the data necessary to do heterochronic and allometric analyses, but not to directly study the proximate causes of ontogeny. Direct knowledge of developmental causation began to come into the picture in the 1980s. Before we discuss the results, let us consider the resistance to developmentalist evolutionary views that came from the adaptationism of the Evolutionary Synthesis.

5. Adaptationism and the Synthesis

From the discussion of allometry and heterochrony it might seem that the developmentalist and the adaptationist explanations were on equal footing. They were not. Adaptationist explanations, even in the absence of direct evidence, possessed much more prestige and scientific plausibility than developmentalist ones during mid-century. Although developmentalists often describe this as an unfair prejudice, it followed upon genuine successes of the adaptationist research program during the middle of the century. Adaptationism was not particularly dominant during the early years of the Synthesis. Genetic drift was considered by many Synthesis thinkers to be the cause of the traits that differentiated between related species. But careful adaptationist studies had revealed that examples of these traits were, surprisingly, adaptive to the species (Cain & Sheppard, 1950). The prestige of adaptationism was well-earned. Nevertheless, it was based on limited data. Adaptationist over-enthusiasm led to conclusions that are now seen as unjustified. Two examples will illustrate the exuberance of adaptationism during this period.

First, it was believed by many adaptationists during the 1950s that major taxonomic groups, such as mammals and birds, were not monophyletic (that is, descended from a single evolutionary ancestor), but polyphyletic (descended from two or more originally distinct ancestors). This means that diverse ancestors had adaptively converged on the common characteristics by which mammals and birds are identified. It was suggested, for example, that all of the anatomical properties that characterized mammals (such as hair, mammary glands, and placental gestation) were coincidental adaptive consequences of the independent evolution of homeothermy (warm-bloodedness) among several pre-mammalian lineages. The power of natural selection to produce adaptive convergence was so highly regarded that it was considered hazardous to infer common ancestry from any degree of anatomical similarity. Extremists were willing to hypothesize that virtually every trait of an organism was there because it served an adaptive purpose to that species. This means that no traits at all should be ascribed to common ancestry (let alone common embryological causation), at least until it was conclusively proven that the traits were not adaptive to that species. Because traits shared between species are assumed to be selectively produced in each species, those who wish to use developmental explanations have virtually nothing to explain.

A second illustration of the consequences of mid-century adaptationism relates to the concept of homologous genes. Mendelian genetics is based on crosses between individuals that have different heritable traits. For practical purposes this means that it is impossible to identify the genetic basis of similar traits within distinct species, simply because it is impossible to crossbreed between species. Given this lack of data, it would seem to be an open question whether or not traits shared between species had the same (or "homologous") genetic causes. Nevertheless, even in the absence of direct genetic evidence, the commitment to adaptationism inclined evolutionists to believe that shared genes were *most likely not* the causes of shared traits. Shared traits were believed to have been independently sculpted in each species in which they appeared. The common commitment of leading evolutionists such as Ernst Mayr and Theodosius Dobzhansky in the 1950s and 1960s was that "If there is only one efficient solution for a certain functional demand, very different gene complexes [in different species] will come up with the same solution, no matter how different the pathway by which it is achieved" (Mayr, 1966, p.609). As we shall see, the adaptationist commitment to convergent rather than conserved similarities was a spectacularly hasty conclusion.

Additional barriers to developmentalist thought came from the articulation of several philosophical and methodological concepts in the form of binary distinctions between classes of phenomena. These were invented to express basic concepts of the MCTH and Synthesis theory, and give the appearance of basic conceptual truths. Nevertheless, they can be taken to imply the irrelevance of development to evolution. We have already used one of these binaries, the distinction between proximate and ultimate causation. Ernst Mayr, the popularizer of this distinction, has used it to argue against developmental evolutionists on the grounds that development is a proximate process while evolution is an ultimate process. Another important but biased binary is the genotype versus phenotype distinction.[2] This distinction predated the MCTH, but was modified from its original meaning in order to reflect the ontological commitments of the MCTH. The updated version is often taken to exhaustively label all organismic factors that are relevant to evolution. If understood in this way, it expresses the black-boxing of embryological development. The genotype is held to "cause" the phenotype, with no reference to the causal activities that take place within the black box of development. If the genotype and the phenotype together provide a complete and adequate account of evolutionary processes, then ontogeny is irrelevant to that account [SEE GENE CONCEPTS] Population genetics deals with the sorting of the traits of populations through evolutionary time. This sorting can be seen in terms of genes or of phenotypic traits, depending on whether one is a gene-selectionist or an individual selectionist. But it cannot be seen as the sorting of ontogenies (or elements of ontogenies), because ontogeny is conceived to be irrelevant to the hereditary causation of traits.

Advocates of developmental evolution are forced to reject the sufficiency of the genotype–phenotype dichotomy. Waddington did so in the 1950s with the proposal to add the *epigenotype* to the genotype/phenotype distinction. The epigenotype is made up of the causal processes in the embryo that mediate between genotype and phenotype. The genotype controls embryological growth (the epigenotype) which in turn builds the phenotype. Waddington was unable to prove to Synthesis evolutionists that his three-part distinction was superior in its explanatory power to the genotype–phenotype binary.

Waddington's situation was typical. An important factor in the persistence of the debates about development is that it was difficult for developmental evolutionists to specify in detail what *they* could explain but population biologists could not. Consider the proposals put forth by developmentalists advocating allometry or heterochrony, for example. Any phenomenon explainable by heterochrony (the similarity between infant chimpanzees and adult humans, for example) was also explainable by adaptation (that infant chimps were subject to different selection pressures than adult chimps, for example – therefore the similarity with adult humans is a mere coincidence). Allometry and heterochrony were too crude to survive a test against the strength of mid-century adaptationism.

6. Direct Debates

During the 1980s the debates began to center around the concept of *constraint*. Developmental evolutionists focused on developmental constraint, but this topic was entangled with several other factors that were also called "constraints." The extreme adaptationism of the 1950s and 1960s had waned somewhat, and adaptationists were now responding to theoretical challenges. Nevertheless, the divergent commitments of the two theoretical orientations resulted in a stalemate in the constraints debates. Let us examine how this happened.

Critics of the Synthesis (including developmentalists) had alleged that adapationists were insensitive to all factors that might limit adaptive perfection, and development

2 Other binaries include germ-line versus soma, and population thinking versus typology. Each of these was used during the 1980s and 1990s to argue against the legitimacy of developmentalist evolutionary theories (Amundson, 2005).

was merely one of those factors. So the term constraint was often used generically to cover a number of non-adaptive causal factors. From the adaptationists' perspective, the challenge was to demonstrate that they took adequate account of factors that might limit adaptive perfection: constraints on adaptation. The critics had asserted that adaptationists neglected these limits. Adaptationists said that such limits were acknowledged (implicitly or explicitly) in every explanation. The foraging behavior of a particular bird was only optimized *under the constraints* of its eyesight and flight range. The allegation that adaptationists ignored constraints seemed absurd to these authors, and they often said so. To them, constraints are merely the background assumptions that frame the stage on which natural selection takes place. Every selective explanation must have those background assumptions, and so the allegation that adaptationists ignored constraints was simply false.

Unfortunately (for the unity of biology) the adaptationist concept of constraints was quite different from the developmentalists' concept. Developmental constraints were conceived not as mere limiting background assumptions, but as visible manifestations of underlying causal processes. The underlying processes were the processes of ontogeny, the building of bodies during embryological development. Ontogeny, and its modifications through evolutionary time, was the focus of developmental interest. The discovery of constraints was a significant part of the purpose of their study. Developmental constraints were seen to constitute direct evidence about the processes that constituted developmental evolution. They pointed toward a certain kind of positive causal activity underlying evolutionary change: modification of the process of ontogenetic development. Ontogenetic processes are themselves productive. They are not restrictive, as the term "constraint" seems to imply. Ontogeny is productive of functioning phenotypes. Ontogenetic processes can be modified, and certain kinds of modifications are more likely than others. Constraints are the shapes of possible or likely changes in ontogeny (Amundson, 1994). The contours of these possibilities are consequences of the ways in which bodies are, and have been, built. Embryologists and (more recently) developmental biologists have traditionally concentrated on the form, the morphology, of organisms. So developmental constraints were considered to be constraints on form, on the possible morphologies of developmental variants. So conceived, constraints (on form) had no direct implications regarding the study of adaptation. They were involved, instead, in the study of how body form had changed through evolution. The relevance of adaptation to this study was simply not a topic of discussion: that was left to the adaptationists.

So we see that the theoretical role of constraints differs greatly between these groups. Adaptationists were concerned with constraints on adaptation, limitations on adaptive perfection. Developmentalists, in contrast, were interested in constraints on *form* – on the possible configurations that bodies can take.[3] Even if adaptationists were to consider

---

3 As an example, consider the universality of four limbs, rather than six or eight, among tetrapods. An adaptationist explanation would concentrate on the relative fitness of variant limb numbers, while a developmentalist would concentrate on the mechanisms by which limbs are constructed in the embryo. The constraints-on-adaptation might be the environmental problems caused by increased limb numbers. The constraints-on-form might be the inflexibility of the embryological processes that produce limb numbers. Although this example is purely imaginary, it illustrates the contrasting interests of the two research orientations.

genuinely developmental constraints (those that are produced by the organization of ontogeny), their relevance would be no different than any other restriction on adaptation, such as a bird's poor eyesight. Constraints would continue to be a background condition for adaptationist explanations, whatever their source. The criteria for success of an adaptationist explanation simply do not require an account of the ontogeny of the trait under consideration; the MCTH made sure of that. Because adaptationist explanations were natural selective explanations, developmentalist accounts of evolution would appear irrelevant to evolution. Developmentalists continued to insist on the importance of constraints. Adaptationists continued to misunderstand this insistence as a complaint against *unconstrained* adaptationism. Adaptationists accepted the need to state the constraining background conditions under which their explanations operated, and saw no point to doing anything further (such as understanding development).

The proximate processes of development began to play a larger role in evolutionary explanations during the 1980s. These studies were based on a detailed understanding of the ontogeny of vertebrate limbs, especially the limbs of the two amphibian groups of urodules (salamanders) and anurans (frogs). David Wake and his colleagues constructed explanations for a number of phylogenetic patterns of limb variation that showed them to result from the mechanisms of limb ontogeny. One was an explanation of a correlation regarding the evolution of digit loss between the two groups. Digit loss occurs frequently, especially in those frogs and salamanders that evolve a miniaturized size. An interesting pattern is that the lost digits differ between anurans and urodeles. Urodeles lose the posterior digits first; anurans lose the anterior digits first. This pattern corresponds with the order in which the digits are ontogenetically produced within the group: urodeles differentiate their digits beginning with the anterior and proceeding to the posterior, and anurans the reverse. So the pattern of digit loss in evolution is the reverse of the pattern of digit differentiation in ontogeny: the last digit developed is the first one lost. Wake and colleagues have argued that a number of similar patterns in the evolution of amphibian limbs can only be understood in terms of the ontogenetic processes by which those limbs are developed (Rienesl & Wagner, 1992; Shubin, Wake, & Crawford, 1995).

These developmentalist explanations of evolutionary patterns conflict with adaptationist standards for a good explanation. Hudson Reeve and Paul Sherman produced an extensive critique of Wake's style of developmentalist explanation from an adaptationist perspective (Reeve & Sherman, 1993). They pointed out that Wake did not examine the ways in which digit loss affected the fitness of the various frog and salamander species. For this reason (they said), Wake had not *explained* the patterns of digit loss, but only re-described them. In order to explain digit loss (by adaptationist standards) one had to demonstrate the effects of fitness on the variant forms. Only if Wake was able to prove that the patterns had not resulted from distinct cases of natural selection for the adaptive benefits of digit loss could he be said to have "explained" digit loss in terms of ontogenetic processes. Adaptationist explanations require an examination of comparative fitnesses of the variants (and do not require an analysis of their comparative ontogenetic sources). From this perspective, developmentalist explanations are no explanations at all.

This is merely one example of the contrast between explanatory standards. From an outsider's point of view, one would think that developmentalists and adaptationists might be equal players in the explanation game. Adaptive and developmental explanations might compete for legitimacy, but each is a possible contender. For most of the twentieth century this was not true. As we see from the Reeve–Sherman critique, developmentalist explanations were at a disadvantage. Adaptationist explanations had exhibited their power ever since the 1940s. As late as the 1990s, developmentalist explanations had no similar track record. Some of the later developmentalist explanations were more appealing, because they were based on known ontogenetic mechanisms rather than such hypothesized causes as heterochrony and allometry. However, few could definitively rule out adaptation as a possible alternative scenario. For this reason, adaptation retained the upper hand. This was all to change by the end of the century.

7. A Torrent of Homologous Genes

T. H. Morgan had foreseen the genetic study of development in the 1920s. That study was delayed for the greater part of the century. The near-total absence of the genetic understanding of development meant that the phenomenalist science of transmission genetics was left alone to form the basis of evolutionary theory. Certain features of mid-century adaptationism must be seen as byproducts of the absence of developmental genetics during this period. This is certainly true of the opinion of Mayr and Dobzhansky that similar traits were probably not due to homologous genes. These authors are not to blame for their lack of data. Transmission genetic analyses were based on genetic crosses, and so the impossibility of crossbreeding species made homologous genes almost impossible to identify. By the turn of the twenty-first century, however, this commitment to the power of natural selection to sculpt similar traits out of diverse genetic resources is open to serious doubt. What has changed? Our understanding of molecular genetics, and the discoveries of a number of extremely deep genetic homologies.

Consider what the adaptationists of the 1970s might have expected from the advancement of developmental genetics. Natural selection was presumed to be able to sculpt common characters from diverse genetic resources. Most similarities among phylogenetically remote species were due to adaptive convergence rather than developmentally conserved traits. Consider two examples of traits that are shared by widely separated groups of animals: eyes, limbs, and bilateral symmetry. Insects and vertebrates are bilaterally symmetrical, and both have limbs and eyes. But the body plans of the two groups are very different, and their common ancestors are lost in evolutionary time. Eyes, limbs, and bilateral body arrangement have obvious selective advantages – they allow the animal to move forward in a search for food. Surely they must be seen as adaptive convergences, not shared development. (It was believed, for example, that eyes had independently evolved about forty different times.) For these reasons, homologous genes were not to be expected except between very closely related species. Natural selection was the causative force in evolution, and selection produced diversity, not commonality. Developmentalists had claimed that perceived commonalities were attributed to shared developmental causes, but they had never been able to prove that the commonalities had *not* arisen from convergent selection. Given the perceived power of selection, there was no reason to expect surprises from the progress of developmental genetics. But surprises there were.

By the early 1990s molecular geneticists were beginning to identify genes on the basis of their molecular composition. The first shock was the discovery that certain genes were shared among nearly all animal groups, from mammals to insects to flatworms. The mere fact of widely shared genes was inconsistent with the expectations of major adaptationists. But the nature of those genes was even more surprising. They acted at the deepest and earliest stages of embryonic development. A gene called *Pax-6* was the first to be identified in widely divergent groups. This gene stimulates the development of eyes (and even primitive eye-spots) in all known taxa. A similar gene exists for the developmental origin of limbs. An entire set of genes, called *Hox* genes, sets up the bilateral body axes (front–back and left–right) and specifies the nature of the various body segments from front to back. The bilateral body plan of virtually all complex animals was invented only once in evolutionary history. The hypothetical ancestor of all of these (all of us) animals is named *Bilateria*. The source of this continuing body plan is embodied in a "toolkit" of developmental genes that remain almost identical, after six hundred million years of evolutionary divergence.

The toolkit genes are quite unlike the genes studied by transmission genetics. Because they are shared by virtually all animal species, their discovery under classical methods of crossbreeding was impossible. The protein products of these genes do not directly affect the phenotype. They control the expression of other genes, and do so in the earliest stages of development.

The Evolutionary Synthesis offered a theory that was based on genes that were conceived to vary in populations, and that had direct effects on the phenotype. This allowed population geneticists to imagine that the sorting of genes in a population was conceptually equivalent to the adaptive sorting of traits – the Darwinian process of natural selection. That analysis led them to doubt that development was relevant to evolution, and to doubt that development had any role to play in explaining such commonalities as the bilateral body plan or the existence of legs and eyes. They were right in their recognition of the diversity of life. But they were wrong in their failure to recognize the commonality that underlay that divergence. Developmental evolutionists had been arguing since the late nineteenth century that evolution could not be understood without understanding development. The invention of transmission genetics (via the MCTH) was a serious challenge to that view: exactly what was it that neo-Darwinian evolutionary theory could not explain, but developmental evolution could? Attempts by developmentalists to explain character distribution by heterochrony and allometry were staved off by arguments that the same phenomena could be explained by ordinary selective processes. But now, with the discovery of the deep homologies, developmentalists had a well-confirmed fact about the unity (not the diversity) of life that tied embryological development deeply into the evolutionary process. Adaptationists like Dobzhansky and Mayr had predicted just the opposite – the absence of any important homologous genes. The deep homologies were not *mere* commonalities, but very early and developmentally important commonalities that tied together shockingly diverse life forms. Metazoa all share their deepest developmental mechanisms. To understand the evolution of *this* group of life forms – the ones here on earth – it is necessary to understand their development.

Evolutionary developmental biology, evo-devo for short, is the new name for the developmental study of evolution. The momentum of the research program of developmental biology is enormous, and a great deal of it has direct implications for evolution. (For an accessible modern introduction see Carroll, 2005.) So developmental evolution is alive and well. However, the reader may be wondering what became of the constraints debates discussed above. Did the developmentalists refute the adaptationists? The answer is no. The debates remain unresolved. The progress of evo-devo has come by way of an explosion of new information from molecular biology. It has not come by way of philosophical and methodological argumentation. Many of the practitioners of evo-devo are not even aware of the old debates, or the conflicts between adaptationist and developmentalist views of evolution that were exposed in those debates. This leaves the philosophical issues in an odd situation – unresolved. The final section of this chapter will discuss this odd situation, and possible future resolutions.

8. What Now?

Philosophers of biology often claim that their field began around 1960. This period was certainly an evolutionary and philosophical watershed. Ernst Mayr's important work of articulating the philosophical and methodological foundations of neo-Darwinian theory began in 1959, corresponding with centennial celebration of Darwin's *Origin of Species*. Mayr criticized most previous philosophers who wrote on biology for failing to recognize population thinking, and introduced other important philosophical concepts such as proximate versus ultimate causation. Mayr's principles were seen by many as the starting point of philosophy of biology. The preponderance of philosophical writing on biology since then fits within the parameters of Evolutionary Synthesis thought. Problems like units of selection, and the proper scientific definitions of concepts like *adaptation*, *fitness*, and *function* fit perfectly well within this framework. This work has continuing value; no one in the evo-devo camp rejects the importance of population thinking and other neo-Darwinian concepts. However, this work does not help us to understand the role of development in evolution. Something other than adaptation and population thinking must be addressed if we are to establish a philosophical understanding of developmental evolution.

For a developmental understanding of evolution, the most productive area of philosophical inquiry has been genetics. Transmission genetics offered no room for the developmental evolutionist. But molecular genetics, and later developmental genetics, made new understandings possible. Philosophical studies of the changing concepts of the gene provide the strongest philosophical transition between the neo-Darwinian style of philosophy of biology and the kind of understanding that will be necessary in the era of evo-devo. However, most of this work does not directly relate developmental genetics to evolution. This is what is needed for a philosophical understanding of evo-devo.

In recent years, as evo-devo has grown, philosophers have gradually begun to recognize the new field of thought, and attribute more explanatory power to developmental concepts. The relation between natural selection explanations and developmental explanations has been explored, and some of the power formerly attributed to natural selection has been challenged. It has even been argued that certain kinds of genetic systems might make it possible for directional evolution to occur even in the absence of natural selection. This kind of evolution happens *only* because of how development (and developmental genetics) is structured in the organism. Such a phenomenon is quite inconsistent with traditional neo-Darwinian thought.

Two sorts of philosophical questions still must be addressed before philosophy of biology can be said to include developmental evolutionary thought. The first are relatively pure philosophical questions. The second are questions of the unity of science, and the relation between evo-devo and neo-Darwinian theory.

The first set of questions is the evo-devo analogs to the kinds of topics addressed by earlier philosophers of biology. What additional philosophical analyses are necessary in order to have the kind of understanding of evo-devo that we now have of neo-Darwinian theory? The neo-Darwinian concepts of adaptation and fitness are not enough. Natural selection plays a relatively small role in evo-devo, and so traditional philosophical issues like the units of selection problem would seem irrelevant. But could aspects of ontogeny be seen as new kinds of "units of selection"? The concept of function, another tradition topic of philosophy of biology, may be given a new reading in an evo-devo context. This work has yet to be done. Other core concepts of evo-devo show a contrast – perhaps even an inconsistency – with neo-Darwinian concepts. The concept of the *Bauplan* or body plan is one such concept, seen as illegitimately typological by purely neo-Darwinian thinkers. Evo-devo is filled with such concepts. For example, the *vertebrate limb* is used in reference to an abstract set of developmental possibilities, not merely the set of all limbs of animals that happen to be vertebrates. I have termed these *developmental type concepts*, and argued that they show a continued tension between population thinking and evo-devo (Amundson, 2005). Could philosophers throw the kind of light on the *Bauplan* that they have thrown on adaptation and fitness? Should Waddington's third-choice concept of the epigenotype be added to the genotype/phenotype distinction? What is to come of Mayr's proximate/ultimate distinction, which had seemed such a barrier to the relevance of development to evolution? Do we need a third choice besides proximate and ultimate to account for developmental evolution? Can development itself be understood as somehow ultimate?

The second set of questions concerns the relation between the scientific disciplines of neo-Darwinism and evo-devo. Some (but by no means all) evo-devo practitioners are concerned about the relation between their field and neo-Darwinism. Unlike the constraint debates of the 1980s, no evo-devo practitioner claims that neo-Darwinism will be overthrown. But many of them (especially those whose early career was in evolutionary biology rather than developmental biology) recognize tensions between the two fields. My comments in the previous paragraph about developmental types show that I agree with these concerns. On the other hand, some evo-devo practitioners expect no special problem in giving a natural selective account of developmental types. (As you can see, I am a skeptic.) A third school of thought seems to be that the two fields will naturally coalesce as evo-devo matures. Until recently the two disciplines have concentrated on different characters and even different organisms. When population geneticists and developmental geneticists begin studying the same characters in the same animals, some accommodation will emerge.

The two decades after 1960 were a formative and exciting period for the philosophy of neo-Darwinian evolutionary biology. A new evolutionary paradigm had emerged, and its concepts offered new ways of thinking for a generation of philosophers. Today is exciting for similar reasons. Evo-devo has finally proven that development is relevant to evolution. The philosophical implications of this new field are only beginning to be addressed.

Acknowledgements

This contribution was much improved by the suggestions of the editors Sahotra Sarkar and Anya Plutynski.

References

Amundson, R. (1994). Two concepts of constraint: adaptationism and the challenge from developmental biology. *Philosophy of Science*, 61, 556–78.

Amundson, R. (2005). *The changing role of the embryo in evolutionary biology: roots of evo-devo*. Cambridge: Cambridge University Press.

Bowler, P. J. (1996). *Life's splendid drama*. Chicago: University of Chicago Press.

Burian, R. M. (2005). *Epistemological papers on development, evolution, and genetics*. Cambridge: Cambridge University Press.

Cain, A. J., & Sheppard, P. M. (1950). Selection in the polymorphic land snail Cepaea nemoralis. *Heredity*, 4, 275–94.

Carroll, S. B. (2005). *Endless forms most beautiful: the new science of evo devo*. New York: Norton.

Gould, S. J. (1977). *Ontogeny and phylogeny*. Cambridge, MA: Harvard University Press.

Hertwig, P. (1934). Probleme der heutigen Vererbungslehre. *Die Naturwissenschaften*, 25, 425–30.

Lillie, F. R. (1927). The gene and the ontogenetic process. *Science*, 66, 361–8.

Mayr, E. (1961). Cause and effect in biology. *Science*, 134, 1501–6.

Mayr, E. (1966). *Animal species and evolution*. Cambridge, MA: Harvard University Press.

Morgan, T. H. (1926). *The theory of the gene*. New Haven, CT: Yale University Press.

Nyhart, L. (1995). *Biology takes form*. Chicago: University of Chicago Press.

Ospovat, D. (1981). *The development of Darwin's theory*. Cambridge: Cambridge University Press.

Reeve, H. K., & Sherman, P. W. (1993). Adaptation and the goals of evolutionary research. *Quarterly Review of Biology*, 68, 1–32.

Rienesl, J., & Wagner, G. P. (1992). Constancy and change of basipodial variation patterns: A comparative study of crested and marbled newts – *Triturus cristatus, Triturus marmoratus* – and their natural hybrids. *Journal of Evolutionary Biology*, 5, 307–24.

Sapp, J. (1987). *Beyond the gene: cytoplasmic inheritance and the struggle for authority in genetics*. New York: Oxford University Press.

Sarkar, S. (1998). *Genetics and reductionism*. Cambridge: Cambridge University Press.

Shubin, N., Wake, D. B., & Crawford, A. J. (1995). Morphological variation in the limbs of *Taricha Granulosa* (Caudata: Salamandridae): evolutionary and phylogenetic mechanisms. *Evolution*, 49, 874–84.

Further Reading

Amundson, R. (2005). *The changing role of the embryo in evolutionary biology: roots of evo-devo*. Cambridge: Cambridge University Press.

Bowler, P. J. (1996). *Life's splendid drama.* Chicago: University of Chicago Press.
Carroll, S. B. (2005). *Endless forms most beautiful: the new science of evo devo.* New York: Norton.
Gould, S. J. (1977). *Ontogeny and phylogeny.* Cambridge, MA: Harvard University Press.
Raff, R. A. (1996). *The shape of life.* Chicago: University of Chicago Press.

Part IV

Medicine

Chapter 15

Self and Nonself

MOIRA HOWES

1. Introduction

Immunology is the science that investigates how organisms defend themselves against infection, harmful substances, and foreign tissue. In order for an organism to defend itself against such threats, however, its immune system presumably must be able to discriminate self from nonself. If the immune system could not make such a discrimination, it might harm the organism it is to defend, rather than the microbes infecting it. Self–nonself discrimination thus appears to be a crucial function of the immune system. Indeed, immunology has been referred to as the "science of self–nonself discrimination" (Klein, 1982).

How self–nonself discrimination is achieved depends, among other things, on whether an organism is an invertebrate or a jawed vertebrate. Self–nonself discrimination is more rudimentary in invertebrates (and jawless fishes) because their immune systems are "innate." Innate immune mechanisms are those that do not change after repeated exposure to a given infectious agent; they do not *learn*. This contrasts with a type of immunity in jawed vertebrates – adaptive immunity – which does change after repeated exposure to pathogens. After the adaptive immune system adapts to a given pathogen, it can target the pathogen with greater precision and eliminate it more rapidly. Because of adaptive immunity, self–nonself discrimination in jawed vertebrates is specialized and precise: vertebrate immune systems are fine-tuned to differences between self and nonself.

So significant do immunologists find the evolution of adaptive immunity – both with respect to enhanced pathogen defense and self–nonself discrimination – that some refer to it as the "immunological Big Bang" (Janeway & Travers, 2005). But while the enhanced precision of the adaptive immune system is unquestionably significant, it does raise a difficult problem with respect to self–nonself discrimination: the precision of the adaptive immune system can be turned against the organism itself. Autoimmune disease occurs when the adaptive immune system targets the self, and the consequences can be disabling and deadly. Thus, one of the key questions in immunology – that of how the immune system avoids harming the organism it protects – gains special force in vertebrate immunology.

Theoretical and empirical research concerning immunological self–nonself discrimination is of interest to philosophers for at least two reasons. First, in immunology and philosophy alike, metaphysical questions exist concerning the nature of the self and its persistence over time. What are the boundaries of the self? How do we define the self? And second, self–nonself discrimination raises philosophical questions concerning explanation and reduction. Are self concepts in immunology genuinely explanatory? Is the investigation of immunology at the molecular level sufficient to explain all immunological phenomena?

In the following, I provide an overview of the different theoretical perspectives of self–nonself discrimination and some of the challenges that have been raised to those perspectives. In this overview, I focus mainly on adaptive immunity in vertebrates, given that most of the debates about self–nonself discrimination concern adaptive immunity; though, as I will suggest later, greater attention to innate immunity may be needed to resolve some of these debates. In part one, I describe three major theoretical perspectives of self–nonself discrimination: these include clonal selection theory and immunological tolerance; three-signal models; and network models. In part two, I examine challenges to contemporary thinking about self–nonself discrimination that complicate the three major theoretical perspectives described. These challenges – including questions about the genetic criterion of selfhood, the viability of the innate–adaptive distinction, and self–nonself discrimination in pregnancy – demonstrate that much conceptual work on self–nonself discrimination remains to be done.

2. Theoretical Perspectives

2.1. *Clonal selection theory, tolerance and self–nonself discrimination*

In the first part of the twentieth century, one of the central puzzles of immunity concerned antibody diversity. Antibodies are large soluble glycoproteins found in the blood and other fluids of the body. Vertebrate organisms develop antibodies to hundreds of millions of substances known as "antigens." Most antigens are protein fragments from microbes or cells of the organism's own body. The puzzle raised by antibody diversity is this: How does the immune system produce antibodies able to interact with such an incredibly diverse array of antigens? What accounts for the diversity of antibody conformations?

In the 1930s and 40s, proposed solutions to the puzzle of antibody diversity focused on the idea that antigens acted as templates for antibody production. In this view, antigens shape antibody structure, somewhat like a mold shapes a form, and this can generate as many different antibody conformations as there are antigens. Niels Jerne, however, showed that the template idea was flawed. Jerne was interested in natural antibodies, which are antibodies that exist in the body prior to exposure to antigens. Natural antibodies thus provided a key reason to reject template theory: if antibodies can exist prior to antigen exposure, then antigens are clearly not involved in their creation. Jerne's interest in natural antibody formation, combined with his interest in Darwinian selection processes, led to a better explanation for antibody diversity: the natural selection theory of antibody formation. In this theory, an antigen selects a

Table 15.1 Some of the main cell types of the immune system

B lymphocyte	Plasma	• Produces antibodies in response to infection • Can recognize and differentiate between different antigens
	Memory	• Circulates the body and remains quiescent until a second encounter with a given pathogen • Can recognize and differentiate between different antigens
T lymphocyte	Helper	• Stimulates B cell growth and differentiation. • Stimulates macrophages. • Can recognize and differentiate between different antigens
	Cytotoxic	• Kills virus-infected cells and tumor cells • Can recognize and differentiate between different antigens
Natural killer cell		• Kills virus-infected cells and tumor cells
Macrophage		• Presents antigen to T helper cells • Ingests and destroys microbes • Activates inflammation
Dendritic cell		• Presents antigen to T helper cells

circulating antibody and the resulting antigen–antibody complex circulates in the body until it is picked up by an antibody-producing cell. The antibody-producing cell then makes more antibodies of that type (Jerne, 1955; Söderqvist, 1994).

The idea of using natural selection to explain antibody formation was significant. But Jerne's assumption that antibody-producing cells could manufacture any configuration of antibody taken up was problematic. Each antibody-producing cell would in principle have to be able to make millions of different conformations of the antibody molecule – an implausible scenario. By 1957, David Talmage and Frank Macfarlane Burnet independently resolved this problem by shifting the selection process from Jerne's antigen–antibody complex to the antibody-producing cells themselves (Taliaferro & Talmage, 1955; Talmage, 1957; Burnet, 1957). Antigens entering the body attach themselves directly to antibody-producing cells having compatible receptors, and in so doing, they select those cells from among others. On the basis of this idea, Burnet developed his *clonal selection theory* of antibody formation. In clonal selection theory, the antibody-producing cell – a B lymphocyte cell – is selected and then proliferates by clonal expansion. (See Table 15.1.) Each of the resulting clones produces only one type of antibody molecule – the same type as the parent cell. This is a much more manageable task for an antibody-producing cell than the generation of innumerable different antibody types. Clonal selection theory thus provided a clear explanation for how vertebrate organisms produce such an incredibly diverse array of antibody conformations.

In the late 1950s, a further refinement of clonal selection theory was made with respect to the question concerning how antibodies confer immunity to pathogens that have previously caused an infection. Gustav Nossal and Joshua Lederberg (1958) found that B cells produce two different types of clones: plasma cells and memory cells. Plasma cells are the cells that generate identical copies of the antibody produced by the parent B cell. Memory cells, however, do not undergo differentiation to become antibody-producing cells. Instead, they remain quiescent in the body and persist long after the initial infection is resolved. If, however, there is a second encounter with the microbe that caused the primary infection, memory cells will initiate a response to the microbe immediately. This fast response eliminates the threat before infection takes hold. Memory cells thus are responsible for the immunity we develop to certain infections after we have fallen ill by them.

Despite the success of clonal selection theory and the discovery of plasma and memory cells, however, fundamental questions about the genetic mechanisms behind the generation of B cell diversity and immunological memory remained. Until the early 1980s, some thought that the genetic diversity responsible for antibody diversity already existed in the germ-line of organisms. Others thought the necessary genetic diversity developed in the organism somatically; on this view, organisms are not born with the necessary genetic diversity, but develop it later.

The latter view turned out to be correct. Susumu Tonegawa (1983) found that as B cells mature into plasma or memory cells, they mix and match genes, add and delete genes, and mutate genes. This recombination, mutation, and addition and deletion of genes explains how such a diverse array of antibody conformations can be created. A similar process also occurs in another type of immune cell – the T helper lymphocyte, a cell that stimulates B cell activity. Receptors on T helper cells are very diverse and they achieve this diversity through genetic recombination (though not through mutation as in B cells). Between different antibodies and T cell receptors, the question of how the immune system recognizes such an enormous variety of antigens was more or less resolved.

The somatic rearrangement of antibody genes is responsible for the precision of the adaptive immune system: it enables the immune system to produce antibodies that are highly specific to any given antigen. Specificity means that if an antibody binds tightly to an antigen from the chicken pox virus, it will not bind well to antigens from anything else: it is specific for chicken pox. As B cells mature, those that best fit the antigen in question will last longer than those that do not, and only those that best fit will survive long enough to become plasma or memory cells. As a result of this process, plasma and memory cells are able to bind to antigens very tightly. The discovery of the somatic rearrangement of antibody genes thus helped to explain the ability of the adaptive immune system to target antigens with a high degree of precision.

But, the precision of adaptive immunity raises a problem for clonal selection theory. In adaptive immune systems, antibodies able to recognize self antigens can develop through somatic rearrangement and mutation. T cell receptors specific for self antigens can also arise through somatic rearrangement. And, we know that T and B cells are capable of mounting immune responses against the self. So, something must normally stop the immune system from targeting self tissues. But what? The principal answer – immunological *tolerance* – became a mainstay of the dominant view of self–nonself discrimination.

Immunological tolerance is a learned unresponsiveness to specific antigens: in short, it is the ability of T and B cells to tolerate or ignore self antigens. Burnet explained tolerance in the context of his clonal selection theory. He argued that B and T cells able to recognize self antigens are selected against – that is, eliminated – early in vertebrate development. Within this early window, an organism can become tolerant of any tissue, including tissue transplanted from other organisms. But if transplantation is attempted after the window closes, the transplant will not be tolerated. Key evidence supporting this developmental account of tolerance was Ray Owen's (1945) observation that cattle that had shared a circulatory system *in utero* did not respond immunologically to each other's blood cell antigens: they were hematopoietic chimeras. Further evidence came from Rupert Billingham, Leslie Brent, and Peter Medawar's 1953 study showing that mice injected with donor cells as pups would accept skin grafts as adults from those same donors. Normally such grafts would be rejected.

It is now well established that the principal means of achieving tolerance involves the elimination of self-reactive immune cells. T cells are eliminated in the thymus if they are able to bind self antigens. An analogous process occurs in B cell development: those cells able to bind self antigens are eliminated in the bone marrow. Tolerance achieved in either of these ways is referred to as "central" tolerance. Through somatic rearrangement and tolerance-inducing mechanisms, the immune system is able to develop cells that specifically bind nonself antigens and eliminate cells that specifically bind self antigens.

The processes creating central tolerance, however, are imperfect: self-reactive cells escape the thymus and bone marrow, and some self antigens are found in tissues that are unavailable for tolerance induction in the thymus. This necessitates a means of achieving tolerance in the periphery of the body. One mechanism for peripheral tolerance is proposed in the two-signal or "associative recognition" model of Peter Bretscher and Melvin Cohn (1968, 1970) and Rod Langman and Melvin Cohn (1993). In the associative recognition model, antigen provides the first signal and this acts as an "off" signal to T cells. This induces tolerance. The second signal, delivered during infection by T helper cells known as effector T helper cells, is an "on" signal that activates the immune system's ability to destroy an infectious agent. The effector T cell recognizes the association between the T cell receiving the first signal and the antigen, hence the "associative recognition" name for the model. The delivery of both signals is thus antigen-dependent. This model is thought to explain self–nonself discrimination because self antigens, which are present continually, will provide a constant source of the first "off" signal, thus inducing tolerance. Only in the occasional instances of infection will a second signal be delivered.

But an important question remains: How is the effector T helper cell that delivers the second "on" signal itself activated? Why does it not remain in an inactive, tolerant state? Cohn (1998) refers to this as "the primer problem." In the two-signal model, this problem is solved by positing an antigen-*in*dependent pathway to T helper cell activation. Bretscher and Cohn's (1968, 1970) antigen-independent pathway holds that if T cells interact with antigen early in their development and in the absence of a stimulatory signal from a T-helper cell, their further differentiation is arrested. Because self antigens are always present, younger T cells will always be exposed to self antigen and, if they are capable of reacting with self antigen, their development will stop and they will pose no threat to the self. If, however, T cells do not react with any available self antigens, their development will slowly continue to the activation stage. Because infections occur sporadically, T cells capable of reacting to them will likely have matured to the activation stage by the time an infectious agent appears on the scene. If correct, the two-signal model offers a relatively simple way to resolve the primer problem within the context of clonal selection theory. It provides a mechanism to distinguish between self and nonself in the periphery of the body.

Self–nonself discrimination – as understood in terms of clonal selection theory – has long been the dominant model in immunology and it continues to have vigorous defenders. However, the reliance upon tolerance to establish a clear self–nonself distinction is problematic for a variety of empirical and conceptual reasons, reasons that call into question the viability of the traditional self–nonself model. Some of these problems are identified and addressed by three-signal models, to which I now turn.

2.2. Three-signal models: the end of the immune self?

One of the difficulties with the traditional model of self–nonself discrimination is that violations of the self–nonself distinction regularly occur and do not appear to cause problems for the organism. Self-reactive immune cells do escape elimination; and self-reactive antibodies known as natural *auto*antibodies exist in individuals showing no signs of autoimmune disease. Moreover, while food is nonself, it is tolerated, as are many airborne substances and species of bacteria. While the gut and respiratory surfaces may be considered the "outside" of the body and introduce the possibility that self and nonself are discriminated spatially (with nonself on the "outside"), many of the nonself substances that engage these surfaces do enter the body. These violations suggest that discrimination between self and nonself is not as straightforward as proponents assume.

A further problem for the traditional self–nonself distinction concerns the antigen-independent development of T cells in the two-signal model. Mature effector T cells could target *newly arising* self proteins in mature organisms. Because a new protein would not be present during T cell development, nothing would stop the development of effector T cells able to recognize it. As Polly Matzinger asks,

> what happens when "self" changes? How do organisms go through puberty, metamorphosis, pregnancy, and aging without attacking newly changed tissues? Why do mammalian mothers not reject their fetuses or attack their newly lactating breasts, which produce milk proteins that were not part of the earlier "self"? (Matzinger, 2002, p.301)

These problems suggest that the boundaries in traditional self–nonself discrimination models may need to be relaxed. Some argue that the immune system only discriminates self from *infectious nonself* (Janeway, 1992) or "*some* self from *some* non-self" (Matzinger, 1994, p.994).

Ephraim Fuchs (1992) and Polly Matzinger's (1994) "danger model," which involves three signals instead of two, is a good example of a more flexible approach to self–nonself discrimination. In the danger model, an antigen-presenting cell provides signals 1 and 2. (See Figure 15.1.) Cellular substances released in response to tissue damage or abnormal cell death emit a third signal – a "danger" signal – which is needed to activate an immune response. Without it, nothing happens. Given that healthy tissues do not emit danger signals, they will not activate the immune system. The danger model thus shifts control of tolerance from the immune system to non-immunological tissues of the body. It is the local health status of tissues, not self–nonself discrimination, that stimulates an immune response.

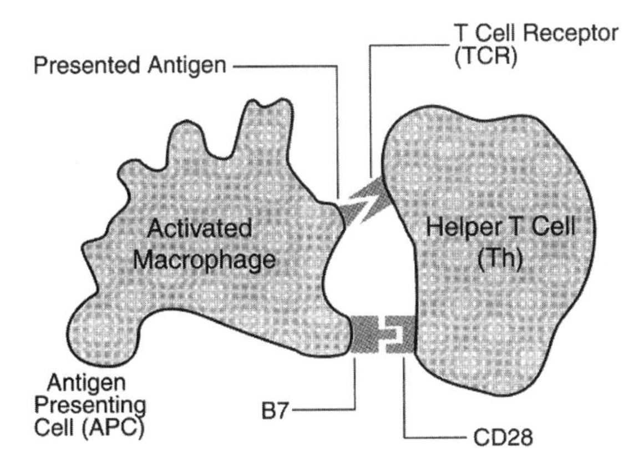

Figure 15.1 A two signal model of T helper lymphocyte activation. The first signal is the antigen presented by the macrophage (in the context of MHC class II) to the T cell receptor. The second signal is a protein (B7 in the diagram) presented to the receptor CD28. In a three signal model, an additional signal is needed to activate an immune response. Heat shock proteins spilled from damaged cells are an example of the sort of additional signal required in three signal models like the danger model. Reprinted from *How the Immune System Works*, L. Sompayrac, Blackwell Science, 1999, with permission from Blackwell

While the addition of a third signal may not seem particularly significant (Cohn, 1998), Matzinger claims that the addition of a danger signal is a small step that "drops us off a cliff, landing us in a totally different viewpoint, in which 'foreignness' of a pathogen is not the important feature that triggers a response, and 'self-ness' is no guarantee of tolerance" (Matzinger, 2002, p.302). When danger is the concern, there is no need for immune mechanisms to distinguish precisely between self and nonself. The danger model also suggests that when immunological investigations are conducted under the rubric of self–nonself discrimination, those investigations – and the treatments for cancer, organ transplantation, pregnancy, and autoimmune disease based thereupon – may target the wrong mechanisms. As Matzinger argues, questions that do not arise in the context of traditional self–nonself discrimination models do arise once selfhood is de-emphasized, including questions such as.

> why liver transplants are rejected less vigorously than hearts; why women seem to be more susceptible than men to certain autoimmune diseases . . . [and] why graft-versus-host disease is less severe in recipients that have had gentle rather than harsh preconditioning treatments . . . (Matzinger, 2002, p.301)

Despite raising these important questions, however, the extent to which the danger model really does depart from traditional self–nonself discrimination theory remains an unsettled matter. The difficulty with thinking that the danger model marks the end of immunological self–nonself discrimination is that some means of distinguishing self from nonself may still be required in the danger model – otherwise, self-reactive T cells could be activated by danger signals with harmful consequences for the organism. And, because the decision to activate the immune response at local sites of infection is not based upon self–nonself discrimination in the danger model, there is no local way to avoid reactions against the self if self-reactive cells are present – one just has to hope that the general system has already eliminated any cells capable of self-reactivity.

Regardless of whether the danger model ultimately spells the end of the immune self, however, the questions it raises strongly suggest that self concepts in immunology require further analysis. The network theoretical perspective discussed next also suggests that analysis of self concepts in immunology remains an important task.

2.3. Network models of immunological self

The debate between clonal selection theory and the network perspective largely concerns how immune activity toward self and nonself is regulated. In clonal selection theory, regulation is achieved by self–nonself discrimination. In network models, regulation is achieved through connections amongst lymphocytes and/or between antibody molecules. Of course, in self–nonself discrimination models it is recognized that immune cells, antibodies, and immune biochemicals form a network of interactions. But in network models, "network" is meant in a more specific sense: it refers to regulatory autoimmunity. Regulatory autoimmunity, as we shall see, has consequences for understanding self–nonself discrimination.

The basis upon which contemporary network views rest is Jerne's (1974) idiotypic theory of the immune system. An idiotype is a lymphocyte antigen receptor whose unique amino acid sequence can be recognized by other lymphocyte receptors, provided they are complementary or "anti-idiotypic." (See Figure 15.2.) Jerne called these recognition interactions between lymphocytes "idiotypic." Given the diversity of lymphocyte receptors and antibodies, there must exist antibodies and lymphocyte receptors that can recognize other antibodies and lymphocyte receptors. If lymphocytes can activate other lymphocytes through idiotypic interactions, a network of interacting lymphocytes, ultimately encompassing the entire immune system, could form. In Jerne's view, this connectivity amongst lymphocytes would then serve to read the state of body and regulate the immune system accordingly, either through activation or suppression. Note that the molecular conformations involved here are all "self" in origin; hence, network perspectives are based on regulatory *auto*immunity.

Antonio Coutinho (1984, 1989) is one of the principal contemporary immunologists associated with the network approach. One interesting way in which Coutinho develops the network hypothesis, beyond Jerne's version, is his division of immune network activities into central and peripheral compartments. Coutinho holds that the immune system consists of a central immune system involving a connected network of lymphocytes that maintain tolerance to self and a peripheral immune system consisting of unconnected lymphocytes that, when stimulated by antigen, begin an immune response.

In Coutinho's model, the immune system does not regulate itself by first discriminating between self and nonself. Self–nonself discrimination is not a property or ability of an individual lymphocyte, such that it is either "turned off" if it can recognize self substances, or "left on" if it can recognize foreign substances. Rather, immune regulation is achieved by discriminating between unperturbed and perturbed states of immune

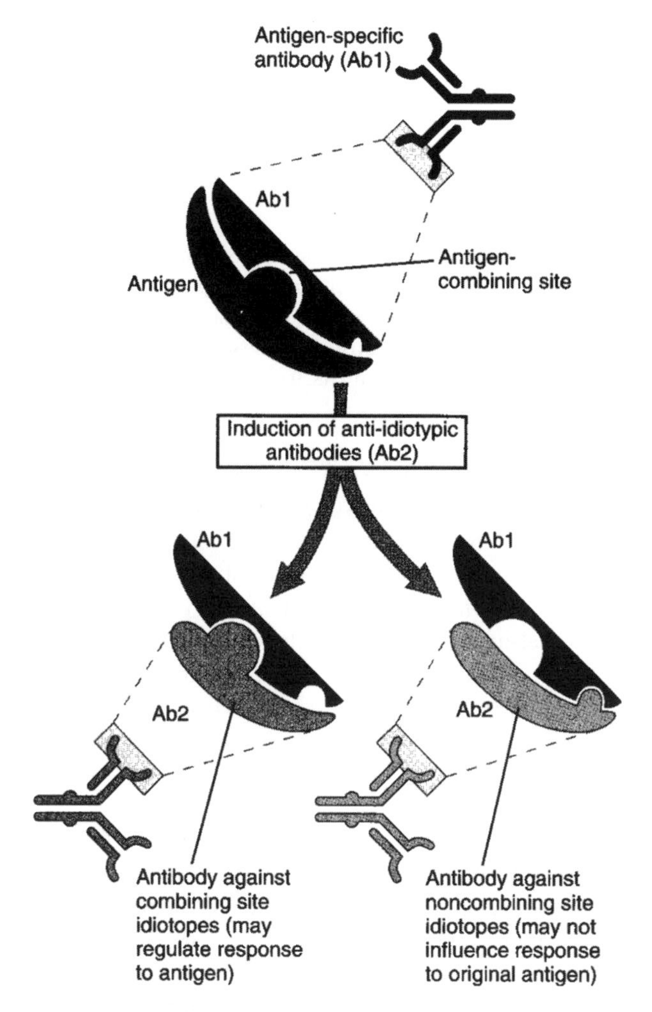

Figure 15.2 Idiotypic interactions between antibodies. Reprinted from *Cellular and Molecular Immunology*, fourth edition, A. Abbas, A. Lichtman and J. Pober, Philadelphia: W. B. Saunders Company, 2000, with permission from Elsevier

connectivity. The immune system is busy interacting with itself and with the body all of the time and the appearance of foreign antigens causes a perturbation of this activity. Because nonself is viewed as a perturbation of the system, it is not really viewed as "nonself" by the immune system. There is only "self" and its perturbations; and hence, we have a theory about how the immune system reacts to the self rather than a theory focusing on immunity to nonself.

Network approaches to the immune self thus depart from the relatively static demarcation between self and nonself found in tolerance views of self–nonself discrimination. The immune self in clonal selection theory is firmly defined: its edges may change, but the core of the self is maintained throughout life. This defined self–nonself distinction is the cause of immune activity (or inactivity, as in the case of tolerance). Unlike traditional self–nonself discrimination, which treats the self as an entity, network views treat the self as a process. The network self does not have a stable core. In network models, self–nonself discrimination is the *outcome* of interactions between lymphocytes and not the starting point for those interactions. Self–nonself discrimination is a consequence, not cause, of immune activity.

In the network perspective,

> self is in no way a well-defined (neither predefined) repertoire, a list of authorized molecules, but rather a set of viable states, of mutually compatible groupings, of dynamical patterns . . . The self is not just a static border in the shape space, delineating friend from foe. Moreover, the self is not a genetic constant. It bears the genetic make-up of the individual and of its past history, while shaping itself along an unforeseen path. (Varela et al., 1988, p.363)

This more dynamic understanding of self also has implications for how experiments are designed in immunology. Because system-wide lymphocyte connectivity is the source of self–nonself discrimination network perspectives, Coutinho argues that *in vitro* experimental investigations of tolerance are limited in what they can tell us. Thus, the evidence provided by *in vitro* experimental studies of tolerance may not apply to naturally occurring tolerance. Similarly, evidence provided by *in vivo* studies using transgenic mice and chimeras (wherein different genetic tissues are mixed in one animal) may also fail to apply to naturally occurring tolerance. If such studies cannot provide adequate support for naturally occurring tolerance, one must return to the organism, to the lymphocyte in its bodily context, to achieve adequate understanding.

Indeed, network perspectives claim to be antireductionistic insofar as they claim that there exist some properties of the immune system that exceed description in terms of the immune system's parts and relations considered in isolation from each other. This means that complete understanding of the immune system will not be achieved by studying component functions in isolation from other immune activities. But given that immunological experimental studies must isolate mechanisms, network accounts have had difficulty finding experimental support. It is simply not possible to replicate experimentally system-wide lymphocyte behavior. Moreover, network models have not yielded much in the way of testable predictions. There are some newer experimental approaches, such as quantitative immunoblotting and multiparametric data analysis, that some immunologists are now using to investigate immune activities in a less isolated manner. However, the extent to which these experiments involve less isolated immune activities remains to be determined.

3. Challenges

Now that the three main theoretical perspectives on self–nonself discrimination have been outlined, I turn to consider several challenges that complicate these perspectives.

3.1. *The major histocompatibility complex: a genetic signature of self?*

The major histocompatibility complex (MHC) genes code for cellular proteins that are unique to each individual vertebrate organism. There are two classes of MHC, each of which is involved in self–nonself discrimination in a different way. MHC class I is found on all nucleated cells of the body. Its function is to sample cellular proteins and display those proteins on the cell surface, where the immune system can see and evaluate them. MHC class II is found only on antigen-presenting cells. It presents fragments of bacterial and viral substances to T helper lymphocytes and thus plays a role in establishing the first signal in lymphocyte activation.

There are at least two other respects in which MHC is relevant to self–nonself discrimination. First, MHC proteins are involved in the acceptance and rejection of transplanted tissue – indeed, they are named for this role. The immune system regards foreign MHC just as it regards viral proteins and when it targets foreign MHC in transplanted tissue, rejection results. Second, MHC plays a role in tolerance induction in the adaptive immune system. MHC is involved in the selection for and against T cells in the thymus. T cells that are aggressively reactive towards self antigens presented in the context of MHC proteins are eliminated. T cells that do not recognize self are retained.

By virtue of its involvement in transplant rejection, tolerance induction, and antigen presentation, MHC appears to provide a secure means of identifying self and nonself. Because of these functions, and because MHC proteins are unique to each individual organism, MHC has been referred to as the "genetic signature" of immunological selfhood (Tauber, 1994). On this view, MHC is a necessary, though not sufficient, element of immunological selfhood (Tauber, 1994).

It would be a mistake to settle for the view that MHC is the "genetic signature" of the self, however, if by this it is meant that the genetic criterion of selfhood is somehow more essential to selfhood than other immune factors contributing to self–nonself discrimination. By way of analogy, the claim that the human genome provides the essence of human selfhood is clearly problematic: the claim ignores biological and social development. Similarly, we should not privilege MHC genes in the development of the immune self. That the MHC contribution is genetic does not afford it some special ontological status. There may be many different routes to self–nonself discrimination, including networks and danger signals. And, and as outlined in the next section, there may also exist innate mechanisms for self–nonself discrimination.

3.2. *Innate immunity: is there self–nonself discrimination without the adaptive immunity?*

In vertebrates, innate immunity provides a first line of defense against infectious organisms. Cells of the innate immune system – such as macrophages and natural killer cells – prevent infection at all points of entry into the body. Macrophages engulf bacteria in a process known as phagocytosis and digest them. Natural killer cells lyse virally infected cells. And, innate immune cells produce biochemicals that stimulate the immune response. It is generally thought, however, that innate immune cells lack the specificity and immunological memory typical of the adaptive immune system and on this basis a firm distinction between innate and adaptive systems is made.

Recent findings in the area of innate immunity suggest that the innate–adaptive distinction may not be so clearly defined after all. Three-signal models, for example, challenge the innate–adaptive distinction, for non-adaptive cells like antigen-presenting cells initiate adaptive immune responses by providing danger signals or their equivalent. Problems with classifying certain immune cells as either innate or adaptive also present a challenge the innate–adaptive distinction. A type of T cell known as the gamma delta (γδ) T cell is a case in point: γδ T cells develop in the thymus like other T cells and have T-cell receptors which suggests they are part of adaptive immunity; however, they are not capable of specificity in the way that other T cells are and so they are more like innate immune cells. Moreover, γδ T cells migrate to epithelial tissues, which is characteristic of innate immune cells; in general, γδ T cells do not circulate to the lymph nodes as do other T cells. The γδ cell thus appears to resist classification as either innate or adaptive.

The classification of the macrophage as an innate immune cell – insofar as it lacks specificity – is also now being questioned. The key challenge to its classification arose during investigations in developmental biology concerning *Toll*, a maternal-effect gene responsible for embryonic dorsal–ventral polarization. A connection between *Drosophila Toll* and immunity was made when it was found that *Toll* mutants had immunological deficiencies (Rich, 2005). Macrophages were found to have many *Toll*-like receptors – "*Toll*-like" because they bear a sequence homology with *Toll* – which can recognize evolutionarily conserved microbial structures. This recognition is not that of adaptive specificity, which must be learned. However, there is some evidence that through *Toll*-like signaling, macrophage receptors gather into clusters. It is suspected that clustering introduces a form of *specificity* into the innate immune response by generating novel molecular receptor configurations. Thus, macrophage functions, long classified as non-specific and innate, may actually include the generation of novel immune specificities. It is also worth noting here that some evidence of immunological memory in invertebrates now exists (Kurtz, 2004). Despite lacking adaptive immune systems, then, invertebrate immune systems may be able to learn. Innate and adaptive systems may thus share features that are commonly used to distinguish them.

Another reason to question the innate–adaptive distinction concerns evidence that tolerance may be achievable in some cases without input from the adaptive immune system. Consider the following example. Epithelial cells lining the intestinal lumen are polarized in their expression of *Toll*-like receptors. *Toll*-like receptors are absent on epithelial surfaces facing the intestinal lumen, but present on the other side. Friendly gut bacteria only come into contact with the cell surfaces and, since there are no *Toll*-like receptors there, no immune response is initiated. But pathogenic bacteria will breach the intestinal epithelium and, in so doing, will encounter the *Toll*-like receptors. This will initiate an immune response. Here, then, tolerance to friendly bacteria, and intolerance to the unfriendly, is achieved without adaptive immunity.

These empirical challenges to the innate–adaptive distinction are intriguing, but there are also philosophical questions that need to be addressed here. Does innateness in immunology resemble notions of innateness at play in cognitive psychology, genetics, and linguistics? What exactly is an *innate* immune phenomenon? The concept of innateness generally involves notions of fixity as well as essentialist views about biological natures, but the extent to which these associations carry over into immunological innateness remains to be determined. Because innate immune recognition is thought to be germ-line encoded, immunological innateness fits these associations quite well. It seems reasonable to suppose, then, that the concept of innateness may be just as problematic for immunology as for other fields. Now that the fixity of innate immunity – its inability to learn – is being challenged, essentialist undertones may prove particularly problematic.

Given the empirical challenges posed to the innate–adaptive distinction by γδ T cells, *Toll*-like receptors, and macrophages, the distinction increasingly appears artificial. And, given the outstanding conceptual issues concerning innateness, what it is to be innate, or adaptive, in the context of immunity requires further analysis. The significance of this conclusion for vertebrate immunology is that it challenges the exclusive role of adaptive immunity and tolerance in the generation of self–nonself discrimination. Research concerning innate immunity may well provide deeper understanding of self–nonself discrimination in immunology (Janeway & Medzhitov, 2002).

3.3. Self–nonself discrimination in pregnancy immunology

Pregnancy has long been described by immunologists as a "paradox" (Medawar, 1953, 1957). Because the fetus has paternally derived MHC proteins, the fetus should appear as nonself, at least partially, to the maternal immune system. From the traditional self–nonself discrimination perspective, the fetus is akin to an organ transplant. Given this, the maternal immune system should try to reject the fetus. The objective of the fetus is presumably to try to prevent this rejection. On this view, then, a constant tension between mother (immunological self) and fetus (immunological nonself) exists at the core of immunity in pregnancy.

Indeed, in immunology, mothers and fetuses are often conceptualized as warring entities battling for control. In order to prevent maternal immune aggression from erupting, Medawar thought that either the fetus must hide from the maternal immune system or the maternal immune system must be suppressed – and updated variations of these ideas exist in the present day. Some evidence appears to support the idea that the fetus hides from the maternal immune system. For example, certain identifying cell markers derived from MHC proteins are either absent or altered in fetal trophoblast cells – the cells that interact most closely with maternal tissues. Other evidence appears to support the idea that certain maternal immune functions are downregulated. Pregnant women have increased vulnerability to certain types of infection and some experience changes in the severity of autoimmune disease.

A number of findings, however, challenge the view that self–nonself discrimination is important in pregnancy immunology and that maternal aggressiveness towards the fetus is the best (or only) way to frame maternal immunology. It may not be appropriate to treat the fetus as a nonself entity always at risk of rejection. This view is supported by the danger model, wherein the maternal immune system only responds to the placenta–fetus if danger signals are present.

Indeed, some reproductive immunologists are now exploring the idea that maternal immune recognition of the fetus is *beneficial* to fetal growth and development. The discovery of lasting microchimerism – the persistence of small numbers of cells from one individual in another – in mothers and their children also suggests that too much has been made of maternal–fetal conflict. There is evidence that maternal immune cells present in children populate sites of infection and may lend immunological assistance (Hall, 2003). There is also evidence that persisting fetal cells contribute to tissue repair in some women long after the birth of their children (Adams & Nelson, 2004). In light of these findings, it is difficult to imagine that maternal–fetal relations should be classified simply in terms of antagonistic self–nonself relations.

Rather than being an immunological paradox or a weakened immunological state, pregnancy is probably a sensible immunological phenomenon and its study may have much to contribute to the development of more adequate models of self–nonself discrimination. Moreover, because viviparity may have been one of the selective pressures driving the evolution of adaptive immunity (Sacks, Sargent, & Redman, 1999), the fact that pregnancy receives little attention may stand in the way not just of immunological understanding, but of evolutionary understanding as well. But in order to envision alternatives to the view that pregnancy is an immunological paradox, different understandings of how selfhood relates to maternal–fetal relationship in pregnancy are needed.

4. Conclusion

As the main theoretical perspectives of self–nonself in immunology and the challenges posed to them illustrate, the issue of the self in immunology is complex and controversial. But recent challenges to immunological self–nonself discrimination should be no cause for despair: though philosophers still lack a satisfactory criterion for self identity, most have not declared the self a useless fiction. Moreover, there is much in biology to suggest that selfhood is important. It therefore seems premature to claim, as some do, that self concepts are no longer useful in immunology (Tauber & Podolsky, 1997, p.377). On the contrary, the question of immunological selfhood appears to be on the cusp of renewed and vigorous inquiry, with revised models of self–nonself relations replacing dated versions. Such revision is especially promising given growing connections between immunology and developmental biology, comparative immunology, neurobiology, and evolutionary biology. The landscape of self–nonself discrimination is changing – and philosophy has a role in coming to understand these changes.

References

Adams, K., & Nelson, J. (2004). Microchimerism: an investigative frontier in autoimmunity and transplantation. *Journal of the American Medical Association*, 291, 1127–31.

Billingham, R., Brent, L., & Medawar, P. (1953). Actively acquired tolerance of foreign cells. *Nature*, 172, 603.

Bretscher, P., & Cohn, M. (1968). Minimal model for the mechanism of antibody induction and paralysis by antigen. *Nature*, 220, 444–8.

Bretscher, P., & Cohn, M. (1970). A theory of self–nonself discrimination. *Science*, 169, 1042–9.

Burnet, F. M. (1957). A modification of Jerne's theory of antibody production using the concept of clonal selection. *Austral Science*, 20, 67–9.

Cohn, M. (1998). A reply to Tauber. *Theoretical Medicine and Bioethics*, 19, 495–504.
Coutinho, A. (1989). Beyond clonal selection and network. *Immunological Reviews*, 110, 63–87.
Coutinho, A., Forni, L., Holmberg, D., Ivars, F., & Vaz, N. (1984). From an antigen-centered clonal perspective to immune responses to an organism-centered, network perspective of autonomous activity in a self-referential immune system. *Immunological Reviews*, 79, 151–68.
Fuchs, E. (1992). Two signal model of lymphocyte activation. *Immunology Today*, 12, 462.
Hall, J. (2003). So you think your mother is always looking over your shoulder? – She may be in your shoulder! *The Journal of Pediatrics*, 142, 233–4.
Janeway, C. (1992). The immune system evolved to discriminate infectious nonself from noninfectious self. *Immunology Today*, 13, 11–16.
Janeway, C., & Medzhitov, R. (2002). Innate immune recognition. *Annual Review of Immunology*, 20, 197–216.
Janeway, C., & Travers, P. (2005). *Immunobiology: the immune system in health and disease.* New York: Garland Science.
Jerne, N. (1955). The natural selection theory of antibody formation. *Proceedings of the National Academy of Science (USA)*, 41, 849–57.
Jerne, N. (1974). Towards a network theory of the immune system. *Annales de l'Institute Pasteur/ Immunology* (Paris), 125C, 373–89.
Klein, J. (1982). *Immunology: the science of self–nonself discrimination.* New York: Wiley.
Kurtz, J. (2004). Memory in the innate and adaptive immune systems. *Microbes and Infection*, 6, 1410–17.
Langman, R., & Cohn, M. (1993). Two signal models of lymphocyte activation? *Immunology Today*, 14, 235–6.
Matzinger, P. (1994). Tolerance, danger, and the extended family. *Annual Review of Immunology*, 12, 991–1045.
Matzinger, P. (2002). The danger model: a renewed sense of self. *Science*, 296, 301–5.
Medawar, P. (1953). Some immunological and endocrinological problems raised by the evolution of viviparity in vertebrates. *Symposia – Society for Experimental Biology*, 44, 320–38.
Medawar, P. (1957). *The uniqueness of the individual.* London: Methuen and Co, Ltd.
Nossal, G., & Lederberg, J. (1958). Antibody production by single cells. *Nature*, 181, 1419–20.
Owen, R. (1945). Immunogenetic consequences of vascular anastomoses between bovine twins. *Science*, 102, 400–1.
Rich, T. (2005). *Toll and toll-like receptors: an immunologic perspective.* New York: Kluwer Academic/ Plenum Publishers.
Sacks, G., Sargent, I., & Redman, C. (1999). An innate view of human pregnancy. *Immunology Today*, 20, 114–18.
Söderqvist, T. (1994). Darwinian overtones: Niels K. Jerne and the origin of the selection theory of antibody formation. *Journal of the History of Biology*, 27, 481–529.
Taliaferro, W., & Talmage, D. (1955). Absence of amino acid incorporation into antibody during the induction period. *Journal of Infectious Diseases*, 97, 88–98.
Talmage, D. (1957). Allergy and immunology. *Annual Review of Medicine*, 8, 239–56.
Tauber, A. (1994). *The immune self: theory or metaphor?* Cambridge: Cambridge University Press.
Tauber, A., & Podolsky, S. (1997). *The generation of diversity.* Cambridge, MA: Harvard University Press.
Tonegawa, S. (1983). Somatic generation of antibody diversity. *Nature*, 302, 575–81.
Varela, F., Coutinho, A., Dupire, B., & Vaz, N. (1988). Cognitive networks: immune, neural, and otherwise. In A. Perelson (Ed.). *Theoretical immunology Part 2* (pp. 359–75). Reading, MA: Addison-Wesley.

Further Reading

Buss, L. (1987). *The evolution of individuality.* Princeton: Princeton University Press.
Cohen, I. (1994). The cognitive principle challenges clonal selection. *Immunology Today,* 13, 441–4.
Dreifus, C. (1998). Blazing an unconventional trail to a new theory of immunity. *The New York Times,* June 16.
Haraway, D. (1991). The biopolitics of postmodern bodies: constitutions of self in immune system discourse. In D. Haraway, *Simians, cyborgs, and women: the reinvention of nature* (pp.203–30). New York: Routledge.
Hunt, J. (Ed.). (1996). *HLA and the maternal–fetal relationship.* Austin, TX: R. G. Landes Company.
Lafferty, K., & Cunningham, A. (1975). A new analysis of allogenic interactions. *Australian Journal of Experimental Biology and Medical Science,* 53, 27–42.
Langman, R. (Ed.). (2000). *Self–nonself discrimination revisited. Seminars in Immunology* 12. Cambridge: Academic Press.
Lederberg, J. (1959). Genes and antibodies. *Science,* 129, 1649–53.
Martin, E. (1994). *Flexible bodies: tracking immunity in American culture from the days of polio to the age of AIDS.* Boston: Beacon Press.
Metchnikoff, E. (1905). *Immunity in infective diseases* (F. Binnie, Trans.). Cambridge: Cambridge University Press.
Moulin, A. (2001). *Singular selves: historical issues and contemporary debates in immunology.* Amsterdam: Elsevier.
Schaffner, K. (1993). *Discovery and explanation in biology and medicine.* Chicago: University of Chicago Press.
Tauber, A. (1991). *Organism and the origins of self.* Dordrecht: Kluwer Academic.
Varela, F., Coutinho, A., Dupire, B., & Vaz, N. (1988). Cognitive networks: immune, neural and otherwise. In A. S. Perelson (Ed.). *Theoretical immunology* (Vol. 3, pp. 359–75). Redwood City: Addison-Wesley Publishing Co.
Weasel, L. (2001). Dismantling the self/other dichotomy in science: towards a feminist model of the immune system. *Hypatia,* 16, 27–44.
Zinkernagel, R., & Doherty, P. (1974). Restriction of *in vitro* T cell-mediated cytotoxicity in lymphocytic choriomeningitis within a syngeneic or semiallogeneic system. *Nature,* 248, 701–2.

Chapter 16

Health and Disease

DOMINIC MURPHY

1. Introduction

The philosophical problem of the definition of disease is mostly a matter of understanding the contributions made to our thinking on the subject by two different sorts of judgments; judgments about the natural functioning of humans and judgments about whether it is bad or good to live a certain way or have a certain property. It is widely agreed among scholars that normative judgments play a role in assessing who is healthy and who is ill, injured, or otherwise unhealthy. Some scholars believe that our normative judgments alone determine who is healthy (e.g., Kennedy, 1983). Others believe that normative judgments must be conjoined with empirical judgments about whether someone's physiology is dysfunctional. This tendency has been dominant in recent philosophical treatments of disease concepts (Bloomfield, 2001; Boorse, 1975; Culver & Gert, 1982; Thagard, 1999). Kitcher (1997, pp.208–9) summarizes the debate as follows:

> Some scholars, *objectivists about disease*, think that there are facts about the human body on which the notion of disease is founded, and that those with a clear grasp of those facts would have no trouble drawing lines, even in the challenging cases. Their opponents, *constructivists about disease*, maintain that this is an illusion, that the disputed cases reveal how the values of different social groups conflict, rather than exposing any ignorance of facts, and that agreement is sometimes even produced because of universal acceptance of a system of values.

In the remainder of this chapter, I will first develop a more elaborate taxonomy based on Kitcher's. Then I will criticize constructivism, introduce objectivism, and discuss some of the difficulties objectivism faces.

2. Objectivism and Constructivism

To begin with, then, we need to recognize that both objectivism and constructivism can take either a *revisionist* or a *conservative* form. One could be a conservative or revisionist objectivist, as well as a conservative or revisionist constructivist. A conservative view says that our folk concept of illness is correct, that we make the right judgments about sickness and health. Since our folk judgments are correct, they should constrain a theoretical picture of health and disease worked out by scientists and clinicians. On this view, when we ask what health and disease are, we are doing conceptual analysis.

A revisionist view says that our concepts of health and disease, though a necessary starting point, should not constrain where the inquiry ends up. It could be that our concepts of health and disease are mistaken. So we have four possible positions. Let's go over them, starting with varieties of objectivism.

First, consider conservative objectivism. Objectivists tend to be conservative, believing that our folk judgments about illness agree with own their stress on underlying bodily malfunction. But objectivists often think that folk concepts set conditions on what counts as health and disease and medicine, and that we should adopt those conditions and look for the processes or states in the world that meet them. A revisionist rejects this understanding of science's relation to common sense.

Revisionist objectivists say that because facts about physiological and psychological functioning obtain regardless of how we think about disease, common sense about disease may get some cases wrong. They regard health and disease as features of the world to be discovered by biomedical investigation, and therefore loosely constrained, at best, by our everyday concepts of health and disease.

Constructivists are usually revisionists. They say that concepts of health and disease (especially when it comes to mental disorder) are used to medicalize behavior that is really just socially deviant or otherwise negatively regarded, and this is normally presented as an unmasking of commonsense assumptions that people really have something objectively wrong with them. A constructivist may accept that diagnoses of ill-health involve objective facts. But the relevant facts, for a constructivist, are not facts about how human minds or bodies work. They are social. Societies share norms, and some people transgress those norms. Some people who violate norms are regarded as immoral, others are called eccentric, and others are regarded as ill. A constructivist can concede that we look for distinguishing features in the biology or psychology of the deviants. But the constructivist says that this search just rationalizes our prior decision to stigmatize something about those people on medical grounds. For a constructivist, we call obesity a disease not because of its effects on health, but because we think fat people are disgusting. That's consistent with our subsequently discovering physical facts about obesity and its relation to poor health outcomes. The crucial constructivist claim is that we look for the medical facts selectively, based on prior condemnations of some people and not others.

Because of this claim that medicine is driven by social norms rather than disinterested inquiry, constructivism tends to be revisionist about our folk concepts. Constructivists usually aim to criticize, rather than explicate, our folk concepts, often for political reasons. But constructivism could be a conservative view, aimed at uncovering our folk theory of health and disease. A constructivist who takes this view says that our folk concept of disease is that of a pattern of behavior or bodily activity that violates social norms.

Objectivism and constructivism could be combined. For example, one could be an objectivist about bodily disease but a constructivist about psychiatry. Claims that we are merely treating conduct we don't like as pathological have much more force in psychiatry than in general medicine. Or one could argue that our folk taxonomy of illness involves both objectivist intuitions about some conditions and constructivist rationalizations about others. You could use this depiction of everyday thought as a premise in an argument for revisionism, on the grounds that our folk concepts are too confused to serve as constraints (Murphy (2006) makes this argument with respect to psychiatry). But constructivism faces several problems, which I will now discuss.

3. Problems for Constructivism

The chief problem constructivism faces is its apparent inability to explain everyday distinctions between the pathological and the merely disapproved of. No constructivist has explained why we call violations of some norms a disease, but not others. The medicalization of unwelcome traits may be growing, but it is not complete, and the constructivist owes us some account of why we are selective about medicalization. We regard physical and behavioral phenomena like stupidity and ugliness in a bad light. But we do not think they are disorders, whereas the claim that obesity is a disorder is increasingly uncontroversial. Constructivists are unable able to distinguish moral and political disputes from medical ones.

We also write the history of medicine in terms of uncovering disordered physical mechanisms and destructive processes, not in terms of changing pictures of norm-violation. And it seems that this way of writing the history lets us explain why medicine is progressing: for instance, we no longer treat hysteria by shoving rat droppings up a woman's nose based on a theory that her womb has wandered away from its proper position and must be induced to return. The germ theory was a great advance because it explained more than predecessors like humoral theories. This view of history lets us criticize some past diagnoses as medically incorrect or politically motivated. It was once argued that American slaves who tried to escape were afflicted with "drapetomania" or the compulsion to flee. Objectivists can say quite straightforwardly that drapetomania was never a disease, regardless of what anyone believed, because it was not caused by malfunctions according to any even moderately correct theory of human biology or psychology. A constructivist thinks that values, not science, drive our judgments of pathology, and the scientific facts are merely marshaled to support prior value judgments. So for a constructivist drapetomania's status as a disease depends on who wins the political battle. Yet drapetomania was never a worthwhile diagnosis. Escape attempts by slaves were not better explained by positing abnormal causal mechanisms instead of a simple desire for freedom.

Constructivism ignores the importance of appropriate causal explanation to our everyday thought about illness. Beliefs about how our bodies normally operate influence our judgments that people are sick: to identify something as a symptom of an illness we need some reason to think the processes underlying it are themselves abnormal. We think that aging is normal but we acknowledge that it brings frailties with it, so our assumptions about normality are sensitive to background conditions. But when aging is bizarrely accelerated, we regard it differently: children with Hutchinson–Gilford progeria syndrome go through all the stages of human aging at an astonishing rate and are nearly always dead by seventeen, in a state of advanced senescence. We are largely ignorant of its etiology, but we think of Hutchinson–Gilford as obviously different to normal aging, and obviously caused by some underlying pathology. In order to be properly applied, the concept of disease, which includes Hutchinson-Gilford but not normal aging, seems to require that a condition have a causal history that involves abnormal physical systems.

Objectivism embodies the important insight that we do not regard disease judgments as unconstrained by the biological or psychological facts. We do in fact think disease judgments depend on appropriate causal explanations. Not just any sort of story about the causes of abnormal behavior will do, of course: if we discovered that a woman was acting strangely due to hypnosis, or because her body was under the control of malicious extraterrestrial scientists, we would not consider her ill. So let's turn to objectivism to see what its prospects are, and whether it should be adopted as a conservative or revisionist position.

4. Objectivism

Objectivism usually comes with a commitment to separating scientific and social assessments of human malfunction which I call *the two-stage picture*: first, we agree on facts about the failure of someone's bodily systems to function properly. When we have decided that, there is still a question about how to think of the person who is malfunctioning. The second stage, of normative judgment, is bypassed by *simple objectivism*, which is the view that all there is to disorder is the failure of someone's psychology or physiology to work normally. A simple objectivist about disease just identifies someone as healthy or disordered depending on the facts about their organismic functioning relative to our best current theory of what human functioning should be. Simple objectivists are very rare, but Szasz (1987) uses simple objectivism about disease in general to anchor his claim that mental disorder is a mythic notion used to justify repression. Szasz is usually read as a constructivist who denies that mental illness exists. But in fact Szasz has a very strict objectivist concept of disease as simply damage to bodily structures, and he concludes that mental disorders cannot exist because they are not the result of tissue damage.

Simple objectivism is wrong. Normative judgments cannot be all there is to the concept of disorder, but they cannot be neglected. They do inform our conclusions about whether it is bad to have an abnormal physical constitution, whether it makes no difference, or even whether it is desirable. Various forms of bodily damage are not regarded as injuries or instances of disease, such as the effects of vaccination, ear-piercing, or childbirth. A recent stir, for example, was made by evidence of a specific brain lesion which turns the patient into a gourmet (Regard & Landis, 1997). Or imagine a skin condition that in some cultures causes the sufferer to be worshipped as an avatar of the divine, or become a sought-after sexual partner. Advocates for the deaf do not deny that their hearing is impaired due to underlying abnormalities. But they deny that the condition is a disease, to be treated as lessening the lives of those who have it. To make sense of these puzzles and disputes we should distinguish between the physical abnormality and the status it confers on the abnormal person. The two-stage picture is designed to capture this idea that whether someone's body is not functioning correctly is a separate question from whether it is bad to be that way.

We have arrived at a generic objectivism that says judgments of illness are sensitive to causal antecedents of the right sort, as well as to value judgments. What are the right causal antecedents? One set of criteria is Culver and Gert's (1982) theoretically minimal requirement of a "nondistinct sustaining cause." In contrast, Boorse and his followers argue that illness necessarily rests on malfunctions in evolved systems that make up a species-specific design plan.

Culver and Gert analyze the concept of a *malady*, which involves suffering an evil, or an increased risk of evil, that depends for its presence on "a condition not sustained by something distinct" from oneself (1982, p.72). A sustaining cause is one with effects that come and go almost simultaneously with the presence or absence of the cause itself. A wrestler's hammerlock is a sustaining cause, but a person trapped in one does not have a malady because the cause (the wrestler) is distinct from the sufferer. If the cause is within the person it is a nondistinct sustaining cause if it is either biologically integrated (like a retrovirus) or if it cannot be removed without difficulty (like a clamp mistakenly left in the body after surgery). The cause can be physical or mental, like a bad childhood (p.87), as long as it is a sustaining cause that is not distinct from the sufferer (p.88).

Culver and Gert recognize the problem of specifying the acceptable causal constraints on illness. They try to solve the problem by the principle of nondistinctness. Anything that is not a distinct cause and which produces suffering or a risk of suffering counts as an acceptable cause of illness. This is an attractively simple solution but it is too inclusive. Someone who is a victim of discrimination could count as having a malady.

Since loss of freedom, opportunity, or pleasure count as evils, according to Culver and Gert (p.71) prolonged unemployment would seem to be an evil. If you can't get a job because you are black, ugly, fat, short, gay, or female then you are unemployed and hence suffering evils, due to nondistinct sustaining aspects of your nature.

The principle of nondistinctness is not enough. People can have distinct causes of wounds and nondistinct properties that are bad for them without being pathological. What Culver and Gert's analysis seems to miss is the idea that a cause of illness produces harms via the distortion of normal processes. To examine both the attractions of this view, and its pitfalls, we should turn to the work of Boorse.

In his influential writings on disease concepts Boorse (1975, 1976) defended what I have called the two-stage picture by distinguishing "disease" from "illness." He understood illness to depend on value judgments about suffering or deviance in addition to the presence of disease, which he saw as the perfectly objective matter of whether someone fails to conform to the "species-typical design" of humans. The species-typical design is a specification, at various levels of analysis, of the component parts of the body and the functions they perform, function being understood in evolutionary terms as failure to contribute to survival and reproduction (1976, pp.62–3). On this view disease is the failure of species-typical design, but "illness" is defined normatively. Boorse argued that a disease only counts as illness if it is undesirable, entitles one to special treatment, or excuses bad behavior. This is the two-stage picture; scientific judgments that a destructive or abnormal bodily or psychological process has occurred, plus, at the second stage, normative judgments about the extent and manner of its impact.

Many objectivist theorists about disease have since employed Boorse's original contention that diseases are malfunctions that cause suffering or justify special treatment. Spitzer and Endicott (1978, p.18) make the point by calling disease categories "calls to action"; they see them as assertions that something has gone wrong with a human organism that has led to negative consequences (see also Papineau, 1994). The qualification of objectivism with avowedly normative criteria of disability or distress is a dominant one, represented more recently by the work of Wakefield's "harmful dysfunction" concept of mental disorder (1992, 1997).

Like Boorse, Wakefield says that our concepts of both mental and physical disorders involve two individually necessary and jointly sufficient components. First, we judge that an internal mechanism is malfunctioning. Wakefield also follows Boorse in understanding dysfunction in evolutionary terms, as a failure of a mechanism to perform the function for which natural selection designed it. Second, we, the surrounding society, judge that the malfunction is harmful. So Wakefield would count as disordered a woman who cannot bear children if her peers regard her as harmed thereby, even if she doesn't want children and would have a hysterectomy if she were fertile. Also like Boorse, Wakefield argues that the same picture applies to mental illness. He assumes that our psychology consists of evolved functional components and that breakdowns in these systems are a necessary condition of mental disorder.

Boorse's appeal to natural function is the most sophisticated explication of the standard objectivist appeal to the intuitive requirement that disorder depends on an appropriate etiology. It avoids some of the conceptual problems that Culver and Gert's view faced. The notion of dysfunction fleshes out the causal requirement more strictly. In addition, the analysis makes room for the idea that conditions like the gourmet lesion raise both a scientific question about normal function and a non-scientific question about whether it is a good or a bad thing to have abnormal functions. However, the harmful dysfunction view also faces difficulties, which illustrate some of the general troubles with objectivism, to which I now turn.

5. Troubles with Objectivism

The harmful dysfunction view is the currently dominant species of objectivism. It is designed to accommodate both our intuitions about the physical causes of disease and our intuitions about the undesirability of certain kinds of existence. As an objectivist view, it requires that human nature has component parts whose failure to work normally we can specify. The picture is of a commonsense concept of disease which bottoms out in a notion of malfunction as the cause of illness, and assumes that once this point is reached we can hand matters over to the sciences. Science tells us what the functional decomposition of the human organism is. Boorse and Wakefield assume that the relevant decomposition is possible and the relevant notion of function is an evolutionary one.

So the assumption is that conceptual analysis can determine the empirical commitments of our disease concepts and then hand the objective determination of structure and function over to the biomedical sciences. There are three problems with this project. First, the stress on a distinctively evolutionary account of function is unattractive, since the biomedical sciences employ a different conception of function. Second, a reliance on functional decomposition as the ultimate justification of judgments of health and disease requires a revisionist, rather than a conservative, account. Third, it may not always be possible to settle contested cases by an appeal to a notion of normal human nature, because that notion is itself contested.

First, why suppose that the relevant concept of function is an adaptive one, and that dysfunction is a failure of a biological system to fulfill its adaptive function? The harmful dysfunction analysis has been developed with little attempt to argue that medicine does in fact use an evolutionary, teleological account of function. In opposition, Schaffner (1993) has argued very convincingly that although medicine might use teleological talk in its attempts to develop a mechanistic picture of how humans work, the teleology is just heuristic. It can be completely dispensed with when the mechanistic explanation of a given organ or process is complete. Schaffner argues that as we learn more about the causal role a structure plays in the overall functioning of the organism, the need for teleological talk of any kind drops out and is superseded by the vocabulary of mechanistic explanation, and that evolutionary functional ascriptions are "necessary, though empirically weak to the point of becoming almost metaphysical" (1993, pp.389–90). For Schaffner, teleological functional ascription is merely heuristic; it focuses our attention on "entities that satisfy the secondary [i.e., mechanistic] sense of function and that it is important for us to know more about" (1993, p.390).

In effect, Schaffner is arguing that the biomedical sciences employ a causal, rather than a teleological, concept of function. This is in the spirit of Cummins's (1975) analysis of function as the causal contribution a structure makes to the overall operation of the system that includes it. Cummins's concept of function is not a historical or evolutionary concept. According to Cummins, a component may have a function even it was not "designed," and, therefore, parts with no selection history can be ascribed a function. In this sense of *function*, Harvey understood the function of the heart two centuries before Darwin. [See FUNCTION AND TELEOLOGY].

Schaffner's skepticism about the general scientific utility of evolutionary accounts of function is overdone, since in some areas of biology functional ascription is indeed teleological. But biomedical ascriptions of function to an organ or structure do not make assertions about adaptedness. Theory-building in medicine only requires that functional structures can be identified and analyzed in terms of their contribution to the overall maintenance of the organism as a living system. Explanation in medicine takes a model of the normal realization of a biological process and uses the model to show how abnormalities stem from the failure of normal relations to apply between components of the model. This requires a non-historical function concept, one that is at home in mechanistic, rather than evolutionary, explanation.

Medical function concepts, then, get their sense from their role in a mechanistic explanation that shows how the operation of a system depends on the contribution of its components. Boorse and his followers have departed from this practice and tied illness conceptually to an evolutionary concept of function. Besides its questionable title to folk usage, this approach faces problems. First, one must show that illnesses are evolutionary fitness-lowerers, which nobody has any idea how to do. The second problem that one faces in tying disease constitutively to evolutionary dysfunction is that one must then conclude that if an illness depends on structures that have no evolved function, it cannot really be an illness. The appendix, for example, appears to have no adaptive role, and hence appendicitis can't be a disease on a strictly evolutionary analysis of psychiatric dysfunction since the appendix is not failing in its teleological function. Other bodily structures might be spandrels or otherwise not a product of natural selection. [See ADAPTATIONISM]. They cannot malfunction in Boorse and Wakefield's sense, and hence cannot be diseased.

Third, it is widely believed that the evolutionary function concept is normative in virtue of being teleological and thus imports normative considerations into the scientific foundations of the two-stage view. I will deal with the ramifications of this objection in a moment, after I argue for the claim that an objectivist conception of disease must be revisionist.

Objections to an evolutionary notion of medical malfunction do not show that there is anything wrong with the general idea of basing judgments of health and disease on a scientifically established picture of normal functional decomposition. However, on this account, it becomes harder to retain the conservative project that looks for the natural phenomena that fall under, and are therefore constrained by, our folk concepts of health and disease. Both sides of Kitcher's objectivist/constructivist divide usually assume that there is a lay concept of disorder that should constrain the scientific understanding of what is or is not a medical disorder. Wakefield, for instance, thinks that some psychiatric diagnoses flout our intuitions by attributing disorder on the basis of behavior alone without looking for malfunctioning mental mechanisms (1997). He is criticizing the scientific, theoretical picture of mental disorder by appeal to folk intuitions. He is searching for necessary and sufficient conditions for the folk concept of mental disorder and assuming that science should search for the psychological processes that fit the concept thus defined. Boorse, too, adduces everyday linguistic usage and commonsense intuitions as evidence, even though he claims to be discussing the clinical concepts of health and disease. But it is one thing to take intuitions as a starting point, and another to say that they are hegemonic.

A revisionist can say that a condition we currently disvalue but do not regard as a disease may turn out to involve malfunction and hence to be a disease, whatever our intuitions say. Objectivists try to resist this. Wakefield argues that our intuitive folk theory of human design means it is "obvious from surface features" when underlying mechanisms are functional or dysfunctional (Wakefield, 1997, p.256). But it's not obvious. It is an empirical issue that could turn out contrary to common sense. Horwitz (2002, p.98), following Wakefield, argues that symptoms of depression are sometimes appropriate responses to stressful events, and hence not evidence of illness. Horwitz points out that it is normal to get depressed in these cases. But to a revisionist it is an open question whether this normal response means that post-bereavement depression is not a disease. We do indeed have expectations about the psychological aftermath of bereavement, but we also think it's normal to get blisters after ingesting mustard gas. That typical response hardly shows that nothing is wrong internally. We expect environmental stressors to have physiological effects, but those effects could nonetheless be mediated by dysfunctional inner mechanisms. For a revisionist it is an empirical question whether one's physiology or psychology is functioning properly, not a conceptual one. Once we hand over the task of uncovering malfunction to the sciences we can no longer constrain the inquiry by making common sense the ultimate arbiter, unless we wish to explicitly import, into the concept of disease, normative considerations derived from folk theories of what normal human nature amounts to. To maintain the separation of science and values that the two-stage picture aspires to, we must be revisionists. However, there is still one issue facing the objectivist who tries to sequester normative judgments in the second part of the two-stage picture. That is the objection, touched on earlier, that ascriptions of function and malfunction are themselves intrinsically normative.

If mechanistic explanation is not normative it is because the criteria for assessing adequate performance by some functional part of the human organism are supplied by nature. The criteria must not be supplied by regulative criteria derived from human goals, but discovered by science. Without reference to human norms, we must be able to ascertain, within acceptable limits of variation, the biological standards that nature has imposed on humans. The goal of finding out how a biological system works is fixed by our interests in health and well-being, but the objectivist assumption is that the goal is met by discovering empirical facts about human biology, not our own, culturally defined, norms. So, we diagnose someone as suffering from mesenteric adenitis not just because they are in discomfort due to fever, abdominal pain and diarrhea, but because the lower right quadrant of the mesenteric lymphatic system displays abnormal inflammation. This thickening of the nodes is not just the objective cause of the discomfort, it is an objective failure of the lymphatic system to play its normal contribution to the overall system. For the objectivist program to work, the biological roles of human organs must be natural facts just as empirically discoverable as the atomic weights of chemical elements.

The view that the correct functional decomposition of humans can be discovered in nature is very strong. It's the view that natural functional standards for human nature exist independently of what people think. Some people will say that since even this view licenses statements about what some biological system ought to be like, medicine is in fact normative in some sense (Bermudez, 1998). But the problem is whether any science is not normative in this sense, since all sciences license expectations about what ought to happen in a normal system. Stars, for example follow a reliable progression through developmental stages, so we can predict what ought to happen to them.

It is possible that norm-free mechanistic explanation is possible in some areas of human nature, but not in others, and this is likely to be especially important for judgments of mental disorder. In cases where such convincing ascription of function to a physiological mechanism is possible it does not seem at all odd to suppose that the standards of good performance are to be found in nature. But that does not mean it will be easy to establish what counts as successful performance, to do it in all cases, nor to establish what counts as normal variation. The idea that only one model of normal human nature exists is too strong, although there is no need to adopt the opposite view, that there just is no reason to expect any general theory of human nature (Dupré, 2001, p.95.)

We can study a biological system and assess its performance relative to a picture of functioning that, making allowance for normal variation, rests on models of the capacity that relate it to natural standards of performance. Variation in biological traits is ubiquitous, and so establishing whether a mechanism is functioning normally is difficult: nonetheless, biologists do it all the time. But not all diagnoses can be tied to a break between normal and abnormal functioning of an underlying mechanism, such as a failure of the kidneys to conserve electrolytes. Nor can we always discover some other abnormality, such as the elevated levels of *helicobacter pylori* bacteria that have been found to be causally implicated in stomach ulcers (discussed in detail by Thagard, 1999). Some conditions, like hypertension or obesity, involve cutting between normal and pathological parts of a continuous variation, even in the absence of clear underlying malfunctions that separate the populations. The more of this we have to do, the more we will have to complicate the analysis by appeal to risk factors and behavioral difficulties rather than natural standards of underlying function. And that raises the worry that the behavioral factors we cite will reflect contested conceptions of human flourishing. This is particularly likely to occur if our concept of health is of not just an absence of disease, but a more positive conception of a flourishing life.

Distinguishing failures to flourish from functional abnormalities will always be a special problem for psychiatry. For example, judgments of irrationality are central to many psychiatric diagnoses, and our standards of rational thought reflect not biological findings but standards derived from normative reflection. The possibility of psychiatric explanation employing the methods and models of physical medicine, then, depends on how much of our psychology is like the visual system – i.e., decomposable into structures to which we can ascribe a natural function (Murphy, 2006). It is not likely that we can decompose the mind/brain into functional components identified through wholly biological, non-normative standards. Within medicine more generally, the prospects for a general objectivism about disease depend on our ability to understand human biology as a set of structures whose functions we can discover empirically, and our capacity to understand disease causally as the product of failures of those structures to perform their natural functions.

References

Bermudez, J. L. (1998). Philosophical psychopathology. *Mind and Language*, 13, 287–307.

Bloomfield, P. (2001). *Moral reality*. New York: Oxford University Press.

Boorse, C. (1975). On the distinction between disease and illness. *Philosophy and Public Affairs*, 5, 49–68.

Boorse, C. (1976). What a theory of mental health should be. *Journal for the Theory of Social Behavior*, 6, 61–84.

Culver, C. M., & Gert, B. (1982). *Philosophy in medicine*. New York: Oxford University Press.

Cummins, R. (1975). Functional analysis. *Journal of Philosophy*, 72, 741–64.

Dupré, J. (2001). *Human nature and the limits of science*. New York: Oxford University Press.

Horwitz, A. V. (2002). *Creating mental illness*. Chicago: University of Chicago Press.

Kennedy, I. (1983). *The unmasking of medicine*. London: Allen & Unwin.

Kitcher, P. (1997). *The lives to come: the genetic revolution and human possibilities* (rev. edn). New York: Simon & Schuster.

Murphy, D. (2006). *Psychiatry in the scientific image*. Cambridge, MA: MIT Press.

Papineau, D. (1994). Mental disorder, illness and biological dysfunction. In A. Phillips Griffiths (Ed.). *Philosophy, Psychology and Psychiatry: Royal Institute of Philosophy Supplement* 37(pp. 73–82). Cambridge: Cambridge University Press.

Regard, M., & Landis, T. (1997). 'Gourmand syndrome': eating passion associated with right anterior lesions. *Neurology*, 48, 1185–90.

Schaffner, K. F. (1993). *Discovery and explanation in biology and medicine.* Chicago: University of Chicago Press.

Spitzer, R. L., & Endicott, J. (1978). Medical and mental disorder: proposed definition and criteria. In R. L. Spitzer & D. F. Klein (Eds). *Critical issues in psychiatric diagnosis* (pp. 15–39). New York: Raven Press.

Szasz, T (1987). *Insanity.* New York: Wiley.

Thagard. P. (1999). *How scientists explain disease.* Princeton: Princeton University Press.

Wakefield, J. (1992). The concept of mental disorder. *American Psychologist*, 47, 373–88.

Wakefield, J. (1997). Diagnosing DSM-IV, part 1: DSM-IV and the concept of disorder. *Behavior Research and Therapy*, 35, 633–49.

Part V

Ecology

Chapter 17

Population Ecology

MARK COLYVAN

1. Introduction

A population is a collection of individuals of the same species that live together in a region. *Population ecology* is the study of populations (especially population abundance) and how they change over time. Crucial to this study are the various interactions between a population and its resources. A population can decline because it lacks resources or it can decline because it is prey to another species that is increasing in numbers. Populations are limited by their resources in their capacity to grow; the maximum population abundance (for a given species) an environment can sustain is called the *carrying capacity*. As a population approaches its carrying capacity, overcrowding means that there are fewer resources for the individuals in the population and this results in a reduction in the birth rate. A population with these features is said to be *density dependent*. Of course most populations are density dependent to some extent, but some grow (almost) exponentially and these are, in effect, density independent. Ecological models that focus on a single species and the relevant carrying capacity are *single species models*. Alternatively, multi-species or community models focus on the interactions of specific species.

The discipline of population ecology holds a great deal of philosophical interest. For a start, we find all the usual problems in philosophy of science, often with new and interesting twists, as well as other problems that seem peculiar to ecology. Some of the former, familiar problems from philosophy of science include the nature of explanation and its relationship to laws, and whether higher-level sciences (like ecology) are reducible to lower-level sciences (like biochemistry). Some of the philosophical problems that arise within population ecology include whether there is a balance of nature and how the uneasy relationship between the mathematical and empirical sides of the discipline might be understood. As we shall see, many of these questions are intricately linked, and providing satisfactory answers is no easy matter. But there is no doubt that there are important lessons for philosophy of science to be gleaned from the study of population ecology.

In what follows I will focus on some of the central questions that are prominent in the recent philosophy of population ecology literature. There are, of course, other questions and problems, some of which the interested reader may pursue in the works listed in the references and further reading. But despite this admittedly less than comprehensive treatment of the philosophical issues in population ecology, those I address will give a sense of the flavor of the philosophical issues that arise in population ecology.

It is worth mentioning that many of the philosophical problems in population ecology are of great importance to working ecologists. For example, the issue of whether there are laws in ecology is seen by many ecologists as an important internal question to their discipline and one that has immediate methodological implications. (If there are no laws, ecologists might settle for a more pragmatic and even pluralist attitude toward their models.) Philosophers have been a little slow to turn their attention to ecology and so working ecologists have had to tackle many of the philosophical issues themselves. As a result a great deal of the philosophical groundwork has been carried out (for the most part, with a high degree of philosophical sophistication) by working ecologists. (See, for example, Ginzburg, 1986; Pimm, 1991; and Turchin, 2001.) But the philosophical problems in population ecology are important in another way. Population ecology itself has a great deal of social and political significance. Conservation management strategies often depend on predictions of population ecology. Where population ecology meets conservation management we find that philosophy of science meets ethics. Typically a great deal more than scientific or philosophical curiosity hangs on the answers to the philosophical and scientific problems faced by population ecology. For example, scientific issues about burden of proof in hypothesis testing have a distinctly ethical dimension. I will say more about such matters in Section 6.

2. Laws in Ecology

It has been claimed that ecology is not law governed (Murray, 1999; O'Hara, 2005). The reasons for denying the existence of laws in ecology are not always clear. Often appeals are made to lack of generality and lack of predictive success, but the complicated nature of ecology seems to feature especially prominently in this debate. We need to be careful not to set the bar too high for lawhood though. Consider the claim that ecology is too complex to submit to general laws. This may well be true but it is not obviously true, and it is certainly not something we can determine *a priori*. After all, we take celestial mechanics to be law governed, even though every massive body in the universe interacts gravitationally with every other massive object. It does not get much more complicated than that! While it is true that populations are affected by a great deal around them – the weather, predators, parasites, resources, fertility, and so on – considerations elsewhere in science show that complexity alone does not disqualify a discipline from being law governed. The complexity might "wash out" (Strevens, 2003), or much of the complexity might be properly ignored in many situations (as we can properly ignore the gravitational influence of Sirius on the Earth when we consider the Earth's orbit around the sun).

A case can be made for accepting that ecology has laws, albeit laws with exceptions. There is a very natural way to think of a highly simplistic and idealized equation like Malthus's equation, $N(t) = N_0e^{rt}$ (where, N is the population abundance, t is time, N_0 is the initial abundance, and r is the population growth rate), as a fundamental law of ecology. After all, this equation can be thought of as analogous to Newton's first law. Each describes what the respective system does in the absence of disturbing influences.

In the ecological case, Malthus's law tells us that populations tend to grow exponentially unless interfered with. Interference can come in the form of density dependence, predators, and so on. Of course there always are disturbing influences, so no population grows exponentially for any significant period of time. But why should this disqualify Malthus's equation from being a law? After all, no massive body in the universe moves with uniform motion, but this does not disqualify Newton's first law. If it is good enough for celestial mechanics, it is good enough for ecology. Malthus's equation can be thought of as a fundamental law of population growth – it describes the default case from which departures are to be explained. Moreover, like Newton's first law, Malthus's equation has considerable empirical support (e.g., the approximate exponential growth of microbial populations in laboratory situations). If we do treat Malthus's equation as a law, analogous to Newton's first law, we are then faced with the project of identifying the "ecological forces" that result in such departures from exponential growth (Ginzburg & Colyvan, 2004).

What of explanation in ecology? On traditional accounts of explanation (e.g., Hempel, 1965), laws are required for explanation. So if ecology does not have laws, there can be no ecological explanation. One response is to deny the traditional account of explanation: ecology has explanations but not laws (Cooper, 2003). Though if what I have suggested above is correct and ecology does have laws, then even on the traditional account of explanation there can be genuinely ecological explanations. Let us focus on the latter response. That is, let us assume that ecology does have laws and ask after the nature of the explanations delivered. There is still a problem for ecological explanation. The laws we are talking about are population-level laws; they are not about the individuals that constitute the populations in question. Consider Malthus's law. It has only initial abundance and the growth rate as parameters, and these both concern properties of the population, not the *individual*. But now here's the problem. Surely the *real* explanation for why a population has the abundance it does will be about births, deaths, immigrations, and emigrations *of individual members*. The law seems to ignore the individual events and yet the latter are what are causally relevant. How can such a law be genuinely explanatory?

I think this argument against ecological laws being explanatory fails. First, note that the argument is very general and, as stated, it would tell against any macro-level explanations of micro-level phenomena. For example, the ideal gas law has only macro-level parameters – the individual properties of gas molecules do not feature in this law – so it would seem that the ideal gas law also falls foul of this line of attack on ecological laws. But, any statistical law – by its very nature – is at the level of ensembles, not of individuals. It would seem that all statistical laws stand or fall together: the ideal gas law, ecological laws, and many others. Surely the argument against ecological laws being explanatory is misguided. I will return to the issue of explanation in ecology in the next section, when I look at mathematical models in ecology.

3. Mathematical Models

Despite being a highly mathematical discipline, ecology has an uneasy relationship with the mathematics it employs. We have already seen that ecology is about assemblages of living organisms and a population grows or declines by adding or subtracting individuals. The details of the population growth or decline will depend entirely on what happens to the individuals that constitute the population in question. But the typical mathematical models of a population ignore the details of individuals. Or rather, all the details about the individuals are packed into a few population-level parameters such as growth rate, carrying capacity, and the like.

In order to focus the discussion, let us consider a couple of simple mathematical models. Recall *Malthus's law* from the previous section. This states that the rate of change of population abundance, with respect to time, is proportional to population abundance. Represented mathematically, this becomes the following simple first-order differential equation:

$$dN/dt = rN,$$

where r is the population growth rate, t is time, and N is the population abundance. Solving this equation yields the familiar exponential growth equation (which we also refer to as Malthus's law):

$$N(t) = N_0 e^{rt},$$

where N_0 is the initial population abundance. Of course populations do not grow exponentially for long (if at all) – eventually their growth is limited by resources. Introducing such considerations into the mathematical model yields the logistic equation:

$$dN/dt = rN(1 - N/K),$$

where r, t, and N are the same as before, and K is the carrying capacity for the population in question. The logistic equation is, arguably, the simplest useful model in population ecology. Despite a number of idealizations (such as ignoring age structure and genetic variation in the population, and treating that carrying capacity as constant), it is a very good description of many populations. Of course there are other refinements one can make, but we won't bother here. The logistic equation will serve as our canonical example of a mathematical model in population ecology.

Now let us turn to the question of the use of mathematical models of population growth. These models are put to at least two different purposes: prediction and explanation. I will return to explanation shortly but for now let us focus on prediction. Most mathematical models are notoriously poor predictors. Of course they can be made to match existing data by suitably adjusting free parameters, but this gives one little confidence in the predictive accuracy of such models. Indeed, models whose parameters are too finely tuned are treated with considerable suspicion. Such models are (pejoratively) called "over-fitted" and are thought to be unrealistically complicated and thus unreliable predictors. So an important question about the predictive reliability of models needs to be addressed: What means are available for guaranteeing that the model will give us the right answers? Or failing such guarantees, how do we go about specifying the degree of confidence in the model?

The kind of uncertainty we are dealing with here is called "model uncertainty" and is notoriously difficult to quantify (Regan et al., 2002). But while a mathematical model may not predict the details, it may preserve gross trends. So, for example, we might find that under any reasonable value of the free parameters (or less commonly, under any reasonable model design) the model gives more or less the same answer. The model thus exhibits a certain robustness, and testing models in this way is called *sensitivity analysis* (Levins, 1966; Morgan & Henrion, 1990, 39–40; Wimsatt, 1987). Of course, a great deal hangs on how "reasonable values of the free parameters" is understood, but in practice, and in at least some cases, ecological theory provides guidance.

One interesting feature of sensitivity analysis is that it gives rise to a supervaluational logic (admittedly, under a non-standard epistemic interpretation of the logic in question). If the population p is deemed to have property Q on all reasonable values of the parameters, then we are confident that p has Q. If p fails to have Q on all reasonable values of the parameters, then we are confident that p does not have Q. But what of the indeterminate cases, where on some reasonable values of the parameters p has Q, while on others p does not have Q? Here it would seem that the right thing to say is that we are neither confident that p has Q nor are we confident that p does not have Q. In short, we assert that p has Q if and only if p has Q on all valuations. The resulting logic is a supervaluational logic and is familiar in the philosophical logic literature as the tool of choice in dealing with vagueness. This logic has interesting features, such as being non-bivalent while preserving the classical law of excluded middle (van Fraassen, 1966; Beall & van Fraassen, 2003). (Strictly speaking we are talking about the logic of the modal operator "confident that . . ." but I will not explore such complications here.)

Validation studies are another way to test a model. Here, one uses part of a data set to construct the model, including the fixing of all free parameters, while withholding another part of the data set. The second, withheld part of the data set is then used to test the model. If the model predicts the withheld data, the model is said to be validated. The problem with such an approach is that it requires large data sets – typically long time-series data of a population – and such data is rarely available. Indeed, the absence of such data is often the motivation for constructing a model in the first place.

The problems concerning model uncertainty are deep and philosophically rich. For a start, such uncertainty does not readily submit to probabilistic treatment (Regan et al., 2002). After all, it is very often impossible to assign values to the probability that the model is correct in every detail. Or at least on standard methods of assigning such probabilities, they will come out to be zero. New methods for dealing with such uncertainty are required. One such approach is non-classical logic. For in the face of serious uncertainty, it is necessary to entertain at least three categories: definitely true, definitely false, and indeterminate. Multi-valued and modal logics may prove fruitful in dealing with uncertainty that resists probabilistic treatment (Regan et al., 2002). There are various questions about the relationship between simplicity and predictive success of models. Can we be more confident in a simple model? This is an old chestnut in the philosophy of science. On the one hand, there are good pragmatic arguments for insisting on simplicity in the models or laws of ecology; thus formulated, the relevant theory will be easier to work with, and generally more tractable. But, on the other hand, what do pragmatic virtues of a theory have to do with truth or even predictive success? Put another way, what is so bad about complex (or over-fitted models)? Interesting work on this problem has been carried out by Forster and Sober (1994), who use a theorem due to Akaike to forge a link between simplicity and predictive success. Mikkelson (2001) applies these ideas specifically to ecology. (See also Colyvan and Ginzburg (2003) for discussion of possible limitations of this approach to simplicity.)

Thus far, I have been focusing on the typical population models that employ population-level properties like carrying capacity, growth rate, and the like. There are extensions that relax some of the assumptions of single aggregated population dynamics. Age- and stage-based models (also known as matrix models; Caswell, 2001) are models in which organisms are differentiated based on their age or morphological features such as size. Each age or stage class then has its own population growth equation that is coupled with other age or stage classes in the model. Meta-population models incorporate space through a population of sub-populations which are separated by a distance (Gotelli, 2001).

These are all population-level models, though, and it is worth saying a little about another kind of model: individual-based models. The latter are models that focus on the properties and behavior of the individuals of a population. The global population-level properties are then derived from the local interactions. Unlike the global population-level models, individual-based models keep track of individual properties and behaviors (DeAngelis & Rose, 1992). They incorporate diversity amongst individuals by representing each individual separately and explicitly specifying attributes such as the individual's age, size, spatial location, gender, energy reserves, etc. Sometimes individual-based models are used to estimate or model population-level parameters (McCauley et al., 1990; Gurney et al., 1990). In a sense, such individual-based models take a bottom-up approach to determining global population-level properties. A familiar example of an individual-based approach is found in various simulations such as "the game of life" and spatialized prisoner's dilemmas. In such simulations, individuals are located in an environment consisting of cells. Individuals are able to take one of a number of states and there are rules about the interactions between neighboring cells (or individuals). Such approaches have been put to good use in shedding light on altruism in populations (Sober & Wilson, 1998) and the evolution of various social structures (Skyrms, 2004).

In population ecology, individual-based models are becoming more widely used. Typically such models are *spatially explicit*. That is, they associate a spatial location with each individual. Such spatially explicit individual-based models are especially useful in modeling species that aren't terribly mobile – otherwise movement rules need to be included and these present serious difficulties. But if the species in question is reasonably sedentary, each individual in the population can be associated with a particular fixed spatial region. These models are particularly suited to plant populations (see, for example, Regan et al., 2003). But with some additional complications individual-based models are also able to be used for animal populations where individuals are allowed to roam over more than one spatial region. Individual-based models are often employed when information about the structure of the population is required. So, for example, spatially explicit individual-based models are very useful for determining forestation patterns – not just the number of individual trees (Deutschman et al., 1997). To some extent at least, individual-based models and the more traditional population-level models are not direct competitors. Very often they are used to answer different questions (Regan, 2002).

Some ecologists take individual-based models to be less problematic than the usual population-level models. For example, individual-based models cannot be accused of ignoring the properties and behavior of the individual members of a population while focusing only on averaged population-level properties. There are still idealizations though. The behavior and properties of the individuals in individual-based models will be highly idealized and often reduced to one of a small number of states. Moreover, the individuals will be restricted to a small number of possible actions. As with other models, the devil is in the details. There is nothing inherently wrong with such idealizations; the question is whether the idealizations at issue are theoretically well motivated and whether they are useful. These are important questions for ecology but they are not, it would seem, questions that will submit to general answers; they must be answered on a case-by-case basis. And it would seem that these questions must be answered for both individual-based models and population-level models.

Another application of mathematical models is to provide understanding and explanation of certain features of the population in question. Here there is less emphasis on getting detailed predictions and instead the focus is on gaining insights into general population trends and the reasons behind them. Such models are rather controversial in ecology. It is thought by some that mathematical models cannot be explanatory, for they either obscure the underlying biological mechanisms or, worse still, they *ignore* the biological mechanisms. After all, if a population is exhibiting periodic behavior, say, the reason for this behavior must have something to do with births, deaths, immigration, and emigration of individual members of the population. The mathematical model, however, typically employs population-level parameters like carrying capacity and growth rate. A mathematical model thus cannot provide explanation because it is not couched in the right terms (or so the argument goes).

The first thing to stress here is that very often the mathematics is just representing the biological facts in a mathematical way. Properly understood, the mathematics neither ignores nor obscures the underlying biological causal mechanisms. Instead of listing all the individuals in a population at different times, for instance, we can summarize this information in terms of equations for the population abundance. The individual organisms might seem to have dropped out of the picture but they have not. All that is relevant about them is represented mathematically in the equation of growth. Consider another example. The constant K in the logistic equation is not just an uninterpreted constant introduced purely for mathematical convenience. As I have already pointed out, K has a very natural *ecological* interpretation as the carrying capacity. (Though, it might be argued that this interpretation is rather abstract and it is mathematically convenient in that the constant K is just a crude summary of the interactions of a population with its environment.)

Next I note that some explanations are more readily drawn from the model than from the biology. For example, the mathematical model may focus attention away from confusing local-level causal interactions and toward higher-level population trends. We see this in the mathematical explanation of why certain populations undergo specific abundance cycles. The explanation in terms of the periodic solutions of coupled differential equations is much clearer than any detailed tracking of specific individual-level interactions.

Finally (and most controversially), it may be that some explanations are best looked upon as essentially mathematical rather than biological. For example, the question of why certain populations are so unstable can be best understood in terms of facts about the instability of the relevant differential equations (May, 1973). Of course this is not to say that there is no biological component to the explanation—just that the mathematics is doing most of the explanatory work. The thought is that the biological system in question is represented by a mathematical model, and there are certain (mathematical or logical) limitations on the way the model can behave. In so far as the mathematical model accurately represents the biological system, then those limitations apply to the biological system as well. The crucial point here is that in many such cases (such as the example above) one cannot reconstruct the explanation in biological terms. No system, biological or otherwise, can violate the laws of mathematics. And sometimes that is all the explanation that is required. (See Colyvan, 2001, ch. 3, for more on non-causal, mathematical explanations.)

The claim (mentioned several paragraphs back) that individual-based models are preferable to population-level models is sometimes turned into an argument for the explanatory superiority of individual-based models. The idea behind this line of thought is that although both kinds of model are typically couched in mathematical language, it is only the basic features of the individual-based models that correspond to non-relational properties of the individual members of the population. While there is no doubt that population-level properties such as growth rate and the like supervene on properties of individuals, it is clear that the growth rate is fully determined by births, deaths, immigration, and emigration of individual members of the population in question. It is only individual-based models that respect the priority of these basic biological events (or so the argument goes). But as I have already suggested, I think it is a mistake to think of population-level models as ignoring these fundamental biological events. The logistic equation, for instance, does not ignore individual births and deaths, it just incorporates all the relevant information about births and deaths into the growth rate. (Of course, in the logistic model, there is the assumption that the growth rate is constant, but that is a different worry.) Moreover, individual-based models cannot claim to have cornered the market on the biologically relevant facts. We can, for example, ask why a particular individual died. Typically, individual-based models need to incorporate probabilities of death in various circumstances, but then these probabilities are just standing proxy for deeper biochemical and ultimately physical causes. If we take this line of reasoning all the way, we might conclude that only physics is explanatory. In which case, providing explanations in ecology might mean performing the reduction of ecology to biology, biology to biochemistry, and biochemistry to physics. Surely something has gone wrong here. Surely there are biological and ecological explanations. The question of ecological explanation will arise again in the next section when we consider a particular ecological phenomenon in need of explanation. For the moment, I just note that the argument against population-level explanations being genuine explanations is unconvincing.

I should mention one final use of mathematical models that I have not yet covered. Ecological models are often used for decision making and ecologists sometimes distinguish such models from both predictive and explanatory models. For example, a decision model might give you insights into the best fire-management policy for a piece of bushland (Richards et al., 1999) by indicating general trends one would expect to find under different management strategies. While such management applications of models are in some ways different from those discussed above, it might be argued that they are properly thought of as a kind of hybrid of the predictive and explanatory models. In these decision models, while exact predictions are not required, ball-park predictions *are* required (for otherwise the model would be of no use for decision making). And while these models may not provide anything so rich as a full explanation of the phenomenon in question, they do need to provide some understanding of the basic relationships spelled out in the model. In any case, I will not discuss these decision models further, although the use of these models (and operations research techniques, more generally) in conservation biology is a very interesting and a relatively new development that deserves further philosophical attention.

4. What is the Reason for Population Cycles?

Population cycles are periodic fluctuations in a population's abundance. Although stable population cycles are relatively rare in nature, they are very important for a number of reasons. First, from an ecological point of view, they are important test cases for various theories of population growth. Very often in science it is useful to turn one's attention to rare cases for insights. (Consider, for example, the importance of understanding the rather rare solar eclipses for our theory of celestial mechanics.) In any case, any decent ecological theory must be able to give a satisfying account of population cycles, rare or not. Second, from a philosophical point of view, population cycles provide some interesting insights into the methodology of population ecology and help shed further light on issues concerning ecological explanation.

Classical population ecology holds that stable population cycles are a result of predator–prey interactions (although some oscillations can be a result of a population overshooting and undershooting carrying capacity). The predator–prey model of population cycles is due to the pioneering work in population ecology by Lotka (1925) and Volterra (1926). This account describes the population of the predator and the prey via two first-order differential equations that explicitly mention the population of the prey (V) and the predator (P), respectively:

$$dV/dt = rV - \alpha VP$$

$$dP/dt = \beta VP - qP$$

where r, q, α, β are constants determined empirically: r is the intrinsic rate of increase in prey population in the absence of predators; q is the per capita death rate of the predator population; α is a measure of *capture efficiency*, which is the effect of a predator on the per capita growth rate of the prey population; and β is a measure of conversion efficiency, which is the ability of the predator to convert prey into per capita predator growth (Gotelli, 2001, pp.126–33).

These equations give rise to a very rich and interesting dynamics. The basic idea of how they produce cycling, though, is rather simple. As the predator population rises, there is more predation and so the prey population declines. As the prey population declines, there is less food for the predators and so the latter's numbers too decline. Once the predator population declines, there is less predation and so the prey population recovers and starts to rise again. The predator population also recovers and on it goes.

There are a number of idealizations made in the standard Lotka–Volterra model. The first is that the predator is a specialist and will starve in the absence of the specific prey in question. It is also assumed that prey population grows exponentially in the absence of predators and that the predators can consume an infinite number of prey. Some of these idealizations can be dropped. For example, functional-response models relax the assumption that individual predators can always increase their prey consumption when the prey population abundance increases (Gotelli, 2001, pp.135–40).

There is little doubt that predator–prey interactions can result in population cycles; the question is whether they are the *only* reason for cycles. The classic example of population cycles due to predator–prey interactions is the Canadian lynx–hare cycles observed by Elton and Nicholson (1942). But there are other examples of population cycles where no known predators exist. But these too can be forced into the predator–prey mould by treating the cycling population as a predator (even if it is a herbivore) and treating the resources (whatever they may be) as prey. So we can think of population–resource models as a generalization of predator–prey models.

There is also another way that cycles might arise. It is a basic assumption (and orthodoxy) throughout population ecology that ecological forces such as predation, limitation of resources, and so on affect the growth rate. But if these ecological forces were to result in a second-order change – affect the rate of change of the growth rate – things might look quite different. The idea here is analogous to forces in mechanics. On the Aristotelian view, forces result in velocities, whereas on the Galilean view, forces result in the second-order quantity, acceleration. The traditional population models are Aristotelian whereas the new second-order proposal is Galilean (Ginzburg & Colyvan, 2004). The second-order model has it that the dynamic state is no longer fully described by population abundance. Since the resulting model is a second-order differential equation, both population abundance and the rate of change of the population abundance are required. This second-order model thus has a time lag built into it. But most importantly, for present purposes, this model can give rise to internally generated population cycles. That is, the model does not need to rely on population interactions for cycling (although such externally driven cycles are still possible); the model is capable of producing stable single-species cycles.

An interesting question arises at this point concerning the mechanism for the cycles and the time lag. (In fact, it is really just the time lag that is in need of a mechanism, because in an important sense the internally generated cycles are just a consequence of the time lag.) It was largely due to the lack of a convincing answer to this question that the second-order theory was given very little attention in the ecological literature until the 1990s. Before I discuss the answer to this question, let me emphasize the importance of providing an answer. After all, you might be tempted to simply dismiss the question. Indeed, this is very close to what happened in the analogous physics case. Why should position depend on both velocity and previous velocities? "That is just the way things are," is the answer. What is the mechanism for two bodies remote from another to have gravitational influence on one another? Again, that is just the way things are. Why not answer the ecological question along similar lines?

Though this response is tempting, to advance it is, I would suggest, to seriously misunderstand the nature of biology and its relationship to physics. Physics, arguably, is the study of the fundamental laws of nature. We all know that explanation must end somewhere and it seems that physics is the appropriate place for it to end. So while we may accept that some basic laws do not admit of further explanation or justification, any such laws should, it would seem, be reserved for physics.

Fortunately for the second-order theory, there is an account of the time lag. A very plausible reason for such time lags (or *inertia*) in population growth is found in the *maternal effect*. This is the phenomenon of "quality" being transferred from mother to daughter. The idea is that a well-nourished and healthy mother produces not only more offspring but also healthier offspring. So, an individual from a healthy mother experiencing a deteriorating environment will do better and be able to continue reproducing longer than individuals, in the same environment, not fortunate enough to have a healthy mother. Similarly, an individual from an unhealthy mother will do poorly despite an improving environment. This means that the population abundance at any time is the product of both the current environment and, to some extent, the environment of the previous generation (Ginzburg & Colyvan, 2004).

The maternal-effect hypothesis provides an elegant answer to the question of the mechanism for the time lags involved in the second-order model of population growth. But other mechanisms are also possible. Predator–prey interactions are still in the mix (though these aren't causes internal to the population). Another possible mechanism is *niche construction*. This is the modification of a population's environment in ways that are beneficial to both the current generation and often to subsequent generations (Sterelny, 2001). A classic example of such niche construction is the building of dams by beavers and the large number of human interventions such as building dwellings that last more than one generation. (In general, niche modifications can last more than one generation and all that is required for the second-order model of cycling to work is a one-generational lag. But time lags of more than one generation can also be accommodated by the theory.) In fact, we can look on the maternal effect as a special case of niche construction.

So what is the cause of population cycles? There may well be more than one cause: predator–prey interactions, maternal effect, and niche construction all seem like plausible candidates – and there may be others. One interesting feature of population cycles is the Calder allometry (Calder, 1984), a correlation between body size of prey and the period of the cycle. Rather surprisingly, the period of the predator–prey cycle does not depend on the size of the predator. This suggests that even in clear cases of predator–prey cycles, the predator might be just along for the ride, with the period of cycling being set by internal (metabolic) properties of the prey. Work continues on the question of the cause of population cycles, and the evidence and arguments cited in this work make for a very interesting case study for philosophers of science.

5. The Balance of Nature Debate

It is often assumed that nature is in balance. The idea is that an ecosystem, if left undisturbed (i.e., without human interference), finds a balance, where all species can coexist.

There are many uses of this idea in ecology, conservation biology, and environmental ethics. I will focus on a couple of these.

The first example of the use of this metaphor is in environmental ethics and conservation management: if an ecosystem is in balance and this is seen as desirable and difficult to obtain, then we ought to avoid any human activity that might disrupt the delicate balance. Such interference would result in a less desirable state for the ecosystem. This line of reasoning is often thought to provide support for conservation efforts to leave ecosystems alone. There are some interesting questions here. What does it mean to say that nature is in balance? Is nature really in balance? Why is balance a desirable state for an ecosystem? Let us take each of these questions in turn.

The idea of the balance of nature, no doubt, springs from various unexpected consequences of human interventions in ecosystems. The introduction of foxes into Australia may have seemed innocent enough at the time but it has had a severe impact on small marsupial populations. Nature, we suppose, was in balance but the introduction of foxes disrupted that balance. But what is the notion of balance that is at work here? I take it that the idea is that balance is to be understood in terms of population abundances not straying too far from some equilibrium value (mean growth rates are zero). Presumably, populations can cycle but abundances do not tend to zero nor do they increase without bound. This is certainly one sense of balance. Another might be that nature is in balance in the sense that, once disturbed, the system returns to some equilibrium state. This tendency is often called *stability*. (There are also other closely related notions such as the speed which the system returns to the original state after a disturbance, and the degree to which the system can be changed by perturbances [Pimm, 1993].)

Now, turn to the question of whether nature is in balance. Obviously the answer to this question will depend on how "balance" is understood. For example, an ecosystem might be in balance in the sense that all the constituent populations have abundances that do not vary greatly, but the ecosystem might still be unstable: a small external interference might result in massive and widespread changes to the ecosystem. On the other hand, an ecosystem might be stable and yet exhibit wild fluctuations in constituent population abundances. Moreover, the timescale is going to be important here. A population abundance that does not change much on one timescale may vary greatly on another. In geological timescales, very few ecosystems can be thought to be balanced in either sense – species become extinct, populations decline and disappear, new species appear in ecosystems. Some of this is driven by climatic and geological change, some by the contingencies of various ecological factors. So let us suppose that we have fixed the timescale to something appropriate and we have decided on the appropriate sense of "balance." Is nature in balance? This is an empirical question and it would be surprising if it submitted to a general answer. It seems plausible that some ecosystems will be in balance while others will not.

Where does this leave us with regard to our final question of why balance is a desirable state for an ecosystem? One (anthropocentric) answer is that we humans require a certain kind of environment for our continuing existence and so we don't want things to change too much from the way they are. Arguably, balance in both senses under discussion is important for this. First, consider balance in the sense of population abundance not varying too wildly. Human survival clearly depends on balance in this sense, at least in those environments humans inhabit. For example, if the population abundance of crucial biotic resources varied wildly it could make human survival in that environment difficult or impossible. (Think of the impacts of droughts on agricultural societies.) Next, consider stability. Life in an unstable environment would be rather tenuous. Any disturbance could, potentially, lead to dramatic and irreversible changes. Moreover, such changes, in general, would be to the detriment of human survival. Although it seems that there is a plausible line of argument from the hypothesis that nature is in balance to the conservation of ecosystems, caution needs to be exercised. For a start, we would hardly want this to be the only case for preserving ecosystems, for surely we would like some reason to preserve changing and unstable environments. Indeed, it might well be argued that unstable environments are more in need of protection from human intervention than stable ones.

The second use of the metaphor of nature in balance is in the complexity–stability hypothesis, which is the hypothesis that the greater the complexity in an ecosystem, the greater its stability. It is well known that the disappearance of so-called *keystone species* can result in loss of stability of an ecosystem and great efforts are directed toward saving species considered keystone. But the complexity–stability hypothesis is much more general than this; it is not restricted to key species upon which many others depend. Again there are issues concerning the meaning of key terms here, most notably: stability and complexity. Do we read complexity in terms of biodiversity, interspecific species interactions, strength of interspecific interactions, or something else? And again there are ethical or conservation management implications: if stability is something to be valued, and a positive feedback between stability and ecological complexity exists, then we ought not to reduce the complexity of ecosystems. But there is also considerable ecological importance for this hypothesis. The complexity–stability hypothesis provides a wonderful example of the kind of debate one finds in ecology between the modelers and more empirically minded ecologists.

On the one hand, modeling work by Robert May (1973) has suggested not only that the complexity–stability hypothesis is false, but that the reverse relationship holds: increased complexity reduces stability. On the other hand, some empirical studies suggest that the complexity–stability hypothesis is true. The shortcomings of the modeling approach we have seen already: the models are idealizations and are quite unlike real ecosystems. Those unsympathetic to modeling are hardly going to reject a plausible piece of ecology, namely the complexity–stability hypothesis, purely as a result of a piece of modeling. But the case for the complexity–stability hypothesis, based on empirical evidence, is also less than convincing. After all, the complexity–stability hypothesis is supposed to be a general result, so appealing to a couple of case studies is not going to win the day. [SEE COMPLEXITY, DIVERSITY, AND STABILITY].

It is worth saying a little about the role of empirical evidence in debates such as this. At the end of the day, empirical evidence is important but it does not, and should not, have the final word. As is well known, it is very difficult (if not impossible) to derive universal generalizations from finite data sets. But even in cases of recalcitrant data, rejecting a hypothesis is no straightforward matter. As Duhem (1954), Lakatos (1970), and Quine (1951) have stressed (in slightly different ways), recalcitrant data do not count against a particular hypothesis. Each hypothesis makes predictions only when combined with a large body of theory (or auxiliary hypotheses). It is the package that is accepted or rejected; a core hypothesis can be protected from recalcitrant data by suitable adjustments elsewhere. The rejection of outliers in data sets is a clear example of such methodology at work. As Robert MacArthur (1972) puts the point:

> Scientists are perennially aware that it is best not to trust theory until it is confirmed by evidence. It is equally true, as Eddington pointed out, that it is best not to put too much faith in facts until they have been confirmed by theory. (p.253)

Of course such considerations in the philosophy of science do not undermine the evidence-based approach to settling issues such as the complexity–stability hypothesis. After all, there are good reasons to be wary of the modeling approach as well. My point is simply to stress that while having evidence on one's side is a good thing, one should not take the high moral ground as a result of this. Like most issues in population ecology, there are no easy answers here.

6. Socio-Political Aspects of Population Ecology

An important application of the theory of population ecology is in *population viability analyses* (PVAs). These are studies of populations under various management regimes and are important for conservation and resource management decisions. PVAs thus have great political importance. To take an example from conservation management, the standard International Union for the Conservation of Nature (IUCN) classification of endangered species relies heavily on estimates of current population numbers and predictions of declines in the near future. Examples of resource management include predictions of fish populations for managing fisheries. While such applications, strictly speaking, belong to conservation biology and natural resource management, there is a close relationship between ecology and these more politically oriented disciplines. Indeed, it can be difficult to disentangle the purely scientific questions about population abundance (which belong to population ecology) from the value-laden decision questions about how best to manage a population (which belong to conservation biology). But there are other ways in which ecology is entangled with socio-political issues. The issue of type I and type II error is perhaps the most striking example.

In standard hypothesis testing in ecology (and elsewhere), one always compares the hypothesis under investigation H with its negation H_0 – the null hypothesis. When making a scientific pronouncement on the matter, there are four possibilities: (i) accepting H when H is false (a false positive or type I error); (ii) failing to reject H_0 when H_0 is false (false negative or type II error); (iii) accepting H when H is true; (iv) failing to reject H_0 when H_0 is true. In standard hypothesis testing, type I error is considered the more serious error and so to guard against making this type of error, a great deal more is required for the acceptance of H. More specifically, we guard against making type I errors by stipulating that we will not accept H unless the evidence for H is overwhelming. That is, we stipulate that H_0 will win the day unless the probability of H_0, given the evidence, is very low. The later probability is the so-called α-level and is somewhat arbitrarily set at 0.05.

The upshot of all this is that standard hypothesis tests in ecology (and elsewhere) are designed to give the benefit of the doubt to H_0: reject H unless H is proven beyond reasonable doubt. Shrader-Frechette (1994) has argued that there is an ethical dimension to hypothesis testing. One sees this most clearly in legal contexts. The principle of "innocent until proven guilty," with its subsequent onus of proof on the prosecution, is clearly an ethical attitude one takes toward uncertainty in law. But setting the α-level at 0.05 is to take a similar stance in scientific hypothesis testing. The thought behind $\alpha = 0.05$ is that, just as in law, less harm is done if we wrongly deny an effect (or wrongly acquit), but it is bad scientific practice to wrongly accept that there is an effect (or wrongly convict). This is clear enough in the legal setting but, again as Shrader-Frechette (1994) points out, it is hard to defend in all scientific contexts. Ecology, it might be argued, presents us with some interesting problem cases.

Consider an ecological hypothesis that is important for conservation management. Take, for example, the hypothesis that a species will suffer a population decline of such proportions that it will warrant being classified as "critically endangered." The null hypothesis will thus be that there will not be such a population decline. Which hypothesis deserves the benefit of the doubt here? Well, that will depend to some extent on your attitude toward the species in question, and environmental issues more generally. But there is a good case to be made for reversing the burden of proof in this case so that we will be inclined to accept that the species is undergoing a population decline unless there is rather compelling evidence that it is *not* so declining. That is, we might set the α-level quite high, 0.95, say. Indeed, one might argue that there is a (non-scientific) value judgment about the choice of the α-level and that this reflects the researcher's attitude toward environmental issues.

There is also a certain amount of arbitrariness about the choice of hypothesis and null hypothesis. If, as in the last example, we take a population decline as *the effect*, the corresponding hypothesis will be that the population is declining, and the null hypothesis will be that there is no decline. As we saw, on standard hypothesis testing (with an α-level of 0.05), it will take some compelling evidence before the null hypothesis is rejected. But what if we were to turn things around and stipulate that the effect is that the population is not declining? Now the hypothesis will be that there is no decline (or if you prefer, that the population is steady or rising) and the null hypothesis will be that the population is declining (or if you prefer, not rising and not steady). Again it will take some compelling evidence before we reject the null hypothesis. But in the absence of any such evidence, in either case we will reject the hypothesis. But what we are accepting will depend on the arbitrariness of how we set up the problem. If you have green sympathies, say, you can set the hypothesis and null hypothesis in such a way that it will be very hard to reject the claim that the population is declining. And similarly, if you do not have such green sympathies you can set the hypothesis and null hypothesis in such a way that it will be very hard to reject the claim that the population is not declining.

It is also worth noting that often in population ecology data is scanty, so α-levels anywhere near the extremes – 0.05 or 0.95 – might be demanding too much. Setting the α-level at the usual 0.05 (or at the other extreme, 0.95) is in effect to always reject the hypothesis (or respectively, accept it), because there will very rarely be enough evidence to get the probabilities in question below 0.05 (or above 0.95). So to sum up, the poverty of data in much of population ecology and the obvious socio-political implications of many ecological hypotheses suggests that the usual hypothesis tests are inappropriate. What we should do about this is not clear. If we allow one to choose one's α-levels depending on one's attitude to the environment, a rather unpalatable relativism about crucial ecological hypotheses looms. After all, if one ecologist sets her α-level at 0.95 because she is an environmentalist and another sets his at 0.05 because he is not, how do we settle the ensuing debate about whether the species in question is declining in numbers? The question of the decline seems to be a scientific question, but allowing value judgments to enter into the scientific process via the choice of α-level (or the arbitrariness of what counts as the effect) undermines the objectivity of science. This unwelcome invasion of ethics also seems to blur the distinction between ecology and politically charged conservation management issues. Be that as it may, the alternative of sticking with the α-level of 0.05 does not solve the problem – it just hides it. Sticking with the traditional α-level of 0.05 is clearly arbitrary, but worse still, such a choice represents a certain bias against the acceptance of any given hypothesis. The result is not objective science; it is just less obviously subjective, because the subjectivity is buried in standard scientific practice.

Issues about uncertainty in population ecology are interesting in their own right, but when one factors in the socio-political importance of a great deal of ecological theory, uncertainty takes on new significance. Indeed, the interaction, on the one hand, of the scientific and statistical questions about uncertainty in ecology and, on the other, the various important management decisions that depend on ecological pronouncements gives population ecology (and ecology more generally) a very unusual place amongst the sciences. (See Mayo forthcoming for more on these issues.)

7. Ecology and Evolution

I will finish by mentioning just a couple of the interesting connections between population ecology and evolutionary theory. Some of these connections go back to the very origins of both disciplines. The first connection is that it is Malthusian growth that drives the struggle for existence, so central to evolutionary theory. After all, populations increase exponentially, yet resources are (eventually) limited. As Charles Darwin (1859) himself pointed out:

> As more individuals are produced than can possibly survive, there must in every case be a struggle for existence, either one individual with another of the same species, or with the individuals of distinct species, or with the physical conditions of life. (p.78)

But the historical connection between evolution and ecology runs even deeper. Although Darwin showed some interest in giving a general account of this struggle for existence, it was Ernst Haeckel who first identified ecology as "the study of all those complex interrelations referred to by Darwin as the conditions of the struggle for existence" (quoted in Cooper, 2003, pp.4–5). More recently, Greg Cooper (2003) has defended this account of ecology and in so doing raises many important issues. Indeed, Cooper argues that many of the issues I have addressed in this chapter (such as the role of models, the question of whether ecology has laws, and whether there is a balance of nature) all arise very naturally in the process of defending the view that ecology is the science of the struggle for existence.

Another related way in which ecology and evolution are connected is via the theory of *r–K*-selection (Pianka, 1970). Intuitively, there are two reproductive strategies organisms might employ: (i) maximize offspring production and (ii) have only a few offspring but ensure a high survival rate. According to the *r–K* selection theory, when a population is maintained at a low density the best reproductive strategy is (i) and the population in question is said to be *r*-selected. On the other hand, populations maintained at high density (i.e., close to carrying capacity) have no advantage in having high offspring reproduction. For these high density populations, the best reproductive strategy is (ii) and such populations are said to be *K*-selected. The names *r-selection* and *K-selection* come from the two parameters in the logistic equation. In *r*-selection, evolution is supposed to favor early semelparous reproduction (reproduction at a single age), large *r*, many offspring with poor survivorship, type III survivorship curve (survival probability is low at early ages but high for the later ages), and small adult body size. (Note that *survival probability* for an age x is the probability that an individual of age x will survive to age $x + 1$.) Mosquitoes are an example of species that are supposed to have evolved under *r*-selection. In *K*-selection, evolution is supposed to favor late, iteroparous reproduction (reproduction at more than one age), small *r*, few offspring with good survivorship, a type I survivorship curve (survival probability is relatively high for early ages and lower for later ages), and large adult body size. Mammals are examples of species that are supposed to have evolved under *K*-selection.

The *r–K* theory, though once popular, now faces serious problems. Some of these problems include: not all species have life history traits that fit the theoretical predictions; attempts to experimentally confirm the theory have failed; and the derivations from the theory did not include age-structured populations (Gotelli, 2001, p.70). One of the more interesting criticisms of the *r–K* theory is that there can be other factors besides population density involved in the evolution of life history traits. For example, according to the *r–K* theory, iteroparous reproduction is supposed to evolve when population density is high and resources are scarce. But such a reproduction strategy might also evolve in volatile environments where there is a risk of losing all offspring (Murphy, 1968). Having more than one shot at reproduction can certainly have its advantages in such environments. Iteroparous reproduction is a way of spreading the risk.

There is much more that could be said about the relationship between population ecology and evolution. Just as in other areas of population ecology, I think that there will be many interesting philosophical issues to emerge.

Acknowledgments

Some of the material in this chapter has been published previously. I'd like to thank Kluwer Academic Publishers for permission to reproduce portions of "Models and Explanation in Ecology," originally published in *Metascience* (Vol. 13, December 2004, pp.334–7) and portions of Colyvan and Ginzburg (2003). Copyright for the material in question remains with Kluwer. I'd also like to thank Lev Ginzburg, Paul Griffiths, Dominic Hyde, Greg Mikkelson, Hugh Possingham, Helen Regan, Sahotra Sarkar, Kim Sterelny, and members of the Ecology Centre at the University of Queensland for fruitful discussions on the issues addressed in this chapter. I'm especially grateful to James Justus, Elijah Millgram, and Helen Regan for comments on earlier drafts. I'd also like to thank the Biohumanities Project at the University of Queensland for research support.

References

Beall, J. C., & van Fraassen, B. C. (2003). *Possibilities and paradox: an introduction to modal and many-valued logics.* Oxford: Oxford University Press.

Calder, W. A. III (1984). *Size, function, and life history.* Mineola, NY: Dover (Dover edition: 1996; first published: 1984).

Caswell, H. (2001). *Matrix population models: construction, analysis and interpretation.* Sunderland, MA: Sinauer Associates.

Colyvan, M. (2001). *The indispensability of mathematics.* New York: Oxford University Press.

Colyvan, M., & Ginzburg, L. R. (2003). The Galilean turn in population ecology. *Biology and Philosophy,* 18(3, June), 401–14.

Cooper, G. J. (2003). *The science of the struggle for existence: on the foundations of ecology.* Cambridge: Cambridge University Press.

Darwin, C. (1859). *The origin of species.* New York: Collier. (Reprint: 1962; original: 1859).

DeAngelis, D. L., & Rose, K. A. (1992). Which individual-based approach is most appropriate for a given problem? In D. L. DeAngelis & L. J. Gross (Eds). *Individual-based models and approaches in ecology: populations, communities and ecosystems* pp. 67–87). New York: Chapman & Hall.

Deutschman, D. H., Levin, S. A., Devine, C., & Buttel, L. A. (1997). Scaling from trees to forests: analysis of a complex simulation model. *Science* (Online), 277 (5332, September 12). URL=<http://www.sciencemag.org/feature/data/deutschman/index.htm>.

Duhem, P. (1954). *The aim and structure of physical theory.* Princeton: Princeton University Press.

Elton, C., & Nicholson, M. (1942). The ten-year cycle in numbers of the lynx in Canada. *Journal of Animal Ecology,* 11(2, November), 215–44.

Forster, M. R., & Sober, E. (1994). How to tell when simpler, more unified, or less ad hoc theories will provide more accurate predictions. *British Journal for the Philosophy of Science,* 45(1, March), 1–35.

Ginzburg, L. R. (1986). The theory of population dynamics: I. back to first principles. *Journal of Theoretical Biology,* 122: 385–99.

Ginzburg, L., & Colyvan, M. (2004). *Ecological orbits: how planets move and populations grow.* New York: Oxford University Press.

Gotelli, N. J. (2001). *A primer of ecology* (3rd edn). Sunderland, MA: Sinauer Press.

Gurney, W. S. C., McCauley, E., Nisbet, R. M., & Murdoch, W. W. (1990). The physiological ecology of *Daphnia*: a dynamic model of growth and reproduction. *Ecology,* 71(2, April), 716–32.

Hempel, C. G. (1965). *Aspects of scientific explanation and other essays in the philosophy of science.* London: Macmillan.

Lakatos, I. (1970). Falsification and the methodology of scientific research programs. In I. Lakatos & A. Musgrave (Eds). *Criticism and the growth of knowledge* (pp. 91–195). Cambridge: Cambridge University Press.

Lawton, J. H. (1999). Are there general laws in biology? *Oikos,* 84, 177–92.

Levins, R. (1966). The strategy of model building in population biology. *American Scientist,* 54(4, July–August), 421–31.

Lotka, A. J. (1925). *Elements of physical biology.* Baltimore: Williams & Wilkins.

MacArthur, R. (1972). Coexistence of species. In J. A. Behnke (Ed.). *Challenging biological problems: directions towards their solution* (pp. 253–9). New York: Oxford University Press.

May, R. M. (1973). *Stability and complexity in model ecosystems.* Princeton: Princeton University Press.

Mayo, D. G. (forthcoming). Uncertainty and values in the responsible interpretation of risk evidence: beyond clean hands vs. dirty hands. *Risk Analysis.*

McCauley, E., Murdoch, W. W., Nisbet, R. M., & Gurney, W. S. C. (1990). The physiological ecology of *Daphnia*: development of a model of growth and reproduction. *Ecology*, 71(2, April), 703–15.

Mikkelson, G. (2001). Complexity and verisimilitude: realism for ecology. *Biology and Philosophy*, 16(4, September), 533–46.

Morgan, M. G., & Henrion, M. (1990). *Uncertainty: a guide to dealing with uncertainty in quantitative risk and policy analysis.* Cambridge: Cambridge University Press.

Murphy, G. I. (1968). Pattern in life history and the environment. *The American Naturalist*, 102(927, September–October), 391–403.

Murray, B. M. Jr. (1999). Is theoretical ecology a science? *Oikos*, 87(3, December), 594–600.

O'Hara, R. B. (2005). An anarchist's guide to ecological theory. Or, we don't need no stinkin' laws. *Oikos*, 110(2, August), 390–3.

Pimm, S. (1991). *The balance of nature? Ecological issues in the conservation of species and communities.* Chicago: University of Chicago Press.

Pianka, E. R. (1970). On r- and K-selection. *The American Naturalist*, 104(940, November–December), 592–7.

Quine, W. V. (1981). Five milestones of empiricism. In *Theories and Things* (pp. 67–72). Cambridge, MA: Harvard University Press.

Richards, S. A., Possingham, H. P., & Tizzard, J. (1999). Optimal fire management for maintaining community diversity. *Ecological Applications*, 9(3, August), 880–92.

Regan, H. M. (2002). Population models: individual based. In R. A. Pastorok, S. M. Bartell, S. Ferson, & L. R. Ginzburg (Eds). *Ecological modeling in risk assessment: chemical effects on populations, ecosystems and landscapes* (pp. 83–95). Boca Raton, FL: Lewis Publishers.

Regan, H. M., Auld, T. D., Keith, D., & Burgman, M. A. (2003). The effects of fire and predators on the long-term persistence of an endangered shrub *Grevillea caleyi*. *Biological Conservation*, 109(1, January), 73–83.

Regan, H. M., Colyvan, M., & Burgman, M. A. (2002). A taxonomy and treatment of uncertainty for ecology and conservation biology. *Ecological Applications*, 12(2, April), 618–28.

Schrader-Frechette, K. (1994). *Ethics and scientific research.* Lanham, MD: Rowman & Littlefield.

Skyrms, B. (2004). *The stag hunt and the evolution of social structure.* Cambridge: Cambridge University Press.

Sober, E., & Wilson, D. S. (1998). *Unto others: the evolution and psychology of unselfish behavior.* Cambridge, MA: Harvard University Press.

Sterelny, K. (2001). Niche construction, developmental systems and the extended replicator. In R. Gray, P. Griffiths, & S. Oyama (Eds). *Cycles of contingency* (pp. 333–50). Cambridge, MA: MIT Press.

Strevens, M. (2003). *Bigger than chaos: understanding complexity through probability.* Cambridge, MA: Harvard University Press.

Turchin, P. (2001). Does population ecology have general laws? *Oikos*, 94, 17–26.

van Fraassen, B. (1966). Singular terms truth value gaps and free logic. *Journal of Philosophy*, 53, 481–95.

Volterra, V. (1926). Variations and fluctuations of the numbers of individuals in animal species living together. (Reprinted in 1931. In R. N. Chapman, *Animal Ecology*. New York: McGraw Hill.)

Wimsatt, W. C. (1987). False models as means to truer theories. In M. Nitecki & A. Hoffmann (Eds). *Neutral models in biology* (pp. 23–55). Oxford: Oxford University Press.

Further Reading

Kingsland, S. E. (1985). *Modelling nature: episodes in the history of population ecology*. Chicago: University of Chicago Press.

Mikkelson, G. (2003). Ecological kinds and ecological laws. *Philosophy of Science*, 70(5, December), 1390–400.

Odenbaugh, J. (2001). Ecology, stability, model building and environmental policy: a reply to some of the pessimism. *Philosophy of Science*, 68:Supp (September), S493–S505.

Odenbaugh, J. (2005). Ecology. In S. Sarkar & J. Pfeifer (Eds). *The Philosophy of Science: an encyclopedia*. New York: Routledge.

Peters, R. (1991). *A critique for ecology*. Cambridge: Cambridge University Press.

Sarkar, S. (2007). Ecology. In E. N. Zalta (Ed.). *Stanford encyclopedia of philosophy (Fall 2007 Edition)*. URL=<http://plato.stanford.edu/archives/fall2007/entries/ecology/>.

Schrader-Frechette, K., & McCoy, E. (1993). *Method in ecology*. Cambridge: Cambridge University Press.

Chapter 18

Complexity, Diversity, and Stability

JAMES JUSTUS

1. Introduction

The stability–diversity–complexity (SDC) debate has persisted as a central focus of theoretical ecology for half a century. The debate concerns the deceptively simple question of whether there is a relationship between the complexity and/or diversity of a biological community and its stability. From 1955, when Robert MacArthur initiated the debate, to the early 1970s, the predominant view among ecologists was that diversity and complexity were important if not the principal causes of community stability. Robert May, a physicist turned mathematical ecologist, confounded this view with analyses of mathematical models of communities that seemed to confirm the opposite, that increased complexity jeopardizes stability. The praise May's work received for its mathematical rigor and the criticisms it received for its seeming biological irrelevance thrust the SDC debate into the ecological limelight, but subsequent analyses have failed to resolve it. Different analyses seem to support conflicting claims and indicate an underlying lack of conceptual clarity about ecological stability, diversity, and complexity.

At a coarse level of description, ecologists disagree little about the concepts of diversity and complexity. A biological community is a set of interacting populations of different species. Its diversity is commonly understood to be positively correlated with the number of species it contains (richness), and how evenly individuals are distributed among these species (evenness) (Pielou, 1975; Margurran, 1988), though other possible components of diversity have been considered. Complexity of a community is positively correlated with its richness, how many of its species interact (connectance), and how strongly they interact. Diversity and complexity are similar properties and may be strongly positively correlated, but they are not identical. Species of a highly diverse community may interact little and therefore exhibit low complexity, and vice versa.

Beyond these relatively uncontested claims, disagreement arises over how the two concepts should be operationalized. Ecologists have proposed several mathematical functions that differ about what properties (richness, evenness, connectance, etc.) are given priority over others in assessing diversity or complexity, and which differ in functional form. Currently, there is little agreement about what operationalizations, especially of diversity, are ultimately defensible (Ricotta, 2005).

Another problematic aspect of the SDC debate is the lack of consensus about how ecological stability should be defined. This reflects uncertainty about what features of a community's dynamics should be considered its stability, and has resulted in studies that suggest conflicting conclusions about the debate based on different senses of ecological stability (e.g., May, 1974; Tilman, 1999; Pfisterer & Schmid, 2002). Ecological stability is not, however, unique in this regard. As McIntosh (1985, p.80) has quipped: "A traditional problem of ecology has been that ecologists, like Humpty Dumpty, often used a word to mean just what they chose it to mean with little regard for what others said it meant." Disagreements about how to define concepts arise in other sciences as well. Careful analysis of the concept of ecological stability (and diversity and complexity) would thus help resolve the SDC debate, as well as illuminate the general problem of finding adequate definitions for concepts in science. Besides providing insights about how problematic scientific concepts should be defined, the SDC debate also has a potential bearing on biodiversity conservation. For most senses of stability, more stable communities are better able to withstand environmental disturbances, thereby decreasing the risk of species extinction. Positive feedback between diversity/complexity and stability would therefore support conservation efforts to preserve biodiversity, assuming biodiversity and ecological diversity/complexity are closely related (see Goodman, 1975; Norton, 1987, chs. 3 and 4).

To better understand the SDC debate, Sections 2–4 trace its history. Section 2 discusses the seminal works that initiated the debate in its current form, and Section 3 examines Lewontin's (1969) analysis of the relationship between mathematical concepts of stability and ecological stability. Section 4 considers May's influential work, which brought greater mathematical rigor and sophistication to the debate and upended the popular slogan among ecologists that "diversity begets stability."

Section 5 presents a comprehensive classification of different senses of ecological stability; argues that the concepts of resistance, resilience, and tolerance jointly define ecological stability adequately; and defends the concept against the charge that it is "conceptually confused" or "inconsistent." Section 6 surveys some common measures of diversity and complexity, and adequacy criteria proposed for them. Section 7 concludes by describing some methodological challenges the evaluation of stability–diversity and stability–complexity relationships confronts.

2. Emergence of the Stability–Diversity–Complexity Debate

Robert MacArthur (1955) published the first precise definition of ecological stability while still a graduate student of Yale ecologist G. E. Hutchinson. To clarify the concept, MacArthur (1955, p.534) first noted that ecologists tended to call communities with relatively constant population sizes stable, and those with fluctuating populations unstable. Stability in this sense denotes *constancy*. He thought, however, that this confused stability with its effects, and offered another account:

> Suppose, for some reason, that one species has an abnormal abundance. Then we shall say the community is unstable if the other species change markedly in abundance as a result of the first. The less effect this abnormal abundance has on the other species, the more stable the community. (1955, p.534)

This account identifies the underlying dynamic responsible for constancy, not constancy itself, as the proper defining property of stability. Stability in this sense depends on how communities respond to disturbance, in this case the abnormal abundance. The smaller the changes in other species abundances, the more stable the community. Although MacArthur did not use the term, this type of stability is a form of *resistance* to disturbance because its attribution to a community is based on the degree that one abnormal abundance changes other species abundances – specifically, the degree the other abundances resist changing – rather than on whether the community returns to equilibrium. For a community at equilibrium, i.e., its populations remain constant if the community is undisturbed, high resistance will ensure relative constancy is retained even if the community is disturbed. Highly resistant communities will therefore usually exhibit approximately constant species abundances through time, which MacArthur believed led many to call them stable. For MacArthur, however, constancy is a consequence of resistance, not equivalent to it.

MacArthur recognized that two properties could account for high resistance: (i) interspecific species interactions, such as predation and competition; and (ii) "intrinsic" properties of species, specifically their physiologies. Focusing on (i), MacArthur (1955, p.534) suggested a "qualitative condition" for stability: "The amount of choice which the energy has in following the paths up through the food web is a measure of the stability of the community." "Measure" in this condition is used in the standard statistical sense to represent the type of relationship exhibited between positively correlated properties, in the same sense that IQ is claimed to measure intelligence, for instance. Thus, the qualitative condition assumes rather than supports the claim that there is a positive correlation between community stability (understood as resistance) and food web structure.

MacArthur justified this assumption with an intuitive argument that a large number of links in a community's food web should make it highly resistant. In a food web where species *S* is atypically abundant, other species abundances are affected less the more widely *S*'s "excess energy" is distributed among different predators. Similarly, a wide variety of alternative prey for *S*'s predators would minimize the effects an abnormally low abundance of *S* would have on them. In either case, the number of links in a food web is positively correlated with community resistance. If correct, it is important to note that this argument only establishes a positive correlation. It does not justify conflating the properties MacArthur used to define ecological stability and those that may be positively correlated with them. Margalef (1958, p.61), for instance, misinterpreted MacArthur's analysis in this way: "In [MacArthur's (1955)] sense, stability *means*, basically, complexity" (emphasis added).

After noting that resistance can be quantified in several ways, MacArthur (1955, p.534) proposed two "intuitive" adequacy conditions for doing so:

(i) resistance should be minimal (e.g., 0) for food webs with exactly one species at each trophic level (food chains); and,
(ii) resistance should increase with the number of food web links.

Conditions (i) and (ii) refer to properties of food webs, not, as MacArthur *defined* resistance, to how communities change after one species becomes abnormally abundant.

What do (i) and (ii) have to do with quantifying how communities change after one species becomes abnormally abundant? What MacArthur was in fact doing was proposing adequacy conditions for quantifying the *measure* of resistance he had just argued for: "the amount of choice which the energy has in following the paths up through the food web." Conditions (i) and (ii) were not intended to help quantify the concept of resistance as MacArthur *defined* it.

This explains how MacArthur could quantify resistance with the Shannon index:

$$-\sum_i p_i \ln p_i; \tag{1}$$

where p_i is the proportion of the community's "food energy" passing through path i in the food web, which does not represent anything about how species abundances in a community are affected by the abnormal abundance of one species. The decision to use this index, which he called "arbitrary," was intended to specify a mathematical function satisfying (i) and (ii), although MacArthur (1955, p.534) noted that it "may be significant" that (1) has the same form as standard measures of entropy and information (cf. Shannon & Weaver, 1949).

With this quantification of resistance, MacArthur described what properties of food webs would maximize it. For m-species communities, (1) is maximized when the species are at m different trophic levels and each level-k species ($k \leq m$) consumes all species at all lower levels. It is minimized when one species consumes the remaining $m - 1$ species, which are all at the same trophic level. If the species consumed per consumer is held constant, moreover, (1) increases with species richness. This fact and the fact that (1) increases with the number of food web links entail either that large numbers of species with restricted diets, or small numbers that consume many different species can produce a particular value of (1). On this basis, MacArthur hypothesized that since species-poor communities only have high values of (1) when consumers eat a wide variety of species, but consumer diets in species-rich communities need not be similarly restricted to attain the same (1) value, species-rich communities will usually be more resistant. This prediction may explain, MacArthur (1955, p.535) suggested, why Arctic communities, which usually contain fewer species than temperate and tropical ones, seemed to exhibit greater population fluctuations.

Compared with other ecological research of the time, MacArthur's analysis was one of the most mathematically sophisticated. Instead of focusing on empirical evidence, his primary concern was to formulate intuitive ideas about food web structure with mathematical precision and explore their implications. Unlike his predominantly data-driven contemporaries, Hutchinson encouraged this approach to ecological questions among his students (Kingsland, 1995). Hutchinson believed that speculative but mathematically rigorous analyses were crucial to stimulating novel approaches to recalcitrant problems. By challenging ecologists to pinpoint their shortcomings, which mathematical clarity helped facilitate, even those later found wanting would stimulate development of improved successors.

Sometimes, however, this kind of speculative research is uncritically accepted and treated as definitive. This was especially true of such work in post-WWII ecology. At that time, many ecologists thought their discipline suffered from a general lack of mathematical precision and the absence of a theoretical basis (Slobodkin, 1953;

Margalef, 1958). It was for this reason that another Hutchinson student, Lawrence Slobodkin, originally encouraged MacArthur to pursue graduate work in biology with Hutchinson after MacArthur finished his master's degree in mathematics in 1953 (Kingsland, 1995). Five years later Slobodkin praised MacArthur's (1955) analysis because he believed that it provided a general method for ranking the stability of different communities based on their qualitative food web structure, and that this would in turn improve ecological theory by helping classify and conceptualize specific mathematical models (Slobodkin, 1958). Slobodkin thought the development of a "unified theory of ecology" required analyses like MacArthur's, and that they would remedy a troubling "trend in theoretical ecology towards each investigator developing his own equations and systems as if he were alone in the field" (1958, p.551). But Slobodkin (1958) also accepted MacArthur's explanation of a positive relationship between ecological stability and food web structure without scrutiny. Hutchinson (1959, p.149) similarly exaggerated and mischaracterized MacArthur's explanation as a "formal proof" based on information theory.

Unfamiliar with MacArthur's work and more wary of ecological theory that was not closely tethered to data, British ecologist Charles Elton (1958) took a more empirical approach to the issue. He was motivated by a cautious skepticism of the biological relevance of Lotka and Volterra's mathematical models of biological communities, and similar approaches to ecological theory. Specifically, for Elton (1958, p.131), "there does not seem much doubt that theories that use the food-chain for an explanation of the regulation of numbers are oversimplified." Elton focused instead on empirical evidence that seemed to show that some communities were more resistant to invasion by exotic species than others, and experienced more population fluctuations than others. Several documented cases of biological invasions and pest outbreaks on islands and in ecosystems "simplified by man" were the main support for his analysis.

Elton's concept of ecological stability had two components: resistance to invasion and constancy of populations (1958, p.145). This differed from MacArthur (1955) in two ways. First, constancy was explicitly part of stability whereas MacArthur thought constancy was a byproduct of stability, and not an appropriate part of its definition. Second, Elton's and MacArthur's concepts of resistance depend upon different types of disturbance. MacArthur's refers to a community's reaction to an abnormal abundance of one species. Elton did not explicitly define his concept of resistance, but it presumably refers to the ability to suppress the establishment, reproduction, and spatial spread of invasive species, i.e., to resist invasion. Ecological stability therefore involves resistance to disturbance for both Elton and MacArthur, but each focuses on different types of disturbance. Consequently, although Elton (1958) and MacArthur (1955) are commonly cited as analyzing the same relationship between stability and diversity (Pimm, 1984; Lehman & Tilman, 2000; McCann, 2000), their analyses presuppose different stability concepts.

Their analyses also study different properties of a community's structure. Elton (1958, p.145) did not define "rich" and "simple" in his claim that "simple communities . . . [rather than] richer ones . . . [are] more subject to destructive oscillations in populations, especially of animals, and more vulnerable to invasions," but species richness, not food web structure as in MacArthur's analysis, was the primary focus of the six kinds of evidence he presented in its support (1958, pp.146–50):

(i) despite his skepticism about their ecological relevance, Elton noted that simple mathematical models of one-predator, one-prey communities predicted fluctuations of population sizes and often mutual extinction, even in the absence of external disturbances;

(ii) experiments on microscopic one-predator, one-prey communities exhibited the same behavior as these mathematical models. Elton (1958) cited Gause (1934), who showed that population fluctuations to the point of extinction were typical in simple protozoan communities;

(iii) small oceanic islands with few species seemed to be more vulnerable to invasion than similar continental areas of the same size and, Elton assumed, more species;

(iv) successful invasions and population explosions of invasive species occurred more often in communities "simplified by man." Elton suggested four types of simplification as potential causes: (a) cultivation of exotic plants without introduction of the fauna normally accompanying them; (b) cultivation of these exotics in partial or complete monocultures; (c) eradication of species that reputedly harm the cultivated plants; and (d) selection of only a few genetic strains for cultivation;

(v) tropical communities, which contain more species and more complicated intra and interspecific dynamics than temperate communities, experienced fewer population explosions, especially of insects; and,

(vi) orchards, which are relatively simple ecological systems, were frequently successfully invaded. Elton suggested that pesticides usually decrease species richness in orchards and eradicate predators of herbivorous insects, which in turn facilitates invasions by exotic species and explosions in natural pest populations.

Since Elton included both constancy and invasion resistance in his concept of ecological stability, (i)–(vi) address different aspects of the SDC debate. (i) and (ii) focus on the lack of constancy, rather than invasibility, of microscopic communities and mathematical models of communities that contain few species. The lack of constancy, not invasibility, of temperate vs. tropical communities is also the focus of (v). Points (iii), (iv), and (vi), on the other hand, concern the greater invasibility of artificially simple or simplified communities – agricultural monocultures like orchards, for instance – and islands which contain relatively few species compared with continental regions of equivalent size.

Along with MacArthur (1955), Elton (1958) has frequently been cited in support of a positive stability–diversity relationship. Elton (1958, p.146) was careful to emphasize, however, the exploratory nature of his analysis, and he explicitly stressed the need for further data collection and study of the issue. An extensive review almost twenty years later (Goodman, 1975), in fact, revealed some of the ways in which subsequent ecological work had failed to support Elton's predictions.

Goodman (1975) first pointed out that (i) and (ii) are only compelling if community models and microscopic communities with higher species richness exhibit less population fluctuation and fewer extinctions than those with fewer species, which was (and remains) unestablished. Second, (iv) and (vi) were not based on controlled experiments and, Goodman further suggested, the simplified ecological systems in question may have achieved highly stable equilibria with their invasive pests were they not continually disturbed by cultivation. Third, (iii) is only compelling if the possibility that island communities are more susceptible to invasion than continental communities, irrespective of species richness, can be excluded as the cause of the pattern. Elton (1958) had not eliminated this possibility and Goodman noted that Preston's (1968) work on the evolution of island species might provide a better explanation of the greater invasibility of island communities than their supposed lower species richness. Fourth, the observations of population fluctuations in temperate regions and relative constancy of tropical populations that constituted Elton's support for (v) had not been borne out by subsequent studies. By the early 1970s, population fluctuations and insect outbreaks in the tropics that rivaled those in temperate regions had been observed (Leigh, 1975). Their apparent preponderance in temperate compared to tropical regions was probably an artifact of the greater attention and resources devoted to the former.

While Elton's (1958) monograph was in press, Cornell entomologist David Pimentel (1961) conducted the first experimental test of a stability–diversity relationship in fallow fields outside Ithaca, New York. During the summers of 1957 and 1958, Pimentel planted wild cabbage (*Brassica oleracea*) in two fields, one containing approximately 300 plant species and another in which he removed all other plants. He then observed differences in the insect and arachnid communities that developed on individual *B. oleracea* plants in the two fields, and found that the densities of a few pest insects increased dramatically in the monoculture and that more herbivores resided on the monoculture plants than on those in the multi-species community. Although Pimentel did not analyze the statistical significance of his results, he and other ecologists (e.g., Connell & Orias, 1964) believed they showed that the insect outbreaks were more severe in the monoculture and thereby confirmed a positive relationship between ecological diversity and stability.

One shortcoming of Pimentel's study is that diversity was narrowly measured as species richness in his experimental design and data analysis. Changes in species richness are relatively easy to measure, but they show nothing about changes in the proportions of individual organisms in each species of a community, i.e., changes in evenness. Evenness is an important component of ecological diversity (see Section 6), so measuring diversity as species richness limits what Pimentel's study, and any other study using this diversity measure, can show about stability–diversity relationships.

Even with this narrow measure, furthermore, it is unclear that the results of the experiment justify Pimentel's (1961, p.84) claim that, "The lack of diversity in . . . [the] single-species [monoculture] planting allowed outbreaks to occur." The problem is that there are two types of diversity in the monoculture field: plant and faunal diversity. In fact, the low plant diversity of the monoculture was accompanied by *increases* in insect and arachnid richness on the monoculture plants (higher than those on *Brassica* plants in the other field) and the latter may have been a more important determinant of the outbreaks. Thus, while Pimentel interpreted his data as evidence of a positive relationship between diversity (measured as plant richness) and stability (absence of pest outbreaks), it could also be interpreted as evidence of a positive relationship between diversity (measured as insect/arachnid richness) and *in*stability (indicated by the outbreaks). Without separating the effects of plant species richness from insect/arachnid richness on outbreak likelihood, Pimentel's results do not provide unequivocal support of a positive stability–diversity relationship.

Another problem is that Pimentel created the monoculture by removing the extant plant community of one field, whereas *Brassica* individuals were comparatively unobtrusively added to the 300 species plant community of the other field. Removing the extant plants undoubtedly initially eliminated the predators of *Brassica* herbivores so that the herbivores already on the *Brassica* that were planted or that immigrated to the monoculture could reproduce unchecked, while herbivores in the other field faced their usual set of predators and did not increase. In other words, the creation of the *Brassica* monoculture eliminated an important component of the extant animal community in that field, whereas the animal community of the other field was relatively undisturbed. The outbreaks may therefore be a consequence of the disturbance that eliminated the predators of *Brassica* herbivores, rather than the low plant-species richness of the monoculture. Pimentel (1961, p.84) recognized this potential confounding effect, and responded:

> the investigator doubts that the time-lag factor [i.e., that fauna had to immigrate to the plants in the monoculture] played a major role in the outbreaks, because wild *Cruciferae* were flourishing adjacent to all plots and provided ample sources of taxa for invasion of the single-species plots.

Whether this response is sound or not, the problem could have been avoided if Pimentel had eliminated all the plants from both fields and then planted a monoculture and a multi-plant-species community.

In discussing his results, Pimentel (1961) proposed an important hypothesis about possible causes of a positive relationship between species richness and constancy of populations. Pimentel may have been the first biologist to recognize that a "portfolio effect" might produce a positive relationship in the same way a diversity of investments usually reduces financial risk (1961, p.84):

> Each host or prey species reacts differently to the same environmental conditions. One host population may decline as another host population increases. This tends to dampen the oscillations of the interacting host and parasite populations and provides greater stability to the system as a whole.

Although isolated one-predator, one-prey communities may fluctuate, this behavior is collectively averaged out in interactions between multiple predator and prey species so that these systems exhibit more constant population sizes overall.

3. Mathematization of Ecological Stability

Although it was probably the first experimental study of the SDC debate, Pimentel's (1961) work received much less attention than the first theoretical analysis of the debate by MacArthur (1955). This likely reflected the transformation of ecology into a more mathematical and theoretical discipline occurring at the time. Largely through the work of Hutchinson and his students (most importantly, MacArthur), mathematical modeling became more sophisticated and prevalent within ecology in the 1960s.

Ecologists became increasingly concerned with formalization and theoretical systematization of ecological concepts (Kingsland, 1995). Hutchinson's (1957) highly abstract set-theoretic definition of the niche as an *n*-dimensional hypervolume is one example, as were attempts to develop precise definitions and measures of ecological stability, diversity, and complexity by mathematically oriented ecologists around the same time. Lewontin (1969, p.13) captured the intellectual shift within ecology: "To many ecologists their science has seemed to undergo a major transformation in the last 10 years, from a qualitative and descriptive science to a quantitative and theoretical one."

In the context of this transformation, the stability theory of linearized differential equations might have seemed to provide an adequate framework for evaluating stability–diversity–complexity relationships. For instance, the dynamics of a biological community near equilibrium could be represented mathematically with such equations, and formal stability criteria developed for them could be used to assess whether the community was ecologically stable. This was one of the methods utilized by Lotka and Volterra, for example, to analyze the stability of biological communities.

A problem with this specific modeling strategy, however, revealed a general difficulty with the new theoretical orientation of ecology: achieving mathematical precision and rigor often made empirical measurement more difficult. For this reason, Patten (1961) criticized the ecological relevance of the stability theory of linear differential equations. He pointed out that representing ecological systems with these equations requires extensive quantitative data about numerous parameters, which are practically impossible to obtain in the field. To illustrate the problem, consider a community represented by:

$$\frac{d\mathbf{x}(t)}{dt} = \mathbf{A}\mathbf{x}(t); \tag{2}$$

where $\mathbf{x}(t)$ is a vector $\langle x_1, x_2, \ldots, x_i, \ldots, x_n\rangle$ $(i = 1, \ldots, n)$ representing the densities of n species, and $\mathbf{A}$ is an $n \times n$ matrix of constant real coefficients $[a_{ij}]$ representing (linear) relationships between species near equilibrium, such as competition, predation, mutualism, etc. (see Justus, 2005). The problem Patten recognized is that the a_{ij} are extremely difficult to measure for complex, natural communities. Thus, "[s]ince it is usually not possible to obtain sufficient data to represent natural ecosystems canonically [as in (2)] and since they are probably not linear, formal stability criteria are not generally available for ecological applications" (Patten, 1961, p.1011).

Challenges like this and the plurality of distinct senses of stability in the ecological literature by the mid-1960s – resistance, resilience, tolerance, and constancy (see Section 5) – convinced many ecologists that a critical assessment of the concept was needed. Ecological stability was not unique in this regard. Many fundamental but problematically unclear ecological concepts were being examined at that time to determine whether they could be reformulated within mathematical frameworks used in other sciences, especially physics (e.g., Kerner, 1957, 1959; Lewontin, 1969). Doing so would specify their meaning clearly and possibly integrate them into a common mathematical framework. Theoretical unification of this kind had proved fruitful in physics, and ecologists had similar aspirations for their discipline. The existence of a well-developed mathematical theory of stability made such a rethinking of ecological stability seem especially promising and in May a symposium at Brookhaven National Laboratory was ostensibly devoted to examining the meaning of stability and diversity (Woodwell & Smith, 1969, p.v). Only one paper seriously addressed this task (Lewontin, 1969), but it profoundly impacted the subsequent development of the SDC debate.

Lewontin (1969) surveyed various mathematical notions of stability and their relation to ecological stability. He began by representing a biological community of n species as a vector $\mathbf{x}(t) = \langle x_1(t), x_2(t), \ldots, x_n(t)\rangle$ in an n-dimensional vector space H where t represents time. Different coordinates of $\mathbf{x}$ were intended to represent different abundances or densities of the n species in the community. A deterministic vector function $\mathbf{T}$ over H, $\mathbf{T}: H \rightarrow H$, represents the mechanisms responsible for the dynamics of the community. $\mathbf{T}$ represents, therefore, density-dependencies, interspecific interactions, gene flow, etc. between the n species and the effects environmental parameters have on these species. $\mathbf{T}$ is often specified, for instance, in matrix form by a mathematical model of the community (e.g., [2] from above). Application of $\mathbf{T}$ to $\mathbf{x}$ usually induces a change in the vector's coordinates. Points in H for which $\mathbf{T} = \mathbf{I}$ where $\mathbf{I}$ is the identity matrix, induce no change in $\mathbf{x}$ and are called equilibrium points. A vector at such a point will not move from it.

Within this framework, Lewontin distinguished "neighborhood" stability (also called local stability) from global stability. Let $\mathbf{x}_q$ be the position vector for some equilibrium point. Following the mathematical theory of stability pioneered by Lyapunov (1892 [1992]), $\mathbf{x}_q$ is neighborhood stable if and only if for any $\mathbf{x}$ *arbitrarily close* to $\mathbf{x}_q$:

$$\lim_{n\rightarrow\infty} \mathbf{T}^n(\mathbf{x}) = \mathbf{x}_q; \tag{3}$$

where $\mathbf{T}^n(x)$ designates n applications of $\mathbf{T}$ to $\mathbf{x}$. The subset of H within which vectors satisfy (3) defines the *domain of attraction* of $\mathbf{x}_q$. Restricting attention to vectors arbitrarily close to $\mathbf{x}_q$ allows approximation of $\mathbf{T}$ by a linear vector function $\mathbf{L}$. In effect, $\mathbf{T}$ behaves as a linear vector function arbitrarily close to $\mathbf{x}_q$. This linearization of $\mathbf{T}$, in turn, allows evaluation of (3) with well-known mathematical techniques (see Hirsch & Smale, 1974). If (3) holds for all of H and not just arbitrarily close to $\mathbf{x}_q$, $\mathbf{x}_q$ is called globally stable.

By representing the perturbation of a biological community as a displacement from $\mathbf{x}_q$ to $\mathbf{x}$, the community's stability can be represented by local and global stability. The set of perturbations (represented by the displaced vectors $\mathbf{x}$) for which the community returns to equilibrium (represented by $\mathbf{x}_q$) determines its attraction domain. In ecological terms, locally and globally stable communities are often informally characterized as those that return to equilibrium after "very small" perturbations – such as slight climatic disturbances perhaps – and those that return after *any* perturbation.

Lewontin (1969, p.16) argued that local stability inadequately defines ecological stability because it only describes system behavior arbitrarily close to a particular point in H. Strictly speaking, therefore, local stability only describes system behavior for *infinitesimal* displacements from $\mathbf{x}_q$. Real-world perturbations, however, are obviously not of infinitesimal magnitude. Any real perturbation will expel a system at a strictly locally stable equilibrium from its infinitesimal stability domain. Besides this, Preston (1969) pointed out in the same Brookhaven symposium that it is fundamentally unclear how infinitesimal displacement can be biologically interpreted (or empirically measured). Local stability therefore says nothing about system response to real-world perturbation. In contrast, Lewontin suggested that the stability of an ecological system depends upon the (non-infinitesimal) size of its attraction domain. If the formally precise notion of "arbitrarily close" is informally construed as "very close," moreover, local stability still only describes system behavior for very small perturbations and thus provides little or no information about attraction domain size. For this reason, local stability poorly defines ecological stability.

Local and global stability also poorly define ecological stability, Lewontin added, because they are dichotomous concepts, whereas biological communities seem to exhibit different degrees of stability. Elton and MacArthur's concepts of ecological stability confirm Lewontin's claim: Elton believed monocultures were *less* stable than "natural" communities, and the Shannon index MacArthur used to operationalize stability obviously takes values other than 1 and 0. Other stability concepts discussed in Section 5, such as resilience and tolerance, are all matters of degree also.

Lewontin drew an important distinction between stability as a perturbation-based concept, which resistance, resilience, and tolerance are, and non-perturbation-based concepts, such as constancy. Constancy, Lewontin (1969, p.21) suggested, "is a property of the actual system of state variables. If the point representing the system is at a fixed position, the system is constant. Stability, on the other hand, is a property of the dynamical space in which the system is evolving." The two concepts are therefore different and not necessarily coextensive. A system in a large, steep domain of attraction, for instance, may be in constant flux due to frequent external perturbations. Conversely, an unperturbed system may be constant at an unstable equilibrium. Partly for this reason, Section 5 argues that ecological stability should not be defined in terms of constancy.

4. The End of the Consensus

By the time of Lewontin's analysis, a strong consensus had emerged that ecological stability is positively associated with diversity and/or complexity (May, 1974; De Angelis, 1975; Pimm, 1991). In a textbook on environmental science, for instance, Watt (1973) deemed the claim that biological diversity promotes population stability a core principle of the discipline. A half-decade later, the consensus had evaporated. The main reason for its demise was the publication of rigorous analyses of mathematical models of communities that seemed to show increased complexity actually decreased stability.

In a one-page *Nature* paper, Gardner and Ashby (1970) initiated the first doubts with an analysis of the relationship between complexity and asymptotic Lyapunov stability in linear models such as (2) from above. Understood as a model of a biological community, the coefficient a_{ij} from (2) represents the effect of species j on species i. Its quantitative value represents the effect's magnitude and its sign represents whether the effect is positive or negative. For these models, complexity was defined in terms of the number of variables (n) and connectance (C). Gardner and Ashby defined connectance as the percentage of nonzero coefficients in $\mathbf{A}$.

Values of the diagonal elements of $\mathbf{A}$ were randomly chosen from the interval $[-1, -0.1]$ to ensure each variable was "intrinsically stable", i.e., self-damped. For given values of n and C, Gardner and Ashby then randomly distributed an equal number of -1 and $+1$ values within the off-diagonal parts of $\mathbf{A}$ in accord with the C value. Whether systems represented by (2) are stable depends upon the eigenvalues of $\mathbf{A}$. These are scalar value roots λ_i of the characteristic polynomial of $\mathbf{A}$, $|\mathbf{A} - \lambda\mathbf{I}| = 0$, where $\mathbf{I}$ is the identity matrix. Lyapunov ([1892] 1992) proved an equilibrium of a system represented by (2) is asymptotically stable iff:

$$\mathrm{Re}\lambda_i(\mathbf{A}) < 0 \text{ for } i = 1, \ldots, n; \tag{4}$$

where $\mathrm{Re}\lambda_i(\mathbf{A})$ designates the real part of λ_i, the i-th eigenvalue of $\mathbf{A}$. Whether (4) holds depends on the pattern of nonzero values within $\mathbf{A}$. Different randomizations specify different patterns, which may produce different stability results as evaluated by (4). The probability of stability for given values of n and C can therefore be approximated with the results from a sufficiently large number of randomizations. Contrary to most expectations, Gardner and Ashby found the probability of local stability was negatively correlated with n, and with C. Interpreted ecologically, their analysis seemed to show that the more species (greater n) and the higher the frequency of species interaction in a biological community (greater C), the less likely it is (locally) stable.

The study that inverted the opinion of most ecologists, however, was published by Robert May three years later (May, 1974). One reason for its influence was that May generalized Gardner and Ashby's analysis by randomly assigning non-diagonal elements values from a distribution with a zero mean and mean square value of s^2 for different values of s. s represents the interaction strength between variables, which had been restricted to $+1$ and -1 by Gardner and Ashby. May then analyzed how the probability of stability changed with different values of n, C, and s; Gardner and Ashby had not analyzed how s affects the probability of stability. The main result was that for systems in which $n >> 1$, there is a sharp transition from high to low probability of stability as s or C exceeds some threshold. He found, for instance, that the probability of stability for these systems is approximately 1 if $s\sqrt{nC} < 1$ and approximately 0 if $s\sqrt{nC} > 1$. His analysis also confirmed Gardner and Ashby's finding that for fixed C and s the probability of local stability decreases with increasing n. In general, these results seemed to demonstrate that high connectance, species richness, or strong species interactions preclude communities from being stable. May defined complexity in terms of connectance, richness, and interaction strength, so the results seemed to confirm a negative stability–complexity relationship.

One compelling feature of May's result was its generality. Besides requiring the entries of the diagonal be -1, no assumption was made about the coefficients of $\mathbf{A}$ from (2). In the parameter space representing all possible linear systems, therefore, May's results seemed to demonstrate that stability is exceedingly rare. It remained possible, May (1974, p.173) recognized, that actual biological communities primarily inhabit a rare stable realm of parameter space:

> Natural ecosystems, whether structurally complex or simple, are the product of a long history of coevolution of their constituent plants and animals. It is at least plausible that such intricate evolutionary processes have, in effect, sought out those relatively tiny and

mathematically atypical regions of parameter space which endow the system with long-term stability.

To address this possibility, May analyzed several common mathematical models of communities. Two patterns emerged. First, generalizations of simple models representing few species, such as one-predator, one-prey Lotka-Volterra models, to more complicated *n*-species models were generally less likely to be stable than the simple models. This underscored Goodman's (1975) criticism of the first type of evidence Elton (1958) cited in favor of a positive stability-diversity relationship (see Section 2) because it showed that multi-species predator–prey models are less, rather than more, stable than predator–prey models of fewer species.

Second, many modifications that made models more realistic also made them less stable. For instance, community models often unrealistically assume an unvarying deterministic environment and thereby set parameters to constant values. May showed that if some or all parameters are allowed to vary stochastically to represent environmental fluctuation, the resulting model is generally less likely to be stable. Similarly, most community models represent birth and death as continuous processes, even though their occurrences are discrete events in nature. May (1974, p.29) found that that more realistic models with discrete variables are likely to be less stable than their continuous counterparts. The disparity between models with discrete and continuous variables also becomes more pronounced as the number of variables increases.

The upshot of May's work was that stability is rare both in the "parameter space" of possible models and for more realistic community models, and that its probability decreases with model complexity. What it showed about actual biological communities, as opposed to models of them, remained unclear. Lewontin (1969) had argued, for instance, that the relationship between local stability and ecological stability is tenuous at best (see Section 3). The worry was that May's analysis may be an interesting mathematical exercise with little or no biological application. For this reason, in fact, May (1974, p.75–6) hedged about the proper interpretation of his results: "the balance of evidence would seem to suggest that, in the real world, increased complexity *is* usually associated with greater stability." Thus, for May, the results "suggest that theoretical effort should concentrate on elucidating the very special and mathematically atypical sorts of complexity which could enhance stability, rather than seeking some (false) 'complexity implies stability' general theorem" (1974, p.77). May recognized that most natural biological communities may have evolved a specific structure (the "atypical complexity") that generates stability. For May, what this structure is should be the focus of ecological modeling.

Despite May's (1974) qualifications and these criticisms, his work was widely accepted (Lawlor, 1978) and taken to overturn the consensus about a positive correlation between community complexity and stability in favor of a negative one (De Angelis, 1975). Earlier work on the SDC debate, however, did not concern the relationship between local stability and complexity, which was May's main focus. MacArthur (1955), Elton (1958), and Pimentel (1961), for instance, whose work set most ecologists' initial expectations about the debate, were concerned with resistance and constancy, not local stability, and focused primarily on species richness rather than complexity. May's analysis only provides insights into the stability–diversity relationship, therefore, if local stability and ecological stability (despite Lewontin's [1969] objections) and ecological diversity and complexity are closely related. Diversity and complexity are related in the weak sense that species richness is a component of both, but this relationship is too weak to guarantee even that a positive correlation between them exists.

5. Contextualization and Classification of Ecological Stability

As the history above indicates, one fundamental obstacle to resolving the SDC debate is finding an adequate definition of ecological stability. Numerous definitions and categorizations of the concept have been proposed (Lewontin, 1969; Orians, 1975; Pimm, 1979, 1984, 1991; Grimm & Wissel, 1997; Lehman & Tilman, 2000) and this plurality is responsible for much of the confusion and lack of progress in resolving the debate. As part of their argument that ecological theory has failed to provide a sound basis for environmental policy – they believe the SDC debate provides a clear example of this failure – Shrader-Frechette and McCoy (1993) have also argued that several proposed definitions of ecological stability are incompatible and that the concept is itself "conceptually confused" or "inconsistent."

A plurality of distinct senses of stability have been used within ecology: resistance, resilience, tolerance, constancy, local and global Lyapunov stability. Before considering these concepts in more detail, it should first be noted that their attribution must be made with respect to two evaluative benchmarks. The first is a system description (M) that specifies how the system and its dynamics are represented. The second is a specified reference state or dynamic (R) of that system against which stability is assessed. In most ecological modeling, M is a mathematical model in which:

(i) variables represent system parts, such as species of a community;
(ii) parameters represent factors that influence variables but are (usually) uninfluenced by them, such as solar radiation input into a community; and,
(iii) model equations describe system dynamics, such as interactions among species and the effect environmental factors have on them.

M therefore delineates the boundary between what constitutes the system, and what is external to it. Relativizing stability evaluations to M is a generalization of Pimm's (1984) relativization of stability to a "variable of interest."

The specification of M partially dictates how R should be characterized. A biological community, for instance, is usually described as a composition of populations of different species. Consequently, R is often characterized in terms of the "normal" population sizes of each species. Since ecological modeling in the late 1960s and 1970s was dominated by the development of mathematically tractable equilibrium models (Chesson & Case, 1986; Pimm, 1991), the "normal" population sizes were often assumed to be those at equilibrium, i.e., constant population sizes the community exhibits unless perturbed. This is not the only possible reference state, however. A community may be judged stable, for instance, with respect to a reference *dynamic* the models exhibit. Common examples are a limit cycle – a closed path C that corresponds to a periodic solution of a set of differential equations and toward which other paths asymptotically approach – or a more complicated attractor dynamic (Hastings et al.,

1993). Ecological stability can also be assessed with respect to some specified range of tolerated fluctuation (Grimm & Wissel, 1997). *R* may also be characterized solely in terms of the presence of certain species. Only extinction would constitute departure from this reference state.

The details of *M* and *R* are crucial because different system descriptions – e.g., representing systems with different variables or representing their dynamics with different functions – may exhibit different stability properties or exhibit them to varying degrees relative to different specifications of *R*. Specifying *R* as a particular species composition vs. as an equilibrium, for instance, can yield different stability results. Similarly, different *M* can produce different assessments of a system's stability properties. Describing a system with difference vs. differential equations is one example (May, 1974a).

The details of *M* and *R* are also important because they may specify the spatial and temporal scales at which the system is being analyzed, which can affect stability assessments. Systems with low resistance but high resilience, for example, fluctuate dramatically in response to perturbation but return rapidly to their reference state *R* (see below). Low resistance is detectable at fine-grained temporal scales, but systems may appear highly resistant at coarser scales because their quick return to *R* prevents detection of fluctuation. Similarly, significant fluctuations in spatially small areas may contribute to relatively constant total population sizes maintained through immigration and emigration in larger regions.

Once (and only once) *M* and *R* are specified, the stability properties of a system can be determined. These properties fall into two general categories, depending on whether they refer to how systems respond to perturbation (relative to *R*) or refer to system properties independent of perturbation response. A perturbation of an ecological system is any discrete event that disrupts system structure, changes available resources, or changes the physical environment (Krebs, 2001). Typical examples are flood, fire, and drought. Perturbations are represented in mathematical models of communities by externally induced temporary changes to variables that represent populations, to parameters that represent environmental factors, and/or to model structure. Many, perhaps most, real-world perturbations of communities should be represented by changes to both variables and parameters. A severe flood, for instance, eradicates individual organisms and changes several environmental factors affecting populations. In the following, let P_v, P_p, and P_{vp} designate perturbations that change only variables, change only parameters, and those that change both, respectively.

There are four plausible adequacy conditions for a definition of ecological stability:

(A1) the ecological stability of a biological community depends upon how it responds to perturbation ([A2]–[A4] specify the form of the required dependency);

(A2) of two communities *A* and *B*, the more ecologically stable community is the one that would exhibit less change if subject to a given perturbation *P*;

(A3) if *A* and *B* are in a pre-perturbation reference state or dynamic *R*, the more ecologically stable community is the one that would most rapidly return to *R* if subject to *P*; and,

(A4) if *A* and *B* are in *R*, the more ecologically stable community is the one that can withstand stronger perturbations and still return to *R*.

(A1) captures the idea that the behavior of a community is a reliable indicator of its ecological stability only if the behavior reflects how perturbation changes the community. If unperturbed, a community may exhibit great constancy throughout some period. It may be, however, that if it had been even weakly perturbed, it would have changed dramatically. Constancy of this community surely does not indicate ecological stability when it would have changed substantially if perturbed slightly. Similarly, variability of a community does not necessarily indicate lack of ecological stability if it is the result of severe perturbations, perturbations that would cause greater fluctuations or even extinctions in less stable communities.

The reason for (A2) is that more stable communities should be less affected by perturbations than less stable ones. Communities that can withstand severe drought, for instance, with little change are intuitively more stable than those modified dramatically. This was the idea, for example, underlying MacArthur's (1955) definition of community stability. The justification for (A3) is that more stable communities should more rapidly return to R following perturbations than less stable ones. This adequacy condition captures the idea that lake communities that return to R quickly after an incident of thermal pollution, for instance, are more stable than those with slower return rates following similar incidents. The ground for the last condition is that communities that can sustain stronger perturbations than others and still return to R should be judged more stable.

Three concepts – resistance, resilience, and tolerance – represent the properties required of ecological stability by (A2)–(A4). Resistance is inversely correlated with the degree a system changes relative to R following perturbation (P_v, P_p, or P_{vp}). Since perturbations vary in magnitude, resistance must be assessed against perturbation strength. Large changes after weak perturbations indicate low resistance; small changes after strong perturbations indicate high resistance. Resistance is thus inversely proportional to perturbation sensitivity.

Depending on M and R, changes in communities can be evaluated in different ways, each of which corresponds to a different measure of resistance. Community resistance is typically measured by the changes in species *abundances* following perturbation. It could, however, be measured by changes in species *composition* following perturbation, or in some other way. Pimm's (1979) concept of species deletion stability, for instance, measures resistance by the number of subsequent extinctions in a community after one species is eradicated.

Different types of perturbations, moreover, yield different measures of resistance. Since evaluating resistance requires considering perturbation strength, strengths of different types of perturbations must be comparable for there to be a single measure of resistance for a system. Such comparisons are sometimes straightforward. If one perturbation eradicates half of species x in a community, for instance, another that eradicates 75 percent of x is certainly stronger. If another perturbation eradicates 25 percent of 3 species or 5 percent of 15 species in the community, however, it is unclear how its strength should be ranked against the perturbation that eradicates 75 percent of x. What criteria could be used to compare strengths of P_v, P_p, or P_{vp} perturbations, to which systems may show differential sensitivity, is even less clear. Systems that are highly resistant to P_v perturbations may be extremely sensitive to even slight P_p. Comparing the resistance of communities is therefore only unproblematic with respect to perturbations of comparable kind.

Resilience is the rate a system returns to R following perturbation (P_v, P_p, or P_{vp}). Like resistance, resilience must be assessed against perturbation strength unless, although unlikely for most types of perturbation, return rate is independent of perturbation strength. Slow return rates after weak perturbations indicate low resilience and rapid rates following strong perturbations indicate high resilience. Systems may not return to R after perturbation, especially following severe perturbation, so, unlike resistance, resilience is only assessable for perturbations that do not prevent return to R. Note that resilience and resistance are independent concepts: systems may be drastically changed by weak perturbations (low resistance) but rapidly return to R (high resilience), and vice versa.

Resilience is commonly measured as the inverse of the time taken for the effects of perturbation to decay relative to R. For a specific mathematical model, this can be determined analytically or by simulation. For the community described by equation (2) above, for instance, resilience to a P_v perturbation that eradicates half of one species could be simply measured by, $|t_{eq} - t_p|$, where t_p is the time at which the community is initially perturbed and t_{eq} is the time at which the community reestablishes equilibrium. Resilience to P_v perturbation is determined by the largest real eigenvalue part for systems modeled by linear differential equations, and analytic methods have been developed to assess resilience to P_v perturbation for nonlinear models. Empirical measurement of resilience for communities in nature, however, is often thwarted by subsequent perturbations that disrupt return to R. This difficulty can be avoided if subsequent perturbations can be evaded with controlled experiments. If the return rate is independent of perturbation strength, estimation of resilience is also more feasible because only the decay rate of the perturbation effects need be measured before the system is further perturbed; measurement of perturbation strength is not required (Pimm, 1984). Like resistance, furthermore, different types of perturbations yield different measures of resilience since return rate to R may depend upon the way in which systems are perturbed. A system may be highly resilient to P_v perturbation and poorly resistant to P_p perturbation, for instance, or more resilience to some P_v or P_p perturbations than others.

Tolerance, or "domain of attraction" stability, is the ability of a system to be perturbed and return to R, regardless of how much it may change and how long its return takes. More precisely, tolerance is positively correlated with the range and strength of perturbations a system can sustain and still return to R. The magnitudes of the strongest perturbations it can sustain determine the contours of this range. Note that tolerance is conceptually independent of resistance and resilience: a system may be severely perturbed and still return to R (high tolerance), even if it changes considerably (low resistance) and its return rate is slow (low resilience), and vice versa.

Similar to resistance and resilience, different kinds of perturbations yield different measures of tolerance. Tolerance to P_v perturbations, for instance, is determined by the maximal changes variables can bear and not jeopardize the system's return to R. With respect to P_v perturbations that affect only one species of a community, for instance, tolerance can be simply measured by the proportion of that species that can be eradicated without precluding the community's return to R. If a nontrivial equilibrium of equation (2) from above is globally stable, for instance, the community described by the equation is maximally tolerant to P_v perturbations relative to this reference state because the community will return to it after any P_v perturbation that does not eradicate one of the species. Variables of a system may be perturbed, however, in other ways. A P_v perturbation may change all the variables, several, or only one; it may change them to the same degree, some variables more severely than others, and so on. How exactly variables are perturbed may affect whether the system returns to R. System tolerance must therefore be evaluated with respect to different types of perturbation. The same goes for assessing tolerance to P_p or P_{vp} perturbations. Note that local asymptotic Lyapunov stability corresponds to tolerance to P_v perturbation in the infinitesimal neighborhood of an equilibrium, and global asymptotic Lyapunov stability corresponds to tolerance to any P_v perturbation (see Section 3).

Although resistance, resilience, and tolerance do not adequately explicate ecological stability individually, they do so collectively. In fact, they constitute jointly sufficient and separately necessary conditions for ecological stability, notwithstanding Shrader-Freschette and McCoy's (1993, p.58) claim that such conditions do not exist. Consider sufficiency first. Since these three concepts represent the properties underlying conditions (A2)–(A4), communities exhibiting them to a high degree would change little after strong perturbations ([A2]), return to R rapidly if perturbed from it ([A3]), and return to R following almost any perturbation ([A4]). As such, these three properties certainly capture ecologists' early conceptions of ecological stability (see Sections 2 and 3), and there does not seem to be any further requirement of ecological stability that a community exhibiting these properties would lack.

Each concept is also necessary. Highly tolerant and resistant communities, for instance, change little and return to R after most perturbations. In regularly perturbing environments, however, even a highly resistant and tolerant community may be iteratively perturbed to the boundary of its tolerance range and "linger" there if its return rate to R is too slow. Subsequent perturbations may then displace it from this range, thereby precluding return to R. If this community rapidly returned to R after most perturbations (high resilience), it would rarely reach and would not linger at its tolerance boundary. In general, low resilience preserves the effects perturbations have on communities for extended, perhaps indefinite durations, which seems incompatible with ecological stability.

Similar considerations show that tolerance and resistance are necessary for ecological stability. A highly resilient and tolerant but weakly resistant community rapidly returns to R following almost any perturbation, but changes significantly after even the slightest perturbation, which seems contrary to ecological stability. The dramatic fluctuation such communities would exhibit in negligibly variable environments is the basis for according them low ecological stability. A highly resilient and resistant but weakly tolerant community changes little and rapidly returns to R when perturbed within its tolerance range, but even weak perturbations displace it from this range and thereby preclude its return to R, which also seems contrary to ecological stability.

Compared with resistance, resilience, and tolerance, constancy is a fundamentally different kind of concept. Unlike them, it is not defined in terms of response to perturbation, and thus violates adequacy condition (A1). Rather, constancy of a biological community is typically defined as a function of the variances and/or covariances in species biomasses. Tilman (1996, 1999; Lehman & Tilman, 2000), for instance, defined ecological stability as "temporal stability" (S_t):

$$S_t(C)=\frac{\sum_{i=1}^{n}\overline{B_i}}{\sqrt{\sum_{i=1}^{n}Var(B_i)+\sum_{\substack{i=1;j=1\\ i\neq j}}^{n}Cov(B_i,B_j)}}; \tag{5}$$

where C designates an n-species communities; B_i is a random variable designating the biomass of species i in an n-species community C assayed during some time period, and let $\overline{B_i}$ designate the expected value of B_i for the time period it is assayed; *Var* designates variance; and *Cov* designates covariance. The motivating intuition for this definition is the idea that if two series of abundances are plotted across time, the more stable one exhibits less fluctuation (Lehman & Tilman, 2000). In particular, if a community becomes more variable as judged by variances and covariances between its biomasses, regardless of what causes the variability, the denominator of (5) increases and its temporal stability decreases. One counterintuitive feature of this definition is that a community could become more stable solely because mean biomasses increase. This problem is easily avoided by measuring constancy strictly in terms of biomass variability [e.g., reformulating (5) with a numerator of 1], but this does not circumvent the fundamental difficulty that defining ecological stability as constancy does not satisfy (A1), irrespective of how constancy is measured. Contrary to the intuition motivating this definition , however, constancy is neither necessary nor sufficient for ecological stability. It is insufficient because a community may exhibit great constancy if unperturbed, but change dramatically if it were even weakly perturbed. Unless constancy is a result of how a community responds to perturbation, it can mask extreme sensitivity to perturbation, which is incompatible with ecological stability. By itself, therefore, constancy is not a reliable indicator of ecological stability.

Constancy is not necessary for ecological stability for two reasons. First, although a highly stable community at equilibrium remains relatively constant, it may fluctuate if subject to severe perturbations, perturbations that would drastically modify or eradicate weakly stable communities. It would be unjustifiable to regard these fluctuations as evidence of low ecological stability. The problem is that because constancy is not defined relative to perturbation response, it cannot distinguish between fluctuations that are a consequence of strong perturbation, which are consistent with ecological stability, and those that reveal susceptibility to weak or moderate perturbations, which are incompatible with ecological stability.

The second reason constancy is not necessary is that a highly stable system will not be constant if R is not an equilibrium reference state. Communities may be highly resistant, resilient, and tolerant with respect to regular limit cycles or more complicated attractor dynamics. In this case, the community changes little relative to the limit cycle or attractor after strong perturbation, rapidly returns to the limit cycle or attractor after strong perturbation, and returns to the limit cycle or attractor even after severe perturbation. Lack of constancy of such a community does not detract from the fact that any adequate conception of ecological stability should judge it highly stable relative to R.

Table 18.1 Different concepts of ecological stability. R designates a reference state or dynamic. P_v designates perturbations to system variables; P_p designates perturbations to parameters; and P_{vp} designates perturbations that affect both variables and parameters

	Type	Definition	Properties
Perturbation-based stability types	Resilience	Rate a system returns to R following P_v, P_p, or P_{vp}	Comparative concept
	Resistance	Inverse of the magnitude a system changes relative to R following P_v, P_p, or P_{vp}	Comparative concept
	Tolerance	Range of P_v, P_p, or P_{vp} a system can sustain and still return to R	Comparative concept
	Local asymptotic stability	A system returns to R following "small" P_v	(i) Dichotomous concept (ii) Special case of tolerance to p_v
	Global asymptotic stability	A system returns to R following any P_v	(i) Dichotomous concept (ii) Special case of tolerance to p_v
Perturbation-independent stability types	Constancy	Inverse of the variability of a system	Comparative concept

Most classifications of ecological stability include an additional non-perturbation-based stability concept, persistence: the time a community remains in R irrespective of whether or not it is perturbed. Retaining a particular species composition or biomasses within delimited ranges are typical reference states for gauging persistence. Persistence is usually measured by how long they do, or are predicted to, exhibit these states. It could, for instance, be measured by the time minimum population levels have been sustained (e.g., nonzero levels), or will be sustained based on predictions from mathematical models (Orians, 1975). As such, persistence is in fact only a special case of constancy measured in terms of the time R has been or will be exhibited, rather than variability with respect to R.

Table 18.1 presents a taxonomy of the stability concepts: resistance, resilience, tolerance, constancy, and local and global Lyapunov stability. The taxonomy classifies these concepts into two general categories, defines each, and lists some of their properties.

It is worth pausing over what the framework for ecological stability presented above shows about the general concept. It certainly shows that ecologists have used the term "stability" to describe several distinct features of community dynamics, although only resistance, resilience, and tolerance adequately define ecological stability. This plurality does not manifest, however, an underlying vagueness, "conceptual incoherence," or "inconsistency" of the concept, as Shrader-Frechette and McCoy (1993, p.57) suggest in their general critique of basic ecological concepts and ecological theories based on them. Two claims seem to ground their criticism. First, that if "stability" is used to designate distinct properties, as it has been in the ecological literature, this indicates the concept is itself conceptually vague and thereby flawed. Although terminological ambiguity is certainly undesirable, most ecologists unambiguously used the term to refer to a specific property of a community and accompanied the term with a precise mathematical or empirical operationalization (see Sections 2–4). Since these were in no sense vague, in no sense was ecological stability "vaguely defined" (Shrader-Frechette & McCoy, 1993, p.40). Ecologists quickly appreciated this terminological ambiguity, moreover, and began explicitly distinguishing different senses of ecological stability with different terms (Odenbaugh, 2001). Lewontin's (1969) review was the first example, and subsequent analyses of the concept did not jettison this insight.

Shrader-Frechette and McCoy's second claim is that, "There is no homogeneous class of processes or relationships that exhibit stability" (1993, p.58). The assumption underlying this claim seems to be that concepts in general, ecological stability in particular, must refer to a homogeneous class to be conceptually unproblematic. That ecological stability does not, and worse, that ecologists have supposedly attributed inconsistent meanings to it, shows that the concept is incoherent, they believe, much like the vexed species concept (1993, p.57).

Shrader-Freschette and McCoy do not offer an argument for this assumption, and it is indefensible as a general claim about what concepts must refer to. Common concepts provide clear counterexamples. The concepts "sibling," "crystal," and "field," for instance, refer to heterogeneous classes, but there is nothing conceptually problematic about them. There is debate about the idea of disjunctive *properties* in work on multiple realization (Fodor, 1974; Kim, 1998; Batterman, 2000), but the criticisms raised there against disjunctive properties do not necessarily apply to disjunctive *concepts*, nor were they intended to. Kim (1998, p.110) emphasizes this point:

> Qua property, dormativity is heterogeneous and disjunctive, and it lacks the kind of causal homogeneity and projectability that we demand from kinds and properties useful in formulating laws and explanations. But [the concept of] dormativity may well serve important conceptual and epistemic needs, by grouping properties that share features of interest to us in a given context of inquiry.

Even if criticisms of disjunctive properties were sound, it therefore would not follow that the disjunctive concepts such as ecological stability are also problematic. The conceptual and epistemic utility of a concept is enhanced, furthermore, if there are clear guidelines for its application. The preceding analysis attempts to provide such guidelines.

Moreover, the definitional statuses of the concepts of ecological stability and species are not analogous. Biologists have proposed plausible, but incompatible competing definitions for the species concept because it is problematically ambiguous (Ereshefsky, 2001). That resistance, resilience, and tolerance have been referred to under the rubric "stability," however, does not show that ecological stability is similarly problematically ambiguous because they are conceptually independent and therefore compatible, as different senses of "species" are not. As classifications of different stability concepts attest (e.g., Lewontin, 1969; Orians, 1975; Pimm, 1984), most ecologists recognized that there are several senses of ecological stability, and individual stability concepts were rarely proposed as *the* uniquely correct definition of ecological stability. Rather, they were and should be understood as distinct features of ecological stability, not competing definitional candidates. Like many scientific concepts, ecological stability is multifaceted, and the distinct referents ecologists attributed to it accurately reflect this. Conceptual multifacetedness alone does not entail conceptual incoherence or inconsistency.

6. Measures of Ecological Diversity and Complexity

Compared to the concept of ecological stability, little attention was devoted to clarifying the concept of ecological diversity or complexity in the early SDC debate. Significant disagreement about how the concept of diversity should be measured, however, emerged during this period (Magurran, 1988). By the early 1970s, moreover, enough attention was being devoted to common indices of diversity to spark criticism, perhaps the most incisive from Hurlbert (1971) (see below).

Conceptually, there was wide agreement among ecologists at the time that diversity has two main components: species richness and evenness (see Section 1). Distinct quantitative diversity indices result from different ways of quantifying and integrating these two notions. For a clearer understanding of these concepts, consider two simple communities, A and B, both composed of two species s_1 and s_2. A and B have the same species richness. If the percentages of individuals distributed among the two species are 0.02 percent and 99.98 percent for A and 50 percent and 50 percent for B, respectively, B seems more diverse than A. This can be represented by a higher evenness value for B. Besides that a diversity index should increase with species richness, therefore, another reasonable adequacy condition seems to be that it should increase with evenness. Many distinct quantitative indices, however, satisfy these two adequacy conditions.

Probably the most popular index of community diversity, then and now, is the Shannon index (H) (see equation [1], Section 2). The index was originally intended to quantify the amount of information in a communicated message (Shannon & Weaver, 1949). Good (1953) and Margalef (1958) were the first to use it as an index of diversity. In the ecological context, p_i designates the proportion of individuals in the i-th species of a community, so that H is at its maximal value for a given species richness n ($H = \ln(n)$) when the individuals are equally distributed among the species.

Another common diversity index (D) is the complement of Simpson's (1949) "measure of concentration":

$$1-\sum_{i=1}^{n} p_i^2. \tag{6}$$

Simpson (1949) explained that his concentration index represents the probability that two individuals chosen at random (with replacement) from a community will belong to the same species, so D represents the probability the two individuals will belong to different species. This probability, like Shannon's index, is at its maximal value for a given species richness n ($D=1-\frac{1}{n}$) when individuals are equally distributed among the species. D is more sensitive to the abundances of species and less sensitive to species richness than Shannon's index (May, 1975; Magurran, 1988). Hurlbert (1971) later proposed, furthermore, that if D is multiplied by $\frac{N}{N-1}$, the resulting index represents the probability of interspecific encounter in the community.

By the late 1960s, a large number of diversity indices had been developed, and numerous empirical studies of different ecological systems were being conducted to estimate diversity using these indices (Pielou, 1975; Magurran, 1988; Sarkar, 2007). In an influential critique of this research agenda, Hurlbert (1971, p.577) argued that, "the term 'species diversity' has been defined in such various and disparate ways that it now conveys no information other than 'something to do with community structure'," and that this indicated a fundamental vagueness of the underlying concept. He thought ecologists had further exacerbated this problem by appropriating statistical measures of diversity developed in nonbiological contexts with dubious ecological relevance. Rather than attempt to rehabilitate the concept by proposing adequacy conditions by which to evaluate the relative merits and weaknesses of different indices, Hurlbert suggested that the search for stability–diversity relationships should be refocused on the relationship between community stability and indices that reflect biologically meaningful properties that might influence community dynamics. His index of the probability of interspecific encounter is one example. Species richness seems to fail this test since it is generally unlikely that extremely rare species (e.g., s_1 in community A above) play an important role in community dynamics. Species richness was and remains, however, the predominant surrogate for diversity in analyses of stability–diversity relationships (e.g., Tilman, 1996, 1999).

Since Hurlbert's critique, ecologists have proposed a multitude of new diversity indices to satisfy plausible adequacy conditions besides those about species richness and evenness (see Ricotta, 2005 for a short review). Diversity indices should increase, for instance, as interspecific taxonomic and functional differences in a community increase. Besides properties of species, spatial properties of their geographical distribution could also be included in a diversity index. Since species distributions are significantly influenced by regional geology and environmental gradients, however, including these properties would expand the scope of diversity indices beyond measuring just the biological properties of communities. Expanded in this way, the "diversity" of the physical environment in which a community resided would also contribute to the value of such indices.

How compatible these additional adequacy conditions are with one another, or with the other conditions is not yet clear. Some conditions appear to be conceptually independent, but some formal diversity indices suggest that others are not. Rao's (1982) "quadratic entropy" diversity index, for instance, which generalizes the Simpson index (Ricotta & Avena, 2003), incorporates interspecific taxonomic and functional differences as well as evenness and species richness into a single quantitative measure. Unlike the Shannon and Simpson indices, however, quadratic entropy violates the adequacy condition that diversity should be maximal for a given species richness when individuals are equally distributed among species (Ricotta, 2005). This is as it should be. If functional or taxonomic information is included in assessments of diversity, then high functional or taxonomic diversity may make a less even community more diverse overall than a more even one. In effect, functional or taxonomic diversity can trump evenness. As new indices are devised, similar incompatibilities between other adequacy conditions may be revealed. Absent a general proof that the conditions themselves are incompatible, however, it remains possible there is a diversity index that satisfies all defensible adequacy conditions.

In evaluating what properties of communities may make them stable, focusing solely on diversity is unjustifiable because intra- and inter-specific dynamics largely determine a community's stability properties. As a function of properties of individual organisms in a community, such as their taxonomic classes, how they are distributed among these classes, etc., even biologically meaningful diversity indices may reveal little about community dynamics. Individuals in species-rich communities with high evenness and taxonomic variety may interact rarely and weakly (intra- and inter-specifically); the former entails nothing about the latter. Hurlbert's (1971) claim that the modified complement of Simpson's index measures the probability of interspecific encounter, for instance, is true only if individuals of different species meet in proportion to their relative abundances. The likelihood may be higher, of course, that species in more even communities will interact more frequently, but the latter cannot be inferred from the former alone. A high or low diversity does not reveal, moreover, anything about how strongly species interact. May (1974) may have focused on complexity rather than diversity for precisely these reasons.

Unlike diversity, complexity is defined in terms of community dynamics. The more species, the more frequently they interact, and the stronger they do, the more complex the community. As a function of intra- and inter-specific dynamics, complexity can only be assessed against a description of these dynamics, usually in the form of a mathematical model (see Section 5). With respect to the simple linear models analyzed by May (1974), for instance, complexity is a function of species richness (n), connectance (C), and mean linear interaction strength (s) (see Section 4). How complexity should be assessed for more complicated nonlinear models, however, is unclear. Determining species richness is obviously unproblematic, and connectance can be determined from functional dependencies between variables in the model. The problem is assessing mean interaction strength. For linear models, the growth rate of each species is a linear function of the abundances of species with which they interact, so s is simply the average of the interaction coefficients. Variables in nonlinear models, however, may interact in disparate ways, and that they may exhibit different functional relationships precludes simply averaging to determine s. Different methods of integrating strengths of distinct types of relationships into a single quantitative complexity value, assuming there is a defensible way of doing this, would beget different measures of complexity.

Restricting complexity to just n, C, and s is also unduly restrictive. The variety of relationships exhibited between variables, the number of parameters, how complicated their relations are with variables, and other properties of community models that represent important features of community dynamics should be part of any defensible measure of ecological complexity. Whether they can be codified into a general complexity index remains to be seen. Without such a codification, however, the question of whether there is a relationship between ecological stability and complexity is poorly formed.

7. Evaluating Stability–Diversity–Complexity Relationships

Given the multitude of ways a biological community can be represented (M), its reference state specified (R), the variety of ecological stability concepts (see Section 5), and the numerous potential operationalizations of ecological diversity and complexity (see Section 6), it is unsurprising that the SDC debate remains unresolved after half a century of ecological research. As might be expected, only a few of the total possible relationships between stability, complexity, and diversity have been analyzed. Specifically, relationships between local stability and complexity of linear models, and between species richness and constancy have been the predominant focus thus far. The tractability of these concepts compared with the general concepts of stability, diversity, and complexity accounts for the selective scrutiny.

Empirically and theoretically oriented ecologists have taken disparate approaches to the debate (e.g., Elton, MacArthur, and May). Theoreticians focus on mathematical models of biological communities, and since these models describe the dynamics of communities, their research has been primarily concerned with the role complexity, rather than diversity, may have in generating or prohibiting stability. Food web models are perhaps the most common type of community model. They usually represent interspecific species interactions by linear relationships between variables, which has two advantages. First, assessing the complexity of these models is straightforward (see Section 6). Second, a well-developed mathematical theory of Lyapunov stability applies to these models, so evaluating their stability seems to be similarly straightforward. A large body of work on food webs since May's (1974) influential monograph has subsequently exploited these facts to uncover properties of community structure that might produce Lyapunov stability (see McCann, 2005 for a short review).

These facts obviously encourage this approach, but at the expense of ecological applicability. The problem is that actual species interactions probably rarely take a linear form, which means food webs poorly represent the dynamics of real-world communities. Species interactions can be treated as linear in the infinitesimal neighborhood of an equilibrium, but this severe restriction precludes inference about how actual communities respond to perturbation (see Section 3). What structural properties of food webs generate or jeopardize Lyapunov stability therefore indicate little about the relation between stability and complexity in actual biological communities. For this reason, Hastings (1988, p.1665) warned, "food web theory is not an adequate approach for understanding questions of stability in nature." It certainly tempers claims that recent advances in food web theory suggest a resolution of the SDC debate (e.g., McCann, 2000).

Evaluating stability–complexity relationships for nonlinear models that more accurately represent community dynamics presents different difficulties. One is the challenge of integrating different types of model properties into a general measure of complexity discussed in Section 6. Another is evaluating the stability of these models. If, as commonly assumed, Lyapunov stability adequately defines ecological stability, the ecological stability of nonlinear community models can be assessed with standard analytic techniques (Hirsh & Smale, 1974). An unequivocal relationship between Lyapunov stability and model complexity (gauged informally), however, has not emerged. Realistic increases in model complexity sometimes decrease the likelihood of stability, but sometimes they increase it. May (1974), for instance, found that including environmental stochasticity and realistic time delays in Lotka–Volterra models of predator–prey communities destabilized them, while incorporating spatial heterogeneity and more realistic predator response to prey were stabilizing (see Section 4). The more realistic and complex the community model, moreover, the less mathematically tractable it and the assessment of its stability properties become. Since no general method for evaluating the Lyapunov stability of nonlinear systems is known, the ultimate verdict on the relationship between Lyapunov stability and complexity remains unclear.

The assumption that ecological stability is adequately defined as Lyapunov stability, however, should be rejected. The reasons it should not are technically involved (see Justus [in press] for details), but the general idea is that Lyapunov stability only considers response to P_v perturbations, whereas ecological stability also concerns response to P_p and P_{vp} perturbations. As such, Lyapunov stability provides only a partial account of the kind of perturbation response ecological stability requires. Unfortunately, a mathematical theory of stability for this more general class of perturbations has not yet been developed within ecology.

Besides these difficulties, estimating parameters of community models with empirical data, especially of more complicated realistic models, is often practically impossible. It is even less likely, furthermore, that the dynamics of a community can be easily discerned from the limited data usually available (see Connell & Sousa, 1983 for a review of methodological problems). For these reasons, some ecologists have decided to focus instead on evaluating stability–diversity–complexity relationships with statistical measures derived from data. This strategy also faces difficulties. The most daunting is the lack of adequate data. Its absence explains why species richness, which is a poor surrogate for ecological diversity, is used in almost all studies of stability–diversity relationships: data on relative species abundances required to estimate the evenness of a community can rarely be collected, whereas the numbers of species often can. For instance, most of David Tilman's analyses of Minnesota grasslands (e.g., Tilman, 1996, 1999; Lehman & Tilman, 2000), which are probably the most spatially and temporally extensive empirical studies of stability–diversity relationships, measure diversity as species richness (a recent study by Tilman et al. [2006] uses the Shannon index to measure the diversity of a Minnesota grassland). His finding of a positive correlation between grass species richness and temporal stability [see (5) above] therefore shows little about the relationship between diversity and ecological stability, because species richness is a poor surrogate for diversity.

Absence of adequate data also explains the prevalence of constancy as the measure of ecological stability in most empirical studies of the SDC debate, temporal stability being the most prominent example. Lehman and Tilman (2000, p.535) suggest that temporal stability is "readily observable in nature," as one reason for using it to measure ecological stability. It only requires, specifically, data on species biomasses. This contrasts, they suggest, with the infeasibility of empirical evaluation of the stability concepts common in mathematical modeling in ecology.

Facile measurability, however, does little to overcome the shortcomings of *defining* ecological stability as constancy (see Section 5). If, alternatively, temporal stability were merely proposed as a *measure* of ecological stability, its suitability depends upon: (i) whether (and how variously) the community is perturbed during the period temporal stability is assessed, so that the measure indicates how the community responds to perturbation; and (ii) whether the reference state (R) is an equilibrium, so that species biomasses will remain constant if unperturbed. Communities exhibiting limit cycles and "strange attractors," for instance, violate (ii); a community that resides in a relatively unvarying, unperturbing environment over the period it is assayed violates (i). Since data are usually collected over short periods given the limited monetary and temporal constraints of ecological research, (i) is rarely satisfied, and it has been suggested that complex dynamics violating (ii) may be widespread. For this and the reasons discussed above, Tilman's (1996, 1999) careful analysis of how species richness can increase temporal stability should probably not be considered as even, "a partial resolution of the long-standing diversity–stability debate" (Lehman & Tilman, 2000, p.548).

This survey of the SDC debate helps explain why it remains unresolved. The concepts of ecological stability, diversity, and complexity are multifaceted and difficult to evaluate, theoretically and empirically, and there is significant disagreement among ecologists about how they should be defined. Consequently, only a few of the total possible stability–diversity and stability–complexity relationships have been analyzed. These analyses, moreover, have focused on relations between tractable, but poor surrogates for the three concepts. The debate's broad scope and potential implications for applied biological fields such as conservation, pest control, and resource management ensure its continued scientific scrutiny. The conceptual and methodological issues it raises merit more philosophical scrutiny than thus far devoted to it.

Acknowledgments

Thanks to Anya Plutynski, Carl Salk, and Sahotra Sarkar for helpful comments.

References

Batterman, R. W. (2000). Multiple realizability and universality. *British Journal for the Philosophy of Science*, 51, 115–145.

Chesson, P. L., & Case, T. J. (1986). Overview: nonequilibrium community theories: chance, variability, history, and coexistence. In J. Diamond & T. J. Case (Eds). *Community ecology* (pp. 229–239). New York: Harper & Row.

Connell, J. H., & Orias, E. (1964). The ecological regulation of species diversity. *American Naturalist*, 98, 399–414.

Connell, J. H., & Sousa, W. P. (1983). On the evidence needed to judge ecological stability or persistence. *American Naturalist*, 121, 789–833.

De Angelis, D. L. (1975). Stability and connectance in food web models. *Ecology*, 56, 238–43.

Elton, C. (1958). *The ecology of invasions by animals and plants.* London: Methuen.

Ereshefsky, M. (2001). *The poverty of the Linnaean hierarchy.* Cambridge: Cambridge University Press.

Fodor, J. (1974). Special sciences, or the disunity of science as a working hypothesis. *Synthese*, 28, 77–115.

Gardner, M. R., & Ashby, W. R. (1970). Connectance of large dynamic (cybernetic) systems: critical values for stability. *Nature*, 228, 784.

Gause, G. F. (1934). *The struggle for existence.* Baltimore: Williams & Wilkins.

Good, I. J. (1953). The population frequencies of species and the estimation of population parameters. *Biometrika*, 40, 237–64.

Goodman, D. (1975). The theory of diversity–stability relationships in ecology. *Quarterly Review of Biology*, 50, 237–66.

Grimm, V., & Wissel, C. (1997). Babel, or the ecological stability discussions: an inventory and analysis of terminology and a guide for avoiding confusion. *Oecolgia*, 109, 323–34.

Hastings, A. (1988). Food web theory and stability. *Ecology*, 69, 1665–8.

Hastings, A., Hom, C. L., Ellner, S., Turchin, P., & Godfray, H. C. J. (1993). Chaos in ecology: is Mother Nature a strange attractor? *Annual Review of Ecology and Systematics*, 24, 1–33.

Hirsh, M., & Smale, S. (1974). *Differential equations, dynamical systems, and linear algebra.* San Diego, CA: Academic Press Inc.

Hurlbert, S. (1971). The nonconcept of species diversity: a critique and alternative parameters. *Ecology*, 52, 577–86.

Hutchinson, G. E. (1959). Homage to Santa Rosalina or why are there so many kinds of animals? *American Naturalist*, 93, 145–59.

Hutchinson, G. E. (1957). Concluding remarks. *Cold Spring Harbor Symposia on Quantitative Biology*, 22, 415–27.

Justus, J. (in press). Ecological and Lyapunov stability. *Philosophy of Science.*

Kerner, E. H. (1957). A statistical mechanics of interacting biological species. *Bulletin of Mathematical Biophysics*, 19, 121–46.

Kerner, E. H. (1959). Further considerations on the statistical mechanics of biological associations. *Bulletin of Mathematical Biophysics*, 21, 217–55.

Kim, J. (1998). *Mind in a physical world: an essay on the mind–body problem and mental causation.* Cambridge, MA: MIT Press.

Kingsland, S. (1995). *Modeling nature.* Chicago: University of Chicago Press.

Krebs, C. (2001). *Ecology.* New York: Benjamin Cummings.

Lawlor, L. R. (1978). A comment on randomly constructed model ecosystems. *American Naturalist*, 112, 445–7.

Lehman, C. L., & Tilman, D. (2000). Biodiversity, stability, and productivity in competitive communities. *American Naturalist*, 156, 534–52.

Leigh, E. G. (1975). Population fluctuations, community stability, and environmental variability. In M. L. Cody & J. M. Diamond (Eds). *Ecology and evolution of communities* (pp. 52–73). Cambridge, MA: Harvard University Press.

Lewontin, R. (1969). The meaning of stability. In G. Woodwell & H. Smith (Eds). *Diversity and stability in ecological systems* (pp. 13–24). Brookhaven, NY: Brookhaven Laboratory Publication No. 22.

Lyapunov, A. ([1892], 1992). *The general problem of the stability of motion.* London: Taylor & Francis.

MacArthur, R. (1955). Fluctuations of animal populations, and a measure of community stability. *Ecology*, 36, 533–6.

Margalef, R. (1958). Information theory in ecology. *General Systems*, 3, 36 –71.

Magurran, A. E. (1988). *Ecological diversity and its measurement.* Princeton, NJ: Princeton University Press.

McCann, K. (2000). The diversity–stability debate. *Nature*, 405, 228–33.

McCann, K. (2005). Perspectives on diversity, structure, and stability. In K. Cuddington & B. Beisner (Eds). *Ecological paradigms lost* (pp. 183–200). New York: Elsevier Academic Press.

McIntosh, R. P. (1985). *The background of ecology: concept and theory.* Cambridge, UK: Cambridge University Press.

May, R. M. (1974). *Stability and complexity in model ecosystems* (2nd edn). Princeton, NJ: Princeton University Press.

Norton, B. G. (1987). *Why preserve natural variety?* Princeton, NJ: Princeton University Press.

Odenbaugh, J. (2001). Ecological stability, model building, and environmental policy: a reply to some of the pessimism. *Philosophy of Science* (Proceedings), 68, S493–S505.
Orians, G. H. (1975). Diversity, stability, and maturity in natural ecosystems. In W. H. van Dobben & R. H. Lowe-McConnell (Eds). *Unifying concepts in ecology* (pp. 139–50). The Hague: W. Junk.
Patten, B. C. (1961). Preliminary method for estimating stability in plankton. *Science*, 134, 1010–11.
Pfisterer, A. B., & Schmid, B. (2002). Diversity-dependent production can decrease the stability of ecosystem functioning. *Nature*, 416, 84–6.
Pielou, E. C. (1975). *Ecological diversity*. New York: John Wiley & Sons.
Pimentel, D. (1961). Species diversity and insect population outbreaks. *Annals of the Entomological Society of America*, 54, 76–86.
Pimm, S. L. (1979). Complexity and stability: another look at MacArthur's original hypothesis. *Oikos*, 33, 351–7.
Pimm, S. L. (1984). The complexity and stability of ecosystems. *Ecology*, 61, 219–25.
Pimm, S. L. (1991). *The balance of nature?* Chicago: University of Chicago Press.
Preston, F. W. (1968). On modeling islands. *Ecology*, 49, 592–4.
Preston, F. W. (1969). Diversity and stability in the biological world. In G. Woodwell & H. Smith (Eds). *Diversity and stability in ecological systems* (pp. 1–12). Brookhaven, NY: Brookhaven Laboratory Publication No. 22.
Rao, C. R. (1982). Diversity and dissimilarity coefficients: a unified approach. *Theoretical Population Biology*, 21, 24–43.
Ricotta, C. (2005). Through the jungle of biological diversity. *Acta Biotheoretica*, 53, 29–38.
Ricotta, C., & Avena, G. C. (2003). An information-theoretical measure of taxonomic diversity. *Acta Biotheoretica*, 51, 35–41.
Sarkar, S. (2007). From ecological diversity to biodiversity. In D. Hull & M. Ruse (Eds). *Cambridge companion to the philosophy of biology*. Cambridge, MA: Cambridge University Press.
Shannon, C. E., & Weaver, W. (1949). *The mathematical theory of communication*. Urbana: University of Illinois Press.
Shrader-Frechette, K. S., & McCoy, E. D. (1993). *Method in ecology*. Cambridge: Cambridge University Press.
Simpson, E. H. (1949). Measurement of diversity. *Nature*, 163, 688.
Slobodkin, L. B. (1953). An algebra of population growth. *Ecology*, 34, 513–19.
Slobodkin, L. B. (1958). Meta-models in theoretical ecology. *Ecology*, 39, 550–1.
Tilman, D. (1996). Biodiversity: population versus ecosystem stability. *Ecology*, 77, 350–63.
Tilman, D. (1999). The ecological consequences of biodiversity: a search for general principles. *Ecology*, 80, 1455–74.
Tilman, D., Reich, P. B., & Knops, J. M. H. (2006). Biodiversity and ecosystem stability in a decade-long grassland experiment. *Nature*, 441, 629–32.
Watt, K. E. F. (1973). *Principles of environmental science*. New York: McGraw-Hill.
Woodwell, G., & Smith, H. (Eds). (1969). *Diversity and stability in ecological systems*. Brookhaven, NY: Brookhaven Laboratory Publication No. 22.

Further Reading

Cooper, G. (2001). Must there be a balance of nature? *Biology and Philosophy*, 16, 481–506.
Egerton, F. N. (1973). Changing concepts of the balance of nature. *Quarterly Review of Biology*, 48, 322–50.

Ives, A. R. (2005). Community diversity and stability: changing perspectives and changing definitions. In K. Cuddington & B. Beisner (Eds). *Ecological paradigms lost* (pp. 159–182). New York: Elsevier Academic Press.

King, A., & Pimm, S. (1983). Complexity and stability: a reconciliation of theoretical and experimental results. *American Naturalist*, 122, 229–39.

May, R. M. (1985). The search for patterns in the balance of nature: advances and retreats. *Ecology*, 67, 1115–26.

McNaughton, S. J. (1977). Diversity and stability of ecological communities: a comment on the role of empiricism in ecology. *American Naturalist*, 111, 515–25.

Sarkar, S. (2005). Ecology. In E. N. Zalta (Ed.). *Stanford encyclopedia of philosophy*. URL=<http://plato.stanford.edu>.

Tilman, D., Lehman, C. L., & Bristow, C. (1998). Diversity–stability relationships: statistical inevitability or ecological consequence? *American Naturalist*, 151, 277–82.

Chapter 19

Ecosystems

KENT A. PEACOCK

The *ecosystem* is the central unifying concept in many versions of the science of ecology, but the meaning of the term remains controversial, and a few authors (e.g., Sagoff, 2003) question whether it marks any clear or non-arbitrary distinction at all. The following definitions will do as a fairly uncontroversial starting point: The terms "ecology" and "economics" themselves come from the Greek root *oikos*, meaning "household." *Ecology* is the branch of biology that deals with the ways in which living organisms organize themselves into dynamic structures that facilitate the exchange of energy, materials, and information between themselves and the larger physical and biological environments in which such structures are situated; while *ecosystems* themselves are, loosely speaking, the structures in question.

This chapter will begin with observations on the meaning and scope of ecology itself. It will then outline the ways in which ecosystems can be understood from a number of perspectives: the ecosystem as the *descriptive unit* of the working field biologist; the *history* of the concept of the ecosystem; the ecosystem as a *dissipative structure*; the ecosystem as *symbiotic association*; the ecosystem in evolutionary theory; and skeptical views according to which the ecosystem is little more than a *descriptive convenience*. The applications of these conceptions to environmental ethics and the problems of ecosystem health and sustainability will then be reviewed. Ecosystems are not merely of theoretical interest, for understanding them may make a critical difference to how successful we humans are in responding to the ecological crisis precipitated by the unprecedented impact we are currently having on the environment. In the end, the practical perspective must serve as our touchstone; the observations of field ecologists, working agronomists, foresters, soil scientists, and conservationists should temper our flights of theoretical fancy.

1. The Scope of Ecology

K. de Laplante (2004) argues that in recent years there have been two major ways to think about ecology itself, a narrowly orthodox approach and what de Laplante calls the "expansive" approach. The orthodox approach holds that ecology should concern itself largely or entirely with nonhuman communities of species, and that the value of ecology is primarily the prediction of the population dynamics of organisms. The expansive approach is more in the spirit of E. Haeckel's original (1869) definition of ecology as "the investigation of the total relations of the animal both to its inorganic and to its organic environment; including above all, its friendly and inimical relations with those plants and animals with which it comes directly or indirectly into contact – in a word, ecology is the study of those complex interrelations referred to by Darwin as the conditions of the struggle for existence" (Haeckel, 1869/1879, quoted in de Laplante, 2004, p.264). Both views accept the fact that even very subtle features of human culture (broadly understood to include our art, science, economy, religion, architecture, technology, and philosophies) could have ecological significance on a planetary scale (Peacock, 1999b); a stock market fluctuation, a change in communications technology, or the promulgation of a novel philosophical doctrine could trigger chains of cause and effect leading to dramatic disruptions of the nonhuman ecosystem. (A fashion trend in Europe in the period 1900–10 led to the extinction of the New Zealand huia bird; Day, 1989.) Because human activities, for better or worse, are so deeply entangled in the present functioning of the planetary system (E. Odum, 1971, p.36, has referred to "man the geological agent"), the notion of an ecological theory that could do even as much as predict animal population distributions and numbers, without taking into account the myriad ways in which human activity impacts nonhuman nature, seems naïve. The essential distinction between the orthodox view and the expansive view is that the latter is concerned not merely with population dynamics, but with all properties of organisms and communities of organisms insofar as they can be understood as consequences or features of their interactions with their physical and biotic environments. In particular, the expansive view understands virtually all aspects of human thought and activity as ecological in nature or implication, and opens the door to a rethinking and redirection of the whole human enterprise on ecological grounds. It is virtually impossible to make sense of most of the notions of the ecosystem that we shall review here without implicitly taking the expansive view of ecology.

It is understandable that many working ecologists have chosen to narrow their focus to matters about which one has a hope of making testable predictions, for a theory that says (as ecology is often taken to say) that "everything connects" and that wholes resist analysis into the interactions of parts risks falling into vacuity. However, it is just a brute fact that the living world is profoundly complex and abounding with interdependencies that resist tractable mathematical description. These barriers to scientific analysis are only compounded by the challenge of scale (some ecosystems span continents or the whole planet itself), the inseparability of the human observer from many biotic systems under study, and the fact that the operations of ecosystems must often be inferred from indirect observations. Thus the challenge for ecologists of all stripes is to arrive at accounts of ecological entities and processes that allow for the complexity, openness, and nonlinearity of ecological systems, but which are at the same time scientifically meaningful.

A number of authors have noted the relative lack of predictive power, especially quantitative predictive power, of ecosystems theory as compared with other branches of science such as chemistry or physics. Predicting the behavior of ecosystems suffers from many of the same difficulties as weather forecasting – nonlinearity, sensitive dependence on initial conditions, complexity, and our lack of full understanding of the dynamics. We cannot accurately model climate without modeling the earth system as a whole, but as with weather and climate forecasting, it is reasonable to hope (as Schneider and Sagan 2005 argue) that the ever-increasing power of computer modeling will allow more effective predictions, both qualitatively (will the icecaps melt?) and quantitatively (when will they melt?). As a quantitative science, ecology is in its infancy, but it has already yielded many qualitative insights that could make a material difference to the probability of the survival of the human species.

2. General Description of Ecosystems

Ecology began as branch of science driven by observations of communities of plants and animals and their interactions with their physical surroundings. In this loose sense even Aristotle was an ecologist. The pre-theoretical aspect of ecology as an observational practice must always remain the essential reference point for ecological theorizing (Odum, 1971). Relatively self-contained entities such as ponds are well studied and their properties suggested that groups of organisms in an approximately bounded physical setting tend to interact in such a way as to define a coherent entity. (See, e.g., Golley, 1993.)

We will first review some generally accepted terminology.

Populations are commonly defined as interbreeding groups of organisms of the same species. A *community* is a group of interacting populations, and it is usually identified relative to a geographical area. However, what counts toward defining communities, ecosystems, and symbiotic associations is causal connectivity, not merely physical proximity; a pod of great whales and their prey may be spread over thousands of square kilometers of ocean, and yet remain connected by underwater sound signals. E. Odum (1964, p.15) emphasized that "coordination at ecological levels involves communication across non-living space."

A *biome* is a grouping of communities in a specific climate region, and characterized usually by plant type; various desert or forest environments (such as the montane cloud forest) are typical biomes. The term *biosphere* has been used in more than one sense. Conventionally it is taken to mean the regions on or in the Earth where life is found, between the lowest ocean depths to the lower atmosphere. The term *biosphere* as introduced by V. I. Vernadsky (1926/1988) is a broader conception: it is, he said, the "surface that separates the planet from the cosmic medium" (1926/1988, p.43), a layer that extends down as far as the lithosphere.

An *ecosystem* or *environment* can be defined loosely as the combination of a community of organisms and the abiotic physical surrounding with which the organisms interact. This leaves open the question of how we identify those features of the abiotic world that constitute the environment for the community in question. This is difficult not only because of the complexity of the causal interactions, both direct and indirect, of life with its surroundings, but because many of the materials in an environment with which organisms interact were once living or are byproducts of life; for instance, carbonate minerals are mostly residues of long-ago marine organisms. The environment proper of a community could include the entire planet, and in this inclusive sense there is, strictly speaking, only one ecosystem.

One non-arbitrary way we can distinguish ecosystems from their surroundings is by the presence or absence of feedback. The CO_2 in the atmosphere both affects and is affected by the biota, but the ultraviolet (UV) flux at the top of the atmosphere can safely be viewed as an external influence because there is no reason to think that the solar output of UV is in any way affected by life on earth. Within the earth system proper, however, it is very difficult to find anything from the top of the stratosphere down to several kilometers into the crust that has not been to some degree affected causally by life.

Ecosystems are open in the sense that they both actively and passively exchange energy, materials, and information with their surroundings in a myriad of ways. (A passive process is driven by gradients such as temperature, pressure, or concentration, while active exchange is driven by expenditure of free energy by the organisms of the system and which can therefore run counter to gradients.) Ecosystems also exhibit periodicities and quasi-periodicities. Long before the term "ecosystem" was coined, biologists noted the phenomenon of *succession*, in which communities develop and apparently reach maturity in a *climax* community which may be approximately stable unless perturbed by outside forces. (Many ecologists today question the existence of climax communities, not only because the notion smacks of teleology, but because it may simply not be the case that ecosystems, particularly vigorous ecosystems, always or even often attain a long-running dynamic equilibrium; Sarkar, 2005a.) There are ecosystems within ecosystems, but many of the characteristic features of ecosystems are, as argued by Odum (1964, 1971), scale-invariant – another factor that makes the concept useful.

Ecosystems are powered by the *autotrophs*, which are photosynthetic or chemosynthetic organisms which derive energy from inorganic sources such as the sun, geothermal sources, or various inorganic chemical reactions. In their relations to the earth system, they are the producers, since they trap the free energy used by all other organisms. The consumers are the *heterotrophs* (including humans), who require organic sources of energy to survive. As will be discussed later in more detail, the sharp distinction between producers and consumers is misleading; it is also tempting to think of the heterotrophs as parasitic upon the autotrophs, but this, too, can be a mistake.

3. History of the Term "Ecosystem"

The term *ecology* predates the term *ecosystem*. As noted above, the discipline of ecology was founded, at least in name, by Haeckel in 1869, though biologists had been practicing ecology for very much longer than that. For some decades following Haeckel, ecologists groped for terms that would capture the sense of the holistic entities, the "quasi-organisms" (Tansley, 1935) that they were studying in nature. The term "biocoenosis" (ecological community) was introduced by K. Möbius in 1877; in 1939 limnologist A. Thienemann used the term *biotope* for the physical environs with which the bioceonosis interacts, and referred to the sum of the bioceonosis and biotope as the *holocoen*, roughly synonymous with our present ecosystem. The term "biogeocoenosis" (synonymous with ecosystem or holocoen) was suggested by the Russian ecologist

V. N. Sukachev in the 1940s. Although more precise than *ecosystem*, this term understandably did not catch on.

The term *ecosystem* was first used in print by botanist A. G. Tansley (1935; Golley, 1993). Tansley defined the ecosystem as

> the whole *system* (in the sense of physics), including not only the organism-complex, but also the whole complex of physical factors forming what we call the environment of the biome – the habitat factors in the widest sense. It is the systems so formed which, from the point of view of the ecologist, are the basic units of nature on the face of the earth. (1935, p.299)

Tansley's ecosystem includes not only the community or communities of organisms, but the physical surroundings – the atmosphere, water, soil, rock – with which they interact. For instance, a body of topsoil considered as an ecosystem includes not only the plants, microorganisms, and numerous other life-forms that inhabit the soil, but also (among other things) the minerals of the soil crumbs, the soil water, and the air which interpenetrates the soil and with which the soil organisms interact.

Viewed this way, the boundaries of ecosystems may seem arbitrary (for instance, soil air is continuous with the entire atmosphere of the Earth); so it must be asked whether it is possible to delineate smaller ecosystems within the biosphere in a non-arbitrary way. Tansley (1935) argued that the distinguishing feature of an ecosystem was that it is a type of *physical* system having an identity defined by a "relative dynamic equilibrium." Although the task remained to explain precisely what this phrase means, Tansley's view implies that subsystems can be picked out from their backgrounds by the presence of cycles of energy, materials, food, or information, in the same sense in which a live electrical circuit could be distinguished from a tangled mass of wiring and components.

There are no truly stable structures in nature, but some structures can be approximately stable (or at least fluctuate around a mean) over thousands or even millions of years, some only over short times. [See Complexity, Diversity, and Stability]. Tansley thought it obvious that natural selection favors ecosystems which tend to be stable. He conceived of ecosystems as founded on plant life, but they could also involve animal and human activity as integral parts. Tansley argued that the "prime task of the ecology of the future" was to investigate the ways in which the components of the ecosystem "interact to bring about approximation to dynamic equilibrium" (1935, p.305). On Tansley's view, communities and biomes are descriptive units, while ecosystems are defined by their underlying dynamics, which may not always be immediately apparent.

In a paper that was to have a strong influence on ecology in the coming decades, R. L. Lindeman (1942, p.400) defined the ecosystem as "the system composed of physical-chemical-biological processes active within a space-time unit of any magnitude, i.e., the biotic community *plus* its abiotic environment." Lindeman's definition is less inclusive than Tansley's, since the latter implicitly points not only to processes within a region of study, but any processes which contribute to dynamic stability. As well, the term "space-time unit," although unclear, suggests physical contiguity, but the type of dynamic coherence indicated by Tansley could be produced by causal interactions acting at quite long range. Lindeman mainly considered the trophic dynamics of ecosystems, which has to do with how energy – primarily in the form of food – is cycled within the ecosystem. This is an important way of accounting for the dynamic stability indicated by Tansley, but is also a narrower conception, since Tansley's definition would in principle allow for any sort of causal interactions (such as amplification and information exchange) that tended to stability. Still, Lindeman and Tansley's conceptions have in common the key notion that the ecosystem is defined in terms of dynamic cycling.

Lindeman's view was adapted by H. T. Odum (1983) and E. P. Odum (1964, 1971) and became the central concept in so-called *systems ecology* or New Ecology (Worster, 1977), the dominant trend in ecology from the 1950s until at least the 1980s. The New Ecology describes the dynamic stability of Tansley as a *homeostasis* in much the same sense in which this term is used in physiology: a quasi-stable state maintained by organisms actively balancing their responses to positive and negative feedbacks. In the systems approach the ecosystem is defined as a *circuit* of energy but it can also be defined in terms of the types of materials in circulation; L. Margulis, for instance, has defined an ecosystem as "the smallest unit that recycles the biologically important elements" (1998, p.105).

It is difficult to define the term "stability" in a non-tendentious way, but it can be loosely defined as resistance to external perturbations and forcing. More precisely, it can be defined as the maintenance, in the face of perturbations, of biophysical parameters within a range suitable for the survival of the life-forms in the system. An important characteristic of homeostatic systems is that they tend to *return* to equilibrium when subjected to perturbations within a certain range of tolerance.

Recent elaborations of the ecosystem concept include ecosystems as complex adaptive systems (Levin 1998) and as self-organizing critical systems (Jørgenson, Mejer, & Neilsen, 1998). What all such conceptions have in common is some notion of a quasi-steady state maintained by cycling of energy, materials, or information.

E. P. Odum was especially influential in defining and promoting the ecosystem as the central unifying concept in ecology. He argued (see, e.g., 1964) that it is necessary to distinguish between physical structure and dynamic function; while cells are structurally very different from forests, there are, on the systems ecology view, key similarities in the way diverse ecosystems at all sizes scales circulate energy, materials, and information. Odum insisted that a purely reductionist approach to biology would lose sight of the emergent structures and properties that appear only at the level of complex systems. By an "emergent property" one means a property that can be meaningfully applied to a complex system as a whole but not the parts of the system. For instance, the sense in which a person may be "healthy" is quite different from the sense in which a cell in that person's body may be said to be healthy. In physical terms emergence takes the form of *synergism*, in which properties of subsystems combine to produce a system-wide effect which is not a linear function of the properties of the parts. (See, e.g., Fath & Patten, 1998.)

Most treatments of the ecosystem in systems ecology focus on the direct interchange and circulation of free energy and nutrients between organisms and their non-living surroundings. However, there are other ways that the dynamic equilibrium cited by Tansley and Odum can be maintained: the circulation of information can be decisive since organisms respond to informational feedbacks from the environment with which they interact; also (and this turns out to be crucial in the discussion of sustainability) organisms can contribute to the energetic synergism of an ecosystem indirectly as well as directly. External energy flows can be steered into the ecosystem by a variety of manipulations; for instance, humans can plant trees and thus promote the input of far more photosynthetic energy than they consume.

In recent years the systems-theory conception of the ecosystem has come under criticism (e.g., Sagoff, 2003) but because it was so influential it must serve as a reference point for ecology for some time to come.

4. Ecosystems as Symbiotic Units

Symbiosis is often taken to be a topic in community ecology, not ecosystems ecology. However, E. Odum (e.g., 1971) frequently stressed the importance of his interpretation of the ecosystem as a kind of *symbiotic* association. This aspect of Odum's view of ecosystems has received relatively little attention, but it is crucial in understanding the possible application of ecosystems theory to sustainability. On this interpretation, a community becomes an ecosystem precisely when it becomes symbiotic.

The term *symbiosis* was introduced by A. de Bary (Paracer & Ahmadjian, 2000). It is often used loosely to suggest cooperation, but as de Bary apparently intended it, and as it is usually used in the professional literature today, it is a more general concept. To say that organisms are symbiotic is to say that in some manner they include each other in their life cycles, but this does not necessarily entail a mutually beneficial interaction; for instance, the malaria parasite is in a symbiotic relationship with its hosts. Some interactions which appear to be parasitic are mutualistic when looked at on a larger scale; predator–prey relationships are typical examples. A wide variety of causal interactions, direct and indirect, can play a role in maintaining a symbiotic state. Fath and Patten (1998) argue for "mutualism as an implicit consequence of indirect interactions and ecosystem organization," and show how indirect interactions contribute to network synergism in ecosystems.

There is a range of symbiosis from pathogenic parasitism to symbiogenesis; this can be defined in terms of increasing degree of cooperation and also in terms of increasing energetic synergism. Pathogenicity occurs when a mutant or emergent parasite overwhelms the defenses of its host and both host and parasite perish; unpleasant medical examples such as metastatic cancer come to mind, but the sort of overpopulation crisis identified by Malthus is also an example of pathogenic parasitism in which the host is the whole biophysical environment exploited by the overpopulating species. There are various degrees of parasitism in which the parasite is partially tolerated by the host. A commensal (such as the human forehead mite *Demodex*) is a parasite which generally cannot survive without the specialized environment provided by its host, but which (usually) neither benefits nor harms its host. Mutualists are organisms which benefit each other in the precise sense that each somehow increases or maintains the other's reproductive success. Mutualistic relations can be facultative (optional) or obligate. In animals such as humans with a complex neurology, mutualistic relations, if they occur, tend to be learned rather than instinctual or biochemically mediated. (In humans, therefore, the maintenance of mutualism is partially a function of culture, broadly understood.) Highly obligate mutualisms sometimes lead to *symbiogenesis*, the creation of a new type of organism. In symbiogenesis, branches of the tree of life occasionally converge, contrary to the classical Darwinian picture where they always keep splitting. L. Margulis (1998) suggests that the formation of symbiotic associations could be a source of evolutionary novelty comparable in importance to mutation, but this view is highly controversial.

Margulis has played a leading role in demonstrating the importance of symbiogenesis in cellular evolution (Margulis, 1998). There is, by now, a large body of evidence supporting serial endosymbiosis, the view that eukaryotic cells are highly obligate mutualistic associations of bacteria.

Margulis and E. Odum (1971) highlighted the importance of the "symbiotic transition" in which an opportunistic parasite can move along the symbiotic scale from parasite, through commensal, to obligate mutualist. Such a transition from parasite to mutualist played an essential role in the evolution of eukaryotic cells, in which parasitic bacteria apparently became organelles of the cells they had originally preyed upon. Symbionts will coevolve even if they do not necessarily become mutualists, because a host will evolve to defend itself from a parasite, while the parasite may evolve to cope with the host's defenses.

There is evidence from cell biology that a transition from parasitism to mutualism will be favored in environments that are closed in a way that leads to resource restriction (Margulis and Sagan, 1995), and this is consistent with Kropotkin's observation (1902/1989) that mutualism is favored over competition in harsher environments. However, the conditions under which mutualism and symbiogenesis are adaptively favored remain unclear, and this remains an important unsolved problem that has much significance for ecology.

5. Ecosystems as Dissipative Structures

The earliest conceptions of the ecosystem defined it in terms of dynamically maintained homeostasis, energy circuits, and feedback loops. Recently, a number of authors have extended this approach by controversially suggesting that the problem of explaining ecosystem stability is the same as the problem of explaining how life itself is thermodynamically possible (Schneider & Sagan, 2005; for a skeptical response, see Farmer, 2005). Schrödinger (1944) noted that any living system apparently violates the Second Law of Thermodynamics within its boundaries, for it maintains a highly ordered or low entropy internal state by the expenditure of energy released through its metabolism. The key to the puzzle, Schrödinger realized, is that living systems shed entropy by actively expelling waste heat. Living organisms and ecosystems belong to the class of *dissipative structures*, far-from-equilibrium, highly ordered states that can only exist where there is a generous externally-applied flow of free energy from a source such as the sun. This suggests a notion of the ecosystem as a dissipative structure – an "eddy," as it were, in the relentless flow of energy down entropic gradients. Paradoxically, an ecosystem's stability is a function of how efficiently it can degrade free energy. Presumably the ecosystem maintains its highly ordered internal cycling of energy because that is the most efficient way for it to produce waste heat.

The dissipative-structure view of ecosystems can be combined with the symbiotic view. An ecosystem proper can be understood as a mutualistic association of organisms, whose mutualistic symbiosis is defined by their thermodynamic relationships (Peacock, 1999). On this view, mutualism involves sharing free energy and thereby implies dynamic coupling between members of a mutualism; the system acts, as it were, as a quasi-rigid body under selective pressure. By combining the non-equilibrium thermodynamic view with the symbiotic understanding of ecosystems, we arrive at the view of ecosystems as quasi-stable dissipative structures, characterized by a circulation of energy, information, and/or materials, in such as way as to confer selective advantage on the association as a whole.

The dissipative-state theory of ecosystems suffers from two related problems. First, it has not so far been shown to have much quantitative predictive power. Second, non-equilibrium statistical mechanics still lacks its Boltzmann, someone who could provide a clear explanation of the principles of the theory in purely statistical terms. In equilibrium thermodynamics it is easily seen that higher-entropy states are more probable. (For instance, air pressure in a closed room is uniform simply because there are enormously more ways for the air molecules to be distributed approximately evenly than unevenly.) But in what sense are there more microstates associated with the vortical motion of a tornado than with turbulent motion? Why would cyclic motions be entropically favored simply because they move material through the system faster? Until such questions can be answered in a rigorous and clear way, the application of dissipative-systems theory to ecosystem dynamics remains an intuitively plausible but still essentially analogical and qualitative hypothesis.

5.1. The Gaia hypothesis

Possibly the most speculative or visionary conception of the ecosystem is the Gaia hypothesis. This is the proposal that it is scientifically meaningful to regard the entire planetary biosphere as a single, self-regulating ecosystem. In its modern form this hypothesis was devised by J. Lovelock, D. Hitchcock, and L. Margulis (Lovelock & Margulis, 1974; Lovelock, 1988; Lovelock, 2003). The Gaia hypothesis was suggested by the observation that many components of the earth's atmosphere are so far from chemical equilibrium that their relative abundance could only be explained by the mediation of life. The atmosphere, Lovelock argues, is a "contrivance" (in the sense that a coral reef or an ant-hill is a contrivance) which maintains temperature, atmospheric composition, and other variables suitable for life by means of an elaborate network of feedbacks. Lovelock's Daisyworld model (1988) demonstrates, apparently, that a sufficiently diverse system of biota could generate its own set-points, so long as the planet remained within a fairly wide range of solar input.

Lovelock has tended to explain Gaia as a biologically mediated control system, but Gaia can also be understood either from a thermodynamic point of view (as a dissipative structure) or the symbiotic point of view (as a planetary-scale mutualism). Opinions differ strongly on whether the Gaia hypothesis is scientifically well founded or arrant speculation. One of its virtues is that it provides a plausible explanation for the maintenance over billions of years of the far-from-equilibrium conditions in the earth favorable to life. However, the Gaia hypothesis has so far been short on predictive power. It is also difficult to square with evolutionary theory; T. M. Lenton (1998) offers a detailed attempt to work out how self-regulation on a planetary scale could be brought about by natural selection.

6. Ecosystems and Evolutionary Biology

The idea that organisms are subject to selective pressure by their biophysical environments is one of the central tenets of Darwinism. What ecology, and ecosystem theory in particular, adds to this is a special emphasis on the fact that organisms can affect their environments as much as the environments affect their organisms. What one might call the "post office" theory of the ecological niche holds that the survival problem for a species consists in adapting itself to a preexistent slot in a much larger backdrop ecosystem. (No one literally believes this any more, but it is a useful approximation when the back-reaction of the organism on its environment is not very important.) The central fact of ecology, however, is that the lines of influence between organism and ecosystem run both ways: organisms adapt to their ecosystems, but they also adapt their ecosystems to themselves, sometimes in ways that are favorable to their future survival, sometimes not. The way in which organisms alter their ecosystems then poses additional survival challenges or opportunities for themselves and other species, and this must be taken into account in any complete picture of the evolution of life.

The existence of self-supportive and cooperative biological systems, which ecosystems are presumed to be on many accounts of ecosystems theory, is a challenge for evolutionary biology. In the late nineteenth century the Russian emigré ecologist P. Kropotkin (1902/1989) criticized the view of T. H. Huxley that "the animal world is on about the same level as a gladiator's show." (Huxley, 1888/1989, p. 330). Kropotkin pointed out that cooperation occurs at many levels in nature, and argued that fitness can just as easily amount to the ability to cooperate as well as to compete, depending upon the demands of the ecological context.

The Kropotkin/Huxley controversy is being replayed today. If it is correct to speak of the persistence of ecosystems as a form of adaptive success (as Tansley and many other ecosystem theorists believed) then that fact might be difficult to understand from the narrow adaptationist/selfish gene point of view, which tends to be skeptical of natural selection acting beyond the level of the individual organism. [See The Units and Levels of Selection]. Could ecosystem stability (which could be read as the tendency of an ecosystem to *survive* over time) be a sign of "group selection"? (Group selection in this context would mean the tendency for organisms to be favored by natural selection partially on the basis of their ability to contribute symbiotically to ecosystem functioning.) The prevalent view in evolutionary theory today is that there is group selection but it is not a dominant factor in evolution. (Sober & Wilson, 1998.) However, this view would have to be revised if ecosystem theorists (in alliance with evolutionary biologists) can succeed in showing that the evolution of many organisms cannot be understood unless their traits were selected for, in important part, on the basis of their ability to contribute to the relative stability and persistence of the ecosystems which support their existence. This remains an open and controversial question. [See Cooperation].

7. Skeptical Critiques of Ecosystem Theory

Ecosystem theory has been criticized from a number of philosophical and scientific directions. American plant ecologist H. Gleason (1882–1975) was an early skeptic about the prevailing theories of F. Clements and others according to which biotic communities were "superorganisms" with definable internal parts, a coherent structure, and law-like behavior (Keller & Golley, 2000). Gleason proposed his "individualistic hypothesis" according to which the plant association was merely a descriptive convenience, and the character of every biotic community in nature was unique and dependent upon statistical variations and the vagaries of individual organisms within it.

M. Sagoff (2003) offers an up-to-date critique of ecosystem theory that is much in the spirit of Gleason. Sagoff points out that there are two complementary trends in many branches of science. Physics usually takes what Sagoff calls a top-down approach in which one attempts to understand the complexities of nature in terms of simple mathematical laws of wide applicability, and from which predictions are derived deductively. Biology perforce tends to use a bottom-up approach which sees nature as irreducibly complex; predictions are made statistically, and every general rule is expected to have exceptions. Sagoff argues that ecosystem ecology is in effect an attempt to turn ecology into a branch of physics, and charges that much of ecosystems theory is circular, vacuous, and incapable of generating testable predictions. While the New Ecologists define ecology as nothing other than the study of ecosystems, Sagoff, in effect, proposes that there is such a thing as ecology without the ecosystem. K. de Laplante and J. Odenbaugh (in press) offer a response to Sagoff's critique. Whether or not Sagoff is entirely correct, this debate should usefully spur ecosystems theorists to a renewed effort to demonstrate the relevance of their model-building to the real world of ponds and people.

For a skeptical view of the notion that nature can be viewed both as a biophysical machine and as a superorganism, see Botkin (1990).

Another approach that is critical of the dominant systems paradigm is non-equilibrium ecology, which charges the New Ecology as exaggerating the degree of stability of ecosystems. These authors insist that real ecosystems such as grasslands (as opposed to idealized mathematical models) are rarely close to equilibrium and cannot be managed effectively were they expected to be such. (Walker & Wilson, 2001; Rohde, 2005.) E. Odum responded that fluctuations within localized systems or even periods of time as long as the glacial epochs should not distract us from the fact that the earth system as a whole has maintained sufficient stability over hundreds of millions of years to permit the continuance of life.

The notion of the ecosystem could also be subjected to the same sort of skeptical critiques that have been directed toward the reality of other scientific entities. The causal workings of ecosystems must often be inferred by indirect evidence; even the descriptive ecosystems of the working field ecologist are to a large extent inferential and theory-laden. However, ecosystems do not seem to have caught the attention of instrumentalists or antirealists within the philosophy of science.

There are also post-modernist and constructivist critiques of ecology (e.g., Evernden, 1992). Keller and Golley (2000, p.13) argue that "scientific ecology . . . is at odds with social constructivism," and defend the view of an "extrasubjective, transcultural meaning in nature which humans can [however imperfectly] discern." If Sagoff is right, however, then ecosystem theory leaves itself open to constructivist criticism by not doing a good enough job of making its concepts operationally meaningful and testable.

8. Ecosystem Integrity and Health

Bodily health in the medical sense can be given a sharp definition as a state of homeostasis (actively maintained equilibrium) that fulfills certain quantitative norms. There is a substantial literature exploring parallels between bodily health and the health or integrity of ecosystems (Costanza, Norton, & Haskell, 1992).

The concept of ecosystem health plays an important role in some conceptions of environmental ethics. Aldo Leopold's influential Land Ethic (1966) elevates the "biotic community" (conceived of as a symbiotic "energy circuit") to an object of ethical regard, and proclaims that a "thing is right when it tends to preserve the integrity, stability, and beauty of the biotic community. It is wrong when it tends otherwise." This is one extreme along a continuum of views about ethical obligations to the environment. One could well have regard for the well-being of the environment that supports human life without subscribing to the theory of the ecosystem as energy circuit. Also, the idea that ecosystems can be treated as objects of ethical regard could not be in itself a *sufficient* basis for environmental ethics (as Leopold's concise wording seems to suggest); whales do not merit protection *merely* because they are parts of an oceanic ecosystem. A further problem is that Leopold freely mixes normative concepts such as "beauty" with descriptive concepts such as "stability." However, Leopold's ideas draw attention to the important notion of ethics as having a crucial role in any human–land symbiosis. For more on the aesthetics and ethics of ecology, see Peacock (1999b) and Schmidtz and Willott (2002). For an up-to-date discussion of the very difficult problems of elucidating the meanings of ecological stability and biodiversity, and the relations between them, see Sarkar (2005a). [See Complexity, Diversity, and Stability].

If anything like E. Odum's conception of ecosystems is correct, the general principles for maintaining ecosystem health would include the preservation (and perhaps judicious repair) of existing pathways of energy and materials. It could be quite important for conservation biology to be able to identify ecosystems in terms of the circuits of energy, information, and materials that define them, and one would want to avoid misguided attempts to "improve" an ecosystem that result in severing those circuits. Certainly both advocates and critics of ecosystems theory would agree that sensitivity and caution are essential in any attempt to apply ecology to real situations where human well-being is at stake.

9. Sustainability from an Ecosystems Point of View

One of the most important applications of ecosystems theory is to help define a possible basis for the sustainability of the global ecology that supports the human species and its complex global civilization. In 1987, the World Commission on Environment and

Development (WCED) published the influential Brundtland Report. This document argued that two interrelated factors constitute the world's current ecological/economic crisis: poverty and the threat to humanity caused by breakdown of "ecosystem services" caused by human over-exploitation of the earth system. It is development (exploitation of the found ecology for human purposes) that presumably is necessary to eradicate poverty, and yet it is precisely exploitive development that undermines the capacity of the earth system to sustain humans indefinitely.

As a solution to the twin imperatives – to advance human prosperity and to respect ecological limitations – the Brundtland Report advocated *sustainable development*, which it defined as "development that meets the needs of the present without compromising the ability of future generations to meet their own needs" (WCED, 1987). Many observers (e.g., Livingston, 1994) have argued that this notion is incoherent. *Prima facie*, the phrase sounds oxymoronic, since the very notion of development seems to imply exploitation of a natural resource for human ends in such a way as to permanently use it up.

The weakness of the Brundtland Report is that it did not define "development" precisely. The Report itself implicitly assumed that development must amount to tapping into the resources and free energy of nonhuman ecosystems: "Development tends to simplify ecosystems and to reduce their diversity of species" (WCED, 1987, p.46).

The question is therefore whether sustainability can amount to anything other than rationing. According to several influential authors, we are in a lifeboat with a finite initial supply of resources which cannot be replenished by any conceivable human action (Georgescu-Roegen, 1977; Daly, 1985; Rees, 1987. For a more nuanced version of the lifeboat picture, see Meadows, Meadows, & Randers, 1992). This neo-Malthusian view is often stated in terms of thermodynamics. H. Daly, for instance, states, "Low entropy is the ultimate resource which can only be used up and for which there is no substitute" (1985, p.90), and according to W. Rees, "The thermodynamic interpretation of the economic process therefore suggests a new definition of sustainable development . . . [as] development that minimizes resource use and the increase in global entropy" (1990, p.19). These authors adduce in support of their lifeboat view of sustainability that version of the Second Law of Thermodynamics which states that entropy can never be decreased in a closed, isolated system. This is not the form of the Second Law that is relevant to ecology, however; the earth system is not thermally isolated, since it is bathed by more solar and geothermal energy than it can possibly use. The lifeboat view of sustainability thus seems to be founded on an elementary misunderstanding of the physics of ecosystems; it is the ecology of the thermos flask. In fact, it is not negentropy (negative entropy, a measure of order), but the capacity of the autotrophic components of the earth system to generate negentropy, which is the "ultimate resource." It is not immediately obvious that humans cannot contribute positively to this in many ways.

There is a sound notion behind the Brundtland definition of sustainable development, despite its unfortunate formulation. The aim is to avoid the *ecological bind*: the tendency of an organism to undermine its own future by the very means that give it a survival advantage in the first place. From the viewpoint of ecosystem theory, it therefore seems natural to define sustainability in this sense in terms of symbiotic concepts, since the biotic relationships that do have the tendency to self-perpetuate are precisely those that are mutualistic, or part of larger symbiotic cycles that are mutualistic. Transitions from parasitical to mutualistic symbioses are frequently noted (Odum, 1971), especially when resources are restricted (Margulis & Sagan, 1995), although this phenomenon requires further scientific study. The achievement of sustainability for human culture would amount to a symbiotic transformation from parasite to facultative mutualist.

This notion of human mutualism is not merely a metaphor; rather, it has the concrete sense that the means by which we garner the resources we need would be also the means by which we sustain the environment. G. A. Whatmough (1996, pp.418–19), citing the horticulturally intensified ecologies of rural England and Japan, observes that "the increase in the density and luxuriance of the whole spectrum of local flora and fauna [was] an entailed consequence of the techniques by which those populations then produced their necessary supplies . . . It can only be by some such means that our species can possibly transform our present parasitic dependence on the found ecology to some kind of symbiotic alternative."

There are two components to sustainability: the conservation and preservation of existing ecosystem function, and (more controversially) the enhancement and intensification of the ecosystem. Although the concept of ecosystem intensification was mooted by A. J. Lotka in 1922, it has received very little discussion. Lotka argued that "suitably constituted organisms [may] enlarge the total energy flux through the system. Whenever such organisms arise, natural selection will operate to preserve and increase them" (1922, p.147). On Lotka's view, an ecosystem may be thought of as a sort of battery that can be charged up by its autotrophs.

At first glance it might seem that humans are inherently incapable of such a mutualism, since we are obligate heterotrophs. However, from a thermodynamic point of view the distinction between autotrophs and heterotrophs is not as sharp as is usually supposed. Consider how an algae cell shunts solar energy into the ecosystem it supports. It has within its body an elegant biochemical mechanism which captures solar energy and uses it to reduce CO_2 and H_2O to carbohydrates and free oxygen. The algae uses a small proportion of the captured energy to support its own metabolism, and the rest is ultimately made available to other organisms in the ecosystem. In effect, the algae acts like a *valve*, diverting part of the external flow of energy into the system and thereby increasing the total circulation of usable energy and materials in the system. What is definitive of this function is the valving capacity. Valves expend far less energy than they can divert or modulate, and there is in general no theoretical limit to an amplification factor.

It is incidental that the mechanism by which the algae diverts energy into the ecosystem is inside its own cellular envelope. A heterotroph (not itself photosynthetic) can do the same thing by manipulations carried on outside its body. Heterotrophic life vastly multiplies the number and kinds of niches within which autotrophic life can operate. Humans can contribute to this process as well: we can, for instance, do things such as plant trees or regenerate topsoil, and if these things are properly done they can divert far more solar energy into the planetary ecosystem than they require for their execution.

Eugene Odum has expressed the problem of sustainability from the ecosystems point of view:

> Obviously it is time for man to evolve to the mutualism stage in his relations with nature . . . if understanding of ecological systems and moral responsibility among mankind can keep pace with man's power to effect changes, the present-day concept of "unlimited exploitation of resources" will give way to "unlimited ingenuity in perpetuating a cyclic abundance of resources." (1971, p.36)

On this view, sustainable development – or more precisely the development of sustainability – amounts to the rearrangement of human affairs so that by means of the techniques we use to survive on this planet, we "pump up" the earth system instead of drawing it down.

Philosophers have an important role to play in helping to define the vision of moral responsibility that could help make this symbiotic transition possible (Norton, 2005). A sense of responsibility usually begins with an awareness of what is required for self-preservation (though it need not end there). As Odum indicates, any such sense of responsibility must be coupled with a sound scientific understanding; the *scientifically informed* sense of ecological moral responsibility called for by Odum is therefore, for humans, nothing other than an indispensable survival tool.

If, on the other hand, scientific and philosophical critiques of ecosystems theory show that Odum's vision is *not* tenable, then humanity needs to know it, and soon, for the study of ecology possesses a particular urgency not shared by most other branches of theoretical science.

Acknowledgments

The author thanks Dawn Collins for research assistance, Bryson Brown, Kevin de Laplante, John Collier, and the editors of this book for perceptive advice, Scott Howell for assistance, and the University of Lethbridge and the Social Sciences and Humanities Council of Canada for support.

References

Botkin, D. B. (1990). *Discordant harmonies: a new ecology for the twenty-first century.* Oxford: Oxford University Press.

Chaffin, T. (1998). Whole-earth mentor: a conversation with Eugene Odum. *Natural History* (October).

Costanza, R., Norton, B. G., & Haskell, B. D. (1992). *Ecosystem health: new goals for environmental management.* Washington, DC: Island Press.

Daly, H. E. (1985). Economics and sustainability: in defense of a steady-state economy. In M. Tobias (Ed.). *Deep ecology.* San Marcos, CA: Avant Books.

Day, D. (1989). *The eco wars: true tales of environmental madness.* Toronto: Key Porter.

de Laplante, K. (2004). Toward a more expansive conception of ecological science. *Biology and Philosophy*, 19, 263–81.

de Laplante, K., & Odenbaugh, J. (in press). What isn't wrong with ecosystem ecology. In Skipper, R. A. Jr., C. Allen, R. Ankeny, C. F. Craver, L. Darden, G. M. Mikkelson, & R. C. Richardson (Eds). *Philosophy across the life sciences.* Cambridge, MA: MIT Press. Preprint at http://www.lclark.edu/~jay/vitae.html.

Evernden, N. (1992). *The social creation of nature.* Baltimore: Johns Hopkins University Press.
Farmer, J. D. (2005). Review of Schneider and Sagan (2005). *Nature*, (436, 4 August), 627–8.
Fath, B. D., & Patten, B. C. (1998). Network synergism: emergence of positive relations in ecological systems. *Ecological Modelling*, 107, 127–43.
Georgescu-Roegen, N. (1977). The steady state and ecological salvation: a thermodynamic analysis. *BioScience*, 27, 266–70.
Golley, F. B. (1993). *A history of the ecosystem concept in ecology: more than the sum of the parts.* New Haven and London: Yale University Press.
Haeckel, E. H. (1869/1879). Über Entwickelungsgang und Aufgabe der Zoologie. Lecture at University of Jena, 1869; in *Gessamelte populäre Vorträge aus dem Gebiete der Entwickelungslehre.* Heft 2. Bonn: Strauss.
Huxley, T. H. (1888/1989). The struggle for existence in human society. The Nineteenth Century, February 1888. Reprint in Kropotkin, P. (1989). *Mutual aid: a factor of evolution.* Montreal: Black Rose Books.
Jørgenson, S. E., Mejer, H., & Neilsen, S. N. (1998). Ecosystem as self-organizing critical systems. *Ecological Modelling*, 111, 261–8.
Keller, D. R., & Golley, F. B. (2000). *The Philosophy of Ecology: From Science to Synthesis.* Athens, GA: University of Georgia Press.
Kropotkin, P. (1902/1989). *Mutual aid: a factor of evolution.* London, 1902. Reprint: Black Rose Books, Montréal, 1989.
Lenton, T. M. (1998). Gaia and natural selection. *Nature*, (394, 30 July), 439–47.
Leopold, A. (1966). The land ethic. In *A Sand County Almanac.* New York: Oxford University Press.
Levin, A. A. (1998). Ecosystems and the biosphere as complex adaptive systems. *Ecosystems*, 1, 431–6.
Lindeman, R. L. (1942). The trophic-dynamic aspect of ecology. *Ecology*, 23(4), 399–417.
Livingston, J. (1994). *Rogue primate: an exploration of human domestication.* Toronto: Key Porter.
Lotka, A. J. (1922). Contribution to the energetics of evolution. *Proceedings of the National Academy of Sciences USA*, 8, 147–51.
Lovelock, J. E. (1988). *The ages of Gaia.* New York: W. W. Norton.
Lovelock, J. E. (2003). The living earth. *Nature*, (426, 18/25 December), 769–70.
Lovelock, J. E., & Margulis, L. (1974). Atmospheric homeostasis by and for the biosphere: the Gaia hypothesis. *Tellus*, 26(1–2), 2–9.
Margulis, L. (1998). *Symbiotic planet: a new look at evolution.* New York: Basic Books.
Margulis, L., & Sagan, D. (1995). *What is life?* New York: Simon & Schuster.
Meadows, D. H., Meadows, D. L., & Randers, J. (1992). *Beyond the limits: confronting global collapse, envisioning a sustainable future.* Toronto: McClelland and Stewart.
Norton, B. G. (2005). *Sustainability: a philosophy of adaptive ecosystem management.* Chicago: University of Chicago Press.
Odum, E. P. (1964). The new ecology. *BioScience*, 14(7), 14–16.
Odum, E. P. (1971). *Fundamentals of ecology* (3rd edn). Orlando, FL: Saunders.
Odum, H. T. (1983). *Systems ecology: an introduction.* New York: Wiley.
Paracer, S., & Ahmadjian, V. (2000). *Symbiosis: an introduction to biological associations.* Oxford and New York: Oxford University Press.
Peacock, K. A. (Ed.). (1996). *Living with the earth: an introduction to environmental philosophy.* Toronto: Harcourt Brace.
Peacock, K. A. (1999a). Staying out of the lifeboat: sustainability, culture, and the thermodynamics of symbiosis. *Ecosystem Health*, 5(2), 91–103.
Peacock, K. A. (1999b). Symbiosis and the ecological role of philosophy. *Dialogue*, 38, 699–717.

Rohde, K. (2005). *Nonequilibrium ecology.* Cambridge: Cambridge University Press.
Sagoff, M. (2003). The plaza and the pendulum: Two concepts of ecological science. *Biology and Philosophy*, 18, 529–52.
Sarkar, S. (2005a). *Biodiversity and environmental philosophy: an introduction.* Cambridge: Cambridge University Press.
Schneider, E. D., & Sagan, D. (2005). *Into the cool: energy flow, thermodynamics, and life.* Chicago & London: University of Chicago Press.
Schrödinger, E. (1944). *What is life? The physical aspect of the living cell.* Cambridge: Cambridge University Press.
Schmidtz, D., & Willott, E. (2002). *Environmental ethics: what really works, what really matters.* New York and Oxford: Oxford University Press.
Sober, E., & Wilson, D. S. (1998). *Unto others: the evolution and psychology of unselfish behavior.* Cambridge, MA: Harvard University Press.
Tansley, A. G. (1935). The use and abuse of vegetational concepts and terms. *Ecology*, 16(3), 284–307.
Ulanowicz, R. (1997). *Ecology: the ascendent perspective.* New York: Columbia University Press.
Vernadsky, V. I. (1926/1998). *The biosphere.* New York: Copernicus, 1998. (Translation by D. B. Langmuir of *Biosfera*, Nauka, Leningrad, 1926.)
Walker, S., & Wilson, J. B. (2001). Tests for nonequilibrium, instability, and stabilizing processes in semiarid plant communities. *Ecology*, 83(3), 809–22.
Whatmough, G. A. (1996). The artifactual ecology: an ecological necessity. In K. A. Peacock (Ed.). *Living with the earth: an introduction to environmental philosophy* (pp. 417–20). Toronto: Harcourt Brace.
World Commission on Environment and Development (WCED) (1987). *Our common future* (The Brundtland Report). New York: Oxford University Press.
Worster, D. (1977). *Nature's economy: a history of ecological ideas.* Cambridge: Cambridge University Press.

Further Reading

Odum, E. P. (1989). *Ecology and our endangered life-support systems.* Sunderland, MA: Sinauer.
Sarkar, S. (2005b). Ecology. *Stanford encyclopedia of philosophy.* http://plato.stanford.edu.
Schneider, S. H., Miller, J. R., & Boston, P. (2004). *Scientists debate Gaia.* Cambridge, MA: MIT Press.
Ulanowicz, R. (1986/2000). *Growth and development: ecosystems phenomenology.* San Jose, CA: toExcel Press.

Chapter 20

Biodiversity: Its Meaning and Value

BRYAN G. NORTON

1. What is Biological Diversity?

Bacterial cells were the first forms of life, beginning about four billion years ago. Since then, life forms have proliferated and, with notable but rare exceptions when diversity has declined catastrophically, life has expanded, evolved, and complexified across time. Today the earth teems with many species arranged in many diverse patterns and relationships spread across varied landscapes. While estimates of the total number of species vary widely, estimates usually cited fall between 13 and 20 million species on earth (Hammond, 1995). As human populations have expanded since the industrial revolution, with technologies becoming more powerful and increasingly capable of pervasive impacts, the diversity of life is again in decline, this time as a result of human activities, especially the fragmentation of forests and other wild habitats. Reversing this trend toward biological simplification has become one of the most urgent of global environmental problems.

By the mid-1980s, participants in the effort to save biological resources began referring to the importance of protecting "biological diversity." Then, as part of the preparations for a symposium organized by the Smithsonian Institution and the National Academy of Sciences, it was suggested that the phrase be contracted, and the term "biodiversity" was born amid the fanfare of a large conference (The National Forum on BioDiversity), a stellar list of speakers, a video, and a traveling Smithsonian road show (Wilson, 1988). This symposium was only one example of a worldwide awakening to perhaps the most distressing of all global environmental threats: the possible destruction of the accumulated diversity of life on earth through habitat transformation and other activities. Scientists have estimated that current extinction rates are at least 1,000 times normal, and they may be as high as 10,000 times normal; they are also believed to be rising.

Because of the urgency of the problem, action is being taken to protect biodiversity, even as scientists and policy makers are only beginning to understand and describe it. This situation of uncertainty favors adaptive management, a promising if underdeveloped approach that encourages learning by doing, and by embedding scientific study within activist efforts to protect species. The current situation, then, is one where there is considerable activity intended to protect biodiversity, but there still exists considerable disagreement about basic concepts and broad theoretical issues. Three questions must be addressed here:

(1) How should we *define* the term, "biodiversity?"
(2) What steps must be taken to protect biodiversity?, and
(3) How should we *characterize* and *measure* the *value* of biodiversity?

Question 2 and 3 will be addressed in Sections 3–5, respectively. In answer to question 1, however, there seems to be a near-consensus answer: there is no generally accepted definition of the key term, "biodiversity." To cover this lexicographic embarrassment, one can offer a negative definition as a starting point: The goal of protecting biodiversity is to interrupt the seemingly inexorable trend toward the impoverishment of the biological world. Providing a positive definition has proven difficult, as will be further discussed in Sections 2 and 3.

It is tempting to think that biological data and theory should determine the definition of biodiversity: once a descriptive definition is developed and adopted, it would then be possible to pose, substantively, the question of what social values are served by natural systems that are characterized by biotic diversity, as defined by biologists. One can also look at this relationship in reverse: biologists study biodiversity in order to save something of value, so the biological definition should be shaped to capture whatever is valued in the diversity of life.

Defining biodiversity thus requires more than a simple act of lexicography. The term, it turns out, must ultimately be defined by the actions of conservationists in protecting biodiversity. It is a term of action, developed to further the normative science of conservation biology. As will be discussed below in Section 2.1, biologists offer quite varied definitions of the term. Proposing and defending a definition is also difficult because the term is used both in scientific contexts, where it is treated mainly as if it were a biological, descriptive term and, at the same time, the term is widely used in deliberations about conservation policy, where it clearly has normative, even honorific status as expressing a widely shared social goal – saving whatever is important in the diversity of life on earth. Anyone who listens to conversations among conservationists and conservation biologists will realize that they are driven by values. Biodiversity is a normatively charged concept, and the science of protecting biodiversity is therefore a normative science.

Assuming for now that the term will continue to be used in both scientific and policy contexts, the dual function of the term requires a definition of biodiversity that fulfills the *broad purposes of policy discourse*, even as we recognize that any definition of this key term *must achieve biological respectability*. A definition that can fulfill these twin purposes can be thought of as a "bridge term," a term that links discourse about public policy goals and objectives to scientific data and theory, all within a public discourse about policy choices. Ideally, a bridge term will express social values in a way that is measurable because it is important, especially in policy and action arenas to have measurable targets, so that the measure(s) can serve to gauge the success of policies. Identifying the goals of biological conservation must involve both good science and an insightful analysis of values associated with a diverse biota. Submitting proposed definitions to a dual criterion may imply compromises: some of the things biologists think are important in some contexts may be ignored because of the need to develop a definition that can improve communication regarding policy goals.

It may not, however, be as difficult as it seems to find a definition that is adequate to both tasks, despite compromises that may be necessary from both a scientific and a communicative viewpoint. One can argue that *all* scientific concepts have a conventional basis, in the sense that they are defined so as to advance the purposes of those who use them. A conventionalist view is especially appropriate when a new term, connected with a new and emerging science (in this case, conservation biology), is introduced, and used in somewhat different ways by scientists and practitioners. In such a context, the conventionalist, experimental attitude, reminiscent of Dewey's call to understand scientific concepts as tools of communication and reason (Dewey, 1910) and Carnap's (1937, 1956) endorsement of the Principle of Tolerance, chooses concepts that serve effective communication in pursuit of shared purposes and goals. According to this doctrine, scientists – and by extension policy discussants – are free to construct languages so as to advance common purposes. Definitions such as the one sought here are thus conventional – they are not chosen because they picture or stand for features of reality – they function rather as tools in human communication, especially communication in pursuit of common goals. The goal of defining biodiversity, therefore, understood in this conventionalist way, should be tied closely to the actions undertaken by scientists, practitioners, and policy makers in efforts to avoid biological impoverishment.

The good news is that communities – such as the large, international activist community that is united by the purpose of protecting biodiversity – can, through discussion, debate, and consensus, test the usefulness of various definitions by discussing and setting goals to protect the perceived social values associated with biodiversity. Proposed definitions will be judged by their usefulness in choosing effective policies. If a community can be defined as a system of interrelated actors whose members share significant interests and purposes, one can – on the present line of reasoning – propose a definition of our key term, biodiversity, that will be useful for communicating shared interests, goals, and purposes of this large community. Environmentalists in general and conservation biologists in particular form the nucleus of such a goal-directed and action-oriented community. These individuals and groups largely agree about what objectives should be pursued. All of this contributes to the emergence of a concept of biodiversity that is a useful communicative tool in the broader political arena. There remains, however, a long way to go in communicating the meaning and importance of biodiversity to policy makers and the public, and to establishing biodiversity protection as a high-priority goal in public policy.

2. The Definition Problem

As noted above, the search for a definition of biodiversity involves a number of factors because the term serves multiple functions: it is a term used by scientists; it clearly represents social values; it is an important term in discussions of environmental policy and management. Before moving forward to discuss these aspects in more detail, this section will survey some of the reasons the term has proved so difficult, and controversial, to define. It will be useful to begin this survey by examining what biologists have said about diversity of life. In subsequent subsections, several issues surrounding the problem of defining "biodiversity" will be addressed.

2.1. Biologists on biodiversity

Diversity has long been an important term in the literature of both biology and ecology, so one might say that, from the perspective of biologists, biodiversity is simply the diversity that exists in the biological world. It turns out, however, that diversity has been given a number of meanings in the literature of the biological disciplines (Pielou, 1975; Magurran, 1988, 2003). For example, there is a longstanding debate about whether diversity is better captured by total species counts or whether some degree of evenness in the comparative size of populations should also be included as an element of diversity. Further, and more profoundly, biodiversity is *multifaceted* – it encompasses diversity at multiple, nested scales of complex systems.

In pioneering work, R.H. Whittaker (1960, 1972, 1975) defined three levels of diversity, and referred to them as Alpha, Beta, and Gamma diversity; subsequently, these have come to be referred to by their more descriptive labels – within-habitat, cross-habitat, and total diversity, respectively. While all three levels are important, Gamma diversity – which is a measure of the overall diversity within a large region, what one might call "geographic-scale" diversity (Hunter, 2002, p.448), should be the goal of biodiversity management, since it incorporates the other two levels (Norton, 1987). Even if this simplification is accepted, however, we must choose what to emphasize: diversity among species, diversity within species, or cross-habitat diversity. All of these aspects of diversity interact dynamically with habitat change, and these aspects may not be highly correlated with each other, so there remain many more choices to make in operationalizing the key term.

David Takacs (1996) asked twenty-one leading biologists to define the term "biodiversity," and transcribed their answers. Anyone reading them cannot help but be impressed by the diversity of the definitions provided, as few of the definitions emphasize exactly the same aspects of biologically diverse systems. One important difference among biologists' definitions of biodiversity is the extent to which they emphasize the more dynamic aspects of biodiversity. Some definitions treat biodiversity as a list of types of entities, which seems to underemphasize process. There is a commonsense distinction, of course, between products and processes, between stocks of diverse entities and the ongoing flow of biological and evolutionary processes. These processes generate and sustain biodiversity, so it is important that they be included either implicitly or explicitly. It can also be argued that unusual ecological phenomena, such as multi-generational migrations of Monarch butterflies, also should be included. When migratory populations of a species are lost, and only sedentary ones survive, something of biological interest is lost (Brower & Malcolm, 1991).

Takacs transcribed the verbal definitions of leading biologists, which exemplify two types, *inventory* definitions and *difference* definitions. Several of the more popular definitions of biodiversity are aptly described as "inventory-style" definitions. E. O. Wilson, for example, provides an inventory when he defines biodiversity as "the variety of life across all levels of organization from genic diversity within populations, to species, which have to be regarded as the pivotal unit of classification, to ecosystems. Each of these levels can be treated, and are treated, independently, or together, to give a total picture. And each can be treated locally or globally" (Takacs, 1996, p.50). Similarly, Daniel Janzen defines biodiversity as "the whole package of genes, populations, species, and the cluster of interactions that they manifest" (Takacs, 1996: p.8). Peter Brussard recognizes the prominence of inventory-type definitions, asserting that the "standard definition" of biodiversity is species diversity, diversity of communities or habitats that species combine into, and the genetic diversity within species (Takacs, 1996, p.46).

Difference definitions, on the other hand, emphasize the differences and associated complexities and interrelations among biological entities. Several of Takacs' scientists reflect this emphasis: "I think of it as fundamentally a measure of difference," says Donald Falk (Takacs, 1996, p.50; Gaston, 1996). Paul Wood (1997, 2000, p.39) says simply that biodiversity is "the sum total of differences among biological entities."

Difference definitions link biodiversity to a "difference" function. Difference definitions emphasize differences among entities rather than inventorying/listing entities that exemplify differences. Wood captures this variation in types of definitions by noting that we can characterize diversity in terms either of "biological entities that are different from one another," or as "differences among biological entities," with the former approach emphasizing the entities involved, while the latter approach focuses on an environmental *condition or state of affairs* relative to the entities" (Wood, 2000, p.39). If applied at the species level, for example, as an illustration for the multiple facets of diversity, a difference definition would emphasize species that are the only one in their genus and would favor genera with no close relatives (see, for example, Weitzman, 1998).

Both inventory definitions and difference definitions can capture the dynamic aspects of diversity. An "inventory" of diverse entities can, in principle, include dynamic aspects. Daniel Janzen, for example, as noted above, includes "the cluster of interactions that they manifest" (Takacs, 1996, p.48) as one element of his inventory. Another definition simply refers to biodiversity as the sum total of the processes creating biodiversity. For example, Terry Erwin defines biodiversity as "the product of organic evolution, that is, the diversity of life in all its manifestations," which emphasizes processes even when the focus is on the products of processes (Takacs, 1996, p.47). While both inventory definitions and difference definitions can include the dynamic aspects of biodiversity, difference definitions more accurately portray the function of processes in maintaining biodiversity (Takacs, 1996, p.50f). While these definitions differ in form and inclusiveness, there may be much less disagreement about what biodiversity means than the variation in verbal definitions indicate.

Difference definitions help us to see what is most valuable in diversity at all levels, because they reveal the role of biodiversity in biological creativity. Whittaker (1975; Norton, 1987; Wood, 2000) hypothesized that "diversity begets diversity," that diverse elements undergoing diverse processes will generate more diversity. This hypothesis also suggests that losses of diversity can create further losses, as species are threatened because their mutualists become rare or disappear. Diversity of all kinds provides options for further creativity – and diversity is important as a contributor to that dynamic. Consider agricultural crops: most production comes from domesticated and even genetically modified seed stock, while wild varieties produce only a tiny portion of the world's food crops. If, however, a major disease or fungus breaks out in domestic lines, the existence of wild varieties, which are more genetically and morphologically diverse, might hold the key to genetic resistance among domestic stocks and re-establish domestic productivity. Differences in biological entities usually contribute thus indirectly to maintaining the flow of goods and services valuable to humans. It has, however, proven very difficult to quantify this contribution or to state a value for it.

Importantly, defining biodiversity in terms of difference supports an interesting extension, the concept of "complementarity," which refers to the contribution a species makes to fully representing features evident in phylogenetic patterns. By concentrating on variation among features (rather than on inventories of species), it is possible to estimate how much additional "evolutionary history, as depicted in the branches of an estimated phylogeny," according to Faith, are represented in a species. He continues: "The degree of complementarity reflects the relative number of additional features depicted in the branches of an estimated phylogeny" (Faith, 2003, 1994; Sarkar, 2004, 2005). Given this conceptualization, it is possible to estimate the degree of complementarity of a species with respect to an existing set, by estimating the number of additional features that would be added by protecting it. Such quantifications can then be used to judge conservation investments, by considering the relative gains that are achieved in protection of feature diversity by the proposed investment. Emphasis on differences within the biological world, rather than on different types of entities, thus points toward useful tools for management decision making. This point will be returned to in Section 4.

2.2. *Can biodiversity be defined?*

Whether difference- or inventory-style definitions are preferred, there exists a deeper problem as to whether, and in what sense, the term "biodiversity" is susceptible to definition at all. It is useful to replace this very general question with two more specific ones:

(1) Is it possible to provide an *index* of biodiversity? (Could there be a single number, reflecting a measurable biological characteristic, that would serve to rank systems according to their degree of diversity?)
(2) Is it possible to operationalize the term, "biodiversity" in particular situations? (Can one provide an operational measure that will usefully characterize those aspects of the natural world that scientists and activists are trying to save?)

A consensus has emerged that question 1 must be answered negatively. While this conclusion is unfortunate – it would be very useful if one could quantitatively rank ecosystems according to their biodiversity, especially if one could measure increments and decrements of biodiversity in a place over time – this outcome opens up interesting possibilities with respect to answering question 2 positively, which will be discussed in Subsection 2.3.

Strong arguments show an *index* that captures all that is legitimately included as biodiversity is not possible. Biodiversity cannot be made a measurable quantity. Wood explains this point as follows: "The main difficulty in defining diversity . . . is its multidimensional character along with the fact that the dimensions are not commensurable. They cannot be reduced to a single, and therefore commensurable, statistic" (Wood, 2000, p.38; also see Sarkar, 2005, p.177). To illustrate Wood's concern, consider the following case: Assume a functioning ecological system made up of **n** species; if one more species invades the system and establishes itself, without losing any species, there is an increment in diversity to the level **n + 1**. Suppose we expand the system by adding another *system*, however. Perhaps a large tree falls in the forest, creating a clearing; this event creates a new micro-habitat and the overall system subsequently has additional species and a range of relationships and functions that were not present in the original system. This would increase "cross-habitat diversity," rather than "within-habitat diversity." To avoid double-counting in an inventory definition duplicative species in the added system should not be counted, but this simple list of species will not capture the real impact on biodiversity that is represented by additions of cross-habitat diversity. Adding cross-habitat diversity also introduces a range of new functions and relationships – genetic and behavioral – that are an important aspect of biodiversity, and these can never be captured as a list of entities.

2.3. *Surrogacy*

The relative consensus that biodiversity cannot be associated with a definitive index has led to a variety of reactions, as there are several options for defining "surrogates" – features of a system that well represent biodiversity in some or all of its aspects. A common reaction, especially in North America where the species-by-species approach to conservation is strong, has been to simplify the problem by treating species counts as the best surrogate for biodiversity as a whole. Sometimes this move is justified by advocates on the grounds that species counts really are the most important aspect, and if species are protected, most other things will be, as well. In other cases, it is justified, apologetically, by the need to communicate in a policy context, even if the key term is not definable in an ideal way, a point that will be returned to in Section 4.

This tendency to equate conservation of biodiversity with species protection, which was reinforced by the heavy emphasis on species in the United States Endangered Species Act of 1973, was a dominant trend in both the science and practice of biodiversity protection in North America. Given this tendency, a small literature has developed on how to quantify *species* diversity. Statisticians have generated algorithms by which one can measure and quantify diversity so as to rank differing arrays of species as more or less diverse. This work exploits the idea behind difference definitions to arrange species lists so as to allow ranking according to the degree of diversity defined by the favored algorithm, and amounts to embracing the simplifying assumption that treats an inventory of species diversity as a surrogate, or "proxy" variable, to represent biological diversity. On this simplifying assumption, it is then possible to operationalize the "diversity" of a set of species by offering a measure of dissimilarity (differences) by aggregating all the differences found in pair-wise comparisons of the species that are elements of the set (Cervigni, 2001). Solow, Polasky, and Broadus (1993) compared sets before and after extinctions occur, and provide a "Preservation-Diversity" (PD) index, which measures the sum of the individual distance of extinct species from the survivors as a negative number. The goal of maximizing diversity, then, would be to push the index as close as possible to zero. Weitzman (1998) suggests an alternative measure of diversity within sets of species by measuring the minimal number of steps to account for its emergence from an evolutionary process.

These approaches offer precise operationalizations of species diversity; as noted, however, they do not solve the problem of the multifaceted nature of biodiversity, because these quantificational rankings presuppose the choice of species counts as the surrogate for the multifaceted concept of biodiversity. Also, while precise, they demand far more data than are ever likely to be available in particular situations (Cervigni, 2001). Perhaps their value is in showing how "difference" definitions can, given important simplifying assumptions, offer a precise measure of diversity for species inventories, but this is a rather narrow application, and it requires the reduction of biodiversity to species diversity.

In a somewhat different reaction to failure of definition, conservationists working in other parts of the world have suggested a more complex view of surrogacy, and view the development of surrogates as an important series of choices in devising conservation plans (Faith, 1992; Pressey et al., 1997; Margules & Pressey, 2000; Sarkar, 2005). One representative of this group, Sarkar, says that, since "true surrogates for biodiversity can only be chosen by convention," it is implicitly assumed "that no satisfactory definition of 'biodiversity' is forthcoming" (Sarkar, 2005, p.177). Sarkar and others see this conclusion as the starting point for operationalization, which involves choosing "surrogates." "Surrogacy," Sarkar says, "is a relation between a 'surrogate' or 'indicator' variable and a 'target' or 'objective' variable." Sarkar then distinguishes true surrogates from target surrogates, with the former representing "general diversity," while various indicator variables are conventionally chosen to measure success and failure in particular situations – the "target" variable, which can be chosen for ease of measurement. The conventions, however, are not arbitrary and they are not immune to criticism, and Sarkar argues that, through these conventions, one can operationalize "biodiversity" in specific situations (Sarkar, 2004), through a process called "place-prioritization." The idea behind this strategy is to identify biodiversity with that which scientists and advocates are trying to save when they engage in an iterative process of improving biodiversity protection for surrogates chosen as indicators. By repeated applications and adjustment of the algorithm, it is possible to identify and measure species and plots that contribute, represented by their degree of complementarity, to conservation goals.

At this point, the recognition of the impossibility of a definitive index of diversity, and the acceptance of the task of choosing surrogates or proxies, leads into some of the most difficult and divisive issues in conservation biology and practice. In Section 4, below, it will be shown how the conservation paradigm that emerged initially in North America diverged from approaches elsewhere, especially in Australia and South Africa, and how the North American paradigm has been reconsidered, leaving the exact status of conservation theory and practice in the US somewhat unsettled.

3. Two Models of Biodiversity Science and Management

Wildlife protection, of course, has a long history – restrictions on hunting and reservation of herds for the "crown" are ancient. These hunting restrictions, which responded to local declines in particular species, gradually expanded, but were almost always directed at single species that were over-hunted for food, fur, or other products. This single-species approach, which expanded over decades and centuries to apply to more and more *game* species, eventually expanded to apply to "non-game wildlife," as many governments struggled to develop broader programs for wildlife protection in the twentieth century. The persistence of the single-species approach, interestingly, shaped the US Endangered Species Act (ESA), as the Act was so written to protect species, subspecies, and populations of listed species, but mentioned habitat only as a *means* to protect species, once listed.

This trend led to the development of a "small populations model" that dominated early work in conservation biology in North America. While authors of the ESA recognized that extinction was only a part of the problem, they nevertheless directed policy at saving vulnerable species and populations from extinction. By contrast, practitioners in Australia and other countries, also accepting the impossibility of defining biodiversity quantitatively, followed a different path by expecting that the conservation effort will begin with a conventional choice of a number of surrogates. The profound effects of these differing choices will be explored in this section.

While biologists all over the world recognized the seriousness of the problem of biological impoverishment, a new – or newly energized – field of conservation biology developed most visibly in North America, as detailed in Sarkar (2005, p.146). These North American biologists gravitated toward the species-by-species approach and early work in the area was mainly directed at small populations of rare species; emphasis in conservation planning was on designing adequate reserves to protect threatened species. Theoretically, advocates of this approach, noting the analogy between "habitat islands" and true islands, endorsed the theory of island biogeography as the guiding theoretical insight in the design of reserve systems (MacArthur & Wilson, 1967; Harris, 1984). Unfortunately, even as this theoretical structure was applied to reserve design, the theory itself came under increasingly severe criticism (Simberloff, 1976; Margules et al., 1982; Margules, 1989; Soulé & Simberloff, 1986).

Despite these criticisms, the ESA-driven focus on designing reserves to protect species with small populations led to the development of new tools of analysis. The central tool of these academic biologists became the population viability analysis (PVA), which sets out to identify minimum viable populations (MVP) for various small-population species. Responding to the management needs of the ESA, biologists studied the likelihood that stochastic fluctuations in populations of rare species might lead to extinction given various levels of protection. For example, researchers commonly defined MVP as a population that is 95 percent likely to survive for one hundred years (Shaffer, 1978). These approaches, however, were based in neo-positivistic science and paid little attention to context: estimation of likelihood of extinction referred to random occurrences, when it is clear that the management context is important in protecting species. Consequently, it was difficult to apply these relationships to actual conservation problems because such problems are always heavily affected by context.

So, while these analyses were useful in developing goals for single-species recovery plans, it became clear that concentration on establishing MVPs for many species left managers always struggling to save species on the brink of extinction, identifying critical habitat, and trying to set aside enough of it to achieve likely long-term survival of the species. The contextless tools of the academic conservation biologists proved inadequate to the highly contextual problems of protecting biodiversity.

The single-species approach also emphasized reserves designed for particular species or guilds of species, which assumed the manager had effective control of critical habitats. As biologists studied species such as predators requiring large ranges, however, it quickly became clear that concentration on reserves would be inadequate. In one influential, though controversial, study, Newmark (1987) presented evidence that, even in large National Parks in North America, significant percentages of extant species have been lost over decades. These realizations led to a broader recognition that successful conservation measures for large predators and other species with large ranges must address not only the content of reserves, but also the larger context – the landscape matrix – if truly representative patterns of wildlife distribution are to persist. This trend led to a protracted debate about the efficacy – and the opportunity costs – of pursuing a policy of connectance through the prioritization of corridor protection (Harris, 1984).

Questions were also raised about the efficacy of corridors, however (Margules et al., 1982): while corridors can link populations and increase gene pools, they can also lead to the transmission of diseases and to other problems. In the end, most theorists and practitioners agreed that saving all species will require attention to the landscape matrix as well as to reserves; it also became clear that referencing all conservation planning decisions to particular species with small populations would eventually swamp the system, as more and more efforts must be concentrated on perhaps doomed small populations as the march of development drives more and more species to the brink of extinction.

One aspect of this model – which was greatly reinforced by the North Americans' almost unique idealization of wilderness (Nash, 1967; Oelschlager, 1991) – was the opportunistic suggestion that, in order to make the reserve system operate like an archipelago, one needs (analogous to a mainland rich in species) a large wilderness area to serve as a font of speciation and new influxes of species. Implicit in this suggestion was the questionable hypothesis that wilderness areas are necessary to support biodiversity, and evidence was presented that in many cases human activities support and increase biodiversity (Nabhan et al., 1982; Sarkar, 2005). Further evidence showed that the goals of wilderness protection and of biodiversity protection diverge in crucial ways (Sarkar, 1999). This became a hotly debated topic in conservation biology, because of ideological commitments of some scientists to the idea of wilderness, as they hoped to leverage commitments to support biodiversity protection into broader support for wilderness protections. This debate, it should be noted, was also part of a broader, cultural debate over the wilderness ideal in American thought (Cronon, 1995; Callicott & Nelson, 1998).

This sometimes bitter debate about conservation management strategies also had an important international aspect. Scientists and development advocates in developing countries reacted with sharp criticisms of the "wilderness" model, which they sometimes associated with North American rhetoric about "intrinsic value" of nonhuman species (which they took as threats to cultural practices using animals) and with Deep Ecology (which they saw as advocating pro-wilderness policies that would limit their development options) (Gadgil, 1987; Gadgil & Guha, 1992, 1995). Some of this work, drawing on efforts in ecological history (Crosby, 1986; Cronon, 1983; Gadgil & Guha, 1987), drew disturbing parallels between the ideologies of colonialism and of the top-down reserve design approach to protecting biodiversity (Crosby, 1986; Gadgil &Guha, 1992, 1995).

Early application of the idea of maintaining MVPs for endangered species in the developing world, coupled with the questionable assumption that wilderness is necessary for biodiversity protection, led to controversial projects which involved forcible relocation of individuals and villages. Projects such as these have been criticized for their insensitivity to needs of local peoples; it has also been learned that, without local support for a preserve, poaching and other misuses cannot be controlled. Fortunately, there is a growing consensus that projects managed, top down, by scientists basing their calculations on positivistic science must be replaced by more context-sensitive projects that take local needs, as well as biodiversity goals, into account in management.

It has been necessary, then, to modify the single-species approach because of practical failures and because the tools developed to deal with small populations at the edge of extinction did not prove helpful enough in managerial situations. While these techniques promised technically sophisticated tools, these tools were made quantifiable, based on general equations, by ignoring many of the contextual aspects that affect the viability of populations. These idealizations have not proved very helpful in planning conservation strategies, which are highly context-dependent. These tools are gradually being replaced with more holistic approaches, as North American conservation biologists become more comfortable working in a post-positivist, problem-oriented, and place-based way to encounter and address the many contextual issues that shape any environmental policy or management action.

Meanwhile, despite the pressure exerted toward a species-by-species policy by the ESA, and despite the theoretical interest of biologists in studying minimal population levels, the practice of biological conservation – led by the Nature Conservancy and other nongovernmental organizations – scientists and managers in the US have gradually moved away from the single-species model toward saving habitat types and endangered "places." Major policy questions that remain open are whether the Endangered Species Act will be modified to encourage more holistic practice in public sector management, or whether alternative policy interventions will be necessary if biodiversity, however defined, is to be protected in the US.

As noted above, conservation practitioners in Australia and elsewhere, not so affected by the species bias and small populations model, also accepting the undefinability of biodiversity, pursued a different approach, referred to as the "declining populations model" (Caughley, 1994; Margules & Pressey, 2000; Sarkar, 2005). This model, which is embodied in a process called systematic conservation planning, represents a promising alternative to the small populations model for conservation decision making. This approach has been well established outside North America, especially in Australia and South Africa, for several decades. Developed more by practitioners than academics, responding to demand for action in response to disappearing habitats, this approach emphasizes systematic conservation planning over analysis of viability of populations, suggesting the importance of non-random, rather than randomized, models as relevant to survival and extinction. Systematic conservation planning is highly contextual and place-based and it does not base its strategy on choosing a single surrogate, species counts, as the US practitioners tended to do. Instead, addressing the complexity of the problem head-on, this approach develops, tests, and revises a list surrogates that can be argued to track the goals of conservationists. It then applies an algorithm that will protect the most, given constraints placed by the current situation and by the limitations set upon the costs or the amount of land that can be managed. This approach does not attempt to define biodiversity, but rather to design a *process* that will save as many important "surrogates" for biodiversity as possible. Surrogates are various measurable attributes, such as species counts, number of trophic levels, etc., that are chosen to represent protection goals. Conventions that identify favored surrogates must emerge from an iterative, learning process.

This "place-prioritization algorithm" is the central operating principle in protecting diversity because it guides systematic conservation planners to develop a "conservation area network" (CAN) which provides protection for all chosen surrogates. It thus avoids the simplifications involved in setting policy mainly to save species and in ignoring real-world constraints. For example, using complementarity and other rules, one could "maximize biodiversity" in an area by applying a place-prioritization algorithm. That maximization is qualified, however, by recognition of context. For example, the political process might set a limit of no more than 10 percent protected land. That constraint will be factored into the prioritization, so the prioritization is in this sense place-sensitive and also sensitive to political context. Even though it is not possible to define biodiversity in the sense of providing a biodiversity index, it may be possible to identify and track conservation targets, accepting that those targets are shaped not just by biological knowledge, but also by social and political values and commitments expressed in a place.

Sarkar (2005) provides the most comprehensive account of this consensus, presenting it as a ten-step process, including partial solutions to some of the problems faced in designing the most effective habitat protection plan, given constraints (including financial limitations, political obstacles, etc.). Sarkar suggests that, if enough care is given to choosing surrogates in the process of place-prioritization, it is possible to operationalize biodiversity as what would be saved by a process such as this. These prioritizations, if they pay proper attention to rarity and "complementarity – understood as representing the distinct features, or 'biodiversity content,' that protecting a place would add to an existing conservation network," can be argued to embody considerable biological information (Sarkar, 2004). Admittedly, non-biological constraints, including economic constraints, shape the definition, which must be taken into account, so the operationalization can be considered an approximation at best (Soulé & Sanjayan, 1998). An operational definition of biodiversity such as this, while certain not to please those who hope for a connection to higher, romantic themes, is consonant with conventionalism, in that definition is treated as an active process in which action – practice – and theory must interact to allow adaptive, experience-based management in difficult situations where social values are threatened.

Advocates of the systematic conservation planning process, and of the associated declining populations model, endorse the general approach of choosing surrogates within a process of systematic conservation planning, and take the additional step of using the outcomes of a place-prioritization algorithm, as applied in a given context, to operationalize biodiversity goals. Biodiversity is operationalized as what is being maximized when the algorithm is executed (Sarkar & Margules, 2002; Sarkar, 2004). So, one creative response to the difficulties in defining "biodiversity" as an easily measured index of diverse taxons or differences among individuals and groups is to focus not on biological theory, but on the *actions* taken or proposed by scientists and activists who are committed to avoid simplification of the biological world. This shift in focus is justified based on the above-noted conventional nature of the scientific measures developed and chosen; conventionalism relates definitions to efforts at communication and cooperative actions, so it is not unreasonable to operationalize the concept of biodiversity by reference to an action-oriented algorithm executed by experts and participants in particular conservation contexts.

The algorithm, which assumes the identification of conservation targets and surrogate indicators, then solves a constrained optimization problem in two steps. The first step in applying a "heuristic place-prioritization algorithm" is to choose an initial conservation area (perhaps corresponding to existing protected areas), and then adding more areas and restrictions until all of the surrogates have adequate protection. Further steps are undertaken to reduce redundancy and to create a protection plan that is as efficient as possible, reducing costs and allowing alternative uses of some lands. Using the idea of complementarity – exploring the minimal plans that will protect all of some chosen targets – in conjunction with other criteria, such as rarity of opportunities to protect target surrogates – thus provides a powerful tool for quantifying conservation goals, even in the absence of a precise definition of "biodiversity." It is noteworthy that the quantified measures developed within the conservation planning process are sensitive to context – emphasis is placed on protecting those targets that do not yet have sufficient protection in the current situation. For more detail and references regarding the development and application of these algorithms, see Sarkar (2004, 2005).

By carefully choosing appropriate surrogates, it is thus possible, through the development of conventions, to provide a context-sensitive algorithm that will allow intelligent decisions to be made in distributing conservation resources, based on an operationalization of place-prioritization. This shifts from attempting to define "biodiversity" in a measurable way, to choosing a non-arbitrary conventional surrogate – an easily measurable variable, or indicator, that is hypothesized to track what scientists and biodiversity advocates are trying to save. While this is somewhat inelegant as a solution, linguistically, it does support a useful operationalization – one that is useful enough to provide an algorithm for sorting conservation plans, an algorithm that can, in principle, guide action.

Implicit in this set of ideas and strategies is a shift away from earlier approaches, developed in North America, which employed island biogeography as the theoretical support for the idea of developing an archipelago of nature parks and preserves as "habitat islands." Recognizing that it will no doubt be impossible to set aside enough reserves to protect all of biodiversity, and recognizing that humans use lands in many ways, biodiversity protection is moving from tools that emphasize the survival of small populations and based in contextless positivistic science, toward more systematic conservation planning that employs post-positivistic and context-sensitive science. This shift is accompanied by increasing acceptance and advocacy of adaptive management, a set of ideas developed by the philosophical forester, Aldo Leopold. Leopold was both an effective wildlife manager and a "theorist" of considerable integrative power. Leopold succeeded in providing a clear-cut alternative – an approach which combined ecology and management in a process of learning – learning by scientists, by managers, and by the public – "social learning." Leopold argued, in 1939, that "ecology is a new fusion point of all the sciences," and he argued for integrating science into management, advocating experimental management and, when that is impossible, careful observation of impacts of human actions on natural systems (Leopold, 1991, pp.266–7). Leopold, then, can be considered the first "adaptive manager," although the term was introduced much later, by C. S. Holling (1978). Leopold anticipated key aspects of adaptive management and he was clearly aware of the need to address systems instead of individual species. A consensus seems to be emerging that, whatever the exact direction of theory and practice in conservation biology, the protection of biodiversity should be understood as a form of adaptive management – learning through doing.

4. Understanding Biodiversity in Public Policy Discourse

Having called into question the possibility of a synoptic index of biodiversity, and having settled for an operationalization of biodiversity based on conventional specification of priorities for further protection, it is difficult to deny that this complex situation is less than ideal for those who hope to advance biodiversity policy. Despite its usefulness in day-to-day conservation work, and despite success in gaining international attention, as in the Biodiversity Convention signed at the Earth Summit in Rio de Janeiro in 1992, with follow-up activities pursued mainly by nongovernmental organizations, the concept of biodiversity has not proven a good vehicle for communicating the importance of biological conservation to policy makers and the general public. Surely, the difficulty and complexity of defining this term contributes to these communication problems.

As noted, however, terms are tools of communication, and tools can be sharpened and improved. A major task of environmentalists and conservation biologists should be to develop better means of communicating biological and ecological information to policy makers and the public (Norton, 1998). In the meantime, however, the use of these imperfect tools is unavoidable if we are to communicate at all. We must use the linguistic tools we have at hand even as we improve them and fashion new ones.

When Thomas Lovejoy was asked by Takacs to define "biodiversity," Lovejoy said "The term is really supposed to mean diversity at all levels of organization. But the way it's most often used is basically relating to species diversity." He continued: "I think for short operational purposes, that *species diversity* is good shorthand. It's not the whole thing, but as you're rushing around trying to do some things, it's the most easily measured, and it's the one at which the measures are the least controversial. But you're really talking about more than that. You're talking about the way species are put together into larger entities and you're talking about genetic diversity within a species" (Takacs, 1996, p.48). Here, then, we have a leading biologist and conservationist advocating a more complex definition for scientific accuracy, but a simpler definition for use as a shorthand in activist and policy contexts.

These problems and discontents have led to speculation that perhaps another term should be chosen when explaining conservation goals to policy makers and members of the public. For example, one might consider "ecological integrity" as a term that expresses both empirical and evaluative content (Angermeier & Karr, 1994). Biologists and ecologists have not generally been comfortable with this term, and it has resisted definition by operationalization or consensus (Norton, 1998). An alternative strategy would be to create a "pairing" of terms, by which the term "biodiversity" – which seems to work well enough among scientists and conservation practitioners – might be paired with another, less technical, term that would be encouraged in discourse with the public and policy makers. Recent opinion research shows, for example, that less than 60 percent of US citizens are familiar with the term "biodiversity," despite significant attempts to spread the word. Focus-group work has also revealed some active hostility toward the term. This focus-group work suggested that the phrase "web of life" resonates much better with respondents than does "biodiversity." This empirical information suggests the possibility of actively cultivating a link between these by, for example, speaking of "the web of life – what scientists call biodiversity" in writing and speaking to the public and policy makers. If these two (or, perhaps other) terms could be linked as synonyms, "biodiversity" could maintain its scientific rigor, while being linked to a more intuitive idea that has broader public appeal.

So, while there remain problems with the concept of biodiversity, especially as it functions in open discourse about what to protect, the conventionalist stance adopted here suggests that the term should be judged in its role in encouraging cooperative action. As work in systematic conservation planning develops heuristic guides that encourage procedures for setting priorities and maximizing chances of protecting desired aspects of diverse communities, a clearer, contextually sensitive, and publicly useful conception of conservation goals may emerge.

5. Identifying and Measuring Values Derived from Biological Diversity

Having discussed the meaning of biodiversity, and having learned in what senses the concept can be defined – and not defined – it is now possible to explore in more detail the *values* associated with biodiversity. Even though the subject of this collection is *science* and biodiversity, the shift to understanding conservation more contextually, especially the operationalization of biodiversity as outcomes of a value-driven process of place-prioritization, inevitably engrains values in the very concepts, language, and measures of conservation biology. Conservation biology, as noted above, is a normative science.

Because biodiversity value is one type of natural value, this exploration must be shaped by the broader discourse of environmental ethics and environmental economics. This literature, often relying on the terminologies and concepts of a "home" discipline, sets out to assess the value according to theories of value prominent in that discipline. Unfortunately, this discourse, suffering from an excess of ideological formulations of problems, is badly polarized, with factions speaking past each other (Norton, 2005). "Ideology," here, refers to beliefs and conceptualizations that are based on commitments that are not resoluble by experiential evidence. Two prominent examples are: (1) the theory that natural values are, or can be, measured as, economic values (Randall, 1986, 1988), a view on which free enterprise is privileged, and the burden lies on those who would interfere with free markets, with economic growth being implicitly accepted as a dominating good; and (2) the theory that nature has human-independent value, whose advocates call for halting or deeply revising growth plans, citing obligations to the natural world as overriding human-oriented values (Taylor, 1986; Rolston, 1994). Proponents on both sides of this ideological debate, rather than trying to use neutral language to describe and to test hypotheses about values, base their arguments on *theories* of value, theories that are not open to refutation by observation or experiment.

Commitment to a theory determines which values are noticed by advocates of the theory, identifies them with a *type* of value, and creates the categories we cite in arguments to protect species and biodiversity. Choosing our biological categories because of *a priori* theories makes it difficult to link science and values together, because terminology and observations, and data sought, will vary according to the *a priori* requirements of non-empirical theories. The present polarization of environmental discourse results from the misleading and counterproductive opposition between anthropocentric economists and advocates of intrinsic value in nature (Norton, 1991, 2005; Norton & Minteer, 2002–3). Because advocates of these two *a priori* theories of environmental values characterize values in incommensurable ways, discourse about environmental values is polarized, characterized by disagreements about the nature of the problem, the goals to be pursued, and the nature of the social values considered worth protecting politically and legislatively.

It is unnecessary to impose these ideologically based constraints on our discourse about environmental values, which is decidedly pluralistic. Environmental values, which are naturally expressed in many ways, need not be – and should not be – forced into the artificial categories necessitated by the economistic assumption that all objects of value can be commodified and given a price. The ideological debate over how to value nature and its diversity shows no signs of progress toward consensus, as economists assume anthropocentrism in their models, while many philosophers reject economic, anthropocentric valuation and attribute intrinsic value to nature.

It can be argued that the most divisive aspect of the polarization of this debate along ideological lines is that both sides insist they have *the* correct theory of value, insisting that all values that count must count in their own accounting framework. This insistence locks both sides into a framework of analysis that expresses their ideological commitment to a theory about the "nature" of environmental value, and makes communication and compromise unlikely. Such theories have been characterized by Christopher Stone (1987, p.199) as "monistic," and defined as the belief that there is a "single coherent and complete set of principles capable of governing all moral quandaries." It may be possible bypass this disciplinary turf war by adopting a pluralistic stance, based on the observable fact that people value nature in many ways, within different worldviews, and that all of these can legitimately be called "human values," expressed in multiple, not-necessarily-commensurable languages.

The point is that saving biodiversity supports a whole range of values. The list is not competitive from a policy standpoint; stronger legislation to protect biodiversity would protect all of these values. In many cases, it is known what should be done to protect biodiversity, and the diversity of reasons given to support such actions accounts for the strong support of biodiversity goals, even if supporters of a popular pro-environmental policy offer very different justifications based in different worldviews, and use very different language to explain protectionist policies. What is important is to motivate governments and communities to address the biodiversity crisis, practically, in the ways we already know work, and we don't have to agree on an abstract characterization of "the value of biodiversity" to agree that actions to protect biodiversity are justified. The same policies – those that protect the most important kinds of biodiversity – are implicated to protect all or most of these values. On the non-ideological, pluralistic view, values associated with protection are additive, not competitive. Saving biodiversity is good policy, however one justifies it.

Value pluralism can be wedded with an experimental spirit ("adaptive management"), according to which participants in the discussion are encouraged to express their values in their own terms, but then to explain and discuss these values with others, a process that encourages comparison and creation of common concepts for expressing values. Over time, by trying out various terms and definitions of value categories, we can assess whether these terms and categories are serving the purpose of communication and enlightened public discourse or not, which occurs naturally as valuation studies are embedded in adaptive management. This process, in other words, at least opens the possibility of consciously developing more effective linguistic tools for characterizing environmental values and threats to them. Defining biodiversity thus becomes a part of this ongoing quest for a definition that captures the values deriving from a diverse biota, one that correctly captures the best science of the day, and also encourages an open discussion of the importance of policies to protect biodiversity.

According to the pluralistic approach favored here, biodiversity is valued in many, probably incommensurable ways, by many different people. These numerous values placed upon, and derived from, wild species – many of which cannot be quantified and factored into cost–benefit accounting, imply that irreversible species loss should not be justified simply if the known costs of saving a species exceed documentable economic benefits. This implication has led some commentators to propose that the appropriate decision rule for determining endangered species policy is "The Safe Minimum Standard of Conservation" (SMS), which states that the resource – species, in this case – should be preserved, *provided the social costs are bearable* (Bishop, 1978; Norton, 1987). This decision rule suggests that those who propose actions that threaten a species' existence must show that the costs of not doing so are unbearable. While this rule is somewhat vague (but see Norton, 2001, pp.223–7), it does capture the common intuition that, given the irreversibility of extinction, and our inability to predict what will prove important or useful in the future, we should act to save species – or other conservation surrogates – except in highly unusual situations. The SMS can be linked with a cost-effectiveness analysis (CEA), economic methods used to estimate the costs of various strategies to save species. In effect, a CEA *assumes* a species is worth saving, and sets out to determine the least-cost method, relieving preservation advocates of the need to quantify the total value of species.

Because biological diversity is a term that must, at least at present, guide social policies of protection, both values and science must be taken into account. The important point stressed here is that, in order to avoid intractable disagreements, ideology must give way to value pluralism, and all of the rich and varied ways that humans use and value nature must be recognized, made explicit, and balanced against each other. Building on this point, it is possible to state a clear, if rather uninformative, requirement for a useful definition for policy contexts: "biodiversity" should be defined so as to refer to those aspects of natural variety that are socially important enough to entail an obligation to protect those aspects for the sake of future generations.

This otherwise uninformative definition has the advantage of linking biodiversity protection to the core theory of sustainability – the idea that each generation should strive to achieve its own goals in such a way as to not destroy the options and opportunities of those who will live in the future (Norton, 2005). Once this linkage is made strong and clear, efforts to protect biodiversity take their place in the larger effort to live sustainably by learning to modify goals, actions, and policies through an ongoing process of social learning and adaptive management.

6. Conclusion

Biodiversity has become one of the central global problems of environmentalism. E. O. Wilson has said that the destruction of genetic and species diversity now being caused by habitat fragmentation will be "the folly our descendants are least likely to forgive us" (Wilson, 1984). The problem will surely become more difficult as the expansion of human settlements, pollution, and the alteration of waters and the atmosphere all increase the vulnerability of species and ecosystems. The multifaceted nature of biodiversity, it was shown, rules out the development of a definitive index of biodiversity, but the practices and procedures of biodiversity advocates are nevertheless adequate to allow communication and cooperative action in the avoidance of biological simplification. Because the term has been embraced more fully among professional biologists than by policy makers and the public, some have advocated development of somewhat different discourses for biologists and the public, but this proposal is controversial.

The question of placing value upon biodiversity has been hampered by ideological formulations of the nature of environmental value. Rather than argue about whether values in nature are economic or non-instrumental, pluralism is suggested as an inclusive alternative whereby the many ways that humans value nature and biodiversity can be considered additive rather than complementary.

The urgent, perceived need to protect biodiversity has encouraged the development of the discipline of conservation biology and a broad set of procedures and various methods have evolved. In the United States, perhaps too much emphasis remains focused on small populations of rare species, while in Australia and other parts of the world, conservation biologists have developed algorithms to design plans for the efficient protection of chosen surrogates. Despite these regional differences, conservation biologists are evolving toward a consensus of protective policies that, first, emphasize protecting habitat and ecological communities that are not adequately protected, with concerns for rare species being integrated into a broader conservation targets approach. Second, this emerging approach, rejecting positivistic, narrowly scientific approaches to understanding impoverishment, is solidly contextual: its advocates take political and social aspects of the problem to be inherent in the process of designing systematic conservation plans. Prospects for the future are highly in doubt, and methods are only now being developed to respond to what is often described as an "extinction crisis," which would only be one result of the biological impoverishment of the earth.

References

Angermeier, P. L., & Karr, J. R. (1994). Biological integrity versus biological diversity as policy directives. *Bioscience*, 44, 690–97.

Bishop, R. (1978). Endangered species and uncertainty: the economics of a safe minimum standard. *American Journal of Agricultural Economics*, 60, 10–18.

Brower, L. P., & Malcolm, S. B. (1991). Animal migrations as endangered phenomena. *American Zoologist*, 31, 265–76.

Callicott, J. B., & Nelson, R., MP. (1998). *The great new wilderness debate*. Athens, GA: University of Georgia Press.

Carnap, R. (1937). *The logical syntax of language*. London: Routledge & Kegan Paul.

Carnap, R. (1956). Empiricism, semantics and ontology. In *Meaning and Necessity*. Chicago: University of Chicago Press.

Caughley, G. (1994). Directions in conservation biology. *Journal of Animal Ecology*, 63, 215–44.

Cervigni, R. (2001). *Biodiversity in the balance: land use, national development, and global welfare*. Cheltenham, UK: Edward Elgar.

Cronon, W. (1983). *Changes in the land: Indians, colonists, and the ecology of New England*. New York: Hill & Wang.

Cronon, W. (Ed.). (1995). *Uncommon ground: toward reinventing nature*. New York: W.W. Norton.

Crosby, A. W. (1986). *Ecological imperialism: the biological expansion of Europe, 900–1900*. Cambridge: Cambridge University Press.

Dewey, J. (1910). *The influence of Darwin on philosophy and other essays in contemporary thought*. New York: Henry Holt & Company, Inc.

Faith, D. P. (1994). Phylogenetic pattern and the quantification of organismal biodiversity. *Philosophical Transactions of the Royal Society of London, Series B*, 345, 45–58.

Faith, D. P. (1992). Conservation evaluation and phylogenetic diversity. *Biological Conservation*, 61, 1–10.

Faith, D. P. (2003). Biodiversity. In Edward N. Zalta (Ed.). *The Stanford encyclopedia of philosophy* (Summer 2003 Edition), URL= <http://plato.stanford.edu/archives/sum2003/entries/biodiversity/>.

Gadgil, M. (1987). Diversity: cultural and biological. *Trends in Ecology and Evolution*, 2, 369–73.

Gadgil, M., & Guha, R. (1992). *This fissured land: an ecological history of India*. Oxford: Oxford University Press.

Gadgil, M., & Guha, R. (1995). *Ecology and equity: the use and abuse of nature in contemporary India*. London: Routledge.

Gaston, K. J. (1996). *Biodiversity: a biology of numbers and difference*. Oxford: Blackwell Science.

Hammond, P. (1995). The current magnitude of biodiversity. In V. H. Heywood & R. T. Watson (Eds). *Global biodiversity assessment* (pp. 113–38). Cambridge: Cambridge University Press.

Harris, L. D. (1984). *The fragmented forest*. Chicago: The University of Chicago Press.

Holling, C. S. (Ed.). (1978). *Adaptive environmental assessment and management.* Chicester: Wiley.

Hunter, M. Jr. (2002). *Fundamentals of conservation biology* (2nd edn). Malden, MA: Blackwell Science.

Leopold, A. (1991). A biotic view of land. In S. Flader & J. B. Callicott (Eds). *The river of the mother of god and other essays by Aldo Leopold.* Madison, WI: University of Wisconsin Press.

MacArthur, R. A., & Wilson, E. O. (1967). *The theory of island biogeography.* Princeton: Princeton University Press.

Magurran, A. E. (1988). *Ecological diversity and its measurement.* Princeton: Princeton University Press.

Magurran, A. E. (2003). *Measuring biological diversity.* Oxford: Blackwell Publishing.

Margules, C. R. (1989). Introduction to some Australian developments in conservation evaluation. *Biological Conservation,* 50, 1–11.

Margules, C. R., Higgs, A. J., & Rage, R. W. (1982). Modern biogeographic theory: are there lessons for nature reserve design? *Biological Conservation,* 24, 115–28.

Margules, C. R., & Pressey, R. L. (2000). Systematic conservation planning. *Nature,* 363, 242–53.

Nabhan, G. P., Rea, A. M., Reichardt, K. L., Melink, E., & Futchinson, E. F. (1982). Papago influences on habitat and biotic diversity: Quitovak Oasis ethnoecology. *Journal of EthnoBiology,* 2, 124–43.

Nash, R. (1967). *Wilderness and the American mind.* New Haven: Yale University Press.

Newmark, W. D. (1987). A land-bridge perspective on mammal extinctions in western North American parks. *Nature,* 325, 430.

Norton, B. G. (1987). *Why preserve natural variety.* Princeton: Princeton University Press.

Norton, B. G. (1991). *Toward unity among environmentalists.* New York: Oxford University Press.

Norton, B. G. (1998). Improving ecological communication. *Ecological Applications,* 8, 350–64.

Norton, B. G. (2001). What do we owe the future? How should we decide? In V. Sharpe, et al. (Eds). *Wolves and human communities.* Washington, DC: Island Press.

Norton, B. G. (2005). *Sustainability: a philosophy of adaptive ecosystem management.* Chicago: University of Chicago Press.

Norton, B. G., & Minteer, B. A. (2002–3). From environmental ethics to environmental public policy: 1970–present. In T. Teintenberg & H. Folmer (Eds). *The international yearbook of environmental & resource economics.* Cheltenham, UK: Edward Elgar Publishing.

Oelschlager, M. (1991). *The idea of wilderness.* New Haven: Yale University Press.

Pielou, E. C. (1975). *Ecological diversity.* New York: John Wiley.

Pressey, R. L., Possingham, H. P., & Day, J. R. (1997). Effectiveness of alternative heuristic algorithms for approximating minimum requirements for conservation reserves. *Biological Conservation,* 80, 207–19.

Randall, A. (1986). Human preferences, economics, and the preservation of species. In B. G. Norton (Ed.). *The preservation of species* (pp. 79–109). Princeton: Princeton University Press.

Randall, A. (1988). What mainstream economists have to say about the value of biodiversity. In B. G. Norton (Ed.). *The preservation of species* (pp. 79–109). Princeton: Princeton University Press.

Rolston, H., III. (1994). *Conserving natural value.* New York: Columbia University Press.

Sarkar, S. (1999). Wilderness preservation and biodiversity conservation – keeping divergent goals distinct. *Bioscience,* 49, 405–12.

Sarkar, S. (2004). Conservation biology. In Edward N. Zalta (Ed.). *The Stanford encyclopedia of philosophy* (Winter 2004 Edition), URL=<http://plato.stanford.edu/archives/win2004/entries/conservation-biology/>.

Sarkar, S. (2005). *Biodiversity and environmental philosophy: an introduction to the issues.* New York: Cambridge University Press, Sections 6.3, 6.5.

Sarkar, S., & Margules, C. R. (2002). Operationalizing biodiversity for conservation planning. *Journal of Biosciences*, 27(S2), 299–308.

Shaffer, M. L. (1978). *Determining minimum viable population sizes: a case study of the grizzly bear.* PhD Dissertation, Duke University.

Simberloff, D. S. (1976). Species turnover and equilibrium island biogeography. *Science*, 194, 572–8.

Solow, A., Polasky, S., & Broadus, J. (1993). On the measurement of biological diversity. *Journal of Environmental Economics and Management*, 24, 60–8.

Soulé, M. E., & Sanjayan, M. A. (1998). Conservation targets: do they help? *Science*, 279, 2060–1.

Soulé, M. E., & Simberloff, D. S. (1986). What do genetics and ecology tell us about the design of nature reserves? *Biological Conservation*, 35, 19–40.

Stone, C. (1987). *Earth and other ethics.* New York: Harper & Row.

Takacs, D. (1996). *The idea of biodiversity: philosophies of paradise.* Baltimore: The Johns Hopkins University Press.

Taylor, P. (1986). *Respect for nature.* Princeton: Princeton University Press.

Weitzman, M. L. (1998). The Noah's ark problem. *Econometrica*, 66, 1279–98.

Whittaker, R. H. (1960). Vegetation of the Siskiyou Mountains, Oregon and California. *Ecological Monographs*, 30, 279–338.

Whittaker, R. H. (1972). Evolution and measurement of species diversity. *Taxon*, 21, 213–51.

Whittaker, R. H. (1975). *Communities and ecosystems* (2nd edn). New York: MacMillan.

Wilson, E. O. (1984). *Biophilia* (p. 121). Cambridge, MA: Harvard University Press. Quotation retrieved July 27, 2007 from <http://www.brainyquote.com/quotes/authors/e/e_o_wilson.html>.

Wilson, E. O. (Ed.). (1988). *Biodiversity.* Washington, DC: National Academy Press.

Wood, P. (1997). Biodiversity as the source of biological resources: a new look at biodiversity values. *Environmental Values*, 6, 251–68.

Wood, P. (2000). *Biodiversity and democracy: rethinking society and nature.* Vancouver, BC: University of British Columbia Press.

Further Reading

Callicott, J. B. (1989). *In defense of the land ethic: essays in environmental philosophy.* Albany: State University of New York Press.

Carson, R. (1962). *Silent spring.* New York: Fawcett Publications.

Costanza, R., d'Arge, R., de Groot, R., Farber, S., Grasso, M., Hannon, B., Limburg, K., Haeem, S., O'Neill, R. V., Paruelo, J., Raskin, R. G., Sutton, P., & van den Belt, M. (1997). The value of the world's ecosystem services and natural capital. *Nature*, 387, 253–60.

Daily, G. C. (1997). *Nature's services: societal dependence on natural ecosystems* Washington, DC: Island Press.

Ehrlich, P. R., & Ehrlich, A. (1981). *Extinction: the causes and consequences of the disappearance of species.* New York: Random House.

Gadgil, M., & Guha, R. (1992). *This fissured land: an ecological history of India.* New Delhi: Oxford University Press.

Goodpaster, K. (1978). On being morally considerable. *Journal of Philosophy*, 75L, 308–25.

Guha, R. (1989). Radical American environmentalism and wilderness preservation: a third world critique. *Environmental Ethics*, 11, 71–83.

Kellert, S. R. (1996). *The value of life: biological diversity and human society.* Washington, DC: Island Press.
Leopold, A. S. (1949). *A Sand County almanac and essays here and there.* Oxford, UK: Oxford University Press.
Marsh, G. P. (1965). *Man and nature: or, physical geography as modified by human action.* Cambridge, MA: Belknap Press of Harvard University Press. (Originally published 1864).
Lawton, J. H., & May, R. M. (Eds). (1995). *Extinction rates.* Oxford: Oxford University Press.
Meffe, G. K., & Carroll, C. R. (1994). *Principles of conservation biology.* Sunderland, MA: Sinauer Associates.
Nabhan, G. P. (1997). *Cultures of habitat: on nature, culture, and story.* Washington, DC: Counterpoint.
Norton, B. G. (1986). *The preservation of species.* Princeton: Princeton University Press.
O'Neill, J. (1993). *Ecology, policy and politics: human well-being and the natural world.* London, UK: Routledge.
Pearce, D. W. (1993). *Economic value and the natural world.* Cambridge, MA: MIT Press.
Pimm, S. L. (1991). *The balance of nature? Ecological issues in the conservation of species and communities.* Chicago: University of Chicago Press.
Pressey, R. L., & Cowling, R. M. (2000). Reserve selection algorithms and the real world. *Conservation Biology*, 15, 275–7.
Primack, R. B. (1993). *Essentials of conservation biology.* Sunderland, MA: Sunderland.
Rolston, H., III. (1988). *Environmental ethics: duties to and values in the natural world.* Philadelphia, PA: Temple University Press.
Scott, M. (forthcoming). *The Endangered Species Act at 30.* Washington, DC: Island Press.
Shafer, M. L. (1981). Minimum population sizes for species conservation. *Bioscience*, 31, 131–4.
Simberloff, D. (1998). Flagships, umbrellas, and keystones: is single-species management passé in the landscape era? *Biological Conservation*, 83, 247–57.
Soulé, M. E. (1985). What is conservation biology? *Bioscience*, 35, 727–34.
Soulé, M. E., & Orians, G. H. (Eds). (2001). Conservation biology: research biology for the next decade. Washington, DC: Island Press.
Stone, C. D. (1974). *Should trees have standing? Toward legal rights for natural objects.* Los Altos, CA: William Kaufmann.
Taylor, P. W. (1986). *Respect for nature: a theory of environmental ethics.* Princeton: Princeton University Press.
Tilman, D. (1999). The ecological consequences of biodiversity: a search for general principles. *Ecology*, 80, 1455–74.
Weitzman, M. L. (1998). The Noah's ark problem. *Econometrica*, 66, 1279–98.

Part VI

Mind and Behavior

Chapter 21

Ethology, Sociobiology, and Evolutionary Psychology

PAUL E. GRIFFITHS

"It is only a comparative and evolutionary psychology that can provide the needed basis; and this could not be created before the work of Darwin."

William McDougall, *Introduction to Social Psychology*, 1908

1. A Century of Evolutionary Psychology

The evolution of mind and behavior was of intense interest to Charles Darwin throughout his life. His views were made public a decade before his death in *The Descent of Man* (e.g., 1981 [1871]) and *The Expression of the Emotions in Man and Animals* (1965 [1872]). Evolutionary psychology has been an active field of research and a topic of public controversy from that time to the present. At least four distinct phases can be distinguished in the development of evolutionary psychology since Darwin and his immediate successor George Romanes. These are: instinct theory, classical ethology, sociobiology, and Evolutionary Psychology, the last of which I capitalize to distinguish it from evolutionary psychology in general.

The instinct theories of Conwy Lloyd Morgan, James Mark Baldwin, William James, William McDougall, and others were an important part of early-twentieth-century psychology (Richards, 1987) but will not be discussed here because no trace of these theories can be discerned in evolutionary psychology today. It was not until the years leading up to World War II that the ethologists Konrad Lorenz and Nikolaas Tinbergen created the tradition of rigorous, Darwinian research on animal behavior that developed into modern behavioral ecology (Burkhardt, 2005). At first glance, research on specifically human behavior seems to exhibit greater discontinuity than research on animal behavior in general. The "human ethology" of the 1960s appears to have been replaced in the early 1970s by a new approach called "sociobiology." Sociobiology in its turn appears to have been replaced by an approach calling itself Evolutionary Psychology. Closer examination, however, reveals a great deal of continuity between these schools. While there have been genuine changes, many of the people, research practices, and ideas of each school were carried over into its successors. At present, while Evolutionary Psychology is the most visible form of evolutionary psychology, empirical and theoretical research on the evolution of mind and behavior is marked by a diversity of ideas and approaches and it is far from clear which direction(s) the field will take in future.

2. The Study of Instinct

In the period immediately following World War I, many psychologists rejected the previously uncontroversial idea of human instinct. This rejection reflected a number of concerns, including the fear that classifying behaviors by their biological function would not create natural psychological or neurological groupings, and the view that "instinct" was a pseudo-scientific substitute for causal explanation (Dunlap, 1919; Kuo, 1921). The concept of instinct was reconstructed in a fresh and more viable form in the mid-1930s, primarily in the work of Konrad Lorenz. In his view, "the large and immeasurably fertile field which innate behaviour offers to analytic research was left unploughed because it lay, as no man's land, between the two fronts of the antagonistic opinions of vitalists and mechanists" (Lorenz, 1950, p.232). Lorenz criticized the behaviorists for reducing the biological endowment of animals to a small number of reflex reactions destined to be assembled into complex adult behaviors by associative learning. But he was also a stern critic of the vitalistic theories of instinct propounded by McDougall and by the leading Dutch comparative psychologist Abraham Bierens de Haan. In his criticism of these authors Lorenz rejected the traditional picture of instincts such as "parenting," which influence the production of many specific behaviors. Instead, Lorenz argued that when a bird "instinctively" feeds its offspring it has no motivation beyond an immediate drive to perform the act of regurgitation in the presence of the stimulus presented by the begging chick. The appearance of an overarching "parenting instinct" is produced by the interaction of a large number of these highly specific instincts and the stimuli (and self-stimuli) which impact the bird in its natural environmental setting. But while traditional instincts were too nebulous for Lorenz, he was convinced that the mechanistic substitutes envisaged by behaviorism – reflexes and tropisms – were inadequate to explain the rich repertoire of instinctive behaviors. Lorenz was committed to the ultimate reduction of instincts to neural mechanism, but such neural mechanisms, he believed, would be far more sophisticated than mere chain reflexes, or tropisms. It is here that we find the significance of Lorenz's famous drive-discharge or "hydraulic" model of instinctual motivation (Figure 21.1). The hydraulic model was complex enough to account for the observed behavior, but simple enough that it might in future be directly mapped onto neural pathways and humoral influences on those pathways.

Drawing on ideas from contemporary neuroscience, Lorenz suggested that the nervous system continuously generates impulses to perform instinctive behaviors, but that the behaviors manifest themselves only when special inhibitory mechanisms are "released" by an external stimulus. Using a mechanical analogy, Lorenz pictured each instinct as a reservoir in which a liquid (R) continually accumulates. The outlet of the reservoir is blocked by a spring-loaded valve (V) which can be opened by presentation of a highly specific sensory "releaser" (Sp). When the valve is opened, the contents of the reservoir ("action specific energy") flow to motor systems and produce the instinc-

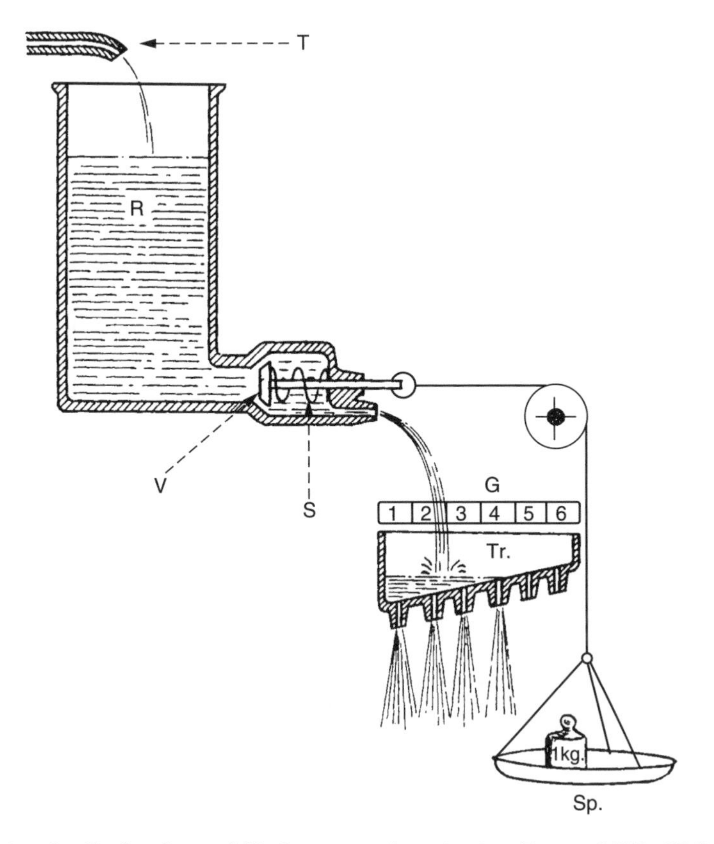

Figure 21.1 The "hydraulic model" of instinctual motivation (Lorenz 1950, 256)

tual behavior pattern. A signal virtue of the hydraulic model, according to Lorenz, is that it captures the apparent spontaneity of some animal behavior: if no releasing stimulus is available, then pressure can accumulate to the point where the valve is forced open and the animal performs instinctive behaviors "in a vacuum." Lorenz regarded the observed phenomenon of "vacuum activities" as one of the most critical clues to the nature of instinctual motivation. Another form of spontaneous behavior to which Lorenz drew attention was "appetitive behavior" – behavior which increases the probability of finding a releasing stimulus for the instinct. He postulated that the accumulation of action specific energy in the reservoir directly causes appetitive behavior. Thus, when external factors such as day-length put it in a suitable hormonal state, a bird will initiate appetitive behaviors that result in it coming into contact with nesting materials which act to release instinctive nest-building behaviors.

A striking feature of Lorenz's instinct theory is that the coordination of instinctive behavior into effective sequences is dependent on the distribution of releasing stimuli in the organism's natural environment. Although each specific instinct – collecting twigs at nesting time, inserting twigs into the nest, and so forth – corresponds to a neural mechanism, the larger structure of instinctual behavior only emerges in the interaction between those mechanisms and the organism's natural environment.

The environment has thus taken over the role of nebulous coordinating forces like the "nesting instinct" postulated by earlier instinct theories. It follows that the study of instinctive behavior requires the observation of the organism in its natural environment.

The program of classical ethology was laid out in Tinbergen's *The Study of Instinct* (1951). The book brought together an impressive body of data and theory concerning animal behavior and showed how far the field had come in recent years, just as Edward O. Wilson's *Sociobiology: The new synthesis* was to do a quarter of a century later (Wilson, 1975). In a striking parallel between the two books, Tinbergen concluded with a more speculative chapter in which the spotlight of the new science was turned on the human mind. But in contrast to the storm that broke over Wilson's head as a result of his chapter on humans, Tinbergen's work was received with general enthusiasm. It played a significant role in the rise of natural history as an entertainment genre in the new medium of television.

The theoretical framework of ethology evolved rapidly in the 1950s and 60s. Three important developments were: 1) the abandonment of Lorenz's identification of instinctive behavior with behavior which is innate as opposed to acquired; 2) the abandonment of the hydraulic model; 3) the integration of ethology with evolutionary ecology, resulting in an increased focus on documenting the adaptive value of behavior.

The eclipse of the Lorenzian concept of innateness in Britain is normally attributed to the influence of American developmental psychobiology in general and Daniel S. Lehrman in particular. Lorenz had denied that instinctive behavior can be "fine-tuned" by experience, as earlier instinct theorists had apparently described in cases such as pecking for grain in chickens. Instead, Lorenz insisted that behavior sequences can always be analyzed to reveal specific components that are innate and other components that are acquired. The innate elements, he thought, were to be explained in terms of the endogenous development of underlying nervous tissue – instincts grow in much the same way as limbs. Lehrman's famous critique documented the fact that endogenous and exogenous influences on behavioral development interact in numerous ways, and that no one pattern of interaction is distinctive of the evolved elements of the behavioral phenotype (Lehrman, 1953). The development of behavior which is instinctive in the sense that it has been designed by natural selection often depends on highly specific environmental influences. Lehrman was also critical of Lorenz's use of the deprivation experiment (raising animals in social isolation and without the ability to practice a behavior) to infer that a behavior is innate *simpliciter*, rather than merely that the factors controlled for in the experiment are not needed for the development of that behavior. Lehrman had been personally acquainted with Tinbergen since before WWII and many of his ideas were incorporated into mainstream ethological theory in Britain (see Tinbergen, 1963b, pp.423–7). Ethological work in the 1960s displayed a sophisticated understanding of the relationship between developmental and evolutionary explanations.

The Cambridge ethologist Robert A. Hinde was probably the first to argue explicitly that the hydraulic model had outlived its usefulness (Hinde, 1956; and see Burkhardt, 2005). Lorenz had created a physical analogy which captured certain observations about instinctive behavior. Tinbergen had already recognized the inadequacy of the original model and had suggested a more complex model along the same lines, with a series of hierarchically organized centers of instinctual motivation influencing one another and, eventually, behavior (Tinbergen, 1951). But Hinde argued that the implications of the fundamental hydraulic analogy had not been borne out by subsequent research. In particular, any empirically adequate model would have to allow "energy" to flow back "uphill" or against the pressure-gradient, thus contradicting the central feature of the analogy. Reliance on the hydraulic model in ethological research was replaced by empirical research on the neurological factors affecting instinctual behavior. Research on the endocrine system was particularly prominent, because this was experimentally tractable at the time.

The third major theoretical development in animal behavior research in the 1950s and 1960s resulted not from external or internal critique, but from the fusion of ethology with a powerful existing British research tradition. Lorenz thought of ethology as the application to behavior of the principles of comparative morphology, the science in which he had been trained as in Vienna as a young man. This led him to reject the orthodox view that behavior is more evolutionarily labile than anatomy: "Such innate, species-specific motor patterns represent characters *that must have behaved like morphological characters in the course of evolution.* Indeed, they must have behaved like *particularly conservative characters*" (Lorenz, 1996 [1948], p.237, his emphases). In British ethology, this emphasis on behaviors as taxonomic characters was replaced by an emphasis on behaviors as adaptations, a change which reflected the greater role of evolutionary ecology in post-synthesis evolutionary biology in Britain, and particularly in Oxford (Burkhardt, 2005). As Tinbergen noted:

> Being a member of the Oxford setup gave me the unique chance to absorb through daily personal contacts, the typical ecology and evolution study-oriented atmosphere of Oxford zoology. Life in this academic community . . . influenced my entire outlook, and the group I now began to build up, from very modest beginnings indeed, began to produce work with a distinctly Oxonian flavour. (Tinbergen, 1985, p.450–1; see also Tinbergen, 1963a)

David Lack, the dominant figure in Oxford ornithology at this time, focused on the ecological functions of bird behavior. A similar emphasis was soon apparent in the work of Tinbergen and his students on the comparative behavior of seabirds, most famously in Esther Cullen's groundbreaking studies of the cliff-nesting adaptations of the Kittiwake. Similarities and dissimilarities between species were interpreted in terms of differing selection regimes as well as, and increasingly instead of, taxonomic relationships.

The mature ideas of the "Tinbergen school" were embodied in the influential programmatic paper "On the aims and methods of ethology" (1963b). Tinbergen began with his favored definition of ethology: "the biology of behavior." Building on previous analyses by Julian Huxley and Ernst Mayr, he argued that the biological study of an organism asks four questions:

(1) Causation
(2) Survival value
(3) Ontogeny
(4) Evolution.

Questions of causation ask what mechanism underlies an observed behavior, such as the collection of nesting materials. The hydraulic model was a hypothesis about causation, albeit an inadequate one.

Questions of survival value ask: "whether any effect of the observed process contributes to survival, if so, how survival is promoted and whether it is promoted better by the observed process than by slightly different processes." This question was the focus of a rich experimental tradition at Oxford, of which H. B. D Kettlewell's studies of industrial melanism in the peppered moth are the most famous example. The mistaken view that survival value cannot be studied by "exact experimentation," Tinbergen argued, reflects "a confusion of the study of natural selection with that of survival value" (1963b: p.418). Even creationists would need to answer questions of survival value: "To those who argue that the only function of studies of survival value is to strengthen the theory of natural selection I should like to say: even if the present-day animals were created the way they are now, the fact that they manage to survive would pose the problem of *how they do this*" (1963b, p.423, my emphasis).

Questions of ontogeny ask how the mechanisms revealed by the study of causation are built. After Lehrman's intervention, work on this question by British ethologists resembled the existing, primarily American, tradition of developmental psychobiology.

Questions of evolution have "two major aims: the elucidation of the course evolution must be assumed to have taken, and the unraveling of its dynamics" (1963b, p.428). The course of evolution is revealed by inferring phylogenies and homologies, as Lorenz had stressed. The dynamics of evolution are revealed by the study of 1) population genetics and 2) survival value (1963b, p.428), studies which correspond to Elliot Sober's (1984) "consequence laws" and "source laws" in evolutionary theory. Source laws explain why one type of organism is fitter than another, while consequence laws tell us what will happen at the population level in virtue of those differences. The study of survival value, Tinbergen notes, can more or less directly demonstrate the "stabilizing" role of particular selection pressures in the evolutionary present, but to infer a larger, "molding" role for those selection pressures in the evolutionary origin of traits we need additional, historical evidence.

Tinbergen's four questions are still used as a framework for research in behavioral biology today (e.g., Manning & Dawkins, 1998).

3. The Triumph of Adaptationism

Lorenz, Tinbergen, and the discoverer of bee language, Karl von Frisch, were awarded a joint Nobel Prize in 1973 for their roles in creating a new science of animal behavior. Ironically, the discipline with which they were so strongly identified – ethology – was on the brink of being eclipsed by a new approach to animal behavior – sociobiology. By the mid-1980s one would have been hard pressed to find a young student of animal behavior who regarded their work as a contribution to ethology, as opposed to behavioral ecology or sociobiology. In their books and journals older ethologists were telling the story of the disappearance of their discipline (e.g., Bateson & Klopfer, 1989). If some of the more polemical writings of early sociobiologists are to be believed, ethology had never risen above the level of descriptive natural history, and had never assimilated the evolutionary biology of the modern synthesis (Barkow, 1979; Barash, 1979). But in reality, the 1970s saw, not the triumph of "sociobiology" over "ethology," but the triumph of adaptationism within English-speaking ethology, so that what can perhaps be most neutrally described as "behavioral ecology" came to dominate animal behavior studies.

By the early 1970s the population genetic models of William D. Hamilton (1964) had created a theoretical tradition that was readily combined with the experimental tradition created by Tinbergen. Behavioral ecologists set out to test the predictions of the new population genetic models through the study of the survival value of different phenotypes in the laboratory and the field. Tinbergen's "survival value" and "evolution" questions came to be seen as the primary questions in animal behavior research. Hence, while the term "sociobiology" was introduced in a revolutionary manner, the research it denoted had come into existence by a far more gradual path. The idea that sociobiology was a break with the past must be primarily credited to Edward O. Wilson, the Harvard biologist who used it as the title of his 1975 book announcing a "new synthesis" in behavioral biology (Wilson, 1975). The term "sociobiology" had been used in various senses since the 1940s (as in the name of the Ecological Society of America's "Section of Animal Behavior and Sociobiology"). Wilson recruited it as a label with which to draw attention to the changes that had occurred in animal behavior research over the previous decade. His book was the subject of public controversy of quite extraordinary intensity, for complex reasons which historians and sociologists of science are only now starting to comprehend (Segerstråle, 2000), and "sociobiology" passed into popular usage as a general term for evolutionary approaches to mind and behavior.

Wilson's book was also the subject of controversy *within* animal behavior studies, for reasons which are easier to comprehend. In a famous amoeba-like diagram, Wilson predicted that sociobiology would ingest and absorb all those parts of behavioral biology that were not ingested and absorbed by an equally voracious cellular neurobiology.

This vision was not welcomed by the existing community of animal behavior researchers, as is evident from the multi-authored review symposium in *Animal Behaviour* (Baerends et al., 1976). At the simplest level, ethologists were reacting in a predictable way to being told that their discipline was outmoded, but that was not all that lay behind their response. Wilson's diagram and the accompanying discussion leave no room for major elements of the research agenda laid out in Tinbergen's "four questions." In effect, Wilson was trying to reduce Tinbergen's quadripartite distinction to Ernst Mayr's equally well-known bipartite distinction between "proximate" and "ultimate" questions in biology. In the process he left out important topics that figured in animal behavior research in the 1960s. Students of behavioral development, for example, did not see themselves fitting into either "cell biology" or "population biology." While their work had a clear role in Tinbergen's ethology, it was not part of Wilson's "new synthesis." As the leading birdsong researcher Peter Slater has written: "E. O. Wilson (1975), in his 'dumb-bell model,' predicted that animal behavior would be swallowed up by neurobiology at one end and sociobiology at the other. As far as song is concerned he has been largely right but only if, as sociobiologists are prone to do, one ignores development" (Slater, 2003).

A distinctive feature of the new behavioral ecology/sociobiology was the conviction that Tinbergen's four questions are not, as he himself had thought, closely interlinked. During the 1960s different ethological research groups had come to focus on different parts of the Tinbergian research program (Durant, 1986, p.1612; see also Burkhardt, 2005). Ethology as a discipline ceased to exist when these groups ceased to see themselves as tackling different aspects of the same problem – the biology of behavior. Researchers like Richard Dawkins, whose favored part of the Tinbergen program was included in the new behavioral ecology, felt no sense of rupture with their earlier work: "My own dominant recollection of [Tinbergen's] undergraduate lectures on animal behavior was of his ruthlessly mechanistic attitude to animal behavior and the machinery that underlay it. I was particularly taken with two phrases of his – 'behavior machinery' and 'equipment for survival'. When I came to write my own first book I combined them into the brief phrase 'survival machine'" (Dawkins, Halliday, & Dawkins, 1991, p.xii). From this perspective *The Selfish Gene* (Dawkins, 1976) differs from *The Study of Instinct* only because of the smooth progress of scientific knowledge. But researchers whose favored Tinbergian questions were "causation" and "ontogeny" found themselves excluded from a new, and highly successful, phase in the study of animal behavior.

It is clear that behavioral ecology/sociobiology and the study of adaptive value and evolutionary origins had a "comparative advantage" over the study of causation and ontongeny during the 1970s. No new discipline comparable to behavioral ecology arose from the other parts of Tinbergen's program, and the rising generation of animal behavior researchers was predominantly attracted to behavioral ecology. Two possible reasons can be advanced for this. First, behavioral ecology made it possible to see particular studies as tests of general hypotheses about the evolutionary process. Behavioral ecology possessed game-theoretic and population-genetic models of a very high degree of generality, and a single, practicable study in the field or the laboratory could constitute a test of the predictions of an entire class of models, such as optimal foraging theory or parental investment theory. With the possible exception of the template theory of song acquisition in passerine birds, the study of causation and behavioral development did not offer general theories of a kind whose adequacy could be meaningfully tested in a single series of experiments. It is not difficult to see why a field in which a practicable series of experiments could test an important theory would be more appealing to young researchers than a field in which in which this appeared impossible. Second, the study of behavioral causation and ontogeny was simply not able to keep up with the study of adaptive value, forcing pragmatic researchers to look for ways to make their research independent of answers to such apparently intractable questions. Studies of causation and ontogeny could, in principle, have contributed to behavioral ecology in a very direct way, by determining a realistic "phenotype set" available for selection to act upon, but in almost all cases those studies were not advanced enough to provide this information. In practice, the phenotype sets of evolutionary models were based on what actually occurs in nature, or on what seemed biologically plausible to the researchers. Hence, instead of developmental biology making a positive contribution to behavioral ecology, it appeared only in the negative role of "developmental constraints" – sets of phenotypes that were inferred on indirect evidence to be in some way unattainable (Maynard Smith et al., 1985). Some researchers argued that the most practicable way to determine the phenotype set was to

		Organism 2	
		Cooperate C	**Defect D**
Organism 1	Cooperate C	a	**b**
	Defect D	**c**	**d**

Figure 21.2 The Prisoner's Dilemma. The values c > a > d > b are the payoff to organism 1 for each possible pair of phenotypes of organisms 1 and 2

build models of optimal adaptation and see when they failed to predict the phenotype observed in nature (Maynard Smith, 1987). One explanation of an organism's failure to manifest the optimal phenotype is that it is not part of the available phenotype set.

4. From Sociobiology to Evolutionary Psychology

Human sociobiology straightforwardly applied the methods of behavioral ecology to the human species. Human behaviors were treated as optimal solutions to adaptive problems, or, more usually, as "evolutionarily stable strategies" in game-theoretic models of competition between organisms (Maynard Smith, 1982). An evolutionarily stable strategy, or ESS, is a phenotype such that, if all members of a population have that phenotype, no mutant phenotype can increase in frequency in the population. The ESS concept is the appropriate conception of an evolutionary equilibrium when selection is "frequency dependent," meaning that the adaptive value of a strategy depends on which strategies are used by the rest of the population. Much human behavioral evolution seems likely to have involved frequency-dependent selection. One of the most prominent topics of research in human sociobiology was the evolution of altruistic behavior, which was seen as the key to the evolution of social behavior more generally. The evolutionary problem posed by the existence of altruistic behavior can be made clear using the game matrix known as "prisoner's dilemma" from the story about two accomplices who are each offered a reduced sentence for betraying the other (Figure 21.2).

The important feature of the prisoner's dilemma is that no matter whether organism 2 has the cooperative phenotype C or the defecting phenotype D, organism 1 will receive a higher payoff if they have the defecting phenotype D, since c > a and d > b. For example, whether or not organism 2 is willing to share food, organism 1 will do better if they are not willing to share food. Hence D is an evolutionary stable strategy: if everyone in an evolving population has phenotype D then any mutant with phenotype C will be selected against. But if both organisms have the defecting phenotype D, they will each only receive payoff d. They would be better off if they could both evolve the cooperative phenotype C, since they would then each receive payoff a > d. But C is not an evolutionarily stable strategy, because in a population of Cs, a mutant with the D phenotype will do better than the Cs and Ds will eventually come to predominate. One well-known solution to this problem proposes that altruistic behaviors can evolve if the competing organisms interact repeatedly during their lifetimes (the "iterated prisoners dilemma") and if they can make their behavior toward other organisms depend on what happened in previous interactions (reciprocation/retaliation) [SEE COOPERATION]. One way to link past interactions to future ones is by recognizing individuals and remembering their behavior, but simpler mechanisms can produce the same outcome. Organisms with the phenotype TFT (tit-for-tat) behave cooperatively in their first encounter with each organism, but in subsequent encounters only cooperate with organisms that cooperated with them in their last encounter. Organisms with the TFT phenotype are "reciprocal altruists" (Trivers, 1971). If the phenotype set contains only the three possibilities C, D and TFT, then TFT is an evolutionarily stable strategy because a population of TFTs cannot be taken over by D mutants. Moreover, under some circumstances TFT mutants can take over in a population composed of Ds. Another solution to the problem of altruism, not necessarily incompatible with the first, draws more directly on Hamilton's work to suggest that altruistic behavior can evolve if the degree of genetic relatedness between the interacting organisms is high enough ("kin selection").

In the late 1980s, sociobiology itself came under attack from a new movement calling itself "Evolutionary Psychology" (Crawford, Smith, & Krebs, 1987; Barkow, Cosmides, & Tooby, 1992). Evolutionary Psychologists argued that the whole project of explaining contemporary human behaviors as a direct result of adaptive evolution was misguided (Symons, 1992). The contemporary environment is so different from that in which human beings evolved that their behavior probably bears no resemblance to the behavior which played a role in human evolution. This problem had been identified by earlier critics of sociobiology (e.g., Kitcher, 1985), but evolutionary psychology followed it up with a positive proposal. Evolutionary theory should be used to predict which behaviors *would have been* selected in postulated ancestral environments. Human behavior today can be explained as the output of mechanisms that evolved to produce those ancestral behaviors when these mechanisms operate in the very different modern environment. Furthermore, the diverse behaviors seen in different cultures may all be manifestations of a single, evolved psychological mechanism operating under a range of local conditions, an idea that originated in an offshoot of sociobiology known as Darwinian anthropology (Alexander, 1979, 1987). Refocusing research on the "Darwinian algorithms" that underlie observed behavior, rather than the behavior itself, lets the evolutionary psychologist "see through" the interfering effects of environmental change and cultural difference to an underlying human nature.

Evolutionary Psychology uses the same population-genetic and evolutionary game-theory models as sociobiology, and there is often little difference in the actual explanations which the two schools offer for human behavior. For example, the classic behavioral ecological explanations of altruistic behavior just discussed are entirely acceptable to Evolutionary Psychologists. Perhaps the best-known experiment in Evolutionary Psychology research was designed to test the hypothesis that humans are reciprocal altruists (Cosmides & Tooby, 1992). In this experiment, Leda Cosmides and John Tooby modified an existing psychological task in which subjects are asked if a conditional rule of the form "If P, then Q" holds in a set of cards, one side of which indicates whether the antecedent (P) of the conditional is true and the other side of which indicates whether the consequent (Q) is true (Figure 21.3). Previous research had shown that many subjects turn over cards whose visible side is marked Q or ~P, despite the fact that these cards are irrelevant to the task, and fail to turn over the card

1. Abstract Problem

Bruce was managing a rural farm in remote Western Australia and told the workers to drench the sheep for parasites. They were also told to mark the sheep with blue dye after drenching. Bruce wants to make sure that they followed the rule:

"If a sheep has been drenched for parasites, then it has been marked with blue dye."

(If P then Q)

The cards below represent sheep. Each card represents one sheep. One side of the card tells whether the sheep was drenched or not, and the other side tells whether the sheep has been marked with blue dye.

Indicate only those card(s) Bruce needs to turn over to see if the rule has been followed:

Drenched	Not Drenched	Blue Dye Mark	No Blue Dye Mark
(P)	*(~P)*	*(Q)*	*(~Q)*

2. Social Exchange Beach-Driving Permit Problem

Sheila was the head ranger for the Byron Bay region in NSW. People that drive on beaches in this area must have a beach driving permit stuck on the left-hand side of their windscreen. Sheila was required to enforce this rule:

"If a person is driving on the beach, then they have displayed a beach driving permit."

(If P then Q)

The cards below have information about vehicles Sheila encounters. Each card represents one vehicle. One side of the card tells if the vehicle is driving on the beach and the other side tells whether the vehicle has a beach driving permit displayed on their windscreen.

Indicate only those card(s) Sheila needs to turn over to see if the rule has been followed:

Driving On Beach	Not Driving On Beach	Beach Driving Permit	No Beach Driving Permit
(P)	(~P)	(Q)	(~Q)

Figure 21.3 Two versions of the Wason card selection task, one an abstract problem and the other a problem concerning social exchange

marked ~Q, despite its relevance. When subjects were given a version of the task in which P and Q were replaced by statements of the general form "If you take the benefit, then you pay the cost" and preceded by descriptions which emphasized what Cosmides and Tooby describe as "social exchange," their performance improved markedly. This result has been used to argue that human psychology has been specifically designed for solving problems to do with "cheating" and "free-riding" in social interactions. This in turn has been taken to confirm the importance of reciprocal altruism in human evolution.

Evolutionary Psychology is a large field (see Buss, 2005 for a representative sample of current work), and it is associated not simply with the methodological approach just described, but also with a number of quite general conclusions about the human mind, conclusions to which many of those who describe themselves as Evolutionary Psychologists subscribe. Some of these conclusions are outlined in the next section.

5. "How the Mind Works"

The classical ethologists based their ideas about mental mechanisms on the neuroscience of the inter-war years, when their program was being formulated. Evolutionary Psychology reflects the state of the sciences of the mind during its own formulation. In particular, the program was influenced by the dominant "classical" school of cognitive science and the idea that the mind is computer software implemented in neural hardware (Marr, 1982; Fodor, 1983). Evolutionary Psychologists argue that the representational, information-processing language of classical cognitive science is ideal for describing the evolved features of the mind. Behavioral descriptions of what the mind does are useless because of the problem of changing environments described above. Neurophysiological descriptions are inappropriate, because behavioral ecology does not predict anything about the specific neural structures that underlie behavior. Models in behavioral ecology predict which behaviors would have been selected in the ancestral environment, but they cannot distinguish between different mechanisms that produce the same behavioral output. Hence, if one accepts the conventional view in cognitive science that many different neural mechanisms could potentially support the same behavior, it follows that behavioral ecology predicts little about the brain except which information-processing functions it must be able to perform:

> When applied to behavior, natural selection theory is more closely allied with the cognitive level of explanation than with any other level of proximate causation. This is because the cognitive level seeks to specify a psychological mechanism's function, and natural selection theory is a theory of function. (Cosmides & Tooby, 1987, p.284)

It is thus slightly confusing that Evolutionary Psychologists talk of discovering psychological "mechanisms," a term which suggests theories at the neurobiological level. What "mechanism" actually refers to in this context is a performance profile – an account of what output the mind will produce given a certain range of inputs.

This fact that evolutionary reasoning yields expectations about the performance profile of the mind fits neatly with the explanatory framework of classical cognitive science. According to the influential account given in David Marr's book *Vision* (1982), explanation in cognitive science works at three, mutually illuminating levels. The highest level concerns the tasks that the cognitive system accomplishes – recovering the shape and position of objects from stimulation of the retina, for example. The lowest level concerns the neurophysiological mechanisms that accomplish that task – the neurobiology of the visual system. The intermediate level concerns the functional profile of those mechanisms, or as it is often described, the computational process that is implemented in the neurophysiology. Hypotheses about the neural realization of the computational level constrain hypotheses about computational processes: psychologists should only propose computational models that can be realized by neural systems. Conversely, hypotheses about computational processes guide the interpretation of neural structure: neuroscience should look for structures that can implement the required computations. Similar relations of mutual constraint hold between the level of task description and the level of computational processes. But there remains something of a puzzle as to how the highest level – the task description – is to be specified other than by stipulation. It seems obvious that the task of vision is to represent things around us, but what makes this true? According to Evolutionary Psychology, claims about task descriptions are really claims about evolution. The overall task of the mind is survival and reproduction in the ancestral environment and the sub-tasks performed by parts of the mind correspond to separate adaptive challenges posed by the ancestral environment. Obviously, it would have been useful for the ancestors of humans to be able to see, so it is predictable that humans will have a visual system. This kind of thinking becomes useful when the function of a psychological mechanism is not as blindingly obvious as in the case of vision. What, for example, is the task description for the emotion system, or for individual emotions such as jealousy or grief? Evolutionary Psychology argues that in such cases it should be evolutionary thinking that sets the agenda for cognitive science, telling it what to look for and how to interpret what it finds.

5.1. *The massive modularity thesis*

One of the best-known claims of Evolutionary Psychology is the "massive modularity thesis" or "Swiss army knife model," according to which the mind contains few if any general-purpose cognitive mechanisms. The mind is a collection of separate "modules" each designed to solve a specific adaptive problem, such as mate-recognition or the enforcement of female sexual fidelity. The flagship example of a mental module is the "Language Acquisition Device" – the mechanism that allows human infants to acquire a language in a way that it is widely believed would not be possible using any general-purpose learning rules (Pinker, 1994). The massive modularity thesis is an example of the kind of evolutionary guidance for cognitive science described in the last section. Evolutionary Psychology argues that evolution would favor multiple modules over domain-general cognitive mechanisms because each module can be fine-tuned for a specific adaptive problem. So cognitive scientists should look for domain-specific effects in cognition and should conceptualize their work as the search for, and characterization of, mental modules.

Evolutionary Psychologists often introduce the idea of modularity with examples from neuropsychology (e.g., Gaulin & McBurney, 2001, pp.24–6). In these examples, "double dissociation" studies, in which clinical or experimental cases show that each of two mental functions can be impaired while the other is performed normally, are used to support the claim that those two functions are performed by separate neural subsystems. But despite their use of these examples, Evolutionary Psychologists are quite clear that the mental modules in which they themselves are interested need not correspond to separate neural subsystems, nor be localized in specific regions of the brain (Gaulin & McBurney, 2001, p.26). The difference between "neural subsystems" and "mental modules" is instructive. The double dissociation experiment is a means for exploring structure–function relationships in the brain. But for the purposes of evolution, what matters is not how the brain is structured, but how it appears to be structured when "viewed" by natural selection. For Evolutionary Psychology, the fact that two functions are dissociated is significant in its own right, and not only as a clue to how those functions are instantiated in the brain. Thus, there are architectures that produce double dissociations but which neuropsychology regards as non-modular, cases where apparent double dissociations are simply misleading (Shallice, 1988, p.250). Evolutionary Psychology, in contrast, would regard these architectures as different ways to produce mental modularity. We might aptly term such mental modules "virtual modules" (Griffiths, in press).

The modularity concept of Evolutionary Psychology derives from that developed in cognitive science of the early 1980s and popularized by Jerry Fodor in *The Modularity of Mind* (1983), but, once again, the differences are instructive. In Fodor's account, the definitive property of a module is "informational encapsulation." A system is informationally encapsulated if there is information unavailable to that system but which is available to the mind for other purposes. For example, in a phobic response the emotional evaluation of a stimulus ignores much of what the subject explicitly believes about the stimulus, suggesting that the emotional evaluation is informationally encapsulated. Fodor lists several other properties of modules, including domain specificity and the possession of proprietary algorithms. A system is domain specific if it only processes information about certain stimuli. It has proprietary algorithms if it treats the same information differently from other cognitive subsystems, something that Evolutionary Psychology identifies with the older idea that the module has "innate knowledge." The leading Evolutionary Psychologists Tooby and Cosmides make it clear that it is these two properties, rather than informational encapsulation, that are the two definitive properties of mental modules. A mental mechanism is simply not a module if "It lacks any a priori knowledge about the recurrent structure of particular situations or problem domains, either in declarative or procedural form, that might guide the system to a solution quickly" (1992, p.104). In the Evolutionary Psychology literature the properties of being domain specific and of having proprietary algorithms are generally referred to simultaneously as "functional specialization." Modules are "complex structures that are functionally organised for processing information" (1992, p.33).

When Evolutionary Psychologists present experimental evidence of "functional specialization" in cognition, it is generally evidence suggesting that information about one class of stimuli is processed differently from information about another class of stimuli – that is, evidence of the use of different proprietary algorithms in the two domains. The interpretation of the Wason card selection task described above exemplifies this pattern of reasoning. In a similar vein, David Buss has argued that people leap to conclusions about sexual infidelity more readily than about other subjects. He uses this to support the view that there is a mental module for dealing with infidelity (Buss, 2000). It seems that Evolutionary psychologists are simply not interested in cases where systems are domain specific but do *not* possess proprietary algorithms (this would be like having two identical PCs running identical software, one for personal use and the other for work). This is presumably because no evolutionary rationale can be imagined for such a neural architecture.

5.2. The monomorphic mind thesis

Tooby and Cosmides have argued strongly for the "monomorphic mind thesis" or "psychic unity of humankind" (1992, p.79). This states that differences in the cognitive adaptations of individual humans or human groups are not due to genetic differences. Instead, such differences are always, or almost always, due to environmental factors that trigger different aspects of the same developmental program. If true, this would make cognitive adaptations highly atypical, since most human traits display considerable individual variation related to differences in genotype. All human beings have eyes, but these eyes exhibit differences in color, size, shape, acuity, and susceptibility to various forms of degeneration over time, all due to differences in genotype. It has been known for half a century that wild populations of most species contain substantial genetic variation, and humans are no exception.

Tooby and Cosmides offer one main argument for the conclusion that the genes involved in producing cognitive adaptations will be the same in all human individuals:

> Complex adaptations necessarily require many genes to regulate their development, and sexual recombination makes it combinatorially improbable that all the necessary genes for a complex adaptation would be together at once in the same individual, if genes coding for complex adaptations varied substantially between individuals. Selection, interacting with sexual recombination, enforces a powerful tendency towards unity in the genetic architecture underlying complex functional design at the population level and usually the species level as well. (Tooby & Cosmides, 1990, p.393)

The authors apply this argument only to psychological adaptations, but its logic extends to all traits with many genes involved in their etiology. The argument seems to overlook the phenomena which the founders of modern neo-Darwinism referred to as "genetic canalization" or "genetic homeostasis" and attributed to the effects of "stabilizing selection" (Schmalhausen, 1949; Dobzhansky & Wallace, 1953; Waddington,

1957). Obviously, evolution will design developmental systems that are robust in the face of environmental variation, but in the middle of the last century new data from the genetics of natural populations indicated that it also designs them to be robust in the face of *genetic* variation. More or less identical "wild type" phenotypes can be generated by a range of genotypes. This is why surprisingly many gene knock-out experiments produce negative results. Development is robust and redundant. Disabling a gene known to be involved in a developmental pathway frequently produces no effect ("null phenotype"), because development contains positive and negative feedback mechanisms that increase transcription of the required gene product from the other allele, initiate transcription from another gene copy, or initiate transcription of a different gene product, and thereby achieve the same ends by different means (Freeman, 2000; Wilkin, 2003).

Pace Tooby and Cosmides, genetics and developmental biology provide no reason to accept the monomorphic mind thesis. Nor is there much direct evidence for the thesis. Behavioral geneticists have documented extensive heritable, individual differences in what are plausibly adaptive characters, such as IQ and personality, and some evolutionary psychologists have put this at the heart of their account of cognitive evolution (e.g., Miller, 2000). One advantage of the thesis is that it makes it impossible to level accusations of racism against Evolutionary Psychology. But this defense is surely unnecessary. If it is assumed that variation in evolved human phenotypes roughly mirrors the known variation in human genotypes (Cavalli-Sforza, Menozzi, & Piazza, 1994), then it follows that the vast majority of adaptive traits are pan-cultural and that any average differences between human groups will be dwarfed by the individual differences within those groups.

6. Evolutionary Psychology Today

Evolutionary Psychology is probably the largest school of evolutionary psychology at the present time, and it is certainly the most prominent in popular science (e.g., Pinker, 1997). However, this particular school has some severe critics (see esp. Buller, 2005; Fodor, 2000) and many other approaches to evolutionary psychology continue to flourish. These are judiciously surveyed in Kevin Laland and Gillian Brown's *Sense and Nonsense: Evolutionary perspectives on human behaviour* (2002; see also Downes, 2001). Two recent collections of papers also emphasize the diversity of ways in which evolution might be thought to inform psychological research (Heyes & Huber, 2000; Scher & Rauscher, 2002). While scientists do not necessarily subscribe to a particular, self-conscious research program like Evolutionary Psychology, many can be classed as engaged in either "human behavioral ecology," "gene–culture coevolution," or "developmental evolutionary psychology."

Human behavioral ecology is a research tradition derived from "Darwinian anthropology," itself an offshoot of sociobiology (Cronk, Chagnon, & Irons, 2000). Human behavioral ecologists continue to believe in the value of testing the predictions of behavioral ecology against contemporary human behavior. The Evolutionary Psychologists' critique of human sociobiology, summarized above, is that the rate of environmental change since the origins of human culture makes it irrational to expect human behavior to maximize reproductive fitness in modern environments. Some behavioral ecologists reply that adaptability is the hallmark of human evolution, so that it is no more irrational to expect humans to maximize their reproductive fitness in a modern city than to expect the rats who live in its sewers to do so (this view can be bolstered further by the "niche-construction" perspective discussed below). Others emphasize the methodological virtues of a paradigm in which hypotheses can be tested directly, as opposed to one in which currently available evidence must be brought to bear on theories about an earlier phase in human evolutionary history. In the light of the discussion of Tinbergen above, we might add that whatever its bearing on evolutionary questions, research in human behavioral ecology could be justified simply by the intrinsic interest of the questions it addresses: how well do human beings survive and reproduce in modern environments and how they achieve this?

Gene–culture coevolution is a flourishing scientific field that has its roots in two major theoretical works from the 1980s, Luca Cavalli-Sforza and Mark Feldman's *Cultural Transmission and Evolution* (1982) and Richard Boyd and Peter Richerson's *Culture and the Evolutionary Process* (1985). These authors developed mathematical models of change in culturally transmitted phenotypic characters in a population as a result of the differential ability of cultural variants to propagate themselves, and of the interaction between this process and genetic change in the same populations. In a flagship example, genetic differences in lactose tolerance in current human populations can be explained as a consequence of the spread of dairy farming, something that is clearly passed from one generation to the next – and from one human population to another – by cultural transmission. These two books have been widely praised for their mathematical sophistication, but criticized for providing "consequence laws" while having no clear program for deriving the matching "source laws" that would be needed to create a genuine evolutionary approach to culture (Sober, 1992). There is, at present, no cultural equivalent of ecology to reveal how the interaction of cultural variants with their environment determines the differential fitness of those variants. During the 1990s gene–culture coevolution was to some extent subsumed under the more general concept of "niche construction" (Odling-Smee, Laland, & Feldman, 1996; Laland, Odling-Smee, & Feldman, 2000; Odling-Smee, Laland, & Feldman, 2003). Conventional evolutionary biology studies how populations change as a consequence of interactions with their environment. Niche-construction studies how environments change as a consequence of interactions with evolving populations. For example, the soil and climate of the Amazon basin are as much a consequence of the biota that has grown up there as of fundamental abiotic parameters such as longitude, topology, and underlying rock strata. On a smaller scale, beavers are exquisitely adapted to life in an environment – the beaver pond – that would not exist if it were not for the dam- and lodge-building activities of beavers.

Human beings can be seen as the "ultimate niche-constructors," in the sense that they modify their environment to a greater extent than any other single species. Rather than seeing humans as having evolved to live in small-scale hunter-gatherer societies and now having to improvise responses to the modern world using unsuitable mental mechanisms, gene–culture coevolution theory sees the relationship between the human mind and the modern world as more like that between the beaver and its dam. A recently published popular book and a collection of classic papers by Boyd and Richerson provide an excellent introduction to this alternative perspective (Boyd & Richerson, 2005; Richerson & Boyd, 2005).

Finally, a "developmentalist" tradition in animal behavior research with its roots in classical ethology and comparative psychology has consistently criticized both sociobiology and Evolutionary Psychology for failing to integrate the evolutionary study of behavior with the study of how behavior develops (Gottlieb, 1997; Bjorklund & Pellegrini, 2002). Accessible introductions to this tradition have been provided by Patrick Bateson and Paul Martin (1999) and by David Moore (2001). Authors closer in orientation to Evolutionary Psychology have also stressed in recent years the importance of integrating evolutionary accounts of the mind with molecular developmental biology and with the neurosciences (e.g., Marcus, 2004). If this trend continues, evolutionary psychology may one day return to Tinbergen's project of constructing a single, integrated "biology of behavior."

References

Alexander, R. (1979). *Darwinism and human affairs.* Seattle: Washington University Press.

Alexander, R. (1987). *The biology of moral systems.* New York: De Gruyter.

Baerends, G. P., Barlow, G. W., Blurton Jones, N. G., Crook, J. H., Curio, E., et al. (1976). Multiple review of Wilson's 'Sociobiology'. *Animal Behaviour,* 24, 698–718.

Barash, D. P. (1979). Human ethology and human sociobiology. *Behavioral and Brain Sciences,* 2 (1), 26–7.

Barkow, J. H. (1979). Human ethology: empirical wealth, theoretical dearth. *Behavioral and Brain Sciences,* 2(1), 27.

Barkow, J. H., Cosmides, L., & Tooby, J. (Eds). (1992). *The adapted mind: evolutionary psychology and the generation of culture.* Oxford: Oxford University Press.

Bateson, P. P. G., & Klopfer, P. H. (Eds). (1989). *Whither ethology?* Vol. 8: *Perspectives in ethology.* New York and London: Plenum.

Bateson, P. P. G., & Martin, P. (1999). *Design for a life: how behavior and personality develop.* London: Jonathan Cape.

Bjorklund, D. F., & Pellegrini, A. D. (2002). *The origins of human nature: evolutionary developmental psychology.* Washington, DC: American Psychological Association.

Boyd, R., & Richerson, P. J. (1985). *Culture and the evolutionary process.* Chicago: Chicago University Press.

Boyd, R., & Richerson, P. J. (2005). *The origin and evolution of cultures.* New York: Oxford University Press.

Buller, D. J. (2005). *Adapting minds: evolutionary psychology and the persistent quest for human nature.* Cambridge, MA: MIT Press/Bradford Books.

Burkhardt, R. W. Jr. (2005). *Patterns of Behavior: Konrad Lorenz, Niko Tinbergen and the founding of ethology.* Chicago: University of Chicago Press.

Buss, D. M. (2000). *The dangerous passion: why jealousy is as essential as love and sex.* New York: Simon & Schuster.

Buss, D. M. (Ed.). (2005). *The handbook of evolutionary psychology.* Hoboken, NJ: Wiley.

Cavalli-Sforza, L. L., & Feldman, M. W. (1982). *Cultural transmission and evolution: a quantitative approach.* Princeton: Princeton University Press.

Cavalli-Sforza, L. L., Menozzi, P., & Piazza, A. (1994). *The history and geography of human genes.* Princeton: Princeton University Press.

Cosmides, L., & Tooby, J. (1987). From evolution to behaviour: evolutionary psychology as the missing link. In J. Dupré (Ed.). *The latest on the best: essays on optimality and evolution* (pp. 277–306). Cambridge, MA: MIT Press.

Cosmides, L., & Tooby, J. (1992). Cognitive adaptations for social exchange. In J. H. Barkow, L. Cosmides, & J. Tooby (Eds). *The adapted mind: evolutionary psychology and the generation of culture* (pp. 163–228). Oxford, New York: Oxford University Press.

Crawford, C., Smith, M., & Krebs, D. (Eds). (1987). *Sociobiology and psychology: ideas, issues and applications.* New York: Lawrence Erlbaum Associates.

Cronk, L., Chagnon, N., & Irons, W. (Eds). (2000). *Adaptation and human behavior: an anthropological perspective.* New York: Aldine de Gruyter.

Darwin, C. (1965 [1872]). *The expression of the emotions in man and animals.* Chicago: University of Chicago.

Darwin, C. (1981 [1871]). *The descent of man and selection in relation to sex.* Facsimile of the first edition. Princeton: Princeton University Press. Original edition, John Murray, London, 1871.

Dawkins, M. S., Halliday, T. R., & Dawkins, R. (Eds). (1991). *The Tinbergen legacy.* London: Chapman & Hall.

Dawkins, R. (1976). *The selfish gene.* Oxford: Oxford University Press.

Dobzhansky, Th., & Wallace, B. (1953). The genetics of homeostasis in *Drosophila. Proceedings of the National Academy of Sciences, USA*, 39(3), 162–71.

Downes, S. M. (2001). Some recent developments in evolutionary approaches to the study of human cognition and behavior. *Biology and Philosophy*, 16(5), 575–95.

Dunlap, K. (1919). Are there any instincts? *Journal of Abnormal Psychology*, 14, 307–11.

Durant, J. R. (1986). The making of ethology: The Association for the Study of Animal Behaviour, 1936–1986. *Animal Behaviour*, 34, 1601–16.

Fodor, J. A. (1983). *The modularity of mind: an essay in faculty psychology.* Cambridge, MA: MIT Press/Bradford Books.

Fodor, J. A. (2000). *The mind doesn't work that way: the scope and limits of computational psychology.* Cambridge, MA: MIT Press.

Freeman, M. (2000). Feedback control of intercellular signaling in development. *Nature*, 408, 313–19.

Gaulin, S. J. C., & McBurney, D. H. (2001). *Psychology: an evolutionary approach.* Upper Saddle River, NJ: Prentice Hall.

Gottlieb, G. (1997). *Synthesizing nature–nurture: prenatal roots of instinctive behavior.* Hillsdale, NJ: Lawrence Erlbaum Assoc.

Griffiths, P. E. (in press). Evo-devo meets the mind: towards a developmental evolutionary psychology. In R. Sansom, & R. N. Brandon (Eds). *Integrating development and evolution.* Cambridge: Cambridge University Press.

Hamilton, W. D. (1964). The genetical evolution of social behaviour, I & II. *Journal of Theoretical Biology*, 7, 1–16, 17–52.

Heyes, C. M., & Huber, L. (2000). *The evolution of cognition. Vienna Series in Theoretical Biology.* Cambridge, MA: MIT Press.

Hinde, R. A. (1956). Ethological models and the concept of "drive". *British Journal for the Philosophy of Science*, 6, 321–31.

Kitcher, P. (1985). *Vaulting ambition: sociobiology and the quest for human nature.* Cambridge, MA: MIT Press.

Kuo, Z. Y. (1921). Giving up instincts in psychology. *Journal of Philosophy*, 18, 645–64.

Laland, K. N., Odling-Smee, F. J., & Feldman, M. W. (2000). Niche construction, biological evolution and cultural change. *Behavioral and Brain Sciences*, 23, 131–57.

Laland, K. N., & Brown, G. R. (2002). *Sense and nonsense: evolutionary perspectives on human behaviour.* Oxford and New York: Oxford University Press.

Lehrman, D. S. (1953). Critique of Konrad Lorenz's theory of instinctive behavior. *Quarterly Review of Biology*, 28(4), 337–63.

Lorenz, K. Z. (1950). The comparative method in studying innate behaviour patterns. *Symposium of the Society of Experimental Biology*, 4 (Physiological mechanisms in animal behaviour), 221–68.

Lorenz, K. Z. (1996). *The natural science of the human species: an introduction to comparative behavioral research. The Russian manuscript (1944–1948)* (R. D. Martin, Trans.). Cambridge, MA: MIT Press.

Manning, A., & Dawkins, M. S. (1998). *An introduction to animal behaviour.* Cambridge: Cambridge University Press.

Marcus, G. F. (2004). *The birth of the mind: how a tiny number of genes creates the complexities of human thought.* New York: Basic Books.

Marr, D. (1982). *Vision.* New York: W.H. Freeman.

Maynard Smith, J. (1982). *Evolution and the theory of games.* Cambridge: Cambridge University Press.

Maynard Smith, J., Burian, R., Kauffman, S., Alberch, P., Campbell, J., Goodwin, B., Lande, R., Raup, D., & Wolpert, L. (1985). Developmental constraints and evolution. *Quarterly Review of Biology*, 60(3), 265–87.

Maynard Smith, J. (1987). How to model evolution. In J. Dupré (Ed.). *The latest on the best: essays on optimality and evolution* (pp. 119–31). Cambridge, MA: MIT Press.

Miller, G. (2000). *The mating mind.* London: Heinemann.

Moore, D. S. (2001). *The dependent gene: the fallacy of "nature versus nurture."* New York: W.H. Freeman/Times Books.

Odling-Smee, F. J., Laland, K. N., & Feldman, M. W. (1996). Niche construction. *American Naturalist*, 147(4), 641–8.

Odling-Smee, F. J., Laland, K. N., & Feldman, M. W. (2003). *Niche construction: the neglected process in evolution* (Vol. 37). Monographs in population biology. Princeton: Princeton University Press.

Pinker, S. (1994). *The language instinct: the new science of language and mind.* UK/US: Allen Lane/William Morrow.

Pinker, S. (1997). *How the mind works.* New York and London: Allen Lane.

Richards, R. J. (1987). *Darwin and the emergence of evolutionary theories of mind and behavior.* Chicago: University of Chicago Press.

Richerson, P. J., & Boyd, R. (2005). *Not by genes alone: how culture transformed human evolution.* Chicago: University of Chicago Press.

Scher, S., & Rauscher, M. (Eds). (2002). *Evolutionary psychology: alternative approaches.* Dordrecht: Kluwer.

Schmalhausen, I. I. (1949). *Factors of evolution: the theory of stabilising selection* (I. Dordick, Trans.). Philadelphia and Toronto: Blakeston.

Segerstråle, U. (2000). *Defenders of the truth: the battle for science in the sociobiology debate and beyond.* Oxford: Oxford University Press.

Shallice, T. (1988). *From neuropsychology to mental structure.* Cambridge: Cambridge University Press.

Slater, P. J. B. (2003). Fifty years of bird song research: a case study in animal behaviour. *Animal Behaviour*, 65, 633–9.
Sober, E. (1984). *The nature of selection: evolutionary theory in philosophical focus*. Cambridge, MA: MIT Press.
Sober, E (1992). Models of cultural evolution. In P. E. Griffiths (Ed.). *Trees of life: essays in the philosophy of biology* (pp. 17–38). Dordrecht: Kluwer Academic Publishers.
Symons, D. (1992). On the use and misuse of Darwinism in the study of human behavior. In J. H. Barkow, L. Cosmides, & J. Tooby (Eds). *The adapted mind: evolutionary psychology and the generation of culture* (pp. 137–59). Oxford: Oxford University Press.
Tinbergen, N. (1951). *The study of instinct*. Oxford: Oxford University Press.
Tinbergen, N. (1963a). The work of the Animal Behaviour Research Group in the Department of Zoology, University of Oxford. *Animal Behaviour*, 11, 206–9.
Tinbergen, N. (1963b). On the aims and methods of ethology. *Zietschrift für Tierpsychologie*, 20, 410–33.
Tinbergen, N. (1985). Watching and wondering. In D. A. Dewsbury (Ed.). *Leaders in the study of animal behavior: autobiographical perspectives* (pp. 431–63). London and Toronto: Associated University Presses.
Tooby, J., & Cosmides, L. (1990). The past explains the present: emotional adaptations and the structure of ancestral environments. *Ethology and Sociobiology*, 11, 375–424.
Tooby, J., & Cosmides, L. (1992). The psychological foundations of culture. In J. H. Barkow, L. Cosmides, & J. Tooby (Eds). *The adapted mind: evolutionary psychology and the generation of culture* (pp. 119–36). Oxford and New York: Oxford University Press.
Trivers, R. L. (1971). The evolution of reciprocal altruism. *Quarterly Review of Biology*, 46(4), 35–57.
Waddington, C. H. (1957). *The strategy of the genes: a discussion of some aspects of theoretical biology*. London: Ruskin House/George Allen & Unwin Ltd.
Wilkin, A. (2003). Canalization and genetic assimilation. In B. K. Hall & W. M. Olson (Eds). *Keywords and concepts in evolutionary developmental biology* (pp. 23–30). Cambridge, MA and London: Harvard University Press.
Wilson, E. O. (1975). *Sociobiology: the new synthesis*. Cambridge, MA: Harvard University Press.

Further Reading

Ethology

Burkhardt, R. W. J. (2005). *Patterns of behavior: Konrad Lorenz, Niko Tinbergen and the founding of ethology*. Chicago: University of Chicago Press.
Bateson, P. P. G., & Klopfer, P. H. (1989). Whither ethology?, *Perspectives in ethology*, Vol. 8. New York and London: Plenum.

Sociobiology

Segerstråle, U. (2000). *Defenders of the truth: the battle for science in the sociobiology debate and beyond*. Oxford: Oxford University Press.
Kitcher, P. (1985). *Vaulting ambition: sociobiology and the quest for human nature*. Cambridge, MA: MIT Press.
Alcock, J. (2001). *The triumph of sociobiology*. Oxford and New York: Oxford University Press.

Evolutionary Psychology

Barkow, J. H., Cosmides, L., & Tooby, J. (1992). *The adapted mind: evolutionary psychology and the generation of culture.* Oxford: Oxford University Press.

Buller, D. J. (2005). *Adapting minds: evolutionary psychology and the persistent quest for human nature.* Cambridge, MA: MIT Press/Bradford Books.

General

Laland, K. N., & Brown, G. R. (2002). *Sense and nonsense: evolutionary perspectives on human behaviour.* Oxford and New York: Oxford University Press.

Chapter 22

Cooperation

J. MCKENZIE ALEXANDER

The Darwinian problem of cooperation is the following: according to the theory of natural selection, behaviors which serve to increase an individual's fitness will be favored over behaviors which decrease an individual's fitness; yet since cooperative behavior generally results in an individual's fitness being lower than what it could have been, had he or she acted otherwise, how is it that cooperative behavior persists? Natural selection, it would seem, should select against cooperative behavior – because of the reduced individual fitness – thereby driving it out of the population and promoting uncooperative behavior.

Closely related to the problem of cooperation is the problem of altruism, which was identified by E. O. Wilson as the "central theoretical problem of socio-biology" (Wilson, 1975, p.3). An altruistic behavior, in the evolutionary sense, causes the donor to incur a fitness cost while conferring a fitness benefit to the recipient (Sober & Wilson, 2000, p.185). According to these definitions, although altruistic behaviors are considered cooperative, the converse need not be true. If all individuals begin with a common baseline fitness and benefits are distributed equally, altruistic individuals have lower fitness than selfish individuals: an altruistic individual incurs both a personal fitness cost (due to his action) while receiving the common fitness benefit (from other altruists in the population), whereas a selfish individual only receives the common fitness benefit. The altruist's fitness is thus lower than what it could have been, had he acted otherwise, and is therefore a cooperative behavior. However, cooperative behavior need not be altruistic because it is possible for a co-operator to fail to maximize his or her individual fitness without incurring an explicit fitness cost. That is, altruistic behavior imposes explicit and actual fitness *penalties* upon individuals, whereas cooperative behavior requires only that the truth of a counterfactual obtain. In the following, this difference between altruistic and cooperative behavior will generally be suppressed.

Historically, attitudes regarding the extent to which evolution is compatible with cooperation have ranged between two extremes represented by Thomas Henry Huxley and Prince Petr Kropotkin in their writings on evolutionary theory in the nineteenth century. Huxley, arguing for the incompatibility of cooperative behavior and evolution, explicitly invoked Hobbesian imagery in his characterization of natural selection:

	Cooperate	Defect
Cooperate	(R,R)	(S,T)
Defect	(T,S)	(P,P)

Figure 22.1 The Prisoner's Dilemma. Payoffs listed for (row, column), where values indicate relative changes in individual fitness, and $T > R > P > S$ and $\frac{T+S}{2} < R$

> the weakest and the stupidest went to the wall, while the toughest and the shrewdest, those who were best fitted to cope with their circumstances, but not the best in any other way, survived. Life was a continuous free fight, and . . . a war of each against all was the normal state of existence. (Huxley, 1888)

Kropotkin, on the other hand, noted how the structures produced by the social insects would have been impossible without a high degree of cooperation:

> The ants and the termites have renounced the "Hobbesian War" and they are the better for it. Their wonderful nests, their buildings superior in size relative to man . . . all of these are the normal outcome of the mutual aid which they practice at every stage of their busy and laborious lives. (Kropotkin, 1902)

The problem of cooperation is compelling because a great deal of cooperative and altruistic behavior clearly exists in nature. Female vampire bats (*Desmodus rotundus*) regurgitate blood obtained during successful feeding runs to other bats that have been less successful in obtaining food (Wilkinson, 1984). Such cooperation is essential to survival, since individual bats can starve to death in 60 hours without food. House sparrows (*Passer domesticus*) emit calls which attract other birds to newly discovered food sources (Summers-Smith, 1963). Indeed, extreme examples of altruistic behavior, such as the existence of sterile workers among the social insects, and the problem they posed for the theory of natural selection, were well known to Darwin. In *The Origin of Species*, he asked, "how is it possible to reconcile this case with the theory of natural selection?" (Darwin 1985 [1859], p.258). The apparent incompatibility, he proposed, "disappears, when it is remembered that selection may be applied to the family, as well as the individual, and may thus gain the desired end" (ibid.: p.258).

In general, the solution to the Darwinian problem of cooperation proceeds by identifying additional features of the evolutionary process which facilitate the emergence and persistence of cooperative behavior, the primary mechanisms being kin selection, reciprocity, and group selection. (One should note that the latter has engendered some controversy [See The Units and Levels of Selection]). Additional mechanisms which have been identified include coercion, mutualism, by-product mutualism, and effects of local interactions.

The most commonly studied model of cooperation is the Prisoner's Dilemma, shown in Figure 22.1, originally developed by Merrill Flood and Melvin Dresher in 1950 while at the Rand Corporation for analyzing strategic conflict during the Cold War. The Prisoner's Dilemma encapsulates the strategic problem underlying the evolution of cooperation produced when individual and collective interests conflict. In the Prisoner's Dilemma, achieving the collectively best outcome – the cooperative outcome – produces a suboptimal result from the point of view of the individual. In this model, each individual faces two courses of action, labeled "Cooperate" and "Defect." If both individuals cooperate, each has a fitness of R, the *reward*. If one individual cooperates and the other defects, the defector has the greatest possible fitness of T, the *temptation* for defecting, while the cooperator earns the lowest possible fitness of S, the *sucker's* payoff. If both individuals defect, each receives a fitness of P, the *punishment* for defecting, which is less than R. (The further condition that $(T+S)/2<R$ is often imposed to insure that, in repeated interactions, cooperative behavior remains more beneficial than alternation of cooperate and defect.) With these particular fitness payoffs, it would seem that natural selection should favor Defect, since it maximizes one's own fitness independent of the behavior of others.

1. Kin Selection

After the modern synthesis, another solution to the problem of cooperation became available. The gene-centered view of evolution (see Dawkins, 1976) recognized that, since it is ultimately genes which are passed from parent to offspring, and individual organisms share portions of their genetic material with other members of the same species, natural selection may favor behaviors that successfully promote the propagation of an individual's genes even if that behavior reduces the number of viable offspring an organism has. [SEE THE UNITS AND LEVELS OF SELECTION]. This view was first given a precise formulation and analysis by Hamilton (1964), who introduced the concept of *inclusive* fitness, which can be thought of as the number of an individual's alleles present in the next generation rather than the actual number of viable offspring of an individual. More precisely, inclusive fitness is the relative representation, in the next generation, of an individual's genes in the overall gene pool. Kin selection is the process of selection which increases the inclusive fitness of the individual.

The theoretical result underlying kin selection is Hamilton's rule, which states that a gene possessed by an individual i increases in frequency whenever $\sum_{j=1}^{n} r_{ij}b_{ij} - c > 0$, where n is the number of individuals affected by the trait the gene encodes, r_{ij} denotes the degree of relatedness between individuals i and j, b_{ij} the benefit conferred by i to j, and c is the associated cost to i of bearing the trait. (The degree of relatedness of two individuals is a real number between 0 and 1 indicating the proportion of genes held in common between the two individuals.) According to Hamilton's rule, cooperative or altruistic acts can evolve provided that the cost/benefit ratio of the act is less than the degree of relatedness between the affected individuals. For example, evolution would favor one sibling sacrificing all of his fitness to help his brother (a degree of relatedness of 0.5) provided that the altruist's act increases his brother's fitness by at least twofold.

A common misinterpretation of Hamilton's rule is that it says organisms are expected to act altruistically toward relatives according to the degree that they are related. (This mistake was made by Dawkins in the first edition of *The Selfish Gene*, corrected in the endnotes to the second edition.) Hamilton's rule states a condition under which altruistic or cooperative behavior toward relatives can evolve; it does not say that evolution is expected to produce an array of behaviors which distribute altruism accordingly across one's relatives.

Hamilton's work, and the idea of kin selection in general, has had great impact upon the field of evolutionary biology for two reasons. The first is that it seemed to provide a more parsimonious account of the evolution of cooperation than Darwin's preferred explanation of group selection. The second is that it provided a theoretical explanation for the haplo-diploid sex determination and eusociality of the social insects. Whereas in most animals sex differentiation occurs through the possession of a different set of sex chromosomes (a heterogametic and homogametic sex), among the social hymenopterans males develop from haploid (unfertilized) eggs and females from diploid (fertilized) eggs. This system of genetic determination of the sexes modifies the degrees of relatedness in such a way so as to strongly favor eusociality. Indeed, eusociality has independently evolved among the social insects no fewer than eleven times.

2. Reciprocity

While kin selection can account for the evolution of cooperation among genetic relatives, it cannot account for the evolution of cooperation among individuals who are not genetically related. Reciprocal altruism, first introduced in an influential paper by Trivers (1971), provides a mechanism through which altruistic or cooperative behavior can evolve even when the individuals who engage in altruistic behavior are not genetically related to one another. Reciprocal altruism is found in a variety of natural environments. Commonly cited examples of this phenomenon include mutual symbioses such as ants and ant-acacias, where the trees provide housing for the ants which, in turn, provide protection for the trees (Janzen, 1966); figs trees and fig-wasps, where the wasps are parasites on the fig flowers but provide the fig trees' method of pollination (Wiebes, 1976; Janzen, 1979); and cleaning symbioses, discussed at length in Trivers' original article. Reciprocal altruism is a robust phenomenon, having independently evolved many times (Trivers notes that it has arisen independently at least three times among shrimp alone).

In Trivers' original model, what promoted the flourishing of cooperative behavior in reciprocal interactions was a common threat from the environment which all faced; engaging in altruistic behavior served to reduce the environmental threat sufficiently so as to be worth each person's incurring the fitness cost imposed by altruistic action. For example, consider the act of saving someone from drowning. Suppose that the probability of dying from drowning is 50 percent if no one attempts a rescue, and that the probability of the rescuer drowning is 5 percent. In addition, assume that the drowning person always dies if his rescuer drowns and the drowning person is always saved if the rescuer does not drown (which is taken to mean that the rescue attempt was successful). If interactions between the drowning person and rescuer were never repeated, then there would be no reason for anyone to attempt to rescue a drowning person. However, if interactions are repeated, so that an individual who was saved from drowning can reciprocate and come to the aid of his rescuer at a later point in time, it is in the interest of each to come to the aid of the other. If every person in the population has the same risk of drowning, people who come to the aid of the other will have, in effect, reduced the original 50 percent chance of dying to only a 10 percent chance. While reduction of risk posed by common threats provides a particularly striking example of the contexts in which reciprocal altruism can arise, the phenomenon is much more widespread, as the examples of mutual symbioses indicate.

Perhaps the most well-known (if somewhat overstated, see Binmore, 1998) example of the evolution of cooperation through reciprocity is the success of Tit-for-Tat in the repeated Prisoner's Dilemma (Axelrod, 1984). Axelrod conducted a computer tournament in which sixty strategies, solicited from many different individuals, were pitted against each other in a "round-robin" competition. Each strategy played five runs of the repeated Prisoner's Dilemma against every other strategy. Each run consisted of the Prisoner's Dilemma being repeated a certain number of times, where the number of repeats was fixed in advance, and common among all strategy pairings.

What Axelrod found, both in the original computer tournament and in a second, larger, tournament held later, was that a very simple strategy favoring cooperative behavior won both tournaments. The strategy, known as Tit-for-Tat, begins by cooperating and then simply mimics the previous play of its opponent in all rounds after the first. If its opponent always cooperates, then Tit-for-Tat will always cooperate. If its opponent defects in the nth stage of the game, Tit-for-Tat will reciprocate by defecting in the $n + 1$st stage of the game; if its opponent should then "apologize" for its nth stage defection with cooperative behavior in the $n + 1$st stage, Tit-for-Tat will accept the apology by cooperating in the $n + 2$nd stage. The simple feedback mechanism employed by Tit-for-Tat is, Axelrod found, remarkably successful at rewarding cooperative behavior and punishing defections in certain environments.

In addition, when Axelrod took the initial strategies and performed an "ecological analysis," modeling a dynamic environment in which more successful strategies became more prolific, Tit-for-Tat still won. This simulation proceeded as follows: initially, each of the submitted strategies was considered to be equally likely in the population. The results from the tournament were assembled into a large payoff matrix specifying how well each strategy did when paired against every other strategy. This matrix was then used to calculate the expected fitness of each strategy in the population, which in the first generation simply equaled the actual fitness earned by each strategy at the end of the original tournament. However, after the first generation, the frequency of each strategy in the population was adjusted according to how well it did at the end of the current generation. From this point on, the expected fitness of each strategy in the population need not necessarily agree with the fitness of each strategy in the original tournament. Even so, within two hundred generations Tit-for-Tat became the most frequently used strategy in the population.

Axelrod identified four beneficial properties of Tit-for-Tat that enabled it to be successful: (1) it was not envious, (2) it was not the first to defect, (3) it reciprocated both cooperation and defection, and (4) it was not too clever (Tit-for-Tat outperformed a strategy which modeled the actions of its opponent as a Markov process, then using Bayesian inference to select which move – Cooperate or Defect – was deemed most likely to maximize its payoff in the next round). He also claimed to provide necessary and sufficient conditions for the collective stability of Tit-for-Tat, where "collectively stable" means that if everyone in the population follows it, no alternative strategy can invade (Axelrod, 1984, p.56). The precise result Axelrod proves is the following proposition:

> Proposition 2. Tit-for-Tat is collectively stable if and only if w is large enough. This critical value of w is a function of the four payoff parameters T, R, P, and S. (Axelrod, 1984, p.59)

The parameter w denotes the probability that both individuals will have another round of interaction in the future, and the critical value which makes Tit-for-Tat collectively stable is $\max\left\{\frac{T-R}{T-P}, \frac{T-R}{R-S}\right\}$.

Unfortunately, Tit-for-Tat's success in Axelrod's tournaments has led some to regard it as *the* solution to the Darwinian problem of cooperation, or as *the* optimal behavior to adopt in the repeated prisoner's dilemma. Tit-for-Tat is not optimal – indeed, it can be proven that in the indefinitely repeated Prisoner's Dilemma no optimal strategy exists. Axelrod himself noted that Tit-for-Tat would not have won the two computer tournaments if two other "natural" competitors had been submitted. One competitor which would have beat Tit-for-Tat is Win–stay, lose–shift (also known as "Pavlov"). Win–stay, lose–shift, like Tit-for-Tat, begins by cooperating on the first move, and then cooperates on future moves if and only if both players adopted the same strategy on the previous move. Suppose that the first individual follows the strategy Win–stay, lose–shift. If both cooperate, he will continue to cooperate on the next move as mutual cooperation is considered to be a "win" and the strategy recommends staying with a win. If both defect, he will switch to cooperating on the next move: mutual defection is considered to be a "loss," so he adopts the other alternative for the next move, which is cooperation. If the first individual defects and the second cooperates, the first individual will continue to defect on the next move, as defection against a cooperator is considered to be a "win." If the first individual cooperates and the second defects, he will switch to defection on the next move, as cooperating against a defector is a "loss," so he switches to the other alternative for the next move, which is in this case defection (Nowak & Sigmund, 1993).

Aside from the fact that Tit-for-Tat would have been beaten in the original tournament by only a marginally simpler strategy, which also does well on the four criteria identified by Axelrod, many other shortcomings of Axelrod's analysis have been identified (Binmore, 1998). Perhaps the most important one is that Tit-for-Tat is not actually immune to being invaded by competing strategies, contrary to Axelrod's claim that it is collectively stable. Lindren and Nordahl (1994) show how, in a model of the infinitely iterated Prisoner's Dilemma with noise and a strategy space which is not bounded in memory length (Tit-for-Tat only has a memory of 1), Tit-for-Tat can be invaded by a variety of other strategies.

Reciprocity promotes cooperation effectively by transforming the structure of the problem from the Prisoner's Dilemma into a different one. Consider what happens in the case where Tit-for-Tat plays against All Defect with the abovementioned payoffs and a probability of future interactions given by w. When Tit-for-Tat plays against Tit-for-Tat, it always cooperates, so the payoffs for the indefinitely iterated interaction are

	Cooperate	Defect
Cooperate	(3,3)	(0,5)
Defect	(5,0)	(1,1)

	Tit-for-Tat	All Defect
Tit-for-Tat	(9,9)	(2,7)
All Defect	(7,2)	(3,3)

Figure 22.2 Reciprocity changes the Prisoner's Dilemma into an Assurance Game. Payoffs listed for (row, column), and $w = \frac{2}{3}$

$$W(\text{TfT}|\text{TfT}) = R + Rw + Rw^2 + Rw^3 + \ldots$$
$$\sum_{i=0}^{\infty} Rw^i = \frac{R}{1-w}.$$

Likewise, the payoffs for the other three possible pairings of Tit-for-Tat and All Defect are as follows:

$$W(\text{TfT}|\text{AllD}) = S + Pw + Pw^2 + Pw^3 + \ldots$$
$$S + \sum_{i=1}^{\infty} Pw^i = S + \frac{Pw}{1-w}.$$
$$W(\text{AllD}|\text{AllD}) = P + Pw + Pw^2 + Pw^3 + \ldots$$
$$\sum_{i=0}^{\infty} Pw^i = \frac{P}{1-w}.$$
$$W(\text{AllD}|\text{TfT}) = T + Pw + Pw^2 + Pw^3 + \ldots$$
$$T + \sum_{i=1}^{\infty} Pw^i = T + \frac{Pw}{1-w}.$$

If the probability of future interactions is sufficiently high, the payoff matrix for choosing between reciprocating cooperative behavior and always defecting becomes that shown in Figure 22.2. Reciprocity can transform the Prisoner's Dilemma into an Assurance Game, or Stag Hunt (Skyrms, 2004).

3. Group Selection

Although the possibility that cooperative behavior might originate through selection acting on levels higher than the individual was first put forward by Darwin in *The Origin of Species*, group selection fell into disrepute when Williams (1966) argued that most alleged instances of group selection could be understood in individualist terms. In recent years, though, Wilson (1980) and Wade (1978) have sought to rehabilitate theories of group selection, arguing for multilevel selection theory. Sober and Wilson (2000) show how group selection can support the emergence and persistence of cooperative behavior under certain conditions.

Whether group selection supports cooperation depends crucially on details of the selection process. For example, Maynard Smith's (1964) "haystack model" of group selection does not support the emergence of cooperation. In this model, field mice live in haystacks, where each haystack is initially populated by a single fertilized female. Each female gives birth in the haystack, which remains populated for several generations. At the end of the first generation, brothers and sisters from the original founding female mate with each other; at the end of the second generation, first cousins mate with first cousins, and so on. After a certain number of generations, all of the haystacks empty, mice mate with randomly chosen partners, and then each fertilized female goes on to found another colony in a new haystack, repeating the process described above. Maynard Smith showed that, under these conditions, cooperation tends to be driven to extinction.

Sober and Wilson's (1998, 2000) model of group selection modifies the process through which groups form. Unlike Maynard Smith's model, where each group (haystack) is initially occupied by a single pregnant female, in the Sober and Wilson model, groups periodically merge into a larger population and re-form by a partitioning of that population into smaller groups. This change, along with the fact that groups may include more than one cooperator at the time of formation, enables cooperation to emerge.

More precisely, suppose that cooperators incur a fitness cost of c and that individuals who receive the benefit of cooperation have their fitness increased by b. In addition, suppose each individual has a baseline fitness of X. If there are n individuals in the group, with p of them being cooperators, then the fitness of a cooperator is

$$W_C = X - c + \frac{b(np-1)}{n-1}$$

since each cooperator has his baseline fitness reduced by c and may possibly receive a benefit from any one of the $np - 1$ other altruists in the group. (The expression $\frac{b(np-1)}{n-1}$ denotes the expected benefit of each altruist in the group.) The fitness of a defector is simply

$$W_D = X + \frac{bnp}{n-1}$$

which exceeds the fitness of a cooperator for two reasons: first, the defector does not incur the fitness cost of cooperating; second, a defector is eligible to receive a benefit from any one of the np cooperators in the group, whereas a cooperator is eligible to receive a benefit from only $np - 1$ cooperators (it is assumed that cooperators cannot bestow benefits to themselves).

Now, suppose we have an initial population consisting of 200 individuals, in which exactly half of the population cooperate. Suppose further that the population divides into two groups of equal size, with the first group containing 20 percent cooperators and the second group contains 80 percent cooperators. The fitness of cooperators and defectors in the first group is then

$$W_C^1 = 10 - 1 + \frac{5(20-1)}{99} = 9.96$$
$$W_D^1 = 10 + \frac{5(20)}{99} = 11.01$$

and the fitness of cooperators and defectors in the second group is

$$W_C^2 = 10 - 1 + \frac{5(80-1)}{99} = 12.99$$
$$W_D^2 = 10 + \frac{5(80)}{99} = 14.04.$$

In both groups, cooperators have lower fitness than defectors, as one would expect given the basic structure of the Prisoner's Dilemma. After reproduction, group one increases in size from 100 to 1,080, with cooperators accounting for only 18.4 percent of the total, and group two increases in size from 100 to 1,320, with cooperators accounting for 78.7 percent of the total. In both groups, the frequency of cooperation has decreased.

However, considering the population as a whole, the total frequency of cooperation has increased. Initially we started with only 200 individuals and a frequency of cooperation of 50 percent. After the first generation, the total population size is 2,400 with the frequency of cooperation being $\frac{0.184 \cdot 1080 + 0.787 \cdot 1320}{2400} = 0.516$. The fact that the frequency of cooperation can decrease in each group individually while increasing in the overall population is an example of Simpson's paradox (see Simpson, 1951; Sober, 1984; and Cartwright, 1978).

4. Coercion

According to coercive theories of cooperation, individuals are coerced into cooperative or altruistic acts by dominant members of the population and face the threat of ejection if they do not comply. Although there is some evidence of coercion in cooperative societies of fish (Balshine-Earn et al., 1998), fairy wrens (Mulder & Langmore, 1993), and naked mole rats (Reeve, 1992), it seems that the majority of forms of cooperation are not coerced.

Closely related to coercive theories of cooperation are retributive theories (Boyd & Richardson, 1992). In this model, groups of size *n* are formed by random sampling from a large population. Within each group, individuals interact in two stages: the first being a cooperative stage where individuals have a choice of either cooperating or defecting (as in the Prisoner's Dilemma), the second being a punishment stage where individuals can punish any member in the group. Boyd and Richardson find that, under certain conditions, retribution-based processes facilitate cooperation in larger groups than is possible with mere reciprocity-based processes. Retribution-based processes can also be a powerful selective and stabilizing force since "moralistic" behaviors, which punish individuals who do not comply with the required behavior, are capable of rendering *any* individually costly behavior evolutionarily stable.

5. Mutualism

For certain animals, the fitness of individual group members tends to increase with group size (Courchamp, Clutton-Brock, & Grenfell, 2000). Mutualist explanations of cooperative behavior point to correlations between group size/success and individual fitness, which thereby reduce the expected gain to individuals by defecting. Kokko, Johnstone, and Clutton-Brock (2001) identify several processes which lead to the creation of these correlations. For example, when greater group size/success leads to greater feeding success in adults, increased success in defending food supplies from competitors, greater efficiency in defending and providing for young, and so on, cooperative group behavior need not be eliminated by defection. While some of the evidence linking group size/success with individual fitness need not differentiate between mutualism and reciprocity, such as when unrelated group members contribute to the common good (Cockburn, 1998), cases where groups accept unrelated immigrants (Piper, Parker, & Rabenold, 1995) or kidnap individuals from other groups (Heinsohn, 1991) seem to favor mutualist accounts over reciprocal altruism.

6. Byproduct Mutualism

Byproduct mutualism occurs when the cooperative behavior benefiting the group coincides with the behavior that maximizes individual fitness. In these cases, the production of beneficial consequences for others through cooperative behavior might be entirely coincidental (Bednekoff, 1997). Note that byproduct mutualism therefore concerns instances of cooperation where the fitness payoffs do not conform to the basic structure of the Prisoner's Dilemma. Hence, there is some question as to whether the behavior deserves the label of "cooperative" in the first place.

Brown (1983) introduced byproduct mutualism by noting that "in many cases of mutualism, CC > DC will be found to prevail rather than DC > CC as required by the prisoner's dilemma." Contrary to the DC > CC > DD > CD ordering of payoffs for the Prisoner's Dilemma, a more likely ordering for species where cooperative activities are more profitable in groups than alone would be "CC > CD > DC = DC" (Brown, 1983, p.30). Figure 22.3 illustrates the payoff matrix for cooperative behavior generated in the context of byproduct mutualism. The structure of the payoff matrix is that of a coordination game, where the choice to Cooperate dominates Defect.

	Cooperate	Defect
Cooperate	(x,x)	(y,w)
Defect	(w,y)	(z,z)

Figure 22.3 The payoff matrix for cooperative behavior generated through byproduct mutualism. Payoffs listed for (row, column), where values indicate relative changes in individual fitness, and $x > y > w \geq z$

Although it is easy to see why natural selection would favor "cooperative" behavior in these instances, part of the interest in byproduct mutualism derives from the fact that, in the study of the evolution of cooperation, it is difficult to determine the payoffs for the acts of Cooperate and Defect. When uncertainty exists as to what the payoffs are, it is an open question as to which payoff matrix best describes the interactive problem. Some experiments with bluejays (Clements & Stephens, 1995) suggest that the observed cooperative behavior is better explained as a result of byproduct mutualism than alternative mechanisms.

7. Local Interactions

Large, panmictic populations that reproduce asexually do not favor the formation of cooperative behavior. One well-known model of this is the *replicator dynamics* by Taylor and Jonker (1978). Suppose we have a large population, where each agent has a certain phenotype σ. For simplicity, assume that there are only finitely many phenotypes $\sigma_1, \ldots, \sigma_m$. Let n_i denote the total number of agents in the population with the phenotype σ_i, with the total size of the population given by $N = \sum_{i=1}^{m} n_i$. For large, panmictic populations, all of the relevant information about the population is contained in the state vector $\vec{s} = \langle s_1, \ldots, s_m \rangle$, where $s_i = \frac{n_i}{N}$ for all i. If the growth rate of the ith phenotype approximately equals the fitness of that phenotype in the population, one can show that the rate of change of the ith phenotype is given by

$$\frac{ds_i}{dt} = s_i (W(i|\vec{s}) - W(\vec{s}|\vec{s}))$$

where $W(i|\vec{s})$ denotes the mean fitness of i in the population and $W(\vec{s}|\vec{s})$ denotes the mean fitness of the population at large. This is continuous replicator dynamics, which assumes that the increase or decrease of the phenotype frequencies occurs without well-defined generational breaks; that is, it assumes there is not a well-defined notion of "next generation" applying to the population (such as biological reproduction in humans).

In a population where p individuals Cooperate and $1-p$ Defect, the expected fitness of Cooperate and Defect are, respectively,

$$W(C|\vec{s}) = p \cdot W(C|C) + (1-p) \cdot W(C|D)$$

and

$$W(D|\vec{s}) = p \cdot W(D|C) + (1-p) \cdot W(D|D).$$

Since $T > R$ and $P > S$, the expected utility of defecting is greater than the expected reward of cooperating, so it follows that $W(D|\vec{s}) > W(\vec{s}|\vec{s}) > W(C|\vec{s})$. From this, it follows that,

$$\frac{ds_D}{dt} = (1-p)(W(D|\vec{s}) - W(\vec{s}|\vec{s})) > 0$$

and

$$\frac{ds_C}{dt} = p(W(C|\vec{s}) - W(\vec{s}|\vec{s})) < 0.$$

Over time, the proportion of the population not defecting will eventually be driven to extinction.

However, if spatial location constrains interaction between individuals, cooperation may emerge. Nowak and May (1992, 1993) show that the spatialized Prisoner's Dilemma favors the evolution of cooperation provided that the fitness payoffs for cooperation lie in a certain range and that there are a certain number of cooperators initially present. In their model, organisms are positioned at fixed locations on a square lattice and interact with their eight nearest neighbors. (In the original paper, all locations on the lattice are occupied and the lattice is considered to wrap at the edges. Although the former assumption is important for their results, the latter is not.) All individuals interact simultaneously and receive a total fitness payoff equaling the sum of all eight interactions. After interacting, behaviors are replicated according to the following rule: if an organism's fitness is lower than the fitness of at least one of his neighbors, that organism will be replaced in the next generation by an offspring from his neighbor who has the highest fitness. (If several neighbors are tied for having the highest fitness, then the neighbor whose offspring replaces the unfit individual is chosen at random.) If an organism's fitness is higher than the fitness of all of his neighbors, that organism's offspring will occupy the same site in the lattice for the next generation.

There are three possible outcomes: cooperation and defection may coexist in stable oscillating patterns, defection may drive cooperation to extinction, or cooperation and defection may coexist in chaotic patterns of mutual territorial invasion. Figures 22.4, 22.5, and 22.6 illustrate each of these possibilities in turn. In Figure 22.4, the case of stable coexistence, the fitness values are $T = 1.1$, $R = 1$, $P = 0$ and $S = -0.1$. In figure 22.5,

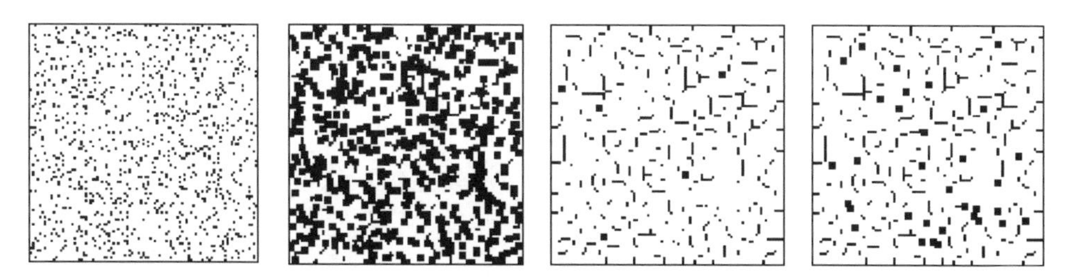

Figure 22.4 The spatial prisoner's dilemma illustrating the evolution of stable cooperative regions. $T = 1.1$, $R = 1$, $P = 0$ and $S = -0.1$

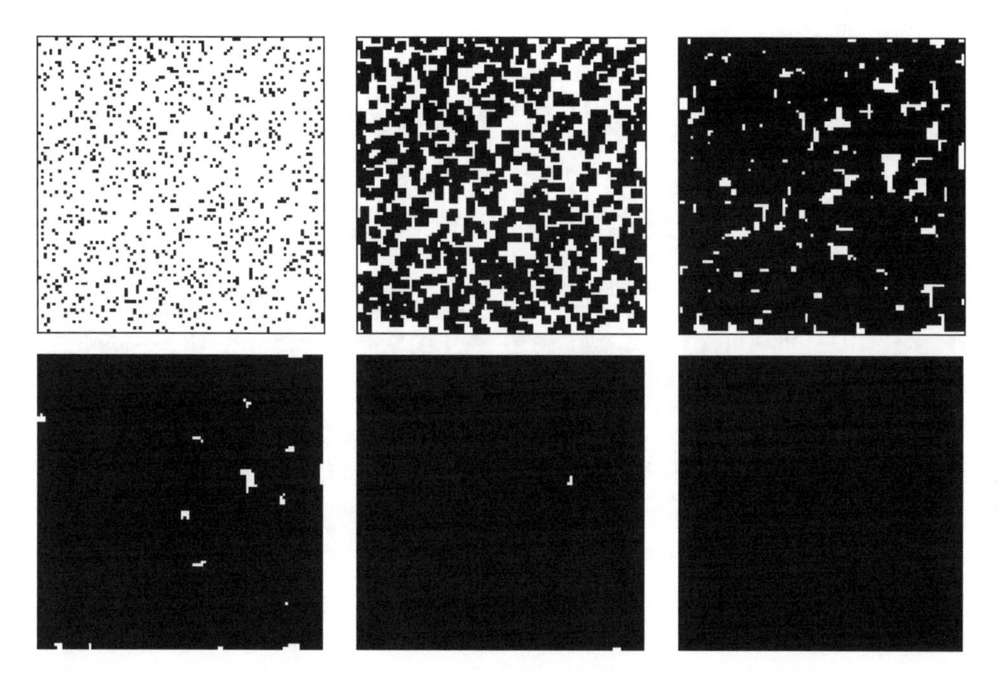

Figure 22.5 The spatial prisoner's dilemma illustrating the evolution of stable cooperative regions. $T = 2.7$, $R = 1$, $P = 0$, $S = -0.1$

with fitness values of $T = 2.7$, $R = 1$, $P = 0$, $S = -0.1$, defectors come to dominate within a relatively short period of time. (Note, though, that these particular fitness values violate the requirement that $(T + S)/2 < R$.) Of particular interest is Figure 22.6, which uses payoff values of $T = 1.6$, $R = 1$, $P = 0$, $S = -0.1$. In this case, the mix of cooperators and defectors in the population fluctuates chaotically. Cooperative regions can be invaded by regions of defectors, and vice versa, without ever settling into a stable evolutionary state.

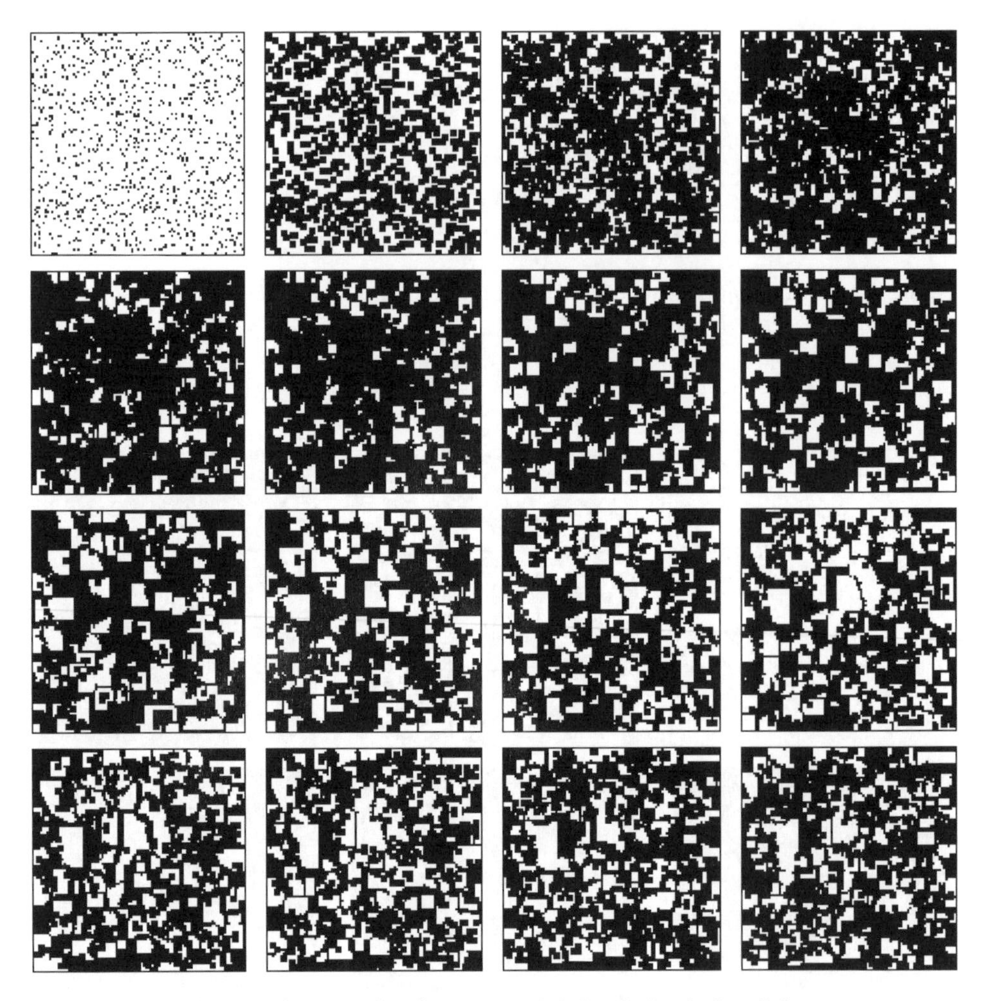

Figure 22.6 The Spatial Prisoner's Dilemma, $T = 1.6$, $R = 1$, $P = 0$, $S = -0.1$

References

Axelrod, R. (1982). *The evolution of cooperation.* New York: Basic Books.

Balshine-Earn, S., Neat, F. C., Reid, H., & Taborsky, M. (1998). Paying to stay or paying to breed? Field evidence for direct benefits of helping behavior in a cooperatively breeding fish. *Behavioral Ecology*, 9, 432–8.

Bednekoff, P. A. (1997). Mutualism among safe, selfish sentinels: A dynamic game. *The American Naturalist*, 150, 373–92.

Binmore, K. (1998). Review of "The Complexity of Cooperation: Agent-Based Models of Competition and Collaboration". *Journal of Artificial Societies and Social Simulation*, 1(1). http://jasss.soc.surrey.ac.uk/1/1/review1.html.

Boyd, R., & Richardson, P. J. (1992). Punishment allows the evolution of cooperation (or anything else) in sizeable groups. *Ethology and Sociobiology*, 13, 171–95.

Brown, J. L. (1983). Cooperation: a biologist's dilemma. In J. S. Rosenblatt (Ed.). *Advances in the study of behavior* (pp. 1–37). New York: Academic Press.

Cartwright, N. (1978). Causal laws and effective strategies. *Noûs*, 13, 419–37.

Clements, R., & Stephens, D. C. (1995). Testing models of non-kin cooperation-mutualism and the prisoner's dilemma. *Animal Behaviour*, 50(2), 527–35.

Cockburn, A. (1998). Evolution of helping behavior in cooperatively breeding birds. *Annual Review of Ecology and Systematics*, 29, 141–77.

Courchamp, F., Clutton-Brock, T., & Grenfell, B. (2000). Multipack dynamics and the Allee effect in African wild dogs *Lycaon pictus*. *Animal Conservation*, 3, 277–86.

Darwin, C. (1985 [1865]). *The origin of species*. London: Penguin.

Dawkins, R. (1976). *The selfish gene*. London: Oxford University Press. Second edition, December 1989.

Hamilton, W. D. (1964). The genetical evolution of social behavior. I and II. *Journal of Theoretical Biology*, 7, 1–52.

Heinsohn, R. G. (1991). Kidnapping and reciprocity in cooperatively breeding white-winged choughs. *Animal Behavior*, 41, 1097–100.

Huxley, T. H. (1888). The struggle for existence in human society. In *Collected Essays* (Vol. 9, pp. 195–236). London: Macmillan.

Janzen, D. H. (1966). Coevolution of mutualism between ants and acacias in Central America. *Evolution*, 20, 249–75.

Janzen, D. H. (1979). How to be a fig. *Annual Review of Ecology and Systematics*, 10, 13–51.

Kokko, H., Johnstone, R. A., & Clutton-Brock, T. H. (2001). The evolution of cooperative breeding through group augmentation. *Proceedings of the Royal Society of London, B*, 268, 187–96.

Kropotkin, P. (1902). *Mutual aid*. London: William Heinemann.

Lindgren, K., & Nordahl, M. (1994). Evolutionary dynamics of spatial games. *Physica, D*, 75, 292–309.

Maynard Smith, J. (1964). Group selection and kin selection. *Nature*, 200, 1145–7.

Mulder, R. A., & Langmore, N. E. (1993). Dominant males punish helpers for temporary defection in superb fairy-wrens. *Animal Behavior*, 45, 830–3.

Nowak, M., & May, R. (1992). Evolutionary games and spatial chaos. *Nature*, 359, 826–9.

Nowak, M., & May, R. (1993). The spatial dilemmas of evolution. *International Journal of Bifurcation and Chaos*, 3, 35–78.

Nowak, M., & Sigmund, K. (1993). A strategy of win–stay, lose–shift that outperforms tit-for-tat in the prisoner's dilemma game. *Nature*, 364, 56–8.

Piper, W. H., Parker, P., & Rabenold, K. N. (1995). Facultative dispersal by juvenile males in the cooperative stripe-backed wren. *Behavioral Ecology*, 6, 337–42.

Reeve, H. K. (1992). Queen activation of lazy workers in colonies of the eusocial naked mole rat. *Nature*, 358, 147–9.

Simpson, E. H. (1951). The interpretation of interaction in contingency tables. *Journal of the Royal Statistical Society B*, 13, 238–41.

Skyrms, B. (2004). *The stag hunt and the evolution of social structure*. London: Cambridge University Press.

Sober, E. (1984). *The nature of selection: evolutionary theory in philosophical focus*. Cambridge, MA: MIT Press.

Sober, E., & Wilson, D. S. (1998). *Unto others: the evolution and psychology of unselfish behavior*. Cambridge, MA: Harvard University Press.

Sober, E., & Wilson, D. S. (2000). Summary of "Unto Others the Evolution and Psychology of Unselfish Behavior". In L. D. Katz (Ed.). *Evolutionary origins of morality: cross disciplinary perspectives* (pp.185–206). Bowling Green: Imprint Academic.

Summers-Smith, J. D. (1963). *The house sparrow*. London: Collins.

Taylor, P. D., & Jonker, L. B. (1978). Evolutionary stable strategies and game dynamics. *Mathematical Biosciences*, 40, 145–56.

Trivers, R. L. (1971). The evolution of reciprocal altruism. *Quarterly Journal of Biology*, 46, 189–226.

Wade, M. J. (1978). A critical review of the models of group selection. *Quarterly Review of Biology*, 53, 101–14.

Wiebes, J. T. (1976). A short history of fig wasp research. *Gardens' Bulletin*, Singapore 29, 207–32.

Wilkinson, G. (1984). Reciprocal food sharing in vampire bats. *Nature*, 308, 181–4.

Williams, G. C. (1966). *Adaptation and natural selection: a critique of some current evolutionary thought*. Princeton: Princeton University Press.

Wilson, D. S. (1980). *The natural selection of populations and communities*. Menlo Park, CA: Benjamin Cummings.

Wilson, E. O. (1975). *Sociobiology: the new synthesis*. Cambridge, MA: Harvard University Press.

Chapter 23

Language and Evolution

DEREK BICKERTON

1. Introduction

Almost all, if not all, species communicate in one form or another. Humans communicate perhaps more than any other species. Although their communications are immensely more complex than those of any other species, and convey an infinitely greater quantity of information, it has seemed to many that human language must have developed out of the communication systems of antecedent species. After all, we evolved as a single species of the primate family, and evolution is normally a gradual process, building on what is already there rather than creating novelties. One might well conclude that human language, different though it might seem from the communication systems of other species, developed out of them by a series of infinitesimal increments, the intermediate forms having been, unfortunately, lost.

However natural such an assumption might appear, there is strong evidence against it. For instance, such basic attributes of language as predication, symbolization, and displacement (the ability to refer to objects and events not physically present) are absent in animal communication systems (ACSs). Further, it is sometimes claimed that the multi-layered nature of modern human language argues against any continuity with ACSs: the basic building blocks of language are phonemes (units of sound meaningless in themselves), which are combined to form morphemes (the smallest meaningful units), which (if they are not in themselves already words) are combined to form words, which can then be combined to form phrases and sentences. But comparing this system with ACSs tell us nothing, since its type of organization may have come relatively late in the development of language. Accordingly, the discussion that follows will refer only to properties found in its most basic and rudimentary forms of language, such as "foreigner talk" (Ferguson, 1971), pidgins (Bakker, 1995), and the like – properties the absence of which would both deprive the word "language" of any meaning, and leave as mysterious as before the means through which those properties did eventually emerge. For instance, without true symbols, it would be impossible to refer to anything that was not physically present, and without predication, which is a semantic relationship before it is a syntactic one, it would not be possible to expand single-unit utterances. As for arguments that the present approach is guided by some anti-scientific or anti-evolutionary agenda, these should be treated as what they are – ways of avoiding inconvenient facts (for arguments against language–ACS continuity see Bickerton, 1990, ch. 1).

2. Fundamental Differences Between Language and ACSs

The crucial differences between ACSs and language are qualitative, not quantitative. One of these involves the difference between symbolic and indexical reference (Deacon, 1997); human language has both, whereas ACSs have only the latter. If a unit is indexical, it can carry reference only if the entity it refers to is physically present. Thus the "leopard" alarm call of the vervet monkey (Cheney & Seyfarth, 1990) is meaningful only in the presence of a leopard; if uttered when no leopard is near, it is either deceptive or meaningless. It is impossible to question or negate a leopard call, since unlike a symbolic unit, an indexical unit does not "stand in place of" its referent but merely "points to" it. A symbolic unit, on the other hand, can be used to make general statements in the absence of any referent ("Leopards have spots") and can be questioned ("Is that a leopard?," or simply, "Leopard?," with a rising inflection) or negated ("No leopards here!"). Deacon (1997) considers the symbolic–indexical distinction to be the major distinction between the language communication systems of our species and others. Certainly it is an absolute, not a scalar one; there cannot, in the nature of things, be any form intermediate between an indexical and a symbolic unit.

Another difference lies in predication. Every linguistic utterance that is not a mere exclamation ("Ouch!" or "Wow!") refers to someone or something (sometimes referred to as the "subject") and then makes a statement about that person or thing (sometimes referred to as the "predicate"). This is true of even the shortest and simplest utterances: "John left," "Time's up!," "Dogs smell." Even imperatives make the same distinction between subject and predicate, although the former is not overtly stated: if I say "Leave!" it is you that are being told to leave, and no one else. If ACSs produce a sequence of calls, each call remains a self-contained unit and its meaning is unaffected by being adjoined to another call: sequences cannot be combined in the way that subject and predicate combine in human language to produce a meaning different from that of either in combination. Even the most primitive forms of language, early-stage pidgins (Bickerton, 1981), employ true combinations where the meaning of the combination is more than the sum of the meanings of its parts. "John" by itself merely refers to a person; "left" by itself merely refers to some action of leaving in the past; but "John left" tells us what a specific person did on a specific occasion. Even the language of children, which initially passes through a one-word stage, already struggles to achieve predication (Scollon, 1974): a child will repeat a word until some grown-up pays attention, then utter another word which expresses some kind of comment on the first.

We are dealing here with another qualitative, not quantitative distinction: either an utterance involves predication, or it does not. The problem for those who believe in continuity between human language and ACSs is to show how predication might have developed from a system that lacked any vestige of predication. At least two scholars have tried to hypothesize intermediate stages between a prior ACS and language. Hockett (Hockett & Ascher, 1964) suggested a possible blending of preexisting calls: for instance, in a situation where food and danger were both present, some hominid might have uttered half the call for "food" together with half the call for "danger." However, sequencing is not the feature that distinguishes human language: predication is. A sequence, even a blending like the example given, is not predication, since "danger" would not constitute a comment (predicate) on a subject ("food").

A more sophisticated proposal has recently been advanced by Wray (1998, 2000). She claims that the earliest forms of language were holophrastic, akin to calls; though they might contain only single and (at least initially) undecomposable units, their meaning would be equivalent to that of a human sentence ("That-animal-is-good-to-eat" or "I-want-to-mate-with-you," for example). According to Wray, such units simply increased in number to a point at which they began to impose an excessive memory load. The holophrases were then decomposed on the basis of phonetic similarities. Wray's own example (2000, p.297) makes the point clearly:

> So if, besides *tebima* meaning *give that to her*, *kumapi* meant *share this with her*, then it might be concluded that *ma* had the meaning *female person + beneficiary*.

There are, however, many problems with this proposal (for a brief review, see Bickerton, 2003, and for a more thorough one, Tallerman, 2004). First, there are clearly only two logical possibilities: either *ma* occurs always and only in holophrases which also contain the meaning "female person + beneficiary," or only some of its occurrences will bear this interpretation while others will not. In the first case, the language would be already synthetic; that is to say, the supposedly undecomposable holophrase would in reality consist of a string of separate (and separable) units combined just as they are combined in the syntaxes of contemporary languages. If it is to be taken seriously, Wray's proposal must assume the second case. But if *ma* also occurred where a female + beneficiary reading was impossible – contexts perhaps as numerous as, or more numerous than, those that could bear such a reading – why would the hearer assume that it referred to a female beneficiary in just those cases where such a reading was possible, and how would that hearer account for the other cases?

But there is an even more basic problem with the holophrase proposal, which involves the tacit assumption that pairs of utterances like *tebima* and *kumapi* could exist in a language that had not already developed the kinds of distinction that only a synthetic language could develop. Hominids developing a holophrastic language would have had to learn that these two different utterances meant two different things. They could do this only by observing differences in the contexts where the two expressions were used. What kind of context would serve to distinguish "Give that to her" from "Share this with her"? Unless the hearer already knew the difference between "give" and "share," and between "this" and "that" (which again assumes the prior existence of a synthetic language in which these would constitute units), the contexts where one or the other expression was appropriate would be virtually identical.

For that matter, the whole proposal depends on there being identity between each holophrase and just one particular synthetic equivalent. But this assumption is quite unrealistic. Suppose there is a holophrastic expression that could be regarded as equivalent to "Don't come near me." It could equally be regarded as equivalent to "Stay away from me," "If you come nearer I'll bite," "Keep your distance," or any of a number of similar expressions. If a phonetic sequence *gu* occurred within this holophrase, how could it be given a unique interpretation? Some might assume it meant "come," others "stay," others "me," others "your," and so on indefinitely. In other words, such a holophrastic language would be highly unlikely ever to decompose into an appropriate set of units. These are by no means the only problems with Wray's proposal, but those cited here should suffice to show that a bridge between any ACS and language is at best extremely difficult and perhaps impossible to construct. To insist on continuity without resolving the problems presented by symbols and predication is simply bad science.

3. Language as Adaptation

Rather than debating the form language first took, it might be more profitable to look at the kinds of selection pressure that might have given rise to it. If language was selected, what was it selected for? Early guesses included communal hunting and the making of tools. Nowadays, few if any evolutionists support these suggestions (see introduction to Hurford et al., 1998). Communal hunting is carried out by a number of species without benefit of language, while tool-making (and even instruction in tool-making) has been found to be performed through observation and imitation, rather than verbally, by the pre-literate hunters and gatherers who, we assume (perhaps even correctly), form the best models for the behavior of our remote ancestors (Ingold & Gibson, 1993). The fact that one might do something better if one had language cannot be a selective pressure – if it were, numerous other species would surely have language too. To break out of the mold of animal communication that had served all other species well since evolution began necessarily required some behavior that was impossible to perform without some language-like system.

Since Humphrey (1976) suggested that the likeliest driving force behind increased cognition and language was intraspecific competition, the search for a selective pressure has focused on the "Machiavellian strategies" (attempts to deceive others to the deceiver's advantage) and high degree of social sophistication found among primates generally, and in particular among the great apes who are our closest relatives. The line of reasoning went as follows: when (presumably among australopithecines) social life grew more complex, intelligence increased to cope with these complexities, until either our ancestors became clever enough to invent language (Donald, 1991) or language spontaneously emerged to satisfy needs for gossip and/or grooming (Dunbar, 1996) or some other social function.

What is striking about the quite extensive literature on the supposed social origins of language is the extent to which it ignores most of what is known about hominid or pre-hominid evolution. All that most writers provide is a straight-line projection from modern ape behavior to modern human behavior, without any reference to particular species or periods of pre-history, and with little if any awareness of the ecology of species antecedent to our own. For instance, it seems to be tacitly assumed that human ancestors had just as much leisure and freedom from predation as modern, forest-dwelling apes have in which to develop and intensify their social lives. Given the size of australopithecines, both absolute and relative to the size of pre-historic predators, and their terrain of open woodland and savanna that was a prime hunting-ground for those predators, this is at best a highly unlikely assumption (Lewis, 1997). The ecological

Table 23.1 Incongruous properties of language and ACSs

Language	ACSs
Symbolism	Indexicality
Mostly objective information	Mostly subjective information
Displacement	No displacement

facts (McHenry, 1994) suggest that there would have been little time for the elaboration of Machiavellian strategies, and a sharply reduced tendency to indulge in them, due to the pressing need for trust, mutual support, and cooperation in the face of predation and the transient, widely scattered nature of food sources.

Moreover, those who claim social pressures as the selective force for language commit what, to many biologists, may seem a cardinal error. As numerous and highly detailed ethological studies have demonstrated (Byrne & Whiten, 1992; Goodall, 1986; Schaller, 1963; Smuts 1987; de Waal, 1982; etc.), apes already have a complex and well-developed social life. If such a life provided a selective pressure for language, how is it that one primate species and one only developed language in (eventually) a highly complex form, while none of the other species developed the least vestige of language? A unique adaptation can only result from a unique pressure. Thus in seeking for the selective pressure that resulted in language, any biologist would look elsewhere than among our closest relatives.

But where to look? The apparent uniqueness of language seems to render the task impossible. However, if instead of treating language as a whole we look at some of its specific properties, there may be a way out of this impasse. One property specific to language is that it conveys objective information – information about things other than the current affective state of the communicator. Indeed, it is almost impossible for a sentence not to convey objective information. Even in flattering someone – "That dress is a perfect match for your eyes" – we cannot avoid conveying the objective information that the dress and the person's eyes are of similar color. In this, language differs from the vast majority of ACSs. Except for warning calls, units in such systems convey only needs, desires, or affective states; interestingly enough, the spontaneous productions of "language"-trained apes are almost all about things they want to eat or do (Terrace et al., 1979). This distinction between language and ACSs is almost certainly linked with the symbolic–indexical distinction. Unless something is a true symbol, it cannot substitute for the physical presence of its referent. However, symbolization is outside the reach of most species (Deacon, 1997) and it may well be that no species can achieve it unless that species has a pressing need to exchange information about things not physically present. Nothing in the life of other primate species provided such a need; only among human ancestors did such a need make itself felt, as will shortly be shown. The capacity kind of information exchange, known as "displacement" (see definition above), forms a basic property of human language. Indeed, three properties of language are tightly linked, and their distribution can be summarized graphically (Table 23.1).

There are a few exceptions, however, to the general rule that ACSs do not convey objective information. These are the "languages" of bees and ants – systems so limited and organisms so phylogenetically remote from humans that researchers have failed to consider any implications they might have for language evolution. However, it may be fruitful to consider them in terms of convergence, a phenomenon familiar to evolutionary biologists (Conway Morris, 2003), on which recent work on niche construction (Odling-Smee, Laland, & Feldman, 2003) has shed much light. The classic example of convergence involves sharks, dolphins, and ichthyosaurs, all of which developed similar fins in response to the pressures of an aquatic existence. Similarly, ant and bee ACSs are adaptations selected for by choice of niche: central-point-based foraging in a fission-fusion mode, with a consequent need for reinforcement. This type of niche puts a premium on exchange of information.

Both bees and ants forage as individuals but recruit conspecifics to exploit transient (and often short-lived) food resources. Using a variety of physical movements (the so-called "round," "waggle," and "vibrating" dances) bees can convey to their fellows the distance, direction, and relative quality of the honey or pollen they have discovered (von Frisch, 1967). Ants also employ a type of "waggle dance" to recruit helpers but lay chemical trails to draw them to the discovered food supply (Sudd & Franks, 1987) – something obviously impossible for bees.

Apes also forage on a fusion-fission basis (Goodall, 1986). In their case, however, food sources are easily accessible, abundant, and (despite seasonal variations) relatively long-lived (von Lawick-Goodall, 1971). They neither need nor create central bases. Such was not the case for *Homo habilis*. Food sources were scattered over a wide area of open woodland and savanna, necessitating much larger day ranges; much food was transient, useless unless exploited within a period of days or even hours; and took a wide variety of forms (tubers, honeycombs, termites, birds' eggs, and, most crucial because most nutritious, the scavenged carcasses of other mammals, see Binford, 1985). An additional problem was posed by predation (Lewis, 1997), which raised serious risks for solitary foragers and favored a central-point strategy (like that of baboons, another ground-dwelling primate, Kummer, 1968) based on a "safe haven" of tall trees or rocks that would serve as night-time protection (as with baboons, such bases may have been sites subject to frequent change rather than permanent or semi-permanent settlements). Under such circumstances, and given a plausible band size of say ~30 individuals, an optimal foraging strategy would consist of dividing the band into several smaller groups to scout resources, returning to the base (or some other pre-determined spot) if recruitment of larger numbers seemed advantageous.

In the course of niche extension (Odling-Smee et al., 2003) *Homo habilis* developed new food-seeking strategies. Their capacity for producing sharp-edged flakes and for using these as well as hammer-stones as tools gave them access to two food sources unavailable to other species. One was the still intact carcasses of megafauna whose skins were too thick to be pierced by the teeth of predators until several hours had elapsed and the skins were ruptured by normal decay processes (Blumenschine, Cavallo, & Capaldo, 1994; Monahan, 1996). The other was the bones of prey at any stage of decomposition, which could be cracked to obtain the rich and highly nutritious marrow within. The first represented a rather narrow window of opportunity, perhaps only a few hours; the second, a considerably longer one, But sources of both types required recruitment, since the first would be attended by major predators, the second by scavengers of all kinds. The scouting group that discovered either would need to recruit the whole band in order for some to fight off the predators while others attacked the hide and butchered the carcass, or carried bones and meat to some more easily defensible site.

But how could recruitment take place? Ants and bees have little individuality and have been programmed by evolution for millions of years to carry out recruitment strategies. Human ancestors had been programmed by evolution for a very different lifestyle – that of the other great apes – and, like any other great ape, had strongly developed individualities. To convince all of them to do the same thing required information far more specific than could be provided by a food call or a scent trail (and in any case the capacity to lay the latter had vanished when still earlier ancestors had selected arboreal niches). Especially if more than one scouting group had found food sources at the same time, specific information about food-type, distance, risks involved, and perhaps other factors was vital for optimal foraging tactics.

Fortunately our ancestors could draw on a capacity widespread among organisms with relatively large brains that originally had nothing to do with communication. This was the capacity to discriminate between a wide variety of natural kinds, in particular other species, as well perhaps as certain types of action, where primate mirror neurons (neurons that fire not only when the subject performs an action but when the subject perceives another performing the same action) may have been helpful (Perrett et al., 1985; Rizzolati, Fogassi, & Gallese, 2001). The resultant categories could have labels applied to them, resulting in a very primitive type of language (nowadays generally referred to as "protolanguage," following Bickerton, 1990). During recent years, it has been shown that chimpanzees (Gardiner & Gardiner, 1969), bonobos (Savage-Rumbaugh, 1986), gorillas (Patterson & Linden, 1982), orangutans (Miles, 1990), dolphins (Herman, 1987), and even sea lions (Schusterman & Krieger, 1984) and African gray parrots (Pepperberg, 1987) can be taught to use simple quasi-linguistic systems that consist of little more than labels attached to concepts/categories. The additional capacity to string such labels together to form elementary propositions seems to have arisen spontaneously and without any explicit training in almost all these animals, suggesting that all they lacked of the prerequisites for protolanguage was a set of labels for preexisting concepts: the rest of the necessary machinery was already in place. This does not mean that ACSs and human language are continuous. Possession of semantic structure (giving rise to a rich set of concepts), sound recognition (enabling hearers to decide whether one sound or set of sounds is the same as, or different from, another), imitative ability and similar capacities may have existed as independent properties in antecedent species, but all these and more had first to be welded together into a single dedicated system before language could begin.

It is sometimes objected that if other species had any kind of language capacity, they would already have deployed it in the wild. Such a belief distorts the way evolution works. Every organism has latent capacities; if this were not so, it would be impossible for species to diversify by extending their niches. However, those latent capacities will never be triggered unless some immediate problem can be resolved or some immediate benefit obtained by exercising them. What would other species have needed language for? With the hindsight that many thousands of years of language development has bestowed, it has seemed to many that language is an adaptive mechanism conferring multiple and unlimited benefits on those who possess it. But we have to imagine not what "language" would confer on a species but what a very small handful of symbolic items would confer. For that is exactly what any language must be at its inception. With a means so limited, there is actually very little one can do (a strong argument against any origin for language in social intercourse, since any kind of social use would presuppose at least a sizeable vocabulary). Apes, for instance, are capable of handling quite complex social lives without language, and of course, no species (least of all our remote ancestors) could have predicted what language might have been able to do for it once the early stages of development were past.

4. The Protolinguistic Adaptation

Accordingly, the most plausible hypothesis for the origin of language is that it developed in the context of extractive foraging by sub-units of small bands, and consisted of a handful of symbolic units used to identify food sources and the location and accessibility of these. The nature of the units remains undetermined, although it has been the subject of some controversy. Some researchers, such as MacNeilage (1998), see language as emerging via the modality of speech from the very beginning. Others, such as Corballis (2002; see also Hewes, 1973) see language as originating in the form of manual gestures. However, there is no reason to regard these choices as mutually exclusive; ant "language" uses chemical, gestural, and tactile modalities, for instance. The most plausible conjecture (and it can be no more than that) is that the first protolanguage users used whatever it took to communicate their message: vocal utterances, gestures, possibly pantomime (Arbib, 2004). The nature of the units is relatively unimportant, so long as they were truly symbolic.

Protolanguage did not supersede the preceding ACS. Humans still have an ACS; the human ACS (which includes sobs, laughter, facial expressions, and manual gestures like fist-shaking and "giving the finger") and language are controlled from different areas of the brain and use different auditory wavelengths, though both are subject to cultural modification (Pinker, 1994). The two systems exist side by side, sometimes augmenting one another but never mixing (a further argument against supposing that one developed out of the other). For reasons discussed above in Section 2.0, its units (whether vocal sounds or manual signs) were most probably discrete and particulate, having much the same kind of referents as modern words – unlike the units of ACSs, whose meanings more closely correspond to those of phrases or sentences. Short propositions ("Dead-mammoth thataway!") could have been produced by simply stringing such units together; perhaps, in the case given, by joining a trumpeting vocalization with a directional gesture. For it would be too much to expect that the symbolism of modern language, with its typically arbitrary associations between signifier and signified, should have emerged full-fledged at the dawn of protolanguage. In all probability, the beginnings of protolanguage included both iconic and indexical units as well as arbitrary, symbolic ones (note that in the example given, a trumpeting sound – iconic – combines with a pointing gesture – indexical – to yield displacement).

Among the misconceptions that have arisen about the nature of protolanguage is that it may have had only a narrow referential domain, and may have required some separate evolutionary development in order to acquire the property, common to all modern languages, of being able to refer to anything one can think of (Jackendoff,

2002; see also Mithen, 1997). Such is, of course, the nature of bee and ant "languages," which can specify food locations and identify outsiders, but little else. However, we must bear in mind that these "languages" are really only ACSs that happen to have acquired, for adaptive reasons, one or two of the properties otherwise found only in language. Like other, less language-like ACSs, they have a specific genetic basis, the result of countless millennia of evolution, and hence are not subject to change or extension by their users. Protolanguage had no genetic basis specific to itself; it simply and opportunistically co-opted the elaborate system of conceptual categorization that had evolved in many of the more advanced mammals and birds. This system was potentially infinite in that its possessors could extend it indefinitely; as experiments by Herrnstein (1979) and associates showed, even pigeons could be trained to recognize fish, which they had certainly never encountered in the wild. Thus, although in its first tentative steps protolanguage was doubtless confined to the domain of foraging, it had built into it from its very beginning the potentiality of reference to anything at all that human ancestors could discriminate.

There can, therefore, be little doubt that once a sufficiently large and varied vocabulary had developed, protolanguage was put to a variety of uses – gossip, alliance-building, planning the group's next moves, and more.

At what stage protolanguage selected the vocal mode must remain a matter for speculation (see Hewes, 1973; McNeilage, 1998, for contrasting views). This, along with other features (the refinement of phonetics and the establishment of a phonemic system, the development of a complex syntactic structure) are things that we know must have happened at some stage between the origin of protolanguage and the emergence of full human language, because all human languages nowadays have such things. We simply do not know, yet, exactly when or even in what sequence these and other related changes took place. The questions most researchers have tried to answer are to what extent these subsequent developments were incorporated into the human genome, and to what extent they merely exploited cognitive and other mechanisms that preexisted language. The section that follows presents some of the approaches that have been made to this still highly controversial issue.

5. Modern Human Language – Innate or Learned?

It is obvious that language cannot be wholly innate, in the way that the songs of certain (though far from all) songbirds are wholly innate. If it were, the species would have only one language (with perhaps minor regional variations), whereas in fact any human infant can learn any of the more than 6,000 (superficially, at least, quite different) human languages. It is obvious that language cannot be wholly learned, since certain aspects of it (its phonology, for instance) are highly determined, and determining factors such as the physical structure of the vocal organs, and even skeletal structure – for example, changes in the degree of basocranial flexion (Lieberman, 1984) – have undergone heavy selection and consequent language-favoring adaptation over the past couple of million years. But between the two indefensible extremes of this section's title, almost every conceivable intermediate position has been defended.

Over the past century, the balance of opinion has undergone at least two major shifts. In the early part of last century, behaviorism was dominant, language was believed to be a purely social construct, and hence if a new language were to be discovered it might differ unpredictably from any previous language. In the second half of the century, however, this view was challenged by generative grammarians (Chomsky, 1957, 1965), who pointed out that all normal humans had similar language abilities, that there were strong structural parallels beneath the apparent diversity of human languages, that the acquisition of language followed an identical course in all normal children, that children frequently produced sentences that they could not have learned through imitation, and that the linguistic input children received was inadequate for any inductive learning of the complex grammatical system underlying that input. This last, known as the "poverty of the stimulus" argument, was believed by generativists to render inevitable the conclusion that most of the syntactic structure of sentences was not learned, but innately specified. Further evidence came from Creole languages, which show a degree of uniformity in their structure that is not predictable from the mix of languages that went into their creation, and is doubly surprising in light of the sparse and conflicting primary data from which their first-generation speakers derived these similarities (Bickerton, 1981; for alternative viewpoints see Lefebvre, 1986; Mufwene, 2003, etc., although none of these satisfactorily accounts for inter-Creole resemblances). Similar phenomena have been observed in the sign languages of Nicaragua (Kegl et al., 1999), where input was even more chaotic and radically reduced.

Belief that syntax was largely innate predominated during the 1960s and early 1970s. However, during the past quarter-century, it has been attacked from a variety of viewpoints. In 1975 the New York Academy of Sciences held the first multidisciplinary conference on the evolution of language (Harnad, Steklis, & Lancaster, 1976), in which Chomsky notoriously dismissed the origin of language as an issue of no more scientific interest than the origin of the heart. Eight years previously, in a work that clearly staked out generative claims in the field of biology, Lenneberg (1967) had professed a similar lack of interest in language evolution; since language left no fossils, the course of that evolution was, he believed, irrecoverable. Future historians of science may well marvel at how the generative movement managed for so long to combine a belief that language was biologically based with a refusal to look at the biological evolution of language (note, however, that recently Chomsky has changed his position, see, e.g., Hauser, Chomsky, & Fitch, 2002). The immediate result was that few linguists, but many scholars from other disciplines who knew little linguistics, concerned themselves with language evolution. Such scholars tended to underestimate the complexity of the data that had to be accounted for. In consequence, while understanding of other aspects of language evolution broadened and deepened, the nature of what had evolved was largely ignored, and the grammars produced by generativists were frequently treated as arcane and convoluted formulations having little to do with the realities of language.

Although prejudice and ignorance played their parts in this opposition, there were legitimate causes for concern. The grammars proposed by generativists and the evolutionary processes known to biologists seemed irreconcilable: it was difficult if not impossible to see how one could have produced the other, hence scholars in the field were overly quick to accept reassurance from non-generative linguists that syntax was simpler than the generativists made it look. Generativists did not help matters by continually changing generative theory. To outsiders, this looked as if they couldn't make up their minds; to insiders it was apparent that each new formulation represented an improvement on its predecessor, although few were rash enough to assume that the latest formulation represented the final truth about syntax. In consequence, there arose a state of mutual incomprehension that has yet to be completely overcome.

Another attack on the generativist/innatist position came from scholars working with models of connectionist networks who carried out computer simulations of language acquisition (Rumelhart & McLelland, 1986). These purported to show that not only could such models acquire particular features of language (such as the English system of past tense), they could even mimic the stages through which, in children, the acquisition process passed. Their claims have been challenged (Marcus, 1996), but other researchers have extended this approach to include computer simulations of how language might have evolved (for a current overview see Briscoe, 2002). As with the acquisition studies, the main thrust of evolutionary simulations has been to show that once linguistic utterances commenced, processes of automatic self-organization would eventually install lexical and syntactic regularities.

There are, however, some problems with this approach. First, although some fairly simple features, such as regularity of word order, have been shown to emerge spontaneously, this has not, with one or two exceptions, been demonstrated for more complex features. Second, the emergence of isolated features, however well these processes are mimicked, is not the same as the emergence of a complex system in which features on many levels are tightly interlocked. Third, some of the researchers have made odd and poorly motivated assumptions about the nature of language. One currently popular view is that languages are "organisms that have had to adapt themselves through natural selection to fit a particular ecological niche: the human brain" (Christiansen & Ellefson, 2002, p.338; see also Deacon, 1997). This, if taken metaphorically, might seem no more than a playful inversion of the innatist view that the structure of the human brain has determined the form that languages take. If taken literally, it is nonsense: languages are not independent entities, like living organisms. How could they "adapt themselves to the brain" unless they had a prior existence outside the human brain, and were delivered to humans (by Martian spacemen, perhaps) ready-made? In fact, the brains to which they supposedly "adapt" can only be what created them in the first place. Moreover, "natural selection" can take place only if there is something to select from. Unless we assume that for every human language there were a dozen or two unfit languages that fell by the wayside, use of the term in this context renders it meaningless.

A fourth and possibly more serious problem with evolutionary simulations lies in the improbable initial conditions that most if not all such programs assume. Improbabilities in the various proposals include, but are not limited to, the following: agents (the term used for the simulated speakers) have access to one another's meanings; agents make a variety of random sounds to express the same meaning; agents employ a mixture of word-equivalents and holophrase-equivalents (see discussion of holophrases in Section 2.0 above). Not one appears to incorporate the most likely initial conditions: speakers know what they mean but their hearers initially don't; speakers pick a single form–meaning combination and stick to it; the referent of the form–meaning combination is a single entity or action, not a state or a situation; meanings are acquired by hearers through observing contexts of use. Until simulations can grapple with plausible real-world scenarios of first-stage language evolution, they will shed little light on it.

The most threatening source of possible counter-evidence to nativist claims comes from brain imaging techniques. Before the introduction of functional magnetic resonance imaging (fMRI), Positron emission tomography (PET) scans, and other means of directly representing neurological processes, neurology, based mainly on aphasia studies, had lent credence to the view that syntax was processed by Broca's area and semantics by Wernicke's area – thus that the brain might indeed contain a localized, discrete analog of Chomsky's "language organ" (Chomsky, 1980). However, scans of actual brains performing various linguistic tasks showed that all of these tasks involved numerous areas of the brain besides the familiar "language areas" – some of them even in the cerebellum, which had previously been believed to be concerned exclusively with non-cognitive functions (Indefrey et al., 2001; Pulvermuller, 2002; Dogil et al., 2002). Note, however, that this evidence rules out only a strictly localist version of innateness. A distributed innateness remains possible. An innate mechanism could consist of a specific wiring plan for the brain, a series of neural connections (linking a variety of areas many of which are also involved in non-linguistic tasks) that are found in human brains but not in those of other species. If so, it remains unclear how such a plan might be instantiated. We need to know more about how brain structure is built up during both pre- and post-natal development. Since neurons in the brain outnumber genes by several orders of magnitude, functions cannot be genetically determined at the cell level. How they might be determined at the level of areas and/or networks remains a profound mystery. Connectionists (people who believe that language and other cognitive capacities do not depend on mental representations, but can be generated by the activities of neural networks alone) would claim that language functions in the brain are not genetically determined at all, but that an equipotential brain is programmed by the input it receives. But of course connectionists have no good explanation for why other ape species, with brains not dissimilar (except in size) to ours, cannot acquire language.

The nativist response to criticism is to state, usually correctly, that the critics simply do not know enough about language (and in particular, about syntax) and in consequence seriously underestimate both the complexity and the task-specificity of the neural machinery required to run it. Typical of the phenomena they invoke are contrasts like the following (an asterisk indicates an ungrammatical sentence):

1a) Who did you think that she saw?
 b) Who did you think she saw?
 c) Who did you think saw her?
 d) *Who did you think that saw her?

If (1a) and (1b) are equally acceptable, why is (1c) acceptable but its equivalent, (1d) not?

2a) Bill needs someone to inspire him.
b) Bill needs someone to inspire.

In (2a) "someone" is to do the inspiring; in (2b), "Bill" is to do it. Why should the presence versus the absence of a pronoun at the end of the sentence change the subject of "inspire," and why should it change them in this direction, rather than the reverse direction?

3a) Bill and Mary wanted a chance to talk to one another.
b) *Bill and Mary wanted Mr Chance to talk to one another.

Why should the switch from a common to a proper noun make (3b) ungrammatical, when its meaning is simple and straightforward – Bill wants Mr Chance to talk to Mary and Mary wants Mr Chance to talk to Bill? How is it that we can't express that meaning unless we spell it out in this way?

4a) Jane is a person that everyone likes as soon as they see her.
b) Jane is a person that everyone likes as soon as they see.
c) *Jane is a person that everyone likes her as soon as they see.

Why is (4) grammatical with a final pronoun or with no pronouns but ungrammatical with a pronoun in the middle?

These are typical of countless puzzling aspects of syntax that are seldom considered by most scholars in the field of evolution. They are not trivial. They represent, not quirks of the English language, but phenomena found across a wide range of languages – perhaps, in one form or another, across all languages. For this to be the case, it is quite implausible that children induced rules that gave the same result in each case in every language. How would you induce a rule involving something that isn't there, as you would have to in inducing anything from examples (1c), (2b), and (4a, b)? It seems much more plausible that examples (1)–(4) do not represent examples of four separate rules, but rather reflect one or more very deep principles that the child could not have induced from data, but that must somehow apply automatically, without any kind of learning being involved. Words have to be learned in every language because they are different in every language, but the so-called "empty categories" such as are found in examples (1c), (2b) and (4a, b) are the same in every language: gaps, where words might be expected to occur and where they can occur, that yield grammatical results in some cases and ungrammatical results in others.

If such phenomena do indeed result from deep principles, then those principles must (somehow) be instantiated both in the human genome and in the human brain, and must have evolved like every other adaptation. In that case, it is irrelevant that linguists still cannot agree what those principles are.

It must be that the brain processes syntax in one particular way, and that such a way is describable. What is known about the capacities of the brain should constrain theories of syntax, at least to the extent that no theory incompatible with such knowledge should be supported. But likewise, what is known about syntax should constrain theories about how the brain generates sentences, to an identical extent. We are still some distance from such a level of interdisciplinary cooperation, but that level must be reached; no theory of language evolution can be complete that does not explain how the basic principles underlying sentence structure came to be the way they are, and not some other way.

Two recent developments in generative grammar have the potentiality to increase chances of arriving at the correct formulation of basic syntactic principles. The first is the Minimalist Program (Chomsky, 1995), whose professed goals are to substitute genuine explanations for mere restatements of problems in other terms, and to bring hypothesized mechanisms down to an irreducible minimum. The second is the Derivational version of that program (Epstein et al., 1998) which, in contrast to earlier versions of generative grammar, builds grammatical structures from the bottom up, instead of first building an entire abstract tree and then inserting lexical items (the Representational approach). However, it is still too early to see where, if anywhere, these developments will lead.

Thus the extent to which syntax is innate remains a highly controversial issue, with no clear signs of a resolution in sight. The question of how syntax evolved, answers to which depend at least in part on the resolution of the innateness issue, is, unsurprisingly, no less confused.

6. The Evolution of Syntax

Was syntax a distinct adaptation, specially selected for? Or was it an exaptation, a mere change in the function of some preexisting capacity? Or did it result, like ice crystals, automatically, due to some hitherto-unstated "law of form"? Or was it the result of a purely fortuitous mutation?

All of these sources have been proposed. The least likely is the fourth (Klein & Edgar, 2002) – that a single mutation could result in all the complexities of syntax. As Pinker and Bloom (1990) noted, these complexities resemble those of the eye, an organ produced by millions of years of natural selection. Consequently, given the gradualness and piecemeal development characteristic of evolution, the first reaction of any biologist would be to suppose that syntax too had evolved in a series of increments, each one somewhat superior to its predecessor, each one specially selected for.

Yet, as so often happens where language is concerned, the straightforward biological solution runs into problems. First there is the problem of time. The eye had tens of millions of years, at least, in which to evolve; syntax has, at most, about two million – unless, contra the balance of the evidence, we are willing to award some degree of syntax to australopithecines. Second, there is the problem of intermediate forms. With the eye, this presents no problems. Countless organisms still survive with eyes in various stages of development. No other organism, however, has anything more language-like than an ACS. In principle, one might partially overcome this deficiency by hypothetically reconstructing intermediate stages. But quite apart form the difficulty of doing this (the flaws in intermediates proposed by Premack, 1985, and Pinker, 1994, are discussed in Bickerton, 1995), there may be a much deeper problem.

The anti-evolution jibe, "What use is 5 percent of an eye?" is easily answered: "More use than 3 percent of an eye." But what use is 5 percent of syntax? The function of syntax is to make utterances automatically processable, hence immediately comprehensible, to the hearer. If what the hearer receives is an utterance in an early-stage pidgin, or the speech of a recent and untutored immigrant, that utterance often can't be quickly and smoothly processed: deprived of the grammatical cues that syntax provides, the hearer frequently has to puzzle over its meaning, must use additional contextual and pragmatic information, and even with the aid of these may still misunderstand the message. In contrast, a message in the hearer's own language will seldom if ever require contextual or pragmatic clues (unless it is structurally ambiguous) and will be understood immediately in the vast majority of cases. Just what, between these two extremes, could 5 percent or 25 percent of syntax do for hearers? While it is easy to see which particular additions to 5 percent of an eye would enable its owner to see more detail, or more colors, or discriminate between more types of object, it remains unclear which particular additions to 5 percent of syntax would improve quantity or quality of understanding – or, if they did, how they would do it.

If there were intermediate grammars, they would have to be individually selected for. Pinker and Bloom (1990) seem to assume that a grammar would consist of large numbers of rules, as in pre-1980 generative grammars; speakers with *n* rules would be replaced by speakers with *n* + 1 rules. Yet at other times they speak as if the units of selection consisted of the constraints on rules (forerunners of the "principles" of more modern grammars) that played an increasing role in grammars from the late 1960s on. One such example they give is Subjacency, a constraint that prevents italicized words from being moved from their original positions (marked by _______) to positions outside the square brackets, as in (5):

5a) [What did Bill deny that he found _______?]
b) *What did Bill deny [the fact that he found_______?]
c) *What did you lose it [and Bill found_______?]
d) *What did Bill tell you [where he had found_______?]

Rules and constraints, however, are equally implausible as targets for selection. In the case of Subjacency, for instance, we would have to make the unlikely assumption that speakers were producing large numbers of sentences like (5b–d) and hearers were failing to understand them until a handful of speakers started limiting their production to sentences like (5a), whereupon members of the second promptly started to have more children than members of the first group.

While a large syntactic increment might secure such a result through female choice (females would mate preferentially with males who controlled a wider variety of syntactic forms and were more readily understandable), it is hard to see how the small increments envisaged by Pinker and Bloom would have any such effect, or would increase fitness in any way sufficient to alter the composition of the gene-pool. Syntacticized language may be adaptive as a whole, once established (it is quite possibly what gave our ancestors the edge over Neanderthals), but in considering how it became established, a long string of adaptations each requiring its own separate selective history hardly seems the likeliest scenario. One therefore has to consider the two remaining alternatives.

The exaptation alternative has been most clearly set forth in Hauser et al. (2002), although this paper appears to be a strange compromise between scholars who previously held diametrically opposed positions – Hauser (1996) affirming continuity with ACSs, Chomsky (1988) equally emphatically denying it. In the 2002 paper, everything in language but recursion – the capacity to expand a linguistic expression without limit, as in *the dog, the black dog, the black dog in the yard, the black dog in the yard that you saw yesterday*, etc. – is regarded as being shared with other species, and the capacity for recursion itself is seen as having been co-opted from some preexisting faculty that originally dealt with other computational problems such as navigation, number quantification, or social relationships. But this proposal is in fact no more than a promissory note. No indication is given as to how the alternatives listed (and doubtless others) would be weighed against one another, what kind of evidence would be sought, or why, given that recursion must consequently exist at least embryonically in apes, apes are quite unable to learn recursion, even though they can learn lexical items with relatively little trouble.

Explanations of the third kind, based on laws of form (Thompson, 1992), Fibonacci numbers, self-organization, and similar factors constitute what is sometimes called a "neo-neo-Darwinist" approach (Piatelli-Palmerini, 1989; Jenkins, 2000). This approach, a reaction to the current "neo-Darwinian" consensus based on the merger of natural selection with post-Mendelian genetics, regards the role of natural selection in current evolutionary theory as being highly exaggerated, and seeks for as many alternative explanations as possible. Such explanations seem more interested in discrediting natural selection than advancing hypotheses specific and coherent enough to be argued about; at least, no such hypotheses have emerged to date. However, an explanation falling into this general class may still prove valid (Calvin & Bickerton, 2000; Bickerton, 2002).

Relative to their size, humans have the largest brains of any animal. It would be strange if this fact and our unique possession of language were unconnected. To some, the connection has seemed to take the form of a causative sequence: big brains → high intelligence → capacity for language. But this cannot be right. If, as seems the likeliest possibility, human language was complete and in place by the early stages of *Homo sapiens sapiens*, it preceded rather than followed the appearance of intelligent behavior; it seems more likely that language itself created human intelligence (McPhail, 1987). However, consider the tasks set for the human brain by the requirements of protolanguage and the requirements of language, respectively. Protolanguage required the brain to send the neural impulses that represent words to the motor areas controlling the organs of speech, one word at a time (represent A, send A, execute A; represent B, send B, execute B, with A, B etc. representing isolated words). Language requires the brain to take the neural impulses that represent a word and then merge it with the neural impulses that represent another word, repeating the process as many times as is necessary to build a complex phrase or sentence (most probably, whatever would come under a single intonation contour) and only then send the entire complex of impulses to the motor areas for execution (represent A, represent B, merge representations of A and B, represent C, merge C with AB . . . send ABC . . . execute ABC). Clearly the second process is far more complex and fraught with problems than the first. Perhaps the most serious problem is to avert message decay – that is, to prevent any parts of the message from becoming garbled during assembly and dispatch. Normal leakage that affects all electrical impulses is worsened by the fact that no given pair of neurons will ever fire with perfect synchronicity. The only way to overcome this is to have the same message sent by large numbers of neurons so that receiving centers average their output. But in all probability, brains smaller than those of humans do not have large enough numbers of neurons that can be spared from other tasks.

The foregoing hypothesis (see Bickerton, 2003 for a fuller discussion) provides an explanation for the ability of many other species to acquire protolanguage and their inability to acquire human language. It would also explain why, even if protolanguage emerged as early as suggested here, true language developed only in our species (and perhaps Neanderthals). It would explain the abruptness with which human intelligence manifested itself (only 40,000 years ago on the conventional wisdom; 90,000 years ago if we recognize recent discoveries in Africa, see McBrearty & Brooks, 2000), if language was indeed the major force in developing human intelligence (and no equally convincing candidate has been proposed).

Until messages could be reliably assembled and transmitted in the linguistic mode, it was safer to use the protolinguistic mode. In other words, even if some true-language ability existed in earlier species, its actual manifestation could have been quite abrupt, as one species (ours) switched entirely to the linguistic mode. A further advantage of the model is that it requires no additional neural machinery over and above what apes come equipped with, apart from an added number of brain cells and some novel connections between these. The phenomena of syntax would then hopefully fall out from the brain's mode of processing and assembling any complex information. But for the moment this remains a very large promissory note, and the model also requires validation from advances in neuroscience and neuroimaging.

7. The "Cultural Evolution" of Language

Is language evolution finished? Some would deny this. In a recent article in *Science*, the authors wrote "Language evolution has not stopped, of course; in fact, it may be progressing more rapidly than ever before" (Culotta & Hanson, 2004, p.1315). This statement reflects a profound misunderstanding. The faculty of language is based, as we have seen, on human biology. For as far back as history will take us, there are no signs of any change in this basic infrastructure, and there is every reason to believe that the languages of 100,000 years ago, though superficially different from ours in many ways, would have the same basic structure. In other words, as far as language is concerned, evolution is at a virtual standstill.

Granted, languages continue to change. Darwin was impressed by the analogy between the way in which languages diverge, diversify, and sometimes die out and the way in which species diverge, diversify, and sometimes go extinct. But this analogy is superficial and leads nowhere. Even the term "cultural evolution" is misleading. Languages change, not in response to cultural developments, but because of either internal or contingent causes. Internally motivated change may result from a variety of factors. For instance, the gradual erosion of sounds at the ends of words, where important grammatical information is often carried, results in the substitution of auxiliary verbs for inflected tenses and pre- or post-positions for case-markers. Change may also result if the increasingly frequent use of a marked word order leads to reanalysis of this as the basic word order. Contingency-motivated change occurs when, through conquest, speakers of one language are dominated by speakers of another language, or when a small language community becomes marginalized and its speakers all die or abandon their native tongue. There is no connection whatsoever between particular types of culture and particular types of language: in the vivid phrase of Sapir (1921), "Alexander walks with the Macedonian swineherd, and Lao Tse with the headhunter of Assam."

In fact, once the biological faculty of language was established, all languages did, or could do, was cycle and recycle through a limited set of possibilities within the narrow envelope that the biological faculty left open for them. Thus to speak of "cultural evolution," at least with respect to language, is a solecism we should learn to avoid.

References

Arbib, M. A. (2004). From monkey-like action recognition to human language: New York: Columbia University Press.

Bakker, P. (1995). Pidgin languages. In J. Arends, P. Muysken, & N. Smith (Eds). *Pidgins and creoles: an introduction.* Amsterdam: John Benjamins.

Bickerton, D. (1981). *Roots of language.* Ann Arbor, MI: Karoma.

Bickerton, D. (1990). *Language and species.* Chicago: University of Chicago Press.

Bickerton, D. (2002). Foraging versus social intelligence in the evolution of language. In A. Wray (Ed.). *The transition to language* (pp. 207–25). Oxford University Press.

Bickerton, D. (2003). Symbol and structure: a comprehensive framework for language evolution. In M. H. Christiansen & S. Kirby (Eds). *Language evolution* (pp. 77–93). Oxford: Oxford University Press.

Binford, L. S. (1985). Human ancestors: changing views of their behavior. *Journal of Anthropological Archaeology,* 4, 292–327.

Blumenschine, R. J., Cavallo, J. A., & Capaldo, S. P. (1994). Competition for carcasses and early hominid behavioral ecology. *Journal of Human Evolution,* 27, 197–214.

Briscoe, T. (Ed.). (2002). *Linguistic evolution through language acquisition: formal and computational models.* Cambridge: Cambridge University Press.

Byrne, R. W., & Whiten, A. (1992). Cognitive evolution in primates; evidence from tactical deception. *Man,* 27, 609–27.

Calvin, W., & Bickerton, D. (2000). *Lingua ex machina: reconciling Darwin and Chomsky with the human brain.* Cambridge, MA: MIT Press.

Cheney, D., & Seyfarth, R. M. (1990). *How monkeys see the world.* Chicago: Chicago University Press.

Chomsky, N. (1988). Language and problems of knowledge. Cambridge, MA: MIT Press.

Chomsky, N. (1957). *Syntactic structures.* The Hague: Mouton.

Chomsky, N. (1965). *Aspects of the theory of syntax.* Cambridge, MA: MIT Press.

Chomsky, N. (1980). *Rules and representations.* New York: Columbia University Press.

Chomsky, N. (1995). *The minimalist program.* Cambridge, MA: MIT Press.

Christiansen, M. H., & Ellefson, M. R. (2002). Linguistic adaptation without linguistic constraint: the role of sequential learning in language evolution. In A. Wray (Ed.). *The transition to language* (pp. 335–58). Oxford University Press.

Christiansen, M. H., & Kirby, S. (2003). *Language evolution*. Oxford: OUP.
Conway Moris, S. (2003). *Life's solution: inevitable humans in a lonely universe*. Cambridge: Cambridge University Press.
Corballis, M. C. (2002). *From hand to mouth: the origins of language*. Princeton: Princeton University Press.
Culotta, E., & Hanson, B. (2004). First words. *Science*, 303, 1315.
Deacon, T. (1997). *The symbolic species*. New York: Norton.
Dogil, G., Ackerman, H., Grodd, W., Haider, H., Kamp, H., Mayer, J., Riecker, A., & Wildgruber, D. (2002). The speaking brain: a tutorial introduction to fMRI experiments in the production of speech, prosody and syntax. *Journal of Neurolinguistics*, 15, 59–90.
Donald, M. (1991). *Origins of the modern mind*. Cambridge, MA: Harvard University Press.
Dunbar, R. I. M. (1996). *Grooming, gossip and the evolution of language*. London: Faber & Faber.
Epstein, S. D., Groat, E. M., Kawashima, R., & Kitahara, H. (1998). *A derivational approach to syntactic relations*. Oxford: Oxford University Press.
Ferguson, C. A. (1971). Absence of copula and the notion of simplicity: a study of normal speech, baby talk, foreigner talk and pidgins. In D. Hymes (Ed.). *The pidginization and creolization of languages* (pp. 141–50). Cambridge: Cambridge University Press.
Frisch, K. von. (1967). Honeybees: do they use direction and distance information provided by their dancers? *Science*, 158, 1072–76.
Gardener, R. A., & Gardener, B. T. (1969). Teaching sign language to a chimpanzee. *Science*, 164, 664–72.
Gibson. K. R., & Ingold, T. (Eds). (1993). *Tools, language and cognition in human evolution*. Cambridge: Cambridge University Press.
Goodall. J. (1986). *The chimpanzees of Gombe: patterns of behavior*. Cambridge, MA: Harvard University Press.
Harnad, S. R., Steklis, H. D., & Lancaster, J. (1976). *Origins and evolution of language and speech*. Annals of the New York Academy of Science, vol. 280.
Hauser, M. D. (1996). *The evolution of communication*. Cambridge, MA: MIT Press.
Hauser, M. D., Chomsky, N., & Fitch, W. T. (2002). The language family: what is it, who has it, and how did it evolve? *Science*, 298, 1569–79.
Herman, L. M. (1987). Receptive competences of language-trained animals. In J. S. Rosenblatt, C. Beer, M. C. Busnel, & P. J. B. Slater (Eds). *Advances in the study of behavior* (Vol. 17; pp. 1–60). San Diego, CA: Academic Press.
Herrnstein, R. J. (1979). Acquisition, generalization, and discrimination reversal of a natural concept. *Journal of Experimental Psychology (Animal Behavior Processes)*, 5, 116–29.
Hewes, G. W. (1973). Primate communication and the gestural origins of language. *Current Anthropology*, 14, 5–24.
Hockett, C. F., & Ascher, R. (1964). The human revolution. *Current Anthropology*, 5, 135–68.
Humphrey, N. K. (1976). The social function of intellect. In P. P. G. Bateson & R. A. Hinde (Eds). *Growing points in ethology* (pp. 303–17). Cambridge: Cambridge University Press.
Indefrey, P., Hagoort, P., Herzog, H., Seitz, R. J., & Brown, C. M. (2001). Syntactic processing in left prefrontal cortex is independent of lexical meaning. *Neuroimage*, 14, 546–55.
Jackendoff, R. (2002). *Foundations of language: brain, meaning, grammar, evolution*. New York: Basic Books.
Jenkins, L. (2000). *Biolinguistics: exploring the biology of language*. Cambridge: Cambridge University Press.
Kegl, J., Senghas, A., & Coppola, M. (1999). Creation through contact: sign language emergence and sign language change in Nicaragua. In. M. DeGraff (ed.). *Language creation and language change: creolization, diachrony and development* (pp. 179–237). Cambridge, MA: MIT Press.
Klein, R. G., & Edgar, B. (2002). *The dawn of human culture*. New York: John Wiley & Sons.

Kummer, H. (1968). *Social organization of hamadryas baboons*. Basel, Switzerland: Karger.

Lawick-Goodall, J. van (1971). *In the shadow of man*. New York: Dell.

Lefebvre, C. (1986). Relexification and creole genesis revisited: the case of Haitian Creole. In P. Muysken, & N. Smith (Eds). *Substrata versus universals in creole genesis* (pp. 279–300). Amsterdam: Benjamins.

Lenneberg, E. (1967). *Biological foundations of language*. New York: Wiley & Sons.

Lewis, M. E. (1997). Carnivorean paleoguilds of Africa: implications for hominid food procurement strategies. *Journal of Human Evolution*, 32, 257–88.

Lieberman, P. (1984). *The biology and evolution of language*. Cambridge, MA: Harvard University Press.

McBrearty, S., & Brooks, A. (2000). The revolution that wasn't: a new interpretation of the origin of modern human behavior. *Journal of Human Evolution*, 39, 453–563.

MacNeilage, P. F. (1998). The frame/content theory of the evolution of speech production. *Behavioral and Brain Sciences*, 21, 499–546.

Marcus, G. F. (1996). Children's overregularization of English plurals: a quantitative analysis. *Journal of Child Language*, 22, 447–59.

McPhail, E. M. (1987). The comparative psychology of intelligence. *Behavioral and Brain Sciences*, 10, 645–95.

McHenry, H. M. (1994). Behavioral ecological implications of early hominid body size. *Journal of Human Evolution*, 27, 77–88.

Miles, H. L. (1990). The cognitive foundations for reference in a signing orangutan. In S. Parker & K. Gibson (eds). *"Language" and intelligence in monkeys and ape: Comparative developmental perspectives* (pp. 511–39). Cambridge: Cambridge University Press.

Mithen, S. (1997). *The prehistory of the mind*. London: Thames & Hudson.

Monahan, C. M. (1996). New zooarchaeological data from Bed II, Olduvai Gorge, Tanzania: implications for hominid behavior in the early Pleistocene. *Journal of Human Evolution*, 31, 93–128.

Mufwene, S. (2003). *The ecology of language evolution*. Cambridge: Cambridge University Press.

Odling-Smee, F. J., Laland, K. N., & Feldman, M. W. (2003). *Niche construction: the neglected process in evolution*. Princeton: Princeton University Press.

Patterson, F., & E. Linden. (1982). *The education of Koko*. New York: Andre Deutsch.

Pepperberg, I. M. (1987). Acquisition of the same/different concept by an African Grey parrot. *Animal Behavior and Learning*, 15, 423–32.

Perrett, D., Smith, P. A. J., Mistlin, A. J., Chitty, A. J., Head, A. S., Potter, D. D., Broenniman, R., Milner, A. P., & Jeeves, M. A. (1985). Visual analysis of body movements by neurones in the temporal cortex of the macaque monkey. *Behavior and Brain Research*, 16 (2–3), 153–70.

Piatelli-Palmerini, M. (1989). Evolution, selection and cognition: from "learning" to parameter setting in biology and the study of language. *Cognition*, 31(1), 1–44.

Pinker, S. (1994). The language instinct. New York: Harper/Collins.

Pinker, S., & Bloom, P. (1990). Natural language and natural selection. *Behavioral and Brain Sciences*, 13, 707–84.

Premack, D. (1985). Gavagai, or the future history of the animal language controversy. *Cognition*, 19, 207–96.

Pulvermuller, F. (2002). A brain perspective on language mechanisms: from discrete engrams to serial order. *Neurobiology*, 574, 1–27.

Rizzolatti, R., Fogassi, L., & Gallese, V. (2001). Neurophysiological mechanisms underlying the understanding and imitation of action. *Nature Reviews Neuroscience*, 2, 661–70.

Rumelhart, D., & McLelland, J. (1986). *Parallel distributed processing, Explorations into the microstructure of cognition*. Cambridge, MA: MIT Press.

Sapir, E. (1921). *Language: an introduction to the study of speech*. New York: Harcourt & Brace.

Savage-Rumbaugh, S. (1986). *Ape language: from conditioned response to symbol.* New York: Columbia University Press.

Schaller, G. B. (1963). *The year of the gorilla.* New York: Ballantyne.

Schusterman, R. J., & Krieger, K. (1984). California sea lions are capable of semantic interpretation. *The Psychological Record,* 34, 3–23.

Scollon, R. (1974). *One child's language from one to two: the origins of construction.* Unpublished doctoral dissertation, University of Hawaii.

Smuts, B. (Ed.). (1987). Primate societies. Chicago: University of Chicago Press.

Sudd, J. H., & Franks, N. R. (1987). *The behavioral ecology of ants.* New York: Chapman & Hall.

Tallerman, M. (2004). *Analyzing the analytic: problems with holistic theories of protolanguage.* Paper presented at the Fifth Biennial Conference on the Evolution of Language. Leipzig, Germany, April.

Terrace, H. S., Pettito, L. A., Sanders, R. J., & Bever, T. G. (1979). Can an ape create a sentence? *Science,* 206, 891–900.

Thompson, D'A. W. (1992). On growth and form. Cambridge: Cambridge University Press.

de Waal, F. B. M. (1982). *Chimpanzee politics: power and sex among apes.* London: Cape.

Wray, A. (1998). Protolanguage as a holistic system for social interaction. *Language and Communication,* 18, 47–67.

Wray, A. (2000). Holistic utterances in protolanguage: the link from primates to humans. In C. Knight, M. Studdert-Kennedy, & J. R. Hurford, (Eds). *The evolutionary emergence of language: social function and the origins of linguistic form* (pp. 285–302). Cambridge: Cambridge University Press.

Wray, A. (Ed.). (2002). *The transition to language.* Oxford: Oxford University Press.

Further Reading

Bickerton, D. (1998). Catastrophic evolution: the case for a single step from proto-language to full human language. In J. R. Hurford, M. Studdert-Kennedy, & C. Knight (Eds). *Approaches to the evolution of language: social and cognitive bases* (pp. 341–58). Cambridge: Cambridge University Press.

Bickerton, D. (1990). *Language and species.* Chicago: University of Chicago Press

Bickerton, D. (in press). *Adam's tongue: how humans made language, how language made humans.* New York: Farrar, Straus & Giroux.

Briscoe, T. (Ed.). (2002). *Linguistic evolution through language acquisition: formal and computational models.* Cambridge: Cambridge University Press.

Christiansen, M. H., & Kirby, S. (Eds.) (2003). *Language evolution.* Oxford: Oxford University Press.

Deacon, T. (1997). *The symbolic species.* New York: Norton.

Hurford J. R., Studdert-Kennedy, M., & Knight, C. (Eds). (1998). *Approaches to the evolution of language: social and cognitive bases.* New York: Cambridge University Press.

Knight, C., Studdert-Kennedy, M., & Hurford, J. R. (Eds). (2000). *The evolutionary emergence of language: social function and the origins of linguistic form.* Cambridge: Cambridge University Press.

McBrearty. S., & Brooks, A. (2000). The revolution that wasn't: a new interpretation of the origin of modern human behavior. *Journal of Human Evolution,* 39, 453–563.

Odling-Smee, F. J., Laland, K. N., & Feldman, M. W. (2003). *Niche construction: the neglected process in evolution.* Princeton: Princeton University Press.

Terrace, H. S., Pettito, L. A., Sanders, R. J., & Bever, T. G. (1979). Can an ape create a sentence? *Science,* 206, 891–900.

Part VII

Experimentation, Theory, and Themes

Chapter 24

What is Life?

MARK A. BEDAU

1. The Fascination of Life

The surface of the Earth is teaming with life, and it is usually easy to recognize. A cat, a carrot, a germ are alive; a bridge, a soap bubble, a grain of sand are not. But it is notorious that biologists have no precise definition of what life is. Since biology is the science of life, one might expect a discussion of the nature of life to figure prominently in contemporary biology and philosophy of biology. In fact, though, few biologists or philosophers discuss the nature of life today. Many think that the definition of life has no direct bearing on current biological research (Sober, 1992; Taylor, 1992). When biologists do say something about life in general, they usually marginalize their discussions and produce something more thought provoking than conclusive. But this is all changing now.

Today the nature of life has become a hot topic. The economic stakes for manipulating life are rising quickly. Biotechnologies like genetic engineering, cloning, and high-throughput DNA sequencing have given us new and unprecedented powers to reconstruct and reshape life. A recent development is our ability to reengineer life to our specifications using synthetic genomics (Gibbs, 2004; Brent, 2004). In this domain attention has fallen on Craig Venter's well-publicized effort to commercialize artificial cells that clean the environment or produce alternative fuels (Zimmer, 2003). The current "wet" artificial life race to synthesize a minimal artificial cell or protocell from scratch in a test tube (Szostak, Bartel, & Luisi, 2001; Rasmussen et al., 2004; Luisi, 2006; Rasmussen et al., 2007) also spotlights life, for the race requires an agreed-upon definition of life, and it must be one that reaches well beyond life's familiar forms. The social and ethical implications of creating protocells will also increase the need for understanding what life is. Current controversies over the origin of life (Oparin, 1964; Crick, 1981; Shapiro, 1986; Eigen, 1992; Morowitz, 1992; Dyson, 1999; Luisi, 1998) and over intelligent design (Pennock, 2001) add more fuel to the fire.

Another recent development that highlights the nature of life is "soft" artificial life attempts to synthesize software systems with life's essential properties (Bedau, 2003a). Soft artificial life has created remarkably life-like software systems, and they seem genuinely alive to some (Langton, 1989a; Ray, 1992), but others ridicule the whole idea of a computer simulation being literally alive (Pattee, 1989).

Further still, recent "hard" artificial life achievements include the first widely available commercial robotic domestic vacuums, Roomba (Brooks, 2002), and the walking robots designed by evolution and fabricated by automated rapid prototyping (Lipson & Pollack, 2000). These robots inevitably raise the question whether a device made only of plastic, silicon, and steel could ever literally be alive. Such scientific developments increase uncertainty about how exactly to demarcate living things.

Biology makes generalizations about the forms life can take, but such generalizations rest on the forms of life that actually exist. Biologists study a number of different model organisms, like *Escherichia coli* (a common bacterium), *Caenohabditis elegans* (a nematode), and *Drosophila melanogaster* (a fruit fly). Picking model organisms that are as different as possible best illustrates the possible forms that life can take, and thus enables the widest generalizations about terrestial life. But all the life on Earth is terrestrial. Thus, these generalizations about life currently hinge on a sample size of one. Maynard Smith (1998) pointed out that artificial life helps mitigate this problem. Natural life comes in an amazing diversity of forms. But they are just a tiny fraction of all possible forms of life. Anytime we can synthesize a system in software, hardware, or wetware that exhibits life's core properties, we have a great opportunity to expand our empirical understanding of what life is.

There are three giants in the history of philosophy who advanced views about life, and their views still echo in contemporary discussion. In the *De Anima* Aristotle expressed the view that life is a nested hierarchy of capacities, such as metabolism, sensation, and motion. This nested hierarchy of capacities corresponds to Aristotle's notion of "soul" or mental capacities, so Aristotle essentially linked life and mind. As part of his wholesale replacement of Aristotelian philosophy and science, Descartes supplanted Aristotle's position with the idea that life is just the operation of a complex but purely materialistic machine. Descartes thought that life fundamentally differed from mind, which he thought was a mode of consciousness. Descartes sketched the details of his mechanistic hypothesis about life in his *Treatise on Man*. Some generations later, Kant's *Critique of Judgement* struggled to square Descartes's materialistic perspective with life's distinctive autonomy and purpose.

Understanding the nature of life is no mere armchair exercise. It involves investigating something real and extremely complex, and with huge potential creativity and power to change the face of the Earth (Margulis & Sagan, 1995). This investigation will by necessity be interdisciplinary, and it will survey an almost astonishing variety of perspectives on life. Interesting and subtle hallmarks like holism, homeostasis, teleology, and evolvability are thought to characterize life. But a precise definition of life remains elusive, partly because of borderline cases such as viruses and spores, and more recently artificial life creations. To add more complication, life figures centrally in a range of philosophical puzzles involving important philosophical issues such as emergence, computation, and mind. So, a diversity of views about life can be expected. Some employ familiar philosophical theories like functionalism. Others use biochemical or genetic explanations and mechanisms. Still others emphasize processes like metabolism and evolvability. The sheer diversity of views about life is itself interesting and deserves an explanation.

2. The Phenomena of Life

Life has various hallmarks and borderline cases, and it presents a variety of puzzles. The rest of this chapter is mainly devoted to explaining these phenomena.

A striking fact of life is the characteristic and distinctive *hallmarks* that it exhibits. These hallmarks are usually viewed as neither necessary nor sufficient conditions for life; they are nonetheless typical of life. Different people provide somewhat different lists of these hallmarks; see, e.g., Maynard Smith, 1986; Farmer & Belin, 1992; Mayr, 1997; Gánti, 2000. But most lists of hallmarks substantially overlap. Another notable point is that the hallmarks itemized on the lists are strikingly heterogeneous. A good illustration is Gánti's hallmarks (or "criteria," as he calls them).

Gánti's hallmarks fall into two categories: real (or absolute) and potential. Real life criteria specify the necessary and sufficient conditions for life in an individual living organism. Gánti's (2003) proposed real life criteria are these:

(1) *Holism.* An organism is an individual entity that cannot be subdivided without losing its essential properties. An organism cannot remain alive if its parts are separated and no longer interact.
(2) *Metabolism.* An individual organism takes in material and energy from its local environment, and chemically transforms them. Seeds are dormant and so lack an active metabolism, but they can become alive if conditions reactivate their metabolism. For this reason, Gánti makes a four-part distinction between things that are alive, dormant, dead, or not the kind of thing that could ever be alive.
(3) *Inherent stability.* An organism maintains homeostatic internal processes while living in a changing environment. By changing and adapting to a dynamic external environment, an organism preserves its overall structure and organization. This involves detecting changes in the environment and making compensating internal changes, with the effect of preserving overall internal organization.
(4) *Active information-carrying systems.* A living system must store information that is used in its development and functioning. Children inherit this information through reproduction, because the information can be copied. Mistakes in information transfer can "mutate" this information, and natural selection can sift through the resulting genetic variance.
(5) *Flexible control.* Processes in an organism are regulated and controlled so as to promote the organism's continued existence and flourishing. This control involves an adaptive flexibility, and can often improve with experience.

In contrast to these "real" criteria, Gánti also proposed "potential" life criteria. An individual living organism can fail to possess life's potential criteria. The defining feature of potential life criteria is that, if enough organisms exhibit them, then life can populate a planet and sustain itself. Gánti proposed three:

(1) *Growth and reproduction.* Old animals and sterile animals and plants are all living, but none can reproduce. So, the capacity to reproduce is neither necessary nor sufficient for being a living organism. But due to the mortality of individual

organisms, a population can survive and flourish only if some organisms in the population reproduce. In this sense, growth and reproduction are what Gánti calls a "potential" rather than "real" life criterion.

(2) *Evolvability.* "A living system must have the capacity for hereditary change and, furthermore, for evolution, i.e. the property of producing increasingly complex and differentiated forms over a very long series of successive generations" (Gánti, 2003, p.79). Since what evolves over time are not individual organisms but populations of them, we should rather say that living systems can be members of a population with the capacity to evolve. It is an open question today exactly which kinds of biological populations have the capacity to produce increasing complexity and differentiation.

(3) *Mortality.* Living systems are mortal. This is true even of clonal asexual organisms, because death can afflict both individual organisms as well as the whole clone. Systems that could never live cannot die, so death is property of things that were alive.

Gánti's life criteria and other lists of life's hallmarks always reflect and express some preconceptions about life. This might seem to beg the question of what life is. Any non-arbitrary list of life's hallmarks was presumably constructed by someone using some criterion to rule examples in or out. But where did this criterion come from, and what assures us it is correct? Why should we be confident that any hallmarks that fit it reveal the true nature of life? Thus, it seems lists of life's hallmarks are not the final word on what life is. As we learn more about life, our preconceptions change, evolve, and mature. So we should expect the same of our lists of life's hallmarks.

Another interesting feature of life is the existence of *borderline cases* that fall between the categories of the living and the nonliving. Familiar examples are viruses and prions, which self-replicate and spread even though they have no independent metabolism. Dormant seeds or spores are another kind of borderline case, the most extreme version of which might be bacteria or insects that are frozen. There are also cases that seem clearly not to be alive but yet possess the characteristic properties of living systems. Hardly anyone considers a candle flame to be alive, but by preserving its form while its constituent molecules are constantly changing, it has something like a metabolism (Maynard Smith, 1986). Populations of microscopic clay crystalites growing and proliferating are another kind of borderline example, especially because they can in appropriate circumstances undergo natural selection (Bedau, 1991). So is a forest fire that is spreading ("reproducing"?) from tree to tree at its edge, somewhat like the edge of a growing population of bacteria. A further kind of borderline case consists of superorganisms, which are groups of organisms, such as eusocial insect colonies, that function like a single organism. Although this is controversial, some biologists think that superorganisms should themselves be thought of as living organisms. Another kind of borderline case consists of soft artificial life creations like Tierra. Tierra is software that creates a spontaneously evolving population of computer programs that reproduce, mutate, and evolve in computer memory. Tierra's inventor thinks that Tierra is literally alive (Ray, 1992). This would radically violate the ordinary concept of life that most of us have. One final category of borderline cases consists of complex adaptive systems found in nature, such as financial markets or the World Wide Web. These exhibit many of the hallmarks of life, and some think that the simplest and most unified explanation of the entire range of phenomena of life is to consider these natural complex adaptive systems to be literally alive (Bedau, 1996, 1998).

3. Puzzles about Life

A third characteristic of life is that it generates a number of puzzles. Seven puzzles are briefly reviewed below. Any account of life should explain the origin of these puzzles; more important, it should resolve the puzzles. Some puzzles might result simply from confusion, but others are open questions about a fundamental and fascinating aspect of the natural world.

Origins. How does life or biology arise from non-life or pure chemistry? What is the difference between a system that is undergoing merely chemical evolution, in which chemical reactions are continually changing the concentrations of chemical species, and a system that contains life? Where is the boundary between living and merely physico-chemical phenomena? How could a naturalistic process bridge the boundary, in principle or in practice? Dennett argues that Darwin's scheme of explanation solves this problem by appealing to "a finite regress, in which the sought-for marvelous property (life, in this case) was acquired by slight, perhaps even imperceptible, amendments or increments" (1995, p.200).

Emergence. How does life involve emergence? B properties are said to *emerge from* A properties when the B properties both depend on, and are autonomous from, the A properties. Different kinds of dependence and autonomy generate different grades of emergence (Bedau, 2003b). One is the "strong" emergence involving in principle irreducible top-down causal powers. An example might be consciousness or qualia in the philosophy of mind (Kim, 1999). If the A and B properties are simultaneous, the emergence of B from A is *synchronic*. It concerns what properties exist at a moment. Those properties might be changing, but the relationship between the A and B properties at an instant are a static snapshot of that dynamic process. By contrast, if the A properties precede the B properties, and the B properties arise over time from the A properties, then the emergence of B from A is *dynamic*. Life is the paradigm case of a dynamic form of "weak" emergence, one that concerns macro properties that are unpredictable or underivable except by observing the process by which they are generated, or by observing a simulation of it (Bedau, 1997, 2003b).

Hierarchy. Various kinds of structural hierarchies characterize life. Each organism has a hierarchical internal organization, and the relative complexity of organizations of different kinds of organisms form another hierarchy. The simplest organisms are prokaryotic cells, which have relatively simple components. More complicated are eukaryotic cells containing complex organelles and a nucleus. Multicellular organisms are even more complicated; they have constituents (individual cells) that also are individual living entities (e.g., they can be kept alive by themselves). In addition, mammals have complex internal organs (such as the heart) that can be harvested and kept alive when an organism dies, and then surgically implanted into another living organism. Two questions arise here. First, why does life tend to generate and encompass such hierarchies? This question applies both to the hierarchy in complexity that spans all organisms together, and also to the organizational hierarchy found within each individual living organism. With regard to the latter, a second question arises. Organisms are our paradigm case of something that is alive, but we also refer to organs and individual cells as alive. For example, apoptosis is an important process by which living cells in an organism undergo programmed death, and hospitals strive to keep certain organs alive after someone dies, so that they are available to be transplanted into someone else. This raises the question whether a mammal, its heart, and the cells therein are each alive in the same or different senses.

Continuum. Can things be more or less alive? Is life a black-or-white Boolean property, or a continuum property with many shades of gray? Common sense leans towards the Boolean view: a rabbit is alive and a rock isn't, end of story. But there are borderline cases like viruses that are unable to replicate without a host. And spores or frozen bacteria remain dormant and unchanging indefinitely but then come back to life when conditions become favorable. Are viruses and spores fully alive? Furthermore, when the original life forms emerged from a pre-biotic chemical soup, they differed very little from their non-living predecessors. Some conclude that there is a continuum of more or less alive things (e.g., Cairns-Smith, 1985; Emmeche, 1994; Dennett, 1995). An alternative is to accept a sharp distinction between life and non-life, but allow that a small step could cross it. The four-fold distinction between things that are (i) inanimate and forever incapable of living, (ii) now living, (iii) dead but formerly living, or (iv) dormant but capable of becoming alive again helps explain away some borderline cases by reclassifying them (e.g., seeds and spores are dormant and not currently living). But it does not fully resolve the continuum puzzle, for there are borderline cases in the four-fold distinction, such as between being dead and alive.

Strong artificial life. Artificial life software and hardware raise the question whether our computer creations could ever literally be alive (Langton, 1989a; Pattee, 1989; Sober, 1992; Emmeche, 1992; Olson, 1997). On the one hand, certain distinctive carbon-based macromolecules play a crucial role in the vital processes of all known living entities; on the other hand, much of artificial life seems to presuppose that life can be realized in a suitably programmed computer. It is important to distinguish two questions here. The first is the philosophically controversial question – *in virtue of what* a computer or a robot could be said to be alive. If this issue were settled, we would face the technical question of whether it is possible to create a software system or hardware device (e.g., a robot) that is literally alive in this sense. The challenge here is whether we could, in fact, realize the processes that were specified in the appropriate materials. The "strong" artificial life position about software is that an instantiation of artificial life software could literally be alive. There is an analogous strong position about "hard" artificial life hardware constructions, and also about "wet" artificial life laboratory constructions. These strong positions contrast with the uncontroversial "weak" positions that computer models, hardware constructions, and wet lab productions are just useful for understanding living systems. And yet, the strong version of wet artificial life is intuitively plausible; we usually accept that something synthesized from scratch in the lab could be literally alive. So the controversy about strong artificial life concerns primarily soft and hard artificial life.

Mind. Another puzzle is whether there is any intrinsic connection between life and mind. Plants, bacteria, insects, and mammals, for example, have various kinds of sensitivity to the environment, various ways in which this environmental sensitivity affects their behavior, and various forms of inter-organism communication (e.g., Dennett, 1997). These are all forms of intelligent behavior, and the relative sophistication of these "mental" capacities seems to correspond to, and explain the relative sophistication of, those forms of life. So it is natural to ask whether life and mind have some deep connection. Evolution creates a genealogical connection between life and mind, of course, but they would be much more deeply unified if Beer is right that "it is adaptive behavior, the . . . ability to cope with the complex, dynamic, unpredictable world in which we live, that is, in fact, fundamental [to intelligence itself]" (Beer, 1990, p.11; see also Maturana & Varela, 1987; Godfrey-Smith, 1994; Clark, 1997). Since all forms of life must cope in one way or another with a complex, dynamic, and unpredictable world, perhaps this adaptive flexibility inseparably connects life and mind.

4. Accounts of Life

There have been various attempts to state the universal characteristics of all forms of life. In this section, I will discuss the main varieties of such accounts of life, indicating some of their motivations, strengths, and weaknesses. I will also note some skeptical positions that deny the usefulness of such accounts.

First, consider the skeptical position that the nature of life is largely irrelevant to biology (Sober, 1992; Taylor, 1992). The reason for this skepticism is that biologists can continue with their biological research whether or not life can be adequately defined, and no matter what view of life prevails in the end. One must admit, though, that recent developments such as attempting to make minimal artificial cells from scratch does require scientists to start to articulate their views about what is essential to life, even if these views fall short of a precise definition. So the issue is no longer irrelevant, if it ever was. For one can set out to construct a minimal form of life only if one has at least a working hypothesis about life's minimally sufficient conditions. Otherwise one would have no idea what to try to make.

A second form of skepticism is the view that life cannot be captured by necessary and sufficient conditions, but instead consists of just a cluster of things sharing only a Wittgenstinian family resemblance. Different forms of life might share various properties or hallmarks, but the individual properties in the cluster each have exceptions. The properties would typically be possessed by living organisms but they would not be strictly necessary or sufficient. Farmer and Belin list eight hallmarks: process; self-reproduction; information storage of self-representation; metabolization; functional interactions with the environment; interdependence of parts; stability under perturbations; and membership in a population with the ability to evolve. They then explain that a cluster conception of life arises from their despair at finding anything more precise than this list of hallmarks.

> There seems to be no single property that characterizes life. Any property that we assign to life is either too broad, so that it characterizes many non-living systems as well, or too specific, so that we can find counter-examples that we intuitively feel to be alive, but that do not satisfy it. (Farmer & Belin, 1992, p.818; see also Taylor, 1992)

The cluster conception amounts to skepticism about the possibility of a unified theory of life.

An advantage of the cluster conception is that it offers a natural explanation for borderline cases. All cluster concepts inevitably have borderline cases. A characteristic of the cluster conception is that it cannot explain why forms of life are unified by one set of hallmarks rather than another. The cluster view must simply accept the hallmarks as given, and then identify the cluster with those hallmarks. Thus, this view can identify life's hallmarks only post hoc; it cannot predict or explain the hallmarks. Those who think that there should be an explanation for life's hallmarks will therefore find the cluster conception unsatisfying.

Another kindred form of skepticism questions the idea that life is a natural kind. Keller (2002) says that life is a human kind, not a natural kind, that is, a distinction created by us, not a distinction in nature. This could explain borderline cases. Since the concept of life changes with the progress of science and technology, one should expect its boundaries to change, thus creating borderline cases. The view also provides some general ammunition against life's puzzles, for a mutable human construct can be expected to spawn puzzles. Keller's argument that life is a human kind suggests that the present presupposition that life has an essence arose only 200 years ago, that the search for life's essence is driven by attempts to make life from non-life (and this tends to dissolve the boundary between life and non-life) and that the new concepts generated by scientific and technological progress violate older taxonomies like the life/non-life distinction (Keller, 2002).

There are problems with all of these arguments. First, all modern scientific concepts like matter and energy arose at some point in human history and have evolved since then. So contingent, datable recent origin does not show that a kind is a human kind, unless it does so at one fell swoop for all scientific concepts. Second, bridging the gap in the laboratory from the non-living to the living need not dissolve the boundary between life and non-life, any more than making the first airplane dissolved the distinction between flying and not flying. Remember that we are seeking the nature of life, not just current conceptions of life.

Now, one answer to the question "what is life?" is simply to give a taxonomy of living things. This is taking the question as a request for an exhaustive list of the kinds of things on the Earth that are alive. This is an interesting historical question, but one riddled with contingencies. The taxonomy is necessarily silent about forms of life that could have existed but did not. This illustrates the taxonomy view's chauvinism in assuming that life as we know it exhausts what life is or could be. Unrelated life forms that exist on an extra-terrestrial site like Europa are absent from all such taxonomies. In any case, we should welcome having our taxonomies adjusted by scientific and technological progress, for that is how we learn.

Some have given a biochemical definition of life. They attempt to specify the biochemical properties that any form of life must have, given the general constraints set by physics and chemistry (Pace, 2001; Benner, Ricardo, & Carrigan, 2004). This includes thermodynamic limits, energetic limits, material limits, and even geographical limits. The features in a biochemical definition are sometimes called life's biochemical "universals." A biochemical definition always presupposes a prior account of life; it states the physical, chemical, and biological possibilities for any biochemical system meeting that prior account of life. The biochemical definitions of Pace (2001) and Benner et al. (2004) presuppose a definition of life based on evolution, so Pace and Benner dwell on the biochemical universals for genetic capacities and emphasize molecules like DNA that can store and transmit information between generations. Biochemical definitions are often myopic and presume that all possible life forms are quite similar to the familiar ones. One could imagine starting with a different conception of life, such as the view based on metabolism, and ending up emphasizing different biochemical universals, such as those that enable open systems to retain their structure in the face of the second law of thermodynamics.

A genetic instance of a biochemical definition of life is Venter's recent genomic definition of life as a minimal genome sufficient to support life (Hutchison et al., 1999). This view inherits the limitations of biochemical definitions. The genomic definition captures the simplest known set of genes sufficient for life. It does not capture genes found in every life form, for the same essential life functions can be achieved by different genes. Many people would question the molecular definition's limitation to genetic properties, on the grounds that life centrally involves much more than genes (Cho et al., 1999).

Everyone in the community of scientists making artificial cells from scratch or "protocells" admits that the nature of life is controversial and contentious, but almost all share the goal of making a self-contained system that metabolizes and evolves (e.g., Rasmussen et al., 2004). That is, an artificial cell is viewed as any chemical system that chemically integrates three processes: The first is the process of assembling some kind of container, such as a lipid vesicle, and living inside it. The second is the metabolic processes that repair and regenerate the container and its contents, and enable the whole system to reproduce. Those chemical processes are shaped and directed by a third chemical process involving encoded information about the system stored in the system ("genes"). Errors ("mutations") can occur when this information is reproduced, so the systems can evolve by natural selection. The integrated-triad view of life requires that the chemical processes of containment, metabolism, and evolution support and enable each other, so that there is functional feedback among all three. This view of protocellular life as an integrated triad of functions accepts any biochemical realization of the triad as genuine life.

The past generation of the philosophy of mind has been dominated by functionalism: the view that mental beings are a certain kind of input–output device and that having a mind is simply having a set of internal states that causally interact (or "function") with respect to each other and with respect to environmental inputs and behavioral outputs in a certain characteristic way. Functionalism with respect to life is the analogous view that being alive is simply realizing a network of processes that interact in a certain characteristic way. Some processes (such as information processing, metabolization, purposeful activity) operate within the organism's lifetime; other processes (such as self-reproduction and adaptive evolution) operate over many generations. These processes are always realized in some material substratum, but the substratum's material nature is irrelevant so long as the *forms* of the processes are preserved. For these reasons, functionalism is an attractive position with respect to life. Chris Langton's defense of artificial life is a classic statement of the case for functionalism with respect to life:

> Life is a property of *form*, not *matter*, a result of the organization of matter rather than something that inheres in the matter itself. (Langton, 1989a, p.41)

> The big claim is that a properly organized set of artificial primitives carrying out the same functional roles as the biomolecules in natural living systems will support a process that is "alive" in the same way that natural organisms are alive. Artificial Life will therefore be genuine life – it will simply be made of different stuff than the life that has evolved here on Earth. (Langton, 1989a, p.33)

We might be unsure about the details of the processes that are definitive of life, and we might wish to reserve judgment about whether artificial life creations are genuinely alive. Nevertheless, it is hard to deny Langton's point that life's characteristic processes like metabolism, information processing, and self-reproduction could be realized in a wide and potentially open-ended range of materials. Thus, the prospects for some form of functionalism with respect to life seem bright.

The main challenge for functionalism with respect to mind concerns consciousness and qualia. It is worth noting that functionalism about life does not face any analogous problems. Another challenge for functionalism with respect to mind is to explain how people's mental states are meaningful or have semantic content. Darwinian natural selection provides a naturalistic explanation of many biological functions of structures in evolved forms of life. This biological functionality gives the internal states of living creatures a kind of meaning or semantic content, so that we can speak of a creature trying to find food for nourishment. Many philosophers are optimistic that the meaning problem in functionalism with respect to mind will be solved by some analogous Darwinian explanation of the biological function of mental states (e.g., Dennett, 1995).

Another apparent threat to functionalism with respect to life is the suggestion that the processes involved in life are, in some relevant sense, unformalizable or non-computational (e.g., Emmeche, 1992). Bedau (1999) thinks that the apparent non-computational quality of life can be explained. Advantageous traits that arise through mutations tend, ceteris paribus, to persist and spread through the population. Furthermore, trait frequencies in the population will tend, ceteris paribus, to change in a way that is generally apt for the population in its exogenously changing environment. These dynamical patterns in trait frequencies emerge as a statistical pattern from the micro-level contingencies of natural selection, mutation, drift, etc. Bedau argues that there is often a special kind of suppleness in these patterns. Such patterns in trait frequencies are not precise and exceptionless universal generalizations, but instead hold only for the most part, only ceteris paribus. Furthermore, those regularities have exceptions that sometimes "prove the rule" in the sense that they are a byproduct of trying to achieve some deeper adaptive goal. For example, Bedau describes a system in which mutation rates can evolve and shows that the mutation rates tend to evolve so as to keep the population's gene pool at the "edge of disorder"; but this regularity has exceptions, some of which are due to the operation of a deeper regularity about mutation rates evolving so as to optimally balance evolutionary "memory" and "creativity" (for details, see Bedau, 1999). In this sort of way, supple regularities reflect an underlying capacity to respond appropriately in an open-ended variety of contexts. This explains a certain kind of unformalizability of life processes, though it also allows life to be captured in appropriate computer models.

Functionalism leaves unanswered exactly which processes play what role in the functional characterization of life. Persisting in the face of the second law of thermodynamics by means of metabolism is the defining process of life according to Schrödinger's influential account:

> When is a piece of matter said to be alive? When it goes on "doing something", moving, exchanging material with its environment, and so forth, and that for a much longer period than we would expect an inanimate piece of matter to "keep going" under similar circumstances . . . It is by avoiding the rapid decay into the inert state of "equilibrium" that an organism appears so enigmatic; . . . How does the living organism avoid decay? The obvious answer is: By eating, drinking, breathing and (in the case of plants) assimilating. The technical term is metabolism . . . (Schrödinger, 1969, pp.74–6)

Metabolism-centered views of life attract many (Margulis & Sagan, 1995; Boden, 1999). They are closely related to views that focus on autopoeisis (Varela, Maturana, & Uribe, 1974; Maturana & Varela, 1987).

The view that metabolism is life's central process has some clear advantages, such as explaining our intuition that a crystal is not alive (there is a metabolic flux of molecules only at the crystal's edge, not inside it). Also, the fact that metabolism is needed to combat entropy implies that metabolism is at least a necessary condition of all physical life forms. Metabolism also naturally explains the four-fold distinction between the non-living, living, dead, and dormant. The non-living cannot metabolize in principle, and the living are now metabolizing. The dead were once living and metabolizing, but now they are decaying. The dormant were once living but now do not metabolize, but they could resume metabolizing given the right circumstances.

The main drawback of metabolism as an all-encompassing account of life is that many metabolizing entities seem intuitively not to be alive or to involve life in any way. Standard examples include a candle flame, a vortex, and a convection cell (Maynard Smith, 1986; Bagley & Farmer, 1992). Such examples by themselves do not prove conclusively that metabolism is insufficient for life, for pre-theoretic intuitive judgments can be wrong. The question is whether on balance metabolism adequately explains life's hallmarks and resolves life's puzzles.

Some think that the central feature underlying all life is the open-ended evolutionary process of adaptation. The central idea is that what distinguishes life is its automatic and open-ended capacity (within limits) to adapt appropriately to unpredictable changes in the environment. From this perspective, what is distinctive of life is the way in which adaptive evolution automatically fashions new and intelligent strategies for surviving and flourishing as local contexts change. Maynard Smith (1975, p.96f; see also Mayr, 1982; Cairns-Smith, 1985) succinctly explains the justification for the view that life crucially depends on the evolutionary process of adaptation:

> We shall regard as alive any population of entities which has the properties of multiplication, heredity and variation. The justification for this definition is as follows: any population with these properties will evolve by natural selection so as to become better adapted to its environment. Given time, any degree of adaptive complexity can be generated by natural selection.

These remarks suggest how the process of adaptive evolution could explain life's hallmarks, borderline cases and puzzles (see Bedau, 1998).

There are a few characteristic criticisms of such evolution-centered views. One is purported counterexamples of creatures that are alive but cannot give birth (mules, old people, etc.) and so cannot contribute to the process of evolution. The typical response is to require that organisms be produced by an evolutionary process, but not that they necessarily can affect further evolution. Another kind of purported counterexample is a clearly non-living system, such as a population of clay crystallites or a free market economy, which evolves by natural selection. Some think that we should accept these unintuitive examples because evolution-centered views provide such a compelling explanation of life's hallmarks, borderline cases, and puzzles (e.g., Bedau, 1998).

Not all of these positions are competing; many are consistent. For example, functionalism is consistent with the protocell integrated-triad account of minimal life. Also, accounts of the nature of life each entail a biochemical characterization of life, and many accounts of life overlap. The problem of understanding life is to identify exactly which of these accounts is true.

5. The Problem of Understanding Life

How should we compare and evaluate accounts of the nature of life? One straightforward answer is simply to see how well each explains the phenomena of life. This amounts to doing three things: explaining life's hallmarks, explaining the borderline cases, and resolving the puzzles about life. The problem of understanding life is the problem of explaining these three things.

One initial difficulty is confusion about what question is at stake. Some investigations think the key test for any account of life is to fit it with our pre-theoretic intuitions about which things are alive and which are not (e.g., Boden, 1999). But one should ask why we should emphasize such intuitions. A good theory of life might make us reconceptualize and recategorize life. This might change our attitudes about exactly which cases are the ones in which life is present. Thus, although they have some weight, our pre-theoretic intuitions are not inviolable.

One could also ask about the *meaning of the word* "life" in today's English. But the stereotypes associated with the term "life" are commonplaces and reflect the lowest common denominator of our current shared picture of life. So we are not likely to learn much about life by relying on what "life" means.

Nor are we likely to learn much by analysis of the *concept* of life. As with the meaning of "life," our current concept of life will reflect our current understanding of life. If we want to learn the real nature of the phenomena with life's hallmarks, borderline cases, and puzzles, we should study the natural phenomena themselves, not our words or concepts. And we should expect our understanding of the phenomena of life to evolve and sometimes improve.

Explaining the phenomena of life involves at least a rough view of life's essence or nature, and perhaps even a rough definition of life. Scientific essentialism, originating from Kripke (1980), is the philosophical view that the essence of natural kinds like water and gold is their underlying causal powers, which are discovered by empirical science (see Bealer, 1987). The essence of substances like water and gold turns out to be their underlying chemical composition. Life, on the other hand, is a certain kind of flexible process, not a fixed chemical substance. So unlike water or gold, life's nature would presumably be captured by the characteristic network of processes (such as metabolism, reproduction, and sensation) that explains its characteristic causal powers. In this regard life is more like heat, which is a certain process in matter (high molecular kinetic energy). A specific temperature (say, 23°C) is a specific kind of process that can occur in all kinds of matter. Life is also a kind of process that can occur in different kinds of material, but unlike temperature not *all* kinds of material can be alive. Mapping the biochemical constraints on the kinds of substances that could instantiate life yields a biochemical definition of life (recall above). Note that scientific essentialism about life might be true, even if contemporary science has reached no consensus about life. Scientific essentialism is a philosophical view about the method by which life's essence would be discovered – it is not a view about the particular content of that essence. The details of the scientific essentialist definition of life might need to await further scientific progress.

It is unclear whether living things have any features that make them essentially alive. In Dennett's opinion, for example, the life/non-life distinction is a matter of degree and life is too "interesting" to have an essence (1995, p.201). In fact, contemporary biology and philosophy of biology thoroughly embrace a Darwinian anti-essentialism according to which species have no essence and their members share no necessary and sufficient properties. Instead, the similarities among the members of a species are only statistical. Species are no more than a cloud or clump in an abstract possible feature space. Although some sub-regions of possible feature space are unoccupied because they are maladaptive, it is an accident exactly which of the acceptable sub-regions are occupied. No sub-regions are any more natural than any other; none are privileged by fixed and immutable Platonic essences. The generalization of this anti-essentialism probably helps account for why so many philosophers are attracted to the cluster concept of life, for that seems like a direct consequence of anti-essentialism.

Darwinian anti-essentialism is directed against a narrow notion of essence that embraces exception-less necessary and sufficient conditions and excludes borderline cases. Borderline cases are one of the hallmarks of life, so the nature of life must be broad and flexible enough to embrace borderline cases. One could embrace Darwinian anti-essentialism but still accept scientific essentialism about life. On this view, the "essence" of life would be whatever process explains the phenomena of life, including life's hallmarks, borderline cases, and puzzles. Life would not be defined by exceptionless conditions but empirically. It is unfortunate that contemporary philosophical terminology obscures that Darwinian anti-essentialism and scientific essentialism about life are compatible.

Clelland and Chyba (2002) argue that it is too early to formulate definitions of life, because our current understanding of life is too limited. They conclude that we should put off formulating definitions until scientists can tell much more about the different forms that life could take. Now might nevertheless be the right time to construct tentative and testable hypotheses about the phenomena of life. These hypotheses will likely be false, but they can aid our search for better theories (Wimsatt, 1987). When we have good theories of life in hand, we can extract their implied definitions of life. So the quest for the definition of life is better recast as the quest for the nature of life.

Life is one of the most fundamental and complex aspects of nature. So accounts of life are rich and interesting, with a complicated structure. They come in many forms, including skepticism, detailed biochemical and molecular descriptions, and abstract functionalism, and they emphasize fundamental biological processes like metabolism and evolution. The criteria for evaluation include their ability to explain life's hallmarks and borderline cases and their ability to resolve the puzzles about life. Many of the main accounts of life still lack substantial development and careful evaluation along a number of these dimensions. Thus, the problem of understanding life is still wide open.

References

Bagley, R., & Farmer, J. D. (1992). Spontaneous emergence of a metabolism. In C. Langton, C. Taylor, J. D. Farmer, & S. Rasmussen (Eds). *Artificial Life II* (pp. 93–140). Redwood City, CA: Addison-Wesley.

Bealer, G. (1987). The philosophical limits of scientific essentialism. In J. Tomberlin (Ed.). *Philosophical perspectives 1* (pp. 289–365). Atascadero: Ridgeway.

Bedau, M. A. (1991). Can biological teleology be naturalized? *The Journal of Philosophy*, 88, 647–55.

Bedau, M. A. (1996). The nature of life. In M. Boden (Ed.). *The philosophy of artificial life* (pp. 332–57). New York: Oxford University Press.

Bedau, M. A. (1997). Weak emergence. In J. Tomberlin (Ed.). *Philosophical perspectives: mind, causation, and world* (vol. 11, pp. 375–99). Oxford: Blackwell.

Bedau. M. A. (1998). Four puzzles about life. *Artificial Life*, 4, 125–40.

Bedau, M. A. (1999). Supple laws in biology and psychology. In V. Hardcastle (Ed.), *Where biology meets psychology: philosophical essays* (pp. 287–302). Cambridge: MIT Press. (The printed version of this paper inadvertently omits a few pages; the full text is available on the web via http://www.reed.edu/~mab. Accessed 11 October 2007)

Bedau, M. A. (2003a). Artificial life: organization, adaptation, and complexity from the bottom up. *Trends in Cognitive Science*, 7(11, November), 505–12.

Bedau, M. A. (2003b). Downward causation and autonomy in weak emergence. *Principia Revista Inernacional de Epistemologica*, 6(1), 5–50.

Beer, R. D. (1990). *Intelligence as adaptive behavior: an experiment in computational neuroethology.* Boston: Academic Press.

Benner, S. A., Ricardo, A., & Carrigan, M. A. (2004). Is there a common chemical model for life in the universe? *Current Opinion in Chemical Biology*, 8, 679–89.

Boden, M. A. (1999). Is metabolism necessary? *British Journal for the Philosophy of Science*, 50, 231–48.

Brent, R. (2004). A partnership between biology and engineering. *Nature Biotechnology*, 22(10), 1211–14.

Brooks, R. (2002). *Flesh and machines: how robots will change us.* New York: Pantheon.

Cairns-Smith, A. G. (1985). *Seven clues to the origin of life.* Cambridge: Cambridge University Press.

Cho, M. K., Magnus, D., Caplan, A. L., McGee, D., & the Ethics of Genomics Group. (1999). Ethical considerations in synthesizing a minimal genome. *Science*, 286(5447), 2087–90.

Clark, A. (1997). *Being there: putting brain, body, and world together again.* Cambridge, MA: MIT Press.
Cleland, C., & Chyba, C. (2002). Defining "life". *Origins of Life and Evolution of the Biosphere*, 32, 387–93.
Crick, F. (1981). *Life itself: its origin and nature.* New York: Simon & Schuster.
Dennett, D. C. (1997). *Kinds of minds: towards an understanding of consciousness.* New York: Basic Books.
Dennett, D. C. (1995). *Darwin's dangerous idea.* New York: Simon & Schuster.
Dyson, F. (1999). *Origins of life* (rev. edn). Cambridge: Cambridge University Press.
Eigen, M. (1992). *Steps toward life.* Oxford: Oxford University Press.
Emmeche, C. (1992). Life as an abstract phenomenon: is artificial life possible? In F. Varela & P. Bourgine (Eds). *Towards a practice of autonomous systems* (pp. 466–74). Cambridge, MA: Bradford Books/MIT Press.
Emmeche, C. (1994). *The garden in the machine: the emerging science of artificial life.* Princeton: Princeton University Press.
Farmer, D., & Belin, A. (1992). Artificial life: the coming evolution. In C. Langton, C. Taylor, J. D. Farmer, & S. Rasmussen (Eds). *Artificial Life II* (pp. 815–40). Redwood City, CA: Addison-Wesley.
Gánti, T. (2003). *The principles of life.* New York: Oxford University Press. With a commentary by James Grisemer and Eörs Szathmáry.
Gibbs, W. W. (2004). Synthetic life. *Scientific American*, 290(April), 75–81.
Godfrey-Smith, P. (1994). Spencer and Dewey on life and mind. In R. Brooks & P. Maes (Eds). *Artificial Life IV* (pp. 80–9). Cambridge, MA: MIT Press/Bradford Books.
Hutchison, C. A., Peterson, S. N., Gill, S. R., Cline, R. T., White, O., Fraser, C. M., Smith, H. O., & Venter, J. C. (1999). Global transposon mutagenesis and a minimal Mycoplasma genome. *Science*, 286(10 December), 2165–9.
Keller, E. F. (2002). *Making sense of life: explaining biological development with models, metaphors, and machines.* Cambridge, MA: Harvard University Press.
Kim, J. (1999). Making sense of emergence. *Philosophical Studies*, 95, 3–36.
Kripke, S. (1980). *Naming and necessity.* Cambridge, MA: Harvard University Press.
Langton, C. (1989a). Artificial life. In C. Langton (Ed.). *Artificial life* (pp. 1–47). Redwood City, CA: Addison-Wesley.
Langton, C. G. (Ed.) (1989b). *Artificial life.* Redwood City: Addison-Wesley.
Lipson, H., & Pollack, J. (2000). Automatic design and manufacture of robotic lifeforms. *Nature*, 406(31 August), 974–8.
Luisi, P. L. (1998). About various definitions of life. *Origins of Life and Evolution of the Biosphere*, 28, 613–22.
Luisi, P. L. (2006). *The emergence of life: from chemical origins to synthetic biology.* Cambridge: Cambridge University Press.
Margulis, L., & Sagan, D. (1995). *What is life?* Berkeley: University of California Press.
Maturana, H. R., & Varela, F. J. (1987). *The tree of knowledge: the biological roots of human understanding* (rev. edn, 1992). Boston: Shambhala.
Maynard Smith, J. (1975). *The theory of evolution* (3rd edn). New York: Penguin.
Maynard Smith, J. (1986). *The problems of biology.* New York: Oxford University Press.
Maynard Smith, J. (1992). Byte-sized evolution. *Nature*, 355(27 February), 772–3.
Mayr, E. (1982). *The growth of biological thought.* Cambridge, MA: Harvard University Press.
Mayr, E. (1997). What is the meaning of "life"? In E. Mayr, *This is biology: the science of the living world* (pp. 1–23). Cambridge, MA: Harvard University Press.
Morowitz, H. J. (1992). *Beginnings of cellular life: metabolism recapitulates biogenesis.* New Haven: Yale University Press.

Olson, E. T. (1997). The ontological basis of strong artificial life. *Artificial Life*, 3, 29–39.
Oparin, A. I. (1964). *Life: its nature, origin, and development* (A. Synge, Trans.). New York: Academic Press.
Pace, N. R. (2001). The universal nature of biochemistry. *Proceedings of the National Academy of Science USA*, 98(January 30), 805–8.
Pattee, H. H. (1989). Simulations, realization, and theories of life. In C. Langton (Ed.). *Artificial life* (pp. 63–78). Redwood City, CA: Addison-Wesley.
Pennock, R. (2001). *Intelligent design creationism and its critics: philosophical, theological & scientific perspectives.* Cambridge, MA: MIT Press.
Rasmussen, S., Chen, L., Deamer, D., Krakauer, D., Packard, N. H., Stadler, P. F., & Bedau, M. A. (2004). Transitions from nonliving to living matter. *Science*, 303(February 13), 963–65.
Rasmussen, S., Bedau, M. A., Chen, L., Deamer, D., Krakauer, D. C., Packard, N. H., Stadler, P. F. (Eds). (2007). *Protocells: bridging nonliving and living matter.* Cambridge, MA: MIT Press.
Ray, T. (1992). An approach to the synthesis of life. In C. Langton, C. Taylor, J. D. Farmer, & S. Rasmussen (Eds). *Artificial life II* (pp. 371–408). Redwood City, CA: Addison-Wesley.
Schrödinger, W. (1969). *What is life? The physical aspect of the living cell.* Cambridge: Cambridge University Press.
Shapiro, R. (1986). *Origins: A skeptic's guide to the creation of life on Earth.* New York: Summit Books.
Sober, E. (1992). Learning from functionalism – prospects for strong artificial life. In C. Langton, C. Taylor, J. D. Farmer, & S. Rasmussen (Eds). *Artificial Life II* (pp. 749–65). Redwood City, CA: Addison-Wesley.
Szostak, J. W., Bartel, D. P., & Luisi, P. L. (2001). Synthesizing life. *Nature*, 409(January 18), 387–90.
Taylor, C. (1992). "Fleshing out" artificial life II. In C. Langton, C. Taylor, J. D. Farmer, & S. Rasmussen (Eds). *Artificial Life II* (pp. 25–38). Redwood City, CA: Addison-Wesley.
Varela, F. G., Maturana, H. R., & Uribe, R. (1974). Autopoiesis: the organization of living systems, its characterization and a model. *Biosystems*, 5, 187–96.
Wimsatt, W. C. (1987). False models as means to truer theories. In M. Niteckiand, & A. Hoffman (Eds). *Neutral modes in biology* (pp. 23–55). Oxford: Oxford University Press.
Zimmer, C. (2003). Tinker, tailor: can Venter stitch together a genome from scratch? *Science*, 299(February 14), 1006–7.

Further Reading

Bedau, M. A., & Packard, N. H. (1992). Measurement of evolutionary activity, teleology, and life. In C. Langton, C. Taylor, J. D. Farmer, & S. Rasmussen (Eds). *Artificial Life II* (pp. 431–61). Redwood City, CA: Addison-Wesley.
Cleland, C. E., & Copley, S. H. (2005). The possibility of alternative microbial life on Earth. *International Journal of Astrobiology*, 4, 165–73.
Code, A., & Moravcsik, J. (1992). Explaining various forms of living. In M. C. Nussbaum & A. O. Rorty (Eds). *Essays on Aristotle's* De Anima (pp. 129–45). Oxford: Clarendon Press.
Emmeche, C. (1994). Is life a multiverse phenomenon? In C. G. Langton (Ed.). *Artificial life III* (pp. 553–68). Redwood City: Addison-Wesley.
Feldman, F. (1992). *Confrontations with the reaper: a philosophical study of the nature and value of death.* New York: Oxford University Press.
Haldane, J. B. S. (1937). What is life? In J. B. S. Haldane, *Adventures of a biologist* (pp. 49–64). New York: Macmillan.

Jonas, H. (1966). *The phenomenon of life: toward a philosophical biology.* New York: Dell.
Keller, E. F. (1995). *Refiguring life: metaphors in twentieth-century biology.* New York: Columbia University Press.
Korzeniewski, B. (2001). Cybernetic formulation of the definition of life. *Journal of Theoretical Biology*, 209, 275–86.
Lange, M. (1996). Life, "artificial life," and scientific explanation. *Philosophy of Science*, 63, 225–44.
Maynard Smith, J., & Szathmáry, E. (1999). *The origins of life: from the birth of life to the origins of language.* New York: Oxford University Press.
Miller, J. G. (1978). *Living systems.* New York: McGraw-Hill.
Murphey, M. P., & O'Neill, L. A. J. (Eds). (1993). *What is life? The next fifty years: speculations on the future of biology.* Cambridge: Cambridge University Press.
Richards, R. J. (2002). *The Romantic conception of life: science and philosophy in the age of Goethe.* Chicago: University of Chicago Press.
Rosen, R. (1991). *Life itself: a comprehensive inquiry into the nature, origin, and fabrication of life.* New York: Columbia University Press.
Sagan, C. (1970). Life. *Encyclopedia Britannica* (15th edn, vol. 10). New York: Macropaedia.
Sterelny, K., & Griffths, P. (1999). What is life? In K. Sterelny & P. Griffiths, *Sex and death: an introduction to philosophy of biology* (pp. 357–77). Chicago: Chicago University Press.

Chapter 25

Experimentation

MARCEL WEBER

1. The Diversity of Experimental Practices in Biology

Experimentation is traditionally understood as the planned production of specific conditions followed by observations in order to gain knowledge about the laws that govern natural phenomena. Its original home as a recognized and distinct category of epistemic activity lies in the physical sciences of the seventeenth century, where it was closely associated with the scientific method itself. By contrast, biology was traditionally viewed as a historical discipline that uses only observation and systematic comparison in order to gain knowledge in its proper domain. However, experimental approaches to studying life were firmly established in the nineteenth century by scientists such as Emil du Bois-Reymond, Claude Bernard, Louis Pasteur, or Gregor Mendel. Unbeknownst to many, Charles Darwin was also a keen experimenter and based his theory of evolution by natural selection partly on experimental results. Today, experimental methods play some role or another in almost all branches of biology.

While there is no generally accepted definition of the terms "experiment" and "experimental," they can be used in a narrow and in a wider sense. In the narrow sense, an experiment involves the production of the antecedent conditions of some causal law in order to observe the consequence. According to this narrow sense, displacing a beehive by some distance in order to observe whether the bees can still find their preferred sources of nectar counts as an experiment, while determining the sequence of a part of the bees' DNA in order to determine its phylogenetic relationship to bumble bees and wasps does not. However, the latter activity may still be viewed as experimental in a wider sense. Sequencing DNA (in this case) involves the controlled manipulation of biological materials in order to make inferences about evolutionary descent and may qualify as experimental on these grounds.

As these simple examples demonstrate, the nature and purpose of experimentation in the life sciences varies considerably. Here, I shall use the terms "experiment" and "experimentation" in the wider sense. I assume that the only constitutive properties of experimentation are, first, that it involves *intervention* with the processes or structures that are being studied and, second, that it requires some kind of *theoretical interpretation*. The first of these properties distinguishes experimentation from observation, while the second property distinguishes it from mere data collection. The latter point will be illustrated later.

It should not be assumed that experiments in biology have been made or are about to be made redundant by recent advances in genomics and bioinformatics. Even though computational approaches (colloquially known as "*in silico* studies") play an increasingly important role in areas such as systems biology, these approaches cannot do without experimental data. For example, when biologists model gene regulatory networks *in silico*, they will need to know which genes interact with each other. This still cannot be predicted from DNA sequence information alone. Furthermore, theoretical models still have to be checked against reality. Thus, biological experimentation is very likely to stay.

While philosophers of science have long neglected experimentation, there is now a considerable body of historical and philosophical literature both on experimentation in general and on biological experimentation specifically. Not all of these studies were pursuing the same questions; however, the following four issues lie behind most of the work that has been done to-date: First, what is the role of so-called "model organisms" in experimental research? Second, is experimental research in biology essentially a process of inventing and testing hypotheses, as traditional philosophers of science such as Karl Popper (1959) held, or does biological experimentation have its own dynamics that is largely independent of specific hypotheses or a theoretical framework? Third, what is the nature of experimental evidence in biology? And fourth, can experiments give scientists access to an objective reality? This chapter tries to show how one could think about these issues.

2. Model Organisms

There is an aspect of modern experimental biology that might seem paradoxical at first sight. On the one hand, most of what we know about genes, cells, and so on is based on research involving a remarkably small number of species: The fruit fly *Drosophila melanogaster* was used to develop the classical methods of genetic mapping (Kohler, 1994). Molecular biologists elucidated some basic genetic mechanisms such as replication (the copying process of DNA) and gene expression (the process by which RNA and proteins are made from genes) in the colon bacillus *Escherichia coli* and its bacteriophages (Cairns, Stent, & Watson, 1992). Yeast has proven to be invaluable for cell biology. The giant nerve cells of squid were helpful to discover the basic mechanism of how neurons transmit information. The mouse is the favorite lab animal of immunologists. At the molecular level, *Arabidopsis thaliana* is the best-understood plant. Finally, the nematode worm *Caenorhabditis elegans*, the fruit fly *Drosophila*, and the see-through zebrafish *Danio rerio* are used with great success to study the molecular basis of embryonic development (the former two also for behavioral genetics).

On the other hand, the knowledge that biologists have acquired with the help of these experimental organisms aspires to be of a *general* nature. Even if Jacques Monod was exaggerating when he (famously) said, "What is true for the colon bacillus is true for an elephant," a large proportion of biological research today is not done just to understand some mechanism in one particular organism, but in a vast number of species including, of course, humans. For this reason, lab organisms such as the mouse or the zebrafish are also known as "model organisms."

How can scientific knowledge gained in research on the model organisms be of a *general* nature, if it is based on research on just a few or even a single species? After all, there probably still exist more than a million species on Earth, while just a couple of dozen have been studied in detail at the molecular level. Is this not too thin an induction base?

Should experimental biologists not widen their induction base by changing their favorite lab organism more often, so as to increase the scope of our biological knowledge? A look at the history of biology shows that, in contrast to botanists and zoologists, they do so quite rarely. Why is this so?

In order to see that the induction base does not always have to be very broad, it is instructive to consider the example of the genetic code. The genetic code relates the sequence of building blocks in DNA (called bases) to the sequence of amino acids that make up protein molecules. (The sequence of bases in DNA determines the sequence of amino acids in proteins). The code was "cracked" in the 1960s, after M. W. Nirenberg and J. H. Matthaei showed that an artificial *in vitro* system for protein synthesis could be programmed to assemble proteins of a certain sequence when fed with a synthetic RNA (ribonucleic acid, a substance that resembles DNA but does usually not occur as a double helix. RNA is used by the cell to make working copies of genes). This RNA contained only the base uracyl (U). Since this RNA caused the *in vitro* system to synthesize a protein made only of the amino acid phenylalanine, this showed that the code word "UUU" specifies "phenylalanine." For theoretical considerations suggested that it must be a triplet of bases, a so-called codon, which specifies a particular amino acid in protein synthesis. Since there are $4^3 = 64$ possible triplets made of the four RNA bases A, U, G, C and only 20 amino acids, the code is redundant, that is, several triplets code for the same amino acid. Molecular biologists have determined all of the 64 assignments, which are canonically known as the genetic code.

While this code was first cracked in the bacterium *Escherichia coli*, a broad variety of species including humans were later found to contain the exact same code. However, there were also deviations found, for example, in the mitochondria. Mitochondria are intracellular organelles whose main function is to provide the cell with energy. They contain DNA and their own genetic system, and this system shows some deviations from the standard genetic code. Furthermore, there are protozoans (single-celled organisms, many of which are pathogenic, e.g., trypanosomes) that have a deviant genetic code.

Since only a very small number of organisms have been analyzed at the molecular level, what entitles molecular biologists to talk about the "standard" genetic code (as they do)? The answer lies in a special kind of *inductive reasoning*. There are different kinds of inductive reasoning, the simplest one being enumerative induction. In enumerative induction, it is inferred from the fact that some individual S_1 has property G and S_2 has G and S_3 has G, and so forth until S_n has G, that *all* S's have property G. Obviously, enumerative induction is highly unreliable, because it is always possible that the next S in the set does not have G, thus making it false that all S have G.

If the universality of the genetic code were based on enumerative induction on the number of species studied molecularly to-date, then molecular biologists would be in trouble. But fortunately, it can be justified with a different kind of reasoning that is more sophisticated and far more reliable than simple enumerative induction. This reasoning involves a so-called "inference to the best explanation." The starting premise for this inference is the fact that the genetic code was found to be identical in a number of organisms that are *not* close relatives in phylogenetic terms. *E. coli*, yeast, *Drosophila*, humans, the plant *Arabidopsis*, the mouse, the zebrafish, and the nematode *C. elegans* all share the same genetic code (in their main genetic system). This list includes species from four different kingdoms of life. Within the highly diverse group of metazoans in the animal kingdom, the standard genetic code is found in taxa that are very remote in terms of ancestry, for example, nematodes, insects, fish, and mammals.

The next premise in the argument states that it is unlikely in the extreme that the genetic codes of these phylogenetically remote organisms should agree if they had arisen independently in each of these lineages. Given the vast number of possible assignments of amino acids to codons, this would amount to an almost cosmic coincidence. Note that there is, so far as we know, no biochemical reason why the genetic code should be as it is. Therefore, it is very safe to conclude that all these diverse organisms have inherited the genetic code from a common ancestor. Given the distribution of this trait in the phylogenetic tree, this common ancestor must be shared by all organisms in existence. But this already means that, unless there is further evolutionary change, the code is universal – which is what we want to show.

Note how we arrived at this conclusion: We started from *particular* facts, namely the agreement of the genetic code in a number of ancestrally remote organisms. We then reasoned that there must be an *explanation* for this agreement; it cannot be attributed to chance or coincidence. The best explanation is the monophyletic origin, which (given the trait's actual distribution) is equivalent to the universality of the code. This is a universal statement. Thus, what we have here is the inductive inference form known as inference to the best explanation.

Thanks to this inference, biologists do not have to check the genetic code for all species in existence; they can infer its universality from what they know about evolution and about the nature of the genetic code. But what about the qualification "unless there is further evolutionary change" that we had to make? Does this not threaten the inference to universality? I think not, as it has been shown that it is very hard for an organism to change its genetic code; this seems to be possible only in some very peculiar genetic systems. Changing the genetic code has been compared to reassigning all the letters and signs on your keyboard at the same time. Thus, we cannot expect many organisms to have changed the genetic code.

Burian (1993, p.366) has called the question of the extent to which experimental findings obtained in one organism can be extrapolated to other organisms "an especially acute version of the traditional philosophical problem of induction." But as I have shown, at least for the case of the genetic code, the problem of induction can actually be solved; for a *justification* can be given for the inductive inference from the few species that have been studied molecularly to all species in existence on Earth. (I take the problem of induction to be the problem of giving a justification for rules of inductive inference, not merely the fact that general statements cannot be deduced from particular premises, which is a basic fact of elementary logic.)

The case of the genetic code may be a special one, because of the extremely low likelihood that it has arisen more than once (in the exact same form) in evolution. In other cases, the inference from the agreement of characters to monophyly and universal distribution may be harder to justify. However, recent advances show that there exist many molecular mechanisms other than the genetic code that are also highly *conserved* in evolution, that is, they stay basically the same even though the organisms of which they are part diverge strongly by evolutionary change. Furthermore, many of these mechanisms have a hierarchical, modular structure with some modules being deployed both in very simple as well as in very complex systems (see Craver & Darden, 2001, for a philosophical account).

An example is provided by the so-called NMDA receptors. These are receptors for the neurotransmitter glutamate, which is the most important excitatory neurotransmitter in the brain. (Neurotransmitters are responsible for the transmission of signals by the synapses in the brain.) The name derives from the fact that these receptors can be activated by the chemical N-methyl-D-aspartate (NMDA). The NMDA receptors can be modulated by the nervous system so as to alter the strength of signal transmission; a phenomenon that is called synaptic plasticity. It was shown that NMDA receptors are involved in long-term potentiation (LTP), where the modifications of the synaptic strength persist for some time. LTP has been studied in invertebrates such as the sea snail *Aplysia*, which show some comparatively simple reflexes. But mounting evidence shows that NMDA receptor-mediated LTP is also involved in learning and memory in animals with a vastly more complex nervous system such as the mouse and probably humans as well. Thus, some molecular mechanisms are found both in very simple systems – where they are more accessible to experimentation – and in more complex ones, making it possible to extrapolate from simpler to more complex organisms (Schaffner, 2001). Thus, it seems to be a fundamental principle of life that there exist modular mechanisms that are deployed by evolution in a large variety of different systems. They are part of a construction kit out of which evolution assembles a large variety of molecular mechanisms in very different organisms. This furnishes part of the explanation why model organisms have proven to be so useful in biology.

My considerations so far show how inferences from model organisms to other organisms can be justified on evolutionary grounds. However, this goes only halfway to explain why biologists do not change their model organisms more often, which is one of the most conspicuous features of modern experimental biology. In order to answer this question in full, some additional aspects of experimental practice in biological science must be taken into account.

Experimental biologists do not only carry out individual experiments; they also prepare and maintain elaborate collections of biological materials such as different strains of organisms carrying certain genetic mutations or, more recently, samples of DNA molecules that can be used for genetically engineering new strains. These collections, which are essential for the scientists' ability to carry out extensive experimental studies, must be carefully documented and properly stored and maintained. Frequently, scientists exchange such materials. It is part of the scientific ethos that research materials are freely provided even to laboratories that directly compete for being first in obtaining certain research results. Biologists organize elaborate networks for the exchange of carefully documented research materials. One of the oldest such networks dates back to the classical school of *Drosophila* genetics founded by Thomas Hunt Morgan (Kohler, 1994). Similar networks were established around other experimental organisms such as the worm *C. elegans* (de Chadarevian, 1998). Thus, complex social networks sometimes form around certain laboratory organisms.

During the last century and continuing, *Drosophila* researchers have built up an enormous collection of fly strains that typically harbor several mutant genes the position of which is known on the genetic map (some of these mutants have occurred spontaneously, while others were induced by chemicals, radiation, or more recently by genetic engineering techniques). Thus, the *Drosophila* system provides a wealth of *resources* for research in genetics. When recombinant DNA technology ("genetic engineering") was developed in the 1970s, it became possible to deploy these resources for molecular studies in *Drosophila*. A fascinating example is provided by the so-called homeobox. This is a genetic element first discovered in *Drosophila*, which is thought to play a central role in regulating gene activity during embryonic development (Gehring, 1998).

The homeobox was found with the help of a technique called "walking on the chromosome," which is an extension of classical gene mapping. This technique, developed in the 1980s, allowed geneticists to isolate and identify DNA fragments from chromosomes. Most importantly, it made it possible to determine the exact location on the chromosome from where the DNA fragments originate. This technique was useful when molecular biologists wanted to isolate the DNA from a gene about which they knew nothing but its chromosomal location. In the 1980s, this was the case for a number of *Drosophila* genes that can give raise to rather bizarre mutations, such as flies with antenna on their heads or flies with an extra pair of wings. Such mutants are termed homeotic (from whence the name "homeobox" derives). The homeobox was discovered once several such genes had been isolated by the technique of chromosomal walking. Remarkably, it was possible to use the DNA fragments isolated from *Drosophila* to identify similar genes in other animals including humans (known as HOX-genes).

The chromosomal walks would not have been possible without the enormous resources that the *Drosophila* system provided. These resources included strains, laboratory techniques, and genetic maps that had been accumulated by several generations of geneticists. An important factor was that these resources from classical, pre-molecular genetics could be *assimilated* into the new molecular technology by creating a hybrid experimental system with an enormous potential for isolating and identifying genes.

This example suggests a partially *economic* explanation for why experimental life scientists rarely change their lab organisms (the following is an adaptation of an original idea due to Kohler, 1991, who worked it out on a different example). Bringing a new organism into the lab and beginning to build up experimental resources from scratch is an enormous investment. Sometimes, scientists make this investment, as Thomas Hunt Morgan did for *Drosophila* in the 1910s and 20s, or as Sydney Brenner did for *C. elegans* in the 1970s. The returns on such an investment can be considerable, as Morgan's or Brenner's Nobel Prizes demonstrate (note that in science – unlike in the economy – returns on investment are not measured in dollars, but in terms of reputation and recognition by peers). But in most situations, scientists can benefit more by using an already established experimental organism, where they can tap into the resources that other scientists have accumulated. So long as a particular organism continues to provide returns in terms of publishable results, scientists will continue to exploit them. Organisms like *E. coli*, *Drosophila*, or *C. elegans* have done so for many decades now, and there is no end in sight yet. Occasions where scientists are forced, as it were, to move into new markets with a newly developed organism – the equivalent of industrial innovation – do exist, but only infrequently. (A recent example is the zebrafish.) This, I suggest, explains why scientists don't change their favorite lab organisms more often.

I am not suggesting that the dynamics of scientific change is *solely* determined by the scientists' selfish interests in advancing their own careers (as some sociologists of science would have it). The epistemic reasons for why model organisms are so useful, which were already discussed, and the economic reasons for why they are being cultivated so extensively are not mutually exclusive. For scientists don't just value their lab organisms and the experimental resources that come with them because they bring them publications, funding, recognition, etc. Even if these things are important elements of science's professional reward system, the model organisms are also valued for epistemic reasons. Ultimately, it is because they have proven to be extremely valuable for producing knowledge that such organisms are cultivated in laboratories. Even if it is true that scientists (like most professionals) try to advance their own careers as much as they can, they can only do so by producing work that is epistemically valuable.

3. Experimental Systems and the "New Experimentalism" in Biology

A traditional view concerning the role of experiments in research has been formulated most succinctly by (Popper, 1959, p.107): "The theoretician puts certain definite questions to the experimenter, and the latter, by his experiments, tries to elicit a decisive answer to these questions, and to no others. All other questions he tries hard to exclude." On this view, theory clearly comes first. For Popper, good scientific theories are always "bold conjectures" and a result of free imagination. The only role that experiments can play in this picture is in determining the truth-value of theories. (Experimental biologists sometimes use the term "theory" in the sense of "unproven conjecture." Following standard practice in the philosophy of science, I use the term in a broader sense, such that it includes any models of biological mechanisms and processes.) [See Models].

Behind this account lies the conviction that, without theory, experimentation (as well as observation) would be "blind," that is, it would senselessly gather data without ever being able to assess the relevance of these data, or even determine when the data collection should terminate. Thus, the "theory first"-view is a legitimate response to naive inductivism, according to which science must start with a theory-free collection of all the relevant facts. But *which* facts are relevant, and how can we know that we have gathered *all* the facts? Only theories can tell us, philosophers such as Popper argued.

More recently, the "theory first"-view has come under attack again. According to a quite heterogeneous school of thought known as "New Experimentalism," experimentation has "a life of its own" (Hacking, 1983, p.150). Although this phrase is obviously metaphorical, it captures well certain aspects of experimental science, including many parts of modern biology. Here are some more specific claims that have been made by New Experimentalists. First, experiments sometimes serve other roles than determining the truth-value of previously conceived theories. For example, they may serve an *explorative* role without thereby being "blind" (Steinle, 1997). In some areas of biology, a large part of experimentation is *preparative*, that is, it aims at producing carefully selected and documented biological materials such as isolated DNA fragments (see the story of the homeobox in the previous section). Both preparative and explorative experimentation do not need preexisting bold conjectures to proceed; specific testable hypotheses may crop up later.

At this point, it may be objected that this kind of work is not experimental, but preparative or exploratory *simpliciter*. Is it not necessary for some research work to count as experimental that it bears on some hypothesis? If this work does not aim at testing specific hypothesis, is it not better described as data collection? Such an objection fails to recognize that the results of such research activities as were described in this and the previous sections are not raw data. A DNA sequence or an identified DNA fragment is already a *theoretical interpretation* of some raw data. This interpretation may even be the result of testing specific hypotheses about what the data say, but this is not essential in my view. What renders research activities such as DNA sequencing or the isolation of DNA fragments experimental (in the wider sense defined in the first section of this chapter) is the involvement of both intervention and theoretical knowledge, where the latter is necessary for interpreting the raw data. This recognition does not mandate a return to the "theory first"-view, as the theoretical knowledge needed to carry out experimental work is not subject to test or confirmation. It is part and parcel of the investigative strategy of a scientific specialty, and it does not aim at conceptual novelty (an example are the laws of gene transmission and recombination in genetics, see Waters, 2004).

Second, many experiments (including preparative and explorative experiments) give rise to *surprises*. This does not just mean that an experiment may violate the experimenters' expectations with respect to outcome, but that sometimes, experimental inquiry may take scientists into totally unexpected directions. Rather than just giving an answer to a question thought up by some theoreticians, experimental inquiries sometimes have the result that the questions themselves change. It is even possible that a series of experiments leads scientists into an altogether different field, for example, from cancer research to molecular biology (see Rheinberger, 1997 and below). In addition, it may occur that an experiment ends up playing a completely different role in the development of science than its authors intended.

Third, experiments can give rise to a kind of knowledge that is historically more stable than theories. An interesting example for this is genetics. While most of the theories about the nature of genes and the physiology of gene action from the early decades of the twentieth century have perished, some of the techniques as well as concepts from classical genetics have survived, even in the era of genomics (see the example of chromosomal walking in the previous section). Thus, experimental knowledge may survive fundamental revisions in theory. This may have implications for the issue of scientific realism (see the final section).

It is not possible here to critically assess all of these claims. But there is a concept that deserves a somewhat closer look: the notion of *experimental system*.

Rheinberger (1997, p.28) characterizes experimental systems as "systems of manipulation designed to give unknown answers to questions that the experimenters themselves are not yet able clearly to ask" and also as "the smallest integral working units of research." Rheinberger argues that research in biology always "begins with the choice of a system rather than with the choice of a theoretical framework" (p.25). Such systems need not be tied to a particular laboratory organism; however, in some cases a model organism may be a central part of an experimental system (see previous section). According to Rheinberger, in contrast to Popper, experimental research in biology does not usually start with a theory, or with well-formulated research questions. Experimental systems may precede the problems that they eventually help to solve. What is more, experimental systems do not just help to *answer* questions, they also help to *generate* them. Where this process leads is impossible to predict; experimental systems give rise to unexpected events. This unpredictable, open-ended character of the research process has not been sufficiently accounted for by traditional philosophy of science.

What Rheinberger is advancing is a theory of the dynamics of scientific change. According to this theory, the development of experimental disciplines in biology is not mainly driven by ideas or theories, as Popper held. Instead, the research process is "driven from behind" by the intrinsic capacities of experimental systems, which the scientists are constantly trying to explore. It should be noted that Rheinberger's notion of experimental system is very broad, as it includes both the material and cognitive resources required to do experiments. Material resources include the biological tissues or cells that are under study, the preparation tools (e.g., centrifuges), and the measurement instruments (e.g., counters for measuring radioactivity). Cognitive resources (my term) include the practical skills required to operate the apparatuses, as well as some theoretical knowledge needed for designing experiments and interpreting the data.

Rheinberger's account is based on a detailed case study of protein synthesis research in the 1940s, 50s and 60s, which eventually led to the cracking of the genetic code. The experimental system used there is an *in vitro* system for protein synthesis. In this system, rat liver cells or *E. coli* cells were broken up by mechanical agitation. The homogenate was then centrifuged in order to remove cellular debris. In order to observe protein synthesis in such so-called "cell-free systems," biochemists added radioactively labeled amino acids and measured the incorporation of this radioactivity into protein. Such *in vitro* systems were developed and refined over many years of research. In the 1960s, they played a central role in the cracking of the genetic code in the elucidation of the steps of protein synthesis. A similar system was used by Nirenberg and Matthaei in their famous poly-U experiment (already mentioned in the previous section).

The story of the *in vitro* system for protein synthesis exemplifies some possibly general features of experimental systems. First, the system moved across traditional disciplinary boundaries. It was first developed for cancer research in order to study the altered metabolism of tumor cells. It was then used by biochemists in order to study the mechanisms of protein synthesis. Eventually, this system helped to solve a fundamental problem in molecular biology. Thus, the practice of several scientific disciplines was substantially transformed by this system.

Second, the *in vitro system* eventually served a purpose for which it was not designed. Nirenberg and Matthaei, when they started to use the *in vitro* system, first wanted to use it for synthesizing specific proteins. Furthermore, the synthetic RNA molecules such as poly-U were not initially added in order to reprogram protein synthesis; they first used it as an inhibitor of nucleases (enzymes that digest DNA or RNA). Later, they used these synthetic RNA molecules as *controls* in order to show that the protein synthesis they observed (with natural mRNA) was template-specific. This is how they came across the effect that poly-U RNA stimulated the incorporation of phenylalanine into protein. Thus, synthetic RNA suddenly moved from the periphery of the experimental system right into its center, and when that happened, the direction of protein synthesis research changed dramatically. Rheinberger (1997, p.212) concludes that Nirenberg and Matthaei's famous poly-U experiment was a direct consequence of "exploring the experimental space of cell-free protein synthesis according to the cutting-edge standards of the biochemical state of the art." This example illustrates how experimental systems – construed, as Rheinberger does, as amalgamations of material objects and practical skills – can generate new, unexpected findings when scientists explore the various things that can be done with the system. Experimental systems act as "generators of surprises."

Thus, it seems that the development of a biological discipline is to a substantial extent governed by the internal dynamics of experimental systems. However, this should not be taken to mean that theoretical ideas play no role whatsoever in the research process. In the 1960s, many molecular biologists and physicists were thinking hard about the so-called "coding problem," i.e., the question of how DNA can specify the amino acid sequence of proteins. Many theoretical schemes were devised and tested; there were even attempts to use cryptographic methods and supercomputers to solve the problem. The modern notions of coding genetic information, transcription and translation, are probably a result of these attempts (see Kay, 2000). However, these theoretical ideas began to affect research on protein synthesis comparatively late. Much of the experimental work done did not serve the purpose of testing preexisting theories or answering questions put forward by theoreticians. However, this should not be taken to mean that there was no theoretical knowledge in this work. For, as I have argued, *some* theoretical interpretation is necessarily involved in an experiment. But there is a clear difference between those theories that are part of the investigative practice of a discipline, and those theories that extend our knowledge into previously unknown territory.

In summary, experimental biology does not seem to conform to the traditional philosophical accounts of the research process, according to which smart theoreticians think up theories for skilful experimenters to test, or ask questions for experimenters to answer. Much biological research can only be understood as an ongoing interaction with experimental systems. A large part of this interaction consists in exploring the space of possible manipulations that an experimental system offers. This activity often does not require a well-formulated research question, or a specific theory to test. Sometimes, it can only be said in retrospect what question an experiment answered; the question might not even have been asked beforehand.

4. The Nature of Evidence

The historical analysis of experimental systems discussed in the previous section provides a radically different explanation for scientific change than traditional philosophical accounts of scientific method such as Karl Popper's (1959). It certainly highlights aspects of scientific practice that philosophers of science have ignored for too long. However, it does not give the full story. In this section, I want to show that inquiries into the nature of experimental evidence, epistemic norms and methodological standards must complement the experimental systems approach. For there are developments in the history of experimental biology that are hard to understand without an appeal to epistemic norms and standards.

A good example is found in the oxidative phosphorylation controversy in biochemistry (*ca.* 1961–77). The issue of this controversy was the mechanism by which the energy released in respiring mitochondria (intracellular organelles that provide the cell with energy) is used to make the energy-rich compound ATP (adenosine triphosphate). ATP powers many biological processes such as muscle contraction or the biosynthesis of small and large molecules needed by the cell. From the 1940s onward, biochemists wanted to understand how the generation of this compound from its dephosphorylated form adenosine diphosphate (ADP) was coupled to respiration (i.e., the breakdown of organic molecules with the help of oxygen). There were two hypothetical schemes proposed for this coupling mechanism, one involving a chemical intermediate, the other a proton concentration gradient across the inner mitochondrial membrane. The latter was also known as the "chemiosmotic" mechanism and had been postulated by the British biochemist Peter Mitchell who later received a Nobel Prize, once the chemiosmotic mechanism was widely acclaimed. It could be argued that this case exemplifies a "theory first" mode of research; however, it can be shown that Mitchell's ideas were also influenced by his involvement with certain experimental systems.

But the question that needs to be addressed here is why biochemists were not able to settle the issue as to which scheme was correct for almost fifteen years. In retrospect, it might look as if Mitchell's opponents were just conservatively clinging to a mechanism that was more familiar to them (i.e., the chemical mechanism). However, it can be shown that the experimental evidence was indeed inconclusive until the mid-1970s (Weber, 2002). Only then did an experimental system become available that allowed biochemists to stage a crucial test, which came out in favor of Mitchell's mechanism. This test involved a new experimental system that contained artificial membranes and purified enzymes from mitochondria and other organisms, developed in the laboratory of Efraim Racker. For the first time, this system allowed biochemists to assemble all the components thought to be necessary for oxidative phosphorylation from purified, isolated components.

In the course of this study, it was possible to show directly that the enzyme that produces ATP from ADP and phosphoric acid can be powered by a proton gradient, as Mitchell had predicted more than a decade ago. One experiment is particularly noteworthy. It was possible to show that the ATPase can also be driven by an enzyme that is devoid of any respiratory function. This enzyme, bacteriorhodopsin, was isolated from a photosynthetic bacterium that lacks respiration (it generates ATP directly with the help of light energy). Today, this experiment is often presented as the crucial experiment that demonstrated the validity of Mitchell's scheme. In other words, this experiment is widely seen as decisive in bringing about a choice between the two main competing explanatory schemes. Even though the possibility of such experiments (which were first envisioned by Francis Bacon) has been doubted by many philosophers of science, I think that something akin to a crucial experiment is possible under certain conditions.

What is important to recognize in the example discussed is that the experiment could play this role only in the context of the whole *series* of experiments involving purified components and artificial membranes. The reason is that it was necessary to show that the respiratory enzymes can actually power the ATPase in the *in vitro*-system. The bacteriorhodopsin experiment was only needed as a *control*, namely to exclude the possibility that a chemical intermediate (as was postulated by the rival, chemical theory) had been co-purified with the respiratory enzymes and that the latter was actually doing the energy coupling. Thus, the conclusion that this experiment served a mainly pedagogical purpose can be avoided (cf. Allchin, 1996). At any rate, one should not speak of a crucial experiment here, but of a crucial experimental *investigation* that involved many experiments and elaborate controls.

This case exemplifies both the strength and the weakness (or incompleteness) of the experimental systems approach discussed in the previous section. For it is true that, even in this apparently theory-driven case, the development of the experimental systems exhibits an internal dynamics that is quite independent of the theoretical controversies that unfolded at conferences and in the scientific literature of the day. For example, Racker's *in vitro* system was a result of research begun around 1960, when Mitchell's theory was either not yet published or hardly noticed by anyone. Until well into the 1970s, Racker was an adherent of the chemical scheme. But most of the experimental work was largely unaffected by this; it simply followed the standard biochemical practice of trying to isolate, purify, and reconstitute the enzymes thought to be necessary for a specific process. (Of course, there were also experiments done directly in order to test Mitchell's theory, in particular in Mitchell's own, private laboratory in Cornwall. However, these experiments, even though they did support the chemiosmotic model, left room for alternative interpretations.) Eventually, the artificial membrane system served a purpose for which it was not designed: a crucial experimental demonstration of the validity of Mitchell's scheme. So far, the New Experimentalism holds out.

But at the same time, this case shows that, in addition to the history of experimental systems, epistemic norms and methodological standards are also required to understand the dynamics of the entire controversy. For the experimental systems approach cannot explain why the *theoretical* controversy could not be resolved before the mid-1970s, nor why it was eventually resolved after all. For this, it is necessary to inquire into the logic of experimental theory testing and to investigate to what extent certain experiments do or do not constitute strong evidence for a theory. Furthermore, such an analysis is required in order to understand why biochemists value certain experimental systems more than others. In biochemistry, two particularly valued features of experimental systems are *simplicity* and *manipulability*, but this may vary from field to field. Many experimental systems (including Rheinberger's *in vitro* system for protein synthesis) are developed toward simplicity and manipulability over time. The reasons have to do with the possibility of ruling out errors or "experimental artifacts" (see the final section), but this brings us already into a methodological ballgame. Thus, the experimental systems approach to explaining scientific change has to be supplemented by a philosophical theory of scientific evidence.

Philosophical theories of scientific evidence try to specify the conditions under which a body of data supports a hypothesis. At present, it is controversial whether there is a *universal* account of scientific evidence. Some think that inferences from evidence can always be modeled as the updating of personal probabilities in accordance with Bayes's theorem (Howson & Urbach, 1989). Others argue that the classical frequentist Neyman–Pearson statistical test, which involves the calculation of error probabilities, provides a universal account of scientific evidence (Mayo, 1996). Both parties to this debate are *monists* with respect to scientific evidence, that is, they hold that a single account of evidence applies to all scientific reasoning. However, given that a diversity of approaches to theory testing, data analysis, and assessment of evidence are, in fact, successfully deployed in different areas of the life sciences (see Taper & Lele, 2004), monists must provide reasons why scientists ought to drop all approaches save one. As no such convincing reasons have been offered, *pluralism* with respect to scientific evidence seems a more attractive option than monism.

For pluralists in matters methodological, the choice of methods and approaches, including the rules of inference, depends on the problems at hand. Bayesian and related methods seem to be useful, for example, in reconstructing phylogeny. These methods require good estimates of the conditional probability of the evidence given the hypotheses under test, which are sometimes available. However, there are many cases where these probabilities can at best be guessed, which renders Bayesian methods useless. Furthermore, many scientists and philosophers object to the Bayesians' use of subjective probabilities in science.

Frequentist statistical tests avoid subjective probabilities, which some take to be an advantage. But Neyman–Pearson methods require good ways of calculating error probabilities, i.e., the frequency of committing a type I or type II statistical error. Such calculations are often possible, for example, in ecology. However, in many cases the error probabilities are spurious because it is not clear in relation to which reference class they should be calculated (frequentist error probabilities always require a reference class).

In areas such as biochemistry or cell biology, statistical tests – Bayesian or frequentist – are sometimes neither helpful nor necessary. This is the case for the example of oxidative phosphorylation discussed above, where statistical methods played no role whatsoever. A qualitative account of scientific evidence is called for in such cases. In other cases, statistical methods are indispensable, but not all statistical methods are equally helpful for all cases. Thus, forcing all scientific practice into the Procrustian bed of some monistic philosophy of scientific evidence seems hardly justifiable. Methodological pluralism is better suited to make sense of the diversity of epistemic practices in the life sciences.

5. Objectivity and Realism

The question of whether there is a sense in which science is objective is one of the thorniest in philosophy, especially if we mean by "objectivity" not just intersubjectivity, but some kind of access to a mind-independent reality. A severe problem for objectivity in this strong sense is the *theory-dependence* (sometimes "theory-ladenness") of observation, championed by Thomas Kuhn (1970) and Paul Feyerabend (1988). Both have argued that any sense-experience that we may have (including, of course, experiments) can be conceptualized in different ways, and no conceptualization is devoid of theoretical assumptions. The thesis of theory-dependence is readily applicable to biological experiments. For example, how do we know that little black bars on an X-ray film can represent a DNA sequence? Only from theoretical knowledge as to how DNA sequencing techniques work. Even microscopic images, be they generated by diffracted electrons or light, require some theoretical interpretation of what it is that we are seeing. I have argued that theoretical interpretations are necessary for some manipulation to count as experimental in the first place.

Harry Collins (1985) has extended the thesis of theory-dependence with an argument that all experimentation is trapped in a so-called "experimenter's regress." This regress arises if some theoretical claim can only be established with certain experimental techniques. In order to do so, of course, these techniques must be reliable. But the judgment of whether the techniques are, in fact, reliable depends on their producing some expected outcome. Yet, Collins argues, what makes us expect this outcome is the very theory in question.

Sylvia Culp (1995) argues against Collins that, at least in experimental biology, the experimenter's regress can be broken. She illustrates this with the example of DNA sequencing. The sequence of nucleotides in a DNA molecule can be determined by at least two different techniques, the so-called chemical and chain-termination techniques. The former proceeds by chemically breaking the DNA with chemicals that are selective for each of the four different bases A, T, G, and C. The latter uses the enzyme DNA polymerase to generate truncated copies of different length. In both techniques, the resulting DNA fragments are separated by electrophoresis, which generates the familiar ladders from which the nucleotide sequence can be read off directly.

Culp's strategy is to grant that both of these techniques are theory-dependent. In other words, much theoretical knowledge is needed to establish that the ladders actually represent a DNA sequence. However, Culp argues that the two techniques are *independently* theory-dependent. This means that, even though both presuppose heavy theoretical assumptions, they do not presuppose the *same* assumptions. The chemical technique presupposes theoretical knowledge about the organic chemistry of nucleic acids, while the chain-termination technique presupposes theories from molecular biology (specifically, the mechanism of DNA replication). These theories are logically independent. Finally, the keystone of Culp's case is that, if nothing goes wrong, both techniques will give the *same* DNA sequence for a well-behaved DNA molecule (there are some sports that don't behave well in the chain termination technique, but I shall ignore this special case).

Culp argues that it would be an improbable coincidence to obtain such an agreement of results by chance. But since the two techniques do not depend on the same theoretical assumptions, it is also not possible to attribute the agreement to shared theoretical presuppositions. Hence, there must be a theory-*in*dependent common cause of the two results. In other words, both of the agreeing experimental results must have been constrained by a causal process from the object studied (i.e., DNA) to the results, that is, the sequences obtained. But this means that the DNA sequences obtained with both techniques must be reliably correlated to the actual objects studied – to the real DNA sequence. Hence, the techniques are objective after all and the experimenter's regress is broken.

In her argument, Culp relies on a principle first suggested by Hans Reichenbach, the *common cause principle*. This principle allows inferences from the conditional probabilities of certain events to the presence of a common cause. However, the principle is known to have its difficulties; for example, it is known to be violated by certain quantum-mechanical phenomena. Furthermore, applications of the principle to a case like Culp's run into the difficulty of how to interpret probability. To make things worse, it could be asked in what sense there is a causal *process* that links two sequencing experiments done with different techniques? After all, the two experiments must necessarily work with numerically different DNA molecules (even if they are taken from the same sample).

Perhaps these problems can be bypassed by avoiding the troublesome common cause principle and using an inference to the best explanation (see above) instead. Specifically, it could be argued that the objectivity of the techniques provides the best explanation for why different sequence determinations typically yield the same result. (This echoes a popular way of arguing for scientific realism.) However, how do we know that this *is* the best explanation? Could there not be some hidden connection between the different experiments that we are missing? After all, the very judgment that the two different techniques are independent presupposes the truth of certain theoretical statements about how these techniques work. But this presupposition is what is open to question in this debate. But then, the purpose of the whole argument is defeated, for it is supposed to convince someone like Collins, who doubts that we can know scientific statements to be true. From his perspective, the whole argument must appear question begging.

It is far beyond the scope of this chapter to resolve this difficult issue. I conclude by pointing out that Collins's claim that observing some expected outcome is the *only* way of checking the reliability of a technique is problematic. A great deal of the work that goes on in laboratories is done in order to check the reliability of some techniques without presupposing some expected outcome. Such an inquiry can come out either way. Sometimes, a technique is found to be unreliable, or, as scientists say, prone to *experimental artifacts*. In other cases, artifacts can be excluded at least with a certainty that matches my certainty that I am not dreaming now. Thus, the reliability of a technique can itself become the *object* of an experimental inquiry. Defenders of the experimenter's regress have not provided any arguments that such inquiries are necessarily doomed for failure.

Acknowledgment

I wish to thank Anya Plutynski for helpful comments on a draft for this chapter.

References

Allchin, D. (1996). Cellular and theoretical chimeras: piecing together how cells process energy. *Studies in the History and Philosophy of Science*, 27(1, March), 31–41.

Burian, R. M. (1993). How the choice of experimental organism matters: epistemological reflections on an aspect of biological practice. *Journal of the History of Biology*, 26(2, Summer), 351–67.

Cairns, J., Stent, G. S., & Watson, J. D. (Eds). (1992). *Phage and the origin of molecular biology* (expanded edn). Cold Spring Harbor: CSHL Press.

Collins, H. M. (1985). *Changing order. Replication and induction in scientific practice.* London: Sage.

Craver, C. F., & Darden, L. (2001). Discovering mechanisms in neurobiology: the case of spatial memory. In P. K. Machamer, R. Grush, & P. McLaughlin (Eds). *Theory and method in the neurosciences* (pp. 112–37). Pittsburgh: Pittsburgh University Press.

Culp, S. (1995). Objectivity in experimental inquiry: breaking data-technique circles. *Philosophy of Science*, 62(3, September), 430–50.

de Chadarevian, S. (1998). Of worms and programmes: Caenorhabditis elegans and the study of development. *Studies in History and Philosophy of Biological and Biomedical Sciences*, 29(1, March), 81–106.

Feyerabend, P. K. (1988). *Against method.* London: Verso.

Gehring, W. J. (1998). *Master control genes in development and evolution: the homeobox story.* New Haven: Yale University Press.

Hacking, I. (1983). *Representing and intervening. Introductory topics in the philosophy of natural science.* Cambridge: Cambridge University Press.

Howson, C., & Urbach, P. (1989). *Scientific reasoning: the Bayesian approach.* La Salle, IL: Open Court.

Kay, L. E. (2000). *Who wrote the book of life? A history of the genetic code.* Stanford: Stanford University Press.

Kohler, R. E. (1991). Systems of production: Drosophila, Neurospora and biochemical genetics. *Historical Studies in the Physical and Biological Sciences*, 22(1, September), 87–129.

Kohler, R. E. (1994). *Lords of the fly. Drosophila genetics and the experimental life.* Chicago: University of Chicago Press.

Kuhn, T. S. (1970). *The structure of scientific revolutions* (2nd edn). Chicago: University of Chicago Press.

Mayo, D. G. (1996). *Error and the growth of experimental knowledge.* Chicago: University of Chicago Press.

Popper, K. R. (1959). *The logic of scientific discovery.* London: Hutchinson Education.

Rheinberger, H.-J. (1997). *Toward a history of epistemic things: synthesizing proteins in the test tube.* Stanford: Stanford University Press.

Schaffner, K. F. (2001). Extrapolation from animal models. Social life, sex, and super models. In P. K. Machamer, R. Grush, & P. McLaughlin (Eds). *Theory and method in the neurosciences* (pp. 200–30). Pittsburgh: University of Pittsburgh Press.

Steinle, F. (1997). Entering new fields: exploratory uses of experimentation. *Philosophy of Science*, 64(4, Supplement to vol. 64), S65–S74.

Taper, M. L., & Lele, S. R. (Eds). (2004). *The nature of scientific evidence. Statistical, philosophical and empirical considerations.* Chicago: University of Chicago Press.

Waters, C. K. (2004). What was classical genetics? *Studies in History and Philosophy of Science*, 35(4, December), 738–809.

Weber, M. (2002). Incommensurability and theory comparison in experimental biology. *Biology and Philosophy*, 17(2, March), 155–69.

Further Reading

Burian, R. M. (1993). Technique, task definition, and the transition from genetics to molecular genetics: aspects of the work on protein synthesis in the laboratories of J. Monod and P. Zamecnik. *Journal of the History of Biology*, 26(3, Fall), 387–407.
Clarke, A. E., & Fujimura, J. H. (Eds). (1992). *The right tools for the job. At work in twentieth-century life sciences.* Princeton: Princeton University Press.
Creager, A. N. H. (2002). *The life of a virus: Tabacco Mosaic virus as an experimental model, 1930–1965.* Chicago: University of Chicago Press.
Culp, S. (1994). Defending robustness: the bacterial mesosome as a test case. In D. Hull, M. Forbes, & R. Burian (Eds). *PSA 1994* (pp. 47–57). East Lansing: Philosophy of Science Association.
Gaudillière, J.-P., & Rheinberger, H.-J. (Eds). (2004a). *Classical genetic research and its legacy: the mapping cultures of twentieth century genetics.* London: Routledge.
Gaudillière, J.-P., & Rheinberger, H.-J. (Eds). (2004b). *From molecular genetics to genomics. mapping cultures of twentieth century genetics.* London: Routledge.
Hagen, J. B. (1999). Retelling experiments: H.B.D. Kettlewell's studies of industrial melanism in peppered moths. *Biology and Philosophy*, 14(1, January), 39–54.
Holmes, F. L. (2001). *Meselson, Stahl, and the replication of DNA: A history of "the most beautiful experiment in biology".* New Haven: Yale University Press.
Hudson, R. G. (1999). Mesosomes: A study in the nature of experimental reasoning. *Philosophy of Science*, 66(2, June), 289–309.
Lederman, M., & Burian, R. M. (1993). Introduction: The right organism for the job. *Journal of the History of Biology*, 26(2, Summer), 235–37.
Rasmussen, N. (2001). Evolving scientific epistemologies and the artifacts of empirical philosophy of science: a reply concerning mesosomes. *Biology and Philosophy*, 16(5, November), 629–54.
Rudge, D. W. (1999). Taking the peppered moth with a grain of salt. *Biology and Philosophy*, 14(1, January), 9–37.
Schaffner, K. F. (1993). *Discovery and explanation in biology and medicine.* Chicago: University of Chicago Press.
Schaffner, K. F. (1994). Interactions among theory, experiment and technology in molecular biology. In D. Hull, M. Forbes, & R. M. Burian (Eds). *PSA 1994* (vol. 2, pp. 192–205). East Lansing: Philosophy of Science Association.
Weber, M. (2005). *Philosophy of experimental biology.* Cambridge: Cambridge University Press.

Chapter 26

Laws and Theories

MARC LANGE

1. Is Biology Like Physics?

Do the concepts figuring in the metaphysical foundations of physics also play important roles in the foundations of biology? Or is biology profoundly different from physics? One respect in which biology and physics may be compared concerns the structure of their theories. In particular, the concept of a law of nature figures prominently in physics. Do biologists also aim to discover laws? Are there distinctively *biological* laws (in contrast, say, to the laws governing DNA's behavior, which belong to physical science)? If so, how do they relate to the laws of physics? If the only laws are the laws of physics, then of what do biological theories consist?

To approach these questions, we must begin with law's role in physics. We will then be in a position to investigate whether the same work needs doing in biology and whether laws are doing it.

2. Laws of Nature: The Standard Picture

Traditionally, there are three kinds of facts. First, there are the logical and metaphysical necessities: facts that absolutely could not have been otherwise. The rest (the "contingent" facts) divide into two classes: the "nomic necessities," which follow from the laws alone (e.g., that all copper objects are electrically conductive), and the "accidents," which do not. Typical accidents are that all of the coins in my pocket today are silver-colored (after Goodman, 1983, p.18) and that all solid gold cubes are smaller than a cubic mile (Reichenbach, 1954, p.10; Hempel, 1966, p.55). What distinguishes laws from accidents?

To begin with, an accident just happens to obtain. A gold cube larger than a cubic mile could have formed, but proper conditions happened never to arise. In contrast, it is no accident that a large cube of uranium-235 never formed, since the laws governing nuclear chain-reactions prohibit it. In short, things *must* conform to the laws of nature – the laws have a kind of *necessity* (weaker than logical or metaphysical necessity) – whereas accidents are just coincidences.

That is to say, had Bill Gates wanted to build a large gold cube, then (I daresay) there would have been a gold cube exceeding a cubic mile. But even if Bill Gates had wanted to build a large cube of uranium-235, all U-235 cubes would still have been less than a cubic mile. In other words, the laws of nature govern not only what actually happens, but also what would have happened under various circumstances that did not actually happen. The laws underwrite "counterfactuals" (Goodman, 1983, pp.8–9), i.e., facts expressed by statements of the form "Had *p* been the case, then *q* would have been the case." An accident would not still have held, had *p* been the case, for some *p* that is "nomically possible" (i.e., consistent with all of the laws' logical consequences).

Counterfactuals are notoriously context-sensitive. In Quine's famous example, the counterfactual "Had Caesar been in command in the Korean War, he would have used the atomic bomb" is correct in some contexts, whereas in others, " . . . he would have used catapults" is correct. What is preserved under a counterfactual supposition, and what is allowed to vary, depends upon our interests in entertaining the supposition. But in any context, the laws would still have held under any nomic possibility *p*. I'll refer to this idea as "nomic preservation."

Because of their necessity, laws have an explanatory power that accidents lack (Hempel, 1966, p.56). For example, a certain powder burns with yellow flames, not another color, because the powder is a sodium salt and it is a law that all sodium salts, when ignited, burn with yellow flames. (More fundamental laws explain that law.) The powder *had* to burn with a yellow flame, considering that it was a sodium salt – and that "had-to-ness" arises from the laws' distinctive kind of necessity. In contrast, we cannot explain why my wife and I have two children by citing the fact that all of the families on our block have two children – since this last is an accident. Were a childless family to try to move onto our block, they would not encounter an irresistible opposing force. (A counterfactual!)

Since we believe that it would be mere coincidence if all of the coins in my pocket today turn out to be silver-colored, our discovery that the first coin I withdraw from my pocket is silver-colored fails to confirm (i.e., justly to raise our confidence in) our hypothesis that the next coin to be examined from my pocket will also turn out to be silver-colored. A candidate law is confirmed differently (Goodman, 1983, p.20): that one sample of a given chemical substance melts at 383 K (under standard conditions) confirms, for every unexamined sample of that substance, that its melting point is 383 K (in standard conditions). This difference in inductive role between laws and accidents seems related to the fact that laws express similarities among things that reflect their belonging to the same "natural kind," whereas accidents do not. The electron, the emerald, and the electromagnetic force are all natural kinds, whereas the families on my block and the gold cubes do not form natural kinds. The natural kinds are, roughly speaking, the kinds recognized by the natural laws (Hacking, 1991). Since we do not expect the coins in my pocket to form a natural kind, we regard as accidental any similarities among them that we have noticed so far. Consequently, we regard a uniformity's having held of examined coins from my pocket as no evidence that it holds of unexamined ones.

That the same claims play these special roles in connection with necessity, counterfactuals, explanations, natural kinds, and inductive confirmations would suggest that scientific reasoning draws an important distinction here, which philosophers characterize as the difference between laws and accidents. (Obviously, this distinction involves what laws *do* rather than which propositions happen to be called "laws"; Heisenberg's uncertainty "principle," the "axioms" of quantum mechanics, and Maxwell's "equations" are laws of physics.) However, it is notoriously difficult to capture the laws' "special roles" precisely. Take counterfactuals. The mathematical function relating my car's maximum speed on a dry, flat road to its gas pedal's distance from the floor is not a law (since it reflects the engine's accidental features). Yet this function supports counterfactuals regarding the car's maximum speed had we depressed the pedal to one-half inch from the floor. This function has "invariance with respect to certain hypothetical changes" (Haavelmo, 1944), though not certain changes to the engine. Indeed, for nearly any accident, there are *some* hypothetical changes with respect to which it is invariant. All gold cubes would still have been smaller than a cubic mile even if I had been wearing a differently colored shirt. Likewise, my car's pedal-speed function can be confirmed inductively and can (together with the road's condition and the pedal's position) explain the car's maximum speed. So although a fact's lawhood apparently makes a difference, it is difficult to identify exactly what difference it makes. This problem's stubbornness – and the difficulty of explaining what a law of nature *is*, in virtue of which it can play these allegedly distinctive roles – has even led some philosophers (van Fraassen, 1989; Giere, 1995) to suggest that it is a mistake to distinguish laws from accidents in reconstructing science.

On the other hand, perhaps by thinking about whether there are laws of biology, we will better understand what laws are and what special roles they play (Lange, 2000).

3. Why Not Laws of Biology? The Problem of Exceptions

Much of biology – "functional biology" – aims to discover various species' "biological properties" (Mayr, 1965): their external morphology, internal anatomical structure, physiology, chemical constitution, environmental tolerances, etc. This is evident from even a casual perusal of journals in physiology, anatomy, medicine, genetics, neurobiology, pathology, psychology, developmental biology, and so on. They contain articles like "Growth of cottontail rabbits (*Sylvilagus floridanus*) in response to ancillary sodium" (McCreedy & Weeks, 1993), "Establishment and maintenance of claw bilateral asymmetry in snapping shrimps (*Alpheus heterochelis*)" (Young, Pearce, & Govind, 1994), "Antibiotic activity of larval saliva of *Vespula* wasps" (Gambino, 1993), and "Learning to discriminate the sex of conspecifics in male Japanese quail (*Coturnix coturnix japonica*)" (Nash & Domjan, 1991). These articles apparently present laws concerning the "biological properties" characteristic of particular species.

It has been objected (Smart, 1963) that such articles fail to present laws since laws, being general, cannot refer to particular locations or individuals, whereas *Vespula* refers implicitly to Earth since a species is defined by its position in the terrestrial evolutionary tree. Perhaps *Vespula* itself is an individual – a chunk of the genealogical nexus – and so there can be no laws about it any more than there could be a separate law of physics about Mars (Hull, 1978, p.353; Ghiselin, 1989; Rosenberg, 1985, p.219). But this argument faces severe obstacles. Even laws of physics may refer to individuals, such as the Big Bang (Dirac, 1938), and if none do, that is a fact discovered about the world rather than built into the concept of what a law *is*. Aristotelian physics posited laws referring to the universe's center and the moon's orbit. Surely that was not a *logical* error (Armstrong, 1983, p.26; Tooley, 1977, p.686). The laws' "generality" apparently has nothing to do with their distinctive relations to induction, explanation, and counterfactuals.

But what are the counterfactual supporting, inductively confirmed, explanatorily potent regularities in these articles? They cannot take the form "All members of species S possess biological property T" because such generalizations either

(1) are not biological (e.g., where T is the property of remaining at less than light's speed), or
(2) have exceptions (e.g., where the regularity is that all human beings – or even just all healthy, uninjured human beings – have ten fingers: Anne Boleyn had eleven fingers), or
(3) lack exceptions merely accidentally. Even if, in fact, every robin's egg is greenish-blue, that is only because a certain mutation happened never to occur; had the mutation occurred, there would have been a robin's egg of a different color. Apparently, "All S's are T" does not function as a law in connection with counterfactuals.

To (3), we might reply that certain traits are necessary for a creature to live; no mutation preventing the embryological development of the human lung would have resulted in a human being living lung-less. So "All [healthy, uninjured . . .] human beings possess lungs" has the requisite invariance under counterfactual perturbations. However, a mutation preventing lung development would not be fatal if (unlikely though this may be) it enabled human beings to (say) produce their own oxygen.

Perhaps "All human beings are mortal" falls under none of the three alternatives above. But what about the generalizations in the various articles? They have exceptions:

> There are no laws about particular species . . . This fact is reflected not only in the role of specimens, but also in the decline of essentialism among biologists: variation, as Mayr has pointed out, is not viewed as a disturbance from some mean property of members of a species which provides its essence; it is viewed as the normal result of recombinations within a lineage. The generalizations about particular species on which taxonomic decisions rest are full of exceptions, and there is no background theory that will enable us to eventually eliminate, reduce, or explain these exceptions. (Rosenberg, 1987, p.195)

The same applies to biological generalizations that aim to cover all species. Mendel's "laws" have exceptions (crossing-over, segregation distortion). (However, Ruse (1970) contends that Mendel's laws – in their modern form – include caveats about linkage and segregation distortion.) Certain important biological generalizations (such as the Hardy–Weinberg "law" and the principle of natural selection, under some interpretations) are exceptionless, but only by being logically necessary and hence non-laws on the traditional picture. (See Brandon, 1997, and Sober, 1997 and 2000, pp.72–4, who accordingly reject the traditional picture of laws as contingent truths.) Some exceptionless, biologically relevant generalizations, such as those concerning the chemical properties of DNA or ATP, are really matters of chemistry rather than biology. The same goes for the "causal regularities" discussed by Waters (1998), such as that blood vessels with . . . chemical constitution under . . . conditions expand with higher internal fluid pressure. Likewise, a generalization that Hull (1974, p.80) suggests might be a biological law – "Any organism with genotype . . . in environment . . . undergoing biochemical reactions . . . will have phenotypic characters . . ." – is just a logical consequence of chemical laws (Beatty, 1995, p.61). Accordingly, many philosophers of biology have concluded that whereas physical science is built around laws, biology involves the application of abstract models on a case-by-case basis. (Among these philosophers are Beatty, 1981, Lloyd, 1988, and Thompson, 1989.) For certain purposes at certain moments, a system may be usefully approximated as a "Mendelian breeding group," for instance, and Mendel's "laws" are trivially true of this model since they define what a "Mendelian breeding group" is. (Physics has been interpreted in this fashion as well; see Cartwright, 1983; Giere, 1995.)

But despite having exceptions, generalizations concerning particular biological species are the goal of much biological research. These generalizations function in biological reasoning – in connection with counterfactuals, explanations, and inductions – much as laws do in the traditional picture. Jane's trachea has cartilaginous rings because Jane is a human being and the human trachea has cartilaginous rings in order to keep it from collapsing between breaths. This individual has eyespots because it is a butterfly of the buckeye species and this species uses eyespots to fool predators. Explanations of this sort are quite ordinary. At the zoo, a child might point to a bird and ask, "Why did he do that?" An adult might properly reply, "That's how pelicans eat." Indeed, biologists employ generalizations of this kind even while describing natural *variation* among conspecifics:

> Within a single species . . . individuals sometimes have the diagnostic characteristics of related species or even genera. The form and number of teeth in mammals are important for classification; yet in a single sample of the deer mouse *Peromyscus maniculatus*, Hooper (1957) found variant tooth patterns typical of 17 other species of *Peromyscus*. Among fossils of the extinct rabbit *Nekrolagus*, Hibbard (1963) found one with the premolar pattern characteristic of modern genera of rabbits; and the *Nekrolagus* pattern is occasionally found in living species. (Futuyma, 1979, p.161).

Even while emphasizing the variation in dentition among conspecific mammals, Futuyma refers to the dentition "characteristic" or "typical" of a given species, and to "the *Nekrolagus* pattern."

But what does "*Nekrolagus* has tooth pattern T" *mean*? It is obviously not a simple statistical generalization. (What particular probability does it assign to T?) My view is that "The S is T" ("S's are characteristically/typically T") specifies a kind of default assumption about S's: If you believe (with justification) that some thing is an S, then you ought to believe it T in the absence of information suggesting that it isn't. This sort of policy may be sufficiently accurate for the purposes of the biological fields (e.g., neurology, physiology, embryology) for which the above journal articles were written, though not for the purposes of a biological discipline that is interested precisely in intraspecific variation, such as population genetics or evolutionary biology. Whether the default inference rule is sufficiently reliable for "The S is T" to be true depends upon how reliable it is – e.g., on how readily available "information to the contrary" is when an S isn't T – and on how tolerant of error we can afford to be, considering the relevant purposes.

This intepretation of "The S is T" explains how it can be that "The lion is tawny" and "The lion with gene A [for albinism] is white" can both be true: when we have no reason to believe that Leo possesses gene A, we ought to believe Leo tawny, but when we have some reason to believe that Leo possesses A, then we ought to withhold judgment, and when we believe that Leo possesses A, then we ought to believe Leo white (in the absence of any further relevant information).

Whatever the merits of this particular view of "The S is T," the key point is that generics (as linguistics call such claims) presumably have a respectable semantics (even if the details are controversial: see Carlson & Pelletier, 1995 – and in the philosophical literature, see Anscombe, 1958, p.14; Clark, 1965; Achinstein, 1965; Molnar, 1967). Whatever their precise truth-conditions, their "exceptions" – emphasized in point (2) above – do not bar their truth and so do not bar their expressing laws of functional biology.

But their relation to counterfactuals – point (3) above – might.

4. Why Not Laws of Biology? The Problem of Accidentalness

Even if "The S is T" is true, that is only because certain mutations have never occurred or certain selection pressures have never been in force. Had evolutionary history been replayed from the same initial conditions, the outcome might well have been radically different, since different mutations might have been introduced, random drift in small populations might have led in another direction, and different selection pressures might consequently have been imposed (Gould, 1989). Had a certain mutation occurred or a certain environmental influence been present, then "The S is T" might not still have held. Therefore, "The S is T" does not have the proper invariance under counterfactual perturbations to qualify as a law. It lacks nomic necessity; it merely expresses "current evolutionary fashions" (Waters, 1998, p.16), a coincidence of natural history (Beatty, 1981, 1995; Rosenberg, 2001a, 2001b, 2001c).

Let's examine this argument more closely. It employs an idea we encountered in Section 2:

> *Nomic Preservation (NP)*: g is nomically necessary if and only if in any context, g would still have held had p obtained, for every p that is logically consistent with every nomic necessity.

It is not the case that "The S is T" would still have held no matter what mutations had occurred or what environmental conditions had obtained. Therefore, by *NP*, "The S is T" is not nomically necessary.

However, this conclusion follows from *NP* only if it is nomically possible for those mutations or environmental conditions to occur under which "The S is T" is not invariant. Undoubtedly, the laws *of physics* (along with the principles of natural selection) are logically consistent with the occurrence of these mutations and environmental conditions. But to argue in this way that "The S is T" is not a law of functional biology – that is, to take as a premise that the laws of physics constitute all of the laws there are – amounts to reasoning in a circle: to arguing that there are no laws of functional biology by presupposing that there are no such laws (Lange, 2000, p.230).

Here is another way to make this point. We can think of *NP* as a general schema. For the laws of physics, it becomes: The laws of physics would still have held under all counterfactual suppositions that are logically consistent with the laws of physics. In other words,

> *NP°*: *g* is one of the laws *of physics* (or a logical consequence of those laws) if and only if in any context, *g* would still have held had *p* obtained, for every *p* that is logically consistent with the laws *of physics*.

But how should *NP°* be extended to determine what it would take for there to be laws of functional biology? The italicized bits in *NP°* might be replaced in either of two ways:

> *g* is one of the laws *of functional biology* (or a logical consequence of those laws) if and only if in any context, *g* would still have held had *p* obtained, for every *p* that is logically consistent with
> *NP′*: the laws *of physics*
> *NP″*: the laws *of functional biology*

NP′ requires that any law of functional biology have the same range of invariance as – and, hence, the same kind of necessity as – the laws of physics. On the other hand, *NP″* permits the range of invariance that *g* must possess, in order to qualify as *necessary* for the purposes of functional biology, to be distinct from the range of invariance characteristic of the laws of physics.

To use *NP′* to argue that there are no laws of functional biology, one must contend that *NP″* fails to capture a kind of invariance associated with a distinctive brand of necessity and explanatory potency. Let us now see whether this is so.

According to *NP*, the laws would all still have held under any counterfactual supposition that is logically consistent with the laws. No accident is always preserved under all of these suppositions. But *NP* alone cannot save our intuition that the nomic necessities possess an especially great power to support counterfactuals. That's because the *range* of counterfactual suppositions under consideration in *NP* (namely, the nomic possibilities) has been designed expressly to suit the nomic necessities.

What if we extend the same courtesy to a set containing accidents, allowing it to pick out a range of counterfactual suppositions especially convenient to itself: those suppositions that are logically consistent with every member of that set? Take a logically closed set of truths that includes the accident that all of the wires on the table are copper but omits the accident that all of the apples on my tree are ripe. Here's a counterfactual supposition that is consistent with every member of this set: had either some wire on the table *not* been made of copper or some apple on the tree *not* been ripe. What would the world then have been like? In many conversational contexts, we would deny that the generalization in the set (the one about the wires) would still have held. (Perhaps it is the case, of neither generalization, that it would still have held.)

The same sort of argument could presumably be made regarding any logically closed set of truths that includes *some* accidents but not *all* of them. Given the opportunity to pick out the range of counterfactual suppositions convenient to itself, the set nevertheless is not invariant under all of those suppositions. (Trivially, every member of the set of *all* truths would still have held under any counterfactual supposition logically consistent with all of them, since *no* counterfactual supposition is so consistent.)

Here, then, is my preliminary suggestion for the laws' distinctive relation to counterfactuals (which in Section 2 we found difficult to specify precisely). Take a set of truths that is "logically closed" (i.e., that includes every logical consequence of its members) and is neither the empty set nor the set of all truths. Call such a set *stable* exactly when every member g of the set would still have been true had p been the case, for each of the counterfactual suppositions p that is logically consistent with every member of the set. My preliminary suggestion: g is a nomic necessity exactly when g belongs to a stable set. (For a more careful discussion, see Lange, 2000.)

What makes the nomic necessities special is their stability: *taken as a set*, they are invariant under as broad a range of counterfactual suppositions as they *could* logically possibly be. *All* of the laws would still have held under *every* counterfactual supposition under which they *could all* still have held. No set containing an accident can make that boast (save for the set of all truths, for which the boast is trivial). Because the set of laws (and their logical consequences) is non-trivially as invariant under counterfactual perturbations as it could be, there is a sense of *necessity* corresponding to it; necessity involves possessing a *maximal* degree of invariance under counterfactual perturbations. No variety of necessity corresponds to an accident, even to one (such as my car's gas pedal-maximum speed function) that would still have held under many counterfactual suppositions. The notion of "stability" allows us to draw a sharp distinction between laws and accidents. It also gives us a way out of the notorious circle that results from specifying the nomic necessities as the truths that would still have held under those counterfactual suppositions consistent with the nomic necessities.

If g is a logical consequence of the laws exactly when g belongs to a stable set, then the laws of physics play no privileged role in picking out the range of counterfactual suppositions under which g must be invariant in order for g to be a logical consequence of the laws. Rather, any set of truths picks out for itself the range of counterfactual suppositions under which that set's members must all be invariant in order for that set to be necessary in the manner of laws. We now have something much more like *NP″* than *NP′*.

To see how the laws of functional biology may fail to be laws of physics, we must go beyond stability *simpliciter* and consider what it would be for a set to be stable *for the purposes of a given scientific field*. Such stability requires, to begin with, that the set's members all be *reliable* – that is, close enough to the truth for that field's purposes. Mill nicely captures the point:

> It may happen that the greater causes, those on which the principal part of the phenomena depends, are within the reach of observation and measurement . . . But inasmuch as other, perhaps many other causes, separately insignificant in their effects, co-operate or conflict in many or in all cases with those greater causes, the effect, accordingly, presents more or less of aberration from what would be produced by the greater causes alone. . . . It is thus, for example, with the theory of the tides. No one doubts that Tidology . . . is really a science. As much of the phenomena as depends on the attraction of the sun and moon . . . may be foretold with certainty; and the far greater part of the phenomena depends on these causes. But circumstances of a local or casual nature, such as the configuration of the bottom of the ocean, the degree of confinement from shores, the direction of the wind, &c., influence in many or in all places the height and time of the tide . . . General laws may be laid down respecting the tides; predictions may be founded on those laws, and the result will in the main . . . correspond to the predictions. And this is, or ought to be meant by those who speak of sciences which are not *exact* sciences. (1961, 6.3.1, pp.552–3)

A "reliable" g must reflect all of the "greater causes." But it may neglect a host of petty influences. For example, classical physics might suffice for the purposes of human physiology or marketing; relativistic corrections are negligible. (Biological controversies often concern the "relative significance" of various factors (Beatty, 1995), and these disputes may be understood as concerning which are the "greater causes" that must figure in biological laws (Sober, 1997, p.S461).)

What range of invariance must the set's members exhibit in order for that set to be stable for the purposes of a given scientific field? Since that field's concerns may be limited, certain claims and counterfactual suppositions lie outside of the field's interests, and the field is irrelevant in certain conversational contexts. A logically closed set is stable *for the purposes of a given science*, and hence its members are nomic necessities *for that field*, if and only if all of its members not only are of interest to the field and reliable for the field's purposes, but also – in every context of interest to the field – would still have been reliable, for the field's purposes, under every counterfactual supposition of interest to the field *and* consistent with the set.

To understand this idea, let's apply it to a real example.

5. A Worked Example: The "Area Law"

Take island biogeography (IB), which deals with the abundance, distribution, and evolution of species living on separated patches of habitat. It has been suggested that ceteris paribus, the equilibrium number S of species of a given taxonomic group on an "island" (as far as creatures of that group are concerned) increases with the island's area A in accordance with a power law: $S = cA^z$. The (positive-valued) constants c and z are specific to the taxonomic group and island group – e.g., Indonesian land birds or Antillean beetles. One theory purporting to explain this "area law" (the "equilibrium theory of IB," developed by MacArthur and Wilson) is roughly that a larger island tends to have larger available habitats for its species, so it can support larger populations of them, making chance extinctions less likely. Larger islands also present larger targets for stray creatures. Therefore, larger islands have larger immigration rates and lower extinction rates, and so tend to equilibrate at higher biodiversity.

Nevertheless, a smaller island nearer the "mainland" may have greater biodiversity than a larger island farther away. This factor is covered by the ceteris paribus qualifier to the "area law." Likewise, a smaller island with greater habitat heterogeneity may support greater biodiversity than a larger, more homogeneous island. This factor is also covered by "ceteris paribus." And there are others. (For more on ceteris-paribus laws in science generally, see Cartwright, 1983, and Earman, Glymour, & Mitchell, 2002.)

But to discover the "area law," ecologists did not need to identify *every* factor that may cause deviations from $S = cA^z$, only those "greater causes" sufficient for the area law to yield predictions good enough for various sorts of applications. These may be practical: the design of nature reserves. Or theoretical: serving as a common starting-point for building more accurate ecological models on a case-by-case fashion, each model incorporating the idiosyncratic features of the particular case for which it is intended. In this role, the area law functions like Hooke's law for springs, for example (Lange, 2000, p.28).

Suppose that the "area law" is indeed reliable. This may not be so. Perhaps only a case-by-case approach makes approximately accurate predictions regarding island biodiversity. I am not trying to prejudge the outcome of scientific research, merely to understand what laws of IB (or "macroecology" (Brown, 1995)) would be. What must the range of invariance of the "area law" be for it to qualify as an IB law – for it to be *necessary* in the relevant sense? There are counterfactual suppositions under which the laws of physics would still have held, but under which the "area law" would not still have held. For example, had Earth always lacked a magnetic field, cosmic rays would have bombarded all latitudes, which might well have prevented life from arising, in which case S would have been zero irrespective of A. Here's another counterfactual supposition: Had evolutionary history proceeded differently so that many species developed with the sorts of flight, orientation, and navigation capacities possessed by actual airplanes. (This supposition, albeit rather outlandish, is nevertheless consistent with physical laws since airplanes exist.) Under this supposition, the "area law" might not still have held, since an island's size as a target for stray creatures might then have made little difference to its immigration rate. (Creatures without the elaborate organs for flight and navigation could have hitched rides on those so equipped.)

The "area law" is not prevented from qualifying as an IB law (i.e., from belonging to a set that is stable for IB purposes) by failing to be preserved under these two counterfactual suppositions, although each supposition is consistent with the laws of physics. The first supposition (concerning Earth's magnetic field) falls outside IB's interests. It twiddles with a parameter that IB takes no notice of or, at least, does not take as a variable. Of course, IB draws on geology, especially paleoclimatology and plate tectonics. Magnetic reversals are crucial evidence for continental drift. But this does not demand that IB be concerned with species distribution had Earth's basic physical constitution been different. Biogeographers are interested in how species would have been distributed had (say) Gondwanaland not broken up, and in how Montserrat's biodiversity would have been affected had the island been (say) half as large. On the other hand, IB is not responsible for determining how species would have been distributed had (say) Earth failed to have had the Moon knocked out of it by an early cataclysm. (Earth's rotation rate would then have been greater, its tides would have been less, and the

CO_2 level in its atmosphere would have been greater.) Biogeographers need not be geophysicists.

The second counterfactual supposition I mentioned (positing many species capable of covering long distances over unfamiliar terrain nearly as safely as short distances over familiar territory) is logically inconsistent with other generalizations that would join the "area law" in forming an IB-stable set. For example, the "distance law" says that ceteris paribus, islands farther from the mainland equilibrate at lower biodiversity. Underlying both the area and distance laws are various constraints – e.g., that creatures travel along continuous paths, that the difficulty of crossing a gap in the creature's habitat increases smoothly with the gap's size (ceteris paribus). These "continuity principles" (MacArthur, 1972, pp.59–60) must join the area and distance laws in the IB-stable set.

The area law might not still have held, had these constraints been violated. Yet the area law's range of invariance under counterfactual suppositions may suffice for it to qualify as an IB law because *other* IB laws, expressing these constraints, make violations of these constraints nomically impossible in IB. Here's an analogy. Take the Lorentz force law: In magnetic field $\boldsymbol{B}$, a point body with electric charge q and velocity $\boldsymbol{v}$ feels a magnetic force $\boldsymbol{F} = (q/c)\boldsymbol{v} \times \boldsymbol{B}$. Presumably, it isn't the case that this law would still have held, had charged bodies been accelerated beyond c. But this law requires no proviso limiting its application to cases where bodies fail to be accelerated beyond c. That's not because there are actually no superluminal accelerations, since a law must hold not merely of the actual world, but also of certain possible worlds. The proviso is unnecessary because *other* laws of physics deem superluminal acceleration to be nomically impossible in fundamental physics. Hence, the Lorentz force law can have the range of invariance demanded of a law of physics without being preserved under counterfactual suppositions positing superluminal accelerations.

A set that is IB stable can omit some laws of physics. The *gross* features of the physical laws captured by constraints like those I've mentioned, along with the other IB laws and the field's interests, may suffice *without the fundamental laws of physics* to limit the relevant range of counterfactual suppositions. The area law would still have held had there been (for example) birds equipped with modest antigravity organs, assisting in takeoffs. The factors affecting species dispersal would then have been unchanged: for example, smaller islands would still have presented smaller targets to off-course birds and so accumulated fewer strays, ceteris paribus. Likewise, the area law would still have held had material bodies consisted of some continuous rigid substance rather than corpuscles. The IB laws's range of stability may in places extend *beyond* the range of stability of the laws of physics; the island-biogeographical laws don't reflect every *detail* of the laws of physics.

This is a crucial point. The IB laws' necessity corresponds to their range of stability. But that range is not wholly contained within the range of stability of the laws of physics (since, as we have just seen, it includes some counterfactual suppositions violating the physical laws). Consequently, the stability of the laws of physics cannot be responsible for the IB laws' stability for IB purposes. The IB laws do not inherit their *necessity* from the physical laws. The kind of necessity characteristic of IB laws is *not possessed* by the physical laws (since the physical laws are not invariant under all of the counterfactual suppositions within the IB laws' range of stability). The approximate *truth* of IB laws might well follow from the physical laws and certain initial conditions that are accidents of physics. The IB laws would then be *reducible* (in an important sense) to physics. Nevertheless, the *lawhood* (as distinct from the *truth*) of IB laws – their stability for IB's purposes – *cannot* follow from physical laws and initial conditions. The IB laws' stability depends on their remaining reliable under certain counterfactual suppositions *violating* physical laws. The physical laws obviously cannot be responsible for the area law's remaining reliable under those counterfactual suppositions.

Hence, if there turned out to be IB laws, IB would have an important kind of *autonomy*. Because the IB laws' lawhood would be irreducible to (various initial conditions and) the lawhood of the fundamental physical laws, IB's nomological explanations (of, for instance, Mauritius's biodiversity) would be irreducible to the explanations of the same phenomena at a more microphysical level.

In other words, there would be two different explanations of why (say) *n* species of land bird currently inhabit Mauritius. One explanation would proceed on the macro level, using IB laws and Mauritius's area, distance from the mainland, and so forth, to explain why there are *n* species rather than many more or fewer. The other explanation would proceed on the micro level, by explaining the fates of various individual creatures that might have migrated to Mauritius and left descendants. (That we could never in practice discover all of these details does not alter the fact that this would be an explanation.) This micro account explains not merely what the macro account explains (why Mauritius is currently inhabited by *n* species rather than *many* more or *far* fewer), but also why Mauritius is currently inhabited by *n* species rather than *one* more or fewer – and, indeed, why Mauritius is inhabited by those particular *n* species rather than a different combination. (Note the differences in contrast classes.) However, it does not follow that the macro account is merely a rough sketch of or promissory note for the micro account. On the contrary, the macro account includes explanatorily relevant information omitted from the micro account, despite its rich detail. In particular, the micro account does not say that Mauritius's biodiversity would have been nearly the same even if, say, the stock of potential migrants (the mainland species of birds) had been very different – indeed, even if some of those species had been made of continuous rigid substance (instead of particles) or had possessed antigravity organs assisting slightly in takeoffs. The IB laws would then still have applied.

As far as IB is concerned, the fact that there are no birds equipped with modest antigravity organs or made of continuous rigid substance is merely an accident of the actual world (like the occurrence of the long-ago storm that deflected a given bird to Mauritius). The macro outcome is insensitive to this accident. The IB explanation of Mauritius's biodiversity uniquely supplies this information.

The MacArthur–Wilson equilibrium theory in IB is typical of many biological models and idealizations. I could just as well have discussed the Hardy–Weinberg law, the logistic equation of population growth, or the Wright/Fisher model of selection. For that matter, I could have discussed macro-level explanations from thermodynamics or economics. All are idealizations that are reliable (for certain purposes), despite including only the "greater causes," and that would still have been reliable under a range of counterfactual suppositions that includes some violations of the laws of fundamental physics.

What would have happened had either some birds possessed modest antigravity organs or the area law been violated? The correct answer is highly context-sensitive. (Compare: What would have happened had Caesar been in command in the Korean War?) In a context concerned with the sorts of things of interest to fundamental physics, the correct answer is that the law of gravity would still have held, and so the area law would have been violated (perhaps because no living things would have evolved). This result does not undermine the IB laws' stability *for IB purposes*, since this context does not matter to IB. Likewise, in a context concerned with the abundance, distribution, and evolution of species living on separated patches of habitat, the correct answer is that the area law would still have held and the law of gravity would not. This result does not undermine the fundamental physical laws' stability *for the purposes of fundamental physics*, since this context is not of interest to fundamental physics. This result does mean, however, that the laws of fundamental physics fail to be stable *simpliciter* (contrary to *NP*°); there is a context where a fundamental physical law would not still have held under a counterfactual supposition *p* that is logically consistent with the fundamental physical laws. These laws are stable for the purposes of fundamental physics just as the area law is stable for IB purposes. It is not the case that the laws of fundamental physics are the *real* laws, whereas the "area law" is a law merely for IB purposes.

6. Evolutionary Accidents as Laws of Functional Biology

Could reliable "The S is T" generalizations form a set that is stable for the purposes of functional biology? If so, then as necessities of functional biology, these generalizations could ground scientific explanations. As with the putative IB laws, these "The S is T" generalizations might exhibit a range of stability under counterfactual suppositions that extends in some respects beyond the range of stability exhibited by the laws of physics. In that case, the explanations supplied by functional biology would be irreducible to explanations in terms of selection operating on organic chemistry.

How, in functional biology, do counterfactuals come to be entertained in the first place? A physician might say that the shooting victim would not have survived even if he had been brought to the hospital sooner, since the bullet punctured his aorta and the human aorta carries all of the body's oxygenated blood from the heart to the systemic circulation. (This fact about the human aorta would still have held, had the victim been brought to the hospital sooner.) Counterfactuals may also arise in connection with functional explanations – e.g., the human trachea has cartilaginous rings in order to make it rigid and so keep it from collapsing between breaths. This explanation depends on the counterfactual "There would be no such rings if they didn't make the trachea rigid." Likewise, that the rings' effect of making the trachea's outer surface white does not explain the rings' presence is bound up with the counterfactual "Were cartilage bright blue instead of white, the human trachea would still have had cartilaginous rings."

Counterfactuals are context-sensitive. Consider "Were cartilage bright blue instead of white." In a context concerned with evolutionary history, it is *incorrect* to say (as we should in a functional biology context) ". . . then the human trachea would still have had cartilaginous rings." For if cartilage had been bright blue instead of white, different selection pressures might have acted upon various creatures of eons past with cartilaginous parts that are visible to predators. Evolutionary history might then have taken a different path, and so human anatomy might have been different; the human trachea might have sported no cartilaginous rings, or the human being might have possessed no trachea at all. Similarly, in a context concerned with molecular structure and the laws of physics, the counterfactual supposition "Were cartilage bright blue instead of white" demands changes of some sort either in the chemical structure of cartilage or in the laws governing light's interaction with molecules. All bets are off as to what the human trachea (if any) would then have been like. In functional biology contexts, though, it is correct to say that were cartilage blue instead of white, the human trachea would still have had cartilaginous rings. The counterfactual supposition, entertained in this context, should not lead us to contemplate how cartilage could have managed to be blue.

In this light, reconsider the argument that "The S is T" expresses an accident of evolution, not a law, since it would not have held had a certain mutation occurred or a certain environmental influence been present. By the argument I have just given, this is not the sort of counterfactual supposition with which functional biology is concerned. Therefore, the failure of reliable "The S is T" generalizations to be preserved under such suppositions does not prevent their forming a set that is stable for the purposes of functional biology.

Rosenberg (2001a, p.158) correctly points out that many different structures could have performed the same function and that natural selection is indifferent between functionally equivalent traits. Therefore, such ecological generalizations as "Allen's rule" – that for any warm-blooded vertebrate species, individuals in cooler climates usually have shorter protruding body parts – are not laws of evolution since selection could instead have prevented heat dissipation by, e.g., thicker fur or feathers (Hull, 1974, p.79; Beatty, 1995, p.57). We might want to explain why the buckeye butterfly has eyespots rather than (say) tasting foul to birds. To answer this why-question, it does not suffice to say "The eyespot discourages predation by birds." Rather, we would need to discover why this particular defense mechanism evolved rather than another – e.g., that eyespots required only a few mutations of a gene already existing for other reasons. Likewise, IB laws fail to explain why one particular combination of n species rather than some other inhabits Mauritius. Nevertheless, IB laws explain why n species rather than many more or far fewer inhabit Mauritius. Likewise, that the eyespot discourages predation explains why the butterfly has the eyespot rather than having no eyespot but otherwise being exactly as it actually is. Indeed, it is only in light of this explanation that it makes sense to ask why the butterfly employs this particular defense mechanism instead of some other.

Had the buckeye butterfly tasted foul to birds, then it might not have sported eyespots. Here we have a counterfactual supposition of interest to functional biology, but under which a reliable "The S is T" generalization is not preserved. However, this result does not undermine the stability (for functional biology) of the set of reliable "The S is T" generalizations. That is because the counterfactual supposition "Had the buckeye butterfly tasted foul to birds" is itself logically inconsistent with some member of the set (namely, that the buckeye butterfly does not taste foul to birds). (Recall the Lorentz force law example.)

Although "The S is T" is an accident rather than a law of evolutionary biology, it can possess *necessity* in functional biology by virtue of belonging to a stable set there. (Brandon, 1997, p.S456 and Schaffner, 1993, pp.121–2; 1995, p.100 appear to be after roughly the same idea in referring to "historical accidentality . . . 'frozen into' a kind of quasi-nomic universality" and thus able, certain contexts, to support counterfactuals in the manner of law.) Take the explanation that the vulture has no feathers on its head and neck because the vulture feeds by sticking its head and neck deep inside the bodies of carrion, so any feathers there would become matted and dirty. This explanation is independent of the details of the laws of physics. Putnam uses a similar example to defend the irreducibility of macro explanations: why a cubical peg, 15/16" on a side, cannot fit into a round hole 1" in diameter. Putnam writes:

> The explanation is that the board is rigid, the peg is rigid, and as a matter of geometric fact, the round hole is smaller than the peg. . . . That is a correct explanation whether the peg consists of molecules, or continuous rigid substance, or whatever. (1975, p.296)

A peg (or vulture) made of continuous rigid substance would violate laws of physics. But the same functional explanation would apply to it. That distinctive range of invariance reflects the irreducibility of this kind of explanation to anything that could be supplied, even in principle, by the laws of physics.

References

Achinstein, P. (1965). "Defeasible" problems. *The Journal of Philosophy*, 62(11, November), 629–33.

Anscombe, G. E. M. (1958). Modern moral philosophy. *Philosophy*, 33(1, January), 1–19.

Armstrong, D. (1983). *What is a law of nature?* New York: Cambridge University Press.

Beatty, J. (1981). What's wrong with the received view of evolutionary theory? In P. D. Asquith & R. N. Giere (Eds). *PSA 1980* (vol. 2, pp. 397–426). East Lansing: Philosophy of Science Association.

Beatty, J. (1995). The evolutionary contingency thesis. In G. Wolters & J. Lennox (Eds). *Theories and Rationality in the Biological Sciences* (pp. 45–81). Pittsburgh: University of Pittsburgh Press.

Brandon, R. (1997). Does biology have laws? The experimental evidence. *Philosophy of Science*, 64(Proc.), S444–57.

Brown, J. H. (1995). *Macroecology*. Chicago: University of Chicago Press.

Carlson, G. N., & Pelletier, F. J. (Eds). (1995). *The generic book*. Chicago: University of Chicago Press.

Cartwright, N. (1983). *How the laws of physics lie*. Oxford: Clarendon.

Clark, R. (1965). On what is naturally necessary. *The Journal of Philosophy*, 62(11, November), 613–25.

Dirac, A. M. (1938). A new basis for cosmology. *Proceedings of the Royal Society (London) Series A*, 165(921), 199–208.

Earman, J. C., Glymour, C., & Mitchell, S. (Eds). (2002). *Erkenntnis*, 57(3), Special Issue on ceteris-paribus laws.

Futuyma, D. (1979). *Evolutionary biology*. Sunderland, MA: Sinauer.
Gambino, P. (1993). Antibiotic activity of larval saliva of *Vespula* wasps. *Journal of Invertebrate Pathology*, LXI(1, January), 110–11.
Ghiselin, M. (1989). Individuality, history and laws of nature in biology. In M. Ruse (Ed.). *What the philosophy of biology is* (pp. 53–66). Dordrecht: Kluwer.
Giere, R. (1995). The skeptical perspective: science without laws of nature. In F. Weinert (Ed.). *Laws of nature* (pp.120–38). Berlin: deGruyter.
Goodman, N. (1983). *Fact, fiction, and forecast* (4th edn). Cambridge, MA: Harvard University Press.
Gould, S. J. (1989). *Wonderful life*. New York: Norton.
Haavelmo, T. (1944). The probability approach to econometrics. *Econometrica*, 12(Suppl.), 1–117.
Hempel, C. G. (1966). *Philosophy of natural science*. Englewood Cliffs, NJ: Prentice-Hall.
Hacking, I. (1991). A tradition of natural kinds. *Philosophical Studies*, 61(1, February), 109–26.
Hull, D. (1974). *Philosophy of biological science*. Englewood Cliffs, NJ: Prentice-Hall.
Hull, D. (1978). A matter of individuality. *Philosophy of Science*, 45(3, September), 335–60.
Lange, M. (2000). *Natural laws in scientific practice*. New York: Oxford University Press.
Lloyd, E. (1988). *The structure and confirmation of evolutionary theory*. Westport, CT: Greenwood Press.
MacArthur, R. (1972). *Geographic ecology*. Princeton: Princeton University Press.
Mayr, E. (1965). *Animal species and evolution*. Cambridge: Harvard University Press.
McCreedy, C. D., & Weeks, H. P., Jr. (1993). Growth of cottontail rabbits in response to ancillary sodium. *Journal of Mammalogy*, LXXIV(1, February), 217–24.
Mill, J. S. (1961). *A system of logic*. London: Longmans Green.
Molnar, G. (1967). Defeasible propositions. *Australasian Journal of Philosophy*, 45(August), 185–97.
Nash, S., & Domjan, M. (1991). Learning to discriminate the sex of conspecifics in male Japanese quail. *Journal of Experimental Psychology: Animal Behavior Processes*, XVII(3, July), 342–53.
Putnam, H. (1975). Philosophy and our mental life. In H. Putnam, *Mind, language, and reality: philosophical papers* (vol. 2, pp. 291–303). Cambridge: Cambridge University Press.
Reichenbach, H. (1954). *Nomological statements and admissible operations*. Dordrecht: North-Holland Press.
Rosenberg, A. (1985). *The structure of biological science*. New York: Cambridge University Press.
Rosenberg, A. (1987). Why does the nature of species matter? *Biology and Philosophy*, 2(2, April), 192–97.
Rosenberg, A. (2001a). Reductionism in a historical science. *Philosophy of Science*, 68(2, June), 135–63.
Rosenberg, A. (2001b). How is biological explanation possible? *British Journal for the Philosophy of Science* 52: 4 (December), 735–60.
Rosenberg, A. (2001c: On multiple realization and the special sciences. *Journal of Philosophy*, 98(7, July), 365–73.
Ruse, M. (1970). Are there laws in biology? *Australasian Journal of Philosophy*, 48(August), 234–46.
Schaffner, K. (1993). *Discovery and explanation in biology and medicine*. Chicago: University of Chicago Press.
Schaffner, K. (1995). Comments on Beatty. In G. Wolters, & J. Lennox (Eds). *Theories and rationality in the biological sciences* (pp. 99–106). Pittsburgh: University of Pittsburgh Press.
Smart, J. J. C. (1963). *Philosophy and scientific realism*. New York: Routledge.

Sober, E. (1997). Two outbreaks of lawlessness in recent philosophy of biology. *Philosophy of Science*, 64(Proc.), S458–67.
Sober, E. (2000). *Philosophy of Biology* (2nd edn). Boulder, CO: Westview Press.
Thompson, P. (1989). *The structure of biological theories*. Albany, NY: SUNY Press.
Tooley, M. (1977). The nature of laws. *Canadian Journal of Philosophy*, 7(December), 667–98.
van Fraassen, B. C. (1989). *Laws and symmetry*. Oxford: Clarendon.
Waters, C. K. (1998). Causal regularities in the biological world of contingent distributions. *Biology and Philosophy*, 13(1, January), 5–36.
Young, R. E., Pearce, J., & Govind, C. K. (1994). Establishment and maintenance of claw bilateral asymmetry in snapping shrimps. *The Journal of Experimental Zoology*, CCLXIX(4, July), 319–26.

Further Reading

Bennett, J. (2003). *A philosophical guide to conditionals*. Oxford: Oxford University Press.
Boyd, R. (1999). Homeostasis, species, and higher taxa. In R. Wilson (Ed.). *Species* (pp. 141–86). Cambridge, MA: Bradford.
Carroll, J. W. (1994). *Laws of nature*. Cambridge: Cambridge University Press.
Carroll, J. W. (Ed.). (2004). *Readings on laws of nature*. Pittsburgh: University of Pittsburgh Press.
Dupré, J. (1981). Natural kinds and biological taxa. *The Philosophical Review*, 90(1, January), 66–90.
Fodor, J. (1974). Special sciences, or: the disunity of science as a working hypothesis. *Synthese*, 28(October), 97–115.
Garfinkel, A. (1981). *Forms of explanation*. New Haven: Yale University Press.
Sober, E. (1980). Evolution, population thinking, and essentialism. *Philosophy of Science*, 47(3, September), 350–83.
Wilkerson, T. E. (1995). *Natural kinds*. Aldershot: Avebury.

Chapter 27

Models

JAY ODENBAUGH

Few terms are used in popular and scientific discourse more promiscuously than "model." (Nelson Goodman, 1976, p.171)

1. Introduction

Philosophical discussions of models and modeling in the biological sciences have exploded in the past few decades. Given three-dimensional models of DNA in molecular genetics, individual-based computer models in population ecology, statistical models in paleontology, diffusion models in population genetics, and remnant models in taxonomy, we clearly should have a philosophical account of such models and their relation to the world. In this essay, I provide a critical survey of the accounts of models provided by philosophers of science and biology including models as analogies, relational structures, partially independent representations, and material objects. However, there is much, much more work to be done.

To understand the importance of models philosophically, we must begin at the proverbial beginning with the "received" view of theories. This Syntactic View has almost no need for talk of models except in the thinnest sense. However, it is here that philosophers began to see the need for some notion of a "model."

2. The Received (Syntactic) View of Theories

Philosophical discussions of models began as a response to the view of theories articulated by the logical empiricists (see Hempel, 1967, pp.182–5 for example). On their account, theories are axiomatic systems given in a formal language. The axioms express purported laws of nature that are true of every object, do not refer to any particular objects, and are necessarily true. From these axioms, in conjunction with particular premises, one can deduce theorems, which may be a description of some particular event or a less general law.

As a formal syntactical system, the theory is just an array of symbols with operations defined on them. If the axiomatic system is to be meaningful as a scientific theory, one must also specify its semantics. The semantics is provided by an interpretation given to the symbols and expressions of the language. Logical empiricists were particularly concerned with the meanings of theoretical terms. How could terms like "electron," "belief," and "gene" be empirically meaningful if their meaning was not directly tied to observation?

The logical empiricists articulated the notion of a *correspondence rule* that would completely or partially define theoretical predicates in terms of observable entities and these would be the extensions given to those predicates. For example, consider the following Hempelian classic, "For all objects x and times t, if x is struck at t, then x breaks at t if and only if x is fragile" (Hempel, 1967, p.109). So, we define the theoretical term "fragility" partially in terms that denote the actual, observable behavior of struck objects. A large literature arose debating the notion of a correspondence rule and whether theories are axiomatic systems (see Suppe, 1977).

The syntactic view of theories does not entail that textbooks, monographs, issues of *Nature*, etc. should include axiomatic systems when they contain scientific theories. This view of theories is consistent with the fact that most theories are not pre-packaged as sets of deductively closed axioms. For example, just because population genetics does not come axiomatized does not show that it could not be (for attempts, see Woodger, 1929, 1952; Williams, 1970, 1973; Ruse, 1973).

The foremost problem with the Received View is that it is simply too distant from scientists' work and affords few insights into how and why theorizing occurs in science (van Fraassen, 1980, p.53–6). The Received View inspired a cottage industry of technical problems which philosophers were eager to solve. However, these problems shed minimal light on the original questions concerning the nature of scientific theories. van Fraassen writes,

> Perhaps the worst consequence of the syntactic approach was the way it focused attention on philosophically irrelevant technical questions. It is hard not to conclude that those discussions of axiomatizability in restricted vocabularies, "theoretical terms," Craig's theorem, "reduction sentences," "empirical languages," Ramsey and Carnap sentences, were one and all off the mark – solutions to purely self-generated problems, and philosophically irrelevant. (1980, p.56)

One of the first critical responses to the Received View invoking the notion of a model was the work of philosopher Mary Hesse (1966; see Achinstein, 1968 as well).

3. Models and Analogies

As we have seen, on the Received View theories are composed of a formal language, a set of axioms, and a set of correspondence rules. According to philosophers like Richard Braithwaite (1962) and Ernest Nagel (1961), partial interpretations *are* models. Here is an example. The following set of uninterpreted formulas might constitute a part of a formal calculus for the kinetic theory of gases (Achinstein, 1968, pp.227–8; Suppes, 1957).

(1) The set P is finite and nonempty,
(2) The set T is an interval of real numbers,
(3) For p in P, s_p is twice differentiable on T,
(4) For p in P, $m(p)$ is a positive real number,
(5) For p and q in P and t in T, $f(p, q, t) = -f(q, p, t)$
(6) For p and q in P and t in T, $f(p,q,t) = -s(p,t) \times f(p,q,t) = -s(q,t) \times f(q,p,t)$
(7) $m(p)D^2 s_p(t) = \sum_{p \in P} f(p,q,t) + g(p,t)$

We can informally interpret the formalism as follows. P designates a class of molecules in a gas, T is a set of elapsed times, s_p is the position of molecule p, $m(p)$ is the mass of p, $f(p, q, t)$ is the force that p exerts on q at time t, and $g(p, t)$ is the resulting external force acting on p at t.

Mary Hesse (1966) argued following N. R. Campbell (1920) that such interpretations would be importantly incomplete since we have ignored much-needed *analogies*. For example, there exists an extremely fruitful analogy between particles in a gas and a set of billiard balls. If we let P designate a set of perfectly elastic billiard balls in a box and we do not change the rest of the interpretation, then one can reinterpret the axioms in more familiar terms. Thus, (5) under the two interpretations are as follows:

(5′) The force exerted by a molecule p on molecule q at time t is equal in magnitude and opposite in direction to that exerted by q on p at time t.
(5″) The force exerted by a billiard ball p on a billiard ball q at time t is equal in magnitude and opposite in direction to that exerted by q on p at time t.

We can use an analogy between the unfamiliar particles and familiar billiard balls for understanding our kinetic theory of gases.

Obviously particles in a gas and billiard balls in a box are not identical. There are properties they knowably share and ones that they do not; moreover, there are properties that they *may* share *unbeknownst* to us. Hesse names these properties *positive analogies*, *negative analogies*, and *neutral analogies* respectively. As a positive analogy, both particles and billiard balls have mass and velocity and obey a principle of conservation of momentum. Of course, particles in a gas do not have numbers written on them though billiard balls do; thus this is a negative analogy. Neutral properties are those that we do not know whether they are shared or not.

Hesse distinguishes between two senses of the term "model." What she calls "model_1" is "the imperfect copy (the billiard balls) *minus the known negative analogy*, . . ." (1966, p.9). She then writes, "Since I shall also want to talk about the second object or copy that includes the negative analogy, let us agree as a shorthand expression to call this 'model_2'" (1966, p.10). Thus, there are two types of models, model_1 and model_2.

Many philosophers of science recognized that scientists reason with models in Hesse's senses. However, following Pierre Duhem (1954) they argued that they were dispensable. In fact, philosophers like Rudolf Carnap would argue that models *should* be dispensed with.

> It is important to realize that the discovery of a model has no more than an aesthetic or didactic or at best heuristic value, but it is not at all essential for a successful application of a physical theory. (1939, p.68)

On Hesse's view (and Campbell's), it was essential to interpret a theory's axioms in terms that were familiar via an analogy. There are essentially two reasons though it will be the second reason that plays the largest part of her rationale. Models are necessary for explanation and for novel prediction.

First, according to Campbell, if a theory is to explain some phenomena, then it must produce understanding in the scientist. The only way to produce such understanding is to provide a model – that is familiar. As Campbell writes,

> The behaviour of moving solid bodies is familiar to every one; every one knows roughly what will happen when such bodies collide with each other or with a solid wall . . . Movement is just the most familiar thing in the world . . . And so by tracing a relation between the unfamiliar changes which gases undergo when their temperature or volume is altered, and the extremely familiar changes which accompany the motions and mutual reactions of solid bodies, we are rendering the former more intelligible; we are explaining them. (Campbell, 1920, p.84)

Hence, if a theory explains some phenomena, thus we must provide models for our theories. However, this argument seems problematic. First, it is not clear what this notion of "understanding" is nor that it is necessary for explanation (Salmon, 1984). Consider the case of mid-1920s quantum mechanics. Given the lack of defensible hidden-variable interpretations or classical models, there was nothing "familiar" in which to interpret it. However, the theory seems to explain a large number of phenomena.

Second, Hesse turns to models and novel predictions. She characterizes theories as either *strongly* or *weakly falsifiable*. A theory is strongly falsifiable if that theory makes *novel* predictions. A theory is weakly falsifiable if it only *accommodates* phenomena. On Hesse's view, theory makes novel predictions by employing models via their neutral analogies though of course this does not guarantee the novel prediction will be confirmed; but the use of models is necessary for novel predictions to occur. As before, quantum mechanical theory has very few if any models in Hesse's sense, but has made novel predictions and been strikingly confirmed. Thus, strongly falsifiable theories need not have Hesse models.

Nonetheless, something like Hesse's approach can be found in areas of biology. For example, there are many analogies between biological systems and physical systems. In modeling predatory–prey systems, we use analogies from statistical mechanics involving laws of mass action – predator and prey interact in proportion to their abundances as would molecules in an ideal gas. Similarly, the diffusion of dye particles due to Brownian motion is analogous to a set of populations at an initial gene frequency p "diffusing" away from that value due to random genetic drift (Roughgarden, 1996, p.69). Lastly, evolutionary biologists have borrowed heavily from microeconomics and created evolutionary game theory where the concept of *fitness* is analogized with *utility* (Maynard Smith, 1983). We even see areas of biology borrowing from other areas. Paleontologist Jack Sepkowski (1976, 1978) argued that paleontological phenomena like speciation and extinction of higher taxa (orders in particular) are very similar to the species' colonization and extinction in archipelagos studied in MacArthur and Wilson's theory of island biogeography (1967). Thus, Hesse's general approach has important applications to the biological sciences. Let's now turn to a more popular alternative to the Received View, the Semantic View of theories.

4. The Semantic View of Theories

The Semantic View of theories is probably the most popular approach to theories and models amongst philosophers of science (van Fraassen, 1980; Suppe, 1989). It has also been endorsed by philosophers of biology (Beatty, 1981; Lloyd, 1988; Thompson, 1988). The Semantic View comes in at least two different varieties. On more "conservative" versions of the Semantic View, the notion of a model is a formal semantic one and the relation between models and empirical systems consists in isomorphisms (van Fraassen, 1980; Lloyd, 1988). On a "liberal" Semantic View, the notion of a model is simply an idealized, abstract structure and the relation between models and empirical systems is similarity (Giere, 1988, 1999).

The Semantic View was developed explicitly as an attempt to address the problems plaguing the Received View. The impetus for the Semantic View comes from model theory. In formal semantics, a model for a set of sentences is an interpretation in which all of the sentences are true. However, in the semantics of formal languages, there are two different ways to construe what a model is. First, models are an interpretation function which assigns objects to names, sets of objects to predicates, and *n*-tuples of objects to relations such that the relevant set of sentences are true. Second, models are a set of objects making the sentences characterizing the theory true (Lloyd, 1988, preface).

It is now apparent why the Semantic View is called the *Semantic* View. Its proponents claim that to understand scientific theories we should primarily focus on their semantics. The relationship of interest is that of satisfaction – models make the theory expressed as a set of sentences true. As the Semantic View has developed, the metalogical notion of models has been questioned (Griesemer, 1990; Downes, 1992; Giere, 1988, 1999). Ironically, the critics have argued that the metalogical concept of a model is excessively removed from the notion found in the sciences. What is the relation between models and empirical systems on the Semantic View?

On the Semantic View of theories, theories are a "family of models." Ronald Giere writes,

> My preferred suggestion, then, is that we understand a theory as comprising two elements: (1) a population of models, and (2) various hypotheses linking those models with systems in the real world. (1988, p.85)

Proponents call these two components of scientific theories the *theoretical definition* and the *theoretical hypothesis*, respectively. But, what are the particular structures of a theory, or "theoretical definition," and what is the relation between these structures and empirical systems?

The most popular answer to the first question is that the models are state spaces.[1] A state space consists in the set of all the possible values of the variables. If each variable is given a determinate value, then there is a particular point in that space which is the *state* of the system. We can represent this state of the system at t as a vector $\mathbf{x}_t$ and its dynamics as a sequence of such states or vectors. Likewise, there are also parameters that mediate the relationships between variables.

There are laws that govern how the system moves in the space. These laws of succession and coexistence are either deterministic or stochastic. For deterministic laws encoded in either differential or difference equations, there is a sequence of states $\langle \mathbf{x}_1, \mathbf{x}_2, \ldots, \mathbf{x}_n \rangle$ such that for each state $\mathbf{x}_i$, there is a single state $\mathbf{x}_j$ such that the system moves from $\mathbf{x}_i$ to $\mathbf{x}_j$. For stochastic laws, there is a sequence of states such that each state has a probability of moving to another state, or remaining in the same state, in the space. There are also laws of coexistence which that determine what regions of the state space the system can occupy. For example, the ideal gas law $PV = nRT$ is a law of coexistence where an ideal gas can only occupy the subspace where the equality is satisfied. It should be apparent that the laws governing a state space need not be metaphysical laws of nature.

The idea of defining a model as a state space should resonate with theoretical biologists. Many of the classic models in theoretical biology are construed in just this way. For example, the Lotka–Volterra interspecific competition models are explicated as state spaces or as "phase portraits." The model for two species is described by the following coupled differential equations.

$$\frac{dN_1}{dt} = \left(\frac{K_1 - N_1 - \alpha_{12} N_2}{K_1} \right)$$
$$\frac{dN_2}{dt} = \left(\frac{K_2 - N_2 - \alpha_{21} N_1}{K_2} \right)$$

The state space is of two dimensions with two variables describing population densities N_1 and N_2. In fact, this space can be depicted as a Cartesian coordinate system where N_1 is the abscissa and N_2 is the ordinate. The parameters of the model are the intrinsic rates of growth r_1 and r_2 of the two populations, and the competition coefficients α_{ij} which describe the per capita effect of an individual of species j on species i. The differential equations are the deterministic laws of succession for the system. Thus, points in the state or phase space represent the joint densities of the populations at a particular time. At equilibrium $dN_i^*/dt = 0$, we have the following laws of coexistence:

$$N_1^* = K_1 - \alpha_{12} N_2^*$$
$$N_2^* = K_2 - \alpha_{21} N_1^*$$

1 One important alternative articulated by Patrick Suppes is that the models are set-theoretic structures defined by set-theoretic predicates. For example, a model of the kinetic theory of gas, mentioned in Section III, can be construed as a structure that is in the extension of the set-theoretic predicate $\langle P, T, s, m, f, g \rangle$. In this essay, I shall focus on the state-space approach for convenience.

The joint values of N_1 and N_2 that satisfy both equations are regions of the state space that the system can occupy at equilibrium (otherwise known as "isoclines").

There are two problems with this notion of model structure. First, if *some* models are not mathematical and state spaces are pieces of mathematics, then some models are not state spaces. Second, biologists and philosophers often agree that models are state spaces, as in our example. However, this agreement may implicitly disguise a substantive disagreement. The metalogical concept of a model on the Semantic View may be distinct from the concept possessed by biologists. We should not confuse the sameness of the particular structures considered to be models with the sameness of the concept of models.

On the Semantic View, a sharp separation is made between theoretical definitions – abstract entities – and theoretical hypotheses – claims about the relationship between models and empirical systems. There is controversy over exactly what this relationship is. The more conservative Semantic View claims that the canonical relation between model and world is that of an isomorphism (see van Fraassen, 1971, pp.107–8, 125–6]).[2]

Isomorphism is a very demanding relation to posit between two structures.[3] A case in point is if the Lotka–Volterra interspecific competition model above is isomorphic to some competing species, then their densities must respond instantaneously to one another, the real-world surrogates of the competition coefficients and carrying capacities must be constant in value, and the density-dependence linear. All of these assumptions are false in many if not all pairs of competing species. If this is true, then there can be no isomorphism between real competing populations qua relational structures and their mathematical counterparts.

There is a way of dealing with this worry suggested by proponents of the conservative Semantic View. First, distinguish those elements of the model that are idealized and those that are not. Second, claim there is an isomorphism, or homomorphism, between the non-idealized elements of the model and the relevant empirical objects, processes, and events. This essentially requires only a partial mapping of structure to structure. van Fraassen (1980) provides one way of doing this (see Suppes, 1962, for a very different way). In essence, he suggests that empirical adequacy consists in a homomorphism, or in his terminology, an embedding between the empirical substructures of the model and the observable parts of the empirical systems. However, if the idealizations concern observables as well, then our model cannot be mapped onto the observational relational structure. For example, the parameters r_i and α_{ij} do not *always* concern unobservables – we can observe some organisms reproduction and competitive interactions.

Ronald Giere's more liberal account of scientific theories and models begins from a different starting point than the Received View or conservative versions Semantic View.

2 An *isomorphism* is a bijective map f such that both f and its inverse f^{-1} are homomorphisms.

3 Isomorphism is an equivalence relation between relational structures ⟨objects, properties⟩. Hence, no isomorphism exists between models and empirical systems since empirical systems are not relational structures. However, if we characterize the empirical system as a relational structure (i.e., model of the data), then such an isomorphism may exist (see Suppes, 1962).

Giere thinks that both of these views are too removed from scientific practice. According to Giere, theories are composed of abstract, idealized models. Moreover, not all models are state spaces or necessarily are mathematical even though some are.

The second part of Giere's account of theories concerns theoretical hypotheses or how models relate to the world. His account does not suggest models are isomorphic or even homomorphic to empirical systems; rather, idealized models are related to the empirical systems through *similarity*. Theoretical hypotheses are propositions of the form, "model M is similar to some designated system S in certain respects and to certain degrees." Moreover, the hypothesis is true just in case M bears that relation to S and false if M does not.

Similarity is a troubling notion. Any two objects x and y will necessarily be similar in *some* respects and to *some* degrees. For a theoretical hypothesis to be nontrivial and non-vacuous, then we must specify the relevant respects and degrees of similarity. As an example, Giere claims,

> The positions and velocities of the Earth and moon in the Earth-moon system are very close to those of a two-particle Newtonian model with an inverse square central force. (1988, p.81)

A more colloquial way of putting this is "The Earth and moon form, to a very high degree of approximation, a two-particle Newtonian gravitational system" (1988, ibid.).[4] Let us now consider the Semantic View's reception by philosophers of biology with respect to evolutionary theory and models. Many argue that it captures the structure of evolutionary theory via models in a powerful and extremely useful way.

Many philosophers of biology (Beatty, 1980, 1981; Lloyd, 1988; and Thompson, 1998) have offered arguments to show how the Received View cannot make sense of biological theories – specifically, evolutionary theory – and how the Semantic View can. Two of the most significant arguments are as follows.[5] First, according to the Received View, a theory is scientific only if it contains laws, but evolutionary theory has no laws. However, the Semantic View does not insist that a theory is scientific only if it has laws. Hence, the Semantic View makes sense of how evolutionary theory is scientific. Second, evolutionary theorists often devise models independently of whether those models are isomorphic to, or fit, empirical phenomena. The Semantic View distinguishes between theory (theoretical models) and empirical applications (theoretical hypotheses) and the Received View cannot. Hence, the Semantic View makes better sense of evolutionary theory than the Received View. Let's consider each argument in turn.

---

4 However, this raises problems for the Semantic View. Theories have various epistemic and semantic properties. They can be the object of propositional attitudes, can be confirmed, and have truth-values. On the Semantic View, theories are pairs of the form ⟨relational structures, propositions⟩. Set-theoretical structures, however, cannot be believed, confirmed, or truth-valued. Hence, theories are not set-theoretical structures. One could adopt a "model-based" propositional view of theories where theories are sets of propositions of the form, "Such-and-such real system S is similar to a designated model M in indicated respects and degrees." This allows us to take seriously our normal scientific practice and grant the fundamental importance of models.

5 My presentation and criticisms of these arguments is indebted to Marc Ereshefsky (1991).

John Beatty has been a forceful proponent of the claim that evolutionary theory lacks laws. A law of nature in his view is a universal or statistical generalization that supports counterfactuals, makes no essential reference to particular objects, and is necessarily true. However, generalizations of the form, "All members of species *S* have trait *T*" fail to be laws. First, they mention particular objects – species (especially if species are individuals). Second, a law of this form would not have counterfactual force. For any trait, evolutionary process like selection, mutation, and drift can eliminate such characters from the species. Beatty writes, "In short, there can be no law of nature to the effect that a genetically based trait is universal within a species or among all species" (1981, p.407).

Philosophers of science like J. J. C. Smart (1966) have drawn the conclusion that evolutionary theory simply is not scientific given its lawlessness. However, Beatty argues *given that* evolutionary theory is scientific, the Semantic View makes much better sense of evolutionary theory and its lack of laws than the Received View. The laws on the Semantic View are laws of succession and coexistence which govern the behavior of abstract objects, not necessarily empirical phenomena. Hence, there need not be any biological laws per se.

This argument can be criticized in at least two ways. First, there may be no laws concerning specific taxa (see Lange, 1995, though); however, this does not entail that there are no evolutionary laws.[6] For example, laws may exist over *evolutionary functional kinds* such as host and parasitoid, predator and prey, *r*-selected and *K*-selected species. Second, suppose that a mathematical model is isomorphic to some empirical phenomena (i.e., data model). Thus, just as the mathematical structure will satisfy the laws of succession and coexistence of the model, so will the empirical phenomena. Hence, the laws will be true of the empirical system of interest. Thus, the claim that evolutionary theory semantically construed has no "metaphysical" laws is false.

One might charge that biological models are idealized and so the mathematical models will not be isomorphic to the empirical system. Thus, the above criticism is moot. However, this will not allow us to dodge the charge that evolutionary theory has ceteris paribus laws – if the system of interest meets the boundary and idealizing conditions, then it would behave in such-and-such a way. For example, if a population at a locus with two alleles *A* and *a* is subject to no mutation, is extremely large, has no migration, etc. then its allele frequencies will change in accordance with the following law of succession:

$$p' = \frac{p^2 w_{AA} + 2pq w_{Aa}}{\bar{w}}$$

Beatty argues against the first criticism by claiming that even evolutionary laws over functional kinds will be false. For example, consider Mendel's first law – we now know of instances where it is false – in cases of meiotic drive. The truth of a law depends on how widespread such cases are. Hence, Mendel's first law is not necessarily true.

6 See Brandon, Beatty, Sober, and Mitchell in *Philosophy of Science*, 1997, Issue 64 for a discussion of evolutionary laws.

However, this argument seems to confuse the boundary and ceteris paribus conditions being satisfied and the law being true. Let's now turn to the second argument for the Semantic View and evolutionary theory.

Biologists often develop evolutionary models and only then attempt to determine if the models have empirical applications. For example, there are a large number of theoretical models to explain the existence of sexual reproduction (Williams, 1975; Maynard Smith, 1978). Likewise, there are models that describe natural selection operating at a variety of units or levels – the allele, gene, chromosome, individual, trait-group, deme, species, ecological communities, etc. (Brandon & Burian, 1984). In both cases, the models are developed independently of empirical application. The Semantic View makes a distinction between theoretical models and hypotheses – it preserves this "division of labor"; however, the Received View does not. Thus, the Semantic View makes better sense of biological practice than does the Received View.

However, the issues are not so clear. Consider the following schematic correspondence rule (where C represents a condition, T a theoretical term, and O an observational term) $\forall x[Cx \rightarrow (Tx \equiv Ox)]$. Correspondence rules do provide *empirical interpretations* but they need not provide *empirical applications*. They tell us, according to the theory, what *would* happen if the relevant set of conditions is met. Of course, whether things are as the theory says is settled by testing the theory. So, both the Received View and the Semantic View can preserve the difference between a theory and a theory's empirical application. As Paul Thompson writes,

> The relationship of a model to phenomena is one of isomorphism, and the establishment of the isomorphism is a complex task not specified by the theory. If the asserted isomorphism is not established, it may be that the theory has no empirical *application*. The theory will nonetheless be empirically *meaningful* . . . in that one knows from the theory what the structure and the behavior of the phenomena would be if the phenomena were isomorphic to the theory. (1989, p.72)

Though the Semantic View has been especially popular amongst philosophers of biology, some philosophers recently have argued that it is deeply flawed. One such group of philosophers is Nancy Cartwright, Margaret Morrison, and Mary Morgan. Philosophers of biology have not picked up on this trend like philosophers of physics and economics have, but it is important to understand their views.

5. Models as Mediators

The "models as mediators" group has provided important criticisms of the Semantic View (Cartwright et al., 1995; Morgan & Morrison, 1999). This new program is not in the business of providing a "theory of models." Rather, it is an attempt to understand how models are constructed and function as they do in mediating between theory and phenomena. Margaret Morrison and Mary S. Morgan write,

> Although we want to argue for some general claims about models – their autonomy and role as mediating instruments, we do not see ourselves as providing a "theory" of models. The latter would provide well-defined criteria for identifying something as model and differentiating models from theories. (1999, p.12)

Nonetheless, the models as mediators group do seem to provide an *implicit* or *functional* characterization of what models are.

> A model is that which is constructed and functions as a representation that allows one to learn about theory and phenomena in a way that is partially independent (autonomous) from both.

In essence, models are *technologies*. They are devices that allow one to connect abstract theory and the phenomena of interest. This approach is a form of instrumentalism – though not of the sort philosophers typically discuss. As Cartwright, Shomar, and Suárez write,

> I want to urge instead an instrumentalist account of science, with theory as one small component. Our scientific understanding and its corresponding image of the world is encoded as much in our instruments, our mathematical techniques, our methods of approximation, the shape of our laboratories, and the pattern of industrial developments as in our scientific theories. My claims is that these bits of understanding so encoded should not be viewed as claims about the nature and structure of reality which ought to have a proper propositional expression that is a candidate for truth or falsehood. Rather they should be viewed as adaptable tools in a common scientific tool box. (1995, p.138)

Why should we believe that models are "mediators" between theory and world? Why should we accept at least the "partial independence" of models from both? One common argument is that theory rarely *applies directly* to phenomena of interest. The conceptual resources of the theory are simply too abstract to characterize actual empirical systems. The only way in which this can be done is through something that mediates between the two – namely, a model. However, models must be at least partially independent of the theory and phenomena. If they are not, then we run into the problem of theory-ladenness *and* phenomena-ladenness; the model will be theory or phenomena laden and hence objectivity is lost. Thus, models must mediate between theory and world and do so in a way that makes them at least partially independent.

Most of the work of the proponents of the models as mediators approach consists in case studies demonstrating how models work and why they are needed (see Morgan & Morrison, 1999, for several case studies). They have also been severe critics of the Semantic View. Cartwright, Shomar, and Suárez write that according to the Semantic View,

> Theories have a belly-full of tiny already formed models buried within them. It takes only the midwife of deduction to bring them forth. On the Semantic View, theories are just collections of models; this view offers then a modern Japanese-style automated version of the covering-law account that does away even with the midwife. (Cartwright et al., 1995, p.139)

They go on to argue that "theories plus auxiliaries do not imply data – or better following Matthias Kaiser's advice in this volume, "phenomena" – even in principle" (1995, p.139). The charge is that the Semantic View is incapable of accommodating the independence and autonomy of models in science. On the Semantic View, theories are in part families of models. Thus, any model must be a member of the family or "derivable" from such a family. However, there are models – often phenomenological models – that are not derivable from theory. Hence, then the Semantic View is false.

The models as mediators group have offered several examples of these sorts of cases. In 1934 Fritz and Heinz London provided a model of superconductivity (Cartwright, Tomar, & Suárez, 1995). Mercury when cooled below 4.2°K will have its electrical resistance drop to near zero so long as it is not in the presence of a strong magnetic field. It turns out that there is a critical phase transition for particular temperatures where it becomes superconductor. One phenomenon of superconductivity that needed to be accounted for is called the *Meissner effect* which occurs when there is a sudden expulsion of magnetic flux from a superconductor when it is cooled below its transition temperature.

The traditional approach was to devise an "acceleration equation" from classical electromagnetic theory. However, London and London realized that the traditional theory-driven account could not account for the Meissner effect. They arrived at model that was independent of (and incompatible with) classical electromagnetic theory, which could account for the Meissner effect. The new model was not simply some "de-idealized" version of the classical theory, nor was it built from "theoretical grounds" provided by the theory. Thus, this episode appears to speak against the Semantic View.

Of course, proponents of the Semantic View have responded to this example. Newton C. A. de Costa and Stephen French (2003) reply,

> Let us suppose it is true that models exist that are developed in a manner that is in some way independent of theory. Still, they can be represented in terms of structures that satisfy certain Suppes predicates . . . Whether a model is obtained deductively from theory or by reflecting on experiment, it can be brought under the wings of the Semantic Approach by representing it in structural terms. And there is a general point here: Surely no one in their right minds would argue that all model development in the sciences proceeds deductively! (2004, p.55)

Thus, the Semanticists' essential claim is that they can concede the partial independence of theory and models. For any model there is some theory to which it belongs; however, there need not be some *single* such family for most models of a domain. In effect, they are claiming that there are *different kinds of theories* – abstract, phenomenological, data, etc. Of course, one might worry that this is a misuse or trivialization of the term "theory".

One problem with applying the models as mediators approach to the biological sciences is that there are few if any *fundamental theories*. Generally speaking, models and claims about them are all we have. There is then no gap between fundamental theory and phenomena. However, one can still make many of the same sorts of proposals given simply a hierarchy of more or less abstract biological models. Let us focus on an example commonly discussed by both the Semanticists and M&Ms – Newton's Second Law of motion and harmonic oscillators. We will then connect the example to biology. Newton's second law can be written as:

$$F = ma = m\left(\frac{d^2x}{dt^2}\right)$$

where m is mass, x is position, and t is time. The force acting on a body is equal to the mass times acceleration. Suppose we want to model a linear oscillator where the force on a particle is proportional to the negative displacement of the particle from its rest position. The second law for this linear restoring force is

$$F = ma = m\left(\frac{d^2x}{dt^2}\right) = -kx$$

where k is the constant of proportionality. So, we have added more detail by adding a specific force function for a linear oscillator.

We can also adjust our model so that it is a model of a simple pendulum. Suppose we have a pendulum of length l subject to a uniform gravitational force, $-mg$. A pendulum moves horizontally and vertically. Let us just consider the horizontal component of the motion. The downward gravitational force $-mg$ is partially balanced by the tension along the string S which has a magnitude of $mg\cos(a)$ where a is the angle of displacement. Let us suppose the horizontal component is $-S\sin(a)$. Since $\sin(a) = x/l$, then the equation of motion for x is

$$m\left(\frac{d^2x}{dt^2}\right) = -mg\cos(a)\sin(a) = -x\left(\frac{mg}{l}\right)\cos(a)$$

So, the force is the negative downward gravitational force divided by the length of the pendulum multiplied by the tension of the string. We have defined a force function for a simple pendulum. We can also offer a convenient approximation at this point. If the angle of displacement a is small enough, then $\cos(a) = 1$. Thus, our new equation for a simple pendulum now is

$$m\left(\frac{d^2x}{dt^2}\right) = -x\left(\frac{mg}{l}\right)$$

Finally, consider the case of a damped linear oscillator. Suppose we have a pendulum for which there is air resistance. Let us assume that the friction is a linear function of velocity. So, we have the equation

$$m\left(\frac{d^2x}{dt^2}\right) = -x\left(\frac{mg}{l}\right) + bv$$

where bv is the friction term. If the friction is significant, then the $x(t)$ cycles will decrease over time (informally, the pendulum's swing decreases with time). We could also add more details; for instance, a driving force that could counteract the friction, but we now can see how the different force functions for the general equation are specified. Moreover, we can see how we can make a simple system like our pendulum into a more complex system by adding things like friction.

What is fundamental in this case is that the assumptions needed to arrive at a linear harmonic oscillator, a simple pendulum, or a damped harmonic oscillator did not follow from Newton's Second Law alone or Newtonian mechanics narrowly construed. We had to make substantive assumptions even some of which were only approximately true given our knowledge of oscillators. Thus, we need mediating models at the interface of fundamental theory and phenomena. Now let's see the same point in the context of population biology.

To see the argument, consider a relatively simple example – the Lotka–Volterra predator–prey model. The model assumes that $dV/dt = f(V, P)$ and $dP/dt = g(V, P)$; the instantaneous rates of change of the prey ("victim") population V and prey population P, respectively, are functions of prey and predator abundances. In this respect, it is like Newton's second law $f = ma$. However, just like Newton's law, we must specify the "acceleration" term or the functional form of the expressions. To derive the model, let us make the following assumptions:

- Growth of prey population is exponential in absence of predators;
- Predator declines exponentially in absence of prey;
- Individual predators can consume an infinite number of prey;
- Predator and prey encounter one another randomly in a homogenous environment;
- Individuals in the predator and prey populations respectively are ecologically and genetically identical;

So, if we let r represent the intrinsic growth rate of the prey, α represent the capture efficiency of the predator, β represent the conversion efficiency of the predator, and q represent the mortality rate of the predator, then we have the following model where V is the prey population and P is the prey population:

$$\frac{dV}{dt} = rV - \alpha VP$$
$$\frac{dP}{dt} = \beta VP - qP$$

In effect, we have used a "law of mass action" in deriving the model (an analogy from chemistry and physics!). The interactions between predator and prey are proportional to their respective abundance. Notice that we are assuming that in the absence of predator, the prey grows exponentially. This is completely unrealistic so we can build into our model density-dependence of the prey population. Thus, we have:

$$\frac{dV}{dt} = rV\left(1 - \frac{V}{K}\right) - \alpha VP$$
$$\frac{dP}{dt} = \beta VP - qP$$

We can also incorporate phenomena like predator satiation since no predator can consume an infinite number of prey. If we let with k be a parameter representing the maximum feeding rate and D is the half-saturation constant, then

$$\frac{dV}{dt} = rV\left(1 - \frac{V}{K}\right) - \left(\frac{kV}{V+D}\right)P$$
$$\frac{dP}{dt} = \beta\left(\frac{kV}{V+D}\right)P - qP$$

What is crucial to realize in each of the cases is that we started with our basic Lotka–Volterra "theory" and we developed models, one with an assumption of logistic growth on the part of the prey and the other incorporating both logistic growth and predator satiation. However, to arrive at these models we had to make substantive empirical assumptions we could not have deduced from our theory. Thus, one might allege that the Semantic View cannot account for partial independence of models and theory – even in the context of biology.

Before the work of the Models as Mediators group, philosopher of biology William Wimsatt provided an account of how biological models are constructed and function. Specifically, he offered an account of the heuristics and biases of model building with one important case study concerning models of group selection. Many models of group selection seemed to demonstrate that it could be significant only rarely. However, Wimsatt found that each of the twelve models surveyed made many assumptions – selection occurred at a single locus, there was a strong form of blending inheritance occurring, group and individual selection operated in opposite directions, and groups did not differ in their reproductive rates – inimical to group selection. He suggested that different researchers made these assumptions because of reductionistic research biases of their modeling strategies, along with assuming that groups were simply "collections of individuals." Thus, the robust conclusion that group selection is generally inefficacious was actually a pseudo-robust claim. Wimsatt's work provides a rich resource for considering the heuristics and biases of model building.

6. Material Models

There is one last topic to consider – material models. Material models take us as far from the traditional concerns of philosophers of science as any we have considered. Jim Griesemer is a philosopher of biology who has proposed ". . . a picture of model-building in biology in which manipulated systems of material objects function as theoretical models" (1990, p.79). One can "abstract" through a material object a structure independent of a propositional representation. Material models provide a *presentative* role in theory development. These models serve theoretical functions in virtue of close connection to the phenomena under investigation.

For example, consider James Watson and Francis Crick's material model of the structure of DNA. Watson learned that the physical chemist Linus Pauling had discovered the structure of a protein molecule α-keratin. Pauling discovered its structure by using physical, scale models of the molecule. Not only did this suggest that DNA would be double helical, but also a methodology for discovering its structure. Watson writes,

> I soon was taught that Pauling's accomplishment was a product of common sense, not the result of complicated mathematical reasoning. Equations occasionally crept into this argument, but in most cases words would have sufficed. The key to Linus' success was his reliance of the simple laws of structural chemistry. The α-helix had not been found by only staring at X-ray pictures; the essential trick, instead, was to ask which atoms like to sit next to each other. In place of pencil and paper, the main working tools were a set of molecular models superficially resembling the toys of preschool children. We could thus see no reason why we should not solve DNA in the same way. All we had to do was to construct a set of molecular models and begin to play – with luck, the structure would be a helix. (1968, pp.50–1)

In their final two-chain model, Watson and Crick modeled DNA molecules with sugar-phosphate "backbones" and the adenine, cytosine, guanine, and thymine bases directed inward with metal plates and wire in a structure that stood six feet tall. This metal model made sense of the amount of water in the open spaces of the molecule, the X-ray diffraction data from Rosalind Franklin, and also Chargraff's rules.

One interesting case study of Griesemer's is that of the remnant models of the naturalist Joseph Grinnell. Grinnell was the first director of the Museum of Vertebrate Zoology at the University of California, Berkeley. Grinnell was particularly interested in expanding evolutionary theory by understanding the "evolution" of the environment. Given California's then "pristine" state, one could inventory the vertebrate fauna and the state could be used as an "ecological-evolutionary laboratory" (1990, p.81).

Grinnell believed environments could be classified according to the causes of the presence and absence of particular species in specified locations. These environments would be so classified according to physiological limits of temperature tolerance of the taxa themselves. Hence, his basic data consisted in the presence or absence of taxa at particular locations and times accompanied by information about the environment. He could construct life-zone maps of patches of homogeneity of ecological factors and thus identify the causes and patterns of selection.

Models for Grinnell then consisted in a remnant model, a specimen of a taxa with identifying tags tying them to a place, their taxonomic status, and a set of recorded environmental data. A theory could be presented by specifying a set of such models at different locations and times accompanied by ecological causal factors placing them in a Grinnellian hierarchy. Thus, models could be preserved in the Museum of Vertebrate Zoology. Griesemer argues that this is significant for the following reasons:

> This is significant because changes of theoretical perspective about the nature of species can be taken into account by pulling the specimens back out of their drawers or off the shelves and reanalyzing the model in terms of a different set of taxonomic designations. This is not possible in the isomorphic *formal* model because once the *information* is recorded that members of a particular taxon were present in a location, there is no recourse – through that information alone – to revise the assessment of specieshood that underlies it, should the theoretical perspective on the nature of species change. (1990, p.820)

Thus, Grinnell pursued a strategy of "vicarious" material model-building. He created an institution of such modeling through the practices of ". . . collecting, note-taking, labeling, cataloging, preserving, and storing" (1990, p.83).

Greisemer's analysis can be seen from the point of view of both the Semantic View and the Models as Mediators programs. First, Grinnell's material models can be used as the basis for theoretical hypotheses about various causal factors shaping taxa. Thus, we have the standard distinction between theoretical models and theoretical hypotheses. However, one wonders what the relation between models and the world is on this account since the models are *part* of the world. From the point of the Models as Mediators program, Grinnell can be understood as working hard both privately and via his home institution to create models that are independent of any theory of the environment.

Recently Newton C. A. de Costa and Steven French (2003) have argued that material models are analog models and analog models can be captured under the Semantic View through the notion of a *partial structure*. In effect, we have relational structures whose domains consist in material objects where each relation R in the structure is actually an ordered triple $\langle R_1, R_2, R_3 \rangle$. Thus, R_1 has in its extension those objects that are known to have the relevant property (positive analogy). R_2 has in its extension those objects known to not have the relevant property (negative analogy). Lastly, R_3 has in its extension the set of objects for which we do not know whether it has the relevant property (neutral analogy). Thus, the notion of an isomorphism, or a less stringent mapping, can capture the relevant similarities between domains. Whether this approach and its notion of "partial truth" will make sense of material models is something left to investigate.

However one ultimately understands material models, they provide resources for reevaluating standard philosophical views.

> Instead of reconstructing theories, the new work aims to interpret a variety of representational practices in parallel with increased attention in cognitive psychology to mental maps and "visual thinking", and in sociology to scientific practice. (2004, p.433)

Griesemer argues ultimately that three-dimensional models will force philosophers to come to terms with the heuristics of model building (2004). In order to understand how a material model represents the world, one must recognize both how the object is made and for what purposes it is made. Thus, an account of material models requires much deeper understanding of scientific practice – one that does not just consider word–world relations.

7. Conclusion

This essay surveyed the work of philosophers of science and biology on models. We have considered models as analogies, relational structures, partially independent representations, and material objects. Whether there is an extant account that can make sense of the bulk of models in biology remains to be seen. However, there is much work to be done. Moreover, we have barely touched on the functions that models play in biology, on how they provide explanations, how they can be tested, and the trade-offs that may exist in model-building.

References

Achinstein, P. (1967). *The concepts of science.* Baltimore: John Hopkins Press.
Beatty, J. (1980). Optimality-design and the strategy of model-building in evolutionary biology. *Philosophy of Science*, 47, 532–61.
Beatty, J. (1981). What is wrong with the received view of evolutionary theory. In P. Asquith & R. Giere (Eds). *PSA* 2 (pp. 397–426). East Lansing, MI: Philosophy of Science Association.
Braithwaite, R. (1962). Models in the empirical sciences. In E. Nagel, P. Suppes, & A. Tarski (Eds). *Logic, methodology, and philosophy of science* (pp. 224–31). Stanford: Stanford University Press.
Brandon, R., & Burian, R. (1984). *Genes, organisms, populations: controversies over the units of selection.* Cambridge, MA: MIT Press.
Campbell, N. R. (1920). *Physics: the elements.* Cambridge: Cambridge University Press.
Carnap, R. (1939). *Foundations of logic and mathematics.* Chicago: University of Chicago Press.
Cartwright, N. Shomar, T., & Suárez, M. (1995). The tool box of science: tools for building of models with a superconductivity example. In W. E. Herfel et al. (Eds). *Theories and models in scientific processes* (pp. 137–49). Amsterdam: Editions Rodopi.
De Costa, N. C. A., & French, S. (2003). *Science and partial truth: a unitary to models and scientific reasoning.* Oxford: Oxford University Press.
Downes, S. (1992). The importance of models in theorizing: a deflationary semantic view. *PSA 1992*, 1, 142–53.
Duhem, P. (1954). *The aim and structure of physical theory.* Princeton: Princeton University Press.
Ereshefsky, M. (1991). The semantic approach to evolutionary theory. *Biology and Philosophy*, 6, 59–80.
Giere, R. (1988). *Explaining science.* Chicago: University of Chicago Press
Giere, R. (1999). *Science without laws.* Chicago: University of Chicago Press.
Goodman, N. (1976). *Languages of art: an approach to a theory of symbols.* Indianapolis: Hackett.
Griesemer, J. (1990). Material models in biology. *PSA 1990* (vol. 2, pp. 79–93). East Lansing, MI: Michigan Philosophy of Science Association.
Griesemer, J. (2004). Three-dimensional models in philosophical perspective. In S. de Chadarevian & N. Hopwood (Eds). *Models: the third dimension of science* (pp. 433–42). Stanford: Stanford University Press.
Hempel, C. (1967). *Philosophy of the natural sciences.* Englewood Cliffs, NJ: Prentice Hall.
Hesse, M. (1966). *Models and analogies in science.* Oxford: Oxford University Press.
Lange, M. (1995). Are there natural laws concerning particular biological species? *The Journal of Philosophy*, XCII, 430–451.
Lloyd, E. (1988). *The structure of evolutionary theory.* Princeton: Princeton University Press.
MacArthur, R. H., & Wilson, E. O. (1967). *The theory of island biogeography.* Princeton: Princeton University Press.
Maynard Smith, J. (1978). *The evolution of sex.* Cambridge: Cambridge University Press.
Maynard Smith, J. (1983). *Evolution and the theory of games.* Cambridge, Cambridge University Press.
Morgan, M., & Morrison, M. (Eds). (1999). *Models as mediators.* Cambridge: Cambridge University Press.
Nagel, E. (1961). *The structure of science.* New York: Harcourt, Brace and World.
Roughgarden, J. (1996). *Theory of population genetics and evolutionary ecology: an introduction.* New York: MacMillan.
Ruse, M. (1973). *The philosophy of biology.* London: Hutchinson and Co. Ltd.

Salmon, W. (1984). *Scientific explanation and the causal structure of the world.* Princeton: Princeton University Press.
Sepkowski, J. (1976). Species diversity in the phanerozoic – species-area effects. *Paleobiology*, 2, 298–303.
Sepkowski, J. (1978). A kinetic model of phanerozoic taxonomic diversity. I. Analysis of marine orders. *Paleobiology*, 4, 223–51.
Smart, J. J. C. (1966). *Philosophy and scientific realism.* London: Routledge & Kegan Paul.
Suppe, F. (1977). *The structure of scientific theories.* Urbana: University of Illinois Press.
Suppe, F. (1989). *The semantic conception of theories and scientific realism.* Urbana: University of Illinois Press.
Suppes, P. (1957). *Introduction to logic.* New York: Van Nostrand.
Suppes, P. (1962). Models of data. In E. Nagel, P. Suppes, & A. Tarski (Eds). *Logic, methodology, and philosophy of science* (pp. 252–61). Stanford: Stanford University Press.
Thompson, P. (1988). *The structure of biological theories.* Albany, NY: SUNY Press.
van Fraassen, B. (1971). *Formal logic and semantics.* New York: Macmillan.
van Fraassen, B. (1980). *The scientific image.* Oxford: Clarendon Press.
Watson, J. (1968). *The double helix.* New York: Atheneum.
Williams, M. B. (1970). Deducing the consequences of evolution. *Journal of Theoretical Biology*, 29, 343–85.
Williams, M. B. (1973). Falsifiable predictions of evolutionary theory. *Philosophy of Science*, 40, 518–37.
Williams, G. C. (1975). *Sex and evolution.* Princeton: Princeton University Press.
Woodger, J. H. (1929). *Biological principles: a critical study.* London: Routledge and Kegan Paul.
Woodger, J. H. (1952). *Biology and language.* Cambridge: Cambridge University Press.

Chapter 28

Function and Teleology

JUSTIN GARSON

1. Introduction

Function statements are used throughout the biological disciplines. For example, it is said that the function of the kidney is to extract waste products from the blood, the function of hemoglobin is the transportation of oxygen to tissue, and the function of myelin sheathing is to promote the efficient conduction of action potentials in the nervous system. In the case of many physical and mental disorders, it is believed that an inner part or process is *malfunctioning* or *dysfunctional* – such as the kidney in glomerulonephritis or myelin in multiple sclerosis – and knowledge of such dysfunctions guides medical research and intervention. Thus, functional language in biology has both theoretical and practical significance.

These examples draw attention to two interesting properties that function statements seem to possess. The first is that they are *explanatory*: to say that the function of myelin is to promote efficient nervous conduction is to say, roughly, *why myelin is there* or why many neural projections are sheathed in myelin. The second is that they are *normative*: the fact that the kidney, in the case of glomerulonephritis – a swelling of the glomeruli which filter the blood – can *fail* to perform its function implies that function statements do not necessarily *describe* what an entity actually does, but they set up a norm that specifies what that entity is *supposed* to do, or "what it is for."

Explanations that purport to explain the existence, form, distribution, or location of an entity by referring to some future state that the entity tends to bring about are referred to as teleological. The term "teleological" derives from the Greek word *telos*, meaning "goal" or "end." Hence function ascriptions are often thought to be a type of teleological explanation. Yet functional explanations seem problematic because they appear to violate the principle that temporally *posterior* events cannot figure into causal explanations for temporally *prior* events. The kidney must *already* be part of the organism in order to filter blood, just as neural projections must already be sheathed in myelin in order to efficiently conduct action potentials. How can a kidney's capacity to filter blood explain why the kidney is there, unless the future is assumed to have some causal influence over the present? (This is often called the problem of "backwards causation.")

The normative status of function statements is also puzzling. It is perfectly clear what one means by saying that an artifact, such as a camera, is *malfunctioning* – namely, that it is incapable of doing what the manufacturer made it for. But what could conceivably be the analog of a "manufacturer" in the biological realm, unless one assumes the existence of a supernatural creator – an assumption commonly deemed to have no place in legitimate scientific explanations? In what sense is the kidney *supposed to* filter the blood, rather than to support hard calcium formations along its inner wall, or to do nothing? Consequently, functional explanations are not only puzzling with respect to what they purport to explain, but they are also suspect of violating important tenets of the modern scientific worldview: the absence of final causes in nature and the illegitimacy of appealing to divine creation or intervention. Nonetheless, they are routinely appealed to throughout the biological disciplines. This suggests that they either ought to be eliminated from biology or analyzed in such a way that the appeal to final causes or supernatural beings is shown to be unnecessary.

One approach to the explication of function statements is simply to accept final causation as a distinct and irreducible type of causation. This is the solution that Aristotle is often thought to have provided. Aristotle's view of causation (*aitia*, which can be translated as "cause" or "reason") involves a rejection of the premise that future events cannot enter into explanations for the existence or form of a trait. His view is that the purpose, or *telos*, for which something exists cannot be eliminated from most biological explanations for the existence or form of a trait. (See his *Physics*, Book II.8 for several central arguments for this claim; also see *Parts of Animals*, Book I.1 for his defense of teleology in the context of biological explanation.)

Of course, to say that reference to future effects cannot be eliminated from an explanation is not to say that such explanations actually refer to a distinct type of causal pathway. Thus, one might interpret Aristotle liberally by suggesting that he was not really advocating the existence of final causes that somehow bring about their own realization, but advocating certain constraints on the nature of good *explanations* (translating *aitia* as "reason," a feature of rational discourse, rather than "cause," a mind-independent feature of the world). This latter reading is more generous, given that modern science has not accepted final causation as a distinct ontological relation. Consequently, supposing that functional language will not be eliminated from biology in the near future, any plausible account of "function" must *either* explain how it can be that the effect produced by a kind of entity can have causal relevance to the existence of the entity, *or* dissolve the misleading appearance that function ascriptions are causal explanations at all. *Etiological* approaches to function adopt the former route; *consequentialist* approaches the latter.

Intuitively, one might motivate either of the two main approaches to function by considering the following question: what distinguishes a *function* of an entity from a mere *effect* that it produces? To take a hackneyed, but simple, example, why is the function of the heart to pump blood rather than to make throbbing sounds? Two different answers present themselves as initially plausible:

(i) according to the etiological view, what distinguishes the *function* of an entity from a mere *effect* is that the capacity of the entity to perform that function explains "why it is there" in that system. For example, it is the capacity of windshield wipers to remove water from windshields that explains why they are on the windshield of a specific car; i.e., why the manufacturers placed them there. Similarly, one could argue that the fact that the heart has been selected for by natural selection because it pumped blood explains why, presently, creatures with hearts exist. Therefore, in conformity with the logic of teleological explanations, it is true to say that the heart's capacity to pump blood explains why hearts currently exist. However, the heart was not selected for because of the beating sounds that it makes, so there is no sense in which the heart "is there" because of its capacity to make such sounds;[1]

(ii) according to the consequentialist view, the function of the heart is to beat, rather than to make noise, because the heart's beating typically contributes to some important activity of the system within which it is contained, and heart sounds do not. In this case, beating contributes to pumping blood and this in turn to the survival of the organism. This solution corresponds to the view that the *function* of an entity consists in a (special sort of) consequence that it produces, and has nothing to do with the cause or origin of the item itself.

The following is composed of two sections. Section 2 will describe the *etiological* (or "backwards-looking") approach, which rejects the premise that function statements refer exclusively to future events. It will enumerate the main conceptual challenges that philosophers have confronted it with, and some of the responses to those challenges. Section 3 will describe several contemporary variants of the consequentialist (or "forward-looking") approach to functions, which rejects the premise that function ascriptions are causal explanations for the form or existence of a trait.

2. Etiological Theories of Function

There are two main versions of the etiological approach: one which refers to the *reasons* that motivate a purposeful being to create a functional object ("representationalism"), and one that refers to the natural history of the functional entity, independently of the notion of representation. (The latter is typically referred to as "etiological," although "etiological," properly speaking, could refer to either view.) These views will be elaborated in turn.

2.1. *Representationalist theories of function*

The first version of the "backwards-looking" approach is standardly employed to explain the sense in which intelligent creatures act for the sake of the future: it is *not* the case

1 There are, of course, exceptional cases in which it can be said that the heart's beating sounds explain why it is there. For example, if the beating sounds made by a person's heart alert a doctor to a life-threatening heart problem that is thereby remedied, then one can say that the heart sounds saved the person's life and therefore they partly explain why the person continues to exist, and hence why the heart continues to be there. Does that mean that that person's heart comes to have the function of making throbbing sounds? These sorts of cases will be described in greater detail below (see Section 2, under "The problem of overbreadth").

that the future effect of one's action (e.g., health as a consequence of exercise) causes the person to act; rather, it is the person's mental *representation* of the future effect, together with her other beliefs and desires, that cause her to act as she does. Thus an indirect reference to the future effect is preserved within the causal explanation for the purposeful action, and hence there is no violation of the normal temporal order of causation.

To the extent that, in order for a "representation" to exist, it must exist within, or have been created by, a mind, then representationalist theories are also *mentalistic* (Bedau, 1990). The assumption that functions are based on mental representations leads to two opposing views about how entities in the natural world come to have functions, the *theological* view and the *eliminativist* view (although the latter might just as appropriately be called the "analogical" view, for reasons that will be discussed below). According to the theological view – most notably advocated by Aquinas (1914 [1269–73]) – biological entities have purposes (e.g., functions) because God make them with those purposes in mind. This assumption is the basis for the fifth argument for the existence of God presented in his *Summa Theologica* (Question 2; Article 3). Roughly, his argument is that since mindless biological entities clearly have purposes, and something can only have a purpose if it has a mind or is controlled by something with a mind, then they must be controlled by something with a mind:

> We see that things which lack intelligence, such as natural bodies, act for an end, and this is evident from their acting always, or nearly always, in the same way, so as to obtain the best result. Hence it is plain that not fortuitously, but designedly, do they achieve their end. Now whatever lacks intelligence cannot move towards an end, unless it be directed by some being endowed with knowledge and intelligence; as the arrow is shot to its mark by the archer. Therefore some intelligent being exists by whom all natural things are directed to their end; and this being we call God. (Ibid., p.27)

Paley's (1839 [1802]) famous design argument for the existence of God rests on a similar perplexity about how things that appear so well formed for a specific purpose could have been products of anything but intelligent design. Contemporary advocates of the theological view of functions include Plantinga (1993, Chapter 11) and Rea (2002, Chapter 5).

Proponents of the *eliminativist* view also believe that functions are based on prior representations, and therefore *if* anything in nature has a function it must have been created for that purpose by an intelligent being. But they argue that appeals to supernatural creation have no place in the context of scientific explanations. Therefore, to the extent that one accepts this stricture on scientific explanation, then one must also accept that biological entities do not "really" have functions (or refuse to countenance them in one's explanations) since they are not typically designed with purposes in mind.[2] Accepting this eliminativist position with respect to the existence of function does not, however, imply that scientists should never *ascribe* functions to biological

2 To say that natural entities are "not created with purposes in mind," excludes, of course, the effects of deliberate human manipulation, such as genetic engineering or artificial selection through breeding. Therefore, terminologically it is probably accurate to distinguish artificial functions and natural (rather than "biological") functions, where "natural" is intended in the sense of "not created or brought about by deliberate or conscious effort."

entities or that it is illegitimate or counterproductive to do so. They may legitimately do so, so long as they recognize that such usage is metaphorical (e.g., it involves examining biological forms "as if" they were created for a purpose) and that it performs a purely heuristic role in stimulating actual scientific theories.

Kant's *Critique of the Power of Judgement* of 1789 contains the classic statement of this eliminativist view. Although he expresses different views on natural ends (*Naturzweck*),[3] one view that he expresses is that biological purposiveness is based on a prior representation: "Here I understand by *absolute purposiveness* [*Zweckmäßigheit*] of natural forms such an external shape as well as inner structure that are so constituted that their possibility must be grounded in an idea of them in our power of judgement" (2000 [1789], p.20; see Section VI of first introduction). In the case of natural ends, then, teleology presupposes the existence of a mind that can represent biological forms prior to creating them. Such a postulate, however, cannot enter into a causal explanation for the existence of such traits, since one of the *a priori* constraints on causal explanation is that both cause and effect must themselves be objects of the natural world. Causality cannot be a relation between the supernatural and the natural world, so long as one is operating within the perspective of natural science, since such a relation is not a possible object of experience: "But purposiveness in nature, as well as the concept of things as natural ends, places reason as cause into a relation with such things, as the ground of their possibility, in a way which we cannot know through any experience" (ibid., p.35; see Section IX of first introduction). Therefore, a function ascription has the status of a heuristic device for scientific research, or what Kant refers calls a *regulative*, rather than constitutive, principle: it can guide the formation of scientific hypotheses or the discovery of new evidence but it does not enter into the content of those hypotheses or the evidential statements (ibid., p.37; also see § 61).

A similar representationalist view, according to which the ascription of functions to the natural world rests upon an analogy to conscious design, is also adopted by the emergentist C. D. Broad (1925, p.82), and it finds more contemporary adherents in Woodfield (1976), Schaffner (1993, pp.403–4), Nissen (1997), and Ruse (1989, p.152). See Bedau (1990) for a critique of mentalistic views.

It was noted above that representationalist theories of function are almost always construed as *mentalistic* theories. Can there also be *non-mentalistic* representational theories of function, where representation is analyzed without appeal to minds? A possible such theory is associated with the distinction between "teleology" and "teleonomy." The term "teleonomy" was coined by the evolutionary biologist Pittendrigh (1958, p.394), to refer to systems that are in some sense "end-directed," but where this end-directedness does not rely on the problematic metaphysical assumptions associated with the word "teleology" – those of final causation or divine creation. However, he does not explicate his use of "end-directedness" or "goals." Mayr (1961, 1974), therefore, should primarily be credited with developing the concept of "teleonomy."

According to Mayr, a process or behavior is "teleonomic" if it is controlled by an internal "program" (1974, p.98). He defines a "program," in turn, as "coded or prear-

3 The following is a very partial account of Kant's view, and neglects his important phenomenological descriptions of the self-organization of living matter (e.g., §65), a phenomenon that he believes to warrant teleological explanation.

ranged information that controls a process (or behavior) leading it toward a given end" (ibid., p.102). Clearly, Mayr's analysis does not eliminate appeal to teleological concepts – such as "being led toward an end." Nonetheless, it does not seem implausible to suggest that the operative concept behind his formulation, like that of the mentalistic view, is that of "representation" – insofar as saying that one thing "carries information about" another thing seems tantamount to saying that the first thing *represents* the second. If this is true, then a teleonomic process might be equivalent to one that tends to develop along a specific trajectory, or into a specific form, by virtue of the fact that it is controlled, in part, by a non-mentalistic *representation* of that trajectory or form. Moreover, he clearly intends that segments of DNA that have been retained by natural selection, as well as neural structures that are shaped in some appropriate way by experience, qualify as containing "coded information." Hence, his analysis would require a naturalistic explication of "information" or "representation" that is appropriate for the biological context and that picks out the structures in question.

The feasibility of providing a naturalistic explication of biological information is defended by Maynard-Smith (2000), Sarkar (2000), and Sterelny (2000), as well as in the context of the "teleosemantic" account of information developed by, e.g., Stampe (1977), Enc (1982), and Millikan (1984) [See Biological Information]. However, most of the analyses depend centrally upon the concept of "function," and consequently cannot be used as part of an explication of "function" itself. The problem of defining a concept of "representation" that does not appeal to "function" is that representation, like function, is often assumed to be a *normative* concept. In other words, a representation can *misrepresent* something, just as a part of a system can *malfunction* (Millikan, 1984, p.17; Dretske, 1986). Hence it is sometimes suggested that the concept of function can be used to explicate the concept of representation, since it may be possible to explain the normative nature of representations by assuming that they have *functions*. A "misrepresentation," on this account, would be something like a sign that fails to perform its function. Moreover, since functions seem to be much more widespread in nature than representations (the heart has the function of pumping blood without being a representation of anything, whereas, plausibly, most representations have the function of guiding behavior), then defining representation in terms of function seems more likely to succeed than defining function in terms of representation.

2.2. Non-representationalist theories of function

Whereas representationalist views resolve the problem of backwards causation by seeking the origin of the functional entity in a prior mental representation, non-representationalist views seek to explain why such entities *currently* exist by appeal to entities of the same type that existed in the *past* and that, by virtue of producing the effect in question, were able to persist over time or to reproduce their kind. On this view, the function of an entity is that effect that entities of its kind produced in the past, which, in turn, contributed to the persistence or reproduction of that entity or type of entity. Thus, non-representationalist theories solve the problem of backwards causation by invoking a cyclical dimension: X did Y at time t_0, and as a consequence, X was able to continue to do Y, or X, by virtue of doing Y, was able to produce entities of the same type as X at time t_1. Such cyclical modes of production are sometimes referred to as "consequence-etiologies" (Wright, 1976, p.116), because one of the consequences that the functional item produces figures into an etiological account of why it continues to exist at a later time.

The most obvious example of a process that generates consequence-etiologies is natural selection, since the reproduction of heritable traits that have higher relative fitness than alternate traits explains the maintenance of the former within a population of reproducing entities. Several biologists throughout the twentieth century drew attention to the connection between teleological statements and natural selection, and stated explicitly that the existence of natural selection can justify the use of teleology in biology.[4] Perhaps the earliest reference comes from the neuroscientist Charles Sherrington, in his *The Integrative Action of the Nervous System* (1906). In that work, Sherrington pauses to reflect on his oft-repeated use of teleological terms such as "purpose," and his considerations suggest strongly that he identifies the purpose of a reflex with what it was selected for:

> That a reflex action should exhibit purpose is no longer considered evidence that a psychical process attaches to it; let alone that it represents any dictate of "choice" or "will." In light of the Darwinian theory every reflex *must* be purposive. We here trench upon a kind of teleology . . . The purpose of a reflex seems as legitimate and urgent an object for natural inquiry as the purpose of the colouring of an insect or a blossom. (Ibid., pp.235–6)

The ethologist Konrad Lorenz makes a similar remark in his 1963 book, *On Aggression*:

> If we ask "What does a cat have sharp, curved claws for?" and answer simply "To catch mice with," this does not imply a profession of any mythical teleology, but the plain statement that catching mice is the function whose survival value, by the process of natural selection, has bred cats with this particular form of claw. Unless selection is at work, the question "What for?" cannot receive an answer with any real meaning. (Lorenz, 1966 [1963], pp.13–4; cited in Griffiths, 1993, p.412)

The evolutionary biologist George Williams also emphasizes this point: "The designation of something as the *means* or *mechanism* for a certain *goal* or *function* or *purpose* will imply that the machinery involved was fashioned by selection for the goal attributed to it" (1966, p.9).[5]

None of these accounts, however, state *why* explanations based on natural selection fit the pattern of teleological explanations–they simply express, as it were, the basic intuition that they do, without articulating a rationale. Perhaps the first attempt to explicitly justify this view is found in the work of the evolutionary biologist Ayala (1968, p.217; 1970, pp.40–1), who points out that in a selectionist explanation, an

4 Lennox (1993) argues that Darwin himself implicitly uses teleological terms such as "end" and "purpose" to describe the outcome of selection processes (ibid., p.415), though Darwin never explicitly states this fact about his usage.

5 It is ironic that the etiological theory was primarily developed by biologists, since one of the main arguments *against* the etiological analysis is that it does not correspond to actual biological usage! (See Section 3).

effect that an entity produces figures into an explanation of why that type of entity currently exists, and this, by definition, constitutes a teleological explanation. Wimsatt (1972) provides a comprehensive philosophical analysis of the logical structure of function statements and argues that insofar as function statements are construed as teleological explanations, selection processes are the only known and plausible way in which such statements can be justified: "[T]he operation of selection processes is not only *not* special to biology, but appears to be at the core of teleology and purposeful activity wherever they occur" (ibid., p.13).[6] More famously, Wright (1973, p.161; also see Wright, 1972) defines "function" in terms of these consequence-etiologies and argues that natural selection can justify function statements (Wright, 1973, p.159).[7]

Several different theories of function stem from this basic insight, and much of the philosophical literature on functions consists in the attempt to ramify, extend, and qualify this viewpoint. Three major challenges to this etiological view, and some of the responses to these challenges, will be presented in order to elucidate the ways in which the position has been developed over time.

(i) *The problem of overbreadth.* The first problem can be understood as a response to Wright's (1973) influential view, although in some form or another it continues to plague etiological theories. According to Wright's explication:

> The function of *X* is *Z* *means*
> (a) *X* is there because it does *Z*,
> (b) *Z* is a consequence (or result) of *X*'s being there. (1973, p.161)

In the artifact context, *X*'s form can be explained by the fact that somebody recognized that form to have a certain capacity (*Z*), and produced it for that reason, thereby fulfilling the first premise. In the biological context, if *X* was selected for by virtue of one of its effects, *Z*, and this selection process partly accounts for its present existence, then it will be true to say that "*X* is there because it does *Z*," thereby also satisfying the first premise. If *X*'s being there allows it to *continue* to do *Z*, then the second will be fulfilled as well. Clearly, the purpose of Wright's fairly general analysis is to present the idea of a cyclical causal process, one that incorporates both natural and artifact functions.

But Wright's general definition is also satisfied by processes that, intuitively, one would not want to ascribe functions to, such as the sort contrived by Boorse (1976) in his critique of Wright. Suppose, for example, that a hose in a laboratory springs a leak, and thereby emits a noxious chemical, and any scientist that attempts to seal the hose gets knocked unconscious by the chemical it emits. Thus it can be said that the leak in the hose contributes to its own persistence by knocking out anyone that comes close enough to fix it (ibid., p.72). But it seems counterintuitive to say that knocking out scientists is the function of the leak, or that the leak has any function at all. Similarly, obesity can contribute to a sedentary lifestyle, which in turn can reinforce obesity. Thus

6 However, he hesitates to build this insight into a *conceptual analysis* of "function," since he comes up with counter-examples that purport to show that being selected for is, strictly speaking, neither necessary nor sufficient for having a teleological function (Wimsatt, 1972, pp.15–16).

7 Wright (1973), like Wimsatt (1972), does not define "function" explicitly in terms of selection, but claims that having been selected for, in fact, suffices for having a function.

it is possible to explain a person's current obesity in terms of one of the consequences his or her obesity produced in the past that contributes to its own persistence (ibid., pp.75–6). Yet, like the hose example, it is seems bizarre to suggest that the function of obesity is to contribute to a sedentary lifestyle.[8]

Boorse's counterexamples have been influential in shaping the development and refinement of etiological theories of function, since they have led many to accept that having been selected for by natural selection, rather than merely having contributed to the continuation of one's present state, is a necessary condition for having a function (see, e.g., Neander, 1983, p.103; Millikan, 1993, pp.34 – 6; Boorse himself (1976, p.76) raises this possibility but rejects it). This view will be referred to as the "selected effects" (SE) theory of function, and some version of it is probably the most widely held theory of function amongst philosophers (Neander, 1983, 1991; Millikan, 1984, 1989a, 1989b, 1993; Brandon, 1990; Griffiths, 1992, 1993; Godfrey-Smith, 1994; Mitchell, 1993, 1995; Allen & Bekoff, 1995a, 1995b). Obesity, though it secures its own persistence by contributing to a sedentary lifestyle, is in no sense *selected over* some other phenotypic trait *because* it contributes to a sedentary lifestyle. Similarly, the leak in the hose is not there because *it*, rather than something else, proved to be more effective in knocking out scientists. This also resolves the problem, noted earlier (fn. 2), of the function of heart sounds – since even if the heart's beating sounds help to protect the heart by alerting physicians to potential heart problems, the heart was not selected for because it makes these sounds. However, by introducing natural selection as a necessary condition on function ascriptions, Wright's theory loses some of its generality, and this is the basis for the second criticism.

(ii) *The problem of conceptual divergence.* The problem of conceptual divergence has two forms. First of all, it is not clear how the SE theory of function adequately explains the functions of artifacts, and hence it entails the existence of a conceptual divergence between artifact "functions" and biological "functions" that is not intuitively obvious. Certainly, *some* types of artifacts undergo a certain selection process, where, over a significant period of time, certain features of its form are replicated, others are modified, and still others are extinguished. Nonetheless, functions are typically ascribed to artifacts on their *first* appearance, and that is because the intention of the designer suffices to give an artifact its function. It does not seem that this can be reconciled with the SE view.

Some philosophers have attempted to lessen this discrepancy by suggesting that the process of *designing* an artifact is akin to natural selection, in that the designer typically *imagines* variations on a given form, and chooses to actualize only that one that is most suitable to his or her purposes. Hence a type of "virtual" selection process takes place (Wimsatt, 1972; Griffiths, 1993). For example, Wimsatt (1972, p.15) raises the possibility of "mental trial and error" in his attempt to assimilate artifact functions to his model of biological functions, and show that some concept of selection over a range of

8 Bedau (1992, p.786) uses the example of a stick floating down a stream that brushes against a rock and gets pinned there by the backwash it creates, and thus is responsible for perpetuating its current position, to make the same point. Clearly, such examples can be multiplied indefinitely.

alternatives underlies both.[9] Another response has been to deny that an accurate explication of the concept of "biological function" must also account for the functions of artifacts (e.g., Godfrey-Smith, 1993, p.347). Perhaps the intuition that there exists a unified concept of function merely reflects the persistence of the "dead metaphor" that biological forms are the product of design (Lewens, 2004, p.13).

Regardless of whether or not the SE theory can successfully assimilate artifact functions, its generality appears to be quite limited in a second way, namely, historically. Harvey, for example, discovered that the function of the heart is to circulate the blood, and he *believed that* he discovered its function: "it is absolutely necessary to conclude that the blood in the animal body is impelled in a circle, and is in a state of ceaseless movement; that this is the act or function which the heart performs by means of its pulse; and that it is the sole and only end of the movement and contraction of the heart" (Harvey, 1894 [1628], p.72). But he did not possess the theory of natural selection. Therefore, if the SE view is accurate, then Harvey meant something altogether different when he spoke of the function of the heart than what modern biologists mean (Frankfurt & Poole, 1966, p.71; Boorse, 1976, p.74; Nagel, 1977, p.284; Enc, 1979, p.346).

One response has been to argue that the SE theory is only intended to be accurate as a conceptual analysis of *modern biological usage* (Neander, 1991, p.176), regardless of whether it captures lay or historical usage. It has also been argued that the goal of explicating "function" is not to provide a *conceptual analysis* at all, but rather, a theoretical definition of "function" (Millikan, 1989b, p.293), in the same way that being H_2O constitutes a theoretical definition of "water." But since theoretical definitions are themselves often tantamount to conceptual analyses of modern scientific usage, the two responses are similar. Schwartz (2004) goes further by emphasizing the stipulative and constructive roles of philosophical definitions of "function," arguing that such definitions constitute *explications* of biological usage, rather than conceptual analyses or theoretical definitions. According to Carnap (1950, see chapters 1 and 2), philosophical explication involves the replacement of a vague concept by a precise one, and hence it often entails making distinctions that did not previously exist in the scientific context in question. It has the character of a proposal, to be accepted or rejected on pragmatic grounds.

The attempt to justify the SE view by appealing to modern biological usage gives rise to a different problem, which is that modern biologists don't always, or even typically, use "function" with *any* etiological import (Amundson & Lauder, 1994; Godfrey-Smith, 1994, p.351; Walsh, 1996, p.558; Schlosser, 1998, p.304; Wouters, 2003, p.658; Sarkar, 2005, p.18; Griffiths, 2005). Although, as noted above, biologists sometimes *do* use "function" more or less synonymously with "adaptation," in many contexts "function" is tied more closely to the current survival value of a trait. For example, as Godfrey-Smith (1994, p.351) points out, according to an influential set of distinctions introduced by Tinbergen (1963), the field of behavioral ethology is largely concerned with four questions concerning behavior: its (proximate) causation, its survival value,

9 However, he also entertains the possibility that an omniscient being, if one exists, might never have to consider a range of alternatives before acting, and yet the actions would nonetheless be purposeful – and thus that it is *conceivable* that the actions of this being could be explained teleologically without being the product of a selection process!

its evolution, and its ontogeny (ibid., p.411). In Tinbergen's usage, "survival value" is synonymous with "function," and explicitly separated from the question of evolution, and in particular, from the selective history of a behavior (ibid., p.423). Mayr (1961), similarly, distinguishes "functional biology" and "evolutionary biology," arguing that the former is concerned with the realm of "proximate causes" and the latter, "ultimate causes" of an entity or process, whereas, according to SE, "function" describes only the realm of ultimate causes.

Even more broadly, "function" is often used to characterize the entire range of activities that a part of a system is capable of performing (e.g., the sense in which "function" is opposed to *structure*). For example, the evolutionary morphologists Bock and Von Walhert (1965, p.274) define the function of an entity simply as "all physical and chemical properties arising from its form," provided that these properties are not relative to the environment, and Amundson and Lauder (1994) argue that this more liberal usage is standard in anatomy, comparative morphology, and physiology. This makes the use of function statements in those disciplines heavily dependent upon the interests of the investigator, since without at least imposing a pragmatic restriction on the appropriate use of function statements, virtually every structure in the natural world can be said to possess a "function." Given these multiple salient uses within biology, the most reasonable attitude to adopt seems to be a pluralistic one (e.g., Millikan, 1989a; Kitcher, 1993; Godfrey-Smith, 1994; Amundson & Lauder, 1994).

(iii) *The problem of vestiges.* A third criticism is that SE does not seem to allow for the possibility of *vestiges*, which are traits such as the human appendix which once possessed functions but have ceased to perform them for so long that they are said to be functionless (Boorse, 1976, p.76; Prior, 1985). The rudimentary ocular cyst of the cave-dwelling fish, *Phreatichthys andruzzii*, is not a dysfunctional eye, but a functionless vestige – even though at some point the organ had been selected for because of sight. But if the vestigial trait had *ever* been selected for, however distantly in the past, then its past contribution to the fitness of ancestral organisms figures into a complete explanation for its present persistence in the population. Therefore, without imposing any temporal restrictions on the explication of "function," it is not clear how that explication can capture the idea that a heritable trait, though it once possessed a function, no longer does, but has been retained because, e.g., the relevant mutations that would have allowed it to atrophy or be replaced never arose.

Another case which supports the need for introducing temporal restrictions on function ascriptions is the case of functional co-optation, in which a trait that initially spread within a population by selection for one of its consequences eventually came to be maintained by selection for something else, or in which a trait that was initially not selected for at all came to be selected for in a new environment. This distinction partly overlaps Gould and Vrba's (1982) well-known distinction between adaptation and exaptation, where a trait is an *adaptation* if it was "built by selection for its current role (ibid., p.6)," and an *exaptation* if it was later "co-opted" for a useful role that it was not originally selected for. For example, plant species of the genus *Dalechampia* probably first used resin secretions as a defense against herbivores; later, they became used as a reward system for pollinators (Armbruster, 1997). Exaptations are ubiquitous in the biological realm and render problematic any simplistic attempt to infer the selective history of a trait from its current contribution to fitness. SE must possess the resources to conceptualize such transitions appropriately.

Perhaps the most widely accepted etiological approach is that which identifies the function of a trait with the effect for which it was selected in the *recent evolutionary past* (Griffiths, 1992, 1993; Godfrey-Smith, 1994). But how should such a temporal unit be defined? Griffiths (1992) defends a version of SE according to which the trait in question must have contributed to its maintenance in a population during the last "evolutionary significant time period" for that trait, and he defines an evolutionarily significant time period for a trait, *T*, as that time period during which, given the mutation rate at the loci controlling *T*, and the population size, one would have expected some regression (atrophy) of *T* were it not making some contribution to fitness (ibid., p.128). Godfrey-Smith (1994), while introducing the expression "modern history theory of functions," leaves the determination of such a unit implicit.

Two other important developments within the structure of etiological views are worth noting before describing consequentialist views. The first is a distinction between "function" and "design," the importance of which is argued for in Allen and Bekoff (1995a, 1995b; also see Kitcher, 1993 and Buller, 2002, who elaborate notions of "design" in relation to which functions are identified). Unlike the concept of function, which can be used broadly to encompass whatever a trait was selected for, the concept of design, they claim, should only be applied to that subset of functions that partly explain the *structural modification of a trait* over time (1995a, p.615). They point out that what something is an "adaptation" for (in Gould and Vrba's sense) is often what it is "designed" for, and that "exaptations" will often correspond to traits which merely have "functions" but were not designed, since they did not undergo any additional structural modification to perform the exapted function.[10]

A second distinction that is useful is that between the "strong" etiological theory and the "weak" etiological theory, which has been implicit in much of the literature but only articulated by Buller (1988, 2002). According to the strong etiological theory, a function of a trait is an effect that, in the past, the trait was selected for (hence it is identical to SE). According to the weak etiological theory, however, the function of a trait is an effect that, in the past, contributed to the fitness of its bearer and thereby contributed to its own reproduction, regardless of whether it was selected for–that is, regardless of whether the requisite variation existed upon which selection could act, or whether existing variation was correlated with differential reproduction. Another way of formulating the distinction is that the strong etiological theory emphasizes the contribution of a trait to *differential* reproduction; the weak etiological theory emphasizes reproduction as such. Both theories, clearly, only ascribe functions to heritable traits.[11]

A simple example drawn from Dover (2000, p.41) can help to clarify the distinction. Suppose that, in a small population, genetic drift carries an allele to fixation at t_0.

10 See Buller (2002), however, who argues that their distinction between "function" and "design" is unprincipled, because whether something is *designed for X*, or merely has the *function of performing X*, often depends upon purely conventional decisions about how selection pressures should be individuated.

11 Buller (2002, pp.230–3) points out that it is not uncommon for philosophers to vacillate between the two forms.

Although that allele has a phenotypic effect, it did not confer any fitness advantage on its possessors. Now suppose that, at t_1, the environment changes in such a way that possession of the allele is necessary for survival. Even though all of the individuals within the population have the allele – so there is no selection for it – they all would have perished at t_1 had any of the alternate alleles gone to fixation at t_0. Thus, at t_2, it can be said that one of the reasons that the allele currently exists is because it produces the effect in question. Consequently, the scenario satisfies the pattern of teleological explanation. But since selection did not enter the scenario, the strong etiological theory does not bestow a function upon the trait, since at t_1, the requisite variation did not exist upon which selection could act, and at t_0, the differential reproduction of alleles was not correlated with differential fitness. Hence, the weak etiological theory is clearly more liberal with respect to the range of evolutionary mechanisms that it considers function bestowing, yet it still permits teleological explanation. Finally, since it only ascribes functions to heritable traits, it avoids the Boorse-style counterexamples described earlier.

3. Consequentialist Theories of Function

Despite the plurality of etiological theories, and despite the attempts to render etiological theories more consistent with modern biological usage, it is often pointed out that typically, when biologists seek to determine the function of an entity, they look to some subset of current dispositions or capacities of that entity rather than to the fossil record. This suggests that despite the modifications that can be imposed on the etiological theory to render it more compatible with biological usage, it does not adequately capture that usage. Thus, some argue, functions, whatever else they may be, must be thought of as current dispositions or consequences of traits, and hence function ascriptions cannot provide causal explanations for the current maintenance of a trait in a population. As noted above, consequentialist theories of function almost invariably conceive of the function of an entity as consisting in its contribution to something else, or its disposition to so contribute. Insofar as functions, in the biological context, are typically ascribed only to *parts* of systems (rather than to the system as a whole), then according to consequentialist theories the function of a trait is typically thought to consist in its contribution to some property or capacity of a more inclusive system – e.g., the contribution of a trait to the fitness of the organism. Hence, in the following, consequentialist theories will be classified according to the *sort* of systemic property or capacity which performance of the function contributes to bringing about or maintaining.[12] In the following, four types of contribution theories will be described:

12 It is not always the case that consequentialist theories define the function of an entity in terms of its contribution to something else. As noted above, according to one liberal biological conception of function, the function of a structure consists of the totality of effects it produces, independently of reference to the environment (Bock & von Wahlert, 1965, p.274). In this theory there is no sense in which a function contributes to anything else, much less a containing system. By the same token, it is not always the case that when a functional entity does contribute to a system, that system is its own inclusive system. This is most obviously true in the case of artifacts, which are typically not "part" of the person who uses them (see Wright, 1973, p.145).

interest-contribution theories; goal-contribution theories; good-contribution theories; and fitness-contribution theories.

3.1. Interest-contribution theories

The most general contribution theory is the interest-contribution theory, according to which the function of an entity consists, roughly, in its contribution to bringing about or maintaining some property of a system that is of interest to an investigator. The most well-known proponent of this theory is Cummins (1975; also see 2002) – so well known, in fact, that such functions are often simply referred to as "Cummins functions" (Millikan, 1989a; Godfrey-Smith, 1993), or even "C-functions" (Walsh & Ariew, 1996). However, as will be elaborated below, Cummins' own view could be appropriately referred to as the "systemic capacity" view, because it restricts functions to the components of complex and hierarchically organized systems.

Cummins (1975) claims that most prior analyses of "function" were flawed because they overlooked the fact that functions refer primarily to a distinctive *style* of explanation ("functional analysis"), and only secondarily to a distinctive *object* of study (e.g., organismic fitness) (ibid., p.756). In keeping with this methodological approach, to ascribe a function to a *part* of a system is to ascribe a capacity to that part, and this capacity is picked out because it plays a salient role in an analytical account of a capacity of the system itself. In this sense, there is nothing mysterious about function ascriptions, since they do not imply that an effect of a trait explains that trait's existence; rather, they merely show how a trait produces the effect in question. This analytical strategy constitutes a special style of scientific explanation, however, because it explains a complex capacity of a system by drawing attention to the simpler capacities of its subsystems and showing how they are organized in such a way as to yield the complex capacity. The more complex the capacity under investigation, the more complex the organization of the system, and the simpler the subsystem capacities, the more interesting such an explanation is. Nonetheless, the appropriateness of function statements is always relative to someone's explanatory interest, even if such ascriptions are not particularly *interesting*.

Hempel (1965 [1959]) and Lehman (1965) appear to hold an early version of the interest-based view. According to Hempel, the function of a system *part* consists, roughly, in its contribution to fulfilling some condition which is necessary for the "adequate, or effective, or proper working order" of the system as a whole (ibid., p.306). Hempel, however, does not attempt a definition of "proper working order"; his view is that each scientific discipline that uses function statements, whether it be biology, psychology, or sociology, must operationalize the notion of "proper working order" in its own terms, and hence his concept of function is explicitly relativized to the explanatory and disciplinary context at hand (ibid., pp.321–2). Similar views that emphasize the explanatory or pragmatic context of function statements are held by Prior (1985), Amundson and Lauder (1994), Hardcastle (1999), Davies (2001), and Craver (2001).

Because of the fact that, according to these views, functions are only limited by the interests – epistemic or pragmatic – of the investigator, they are often accused of overbreadth. On the one hand, "functions" could be ascribed throughout the non-organic world. For example, a particular arrangement of rocks can have the "function" of contributing the widening of a river delta downstream from it (Kitcher, 1993, p.390), and clouds can have the function of promoting vegetation growth (Millikan, 1989b, p.294). On the other hand, functions can be applied to entities that are clearly malfunctioning or maladaptive; as Cummins himself points out, the appendix keeps people vulnerable to appendicitis but it sounds strange to call this one of its functions (Cummins, 1975, p.752) – even though medical researchers are clearly interested in providing an analytical account of how this takes place! Yet these criticisms seem to misconstrue Cummins' insistence on the methodological, rather than substantive, character of functional analysis. Certainly, if, on the systemic capacity theory, function ascriptions were primarily intended to perform the substantive role of delineating a special type of system, then the liberality objection would be well taken, since such ascriptions would be vacuous. But since functional analysis is held to mark a style of explanation, then the liberality objection does not hold – it would be tantamount to suggesting that "conceptual analysis" is too liberal because, in principle, it applies to any concept!

3.2. Goal-contribution theories

According to goal-contribution theories, the function of a part of a system consists in its contribution to a *goal* of that system. The notion of a "goal" or of a "goal-directed system" occupied a significant place in philosophical approaches to teleology from the 1940s through the early 1970s (Rosenblueth et al., 1943; Sommerhoff, 1950, 1969; Braithwaite, 1953; Nagel, 1953, 1961; Beckner, 1969; Manier, 1971). However, it largely fell out of favor among philosophers of biology in the early 1970s, partly owing to the predominance of evolutionary considerations within that tradition and partly owing to internal conceptual shortcomings (Wimsatt, 1972; Ruse, 1973; Hull, 1973). In short, a goal-directed system is one that exhibits a capacity to attain a specific value for some system variable, or to maintain the variable within a range of values, in the face of environmental perturbation, via the existence of compensatory activity operating amongst the system's parts. The maintenance or attainment of a given value for the system variable is considered the *goal* of the system, and the specific contribution of a part of the system to that goal is considered to be the *function* of that part (Boorse, 1976, p.77; Nagel, 1977, p.297). Thus any system may have several goals; additionally, any sufficiently complex system can be analyzed as a hierarchy of goal-directed systems. Boorse (1976, 2002) advocates a goal-contribution theory and claims that individual survival and reproduction constitute the "apical goals" of the organism (2002, p.76); hence his general theory of function is largely coextensive with the fitness-contribution view when instantiated in the biological context.

Two paradigmatic cases of "natural" or "mechanical" purposiveness largely inspired this approach to teleology: homeostatic mechanisms drawn from physiology and servomechanisms that constitute the subject matter of cybernetics. As an example of the first type of mechanism, the percentage of water in the blood remains at around 90 percent throughout an individual's lifetime. This is because if it drops far below this level, the muscles increase the rate at which they infuse the blood with water; if it rises far above this level, the kidneys increase the rate at which they extract water from the blood. In this manner, the constancy of the water level of the blood is not a static phenomenon; it is actively maintained via compensatory mechanisms that operate throughout the body in the face of perturbation. Servomechanisms, such as heat-seeking missiles, exhibit a similar capacity to actively maintain a specific trajectory in the face of perturbation, and to adapt that trajectory to the moving position of the target. The oft-repeated slogan that goal-directed systems exhibit "plasticity" and "persistence" (e.g., Nagel, 1977, p.272; Enc & Adams, 1992, p.650) captures two central features of the concept of goal-directedness. On the one hand, such systems exhibit *plasticity* in that the same effect can be reached from a number of initial systemic configurations and by virtue of a number of different mechanisms or pathways. On the other hand, such systems *persist* in their course of action to the extent that they have the ability to attain or maintain a course of action in the face of environmental perturbation.[13]

Since negative feedback systems are capable of exhibiting self-regulation, the concept of goal-directedness has often been analyzed narrowly in terms of *negative feedback* (Rosenbleuth et al., 1943; Manier, 1971; Adams, 1979; Faber, 1984; but see Wimsatt, 1971, for criticism of the concept of "feedback"). However, theories of goal-directedness that emphasize the compensatory and self-regulatory activity of systems are not necessarily tied to negative feedback. Hull (1973) points out that a system can exhibit the plasticity required to be goal-directed without being guided by negative feedback. For example, if the kidney does not succeed in ridding the body of excess water, then sweating may do so, but the different responses are not clearly regulated by a single negative feedback system (ibid., pp.110–11). (Nagel, 1953, p.211, Sommerhoff, 1969, pp.198–9, and Schlosser, 1998, p.309 also point out limitations of the negative feedback model for analyzing goal-directedness.)

Recently, Schlosser (1998) adopted some of the basic insights from the goal-supporting theory while rejecting its association with negative feedback (ibid., p.309) – although, strictly speaking, his theory should not be conflated with a goal-contribution view. According to his view, if a state or property of a system has a function then there exists a set of circumstances under which it is necessary for its own "reproduction" – that is, its trans-generational reproduction or intra-generational persistence (ibid., p.326). However, in order to avoid the Boorse-type counterexamples described above, he stipulates that the system in question must be capable of *complex* self-reproduction–that is, the system must be capable of reproducing the state in different ways, depending upon the environmental circumstances (ibid., p.312). Hence his view incorporates the plasticity criterion associated with goal-supporting theories while leaving fairly open the mechanisms by which this plasticity is realized.

Two main problems afflict goal-contribution theories, the "problem of vacuousness" and the "problem of goal-failure." The problem of vacuousness stems from the fact that the standard characterization of a goal-directed system as one that exhibits "plasticity

13 It is sometimes argued that the goal-supporting account does not allow one to determine a system *goal*, and consequently, that this goal must be arbitrarily stipulated (Wimsatt, 1972, pp.20–2; Schaffner, 1993, pp.367–8; Schlosser, 1998, p.327). However, the above examples show this claim to be inaccurate. In the homeostatic case, *that* maintaining the water content of the blood at around 90% qualifies as a "goal" of the system is a consequence of the definition of "goal" and a rudimentary understanding of physiology, and need not be arbitrarily stipulated.

and persistence" with respect to a given end is not sufficient for imposing a substantive distinction between different types of systems, for almost all systems can be described as seeking an equilibrium state which can be reached from different initial states and in different ways (Wimsatt, 1971; Woodfield, 1976; Nissen, 1980–1; Bedau, 1993). A pendulum swinging to a state of rest, a ball rolling from the top of a bowl to the bottom, and an elastic solid returning to its original condition after the imposition of tension would all represent goal-directed systems. Consequently, unless one specific mechanism, such as negative feedback, is included within the definition, it is difficult to exclude such counterexamples. Sommerhoff (1950, p.86), and Nagel (1977, p.273), attempt to exclude such systems by imposing an independence condition on the variables, which roughly states that all of the controlling variables must be independently manipulable.

The problem of goal-failure stems from the fact that most explications of goal-directedness have tacitly or explicitly assumed that the supposed goal-directed behavior is successful, and as a consequence it is not clear how to explain the intuition that a non-conscious entity can have a goal and yet fail to satisfy it (Scheffler, 1959; Beckner, 1959; Hull, 1973). Manier (1971, p.234) and Adams (1979, p.506) address this problem by arguing that what makes a negative feedback system "goal-directed" is not that it actually achieves its goal, but that it is governed by an internal representation of the goal-state. (This brings the goal-contribution theory closer to an etiological theory such as Mayr's (1961, 1974), as described above.) This, however, raises the additional onus of providing a naturalistic account of "representation" that does not itself appeal to function.

3.3. *Good-contribution theories*

The core idea behind good-contribution theories of function is that in order for an entity to possess a function, performance of that function must (usually or typically) have a *beneficiary*. It must be useful for, beneficial for, or otherwise represent a "good" for some agent or system. This type of teleology is fairly evident in the world of artifacts, because artifacts are produced for a purpose and hence for an end deemed useful or beneficial by someone. Consequently, the good-contribution view is closely associated with the mentalistic view described above. However, this doctrine is not identical with mentalism, because it is not incoherent to ascribe "interests" or "goods" to biological entities that cannot be said to possess the sort of mental life required by mentalism.

Canfield (1964), for example, defines the function of an entity simply as some useful contribution it makes to a system: "A function of *I* (in *S*) is to do *C means I* does *C* and that *C* is done is useful to *S*" (ibid., p.290). In the biological context, he argues, the "usefulness" of a trait can be identified with its making a contribution to the survival or reproductive capacity of its bearer (ibid., p.292). Sorabji (1964) also expounds a good-contribution theory, and he argues that Plato and Aristotle hold this view. Ayala (1970) amends his etiological analysis by incorporating the concept of "utility" into his account: a feature of a system is "teleological" if it possesses "utility for the system in which it exists and such utility explains the presence of the feature in the system" (ibid., p.45). Thus, although strictly speaking, Ayala's position is an etiological one, it also incorporates the concept of benefit. Bedau (e.g., 1991, 1992, 1993) is the most prominent current defender of the good-contribution theory; also see McLaughlin (2001; especially chapter 8) for a recent defense of the view that any adequate theory of function must incorporate such a "welfare" provision.

Presumably, one of the main advantages of such a view is that it appears to bridge the divide between natural functions and artifact functions, for, whereas artifact functions are "useful" by virtue of conscious design, natural functions are "useful" by virtue of their fitness contribution. In other words, the same concept is instantiated differently depending on the context, and hence there is no deep conceptual divergence between the usages. Moreover, as Bedau (1992) points out, this solution would resolve some of the Boorse-style counterexamples described in relation to the etiological view – for example, the stick that is pinned to the rock because of the backwash it creates does not have the "function" of creating the backwash, and that is because being pinned to the rock is not "good for anything" (ibid., p.787).

However, a significant problem with the good-contribution view is that it does not allow functions to be distinguished from "fortuitous benefits" or "lucky accidents." Frankfurt and Poole (1965), for example, criticize Canfield (1964) because heart sounds sometimes *do* have good consequences for fitness by alerting a physician to a potential life-threatening ailment, yet it does not have this as a function. (Wright, 1973, pp.145–6 and Bedau, 1992, p.787 also raise this problem.) One solution to this would be to incorporate a statistical component: in order to have a function, the activity in question must usually, or typically, contribute to some good. But as Millikan (1984, p.29) famously points out, statistical normalcy is not a reliable guide to functionality, since the probability that a given sperm will actually fertilize an egg is extremely low, yet fertilization is without doubt the function of sperm. Most sperm are quite literally good for nothing. Finally, of course, accepting something like the good-contribution view would most likely spell the death of the project of "naturalizing teleology," since the ascription of function would be explicitly value-relative, and values are notoriously difficult to situate within the natural world.

Bedau (1992, p.794), like Ayala (1970), suggests the possibility of a theory of biological teleology that conjoins the etiological view and the good consequence view and that would ameliorate the problem of fortuitous benefits. According to this view, a trait would come to possess a function because its persistence is partly explained by its past contribution to a *beneficial consequence* (e.g., increased fitness). However, he does not go so far as to offer an unqualified endorsement of this view, since the *goodness* of the result (increased fitness) does not itself perform an essential explanatory role in the etiology of the trait, but is only, as it were, externally linked to that explanation (ibid., pp.801–2). McLaughlin (2001), however, develops a similar view according to which, in order to have a function, a trait must have produced a beneficial consequence that contributed to its own persistence or reproduction (ibid., p.168).

3.4. Fitness-contribution theories

The basic, unqualified idea behind fitness-contribution theories is that the function of a trait consists in its contribution to the fitness of the organism (or, more generally, to the fitness of the biological system of which it is a part). Thus, according to this view, the ascription of a function to a trait does not explain why that trait currently exists, although ascription of a function to *ancestral* tokens of a trait can play a role in an explanation for the *current* persistence of that trait. Fitness-contribution views are proposed by Canfield (1964), Lehman (1965), Ruse (1971, 1973), Bechtel (1986), Bigelow and Pargetter (1987), Horan (1989), Walsh (1996), and Wouters (2003, 2005) (although, as pointed out above, Canfield (1964) accepts this view insofar as he defines "function" in terms of utility and believes that the fitness contribution made by a trait is "useful" to the organism). Sarkar (2005, p.18) presents a generalization of this view, according to which a part of a system must merely contribute to the persistence of its containing system in order to have a function, and not necessarily to the reproduction of that system. This would allow functions to be assigned to the parts of, e.g., sterile organisms.

One problem with this unqualified view is that, in principle, fitness assignments can vary wildly depending upon fluctuations in the current environment, but function assignments tend to be relatively stable. For example, one can create an abnormal, transient environment in which a trait that is usually maladaptive possesses survival value, but it seems counterintuitive to say that the trait comes to possess a new function in that environment. Moreover, even traits that are, on average, adaptive in a given environment can, in certain environments, become maladaptive. But it is not said that in such an environment the trait no longer has a function, but that it is unable to perform its function.

Such counterexamples suggest that such function ascriptions should be relativized to a "normal" or "average" environment, in order to exclude abnormal or transient ones. This recognition led Bigelow and Pargetter (1987) to propose that a trait has a function when it bestows a survival-enhancing propensity on the organism that possesses it, in that organism's natural habitat (ibid., p.192). Thus, their definition of function introduces a counterfactual element – if the trait *were* in its natural habitat, then it would, ceteris paribus, contribute to the fitness of its bearer. Yet this introduces further problems. Obviously, the "natural habitat" for an organism is not necessarily the organism's *current* habitat. But if not, then what constitutes an organism's natural habitat? One candidate for the natural habitat of an organism is that habitat in which it has, historically, flourished (Millikan, 1989b, p.300; Mitchell, 1993, pp.258–9; Godfrey-Smith, 1994, p.352; Walsh, 1996, p.562). But then the propensity theory of functions is rendered perilously close to some version of the etiological theory, since its incorporation of a historical component violates the spirit of the "forward-looking" view they endorse. Walsh (1996; also see Walsh & Ariew, 1996) attempts to eliminate the problem of defining the organism's "natural habitat" by proposing a relational theory of function, according to which the function statement must be relativized to a specific "selective regime," which may have occurred in the past or the present. Hence, in his view, there are no functions *simpliciter*; in order to assign a function one must state precisely the nature of the environment within which the trait contributes to fitness.

A similar problem stems from the following consideration. In order to estimate the contribution of a trait to fitness, one must compare the average fitness of organisms that possess the trait with the average fitness of those that do not. But if no variation for that trait currently exists–such as the human kneecap – then it is not clear what to compare its performance with (Frankfurt & Poole, 1965, pp.71–2; Wimsatt, 1972, pp.55–61; Millikan, 1989a; Godfrey Smith, 1994, p.352). One possibility would be to compare it with the variation that existed at an *earlier* time. But again, this brings the propensity theory closer in spirit to the etiological view.

Wouters (2003, 2005) proposes a version of the fitness-contribution view according to which, in order to have a function, a trait must confer a biological advantage upon its possessor, relative to some actual or counterfactual set of variants. This resolves the problem insofar as one must explicitly stipulate the range of variation in question. Moreover, he argues that this reflects standard practice within some fields of biology. In optimality models of adaptation, for example, the relevant range of alternatives (the "phenotype set") is typically derived from biologically informed assumptions about what is physically, ecologically, or physiologically possible (Parker & Maynard Smith, 1990, p.27; also see Wouters, 2005, p.43). However, merely stipulating the range of variants in question seems to introduce an element of arbitrariness into function ascriptions. Relative to one hypothetical set of variants, a trait has a function; relative to another set, it does not. Clearly, something more substantive should be said about how this range of variation can be non-arbitrarily determined.

As noted above, the main advantage of contribution-based theories is that they are more consistent with the majority of biological usage. Moreover, given the difficulty of inferring the evolutionary history of a trait from its current activity, it makes the practice of ascribing functions much more amenable to empirical testing. However, these theories appear to deprive functions of two of the properties that have, historically, been associated with their use and that continue to be associated with them. The first is the notion that they are explanatory in the sense that they specify an efficient cause for the current existence of the trait. What this means is not that the fact that a trait *had a function* in the past explains its current existence, but a trait's *having a function* explains its current existence. The second is that they are normative. On the etiological view, the distinction between functioning properly, malfunctioning, and inability to function due to an abnormal environment is rendered tolerably clear: because of the fact that function is a historical concept, something can have a function without being able to perform it. It is controversial whether these distinctions can be drawn clearly within consequentialist theories, though it has been argued that consequentialist views can sustain normative interpretations of function (Wimsatt, 1972, p.47; Walsh, 1996, p.568; Schlosser, 1998, p.327).

Such considerations reinforce the value of adopting a pluralistic and context-dependent approach to analyzing "function." In other words, in order to evaluate the meaning of a particular usage of "function" in a biological context, one must first identify the particular explanatory or pragmatic context in which that usage is embedded. If, for example, the ascription is intended to support an inference about how a trait evolved, or, perhaps, to make a normative claim about how the trait ought to behave, then an etiological concept of function may be implied. Alternatively, if the ascription is intended to sketch a prediction about the future survival value of a trait, or simply a prediction about what sort of behavior one ought to expect the trait to produce under well-defined circumstances, then a fitness-contribution theory, or an interest-based view, may be sufficient. What is crucial, then, is that different concepts of function allow one to articulate precisely the ontological and epistemological commitments that are implied by a given usage, and to ensure either that those commitments are satisfied in that context, or that the conditions under which the function ascription would be warranted can be explicitly stated.

Acknowledgment

The author wishes to express his gratitude to David J. Buller, Paul Griffiths, Anya Plutynski, Sahotra Sarkar, Gerhard Schlosser, and Arno Wouters for comments and criticism on earlier drafts of this article.

References

Adams, F. (1979). A goal-state theory of function attributions. *Canadian Journal of Philosophy*, 9, 493–518.

Allen, C., & Bekoff, M. (1995a). Biological function, adaptation, and natural design. *Philosophy of Science*, 62, 609–22.

Allen, C., & Bekoff, M. (1995b). Function, natural design, and animal behavior: philosophical and ethological considerations. In N. S. Thompson (Ed.). *Perspectives in ethology* (Vol. 11): *Behavioral design* (pp. 1–46). New York: Plenum Press.

Amundson, R., & Lauder, G. V. (1994). Function without purpose: the uses of causal role function in evolutionary biology. *Biology and Philosophy*, 9, 443–69.

Aquinas, S. Thomas (1914 [1269–73]). *The "Summa Theologica" of St. Thomas Aquinas. Part I. QQ. I.–XXVI.* London: Burns Oates & Washbourne.

Armbruster, W. S. (1997). Exaptations link evolution of plant–herbivore and plant–pollinator interactions: a phylogenetic inquiry. *Ecology*, 78, 1661–72.

Ayala, F. J. (1968). Biology as an autonomous science. *American Scientist*, 56, 207–21.

Ayala, F. J. (1970). Teleological explanations in evolutionary biology. *Philosophy of Science*, 37, 1–15.

Bechtel, W. (1986). Teleological function analyses and the hierarchical organization of nature. In N. Rescher (Ed.). *Current issues in teleology* (pp. 26–48). Lanham, MD: University Press of America.

Beckner, M. (1959). *The biological way of thought.* New York: Columbia University Press.

Beckner, M. (1969). Function and teleology. *Journal of the History of Biology*, 2, 151–64.

Bedau, M. (1990). Against mentalism in teleology. *American Philosophical Quarterly*, 27, 61–70.

Bedau, M. (1991). Can biological teleology be naturalized? *Journal of Philosophy*, 88, 647–55.

Bedau, M. (1992). Where's the good in teleology? *Philosophy and Phenomenological Research*, 52, 781–805.

Bedau, M. (1993). Naturalism and teleology. In S. J. Wagner & R. Warner (Eds). *Naturalism: a critical appraisal* (pp. 23–51). Notre Dame, IN: University of Notre Dame.

Bigelow, J., & Pargetter, R. (1987). Functions. *Journal of Philosophy*, 84, 181–96.

Bock, W., & von Wahlert, G. (1965). Adaptation and the form–function complex. Evolution, 19, 269–99.

Boorse, C. (1976). Wright on functions. *Philosophical Review*, 85, 70–86.

Boorse, C. (2002). A rebuttal on functions. In A. Ariew, R. Cummins, & M. Perlman (Eds). *Functions: new essays in the philosophy of psychology and biology* (pp. 63–112). Oxford: Oxford University Press.

Braithwaite, R. B. (1953). *Scientific explanation.* Cambridge: Cambridge University Press.

Brandon, R. N. (1990). *Adaptation and environment.* Princeton: Princeton University Press.

Broad, C. D. (1925). *The mind and its place in nature.* New York: Harcourt, Brace, & Company.

Buller, D. J. (1998). Etiological theories of function: a geographical survey. *Biology and Philosophy,* 13, 505–27.

Buller, D. J. (2002). Function and design revisited. In A. Ariew, R. Cummins, & M. Perlman (Eds). *Functions: new essays in the philosophy of psychology and biology* (pp. 222–43). Oxford: Oxford University Press.

Canfield, J. (1964). Teleological explanations in biology. *British Journal for the Philosophy of Science,* 14, 285–95.

Carnap, R. (1950). *Logical foundations of probability.* Chicago: University of Chicago Press.

Craver, C. F. (2001). Role functions, mechanisms, and hierarchy. *Philosophy of Science,* 68, 53–74.

Cummins, R. (1975). Functional analysis. *Journal of Philosophy,* 72, 741–65.

Cummins, R. (2002). Neo-teleology. In A. Ariew, R. Cummins, & M. Perlman (Eds). *Functions: new essays in the philosophy of psychology and biology* (pp. 157–72). Oxford: Oxford University Press.

Davies, P. S. (2001). *Norms of nature: naturalism and the nature of functions.* Cambridge, MA: MIT Press.

Dover, G. (2000). *Dear Mr Darwin: letters on the evolution of life and human nature.* Berkeley: University of California Press.

Dretske, F. (1986). Misrepresentation. In R. Bogdan (Ed.). *Belief: form, content, and function* (pp.17–36). Oxford: Clarendon Press.

Enc, B. (1982). Intentional states of mechanical devices. *Mind,* 91, 161–82.

Enc, B., & Adams, F. (1992). Functions and goal-directedness. *Philosophy of Science,* 59, 635–54.

Faber, R. J. (1984). Feedback, selection, and function: a reductionistic account of goal-orientation. In R. S. Cohen & M. W. Wartofsky (Eds). *Methodology, metaphysics and the history of science* (pp. 43–135). Dordrecht: D. Reidel.

Frankfurt, H. G., & Poole, B. (1966). Functional analyses in biology. *British Journal for the Philosophy of Science,* 17, 69–72.

Godfrey-Smith, P. (1993). Functions: consensus without unity. *Pacific Philosophical Quarterly,* 74, 196–208.

Godfrey-Smith, P. (1994). A modern history theory of functions. *Noûs,* 28, 344–62.

Gould, S. J., & Vrba, E. S. (1982). Exaptation: a missing term in the science of form. *Paleobiology,* 8, 4–15.

Griffiths, P. E. (1992). Adaptive explanation and the concept of a vestige. In P. Griffiths (Ed.). *Trees of life: essays in philosophy of biology* (pp. 111–31). Dordrecht: Kluwer.

Griffiths, P. E. (1993). Functional analysis and proper function. *British Journal for the Philosophy of Science,* 44, 409–22.

Griffiths, P. E. (2005). Function, homology, and character individuation. *Philosophy of Science,* forthcoming.

Hardcastle, V. G. (1999). Understanding functions: a pragmatic approach. In V. G. Hardcastle (Ed.). *Where biology meets psychology: philosophical essays* (pp. 27–43). Cambridge, MA: MIT Press.

Harvey, W. (1894 [1628]). *An anatomical dissertation upon the movement of the heart and blood in animals: being a statement of the discovery of the circulation of the blood.* Canterbury: G. Moreton.

Hempel, C. G. (1965 [1959]). The logic of functional analysis. In C. G. Hempel (Ed.). *Aspects of scientific explanation* (pp. 297–330). New York: Free Press.

Horan, B. (1989). Functional explanations in sociobiology. *Philosophy and Biology*, 4, 131–58.

Hull, D. (1973). *Philosophy of biological science*. Englewood Cliffs, NJ: Prentice-Hall.

Kant, I. (2000 [1789]). *Critique of the power of judgement*. Cambridge: Cambridge University Press.

Kitcher, P. (1993). Function and design. *Midwest Studies in Philosophy*, 18, 379–97.

Lehman, H. (1965). Functional explanation in biology. *Philosophy of Science*, 32, 1–20.

Lennox, J. G. (1993). Darwin *was* a teleologist. *Biology and Philosophy*, 8, 409–21.

Lewens, T. (2004). *Organisms and artifacts: design in nature and elsewhere*. Cambridge, MA: MIT Press.

Lorenz, K. (1966 [1963]). *On aggression*. New York: Harcourt, Brace & World.

Manier, E. (1971). Functionalism and the negative feedback model in biology. *Boston Studies in the Philosophy of Science*, 8, 225–40.

Maynard-Smith, J. (2000). The concept of information in biology. *Philosophy of Science*, 67, 177–94.

Mayr, E. (1961). Cause and effect in biology. *Science*, 134, 1501–6.

Mayr, E. (1974). Teleological and teleonomic: a new analysis. *Boston Studies in the Philosophy of Science*, 14, 91–117.

McLaughlin, P. (2001). *What functions explain: functional explanation and self-reproducing systems*. Cambridge: Cambridge University Press.

Millikan, R. G. (1984). *Language, thought, and other biological categories*. Cambridge, MA: MIT Press.

Millikan, R. G. (1989a). An ambiguity in the notion "function". *Biology and Philosophy*, 4, 172–6.

Millikan, R. G. (1989b). In defense of proper functions. *Philosophy of Science*, 56, 288–302.

Millikan, R. G. (1993). *White queen psychology and other essays for Alice*. Cambridge, MA: MIT Press.

Millikan, R. G. (2002). Biofunctions: two paradigms. In A. Ariew, R. Cummins, & M. Perlman (Eds). *Functions: new essays in the philosophy of psychology and biology* (pp. 113–43). Oxford: Oxford University Press.

Mitchell, S. D. (1993). Dispositions or etiologies? A comment on Bigelow and Pargetter. *Journal of Philosophy*, 90, 249–59.

Mitchell, S. D. (1995). Function, fitness, and disposition. *Biology and Philosophy*, 10, 39–54.

Nagel, E. (1953). Teleological explanation and teleological systems. In S. Ratner (Ed.). *Vision and Action* (pp. 537–58). New Brunswick, NJ: Rutgers University Press.

Nagel, E. (1961). *The structure of science*. New York: Harcourt, Brace and World.

Nagel, E. (1977). Teleology revisited: goal-directed processes in biology and functional explanation in biology. *Journal of Philosophy*, 74, 261–301.

Neander, K. (1983). *Abnormal psychobiology*. PhD Diss., La Trobe.

Neander, K. (1991). Functions as selected effects: the conceptual analyst's defense. *Philosophy of Science*, 58, 168–84.

Nissen, L. (1980–1). Nagel's self-regulation analysis of teleology. *The Philosophical Forum*, 12, 128–38.

Nissen, L. (1997). *Teleological language in the life sciences*. Lanham, MD: Rowman and Littlefield.

Paley, W. (1839 [1802]). *Natural theology*. New York: Harper.

Parker, G. A., & Maynard Smith, J. (1990). Optimality theory in evolutionary biology. *Nature*, 348, 27–33.

Pittendrigh, C. S. (1958). Adaptation, natural selection, and behavior. In A. Roe & G. G. Simpson (Eds). *Behavior and evolution* (pp. 390–416). New Haven: Yale University Press.

Plantinga, A. (1993). *Warrant and proper function*. Oxford: Oxford University Press.

Prior, E. W. (1985). What is wrong with etiological accounts of biological function? *Pacific Philosophical Quarterly*, 66, 310–28.

Rea, M. (2002). *World without design: the ontological consequences of naturalism*. Oxford: Clarendon Press.

Rosenblueth, A., Wiener, N., & Bigelow, J. (1943). Behavior, purpose and teleology. *Philosophy of Science*, 10, 18–24.

Ruse, M. (1971). Functional statements in biology. *Philosophy of Science*, 38, 87–95.

Ruse, M. (1973). *The philosophy of biology*. London: Hutchinson University Library.

Ruse, M. (1989). *The Darwinian paradigm: essays on its history, philosophy and religious implications*. London: Routledge.

Sarkar, S. (2000). Information in genetics and developmental biology: comments on Maynard-Smith. *Philosophy of Science*, 67, 208–13.

Sarkar, S. (2005). *Molecular models of life: philosophical papers on molecular biology*. Cambridge, MA: MIT Press.

Schaffner, K. F. (1993). *Discovery and explanation in biology and medicine*. Chicago: University of Chicago Press.

Scheffler, I. (1959). Thoughts on teleology. *British Journal for the Philosophy of Science*, 9, 265–84.

Schlosser, G. (1998). Self-reproduction and functionality: a systems-theoretical approach to teleological explanation. *Synthese*, 116, 303–54.

Schwartz, P. (2004). An alternative to conceptual analysis in the function debate. *The Monist*, 87, 136–53.

Sherrington, C. S. (1906). *The integrative action of the nervous system*. New Haven: Yale University Press.

Sommerhoff, G. (1950). *Analytical biology*. London: Oxford University Press.

Sommerhoff, G. (1969). The abstract characteristics of living systems. In F. E. Emery (Ed.). *Systems thinking* (pp. 147–202). Harmondsworth: Penguin.

Sorabji, R. (1964). Function. *Philosophical Quarterly*, 14, 289–302.

Stampe, D. (1977). Towards a causal theory of linguistic representation. *Midwest Studies in Philosophy*, 2, 42–63.

Sterelny, K. (2000). The "genetic program" program: a commentary on Maynard Smith on information in biology. *Philosophy of Science*, 67, 195–201.

Tinbergen, N. (1963). On the aims and methods of ethology. *Zeitschrift für Tierpsychologie*, 20, 410–29.

Walsh, D. M. (1996). Fitness and function. *British Journal for the Philosophy of Science*, 47, 553–74.

Walsh, D. M., & Ariew, A. (1996). A taxonomy of functions. *Canadian Journal of Philosophy*, 26, 493–514.

Williams, G. C. (1966). *Adaptation and natural selection: a critique of some current evolutionary thought*. Princeton: Princeton University Press.

Wimsatt, W. C. (1971). Some problems with the concept of feedback. *Boston Studies in the Philosophy of Science*, 8, 241–56.

Wimsatt, W. C. (1972). Teleology and the logical structure of function statements. *Studies in the History and Philosophy of Science*, 3, 1–80.

Woodfield, A. (1976). *Teleology*. Cambridge: Cambridge University Press.

Wouters, A. (2003). Four notions of biological function. *Studies in History and Philosophy of Biological and Biomedical Sciences*, 34, 633–68.

Wouters, A. (2005). The functional perspective in organismic biology. In T. A. C. Reydon & L. Hemerik (Eds). *Current themes in theoretical biology* (pp. 33–69). Dordrecht: Springer.

Wright, L. (1972). A comment on Ruse's analysis of function statements. *Philosophy of Science*, 39, 512–14.
Wright, L. (1973). Functions. *Philosophical Review*, 82, 139–68.
Wright, L. (1976). *Teleological explanations: an etiological analysis of goals and functions*. Berkeley: University of California Press.

Further Reading

Allen, C., Bekoff, M., & Lauder, G. (Eds). (1998). *Nature's purposes*. Cambridge, MA: MIT Press.
Ariew, A., Cummins, R., & Perlman, M. (2002). *Functions: new essays in the philosophy of psychology and biology*. Oxford: Oxford University Press.
Buller, D. J. (Ed.). (1999). *Function, selection, and design*. Albany, NY: SUNY Press.
Wouters, A. (2005). The function debate in philosophy. *Acta Biotheoretica*, 53, 123–51.

Chapter 29

Reductionism in Biology

ALEX ROSENBERG

Biological Reductionism holds that all facts, including all the non-macromolecular biological facts, are fixed by the facts of molecular biology. Accordingly, nonmolecular biological explanations need to be completed, corrected, made more precise or otherwise deepened by more fundamental explanations in molecular biology. Antireductionism does not dispute reductionism's metaphysical claim about the fixing of biological facts by macromolecular ones, but denies it has implications either for explanatory strategies or methodological morals. The antireductionists hold that explanations in functional biology need not be corrected, completed, or otherwise made more adequate by explanations in terms of molecular biology.

1. Reduction as Relation between Theories: Historical Considerations

Reduction was supposed by the post-Logical Positivists (or Logical Empiricists, as some preferred to call themselves) to be a relation between theories. In Ernest Nagel's *Structure of Science* (1961), reduction is characterized by the deductive derivation of the laws of the reduced theory from the laws of the reducing theory. The deductive derivation requires that the concepts, categories, explanatory properties, or natural kinds of the reduced theory be captured in the reducing theory. To do so, terms that figure in both theories must share common meanings. Though often stated explicitly, this second requirement is actually redundant as valid deductive derivation presupposes that the terms in which the theories are expressed are interdefinable. However, as exponents of reduction noted, the most difficult and creative part of a reduction is establishing these connections of meaning, i.e., formulating "bridge principles," "bi-lateral reduction sentences," "coordinating definitions" that link the concepts of the two theories. Thus it was worth stating the second requirement explicitly. Indeed, early and vigorous opponents of reduction as the pattern of scientific change and theoretical progress argued that the key concepts of successive theories are in fact incommensurable in meaning, as we shall see immediately below.

Within a few years after Watson and Crick uncovered the structure and function of DNA, reductionists began to apply their analysis to the putative reduction of Mendelian or population genetics to molecular genetics. The difficulties they encountered in pressing Watson and Crick's discovery into the mold of theoretical reduction became a sort of "poster-child" for antireductionists. In an early and insightful contribution to the discussion of reduction in genetics, Kenneth Schaffner (1967) observed that reduced theories are usually less accurate and less complete in various ways than reducing theories, and therefore incompatible with them in predictions and explanations. Accordingly, following Schaffner, the requirement was explicitly added that the reduced theory needs to be "corrected" before its derivation from the reducing theory can be effected. This raised a problem which became nontrivial in the fall-out from Thomas Kuhn's *Structure of Scientific Revolutions* (1962), and Paul Feyerabend's "Reduction, Empiricism and Laws" (1964). It became evident in these works that "correction" sometimes resulted in an entirely new theory, whose derivation from the reducing theory showed nothing about the relation between the original pair. Feyerabend's examples were Aristotelian mechanics, Newtonian mechanics, and Relativistic mechanics, whose respective crucial terms, "impetus" and "inertia," "absolute mass" and "relativistic mass" could not be connected in the way reduction required. No one has ever succeeded in providing the distinction that reductionism required between "corrections" and "replacements."

More fundamentally, reductionism as a thesis about formal logical relations among theories was undermined by the increasing dissatisfaction among philosophers of science with the powers of mathematical logic to illuminate interesting and important methodological matters such as explanation and theory testing. Once philosophers of science began to doubt whether deduction from laws was always sufficient or necessary for explanation, the notion that inter-theoretical explanation need take the form of reduction was challenged. Reductionism is closely tied to the axiomatic or so-called syntactic approach to theories, an approach which explicates logical relations among theories by treating them as axiomatic systems expressed in natural or artificial languages [SEE MODELS]. Indeed, "closely tied" may be an understatement, since deduction is a syntactic affair, and is a necessary component of reduction. But for a variety of reasons, the syntactic approach to theories has given way among some philosophers of biology to the so-called "semantic" approach to theories [SEE MODELS]. The semantic approach treats theories not as axiomatic systems in artificial languages, but as sets of closely related mathematical models. On the semantic view the reduction of one theory to another is a matter of employing (one or more) model(s) among those which constitute the more fundamental theory to explain why each of the models in the less fundamental theory are good approximations to some empirical processes, showing where and why they fail to be good approximations in other cases. The models of the more fundamental theory can do this to the degree that they are realized by processes that underlie the phenomena realized by the models of the less fundamental or reduced theory. There is little scope in this sort of reduction for satisfying the criteria for post-positivist reduction.

To the general philosophical difficulties that the post-positivist account of reduction faced, biology provided further obstacles. David Hull (1973) was the first to notice it is difficult to define the term "gene" as it figures in functional biology by employing only concepts from molecular biology. The required "bridge principles" between the concept of gene as it figures in population biology, evolutionary biology, and elsewhere in functional biology and as it figures in molecular biology could not be constructed. And all the ways philosophers contrived to preserve the truth of the claim that the gene is nothing but a (set of) string(s) of nucleic acid bases could not provide the systematic link between the functional "gene" and the macromolecular "genes" required by a reduction. There is of course less trouble identifying "tokens" – particular bits of matter we can point to – of the population biologist's genes with "tokens" of the molecular biologist's genes. But token-identities won't suffice for reduction. The second problem facing reductionism in biology is the absence of laws, either at the level of the reducing theory or of the reduced theory, or between them. If there aren't any laws in either theory, there is no scope for reduction at all. [See Laws].

Initially, antireductionists were able to show that the criterion of connectability with respect to the Mendelian and the molecular gene was not fulfilled as the two theories were in fact stated. But a stronger conclusion can be sustained: the criterion required by post-positivist reductionism cannot be satisfied as a fundamental matter of biological process. Individuation of types in biology is almost always via function: to call something a wing, a fin, or a gene is to identify it in terms of its function. But biological functions are as a matter of fact all naturally-selected effects. And natural selection for adaptations – i.e., environmentally appropriate effects – is blind to differences in physical structure that have the same or roughly similar effects. Natural selection "chooses" variants by *some of their effects*, those which fortuitously enhance survival and reproduction. When natural selection encourages variants to become packaged together into larger units, the adaptations become functions. Selection for adaptation and function kicks in at a relatively low level in the organization of matter. Accordingly, the structural diversity of the tokens of a given Mendelian or classical or population biological or generally "functional" gene will mean that there is no single molecular structure or manageably finite number of sets of structures that can be identified with a single functional gene. [See Gene Concepts].

Philosophers will recognize the relationship between the functional gene and the DNA sequence as one of "multiple realization" common to the relation functionalism in the philosophy of psychology alleges to obtain between psychological states and neural ones. The blindness of selection for effects to differences in structure provides the explanation for why multiple realization obtains between genes and polynucleotide molecules. Indeed almost every functional kind in biology will be multiply realized, owing to the fact that the kind has an evolutionary etiology.

Functional biology tells us that there is a hemoglobin gene, and yet there is no unique sequence of nucleic acids that is identical to this hemoglobin gene – nothing that could provide a macromolecular definition of the hemoglobin gene of functional biology. Of course there is some ungainly disjunction of all the actual ways nucleic acid sequences nowadays do realize or in the past have realized the hemoglobin gene – i.e., all the sequences that can be translated, and transcribed into RNA which in a local ribosome will code for one or another of the different types – fetal, adult, or the varying defective hemoglobin protein sequences. But this ungainly disjunction, even if we knew it (and we don't), won't serve to define the functional hemoglobin gene. The reason is obvious to the molecular biologist. An even vaster disjunction of nucleic acid sequences than the actual sequence will work just as well, or indeed just as poorly, to constitute the functional hemoglobin gene (and probably will do so in the future, given environmental contingencies and mutational randomness). Just think of the alternative introns that could separate exon regions of the sequence (and may do so in the future, given mutation and variation). Then there are all the promoter and repressor genes, and their alternative sequences, not to mention the genes for producing the relevant ribosomal protein-synthesizing organelles, all equally necessary for the production of the hemoglobin protein, and so claiming as much right to be parts of the functional hemoglobin gene as the primary sequence of the coding region of structural gene itself. Just as the actual disjunction is too complex to state, and yet not biologically exhaustive of the ways to code for a working hemoglobin protein, so also all these other contributory sequences don't exhaust the actual biological alternatives, and so make the macromolecular definition of the functional hemoglobin gene a will-o-the wisp.

In other words, on this view, being a hemoglobin molecule "supervenes," in the philosopher's term, on being a particular sequence of amino acids, even though there is no complete specification possible or scientifically fruitful of all the alternative particular sequences of amino acids that could constitute (i.e., realize) the function of the hemoglobin molecule in oxygen transport. Roughly, a biological property, P, supervenes on a (presumably complex) physical and/or chemical property Q if and only if when any thing has property P, it has some physical/chemical property Q, and anything else that has physical/chemical property Q, must also have biological property P. (See Rosenberg, 1978.) There is among philosophers a fairly sustained debate about the force of the "must" in this formulation. Does the supervenience of the biological on the physical/chemical have to obtain in virtue of natural laws, or even some stronger sort of metaphysical necessity? As many philosophers hold, biological properties are "local." Such properties make implicit but ineliminable reference to a particular place and time (the Earth since 3.5 billion years ago). Thus, it may be that biological properties are only locally supervenient, a much weaker thesis than one which makes it a matter of general law everywhere and always in the universe. (See "Concepts of Supervenience" in Kim, 1992).

When a biological property is supervenient on more than one complex physical/chemical property, then it is also a multiply realized property. The supervenience of the biological on the physical is a way of expressing the thesis of physicalism. The blindness of natural selection to differences in structure is what turns the supervenience of the biological on the physical into the multiple realization of the biological by the physical. This structural diversity explains why no simple identification of molecular genes with the genes of population genetics of the sort post-positivist reduction requires is possible. More generally, the reason there are no laws in biology is thus the same reason there are no bridge-principles of the sort post-positivist reduction requires (This result will be even less surprising in light of the post-positivist realization that most bridge principles in science will be laws, not definitions.)

The unavoidable conclusion is that as far as the post-positivist or "layer-cake" model of intertheoretical reduction is concerned, none of its characteristic preconditions are to be found in theories of functional biology, theories of molecular biology, or for that matter in any future correction of one or the other of these theories.

2. Antireductionism about Intertheoretical Relations

If antireductionism were merely the denial that post-positivist reduction obtains among theories in biology, it would be obviously true. But recall, antireductionism is not merely a negative claim. Most antireductionists add to the negative claim some or all of the following four theses: a) there are generalizations at the level of functional biology, b) these generalizations are explanatory, c) there are no further generalizations outside of functional biology which explain the generalizations of functional biology, and d) there are no further generalizations outside functional biology which explain better, more completely, or more fully, what the generalizations of functional biology explain. (For examples of such arguments cf. either Kitcher, 1984, or Kitcher, 1999.)

All four components of antireductionism are daunted by at least some of the same problems that vex reductionism: the lack of laws in functional biology and the problems facing an account of explanation in terms of derivation from laws. If there are no laws and/or explanation is not a matter of subsumption, then antireductionism is false too. But besides the false presuppositions antireductionism may share with reductionism, it has distinct problems of its own. Indeed, these problems stem from the very core of the antireductionists' argument, the appeal to ultimate explanations underwritten by the theory of natural selection.

To see the distinctive problems that an appeal to the ultimate/proximate distinction raises for biology's autonomy, consider a paradigm of putative irreducible functional explanation advanced by antireductionists. The example is due to one of the most influential of antireductionist physicalists, Phil Kitcher. It is one that has gone largely unchallenged in the almost two decades between the first and the latest occasion in which it has been invoked in his rejection of reductionism. The example is the biologist's explanation of independent assortment of functional genes:

> The explanadum is
> Genes on different chromosomes, or sufficiently far apart on the same chromosome, assort independently.

According to Kitcher, the functional biologist proffers an explanans for (G), which we shall call (PS):

> (PS) Consider the following kind of process, a *PS*-process (for *pairing* and *separation*). There are some basic entities that come in pairs. For each pair, there is a correspondence relation between the parts of one member of the pair and the parts of the other member. At the first stage of the process, the entities are placed in an *arena*. While they are in the arena, they can exchange segments, so that the parts of one member of a pair are replaced by the corresponding parts of the other members, and conversely. After exactly one round of exchanges, one and only one member of each pair is drawn from the arena and placed in the *winners box*. In any PS-process, the chances that small segments that belong to members of different pairs or that are sufficiently far apart on members of the same pair will be found in the winners box are independent of one another. (G) holds because the distribution of chromosomes to games at meiosis is a PS-process.

Kitcher writes, "This I submit is a full explanation of (G), and explanation that prescinds entirely from the stuff that genes are made of" (Kitcher, 1999, pp.199–200).

Leave aside for the moment the claim that (PS) is a full explanation of (G), and consider why, according to the antireductionist, no molecular explanation of (PS) is possible. The reason is basically the same story we learned above about why the kinds of functional biology cannot be identified with those of molecular biology. Because the same functional role can be realized by a diversity of structures, owing to the blindness of selection for differences in structure, the full macromolecular explanation for (PS) or for (G) will have to advert to a range of physical systems that realize independent assortment in many different ways. These different ways will be an unmanageable disjunction of alternatives so great that we will not be able to recognize what they have in common, if indeed they do have something in common beyond the fact that each of them will generate (G). Even though we all agree that (G) obtains in virtue only of macromolecular facts, nevertheless, we can see that because of their number and heterogeneity these facts will not explain (PS), still less supplant (PS)'s explanation of (G), or for that matter supplant (G)'s explanation of particular cases of genetic recombination. This is supposed to vindicate antireductionism's theses that functional explanations are complete and that functional generalizations cannot be explained by non-functional ones, nor replaced by them.

But this argument leaves several hostages to fortune. Begin with (G). If the argument of the previous section is right, (G) is not a law at all, but the report of a conjunction of particular facts about a spatiotemporally restricted kind, "chromosomes," of which there are only a finite number extant over a limited time period at one spatio-temporal region (the Earth). Accordingly, (G) is not something which we can expect to be reduced to the laws of a more fundamental theory, and the failure to do so constitutes no argument against reductionism classically conceived, nor is the absence or impossibility of such a reduction much of an argument *for* antireductionism.

The antireductionist may counter that regardless of whether (G) is a generalization, it has explanatory power and therefore is a fit test-case for reduction. This, however, raises the real problem which daunts antireductionism. Antireductionism requires an account of explanation to vindicate its claims. Biologists certainly do accord explanatory power to (G). But how does (G) explain? And the same questions are raised by the other components of the antireductionist's claims. Thus, what certifies (PS) – the account of PS-processes given above – as explanatory, and what prevents the vast disjunction of macromolecular accounts of the underlying mechanism of meiosis from explaining (PS), or for that matter from explaining (G) and indeed whatever it is that (G) explains?

There is one tempting answer to this question due to Putnam (1975, pp.295–8) and Garfinkel (1981), which is widely popular among antireductionists. This is the square peg–round hole argument. On this view, explanations of why a particular square peg does not go through the round hole in a board based on considerations from geometry are superior to explanations of the same event that advert to quantum mechanics; the former explanations are entirely adequate and correct, and require no supplementation, correction, or deepening by more fundamental considerations about the material composition of the peg and board, or laws and generalizations that they instantiate.

Reductionists are inclined to argue that explanations of why square pegs don't go through round holes which advert to geometry only are either seriously incomplete or false: We need to add information that assures us of the rigidity of the materials under the conditions that obtain when the peg is pushed through the hole, and once we begin trying to make our explanation complete and correct, the relevance of the more fundamental physical facts and laws governing them becomes clearer. Sober (1999) advances a slightly different argument against Putnam's conclusion that the geometrical explanation is superior, which, however, has a conclusion similar to the reductionist's. He notes that Putnam's argument begins by conceding that both explanations are correct, or at least equally well supported. Accordingly he infers that the only reason Putnam can offer for preferring the broader, geometrical explanation to the deeper physical one is our "subjective" interests. Putnam would be better advised simply to deny that the quantum theoretical description of the causal process instantiated by the peg and hole is explanatory at all. But it is hard to see how one could disqualify the quantum story as not explanatory, even if it were guilty of irrelevant detail and silence about an objective pattern instantiated by this and other peg-and-hole cases.

Why is a macromolecular explanation of (PS) not in the cards? One answer is presumably that it is beyond the cognitive powers of any human contemplating the vast disjunction of differing macromolecular processes, each of which gives rise to meiosis, to recognize that conjoined they constitute an explanation of (PS). Or similarly, it is beyond the competence of biologists to recognize how each of these macromolecular processes gives rise to (G). That the disjunction of this set of macromolecular processes implements PS-processes and thus brings about (PS) and (G) does not seem to be at issue. Only someone who denied the thesis of physicalism – that the physical facts fix all the biological facts – could deny the causal relevance of this vast motley of disparate macromolecular processes to the existence of (PS) and the truth of (G).

In fact, there is something that the vast disjunction of macromolecular realizations of (PS) have in common that would enable the conjunction of them to fully explain (PS) to someone with a good enough memory for details. Each was selected for because each implements a PS process and PS processes are adaptive in the local environment of the Earth from about the onset of the sexually reproducing species to their extinction. Since selection for implementing PS processes is blind to differences in macromolecular structures with the same or similar effects, there may turn out to be nothing else completely common and peculiar to all macromolecular implementations of meiosis besides their being selected for implementing PS processes. But this will be a reason to deny that the conjunction of all these macromolecular implementations explain (PS) and/or (G), only on a Protagorian theory of explanation.

Antireductionists who adopt what is called an erotetic account of explanation (for example, Sober, 1993, p.25), in preference to a unification account, a causal account or the traditional D-N account of explanation, will feel the attractions of the Putnam/Garfinkel approach. For the erotetic account of explanations treats them as answers to "why questions" posed about a particular occurrence or state of affairs, which are adequate – i.e., explanatory – to the degree they are appropriate to the background information of those who pose the why-question and to the degree that the putative explanation excludes competing occurrences or states of affairs from obtaining. Since it may be that we never know enough for a macromolecular answer to the question of why does (G) obtain, no macromolecular explanation of why (G) obtains will be possible. Similarly, we may never know enough for a macromolecular explanation of (PS) to be an answer to our question "Why do PS processes occur?" But this seems a hollow victory for antireductionism, even if we grant the tendentious claim that we will never know enough for such explanations to succeed. What is worse, it relegates antireductionism to the status of a claim about biologists, not about biology. Such philosophical limitations on our epistemic powers have been repeatedly breeched in the history of science.

Antireductionists wedded to alternative, non-erotetic accounts of explanation cannot adopt the gambit of a Putnam/Garfinkel theory of explanation in any case. They will need a different argument for the claim that neither (G) nor (PS) can be explained by its macromolecular supervenience base (see fn. 2 above), and for the claim that (PS) does explain (G) and (G) does explain individual cases of recombination. One argument such antireductionists might offer for the former claim rests on a metaphysical thesis: that there are no disjunctive properties or that if there are, such properties have no causal powers. Here is how the argument might proceed: The vast motley of alternative macromolecular mechanisms that realize (PS) have nothing in common. There is no property – and in particular no property with the causal power to bring about the truth of (G) – which they have in common. Physicalism (which all antireductionists party to this debate embrace) assures us that whenever PS obtains, some physical process, call it $P_{i,}$ obtains. Thus we can construct the identity (or at least the bi-conditional) that

$$(R)\ PS = P_1,\ v\ P_2\ v \ldots v\ P_i,\ v \ldots v\ P_m,$$

where m is the number, a very large number, of all the ways macromolecular processes can realize PS processes.

The Putnam/Garfinkel theory of explanation tells us that (R) is not explanatory roughly because it's too long a sentence for people to keep in their minds. A causal theory of explanation might rule out R as explaining PS on the ground that the disjunction, P_1, v P_2 v . . . v P_i, v . . . v P_m, is not the *full* cause. This might be either because it was incomplete – there is always the possibility of still another macromolecular realization of PS arising, or because disjunctive properties just aren't causes, have no causal powers, perhaps aren't really properties at all. A unificationist theory of explanation (or for that matter a D-N account) might hold that since the disjunction cannot be completed, it will not effect deductive unifications or systematizations. Thus (PS) and (G) are the best and most complete explanations biology can aspire to. Antireductionist versions of all three theories, the causal, the unificationist, and the Protagorian, need the disjunction in (R) to remain uncompleted in order to head off a reductionist explanation of (PS) and/or (G).

Consider the first alternative, that (R) is not complete, either because some disjuncts haven't occurred yet or perhaps that there are an indefinite number of possible macromolecular implementations for (SP). This in fact seems to me to be true, just in virtue of the fact that natural selection is continually searching the space of alternative adaptations and counter-adaptations, and that threats to the integrity and effectiveness of meiosis might in the future result in new macromolecular implementations of (PS) being selected for. But this no concession to antireductionism. It is part of an argument that neither (PS) nor (G) reports an explanatory generalization, that they are in fact temporarily true claims about local conditions on the Earth.

On the second alternative, (R) can be completed in principle, perhaps because there are only a finite number of ways of realizing a (PS) process. But the disjunction is not a causal or a real property at all. Therefore it cannot figure in an explanation of either (PS) or (G). There are several problems with such an argument. First, the disjuncts in the disjunction of P_1, v P_2 v . . . v P_i, v . . . v P_m, do seem to have at least one or perhaps even two relevant properties in common: each was selected for implementing (PS) and causally brings about the truth of (G). Second, we need to distinguish predicates in languages from properties in objects. It might well be that in the language employed to express biological theory, the only predicate we employ that is true of every P_i is a disjunctive one, but it does not follow that the property picked out by the disjunctive predicate is a disjunctive property. Philosophy long ago learned to distinguish things from the terms we hit upon to describe them.

How might one argue against the causal efficacy of disjunctive properties? One might hold that disjunctive properties will be causally efficacious only when their disjuncts subsume similar sorts of possible causal processes. If we adopt this principle, the question at issue becomes one of whether the disjunction of P_1, v P_2 v . . . v P_i, v . . . v P_m subsumes similar sorts of causal processes. The answer to this question seems to be that the disjunction shares in common the features of having been selected for resulting in the same outcome – PS processes. Thus, the disjunctive predicate names a causal property, a natural kind. Antireductionists are hard pressed to deny the truth and the explanatory power of (R).

Besides its problems in undermining putative macromolecular explanations of (PS), (G) and what (G) explains, antireductionism faces some problems in substantiating its claims that (PS) explains (G) and (G) explains individual cases of genetic recombination. The problems of course stem from the fact that neither (PS) nor (G) are laws, and therefore an account is owing of how statements like these can explain.

3. Reductionism as a Thesis about Explanations in Biology

Both the "layer-cake" reductionism of post-positivist philosophers of science and its antireductionist rejection are irrelevant to the real issue about the relation between macromolecular biology and molecular biology. If there is a real dispute here, it is not about the derivability or underivability of laws in functional biology from laws in molecular biology, as there are no laws in either subsdicipline. Nor can the real dispute turn on the relationship between theories in molecular and functional biology. There is only one general theory in biology, the theory of natural selection. And it is equally indispensable to functional and molecular biology. Once this conclusion is clear, the question of what was reductionism in the post-positivist past can be replaced by the question of what reductionism is now. For the obscalesence of the post-positivist model of reduction hardly makes the question of reductionism or its denial obsolete. The accelerating pace of developments in molecular biology makes this question more pressing than ever. But it is now clear that the question has to be reformulated if it is to make contact with real issues in biology.

Biology is unavoidably about Earthly phenomena. Its explanatory resources are spatiotemporally restricted in their meanings. Thus, the debate between reductionism and antireductionism will have to be one about the explanation of particular historical facts, some obtaining for longer than others, but all of them ultimately the contingent results of general laws of natural selection operating on boundary conditions. Reductionism needs to claim that the most complete, correct, and adequate explanations of historical facts uncovered in functional biology are given by appeal to other historical facts uncovered in molecular biology, plus some laws that operate at the level of molecular biology. Antireductionism must claim that there are at least some explanations in functional biology that cannot be completed, corrected, or otherwise improved by adducing wholly non-functional considerations from molecular biology. One way to do this would be to show that there are some functional biological phenomena that cannot in principle be decomposed or analyzed into component molecular processes. But such a demonstration would threaten the antireductionist's commitment to physicalism. A more powerful argument for antireductionism would be one that shows that even in macromolecular explanations, there is an unavoidable commitment to ultimate explanation by (implicit) appeal to irreducible functional – i.e., evolutionary, laws.

Reductionists can provide a strong argument for their view and rebut antireductionist counterargument effectively. But to do so they need to show that ultimate explanations in functional biology are unavoidably inadequate, and inadequate in ways that can only be improved by proximate explanations from molecular biology. This would indeed refute antireductionism. Or it would do so if the reductionist can show that these proximate explanations are not just disguised ultimate explanations themselves. What the reductionist must ultimately argue is that the laws of natural selection, to which even their most macromolecular explanations implicitly advert, are reducible to laws of physical science. This second challenge is the gravest one that reductionism faces. For if at the basement level of molecular biology there is to be found a general law not reducible to laws of physics and chemistry, then antireductionism will be vindicated at the very core of the reductionist's favored subdiscipline.

Let us consider the first challenge – that of showing what makes ultimate explanations in functional biology inadequate in ways only proximate molecular explanations can correct. Recall Mayr's (1982) distinction between proximate and ultimate explanation. Consider the question why do butterflies have eyespots on their wings. This question may express a request for an adaptationist explanation that accords a function, in camouflage for instance, to the eyespot on butterfly wings, or it may be the request for an explanation of why at a certain point in development eyespots appear on individual butterfly wings and remain there throughout their individual lives. The former explanation is an ultimate one, the latter is a proximate one. Reductionism in biology turns out to be the radical thesis that ultimate explanations must give way to proximate ones and that these latter will be molecular explanations.

To expound this thesis about explanations, reductionism adduces another distinction among explanations. The distinction is between what are called "how-possibly explanations" and "why-necessary explanations." How-possible explanations show how something could have happened, by adducing facts which show that there is after all no good reason for supposing it could not have happened. A why-necessary explanation shows that its explanandum had to have happened. As we will see, the how-possible/why-necessary distinction is quite different from the proximate/ultimate distinction. In particular, there can be both how-possible ultimate explanations and why-necessary ultimate explanations.

There is an important asymmetry between how-possible and why-necessary explanations that philosophers of history recognized (cf., for instance, Dray, 1957). Once a how-possible explanation has been given, it makes perfect sense to go on and ask for a why-necessary explanation. But the reverse is not the case. Once a why-necessary explanation has been given, there is no point asking for a how-possible explanation. For in showing why something had to happen, we have removed all obstacles to its possibly happening. Some philosophers of history went on to suggest that why-necessary explanations are "complete." But this is a notion hard to make clear in the case of, say, causal explanations, in which it is impossible to describe all the conditions, positive and negative, individually necessary and jointly sufficient for the occurrence of an event which we seek to explain. For our purposes all that will be required is the observation that a why-necessary explanation is more complete than a how-possible explanation, and that is the source of the asymmetry between them.

Consider the ultimate explanation for eyespots in the buckeye butterfly species *Precis coenia*. Notice, to begin with, that there is no scope for explaining the law that these butterflies have eyespots, or patterns that may include eyespots, scalloped color patterns, or edge-bands, even though almost all of them do have such markings. There is no such law to be explained, as there are no laws about butterflies, still less any species of them. That the buckeye butterfly has such eyespots is, however, a historical fact to be explained.

The ultimate explanation has it that eyespots on butterfly and moth wings have been selected for over a long course of evolutionary history. On some butterflies these spots attract the attention and focus the attacks of predators onto parts of the butterfly less vulnerable to injury. Such spots are more likely to be torn off than more vulnerable parts of the body, and this loss does the moth or butterfly little damage, while allowing it to escape. On other butterflies, and especially moths, wings and eyespots have also been selected for, taking the appearance of an owl's head, brows, and eyes. Since the owl is a predator of those birds that consume butterflies and moths, this adaptation provides particularly effective camouflage.

Here past events help to explain current events via principles of natural selection. Such ultimate explanations have been famously criticized as "just-so" stories, allegedly too easy to frame and too difficult to test (Gould & Lewontin, 1979); though its importance has been exaggerated, there is certainly something to this charge. Just because available data or even experience shows that eyespots are widespread does not guarantee that they are adaptive now. Even if they are adaptive now, this is by itself insufficient grounds to claim they were selected because they were the best available adaptation for camouflage, as opposed to some other function, or for that matter that they were not selected at all but are mere "spandrels," or traits riding piggy-back on some other means of predator avoidance or some other adaptive trait.

Reductionists will reply to this criticism that adaptationist ultimate explanations of functional traits are "how-possibly" explanations, and the "just-so story" charge laid against ultimate explanations on these grounds mistakes incompleteness (and perhaps fallibility) for untestability. The reductionist has no difficulty with the ultimate functional how-possibly explanation, as far as it goes. For its methodological role is partly one of showing how high fitness *could* in principle be the result of purely non-purposive processes. More importantly, on the reductionist's view, such a how-possibly explanation sets the research agenda which seeks to provide why-necessary explanations. It is these why-necessary explanations which cash in the promissory notes offered by the how-possibly explanation. But if we are not already convinced reductionists we may well ask, why must such why-necessary explanations be macromolecular? The reason is to be found in a limitation on ultimate explanations recognized by many: its silence about crucial links in the causal chains to which it adverts.

The how-possibly explanation leaves unexplained several biologically pressing issues, ones which are implicit in biologically well-informed requests for an ultimate explanation. These are the question of what alternative adaptive strategies were available to various lineages of organisms, and which were not, and the further question of how the feedback from adaptedness of functional traits – like the eyespot – to their greater subsequent representation in descendants was actually effected. The most disturbing lacuna in a how-possibly explanation is its silence on the causal details of exactly which feedback loops operate from fortuitous adaptedness of traits in one or more distantly past generations to improved adaptation in later generations and how such feedback loops approach the biological fact to be explained as a locally constrained optimal design. Dissatisfaction with such explanations, as voiced by those suspicious of the theory of natural selection, those amazed by the degree of apparent optimality of natural design, as well as the religious creationist, all stem from a single widely shared and very reasonable scientific commitment. It is the commitment to complete causal chains, along with the denial of action at a distance, and the denial of backward causation. Long before Darwin, or Paley for that matter, Spinoza diagnosed the defect of purposive or goal-directed explanation: it "reverses the order of nature," and makes the cause the effect. Natural selection replaces goal-directed processes. But natural selection at the functional level is silent on the crucial links in the causal chain which convert the appearance of goal-directedness into the reality of efficient causation. Therefore, explanations that appeal to it sometimes appear to be purposive or give hostages to fortune, by leaving too many links in their causal chains unspecified. Darwin's search for a theory of heredity reflected his own recognition of this fact.

The charge that adaptational explanations are unfalsifiable or otherwise scientifically deficient reflects the persistent claim by advocates of the adequacy of ultimate explanations that their silence on these details is not problematic.

Only a molecular account of the process could connect all the links in the causal chain. Such an account would itself also be an adaptational explanation: it would identify strategies available for adaptation by identifying the genes (or other macromolecular replicators) which determine the characteristics of *Lepidopterans* evolutionary ancestors, and which provide the only stock of phenotypes on which selection can operate to move along pathways to alternative predation-avoiding outcomes – leaf color camouflage, spot-camouflage, or other forms of Batesian mimicry, repellant taste to predators, Mullerian mimicry of bad-tasting species, etc. The reductionist's "why-necessary explanation" would show how the extended phenotypes of these genes competed and how the genes which generated the eyespot eventually become predominant, i.e., are selected for. In other words, the reductionist holds that a) every functional ultimate explanation is a how-possibly explanation, and b) there is a genic and biochemical pathway selection process underlying the functional how-possibly explanation. As we shall see below, reduction turns the merely how-possible scenario of the functional ultimate explanation into a why-necessary proximate explanation of a historical pattern. Note that the reductionist's full explanation is still a historical explanation in which further historical facts – about genes and pathways – are added, and are connected together by the same principles of natural selection that are invoked by the ultimate functional how-possibly explanation. But the links in the causal chain of natural selection are filled in to show how past adaptations were available for and shaped into today's functions.

Antireductionists will differ from reductionists not on the facts but on whether the initial explanation was merely an incomplete one or just a how-possibly explanation. Antireductionists will agree that the macromolecular genetic and biochemical pathways are causally necessary to the truth of the purely functional ultimate explanation. But they don't complete an otherwise incomplete explanation. They are merely further facets of the situation that molecular research might illuminate (Kitcher 1999: p.199). The original ultimate answer to the question, why do butterflies have eyespots, does provide a complete explanatory answer to a question. Accordingly, how-possibly explanations are perfectly acceptable ones, or else the ultimate explanation in question is something more than a mere how-possibly explanation.

Who is right here?

4. Reductionism and Explanation in Evolutionary Biology

On an erotetic view, how-possible and why-necessary explanations may be accepted as reflecting differing questions expressed by the same words. The reductionist may admit that there are contexts of inquiry in which how-possible answers to questions satisfy explanatory needs. But the reductionist will insist that in the context of advanced biological inquiry, as opposed, say, to secondary school biology instruction, for example, the how-possible question either does not arise, or having arose in a past stage of inquiry, no longer does. How-possible questions do not arise where the phenomena to be explained are not adaptations at all, for instance constraints, or spandrels, and the only assurance that in fact how-possible explanations make true claims is provided by a why-necessary explanation that cashes in their promissory notes by establishing the adaptive origins of the functional traits in molecular genetics. This will become clearer as we examine proximate explanation in biology.

Consider the proximate explanation from the developmental biology of butterfly wings and their eyespots. Suppose we observe the development of a particular butterfly wing, or for that matter suppose we observe the development of the wing in all the butterflies of the buckeye species, *Precis coenia*. Almost all will show the same sequence of stages, beginning with a wing imaginal disk eventuating in a wing with such spots, and a few will show a sequence eventuating in an abnormal wing or one without the characteristic eyespot maladapted to the butterfly's environment. Rarely one may show a novel wing or markings fortuitously better adapted to the environment than the wings of the vast majority of members of its species.

Let's consider only the first case. We notice in one buckeye caterpillar (or all but a handful) that during development an eyespot appears on the otherwise unmarked and uniform epithelium of the emerging butterfly wing. If we seek an explanation of the sequence in one butterfly, the general statement, that in all members of its species development results in the emergence of an eyespot on this part of the wing, is unhelpful. First, because examining enough butterflies in the species shows it is false. Second, even with an implicit ceteris paribus clause, or a probabilistic qualification, we know that the "generalization" simply describes a distributed historical fact about some organisms on this planet around the present time and for several million years in both directions. One historical fact cannot by itself explain another, especially not if its existence *entails* the existence of the fact to be explained. That all normal wings develop eyespots does not go very far in explaining why one does.

Most nonmolecular generalizations in developmental biology are of this kind. That is, they may summarize sequences of events in the lives of organisms of a species or for that matter in organisms of higher taxa than species. Here is an example of typical generalizations in developmental biology from Wolpert et al. (1998, p.320):

> Both leg and wing discs [in *Drosophila*] are divided by a compartmental boundary that separates them into anterior and posterior developmental region. In the wing disc, a second compartment boundary between the dorsal and ventral regions develops during the second larval instar. When the wings form at metamorphosis, the future ventral surface folds under the dorsal surface in the distal region to form the double layered insect wing.

Despite its singular tone, this is a general claim about all (normal) drosophila embryo, and their leg- and wing-imaginal discs. And it is a purely descriptive account of events in a temporal process recurring in all (normal) *Drosophila* larva. For purposes of proximate explanation of why a double layer of cells is formed in any one particular embryo's imaginal disc, this statement is no help. It simply notes that this happens in them all, or that it does so "in order" to eventually form the wing, where the "in order to" is implicit in the small word "to."

How is the pattern of eyespot development described in the extract from Wolpert in fact to be proximally explained? The details of a developmental explanation will show its special relevance to the proximate/ultimate distinction. Having identified a series of genes that control wing development in *Drosophila*, biologists then discovered homologies between these genes and genes expressed in butterfly development, and that whereas in the fruit fly they control wing formation, in the butterfly they also control pigmentation. The details are complex but following out a few of them shows us something important about how proximate why-necessary explanations can cash in the promissory notes of how-possible explanations and in principle reduce ultimate explanations to proximate ones.

In the fruit fly, the wing imaginal disk is first formed as a result of the expression of the gene *wingless* (so called because its deletion results in no wing imaginal disk and no wing) which acts a position signal to cells directing specialization into the wing disc-structure. Subsequently, the homeotic selector gene *apterous* is switched on and produces apterous protein only in the dorsal compartment of the imaginal disk control formation of the dorsal (top) side of the wing and activates two genes, *fringe* and *serrate*, which form the wing margin or edge. These effects were discovered by preventing dorsal expression of *apertous*, which results in the appearance of ventral (bottom) cells on the dorsal wing, with a margin between them and other (nonectopic) dorsal cells. Still another gene, *distal-less*, establishes the fruit fly's wing tip. Its expression in the center of the (flat) wing imaginal disk specifies the proximo-distal (closer to body/further from body) axis of wing development. It is the order in which certain genes are expressed, and the concentration of certain proteins in the ovum, which explain the appearance of eyespot development in the buckeye butterfly.

Once these details were elucidated in *Drosophila* it became possible to determine the expression of homologous genes in other species, in particular in *Precis coenia*. To begin with, nucleic acid sequencing showed that genes with substantially the same sequences were to be found in both species. In the butterfly these homologous genes were shown to also organize and regulate the development of the wing, though in some different ways. For instance, in the fruit fly *wingless* organizes the pattern of wing margins between dorsal and ventral surfaces, restricts the expression of *apterous* to dorsal surfaces, and partly controls the proximo-distal access where *distal-less* is expressed. In the butterfly, *wingless* is expressed in all the peripheral cells in the imaginal disk which will not become parts of the wing, where it programs their death (Nijhout, 1994, p.45). *Apterous* controls the development of ventral wing surfaces in both fruit flies and butterflies, but the cells in which it is expressed in the *Drosophila* imaginal disk are opposite those in which the gene is expressed in *Precis* imaginal disks. As Nijhout describes the experimental results:

> The most interesting patterns of expression are those of *Distal-less*. In *Drosophila Distal-less* marks the embryonic premordium of imaginal disks and is also expressed in the portions of the larval disk that will form the most apical [wing-tip] structures . . . In *Precis* larval disks, *Distal-less* marks the center of a presumptive eyespot in the wing color pattern. The cells at this center act as inducers or organizers for development of the eyespot: if these cells are killed, no eyespot develops. If they are excised, and transplanted elsewhere on the wing, they induce an eyespot to develop at an ectopic location around the site of implantation . . . the pattern of *Distal-less* expression in *Precis* disks changes dramatically in the course of the last larval instar [stage of development]. It begins as broad wedge shaped patters centered between wing veins. These wedges gradually narrow to lines, and a small circular pattern of expression develops at the apex of each line . . .
>
> What remains to be explained is why only a single circle of *Distal-less* expression eventually stabilizes on the larval wing disks. (Nijhout, 1994, p.45)

In effect, the research program in developmental molecular biology is to identify genes expressed in development, and then to undertake experiments – particularly ectopic gene-expression experiments – which explain the long-established observational "regularities" reported in traditional developmental biology. The *explanantia* uncovered are always "singular" boundary conditions insofar as the explananda are spatiotemporally limited patterns, to which there are always exceptions of many different kinds. The reductionistic program in developmental molecular biology is to first explain the wider patterns, and then explain the exceptions – "defects of development" (if they are not already understood from the various ectopic and gene deletion experiments employed to formulate the why-necessary explanation for the major pattern).

The developmental molecular biologists who reported the beginnings of the proximal explanation sketched above, S. B. Carroll and colleagues, eventually turned their attention to elucidating the ultimate explanation. Carroll et al. write:

> The eyespots on butterfly wings are a recently derived evolutionary novelty that arose in a subset of the Lepidoptera and play an important role in predator avoidance. The production of the eyespot pattern is controlled by a developmental organizer called the focus, which induces the surrounding cells to synthesize specific pigments. The evolution of the developmental mechanisms that establish focus was therefore the key to the origin of butterfly eyespots. Carroll (Keys et al., p.532)

What Carroll's team discovered is that the genes and the entire regulatory pathway that integrates them and which control anterior/posterior wing development in the *Drosophila* (or its common ancestor with butterflies) have been recruited and modified to develop the eyespot focus. This discovery of the "facility with which new developmental functions can evolve . . . within extant structures" (p.534) would have been impossible without the successful why-necessary answer to the proximate question of developmental biology.

Of course, the full why-necessary proximate explanation for any particular butterfly's eyespots is not yet in, nor is the full why-necessary proximate explanation for the development of the *Drosophila*'s (or its ancestor's) wing. But once they are in, the transformation of the ultimate explanation of why butterflies have eyespots on their wings into a proximate explanation can begin. This fuller explanation will still rely on natural selection. But it will be one in which the alternative available strategies are understood and the constraints specified, the time and place and nature of mutations narrowed, in which adaptations are unarguably identifiable properties of genes – their immediate or mediate gene products (in Dawkins's terms, their extended phenotypes), and in which the feedback loops and causal chains will be fully detailed, and the scope for doubt, skepticism, questions, and methodological critique that ultimate explanations are open to will be much reduced.

The macromolecular reductionist holds that why-necessary explanations can only be provided for by adverting to the macromolecular states, processes, events, and patterns that these nonmolecular historical events and patterns supervene on. Any explanation that does not do so cannot claim to be an adequate, complete why-necessary explanation. The reductionist does not claim that biological research or the explanations it eventuates in can dispense with functional language or adaptationism. Much of the vocabulary of molecular biology is thoroughly functional. Nor is reductionism the claim that all research in biology must be "bottom up" instead of "top down" research. So, far from advocating the absurd notion that molecular biology can give us all of biology, the reductionists' thesis is that we need to identify the patterns at higher levels because they are the explananda that molecular biology provides the explanantia for. What the reductionist asserts is that functional biology's explanantia are always molecular biology's explananda.

So, why isn't everyone a reductionist, why indeed, does antireductionism remain the ruling orthodoxy among philosophers of biology and even among biologists? Because, in the words of one antireductionist, reductionism's alleged "mistake consists in the loss of understanding through immersion in detail, with concomitant failure to represent generalities that are important to 'growth and form'" (Kitcher: 206). The reductionist rejects the claim that there is a loss of biological understanding in satisfying reductionism's demands on explanation, and denies that there are real generalities to be represented or explained. In biology there is only natural history – the product of the laws of natural selection operating on macromolecular initial conditions.

Reductionism accepts that selection obtains at higher levels, and that even for some predictive purposes, focus on these levels often suffices. But the reductionist insists that the genes, and proteins they produce, are still the "bottleneck" through which selection, among other vehicles, is channeled. Without them, there is no way to improve on the limited explanatory power to be found in functional biology. Insofar as science seeks more complete explanation for historical events and patterns on this planet, with greater prospects for predictive precision, it needs to pursue a reductionistic research program. That is, biology can nowhere remain satisfied with how-possible ultimate explanations, it must seek why-necessary proximate explanations, and it must seek these explanations in the interaction of macromolecules.

But this argument leaves a hostage to fortune for reductionism about biology, one large enough to drive home a decisive antireductionist objection. Although the reductionism here defended claims to show that the how-possible ultimate explanations must be cashed in for why-necessary ultimate explanations, these explanations are still ultimate, still evolutionary – they still invoke the principle of natural selection. And until this principle can be reduced to physical law, it remains open to say that even at the level of the macromolecules, biology remains independent from physical science. Thus, the reduction of molecular biology to physical science remains an agenda item for physicalism.

References

Dray, W. (1957). *The function of general laws in history*. Oxford: Clarendon Press.

Feyerabend, P. (1964). Reduction, empiricism and laws. *Minnesota Studies in the Philosophy of Science, v. III*. Minneapolis: University of Minnesota Press.

Garfinkel, A. (1981). *Forms of explanation*. New Haven: Yale University Press.

Gould, S. J., & Lewontin, R. (1979). The spandrels of St. Marco and the Panglossian paradigm. *Proceedings of the Royal Society of London*, B205, 581–98.

Hull, D. (1973). *Philosophy of biological science*. Englewood Cliffs, NJ: Prentice Hall.

Keys, D. N., Lewis, D. L., Selegue, J. E., Pearson, B. J., Goodrich, L. V., Johnson, R. L., Gates, J., Scott, M. P., & Carroll, S. B. (1999). Recruitment of a *Hedgehog* regulatory circuit in butterfly eyespot evolution. *Science*, 283, 532–4.

Kim, J. (1992). Downward causation. In A. Beckermann, H. Flohr, & J. Kim (Eds). *Emergence or reduction?* (pp. 119–38). Berlin: de Gruyter.

Kitcher, P. (1984). 1953 and all that: a tale of two sciences. *Philosophical Review*, 93, 353–73.

Kitcher, P. (1999). The hegemony of molecular biology. *Biology and Philosophy*, 14, 195–210.

Kuhn, T. (1962). *The structure of scientific revolutions*. Chicago: University of Chicago Press.

Mayr, E. (1982). *The growth of biological thought*. Cambridge, MA: Belnap Press.
Nagel, E. (1961). *Structure of science: problems in the logic of scientific explanation*. New York: Harcourt, Brace & World.
Nijhout, F. H. (1994). Genes on the wing. *Science*, 265, 44–5.
Putnam, H. (1975). *Mind, language, and reality*. New York: Cambridge University Press.
Schaffner, K. (1967). Approaches to reductionism. *Philosophy of Science*, 34, 137–47.
Sober, E. (1993). *The philosophy of biology*. Boulder, CO: Westview Press.
Sober, E. (1999). Multiple realizability arguments against reduction? *Philosophy of Science*, 66, 542–64.
Wolpert, L., Beddington, R., Jessell, T., Lawrence, P., Meyerowitz, E., & Smith, J. (1998). *Principles of development* (2nd edn). Oxford: Oxford University Press.

Further Reading

McShea, D., & Rosenberg, A. (2007). *Philosophy of biology: a contemporary approach*. London: Routledge, ch. 6.
Kitcher, P. (1999). Hegemony of molecular biology. *Biology and Philosophy*, 14(2, April), 195–210.
Rosenberg, A. (1985). *Structure of biological science*. Cambridge: Cambridge University Press.
Rosenberg, A. (2006). *Darwinian reduction or how to stop worrying and love molecular biology*. Chicago: University of Chicago Press.
Sarkar, S. (1998). *Genetics and reductionism*. Cambridge, UK: Cambridge University Press.
Schaffner, K. (1993). *Discovery and explanation in biology and medicine*. Chicago: University of Chicago Press.
Sober, E. (1999). Multiple realizability arguments against reduction. *Philosophy of Science*, 66, 542–64.
Waters, C. K. (1990). Why the anti-reductionist consensus won't survive: the case of classical Mendelian genetics. *PSA 1990* (vol. 1, pp. 125–39). East Lansing, MI: Philosophy of Science Association.
Wimsatt, W. C. (1976). Reductionism and levels of organization. In G. G. Globus, G. Maxwell, & I. Savodnik (Eds). *Consciousness and the brain: a scientific and philosophical inquiry* (pp. 205–67). New York: Plenum.

Index